SIXTH EDITION

SOCIOLOGY

[THE ESSENTIALS]

Margaret L. Andersen

University of Delaware

Howard F. Taylor

Princeton University

 WADSWORTH
CENGAGE Learning™

Australia • Brazil • Japan • Korea • Mexico • Singapore • Spain • United Kingdom • United States

Sociology: The Essentials, **Sixth Edition**
Margaret L. Andersen and Howard F. Taylor

Senior Publisher: Linda Schreiber

Associate Developmental Editor: Dan Moneypenny

Assistant Editor: Erin Parkins

Editorial Assistant: Rachael Krapf

Media Editor: Lauren Keyes

Marketing Manager: Andrew Keay

Marketing Assistant: Jillian Myers

Marketing Communications Manager: Laura Localio

Content Project Manager: Cheri Palmer

Creative Director: Rob Hugel

Art Director: Caryl Gorska

Print Buyer: Linda Hsu

Rights Acquisitions Account Manager, Text: Roberta Broyer

Rights Acquisitions Account Manager, Image: Don Schlotman

Production Service: TBC Project Management

Text Designer: Riezebos Holzbaur Design Group

Photo Researcher: Kelly Franz

Copy Editor: Jane Loftus

Illustrator: Macmillan Publishing Solutions

Cover Designer: Lee Friedman

Cover Image: © Michael Christopher Brown/ Corbis

Compositor: Macmillan Publishing Solutions

Library of Congress Control Number: 2009936913

ISBN-13: 978-0-495-81223-4

ISBN-10: 0-495-81223-4

Wadsworth
20 Davis Drive
Belmont, CA 94002-3098
USA

Cengage Learning is a leading provider of customized learning solutions with office locations around the globe, including Singapore, the United Kingdom, Australia, Mexico, Brazil, and Japan. Locate your local office at **www.cengage.com/global.**

Cengage Learning products are represented in Canada by Nelson Education, Ltd.

To learn more about Wadsworth, visit **www.cengage.com/wadsworth**

Purchase any of our products at your local college store or at our preferred online store **www.ichapters.com.**

To Richard and Pat, with love

Printed in the United States of America
2 3 4 5 6 7 13 12 11 10

BRIEF CONTENTS

Part One INTRODUCING THE SOCIOLOGICAL IMAGINATION

1. The Sociological Perspective 3

Part Two STUDYING SOCIETY AND SOCIAL STRUCTURE

2. Culture and the Media 25
3. Doing Sociological Research 55
4. Socialization and the Life Course 73
5. Social Interaction and Social Structure 101
6. Groups and Organizations 127
7. Deviance and Crime 151

Part Three SOCIAL INEQUALITIES

8. Social Class and Social Stratification 179
9. Global Stratification 211
10. Race and Ethnicity 235
11. Gender 263
12. Sexuality 289

Part Four SOCIAL INSTITUTIONS

13. Families and Religion 313
14. Education and Health Care 345
15. Politics and the Economy 371

Part Five SOCIAL CHANGE

16. Population, the Environment, and Social Change 401

CONTENTS

PART ONE   *Introducing the Sociological Imagination*

The Sociological Perspective 3

What Is Sociology? 4

The Sociological Perspective 5
Discovering Unsettling Facts 7
Debunking in Sociology 7
Establishing Critical Distance 9

The Significance of Diversity 10
Defining Diversity 10
Society in Global Perspective 12

The Development of Sociological Theory 13
The Influence of the Enlightenment 13
Classical Sociological Theory 14
Sociology in America 17

Theoretical Frameworks in Sociology 18
Functionalism 18
Conflict Theory 20
Symbolic Interaction 20
Diverse Theoretical Perspectives 21

Chapter Summary 22
DOING SOCIOLOGICAL RESEARCH: *Debunking the Myths of Black Teenage Motherhood* **8**
UNDERSTANDING DIVERSITY: *Become a Sociologist* **10**
WHAT WOULD THEY SAY NOW? *Classical Theorists Reflect on the War in Iraq* **15**
CAREERS IN SOCIOLOGY **22**

PART TWO   *Studying Society and Social Structure*

Culture and the Media 25

Defining Culture 27
Characteristics of Culture 28

The Elements of Culture 31
Language 31
Norms 34
Beliefs 36
Values 36

Cultural Diversity 37
Dominant Culture 38
Subcultures 39
Countercultures 40
Ethnocentrism and Cultural Relativism 40
The Globalization of Culture 41

The Mass Media and Popular Culture 42
The Organization of Mass Media 42
The Media and Popular Culture 43
Race, Gender, and Class in the Media 44

Theoretical Perspectives on Culture and the Media 46
Culture and Group Solidarity 47
Culture, Power, and Social Conflict 47
Symbolic Interaction and the Study of Culture 48

Cultural Change 49
Culture Lag 50
Sources of Cultural Change 50

Chapter Summary 52
DOING SOCIOLOGICAL RESEARCH: *Tattoos: Status Risk or Status Symbol?* **30**
UNDERSTANDING DIVERSITY: *The Social Meaning of Language* **34**
WHAT WOULD THEY SAY NOW? *Classical Theorists on Hip-Hop!* **49**
A SOCIOLOGICAL EYE ON THE MEDIA: *The Blogging Culture* **51**

Doing Sociological Research 55

The Research Process 56
Sociology and the Scientific Method 56
Research Design 57

The Tools of Sociological Research 63
The Survey: Polls, Questionnaires, and Interviews 63
Participant Observation 65
Controlled Experiments 66
Content Analysis 67
Historical Research 67
Evaluation Research 68

Research Ethics: Is Sociology
  Value Free? 68

Chapter Summary 70
A SOCIOLOGICAL EYE ON THE MEDIA: *Research and the Media* **60**

STATISTICS IN SOCIOLOGY 64

DOING SOCIOLOGICAL RESEARCH: *A Cop in the Hood: Participant Observation* **70**

Chapter 4

Socialization and the Life Course 73

The Socialization Process 74
  The Nature–Nurture Controversy 76
  Socialization as Social Control 76
  Conformity and Individuality 77
  The Consequences of Socialization 77

Agents of Socialization 78
  The Family 78
  The Media 79
  Peers 79
  Religion 80
  Sports 81
  Schools 81

Theories of Socialization 82
  Psychoanalytic Theory 82
  Social Learning Theory 83
  Functionalism and Conflict Theory 84
  Symbolic Interaction Theory 85

Growing Up in a Diverse Society 87

Aging and the Life Course 88
  Childhood 88
  Adolescence 88
  Adulthood 90
  Age and Aging 91
  Rites of Passage 95

Resocialization 96
  The Process of Conversion 97
  The Brainwashing Debate 97

Chapter Summary 98
UNDERSTANDING DIVERSITY: *My Childhood (Bong Hwan Kim)* **75**

DOING SOCIOLOGICAL RESEARCH: *Children's Understanding of Race* **89**

Chapter 5

Social Interaction and Social Structure 101

What Is Society? 102
  Macro- and Microanalysis 102
  Social Institutions 103
  Social Structure 104

What Holds Society Together? 104
  Mechanical and Organic Solidarity 104
  Gemeinschaft and Gesellschaft 105

Types of Societies 105
  Preindustrial Societies 106
  Industrial Societies 107
  Postindustrial Societies 109

Social Interaction and Society 109
  Groups 109
  Status 110
  Roles 111
  Everyday Social Interaction 112
  Interpersonal Attraction 114

Theories About Analyzing Social
  Interaction 116
  The Social Construction of Reality 117
  Ethnomethodology 117
  Impression Management and Dramaturgy 118
  Social Exchange and Game Theory 119

Interaction in Cyberspace 120

Chapter Summary 124
DOING SOCIOLOGICAL RESEARCH: *Doing Hair, Doing Class* **120**

[**seeing society in everyday life**

Social Interaction in an Age
of Technology   122

Chapter 6

Groups and Organizations 127

Types of Groups 128
  Dyads and Triads: Group Size Effects 129
  Primary and Secondary Groups 129
  Reference Groups 131
  In-Groups and Out-Groups 132
  Social Networks 133

Social Influence in Groups 134
  The Asch Conformity Experiment 135
  The Milgram Obedience Studies 135
  The Iraqi Prisoners at Abu Ghraib: Research
    Predicts Reality? 137
  Groupthink 137
  Risky Shift 138

Formal Organizations and
  Bureaucracies 139
  Types of Organizations 140
  Bureaucracy 141
  Bureaucracy's Other Face 141
  Problems of Bureaucracies 142
  The McDonaldization of Society 144

Diversity: Race, Gender, and Class in
  Organizations 145

Functional, Conflict, and Symbolic
  Interaction: Theoretical Perspectives 146

Chapter Summary 147
DOING SOCIOLOGICAL RESEARCH: *Sharing the Journey* 130

Chapter 7

Deviance and Crime 151

Defining Deviance 152
  Sociological Perspectives on Deviance 152
  The Medicalization of Deviance 154

Sociological Theories of Deviance 155
  Functionalist Theories of Deviance 155
  Conflict Theories of Deviance 159
  Symbolic Interaction Theories of Deviance 161

Forms of Deviance 164
  Mental Illness 164
  Social Stigmas 164
  Substance Abuse: Drugs and Alcohol 165

Crime and Criminal Justice 165
  Measuring Crime: How Much Is There? 166
  Organized Crime 169
  Corporate Crime and Deviance: Doing Well,
    Doing Time 169
  Race, Class, Gender, and Crime 170
  The Criminal Justice System: Police, Courts,
    and the Law 172

Terrorism as International Crime:
  A Global Perspective 175

Chapter Summary 175
WHAT WOULD THEY SAY NOW? *Classical Theorists Reflect on School
  Shootings* **158**

A SOCIOLOGICAL EYE ON THE MEDIA: *Images of Violent Crime* **167**
UNDERSTANDING DIVERSITY: *The Rich Get Richer and the Poor Get
  Prison* **171**

PART THREE   *Social Inequalities*

Chapter 8

Social Class and Social Stratification 179

Social Differentiation and Social
  Stratification 180
  Estate, Caste, and Class 181
  Defining Class 182

The Class Structure of the United
  States 183
  Layers of Social Class 187
  Class Conflict 189
  Growing Inequality and the Distribution of
    Wealth and Income 190
  Diverse Sources of Stratification 194

seeing society in everyday life
Social Inequality   184

Social Mobility: Myths and Realities 195
  Defining Social Mobility 196
  The Extent of Social Mobility 196
  Class Consciousness 196

Why Is There Inequality? 197
  Karl Marx: Class and Capitalism 197
  Max Weber: Class, Status, and Party 198
  Functionalism and Conflict Theory: The Continuing
    Debate 199

Poverty 202
  Who Are the Poor? 202
  Causes of Poverty 204
  Welfare and Social Policy 206

Chapter Summary 208
DOING SOCIOLOGICAL RESEARCH: *The Fragile Middle Class* **193**
A SOCIOLOGICAL EYE ON THE MEDIA: *Reproducing Class
  Stereotypes* **197**
WHAT WOULD THEY SAY NOW? *Classical Theorists Reflect on Social
  Class and Sports* **201**

Chapter 9

Global Stratification 211

Global Stratification 212
Rich and Poor 212
Global Networks of Power and Influence 215
Race and Global Inequality 216

Theories of Global Stratification 217
Modernization Theory 217
Dependency Theory 218
World Systems Theory 219

Consequences of Global Stratification 221
Population 221
Health and Environment 222
Education and Illiteracy 223
Gender Inequality 223
War and Terrorism 223

World Poverty 225
Who Are the World's Poor? 226
Women and Children in Poverty 227
Poverty and Hunger 227
Causes of World Poverty 229

Globalization and Social Change 230

Chapter Summary 231
DOING SOCIOLOGICAL RESEARCH: *Servants of Globalization: Who Does the Domestic Work?* **216**
UNDERSTANDING DIVERSITY: *War, Childhood, and Poverty* **228**

Chapter 10

Race and Ethnicity 235

Race and Ethnicity 236
Ethnicity 236
Race 236
Minority and Dominant Groups 239

Racial Stereotypes 240
Stereotypes and Salience 240
The Interplay Among Race, Gender, and Class Stereotypes 240

Prejudice, Discrimination, and Racism 241
Prejudice 241
Discrimination 242
Racism 243

Theories of Prejudice and Racism 244
Psychological Theories of Prejudice 245
Sociological Theories of Prejudice and Racism 245

Diverse Groups, Diverse Histories 246
Native Americans: The First of This Land 247
African Americans 248
Latinos 249
Asian Americans 250
Middle Easterners 252
White Ethnic Groups 252

Attaining Racial and Ethnic Equality: The Challenge 254
The White Immigrants Made It: Why Can't They? 254
Segregation and the Urban Underclass 255
The Relative Importance of Class and Race 256
The Civil Rights Strategy 256
Radical Social Change 258
Affirmative Action 258

Chapter Summary 260
DOING SOCIOLOGICAL RESEARCH: *American Apartheid* **255**
UNDERSTANDING DIVERSITY: *Race and Hurricane Katrina* **257**

Chapter 11

Gender 263

The Social Construction of Gender 264
Defining Sex and Gender 265
Sex Differences: Nature or Nurture? 265

Gender Socialization 267
The Formation of Gender Identity 267
Sources of Gender Socialization 270
The Price of Conformity 272
Race, Gender, and Identity 272
Gender Socialization and Homophobia 274
The Institutional Basis of Gender 274

Gender Stratification 275
Sexism and Patriarchy 275
Women's Worth: Still Unequal 276
The Devaluation of Women's Work 279
Balancing Work and Family 280

Theories of Gender 281
The Frameworks of Sociology 281
Feminist Theory 282

Gender in Global Perspective 283

Gender and Social Change 284
Contemporary Attitudes 284
Legislative Change 285

Chapter Summary 286

A SOCIOLOGICAL EYE ON THE MEDIA: *Cultural Gatekeepers and the Construction of Femininity* **266**

DOING SOCIOLOGICAL RESEARCH: *Eating Disorders: Gender, Race, and the Body* **273**

[seeing society in everyday life

Gender, Bodies, and Beauty 268

Chapter 12

Sexuality 289

Sex and Culture 290

Sex: Is It Natural? 290
The Social Basis of Sexuality 291

Contemporary Sexual Attitudes and Behavior 292

Changing Sexual Values 292
Sexual Practices of the U.S. Public 294

Sexuality and Sociological Theory 294

Sex: Functional or Conflict-Based? 294
Symbolic Interaction and the Social Construction of Sexual Identity 296

Sex, Diversity, and Inequality 297

Sexual Politics 297
The Influence of Race, Class, and Gender 297
A Global Perspective on Sexuality 299
Understanding Gay and Lesbian Experience 300

Sex and Social Issues 302

Birth Control 302
New Reproductive Technologies 303
Abortion 303
Pornography and the Sexualization of Culture 304
Teen Pregnancy 305
Sexual Violence 307

Sex and Social Change 308

The Sexual Revolution: Is It Over? 308
Technology, Sex, and Cybersex 308
Commercializing Sex 308

Chapter Summary 309

A SOCIOLOGICAL EYE ON THE MEDIA: *Publicity Traps: Sex on TV Talk Shows* **293**

DOING SOCIOLOGICAL RESEARCH: *Teens and Sex: Are Young People Becoming More Sexually Conservative?* **295**

WHAT WOULD THEY SAY NOW? *Classical Theorists Reflect on Sex and Popular Culture* **298**

UNDERSTANDING DIVERSITY: *Gay-Bashing and Gender Betrayal* **301**

PART FOUR   Social Institutions

Chapter 13

Families and Religion 313

Defining the Family 314

Extended Families 315
Nuclear Families 316

Sociological Theory and Families 317

Functionalist Theory and Families 318
Conflict Theory and Families 318
Feminist Theory and Families 319
Symbolic Interaction Theory and Families 319

Diversity Among Contemporary American Families 319

Female-Headed Households 320
Married-Couple Families 321
Stepfamilies 322
Gay and Lesbian Households 322
Single People 323

Marriage and Divorce 324

Marriage 324
Divorce 326
Family Violence 327

Changing Families in a Changing Society 328

Global Changes in Family Life 328
Families and Social Policy 329

Defining Religion 331

The Significance of Religion in the United States 332

The Dominance of Christianity 332
Measuring Religious Faith 333
Forms of Religion 333

Sociological Theories of Religion 334

Emile Durkheim: The Functions of Religion 335
Max Weber: The Protestant Ethic and the Spirit of Capitalism 335
Karl Marx: Religion, Social Conflict, and Oppression 335
Symbolic Interaction: Becoming Religious 336

Diversity and Religious Belief 336
The Influence of Race and Ethnicity 336
Religious Extremism 338

Religious Organizations 339

Religion and Social Change 340

Chapter Summary 341
UNDERSTANDING DIVERSITY: *Interracial Dating and Marriage* **317**
A SOCIOLOGICAL EYE ON THE MEDIA: *Idealizing Family Life* **320**
DOING SOCIOLOGICAL RESEARCH: *Men's Caregiving* **324**

Education and Health Care 345

Schooling and Society: Theories of Education 346
The Rise of Education in the United States 346
The Functionalist View of Education 347
The Conflict View of Education 348
The Symbolic Interaction View of Education 348

Does Schooling Matter? 348
Effects of Education on Occupation and Income 348
Effects of Social Class Background on Education and Social Mobility 349
Education, Social Class, and Mobility Seen Globally 350

Education and Inequality 352
Cognitive Ability and Its Measurement 352
Ability and Diversity 353
Genetics and Intelligence: *The Bell Curve* Debate 353
Tracking and Labeling Effects 354
Teacher Expectancy Effect 355
Schooling and Gender 355
Stereotype Threat 356

Health Care in the United States 357
The Emergence of Modern Medicine 357
Health, Diversity, and Social Inequality 358
Race and Health Care 359
Social Class and Health Care 360
Gender and Health Care 362
AIDS and Sexually Transmitted Diseases (STDs) 362

Theoretical Perspectives on Health Care 363
The Functionalist View of Health Care 363
The Conflict Theory View of Health Care 364
Symbolic Interactionism and Health Care 365

The Health Care Crisis in the United States 366
The Cost of Health Care 366
Medical Malpractice 367
A Response to the Problem: HMOs 367
The Universal Health Care Debate 367

Chapter Summary 368
UNDERSTANDING DIVERSITY: *The Americans with Disabilities Act* **359**

Politics and the Economy 371

Power, Politics, and the State 372
The State and Social Order 372
Global Interdependence and the State 373

Power and Authority 373
Types of Authority 373
The Growth of Bureaucratic Government 374

Theories of Power 374
The Pluralist Model 375
The Power Elite Model 375
The Autonomous State Model 376
Feminist Theories of the State 377

Government: Power and Politics in a Diverse Society 377
Diverse Patterns of Political Participation 377
Political Power: Who's in Charge? 379
Women and Minorities in Government 381
The Military 381

Economy and Society 385
The Industrial Revolution 385
Comparing Economic Systems 385

The Changing Global Economy 386
A More Diverse Workplace 387
Deindustrialization 387
Technological Change 387
Immigration 388

Sociological Perspectives on Work 389
The Division of Labor 389
Functionalism, Conflict Theory, and Symbolic Interaction 390

Characteristics of the Labor Force 391
Unemployment and Joblessness 391
Diversity in the U.S. Occupational System 393

Power in the Workplace 394
Sexual Harassment 394
Gays and Lesbians in the Workplace 395
Disability and Work 396

Chapter Summary 397
UNDERSTANDING DIVERSITY: *Diversity in the Power Elite* **378**
DOING SOCIOLOGICAL RESEARCH: *All in the Family: Children of Immigrants and Family Businesses* **392**

PART FIVE *Social Change*

Chapter 16
Population, the Environment, and Social Change 401
Demography and the U.S. Census 402

Diversity and the Three Basic Demographic Processes 403
Birthrate 403
Death Rate 404
Migration 404

Population Characteristics 405
Sex Ratio and the Population Pyramid 406
Cohorts 407

Theories of Population Growth Locally and Globally 407
Malthusian Theory 407
Demographic Transition Theory 408
The "Population Bomb" and Zero Population Growth 409

Checking Population Growth 409
Family Planning and Diversity 409

Urbanization 410
Urbanization as a Lifestyle 410
The New Suburbanites 410

Ecology and the Environment 411
Vanishing Resources 411
Environmental Pollution 412
Environmental Inequality 413
Feminism and the Environment 415
Environmental Policy 415

seeing society in everyday life
The Hurting Environment 416

What Is Social Change? 418

Theories of Social Change 419
Functionalist and Evolutionary Theories 419
Conflict Theories 420
Cyclical Theories 420

Global Theories of Social Change 421
Modernization Theory 421
Dependency Theory 422

Modernization 422
From Community to Society 422
Social Inequality, Powerlessness, and the Individual 423

The Causes of Social Change 423
Revolution 423
Cultural Diffusion 423
Inequality and Change 424
War and Social Change 424
Technological Innovation and the Cyberspace Revolution 425
Mobilizing People for Change 426

Collective Behavior and Social Movements 426
Characteristics of Collective Behavior 426
The Organization of Social Movements 427
Origins of Social Movements 428
Theories of Social Movements 429

Diversity, Globalization, and Social Change 431

Chapter Summary 432

Glossary 435
References 445
Photo Credits 477
Name Index 479
Subject Index 486

BOXES

Doing Sociological Research

Debunking the Myths of Black Teenage
    Motherhood 8
Tattoos: Status Risk or Status Symbol? 30
A Cop in the Hood: Participant Observation 70
Children's Understanding of Race 89
Doing Hair, Doing Class 120
Sharing the Journey 130
The Fragile Middle Class 193
Servants of Globalization: Who Does the Domestic
    Work? 216
American Apartheid 255
Eating Disorders: Gender, Race, and the Body 273
Teens and Sex: Are Young People Becoming More
    Sexually Conservative? 295
Men's Caregiving 324
All in the Family: Children of Immigrants and Family
    Businesses 392

Understanding Diversity

Become a Sociologist 10
The Social Meaning of Language 34
My Childhood (Bong Hwan Kim) 75
The Rich Get Richer and the Poor Get Prison 171
War, Childhood, and Poverty 228

Race and Hurricane Katrina 257
Gay-Bashing and Gender Betrayal 301
Interracial Dating and Marriage 317
The Americans with Disabilities Act 359
Diversity in the Power Elite 378

A Sociological Eye on the Media

The Blogging Culture 51
Research and the Media 60
Images of Violent Crime 167
Reproducing Class Stereotypes 197
Cultural Gatekeepers and the Construction
    of Femininity 266
Publicity Traps: Sex on TV Talk Shows 293
Idealizing Family Life 320

What Would They Say Now?

Classical Theorists Reflect on the War in Iraq 15
Classical Theorists on Hip-Hop! 49
Classical Theorists Reflect on School Shootings 158
Classical Theorists Reflect on Social Class and
    Sports 201
Classical Theorists Reflect on Sex and
    Popular Culture 298

MAPS

Mapping America's Diversity

MAP 2.1   Language as Evidence of Cultural
           Diversity 38

MAP 5.1   Population Density and Social
           Interaction 108

MAP 7.1   Suicide Rates by State 157

MAP 8.1   Median Income in the United States 188

MAP 8.2   Poverty in the United States 204

MAP 10.1   The New Immigration 247

MAP 13.1   Religiosity Across the Nation 339

MAP 14.1   Persons Not Covered by Health
           Insurance 361

MAP 15.1   Youth and the Obama Election 380

MAP 16.1   Total Number of Waste Sites
           by State 414

Viewing Society in Global Perspective

MAP 3.1   The Human Development Index 62

MAP 4.1   The World's Children 90

MAP 9.1   Rich and Poor 213

MAP 9.2   World Poverty 225

MAP 16.2   Technological Penetration 425

Preface

Studying sociology opens new ways of looking at the world. It is a perspective that is grounded in careful observation of social facts, as well as analyses of how society operates. For students and faculty alike, studying sociology can be exciting, interesting, and downright fun, even though it also deals with sobering social issues, such as inequality, racism, and crime, to name a few examples. In this book, we try to capture the excitement of the sociological perspective, while also introducing students to how sociologists do research and explain social interaction and social structure. We try to do so in a way that is engaging and accessible to undergraduate readers, while also preserving the integrity of sociological research and theory. Our experience in teaching introductory students shows us that students can appreciate the revelations of sociological research and theory if presented in a way that engages them and connects to their lives. We have kept this in mind throughout this revision and have focused on material that students can understand and apply to their own social worlds.

With each new edition, we think about the new students who will be reading the book. Although the basic perspective of sociology persists over time, the issues that new generations of learners face change. We find that keeping up with this generation is invigorating, and we hope our excitement and interest in teaching them sociology is apparent in the content and tone of the book. While we always retain a strong focus on core sociological concepts and ideas, we try to present our book in such a way that students will continue to tell us that they learn from it. As teachers and authors, one of our most rewarding moments is when a student tells us something about an insight they gained from *Sociology: The Essentials*.

We have added a particular new feature to this book with students in mind—the boxes titled **What Would They Say Now?** Each of these boxes (found in Chapters 1, 2, 7, 8, and 12) takes a very current topic of interest to students (the war in Iraq, hip-hop music, school shootings, sports and social class, sex and popular culture) and shows how the classical theorists within sociology would interpret these different subjects. We think students will find these boxes engaging and that they will "bring theory to life" in a way that shows the continuing importance of different sociological theories.

Each time we revise the book, we are also surprised by how many new topics for sociological analysis emerge and how much new research in sociology has developed, even in a short period of time. Events in the world also continue to shape the examples and explanations that we provide for our readers. Our book is known for its strong focus on sociological research and theory and for being deeply grounded in the inclusion of diversity. We have retained these focal points in the new edition.

An Approach That Reflects the Significance of Sociology

Diversity

With an increasingly diverse population in the United States, the study of diversity is central to our book. Unlike other introductory texts that add diversity to a preexisting approach to sociology, diversity has been part of the texture of this book since the first edition. And, with new research emerging on topics such as globalization, immigration, and changing race and ethnic patterns, diversity is just as important as ever. Indeed, the election of President Barack Obama since our last edition signals an increased interest, especially among young people, in forging a new society. Yet, social inequalities remain critical in how society is organized, how inequality shapes the experiences of different groups, and how diversity is shaping (and is shaped by) contemporary social changes. Thus, in this book, not only do we pay attention to diverse social groups, but we also maintain a core focus on the dynamics of inequality in society.

We define diversity to include the differences in experience created by social factors, including race, ethnicity, class, gender, age, religion, sexual orientation, and region of residence. We do not think of diversity as just the study of victims, although systems of disadvantage are clearly part of society. We stress the positive aspects of a diverse society, as well as its problems. We are pleased that our thorough integration of diversity has led reviewers to comment that our book provides the most comprehensive coverage of diversity of any book on the market.

Current Theory and Research

According to readers and reviewers of this book, our presentation of current theory and research is one of its strongest features. We use the most current research throughout the book to show students the value of a sociological education. The box feature **Doing Sociological Research** showcases different research studies. We include this so that students will understand not just the concepts and procedures of research methods but also how sociologists do their work and how the questions they ask are linked to the methods of inquiry that they use. Each box begins by identifying the question asked by the researcher. Then it describes the method of research, presents the findings and conclusions, and briefly discusses the implications. These boxes also show students the diverse ways that sociologists conduct research and thus feature the rich and varied content of the discipline.

We also help students understand how different theoretical frameworks in sociology interpret different topics by including tables in every chapter that concisely compare different theoretical viewpoints, showing how each illuminates certain questions and principles. We think this comparison of theories helps students understand an important point: Starting from a different set of assumptions can change how you interpret different social phenomena.

Critical Thinking and Debunking

We use the theme of *debunking* in the manner first developed by Peter Berger (1963) to look behind the facades of everyday life, challenging the ready-made assumptions that permeate commonsense thinking. Debunking is a way for students to develop their critical thinking, and we use the debunking theme to help students understand how society is constructed and sustained. This theme is highlighted in the **Debunking Society's Myths** feature found throughout every chapter.

In this edition, we also include a feature to help students see the relevance of sociology in their everyday lives. The box feature **See for Yourself** allows students to apply a sociological concept to an observation from their own lives, thus helping them develop their critical abilities and understand the importance of the sociological perspective.

Critical thinking is a term widely used but often vaguely defined. We use it to define the process by which students learn to apply sociological concepts to observable events in society. Throughout the book, we ask students to use sociological concepts to analyze and interpret the world they inhabit. This is reflected in the **Thinking Sociologically** feature that is also present in every chapter.

Because contemporary students are so strongly influenced by the media, we also encourage their critical thinking through the box feature **A Sociological Eye on the Media.** These boxes examine sociological research that challenges some of the ideas and images portrayed in the media. This not only improves students' critical thinking skills but also shows them how research can debunk these ideas and images.

Social Change

The sociological perspective helps students see society as characterized both by constant change and social stability. How society changes and the events—both dramatic and subtle—that influence change are analyzed throughout this book. For example, the **Seeing Society in Everyday Life** photo essay in Chapter 5 discusses the impact of new technologies on social interaction, a topic that will be of special interest to students very much affected by these changes. New material added throughout on the impact of the economic recession in 2009 also shows students how their lives—seemingly individual—are greatly influenced by social structures beyond their control.

Global Perspective

One of the main things we hope students learn in an introductory course is how broad-scale conditions influence things as seemingly individualized as everyday lives. Understanding this idea is a cornerstone of the sociological perspective and one of the main lessons learned in introductory courses. One way to see this is to help students understand how the increasingly global character of society affects day-to-day realities. Thus, we use a global perspective to examine how global changes are affecting all parts of life within the United States, as well as other parts of the world. This means more than including cross-cultural examples. It means, for example, examining phenomena such as migration and immigration or helping students understand that their own consumption habits are profoundly shaped by global interconnections. The availability of jobs, too, is another way students can learn about the impact of an international division of labor on work within the United States. Our global perspective is found in the research and examples cited throughout the book, as well as in various chapters that directly focus on the influence of globalization on particular topics, such as work, culture, and crime. The map feature **Viewing Society in Global Perspective** also brings a global perspective to the subject matter.

New to the SIXTH Edition

We have made various changes to the sixth edition to make it current and to reflect new developments in sociological research. Taken together, these changes should make the sixth edition easier for instructors to teach and even more accessible and interesting for students.

At the suggestion of reviewers, we have added a separate chapter on sociological research methods (Chapter Three, "Doing Sociological Research"). We believe this strengthens the research focus in the book and allows instructors to give more attention to the tools of sociological inquiry. Yet, we also know that students will not grasp the importance of sociological research until they have some grounding in the interesting topics of sociology. For this reason, we have placed the research chapter as the third chapter in the book, following Chapter 1, which introduces the core sociological perspective, and Chapter 2, which discusses culture. We think this ordering of chapters will tweak students' interest in the insights of our discipline prior to learning about the tools of sociological research. Of course, instructors who would prefer introducing students to research first can simply reverse the order in which they teach the chapters.

Because of adding this new chapter, we have then combined two of the final chapters from the fifth edition. The final chapter is now "Population, the Environment, and Social Change," a chapter that now combines population, urbanization, and the environment with social change and social movements. Population expansion and change, as well as challenging changes in the physical and social environments of our society, are issues of tremendous importance, and these changes have been the causes of a number of social movements and thus social and cultural change.

Sociology: The Essentials is organized into five major parts: "Introducing the Sociological Imagination" (Chapter 1); "Studying Society and Social Structure" (Chapters 2 through 7); "Social Inequalities" (Chapters 8 through 12); "Social Institutions" (Chapters 13 through 15); and "Social Change" (Chapter 16).

Part I, "Introducing the Sociological Imagination," introduces students to the unique perspective of sociology, differentiating it from other ways of studying society, particularly the individualistic framework students tend to assume. Within this section, **Chapter 1, "The Sociological Perspective,"** introduces students to the sociological perspective. The theme of debunking is introduced, as is the sociological imagination, as developed by C. Wright Mills. This chapter briefly reviews the development of sociology as a discipline, with a focus on the classical frameworks of sociological theory, as well as contemporary theories, such as feminist theory and postmodernism. We have added Iba B. Wells Barnett to the discussion of sociological theory, given her influence and insight on race and society in the early history of sociology. We have updated examples in this chapter, added First Lady Michelle Obama to the list of sociology majors, and included the new box, "What Would They Say Now: Classical Theorists on the War in Iraq."

In **Part II, "Studying Society and Social Structure,"** students learn some of the core concepts of sociology. It begins with the study of culture in

Chapter 2, "Culture and the Media," a new title that reflects the added attention to the media, given how important the mass media are in students' lives. There is a new section on how the media are organized, something students rarely think about even while they are influenced by media images. We reorganized the chapter somewhat to include discussion of ethnocentrism and cultural relativism in the section on diversity. We include one of the new "What Would They Say Now" boxes on the topic of hip-hop—important in youth culture. We have included numerous new examples and updated graphics throughout.

Chapter 3, "Doing Sociological Research," contains an updated discussion of the research process and the tools of sociological research—the survey, participant observation, controlled experiments, content analysis, historical research, and evaluation research. A new participant observation study—"A Cop in the Hood"—is included as a box. Included also are new discussions and recent participant observation studies, as well as recent studies in controlled experiments.

Chapter 4, "Socialization and the Life Course," contains new material on socialization theory and research, including agents of socialization such as the media, research on children's understanding of race, and a comparison of theories of socialization. The chapter also includes discussion of aging and the life course, as well as the conversion process and "brainwashing."

Chapter 5, "Social Interaction and Social Structure," emphasizes how changes in the macrostructure of society influence the microlevel of social interaction. We do this by focusing on technological changes that are now part of students' everyday lives and making the connection between changes at the societal level in the everyday realities of people's lives. The chapter also includes a discussion and comparison of newer theories about analyzing social interaction, such as dramaturgical analysis and game theory. The chapter includes new material on interaction in cyberspace, Internet dating, and an essay on how technology has affected social interaction.

In **Chapter 6, "Groups and Organizations,"** we study social groups and bureaucratic organizations, using sociology to understand the complex processes of group influence, organizational dynamics, and the bureaucratization of society. The chapter includes updates of research on juries, social networks, and the torture of Iraqi prisoners by Americans at the Abu Ghraib Prison in Iraq. New research on race, class and gender in bureaucratic organizations is included.

Chapter 7, "Deviance and Crime," includes a look at the sociological theories of deviance with attention to labeling theory; corporate crime and deviance; and the effects of race, class, and gender on arrest rates. The core material is illustrated with contemporary events, such as school rampage shootings, and updates on corporate and white-collar crime, including a discussion of Ponzi schemes such as that perpetrated by Bernard Madoff. Included in this chapter is recent research on what has become *mass racialized incarceration* of Blacks and Hispanics in U.S. prisons.

In **Part III, "Social Inequalities,"** each chapter explores a particular dimension of stratification in society. Beginning with the significance of class, **Chapter 8, "Social Class and Social Stratification,"** provides an overview of basic concepts central to the study of class and social stratification. An important revision to this chapter is material on the economic crisis revealed throughout 2009, including new material on housing foreclosures, unemployment, and current material on mobility—upward and downward! There is a new "What Would They Say Now" box on sports and social class, since sports are familiar to students and a good way to reveal different theoretical analyses of how class works. Updated data on income, wealth, and poverty is included throughout.

Chapter 9, "Global Stratification," follows, with a particular emphasis on understanding the significance of global stratification, the inequality that has developed among, as well as within, various nations. There is an expanded discussion of the impact of the war on poverty, given how important this topic is in current affairs. We also revised the section on globalization and social change to better analyze the impact of corporate globalization and cultural changes stemming from an increasingly global world.

Chapter 10, "Race and Ethnicity," is a comprehensive review of the significance of race and ethnicity in society, plus discussion of very recent studies of effects of Latino/a immigration, of skin color gradation in both Black and Hispanic communities, and the relevance of net worth as opposed to annual income in Black communities. Included are topics such as an update on forms and types of racism, the relevance of race in the devastation caused by Hurricane Katrina, a new discussion of the combined effects of race and class, a new look at ways to attain racial equality, and speculation about the effects of the Obama presidency upon race in the United States.

Chapter 11, "Gender," focuses on gender as a central concept in sociology closely linked to systems of stratification in society. The chapter links the social construction of gender to issues of homophobia, then is followed by the separate chapter on sexuality. This edition maintains a separate chapter on sexuality (Chapter 12), reflecting the increased attention in sociological research to sexuality as a component of social stratification. The new edition strengthens the discussion of violence against women, adds a discussion of the so-called mommy tax, and clarifies that sexism is both an individual belief and an institutional process.

Chapter 12, "Sexuality," treats sexuality as a social construction and a dimension of social stratification and inequality. The revised chapter includes new information on teen pregnancy, sexual attitudes, and more international material on abortion, sexual activity, and contraception. New research studying the influence of abstinence-only sex education on sexual behaviors is included, as is new material on the "hooking up" phenomenon and new research on the so-called pornification of mainstream culture. There is also expanded international coverage on rape as a consequence of war and new material on sex tourism and sex trafficking.

Part IV, "Social Institutions," includes three chapters, each focusing on basic institutions within society. **Chapter 13, "Families and Religion,"** maintains its inclusion of important topics in the study of families, such as interracial dating, debates about same-sex marriage, fatherhood, gender roles within families, and family violence. The section on religion includes much more research on Muslims in the United States and more material on megachurches, which are increasingly influential. There is also new material on racial–ethnic diversity and religious belief.

Chapter 14, "Education and Health Care," discusses the institutions of education and health care and gives an overview of education in the United States, including recent developments in education and inequality, plus a look at health and sickness seen globally. It includes theoretical perspectives on health care and the health care crisis in America, including difficulties in access to health care by race, gender, and social class. New to this chapter is recent research and data on inequality and standardized testing, on the "cognitive elite" and the nature-versus-nurture argument, the effects of schooling and gender on income and individual mobility, and the effects of tracking in schools. The chapter also contains an update of the health care crisis in America and the issue of universal health care.

Chapter 15, "Politics and the Economy," analyzes the state, power, and authority and bureaucratic government. It also contains a detailed discussion of theories of power in addition to coverage of the economy seen globally and characteristics of the labor force. This chapter has been substantially revised to reflect the presidential election of 2008 as well as to reflect the economic problems that have emerged since the last edition. There is new material on politics on the historic participation of people of color in the presidential election, as well as a discussion of youth voters. The section on the military includes updated material on women and minorities in the military, including family strains from separation, and new material on the privatization of the military. New research on sexual harassment and assault in the military is also included.

Part V, "Social Change," includes **Chapter 16, "Population, the Environment, and Social Change,"** which is an all-new chapter combining population, environmental issues, social movements, and social change. Outlined are the basic demographic processes, urbanization, and theories of population growth, as well as research on pollution and depletion of the physical environment. Included is a new photo essay on the very real effects of environmental pollution, global warming, and the role of new ("green") technologies. New data are included on racial–ethnic migration and the role of environmental racism. Also addressed in this new chapter are social change and social movements, with detailed coverage on theories of social change, the causes of social change, and the nature and types of social movements. New discussions include inequality, powerlessness and the individual, as well as a broader discussion of social movements.

Features and Pedagogical Aids

The special features of this book flow from its major themes: diversity, current theory and research, debunking and critical thinking, social change, and a global perspective. The features are also designed to help students develop critical thinking skills so that they can apply abstract concepts to observed experiences in their everyday life and learn how to interpret different theoretical paradigms and approaches to sociological research questions.

Critical Thinking Features

The feature **Thinking Sociologically** takes concepts from each chapter and asks students to think about these concepts in relationship to something they can easily observe in an exercise or class discussion. The feature **Debunking Society's Myths** takes certain common assumptions and shows students how the sociological perspective would inform such assumptions and beliefs.

See for Yourself

Each chapter includes a feature, **See for Yourself,** intended to provide students with the chance to apply sociological concepts and ideas to their own observations. This feature can also be used as the basis for writing exercises, helping students improve both their analytic skills and their writing skills.

Seeing Society in Everyday Life

These photo essays help students understand concepts more clearly. Essays include "Social Interaction in an Age of Technology" in Chapter 5; "Social Inequality" in Chapter 8; "Gender, Bodies, and Beauty" in Chapter 11; and "The Hurting Environment" in Chapter 16.

Unparalleled Integration of Web-Based Resources

Instructors will find that several technology-based teaching enhancements are integrated throughout the book, making this book the best conceived in using Internet tools for teaching and learning introductory sociology.

The *Sociology: The Essentials* companion website has a multitude of resources for students. Links to the student website are provided at the end of every chapter. On the companion website, students will find suggested readings, web links, a MicroCase® Online feature that teaches students how to research society, learning objectives, Internet exercises, quizzes, and flash cards—available at **www.cengage.com/sociology/andersen**.

In addition, **CengageNOW™,** a self-study assessment, web-based assessment, and study tool, is integrated in every chapter. (See the Supplements section later in this preface for more details on this robust study tool.)

An Extensive and Content-Rich Map Program

We use the map feature that appears throughout the book to help students visualize some of the ideas presented, as well as to learn more about regional and international diversity. One map theme is **Mapping America's Diversity** and the other is **Viewing Society in Global Perspective.** These maps have multiple uses for instructional value, beyond instructing students about world and national geography. The maps have been designed primarily to show the differentiation by county, state, and/or country on key social facts. For example, in Chapter 4 we show the dispersion of the population under five years of age, both nationally and worldwide. Students can use this information to ask questions about how the age distribution of the population might be related to immigration, poverty, or global stratification.

We have included a critical thinking component to the maps to integrate them more effectively with the chapter material. Thus, each map includes **critical thinking questions** that ask students to interpret the map data within the context of concepts and ideas from the chapter.

High-Interest Theme Boxes

In addition to new **What Would They Say Now?** boxes described earlier, we use three high-interest themes for the box features that embellish our focus on diversity and sociological research throughout the text. **Understanding Diversity** boxes further explore the approach to diversity taken throughout the book. In most cases, these box features provide personal narratives or other information designed to teach students about the experiences of different groups in society.

Because many are written as first-person narratives, they can invoke students' empathy toward groups other than those to which they belong—something we think is critical to teaching about diversity. We hope to show students the connections between race, class, and other social groups that they otherwise find difficult to grasp.

The box feature **Doing Sociological Research** is intended to show students the diversity of research questions that form the basis of sociological knowledge and, equally important, how the question a researcher asks influences the method used to investigate the question.

We see this as an important part of sociological research—that how one investigates a question is determined as much by the nature of the question as by allegiance to a particular research method. Some questions require a more qualitative approach, others, a more quantitative approach. In developing these box features, we ask, What is the central question sociologists are asking? How did they explore this question using sociological research methods? What did they find? and What are the implications of this research? We deliberately selected questions that show the full and diverse range of sociological theories and research methods, as well as the diversity of sociologists. Each box feature ends with **Questions to Consider** to encourage students to think further about the implications and applications of the research.

The feature **A Sociological Eye on the Media,** found in several chapters, examines some aspect of how the media influence public understanding of some of the subjects in this book. We think this is important because sociological research often debunks taken-for-granted points of view presented in the media, and we want students to be able to look at the media with a more critical eye. Because of the enormous influence of the media, we think this is increasingly important in educating students about sociology. In addition to the features just described, there is an entire set of learning aids within each chapter that promotes student mastery of the sociological concepts.

In-Text Learning Aids

Chapter Outlines. A concise chapter outline at the beginning of each chapter provides students with an overview of the major topics to be covered.

Key Terms. Key terms and major concepts appear in bold when first introduced in the chapter. A list of the key terms is found at the end of the chapter, which makes study more effective. Definitions for the key terms are found in the glossary.

Theory Tables. Each chapter includes a table that summarizes different theoretical perspectives by comparing and contrasting how these theories illuminate different aspects of different subjects.

Chapter Summary in Question-and-Answer Format. Questions and answers highlight the major points in each chapter and provide a quick review of major concepts and themes covered in the chapter.

A **Glossary** and complete **References** for the whole text are found at the back of the book.

Supplements

Sociology: The Essentials, Sixth Edition, is accompanied by a wide array of supplements prepared to create the best learning environment inside as well as outside the classroom for both the instructor and the student. All the continuing supplements for *Sociology: The Essentials,* Sixth Edition, have been thoroughly revised and updated. We invite you to take full advantage of the teaching and learning tools available to you.

For the Instructor

Instructor's Resource Manual. This supplement offers the instructor brief chapter outlines, student learning objectives, ASA recommendations, key terms and people, detailed chapter lecture outlines, lecture/discussion suggestions, student activities, chapter worksheets, video suggestions, video activities, and Internet exercises. Also included is a listing of resources offered with the text, as well as an assortment of useful appendices, which include activities that help integrate ASA task force recommendations in your course.

Test Bank. This test bank consists of a myriad of multiple-choice, true/false, short-answer and essay questions for each chapter, all with page references to the text. Each multiple-choice item has the question type (factual, applied, or conceptual) indicated, and all test questions will be mapped to a learning objective for the chapter. All questions are also labeled as new, modified, or pick-up so instructors know if the question is new to this edition of the test bank, modified but picked up from the previous edition of the test bank, or picked up straight from the previous edition of the test bank.

PowerLecture with JoinIn™ and ExamView® CD-ROM. This easy-to-use, one-stop digital library and presentation tool includes the following:

- Preassembled Microsoft® PowerPoint® lecture slides with graphics from the text, making it easy for you to assemble, edit, publish, and present custom lectures for your course.

- The PowerLecture CD-ROM includes video-based polling and quiz questions that can be used with the JoinIn™ on TurningPoint® personal response system.

- PowerLecture also features ExamView® testing software, which includes all the test items from the printed Test Bank in electronic format, enabling you to create customized tests of up to 250 items that can be delivered in print or online.

Classroom Activities for Introductory Sociology. Made up of contributions from introductory sociology instructors from around the country, this print supplement will be offered free to adopters of Andersen/Taylor's *Sociology: The Essentials,* Sixth Edition. The booklet features classroom activities, student projects, and lecture ideas to help instructors make topics fun and interesting for students. With general teaching tactics as well as topic-focused activities, it has never been easier to find a way to integrate new ideas into your classroom.

Introduction to Sociology Group Activities Workbook. This supplement is both a workbook for students and a repository of ideas for instructors.

The workbook contains both in- and out-of-class group activities on activity sheets that students can complete, tear out, and turn in to the instructor. Also included are teaching ideas for video clips to anchor group discussions, maps, case studies, group quizzes, ethical debates, group questions, group project topics, and ideas for outside readings.

Extension: Wadsworth's Sociology Reader Collection. Create your own customized reader for your sociology class, drawing from dozens of classic and contemporary articles found on the exclusive Cengage Wadsworth TextChoice database. Using the TextChoice website (**www.TextChoice.com**) you can preview articles, select your content, and add your own original material. TextChoice will then produce your materials as a printed supplementary reader for your class.

Classroom Presentation Tools for the Instructor

Videos. Adopters of *Sociology: The Essentials,* Sixth Edition, have several different video options available with the text. Please consult with your Cengage Learning sales representative to determine if you are a qualified adopter for a particular video.

Wadsworth's Lecture Launchers for Introductory Sociology. An exclusive offering jointly created by Cengage Wadsworth and DALLAS Tele-Learning, this video contains a collection of video highlights taken from the *Exploring Society: An Introduction to Sociology Telecourse* (formerly *The Sociological Imagination*). Each three- to six-minute video segment has been specially chosen to enhance and enliven class lectures and discussions of twenty key topics covered in the Introduction to Sociology course. Accompanying the video is a brief written description of each clip, along with suggested discussion questions to help effectively incorporate the material into the classroom. Available on VHS or DVD.

Sociology: Core Concepts Video. Another exclusive offering jointly created by Cengage Wadsworth and DALLAS TeleLearning, this video contains a collection of video highlights taken from the *Exploring Society: An Introduction to Sociology Telecourse* (formerly *The Sociological Imagination*). Each fifteen- to twenty-minute video segment will enhance student learning of the essential concepts in the introductory course and can be used to initiate class lectures, discussion, and review. The video covers topics such as the sociological imagination, stratification, race and ethnic relations, social change, and more. Available on VHS or DVD.

ABC® Video Series, Volumes I–IV. This series of videos, comprising footage from ABC broadcasts, is specially selected and arranged to accompany your introductory sociology course. The segments may be used in conjunction with Wadsworth's introductory sociology texts to help provide a real-world example to illustrate course concepts or to instigate discussion.

Wadsworth Sociology Video Library. Bring sociological concepts to life with videos from Wadsworth's Sociology Video Library, which includes thought-provoking offerings from *Films for the Humanities* as well as other excellent educational video sources. This extensive collection illustrates important sociological concepts covered in many sociology courses.

Supplements for the Student

Study Card. This handy card provides all the important sociological concepts highlighted in Andersen/Taylor, broken down by chapter. Providing a large amount of information at a glance, this study card is an invaluable tool for a quick review.

CengageNOW™. This online tool provides students with a customized study plan based on a diagnostic pretest that students take after reading each chapter. The study plan provides interactive exercises, learning modules, animations, video exercises, and other resources to help students master concepts that are central to sociology. After the study plan has been reviewed, students can take a posttest to monitor their progress in mastering the chapter concepts. Instructors may bundle this product for their students with each new copy of the text for free! If your instructor did not order the free access code card to be packaged with your text, or if you have a used copy of the text, you can still obtain an access code for a nominal fee. Just visit the Cengage Wadsworth E-Commerce site at **www.ichapters.com**, where easy-to-follow instructions help you purchase your access code.

Study Guide. This student study tool has been significantly revised from the previous edition in order to better suit your study needs. Each chapter contains a brief chapter outline, 10–15 questions to guide your reading, key terms and key people matching exercises with page references to the text, a chapter review where you can fill in the blanks to complete the summaries of each section, Internet and InfoTrac College Edition exercises, and practice tests consisting of multiple-choice, true/false, and short answer questions, as well as essay suggestions. Each chapter has an answer key in the back of the study guide, including answers to the key terms, key people, chapter review, and practice test sections.

Practice Tests. This collection of practice tests helps students adequately prepare for exams by presenting them with multiple-choice, true/false, short-answer, and essay questions that are similar in quality to the test bank questions. Multiple-choice and true/false answers are included, and page references are provided for all questions.

Researching Society with MicroCase® Online Booklet. This supplement contains MicroCase exercises for each chapter. Students can see the results of actual research by using the Wadsworth MicroCase Online feature available from the Wadsworth website. This feature allows students to look at some of the results from national surveys, census data, and other data sources. They can either explore this easy-to-use feature on their own or use the examples provided. Previously in the textbook, these MicroCase exercises are now in their own supplement.

Internet-Based Supplements

InfoTrac® College Edition. Available as an option with newly purchased texts, InfoTrac College Edition gives instructors and students four months of free access to an extensive online database of reliable, full-length articles (not just abstracts) from thousands of scholarly and popular publications, going back as much as twenty-two years. Among the journals available are *American Journal of Sociology, Social Forces, Social Research,* and *Sociology.*

WebTutor™ on WebCT® and Blackboard®. This web-based software for students and instructors takes a course beyond the classroom to an anywhere, anytime environment. Students gain access to a full array of study tools, including chapter outlines, chapter-specific quizzing material, interactive games and maps, and videos. With WebTutor Advantage, instructors can provide virtual office hours, post syllabi, track student progress with the quizzing material, and even customize the content to suit their needs.

Wadsworth's Sociology Home Page at www .cengage.com/sociology. Combine this text with the exciting range of web resources on Wadsworth's Sociology Home Page and you will have truly integrated technology into your learning system. Wadsworth's Sociology Home Page provides instructors and students with a wealth of information and resources, such as *Sociology in Action, Census 2000: A Student Guide for Sociology,* Research Online, Sociology Timeline, a Spanish glossary of key sociological terms and concepts, and more.

Companion Website for *Sociology: The Essentials.* The book's companion website includes chapter-specific resources for instructors and students. For instructors, the site offers a password-protected instructor's manual, PowerPoint presentation slides, and more. For students, there is a multitude of text-specific study aids, including the following:

- Tutorial practice quizzes that can be scored and emailed to the instructor
- Web links
- Glossary
- Flash cards
- MicroCase Online data exercises
- Crossword puzzles
- Virtual explorations
- And much more!

You can visit the companion site for *Sociology: The Essentials* at **www.cengage.com/sociology/ andersen.**

Acknowledgments

We relied on the comments of many reviewers to improve the book, and we thank them for the time they gave in developing very thoughtful commentaries on the different chapters. Thanks to Megan L. Duesterhaus, University of Central Florida; Catherine M. Felton, Central Piedmont Community College; Jennifer Holz, University of Akorn–Wayne College; Dana Hysock, Macon State College; Gina Luby, DePaul University, Joanna Maatta, The Pennsylvania University; Aurea Osgood, Winona State University; and KC Williams, Coastal Carolina Community College.

We would also like to extend our thanks to those who reviewed the first five editions of *Sociology: The Essentials.* Brenda N. Bauch, Jefferson College; E. M. Beck, University of Georgia; Alessandro Bonanno, Sam Houston State University; G. M. Britten, Lenoir Community College (North Carolina); James E. Coverdill, University of Georgia; Susan Crafts, Niagara Community College; Jean E. Daniels, California State University; Angela D. Danzi, Farmingdale State University, Northridge; Ione Y. DeOllos, Ball State University; Marlese Durr, Wright State University; Lois Easterday, Onondaga Community College; Cynthia K. Epperson, St. Louis Community College at Meramec; Lynda Ann Ewen, Marshall University; Grant Farr, Portland State University; Irene Fiala, Kent State University–Ashtabula; James Fillman, Bucks County Community College; Lorna Forster, Clinton Community College; Darlaine Gardetto, St. Louis Community College–Meramec; Patricia Gibbs, Foothill College; Bethany Gizzi, Monroe Community College; Jennifer Hamer, Wayne State University; Sara E. Hanna, Oakland Community College, Highland Lakes Campus; Shannon Houvouras, University of West Georgia; James R. Hunter, Indiana University Purdue University Indianapolis; Jon Iannitti, SUNY College of Agriculture and Technology at Morrisville; Carol A. Jenkins, Glendale Community College; Diane E. Johnson, Kutztown University of Pennsylvania; Katherine Johnson, Niagara Community College; Anna Karpathakis, Kingsborough Community College, CUNY; Alice Abel Kemp, University of New Orleans; Keith Kirkpatrick, Victoria College (Texas); Tim Kubal, California State University, Fresno; Elizabeth D. Leonard, Vanguard University; James Lindberg, Montgomery College, Rockville Campus, Maryland; Martha O. Loustaunau, New Mexico State University; Brad Lyman, Baltimore City Community College; Susan A. Mann, University of New Orleans; Brian L. Maze, Franklin University; Leland C. McCormick, Minnesota State University–Mankato; Christine Monnier, College of DuPage; Brian Moss, Oakland Community College–Waterford Campus (MI); Timothy Owens, Purdue University; Tara Perrello, Fordham University; Linda L. Petroff, Central Community College; David L. Phillips, Arkansas State University; Billie Joyce Pool, Homes Community College; Ralph Pyle, Michigan State University; David

Redburn, Furman University; Lesley Williams Reid, Georgia State University; Lisa Riley, Creighton University; Fernando Rivera, University of Central Florida; Michael C. Smith, Milwaukee Area Technical College; Beverly Stiles, Midwestern State University; Tracey Steele, Wright State University; Melvin Thomas, North Carolina State University; Richard Sullivan, Illinois State University; Judith Warner, Texas A&M International University; Stephanie Williams, Arizona State University; Sheryline A. Zebroski, St. Louis Community College, Florissant; Carl W. Zeigler, Elgin Community College; and Brenda S. Zicha, Mott Community College.

We also thank the following people, each of whom provided critical support in different, but important ways: Alison Bianchi, Cindy Gibson, Linda Keen, Nancy Quillen, and Judy Watson.

We are fortunate to be working with a publishing team with great enthusiasm for this project. We thank all of the people at Wadsworth who have worked with us on this and other projects, but especially we thank Chris Caldeira, Dan Moneypenny, and Cheri Palmer for their efforts on behalf of our book and the guidance and advice, not to mention the hard work, they have given to this project. We especially thank Dusty Friedman of TBC Project Management for her extraordinary attention to detail; we appreciate enormously her talent and perseverance. We thank Jane Loftus for her very careful copyediting of the manuscript and Kelly Franz for her work on the photo research. Finally, our special thanks also go to our spouses Richard Morris Rosenfeld and Patricia Epps Taylor for their ongoing support of this project.

About the Authors

Margaret L. Andersen, raised in Oakland, California; Rome, Georgia; and Boston, is Edward F. and Elizabeth Goodman Rosenberg Professor of Sociology at the University of Delaware. She received her Ph.D. from the University of Massachusetts, Amherst, and her B.A. from Georgia State University. She is the author of *Thinking about Women: Sociological Perspectives on Sex and Gender* (Allyn and Bacon) and the best-selling Wadsworth text, *Race, Class, and Gender: An Anthology* (with Patricia Hill Collins). She is also the author of *On Land and On Sea: A Century of Women in the Rosenfeld Collection* and *Living Art: The Life of Paul R. Jones, African American Art Collector.* She has recently served as Vice President of the American Sociological Association from which she has also received the prestigious Jessie Bernard Award. She has also been awarded the SWS Feminist Lecturer Award, given annually by SWS (Sociologists for Women in Society) to a social scientist whose work has contributed to improving the status of women in society. She currently serves as Chair of the National Advisory Board of the Center for Comparative Studies in Race and Ethnicity at Stanford University. She has served as the Interim Dean of the College of Arts and Science and Vice Provost for Academic Affairs at the University of Delaware, where she has also won the University's Excellence in Teaching Award. She lives on the Elk River in Maryland with her husband Richard Rosenfeld.

Howard F. Taylor was raised in Cleveland, Ohio. He graduated Phi Beta Kappa from Hiram College and has a Ph.D. in sociology from Yale University. He has taught at the Illinois Institute of Technology, Syracuse University, and Princeton University, where he is presently professor of sociology and former Director of the African American Studies Center. He has published over fifty articles in sociology, education, social psychology, and race relations. His books include *The IQ Game* (Rutgers University Press), a critique of hereditarian accounts of intelligence; *Balance in Small Groups* (Van Nostrand Reinhold), translated into Japanese; and the forthcoming *Race and Class and the Bell Curve in America.* He has appeared widely before college, radio, and TV audiences, including ABC's *Nightline.* He is past President of the Eastern Sociological Society, and a member of the American Sociological Association and the Sociological Research Association, an honorary society for distinguished research. He is a winner of the DuBois-Johnson-Frazier Award, given by the American Sociological Association for distinguished research in race and ethnic relations, and the President's Award for Distinguished Teaching at Princeton University. He lives in Pennington, New Jersey, with his wife, a corporate lawyer.

1

The Sociological Perspective

What Is Sociology?

The Sociological Perspective

The Significance of Diversity

The Development of Sociological Theory

Theoretical Frameworks in Sociology

Chapter Summary

[

Imagine that you had been switched with another infant at birth. How different would your life be? What if your accidental family was very poor ... or very rich? How might this have affected the schools you attended, the health care you received, the possibilities for your future career? If you had been raised in a different religion, would this have affected your beliefs, values, and attitudes? Taking a greater leap, what if you had been born another sex or a different race? What would you be like now?

We are talking about changing the basic facts of your life—your family, social class, education, religion, sex, and race. Each has major consequences for who you are and how you will fare in life. These factors play a major part in writing your life script. Social location (meaning a person's place in society) establishes the limits and possibilities of a life.

Consider this:

- The pay gap between women and men, which had been declining since the 1980s, has recently *increased* between college-educated women and men (Blau and Kahn 2004; Leonhardt 2006).

continued

Jeff Greenberg/The Image Works

CHAPTER ONE
Chapter one

- Black Americans who kill Whites are much more likely to face the death penalty than Blacks who kill other Black people, and given the death penalty, Black Americans are more likely than Whites to actually be executed (Paternoster et al. 2003).

These conclusions, drawn from current sociological research, describe some consequences of particular social locations in society. Although we may take our place in society for granted, our social location has a profound effect on our chances in life. The power of sociology is that it teaches us to see how society influences our lives and the lives of others, and it helps us explain the consequences of different social arrangements.

Sociology is the study of human behavior, including the significance of diversity.

Sociology also has the power to help us understand the influence of major changes on people. Currently, rapidly developing technologies, increasing globalization, a more diverse population in the United States, and changes in women's roles are affecting everyone in society, although in different ways. How are these changes affecting your life? Perhaps you rely on a cell phone to keep in touch with friends, or maybe your community is witnessing an increase in immigrants from other places, or maybe you see women and men trying hard to balance the needs of both work and family life. All of these are issues that guide sociological questions. Sociology explains some of the causes and consequences of these changes.

Although society is always changing, it is also remarkably stable. People generally follow established patterns of human behavior, and you can generally anticipate how people will behave in certain situations. You can even anticipate how different social conditions will affect different groups of people in society. This is what sociologists find so interesting: *Society is marked by both change and stability.* Societies continually evolve, creating the need for people to adapt to change while still following generally established patterns of behavior.

What Is Sociology?

Sociology is the study of human behavior in society. Sociologists are interested in the study of people and have learned a fundamental lesson: *All human*

behavior occurs in a societal context. That context—the institutions and culture that surround us—shapes what people do and think. In this book, we will examine the dimensions of society and analyze the elements of social context that influence human behavior.

Sociology is a scientific way of thinking about society and its influence on human groups. Observation, reasoning, and logical analysis are the tools of the sociologist, coupled with knowledge of the large body of theoretical and analytical work done by previous sociologists and others. Sociology is inspired by the fascination people have for the thoughts and actions of other people, but it goes far beyond casual observations. It attempts to build on observations that are objective and accurate to create analyses that are reliable and that can be validated by others.

Every day, the media in their various forms (television, film, video, digital, and print) bombard us with social commentary. Whether it is Oprah Winfrey or Ellen DeGeneres, media commentators provide endless opinion about the various and sometimes bizarre forms of behavior in our society. Sociology is different. Sociologists may appear in the media, and they often study the same subjects that the media examine, such as domestic violence or crime, but sociologists use specific research techniques and well-tested theories to explain social issues. Indeed, sociology can provide the tools for testing whether the things we hear about society are actually true. Much of what we hear in the media and elsewhere about society, although delivered with perfect earnestness, is misstated and sometimes completely wrong, as you will see in some of the "Debunking Society's Myths" examples featured throughout this book.

Question:

What do the following people have in common?

Michelle Obama

Robin Williams (actor, comedian)

Ronald Reagan

Rev. Martin Luther King, Jr.

Debra Winger (actress)

Regis Philbin (TV personality)

Rev. Jesse Jackson

Dr. Ruth Westheimer (the "sex doctor")

Saul Bellow (novelist; Nobel Prize recipient)

Joe Theismann (former football player)

Congresswoman Maxine Waters (from California)

Senator Barbara Mikulski (from Maryland)

Answer:

They were all sociology majors!

Source: Compiled by Peter Dreier, Occidental College. Full list available on the home page of the American Sociological Association **(www.asanet.org/page.ww?section=Students&name= Famous+Sociology+Majors).**

The subject matter of sociology is everywhere. This is why people sometimes wrongly believe that sociology just explains the obvious. But sociologists bring a unique perspective to understanding social behavior and social change. Even though sociologists often do research on familiar topics, such as youth cultures or relations between women and men—they do so using particular research tools and specific frames of analysis (known as sociological theory). Psychologists, anthropologists, political scientists, economists, social workers, and others also study social behavior, though each has a different perspective or "angle" on people in society. Together, these fields of study (also called disciplines) make up what are called the social sciences.

The Sociological Perspective

Think back to the opening of this chapter where you were asked to imagine yourself growing up under completely different circumstances. Our goal in that passage was to make you feel the stirring of the *sociological perspective*—the ability to see the societal patterns that influence individual and group life. The beginnings of the sociological perspective can be as simple as the pleasures of watching people or wondering how society influences people's lives. Indeed, many students begin their study of sociology because they are "interested in people." Sociologists convert this curiosity into the systematic study of how society influences different people's experiences within it.

C. Wright Mills (1916–1962) was one of the first to write about the sociological perspective in his classic book, *The Sociological Imagination* (1959). He wrote that the task of sociology was to understand the relationship between individuals and the society in which they live. He defined the **sociological imagination** as the ability to see the societal patterns that influence the individual as well as groups of individuals. Sociology should be used, Mills argued, to reveal how the context of society shapes our lives. He thought that to understand the experience of a given person or group of people, one had to have knowledge of the social and historical context in which people lived.

Think, for example, about the time and effort that many people put into their appearance. You might ordinarily think of this merely as personal grooming or an individual attempt to "look good," but there are significant social origins of this behavior. When you stand in front of the mirror, you are probably not thinking about how society is present in your reflection. But as you look in the mirror, you are seeing how others see you and are very likely adjusting your appearance with that in mind, even if not consciously. Therefore, this seemingly individual behavior is actually a very social act. If you are trying to achieve a particular look, you are likely doing so because of social forces that establish particular ideals, which are produced by industries that profit enormously from the products and services that people buy, even when they do so believing this is an individual choice.

Some industries suggest that you should be thinner or curvier, your pants should be baggy or straight, your breasts should be minimized or maximized—either way you need more products. Maybe you should have a complete makeover! Many people go to great lengths to try to achieve a constantly changing beauty ideal, one that is probably not even attainable (such as flawless skin, hair never out of place, perfectly proportioned body parts). Sometimes trying to meet these ideals can even be hazardous to your physical and mental health.

The point is that the alleged standards of beauty are produced by social factors that extend far beyond an individual's concerns with personal appearance. Beauty ideals, like other socially established beliefs and practices, are produced in particular social and historical contexts. People may come up with all kinds of personal strategies for achieving these ideals: they may buy more products, try to lose more weight, get a Botox treatment, or even become extremely depressed and anxious if they think their efforts are failing. These personal behaviors may seem to be only individual issues, but they have basic social causes. That is, the origins of these behaviors exist beyond personal lives. The sociological imagination permits us to see that something as seemingly personal as how you look arises from a social context, not just individual behavior.

Sociologists are certainly concerned about individuals, but they are attuned to the social and historical context that shapes the experiences of individuals and groups. A distinction made by the sociological imagination is that between *troubles* and *issues*. **Troubles** are privately felt problems that spring from events or feelings in a person's life. **Issues** affect large numbers of people and have their origins in the institutional arrangements and history of a society (Mills 1959). This distinction is the crux of the difference between individual experience and **social structure,** defined as the organized pattern of social relationships and social institutions that together constitute society. Issues shape the context within which troubles arise. Sociologists employ the sociological perspective to understand how issues are shaped by social structures.

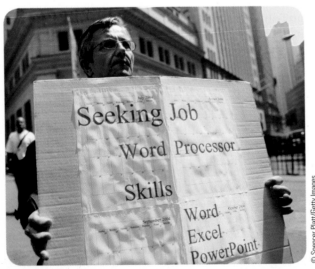

Personal troubles are felt by individuals who are experiencing problems; social issues arise when large numbers of people experience problems that are rooted in the social structure of society.

SEE *FOR YOURSELF*
TROUBLES AND ISSUES

Personal troubles are everywhere around us: alcohol abuse or worries about money or even being upset about how you look. At the individual level, these things can be deeply troubling, and people sometimes need personal help to deal with them. But most personal troubles, as C. Wright Mills would say, also have their origins in societal arrangements. Take the example of alcohol abuse or money problems—or, perhaps, another personal trouble with which you are familiar.

What are some of the things about society—not just individuals—that might influence this personal trouble? Make a list of things in society that might affect this personal trouble. How does this help you understand the distinction that Mills makes between *personal troubles* and *social issues*. Then explain what Mills means by saying that personal biographies are linked to the structure of society.

Mills used the example of unemployment to explain the meaning of troubles versus issues. When a person becomes unemployed, he or she has a personal trouble. In addition to financial problems, the person may feel a loss of identity, may become depressed, may lose touch with former work associates, or may have to uproot a family and move. The problem of unemployment, however, is deeper than the experience of one person. Unemployment is rooted in the structure of society; this is what interests sociologists. What societal forces cause unemployment? Who is most likely to become unemployed at different times? How does unemployment affect an entire community (for instance, when a large plant shuts down) or an entire nation (such as during the economic downturn of recent years)? Sociologists know that unemployment causes personal troubles, but understanding unemployment is more than understanding one person's experience. It requires understanding the social structural conditions that influence people's lives.

The specific task of sociology, according to Mills, is to comprehend the whole of human society—its personal and public dimensions, historical and contemporary—and its influence on the lives of human beings. Mills had an important point: People often feel that things are beyond their control, meaning that they are being shaped by social forces larger than their own individual life. Social forces influence our lives in profound ways, even though we may not always know how. Consider this. Most likely you remember what you were doing on September 11, 2001, when you first heard that terrorists had flown planes into the World Trade Center in New York City. Obviously, this affected people's personal lives, but its impact and its causes go beyond the personal troubles it produced. The sociological perspective explains many dimensions of this event and its aftermath, including, as one example, the significance of cultural symbols that have emerged in the aftermath of 9/11. Think of the T-shirts that many people now wear honoring New York's firefighters. Of course, the social forces that influence people's lives are not always that drastic and include the ordinary events of everyday life.

Sociology is an **empirical** discipline. This means that sociological conclusions are based on careful and systematic observations, as we will see in Chapter 3 on sociological research methods. In this way, sociology is very different from ordinary common sense. For empirical observations to be useful to other observers, they must be gathered and recorded rigorously. Sociologists are also obliged to reexamine their assumptions and conclusions constantly. Although the specific methods that sociologists use to examine different problems vary, as we will see, the empirical basis of sociology is what distinguishes it from mere opinion or other forms of social commentary.

Discovering Unsettling Facts

In studying sociology, it is crucial to examine the most controversial topics and to do so with an open mind, even when you see the most disquieting facts. The facts we learn through sociological research can be "inconvenient" because the data can challenge familiar ways of thinking. Consider the following:

- Despite the widespread idea promoted in the media that well-educated women are opting out of professional careers to stay home and raise children, the proportion of college-educated White women who stay home with children has actually declined; those who do opt out do so more typically because of their frustrations with workplaces (Stone 2007).

- Same-sex couples are more likely to be interracial than are heterosexual couples (Rosenfeld and Kim 2005).

- The number of women prisoners has increased at almost twice the rate of increase for men; two-thirds of women and half of men in prison are parents (Sabol and Couture 2008; Glaze and Maruschak 2008).

These facts provide unsettling evidence of persistent problems in the United States, *problems that are embedded in society, not just in individual behavior.* Sociologists try to reveal the social factors that shape society and determine the chances of success for different groups. Some never get the chance to go to college; others are unlikely to ever go to jail. These divisions persist because of people's placement within society.

Sociologists study not just the disquieting side of society. Sociologists may study questions that affect everyday life, such as how young boys and men are affected by changing gender roles (Kimmel 2008), how children of immigrants fare (Park 2005), or how racism has changed in recent years (Zuberi 2001; Bonilla-Silva 2001). There are also many intriguing studies of unusual groups, such as cyberspace users (Kendall 2002), strip club dancers (Barton 2006), or heavily tattooed people, known as collectors (Irwin 2001). The subject matter of sociology is vast. Some research illuminates odd corners of society; other studies address urgent problems of society that may affect the lives of millions.

Debunking in Sociology

The power of sociological thinking is that it helps us see everyday life in new ways. Sociologists question actions and ideas that are usually taken for granted. Peter Berger (1963) calls this process debunking. **Debunking** refers to looking behind the facades of

© William Thomas Cain/Getty Images

© Lindsay Hebberd/Corbis

Cultural practices that seem bizarre to outsiders may be taken for granted or defined as appropriate by insiders.

everyday life—what Berger called the "unmasking tendency" of sociology (1963: 38). In other words, sociologists look at the behind-the-scenes patterns and processes that shape the behavior they observe in the social world.

Take schooling, for example: We can see how the sociological perspective debunks common assumptions about education. Most people think that education is primarily a way to learn and get ahead. Although this is true, a sociological perspective on education reveals something more. Sociologists have concluded that more than learning takes place in schools; other social processes are at work. Social cliques are formed where some students are "insiders" and others are excluded "outsiders." Young schoolchildren acquire not just formal knowledge but also the expectations of society and people's place within it. Race and class conflicts are often played out in schools (Lewis 2003). Relative to boys, girls are often shortchanged by the school system—receiving less attention and encouragement, less interaction with teachers, less instruction in the sciences, and many other deficits disproportionately forced upon them (American Association of University Women 1998; Sadker and Sadker 1994). Poor children seldom have the same resources in schools as middle-class or elite children, and they are often assumed to be incapable of doing schoolwork and are treated accordingly. The somber reality is that schools may actually stifle the opportunities of some children rather than launch all children toward success.

Debunking is sometimes easier to do when looking at a culture or society different from one's own. Consider how behaviors that are unquestioned in one society may seem positively bizarre to an outsider. For a thousand years in China, it was usual for the elite classes to bind the feet of young girls to keep the feet from growing bigger—a practice allegedly derived from a mistress of the emperor. Bound feet were a sign of delicacy and vulnerability. A woman with large feet (defined as more than 4 inches long!) was thought to bring shame to her husband's household. The practice was supported by the belief that men were highly aroused by small feet, even though men never actually saw the naked foot. If they had, they might have been repulsed, because a woman's actual foot was U-shaped and often rotten and covered with dead skin (Blake 1994). Outside the social, cultural, and historical context in which it was practiced, foot binding seems bizarre, even dangerous. Feminists have pointed out that Chinese women were crippled by this practice, making them unable to move about freely and more dependent on men (Chang 1991).

> doing sociological research

Debunking the Myths of Black Teenage Motherhood

Research Question: Sociologist Elaine Bell Kaplan knew that there was a stereotypical view of Black teen mothers, that they had grown up in fatherless households where their mothers had no moral values and no control over their children. The myth of Black teenage motherhood also depicts teen mothers as unable to control their sexuality, as having children to collect welfare checks, and as having families who condone their behavior. Is this true?

Research Method: Kaplan did extensive research in two communities in the San Francisco Bay Area—East Oakland and Richmond—both communities with a large African American population and typical of many inner-city, poor neighborhoods. Once thriving Black communities, East Oakland and Richmond are now characterized by high rates of unemployment, poverty, inadequate schools, crime, drug-related violence, and high numbers of single-parent households. Having grown up herself in Harlem, Kaplan knew that communities like those she studied have not always had these problems, nor have they condoned teen pregnancy. She spent several months in these communities, working as a volunteer in a community teen center that provided educational programs, day care, and counseling to teen parents, and "hanging out" with a core group of teen mothers. She did extensive interviews with thirty-two teen mothers, supplementing them when she could with interviews with their mothers and, sometimes, the fathers of their children.

Research Results: Kaplan found that teen mothers adopt strategies for survival that help them cope with their environment, even though these same strategies do not help them overcome the problems they face. Unlike what the popular stereotype suggests, she did not find that the Black community condones teen pregnancy; quite the contrary, the teens felt embarrassed and stigmatized by being pregnant and experienced tension and conflict with their mothers, who saw their pregnancy as disrupting the hopes they had for their daughters' success. When the women

This is an example of outsiders debunking a practice that was taken for granted by those within the culture. Debunking can also call into question practices in one's own culture that may normally go unexamined. Strange as the practice of Chinese foot binding may seem to you, how might someone from another culture view wearing shoes that make it difficult to walk? Or piercing one's tongue or eyebrow? These practices of contemporary U.S. culture are taken for granted by many, just as was Chinese foot binding. Until these cultural processes are debunked, seen as if for the first time, they might seem normal.

Debunking Society's Myths

Myth: Violence is at an all-time high.

Sociological research: Although rates of violence are high in the United States, at least twice before (in 1930 and 1980) the murder rate was at a comparable level. It is also true, however, that rates of violence in the United States are higher than in any other industrialized nation (Levine and Rosich 1996; Best 1999).

Establishing Critical Distance

Debunking requires critical distance—that is, being able to detach from the situation at hand and view things with a critical mind. The role of critical distance in developing a sociological imagination is well explained by the early sociologist **Georg Simmel** (1858–1918). Simmel was especially interested in the role of *strangers* in social groups. Strangers have a position both inside and outside social groups; they are part of a group without necessarily sharing the group's assumptions and points of view. Because of this, the stranger can sometimes see the social structure of a group more readily than people who are thoroughly imbued with the group's worldview. Simmel suggests that the sociological perspective requires a combination of nearness and distance. One must have enough critical distance to avoid being taken in by the group's definition of the situation, but be near enough to understand the group's experience.

Sociologists are not typically strangers to the society they study. You can acquire critical distance through a willingness to question the forces that shape social behavior. Often, sociologists become interested in things because of their own experiences. The biographies of sociologists are rich with examples of how their personal lives informed the questions

had to go on welfare, they also felt embarrassed, often developing strategies to hide the fact that they were dependent on welfare to make ends meet. These conclusions run directly counter to the public image that such women do not value success and live in a culture that promotes welfare dependency.

Conclusions and Implications: Instead of simply stereotyping these teens as young and tough, Kaplan sees them as struggling to develop their own gender and sexual identity. Like other teens, they are highly vulnerable, searching for love and aspiring to create a meaningful and positive identity for themselves. But failed by the educational system and locked out of the job market, the young women's struggle to develop an identity is compounded by the disruptive social and economic conditions in which they live.

Kaplan's research is a fine example of how sociologists debunk some of the commonly shared myths that surround contemporary issues. Carefully placing her analysis in the context of the social structural changes that affect these young women's lives, Kaplan provides an excellent example of how sociological research can shed new light on some of our most pressing social problems.

Questions to Consider

1. Suppose that Kaplan had studied middle-class teen mothers. What similarities and differences would you predict in the experiences of middle-class and poor teen mothers? Does race matter? In what ways does your answer *debunk* myths about teen pregnancy?

2. Make a list of the challenges you would face were you to be a teen parent. Having done so, indicate those that would be considered personal troubles and those that are social issues. How are the two related?

Source: Kaplan, Elaine Bell. 1996. *Not Our Kind of Girl: Unraveling the Myths of Black Teenage Motherhood.* Berkeley, CA: University of California Press.

understanding diversity

Become a Sociologist

Individual biographies often have a great influence on the subjects sociologists choose to study. The authors of this book are no exception. Margaret Andersen, a White woman, now studies the sociology of race and women's studies. Howard Taylor, an African American man, studies race, social psychology, and especially race and intelligence testing. Here, each of them writes about the influence of their early experiences on becoming a sociologist.

Margaret Andersen

As I was growing up in the 1950s and 1960s, my family moved from California to Georgia, then to Massachusetts, and then back to Georgia. Moving as we did from urban to small-town environments and in and out of regions of the country that were very different in their racial character, I probably could not help becoming fascinated by the sociology of race. Oakland, California, where I was born, was highly diverse; my neighborhood was mostly White and Asian American. When I moved to a small town in Georgia in the 1950s, I was ten years old, but I was shocked

Courtesy of Richard Rosenfeld

by the racial norms I encountered. I had always loved riding in the back of the bus—our major mode of transportation in Oakland—and could not understand why this was no longer allowed. Labeled by my peers as an outsider because I was not southern, I painfully learned what it meant to feel excluded just because of "where you are from."

When I moved again to suburban Boston in the 1960s, I was defined by Bostonians as a southerner and ridiculed. Nicknamed "Dixie," I was teased for how I talked. Unlike in the South, where despite strict racial segregation Black people were part of White people's daily lives, Black people in Boston were even less visible. In my high school of 2500 or so students, Black students were rare. To me, the school seemed not much different from the strictly segregated schools I had attended in Georgia. My family soon returned to Georgia, where I was an outsider again; when I later returned to Massachusetts for graduate school in the 1970s, I worried about how a southerner would be accepted in this "Yankee" environment. Because I had acquired a southern accent, I think many of my teachers stereotyped me and thought I was not as smart as the students from other places.

These early lessons, which I may have been unaware of at the time, must have kindled my interest in the sociology of race relations. As I explored sociology, I wondered how the concepts and theories of race relations applied to women's lives. So much of

they asked. Among sociologists are former ministers and nuns now studying the sociology of religion, women who have encountered sexism who now study the significance of gender in society, rock-and-roll fans studying music in popular culture, and sons and daughters of immigrants now analyzing race and ethnic relations (see the box "Understanding Diversity: Becoming a Sociologist").

The Significance of Diversity

The analysis of diversity is one central theme of sociology. Differences among groups, especially differences in the treatment of groups, are significant in any

society, but they are particularly compelling in a society as diverse as that in the United States.

Defining Diversity

Today, the United States includes people from all nations and races. In 1900, one in eight Americans was not White; today, racial and ethnic minority groups, including African Americans, Latinos, American Indians, Native Hawaiians, Asian Americans, and people of more than one race represent 35 percent of Americans, and that proportion is growing (U.S. Census Bureau 2009; see Figure 1.1; see also Table 1.1). These broad categories themselves are internally diverse, including, for example, those with long-term roots in the United States, as well as Cuban Americans, Salvadorans, Cape Verdeans, Filipinos, and many others.

what I had experienced growing up as a woman in this society was completely unexamined in what I studied in school. As the women's movement developed in the 1970s, I found sociology to be the framework that helped me understand the significance of gender and race in people's lives. To this day, I write and teach about race and gender, using sociology to help students understand their significance in society.

Howard Taylor

Courtesy of Howard Taylor

I grew up in Cleveland, Ohio, the son of African American professional parents. My mother, Murtis Taylor, was a social worker and the founder and then president of a social work agency called the Murtis H. Taylor Multi-Service Center in Cleveland, Ohio. She is well known for her contributions to the city of Cleveland and was an early "superwoman," working days and nights, cooking, caring for her two sons, and being active in many professional and civic activities. I think this gave me an early appreciation for the roles of women and the place of gender in society, although I surely would not have articulated it as such at the time. My father was a businessman in a then all-Black life insurance company. He was also a "closet scientist," always doing experiments and talking about scientific studies. He encouraged my brother and me to engage in science, so we were always experimenting with scientific studies in the basement of our house. In the summers, I worked for my mother in the social service agency where she worked, as a camp counselor, and in other jobs. Early on, I contemplated becoming a social worker, but I was also excited by science. As a young child, I acquired my father's love of science and my mother's interest in society. In college, the one field that would gratify both sides of me, science and social work, was sociology. I wanted to study human interaction, but I also wanted to be a scientist, so the appeal of sociology was clear.

At the same time, growing up African American meant that I faced the consequences of race every day. It was always there, and like other young African American children, I spent much of my childhood confronting racism and prejudice. When I discovered sociology, in addition to bridging the scientific and humanistic parts of my interests, I found a field that provided a framework for studying race and ethnic relations. The merging of two ways of thinking, coupled with the analysis of race that sociology has long provided, made sociology fascinating to me.

Today, my research on race, class, gender, and intelligence testing seems rooted in these early experiences. I do quantitative research in sociology and see sociology as a science that reveals the workings of race, class, and gender in society.

Perhaps the most basic lesson of sociology is that people are shaped by the social context around them. In the United States, with so much cultural diversity, people will share some experiences, but not all. Experiences not held in common can include some of the most important influences on social development, such as language, religion, and the traditions of family and community. Understanding diversity means recognizing this diversity and making it central to sociological analyses.

In this book, we use the term *diversity* to refer to the variety of group experiences that result from the social structure of society. **Diversity** is a broad concept that includes studying group differences in society's opportunities, the shaping of social institutions by different social factors, the formation of group and individual identity, and the process of social change.

Diversity includes the study of different cultural orientations, although diversity is not exclusively about culture.

Understanding diversity is crucial to understanding society because fundamental patterns of social change and social structure are increasingly patterned by diverse group experiences. There are numerous sources of diversity, including race, class, gender, and others as well. Age, nationality, sexual orientation, and region of residence, among other factors, also differentiate the experience of diverse groups in the United States. And as the world is increasingly interconnected through global communication and a global economy, the study of diversity also encompasses a global perspective—that is, an understanding of the international connections existing across national borders and the impact of such connections on life throughout the world.

Robert Brenner/PhotoEdit

Understanding diversity is important in a society comprising so many different groups, each with unique, but interconnected, experiences.

*thinking*sociologically

What are some of the sources of *diversity* on your campus? What questions might a sociologist pose about the significance of diversity in this environment?

Society in Global Perspective

No society can be understood apart from the global context that now influences the development of all societies. The social and economic system of any one society is increasingly intertwined with those of other nations. Coupled with the increasing ease of travel and telecom-

munication, this means that a global perspective is necessary to understand change both in the United States and in other parts of the world.

To understand globalization, you must look beyond the boundaries of your own society to see how patterns in any given society are increasingly being shaped by the connections between societies. Comparing and contrasting societies across different cultures is valuable. It helps you see patterns in your own society that you might otherwise take for granted, and it enriches your appreciation of the diverse patterns of culture that mark human society and human history. A global perspective, however, goes beyond just comparing different cultures; it also helps you see how events in one society or community may be linked to events occurring on the other side of the globe.

For instance, return to the example of unemployment that C. Wright Mills used to distinguish troubles and issues. One man may lose his job in Peoria, Illinois, and a woman in Los Angeles may employ a Latina domestic worker to take care of her child while

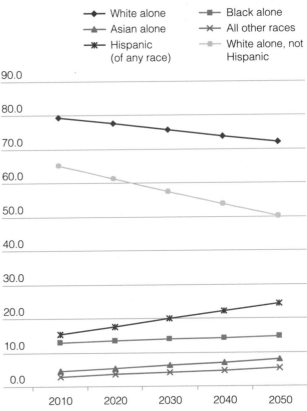

FIGURE 1.1 *Percent of U.S. Population by Race and Ethnicity*

It is a social fact that by 2050 White, non-Hispanics will make up half the U.S. population. Using the information above, what will the population percentages for different racial–ethnic groups be when you are thirty? Fifty? How do you think this will affect your life? How does this illustrate how social facts influence the lives of individual people and diverse groups in society?

Source: U.S. Census Bureau. 2004. "U.S. Interim Projections by Age, Sex, Race, and Hispanic Origin." **www.census.gov/pc/www/usinterimproj/**

Warren Miller/The New Yorker Collection/cartoonbank.com

"Actually, Lou, I think it was more than just my being in the right place at the right time. I think it was my being the right race, the right religion, the right sex, the right socioeconomic group, having the right accent, the right clothes, going to the right schools . . ."

table1.1 Minorities in the U.S. Population					
Percent of Total Population	2010	2020	2030	2040	2050
White alone[1]	79.3%	77.6%	75.8%	73.9%	72.1%
Black alone	13.1	13.5	13.9	14.3	14.6
Asian alone	4.6	5.4	6.2	7.1	8.0
All other races[2]	3.0	3.5	4.1	4.7	5.3
Hispanic (may be of any race)[3]	15.5	17.5	20.1	22.3	24.4
White alone, not Hispanic	65.1	61.3	57.5	53.7	50.1

Note: The U.S. census counts race and Hispanic ethnicity separately. Thus Hispanics may fall into any of the race categories. As a result, only the percentages for White alone, Black alone, Asian alone, and all other races add to 100%.

[1]The U.S. Census Bureau uses this designation to refer to those who indicated on the census form that they were only of one race.
[2]Includes American Indian and Alaska Natives alone, Native Hawaiian and other Pacific Islanders alone, and those of two or more races.
[3]See Note above.

Source: U.S. Census Bureau. 2004. "U.S. Interim Projections by Age, Sex, Race, and Hispanic Origin." **www. census.gov/pc/www/usinterimproj/**

she pursues a career. On the one hand, these are individual experiences for all three people, but they are linked in a pattern of globalization that shapes the lives of all three. The Latina domestic may have a family whom she has left in a different nation so that she can afford to support them. The corporation for which the Los Angeles woman works may have invested in a new plant overseas that employs cheap labor, resulting in the unemployment of the man in Peoria. The man in Peoria may have seen immigrant workers moving into his community, and one of his children may have made a friend at school who speaks a language other than English.

Such processes are increasingly shaping many of the subjects examined in this book—work, family, education, politics, just to name a few. Without a global perspective, you would not be able to fully understand the experience of any one of the people just mentioned much less how society is being shaped by these processes of change and global context. Throughout this book, we will use a global

Globalization brings diverse cultures together, but it is also a process by which Western markets have penetrated much of the world.

AP Images/Eugene Hoshiko

perspective to understand some of the developments shaping contemporary life in the United States.

The Development of Sociological Theory

Like the subjects it studies, sociology is itself a social product. Sociology first emerged in western Europe during the eighteenth and nineteenth centuries. In this period, the political and economic systems of Europe were rapidly changing. Monarchy, the rule of society by kings and queens, was disappearing in western Europe. These changes generated new ways of thinking. Religion as the system of authority and law was giving way to scientific authority. At the same time, capitalism grew. Contact between different societies increased, and worldwide economic markets developed. The traditional ways of the past were giving way to a new social order. The time was ripe for a new understanding.

The Influence of the Enlightenment

The **Enlightenment** in eighteenth- and nineteenth-century Europe had an enormous influence on the development of modern sociology. Also known as the Age of Reason, the Enlightenment was characterized by faith in the ability of human reason to solve society's problems. Intellectuals believed that there were natural laws and processes in society to be discovered and used for the general good. Modern science was gradually supplanting traditional and religious explanations for natural phenomena with theories confirmed by experiments.

The earliest sociologists promoted a vision of sociology grounded in careful observation. **Auguste Comte** (1798–1857), a French philosopher who coined the term

As one of the earliest observers of American culture, Harriet Martineau used the powers of social observation to record and analyze the social structure of American society. Long ignored for her contributions to sociology, she is now seen as one of the founders of early sociological thought.

© Culver Pictures

sociology, believed that just as science had discovered the laws of nature, sociology could discover the laws of human social behavior and thus help solve society's problems. This approach is called **positivism,** a system of thought, still prominent today, in which scientific observation and description is considered the highest form of knowledge, as opposed to, say, religious dogma or poetic inspiration. The modern scientific method, which guides sociological research, grew out of positivism.

Alexis de Tocqueville (1805–1859), a French citizen, traveled to the United States as an observer beginning in 1831. Tocqueville thought that democratic values and the belief in human equality positively influenced American social institutions and transformed personal relationships. Less admiringly, he felt that in the United States the tyranny of kings had been replaced by the "tyranny of the majority." He was referring to the ability of a majority to impose its will on everyone else in a democracy. Tocqueville also felt that despite the emphasis on individualism in American culture, Americans had little independence of mind, making them self-centered and anxious about their social class position (Collins and Makowsky 1972).

Another early sociologist is **Harriet Martineau** (1802–1876). Like Tocqueville, Martineau, a British citizen, embarked on a long tour of the United States in 1834. She was fascinated by the newly emerging culture in America. Her book *Society in America* (1837) is an analysis of the social customs that she observed. This important work was overlooked for many years, probably because the author was a woman. It is now recognized as a classic. Martineau also wrote the first sociological methods book, *How to Observe Morals and Manners* (1838), in which she discussed how to observe behavior when one is a participant in the situation being studied.

Classical Sociological Theory

Of all the contributors to the development of sociology, the giants of the European tradition were Emile Durkheim, Karl Marx, and Max Weber. They are classical thinkers because the analyses they offered more than 150 years ago continue to enlighten our understanding of society, not just in sociology but in other fields as well (such as political science and history).

Emile Durkheim. During the early academic career of the Frenchman **Emile Durkheim** (1858–1917), France was in the throes of great political and religious upheaval. Anti-Semitism (hatred of Jews) was being expressed, along with ill feeling among other religions, as well. Durkheim, himself Jewish, was fascinated by how the public degradation of Jews by non-Jews seemed to calm and unify a large segment of the divided French public. Durkheim later wrote that public rituals have a special purpose in society, creating social solidarity, referring to the bonds that link the members of a group. Some of Durkheim's most significant works explore the question of what forces hold society together and make it stable.

According to Durkheim, people in society are glued together by belief systems (Durkheim 1947/1912). The rituals of religion and other institutions symbolize and reinforce the sense of belonging. Public ceremonies create a bond between people in a social unit. Durkheim thought that by publicly punishing people, such rituals sustain moral cohesion in society. Durkheim's views on this are further examined in Chapter 7, which discusses deviant behavior.

Durkheim also viewed society as an entity larger than the sum of its parts. He described this as society *sui generis* (which translates as "thing in itself"), meaning that society is a subject to be studied separately from the sum of the individuals who compose it. Society is external to individuals, yet its existence is internalized in people's minds—that is, people come to believe what society expects them to believe. Durkheim conceived of society as an integrated whole—each part contributing to the overall stability of the system. His work is the basis for *functionalism,* an important theoretical perspective that we will return to later in this chapter.

One contribution from Durkheim was his conceptualization of the *social.* Durkheim created the term **social facts** to indicate those social patterns that are *external* to individuals. Things such as customs and social values exist outside individuals, whereas psychological drives and motivation exist inside people. Social facts, therefore, are not to be explained by

Emile Durkheim established the significance of society as something larger than the sum of its parts. Social facts stem from society and have profound influence on the lives of people within society.

© Bettmann/Corbis

what would they say now?

Classical Theorists Reflect on the War in Iraq

Suppose some of the classical theorists of sociology were suddenly brought back to life. What might they say about the contemporary war in Iraq?

Emile Durkheim: War, though involving human tragedy, also promotes a sense of group and national identity. Patriotic symbols, such as bumper stickers saying Support Our Troops, promote a collective sense of national identity.

Max Weber: War involves multiple dimensions of society. Wars are often fought for political reasons, but also involve economic forces (such as the need for oil in Western societies). Cultural forces, such as the conflict between western values and Islamic value systems, are also a major social force.

Karl Marx: Wars must be understood as the result of the dynamics of capitalism. Not only do they involve economic greed (such as the profits garnered from the pursuit of oil), but private, economic interests are paramount in understanding war, such as the businesses that profit from them.

Jane Addams: As a sociologist who won the Nobel Peace Prize, I must advocate for peaceful resolutions to world conflicts. Were the more pacifist values of women to guide international affairs, we would not experience as many conflicts as we do in a more masculine-based political culture.

W.E.B. Du Bois: War has a disproportionate impact on people of color. Africans Americans (and, he would now note, Latinos) are most likely to be recruited to fight wars, and thus there is a disproportionate impact of tragedy on these as well as low-income, working-class families.

Ida B. Wells: War has a profound influence on the lives of women and people of color, often resulting in rape and other atrocious human rights violations by occupying armies.

biology or psychology but are the proper subject of sociology; they are its reason for being.

A striking illustration of this principle was Durkheim's study of suicide (Durkheim 1951/1897). He analyzed rates of suicide in a society, as opposed to looking at individual (psychological) causes of suicide. He showed that suicide rates varied according to how clear the norms and customs of the society were, whether the norms and customs were consistent with each other and noncontradictory. Anomie (normlessness) existed where norms were either grossly unclear or contradictory; the suicide rates were higher in such societies or such parts of a society. It is important to note that this condition was external, thus outside of each individual taken singly. In this sense such a condition is truly societal.

Durkheim held that social facts, though they exist outside individuals, nonetheless pose constraints on individual behavior. Durkheim's major contribution was the discovery of the social basis of human behavior. He proposed that society could be known through the discovery and analysis of social facts. This is the central task of sociology (Bellah 1973; Coser 1977; Durkheim 1950/1938).

Karl Marx. It is hard to imagine another scholar who has had as much influence on intellectual history as **Karl Marx** (1818–1883). Along with his collaborator, Friedrich Engels, Marx not only changed intellectual history but world history too.

Marx's work was devoted to explaining how capitalism shaped society. He argued that capitalism is an economic system based on the pursuit of profit and the sanctity of private property. Marx used a class analysis to explain capitalism, describing capitalism as a system of relationships among different classes, including capitalists (also known as the bourgeois class), the proletariat (or working class), the petty bourgeoisie (small business owners and managers), and the *lumpenproletariat* (those "discarded" by the capitalist system, such as the homeless). In Marx's view, profit, the goal of capitalist endeavors, is produced through the exploitation of the working class. Workers sell their labor in exchange for wages, and capitalists make certain that wages are worth less than the goods the workers produce. The difference in value is the profit of the capitalist. In the Marxist view, the capitalist class system is inherently unfair because the entire system rests on workers getting less than they give.

Marx thought that the economic organization of society was the most important influence on what humans think and how they behave. He found that the beliefs of the common people tended to support the interests of the capitalist system, not the interests of the workers themselves. Why? Because the capitalist class controls not only the production of goods but also the production of ideas. It owns the publishing companies, endows the universities where knowledge is produced, and controls information industries.

Karl Marx analyzed capitalism as an economic system with enormous implications for how society is organized, in particular how inequality between groups stems from the economic organization of society.

© Bettmann/Corbis

Marx considered all of society to be shaped by economic forces. Laws, family structures, schools, and other institutions all develop, according to Marx, to suit economic needs under capitalism. Like other early sociologists, Marx took social structure as his subject rather than the actions of individuals. It was the *system* of capitalism that dictated people's behavior. Marx saw social change as arising from tensions inherent in a capitalist system—the conflict between the capitalist and working classes. Marx's ideas are often misperceived by U.S. students because communist revolutionaries throughout the world have claimed Marx as their guiding spirit. It would be naive to reject his ideas solely on political grounds. Much that Marx predicted has not occurred—for instance, he claimed that the "laws" of history made a worldwide revolution of workers inevitable, and this has not happened. Still, he left us an important body of sociological thought springing from his insight that society is systematic and structural and that class is a fundamental dimension of society that shapes social behavior.

Max Weber. Max Weber (1864–1920; pronounced "Vay-ber") was greatly influenced by Marx's work and built upon it. But, whereas Marx saw economics as the basic organizing element of society, Weber theorized that society had three basic dimensions: political, economic, and cultural. According to Weber, a complete sociological analysis must recognize the interplay between economic, political, and cultural

Max Weber used a multidimensional approach to analyzing society, interpreting the economic, cultural, and political organization of society as together shaping social institutions and social change.

© AKG/London

institutions (Parsons 1947). Weber is credited with developing a *multidimensional* analysis of society that goes beyond Marx's more one-dimensional focus on economics.

Weber also theorized extensively about the relationship of sociology to social and political values. He did not believe there could be a value-free sociology, because values would always influence what sociologists considered worthy of study. Weber thought sociologists should acknowledge the influence of values so that ingrained beliefs might not interfere with objectivity. Weber professed that the task of sociologists is to teach students the uncomfortable truth about the world. Faculty should not use their positions to promote their political opinions, he felt; rather, they have a responsibility to examine all opinions, including unpopular ones, and use the tools of rigorous sociological inquiry to understand why people believe and behave as they do.

An important concept in Weber's sociology is *verstehen* (pronounced "ver-shtay-en"). **Verstehen,** a German word, refers to understanding social behavior from the point of view of those engaged in it. Weber believed that to understand social behavior, one had to understand the meaning that a behavior had for people. He did not believe sociologists had to be born into a group to understand it (in other words, he didn't believe "it takes one to know one"), but he did think sociologists had to develop some subjective understanding of how other people experience their world. One major contribution from Weber was the definition of *social action* as a behavior to which people give meaning (Gerth and Mills 1946; Parsons 1951b; Weber 1962), such as placing a bumper sticker on your car that states pride in U.S. military troops.

Sociology in America

American sociology was built on the earlier work of Europeans, but unique features of U.S. culture contribute to its distinctive flavor. Less theoretical and more practical than their European counterparts, early American sociologists believed that if they exposed the causes of social problems, they could alleviate some of the consequences, which are measured in human suffering.

Early sociologists in both Europe and the United States conceived of society as an organism, a system of interrelated functions and parts that work together to create the whole. This perspective is called the **organic metaphor.** Sociologists saw society as constantly evolving, like an organism. The question many early sociologists asked was to what extent humans could shape the evolution of society.

Many were influenced in this question by the work of British scholar **Charles Darwin** (1809–1882), who revolutionized biology when he identified the process termed *evolution,* a process by which new species are created through the survival of the fittest. **Social Darwinism** was the application of Darwinian thought to society. According to the social Darwinists, the "survival of the fittest" is the driving force of social evolution as well. They conceived of society as an organism that evolved from simple to complex in a process of adaptation to the environment. They theorized that society was best left alone to follow its natural evolutionary course. Because social Darwinists believed that evolution always took a course toward perfection, they advocated a *laissez-faire* (that is, "hands-off") approach to social change. Social Darwinism was thus a conservative mode of thought; it assumed that the current arrangements in society were natural and inevitable (Hofstadter 1944).

Most other early sociologists in the United States took a more reform-based approach. Nowhere was the emphasis on application more evident than at the University of Chicago, where a style of sociological thinking known as the Chicago School developed. The Chicago School is characterized by thinkers who were interested in how society shaped the mind and identity of people. We study some of these thinkers, such as George Herbert Mead and Charles Horton Cooley, in Chapter 4. They thought of society as a human laboratory where they could observe and understand human behavior in order to be better able to address human needs, and they used the city in which they lived as a living laboratory.

Robert Park (1864–1944), from the University of Chicago, was a key founder of sociology. Originally a journalist who worked in several midwestern cities, Park was interested in urban problems and how different racial groups interacted with each other. He was also fascinated by the sociological design of cities, noting that cities were typically sets of concentric circles. At the time, the very rich and the very poor lived in the middle, ringed by slums and low-income neighborhoods (Collins and Makowsky 1972; Coser 1977; Park and Burgess 1921). Park would still be intrigued by how boundaries are defined and maintained in urban neighborhoods. You might notice this yourself. A single street crossing might delineate a Vietnamese neighborhood from an Italian one, an affluent White neighborhood from a barrio. The social

Jane Addams, the only sociologist to win the Nobel Peace Prize, used her sociological skills to try to improve people's lives. The settlement house movement provided social services to groups in need, while also providing a social laboratory in which to observe the sociological dimensions of problems such as poverty.

© Bettmann/Corbis

Ida B. Wells-Barnett is now well-known for her brave campaign against the lynching of African American people. Less known are her early contributions to sociological thought.

Schomburg, New York Public Library

structure of cities continues to be a subject of sociological research.

Many early sociologists of the Chicago School were women whose work is only now being rediscovered. **Jane Addams** (1860–1935) was one of the most renowned sociologists of her day. She was a leader in the settlement house movement, which provided community services and did systematic research designed to improve the lives of slum dwellers, immigrants, and other dispossessed groups. Addams, the only practicing sociologist ever to win a Nobel Peace Prize (in 1931), never had a regular teaching job. Instead, she used her skills as a research sociologist to develop community projects that assisted people in need (Deegan 1988).

Another early sociologist, widely noted for her work in the anti-lynching movement, was **Ida B. Wells-Barnett** (1862–1931). Born a slave, Ida B. Wells learned to read and write at Rust College, a school established for freed slaves, later receiving her teaching credentials at Fisk University. She wrote numerous essays on the status of African Americans in the United States and was an active crusader against lynching and for women's rights, including the right to vote. Because she was so violently attacked—in writing and in actual threats—because of her passionate work, she often had to write under an assumed name. Until recently, her contributions to the field of sociology have been largely unexamined. Interestingly, her grandson, Troy Duster (b. 1936), now a faculty member at New York University and the University of California, Berkeley, became the president of the American Sociological Association

An insightful observer of race and culture, W.E.B. Du Bois was one of the first sociologists to use community studies as the basis for sociological work. His work, long excluded from the "great works" of sociological theory, is now seen as a brilliant and lasting analysis of the significance of race in the United States.

Courtesy of University of Massachusetts at Amherst

in 2004 (Lengermann and Niebrugge-Brantley 1998; Henry 2008; Giddings 2008).

W.E.B. Du Bois (1868–1963; pronounced "due boys") was one of the most important early sociological thinkers in America. Du Bois was a prominent Black scholar, a cofounder in 1909 of the NAACP (National Association for the Advancement of Colored People), a prolific writer, and one of America's best minds. He received the first Ph.D. ever awarded to a Black person in any field (from Harvard University), and he studied for a time in Germany, hearing several lectures by Max Weber. Du Bois was deeply troubled by the racial divisiveness in society, writing in a classic essay published in 1901 that "the problem of the twentieth century is the problem of the color line" (Du Bois 1901: 354). Like many of his women colleagues, he envisioned a community-based, activist profession committed to social justice (Deegan 1988); he was a friend and collaborator with Jane Addams. He believed in the importance of a scientific approach to sociological questions, but also thought that convictions always directed one's studies.

Theoretical Frameworks in Sociology

The founders of sociology have established theoretical traditions that ask basic questions about society and inform sociological research. The idea of theory may seem dry to you because it connotes something that is only hypothetical and divorced from "real life"; however, sociological theory is one of the tools that sociologists use to interpret real life. Sociologists use theory to organize their observations and apply them to the broad questions sociologists ask, such as: How are individuals related to society? How is social order maintained? Why is there inequality in society? How does social change occur?

Different theoretical frameworks within sociology make different assumptions and provide different insights about the nature of society. In the realm of *macrosociology* are theories that strive to understand society as a whole. Durkheim, Marx, and Weber were macrosociological theorists. Theoretical frameworks that center on face-to-face social interaction are known as *microsociology*. Some of the work derived from the Chicago School—research that studies individuals and group processes in society—is microsociological. Although sociologists draw from diverse theoretical perspectives to understand society, three broad traditions form the major theoretical perspectives that they use: functionalism, conflict theory, and symbolic interaction.

Functionalism

Functionalism has its origins in the work of Durkheim, who you will recall was especially interested in how social order is possible or how society remains

Key Sociological Concepts

As you build your sociological perspective, you must learn certain key concepts in order to begin understanding how sociologists view human behavior. *Social structure*, *social institutions*, *social change*, and *social interaction* are not the only sociological concepts, but they are fundamental to grasping the sociological perspective.

Social Structure. Earlier, we defined social structure as the organized pattern of social relationships and social institutions that together constitute society. Social structure is not a "thing," but refers to the fact that social forces not always visible to the human eye guide and shape human behavior. Acknowledging that social structure exists does not mean that humans have no choice in how they behave, only that those choices are largely conditioned by one's location in society.

Social Institutions. In this book, you will also learn about the significance of **social institutions,** defined as established and organized systems of social behavior with a particular and recognized purpose. The family, religion, marriage, government,

and the economy are examples of major social institutions. Social institutions confront individuals at birth and they transcend individual experience, but they still influence individual behavior.

Social Change. As you can tell, sociologists are also interested in the process of **social change,** the alteration of society over time. As much as sociologists see society as producing certain outcomes, they do not see society as fixed, nor do they see humans as passive recipients of social expectations. Sociologists view society as stable but constantly changing.

Social Interaction. Sociologists see **social interaction** as behavior between two or more people that is given meaning. Through social interaction, people react and change, depending on the actions and reactions of others. Since society changes as new forms of human behavior emerge, change is always in the works. As you read this book, you will see that these key concepts—social structure, social institutions, social change, and social interaction—are central to the sociological imagination.

relatively stable. **Functionalism** interprets each part of society in terms of how it contributes to the stability of the whole. As Durkheim suggested, functionalism conceptualizes society as more than the sum of its component parts. Each part is "functional" for society—that is, contributes to the stability of the whole. The different parts are primarily the institutions of society, each of which is organized to fill different needs and each of which has particular consequences for the form and shape of society. The parts each then depend on one another.

The family as an institution, for example, serves multiple functions. At its most basic level, the family has a reproductive role. Within the family, infants receive protection and sustenance. As they grow older, they are exposed to the patterns and expectations of their culture. Across generations, the family supplies a broad unit of support and enriches individual experience with a sense of continuity with the past and future. All these aspects of family can be assessed by how they contribute to the stability and prosperity of society. The same is true for other institutions.

The functionalist framework emphasizes the consensus and order that exist in society, focusing on social stability and shared public values. From a functionalist perspective, disorganization in the system, such as deviant behavior and so forth, leads to change because societal components must adjust to achieve stability. This is a key part of functionalist theory—that when

one part of society is not working (or is *dysfunctional,* as they would say), it affects all the other parts and creates social problems. Change may be for better or worse; changes for the worse stem from instability in the social system, such as a breakdown in shared values or a social institution no longer meeting people's needs (Collins 1994; Eitzen and Baca Zinn 2006; Turner 1974).

thinkingsociologically

What are the *manifest functions* of grades in college? What are the *latent functions*?

Functionalism was a dominant theoretical perspective in sociology for many years, and one of its major theorists was **Talcott Parsons** (1902–1979). In Parsons's view, all parts of a social system are interrelated, with different parts of society having different basic functions. Functionalism was further developed by **Robert Merton** (1910–2003). Merton saw that social practices often have consequences for society that are not immediately apparent, not necessarily the same as the stated purpose. He suggested that human behavior has both manifest and latent functions. *Manifest functions* are the stated and intended goals of social behavior.

Critics of functionalism argue that its emphasis on social stability is inherently conservative and that it understates the roles of power and conflict in society. Critics also disagree with the explanation of inequality offered by functionalism—that it persists because social inequality creates a system for the fair and equitable distribution of societal resources. Functionalists would, for example, argue that it is fair and equitable that the higher social classes earn more money since they, so it is argued, are more important (functional) to society. Critics of functionalism argue that functionalism is too accepting of the status quo. Functionalists would counter this argument by saying that, regardless of the injustices that inequality produces, inequality serves a purpose in society: It provides an incentive system for people to work and promotes solidarity among groups linked by common social standing.

Conflict Theory

Conflict theory emphasizes the role of coercion and power, a person's or group's ability to exercise influence and control over others, in producing social order. Whereas functionalism emphasizes cohesion within society, conflict theory emphasizes strife and friction. Derived from the work of Karl Marx, conflict theory pictures society as fragmented into groups that compete for social and economic resources. Social order is maintained not by consensus but by domination, with power in the hands of those with the greatest political, economic, and social resources. When consensus exists, according to conflict theorists, it is attributable to people being united around common interests, often in opposition to other groups (Dahrendorf 1959; Mills 1956).

According to conflict theory, inequality exists because those in control of a disproportionate share of society's resources actively defend their advantages. The masses are not bound to society by their shared values but by coercion at the hands of the powerful. In conflict theory, the emphasis is on social control, not consensus and conformity. Groups and individuals advance their own interests, struggling over control of societal resources. Those with the most resources exercise power over others; inequality and power struggles are the result. Conflict theory gives great attention to class, race, and gender in society because these are seen as the grounds of the most pertinent and enduring struggles in society.

Whereas functionalists find some benefit to society in the unequal distribution of resources, conflict theorists see inequality as inherently unfair, persisting only because groups who are economically advantaged use their social position to their own betterment. Their dominance even extends to the point of shaping the beliefs of other members of the society by controlling public information and having major influence over institutions such as education and religion. From the conflict perspective, power struggles between conflicting groups are the source of social change. Typically, those with the greatest power are able to maintain their advantage at the expense of other groups.

Conflict theory has been criticized for neglecting the importance of shared values and public consensus in society while overemphasizing inequality. Like functionalist theory, conflict theory finds the origins of social behavior in the structure of society, but it differs from functionalism in emphasizing the importance of power.

Symbolic Interaction

The third major framework of sociological theory is **symbolic interaction theory.** Instead of thinking of society in terms of abstract institutions, symbolic interactionists consider immediate social interaction to be the place where "society" exists. Because of the human capacity for reflection, people give meaning to their behavior, and this is how they interpret the different behaviors, events, or things that are significant for sociological study.

Because of this, symbolic interaction, as its name implies, relies extensively on the symbolic meaning that people develop and rely upon in the process of social interaction. Symbolic interaction theory emphasizes face-to-face interaction and thus is a form of microsociology, whereas functionalism and conflict theory are more macrosociological.

Derived from the work of the Chicago School, symbolic interaction theory analyzes society by addressing the subjective meanings that people impose on objects, events, and behaviors. Subjective meanings are given primacy because, according to symbolic interactionists and according to Thomas's dictum mentioned earlier, people behave based on what they *believe,* not just on what is objectively true. Thus, society is considered to be socially constructed through human interpretation (Berger and Luckmann 1967; Blumer 1969; Shibutani 1961). Symbolic interactionists see meaning as constantly modified through social interaction. People interpret one another's behavior, and it is these interpretations that form the social bond. These interpretations are called the "definition of the situation." For example, why would young people smoke cigarettes even though all objective medical evidence points to the danger of doing so? The answer is in the definition of the situation that people create. Studies find that teenagers are well informed about the risks of tobacco, but they also think that "smoking is cool," that they themselves will be safe from harm, and that smoking projects an image—a positive identity for boys as a "tough guy" and for girls as fun-loving, mature, and glamorous. Smoking is also defined by young women as keeping you thin—an ideal constructed through dominant images of beauty. In other words, the symbolic meaning of smoking overrides the actual facts regarding smoking and risk (Stjerna et al. 2004).

thinkingsociologically

Think about the example given about smoking, and using a *symbolic interaction* framework, how would you explain other risky behaviors, such as steroid use among athletes or eating disorders among young women?

Symbolic interaction interprets social order as constantly negotiated and created through the interpretations people give to their behavior. In observing society, symbolic interactionists see not simply facts but "social constructions," the meanings attached to things, whether those are concrete symbols (like a certain way of dress or a tattoo) or nonverbal behaviors. To a symbolic interactionist, society is highly subjective—existing in the minds of people, even though its effects are very real.

Functionalism, conflict theory, and symbolic interaction theory are by no means the only theoretical frameworks in sociology. For some time, however, they have provided the most prominent general explanations of society. Each has a unique view of the social realm. None is a perfect explanation of society, yet each has something to contribute. Functionalism gives special weight to the order and cohesion that usually characterizes society. Conflict theory emphasizes the inequalities and power imbalances in society. Symbolic interaction emphasizes the meanings that humans give to their behavior. Together, these frameworks provide a rich, comprehensive perspective on society, individuals within society, and social change (see Table 1.2).

Diverse Theoretical Perspectives

Sociological theory is in transition. In addition to the three frameworks just discussed, other theories are also used within sociology. Contemporary sociological theory has been greatly influenced by the development of **feminist theory,** which analyzes the status of women and men in society with the purpose of using that knowledge to better women's lives. Feminist theory has created vital new knowledge about women and transformed what is understood about men. Feminist scholarship in sociology, by focusing on the experiences of women, provides new ways of seeing the world and contributes to a more complete view of society. We examine feminist theory throughout this book in the context of particular topics, such as family (see Chapter 13) and politics (see Chapter 15).

Many contemporary theorists are increasingly influenced by postmodernism—a strand of thinking now influencing many disciplines. **Postmodernism** is based on the idea that society is not an objective thing. Instead, it is found in the words and images that people use to represent behavior and ideas. Postmodernists think that images and texts reveal the underlying ways that people think and act. Postmodernist studies typically involve detailed analyses of images, words, film, music, and other forms of popular culture. In a civilization such as that in the United States, saturated by the imagery of the mass media, postmodernist analysis illuminates much about society. Postmodernist thinkers see contemporary life as involving multiple experiences and interpretations, but they avoid categorizing human experience into broad and abstract concepts such as institutions or society (Rosenau 1992).

Whatever the theoretical framework used, theory is evaluated in terms of its ability to explain observed social facts. The sociological imagination is not a single-minded way of looking at the world. It is the ability to observe social behavior and interpret that behavior in light of societal influences.

table 1.2 *Three Sociological Frameworks*

Basic Questions	Functionalism	Conflict Theory	Symbolic Interaction
What is the relationship of individuals to society?	Individuals occupy fixed social roles.	Individuals are subordinated to society.	Individuals and society are interdependent.
Why is there inequality?	Inequality is inevitable and functional for society.	Inequality results from a struggle over scarce resources.	Inequality is demonstrated through the importance of symbols.
How is social order possible?	Social order stems from consensus on public values.	Social order is maintained through power and coercion.	Social order is sustained through social interaction and adherence to social norms.
What is the source of social change?	Society seeks equilibrium when there is social disorganization.	Change comes through the mobilization of people struggling for resources.	Change evolves from an ever-evolving set of social relationships and the creation of new meaning systems.
Major Criticisms			
	This is a conservative view of society that underplays power differences among and between groups.	The theory understates the degree of cohesion and stability in society.	There is little analysis of inequality, and it overstates the subjective basis of society.

Careers in Sociology

Now that you understand a bit more what sociology is about, you may ask, What can I do with a degree in sociology? This is a question we often hear from students. There is no single job called "sociologist" like there is "engineer" or "nurse" or "teacher," but sociology prepares you well for many different kinds of jobs, whether with a bachelor's degree or a postgraduate education. The skills you acquire from your sociological education are useful for jobs in business, health care, criminal justice, government agencies, various nonprofit organizations, and other job venues.

For example, the research skills one gains through sociology can be important in analyzing business data or organizing information for a food bank or homeless shelter. Students in sociology also gain experience working with and understanding those with different cultural and social backgrounds; this is an important and valued skill that employers seek. Also, the ability to dissect the different causes of a social problem can be an asset for jobs in various social service organizations.

Some sociologists have worked in their communities to deliver more effective social services. Some are employed in business organizations and social services where they use their sociological training to address issues such as poverty, crime and delinquency, population studies, substance abuse, violence against women, family social services, immigration policy, and any number of other important issues. Sociologists also work in the offices of U.S. representatives and senators, doing background research on the various issues addressed in the political process.

These are just a few examples of how sociology can prepare you for various careers. A good way to learn more about how sociology prepares you for work is to consider doing an internship while you are still in college.

For more information about careers in sociology, see the booklet, *Careers in Sociology,* available through the American Sociological Association. You can find it online at either **www.asanet.org/student/career/homepage .html** or **http://sociology.wadsworth.com**

Critical Thinking Exercise

1. Read a national newspaper over a period of one week and identify any experts who use a sociological perspective in their commentary. What does this suggest to you as a possible career in sociology? What are some of the different subjects about which sociologists provide expert information?

2. Identify some of the students from your college who have finished degrees in sociology. What different ways have they used their sociological knowledge?

Chapter Summary

- ***What is sociology?***

  Sociology is the study of human behavior in society. The *sociological imagination* is the ability to see societal patterns that influence individuals. Sociology is an *empirical* discipline, relying on careful observations as the basis for its knowledge.

- ***What is debunking?***

  Debunking in sociology refers to the ability to look behind things taken for granted, looking instead to the origins of social behavior.

- **Why is diversity central to the study of sociology?**

  One of the central insights of sociology is its analysis of social diversity and inequality. Understanding *diversity* is critical to sociology because it is necessary to analyze *social institutions* and because diversity shapes most of our social and cultural institutions.

- **When and how did sociology emerge as a field of study?**

  Sociology emerged in western Europe during the *Enlightenment* and was influenced by the values of critical reason, humanitarianism, and positivism. *Auguste Comte,* one of the earliest sociologists, emphasized sociology as a positivist discipline. *Alexis de Tocqueville* and *Harriet Martineau* developed early and insightful analyses of American culture.

- **What are some of the basic insights of classical sociological theory?**

  Emile Durkheim is credited with conceptualizing society as a social system and with identifying *social facts* as patterns of behavior that are external to the individual. *Karl Marx* showed how capitalism shaped the development of society. *Max Weber* sought to explain society through cultural, political, and economic factors.

- **What are the major theoretical frameworks in sociology?**

  Functionalism emphasizes the stability and integration in society. *Conflict theory* sees society as organized around the unequal distribution of resources and held together through power and coercion. *Symbolic interaction* theory emphasizes the role of individuals in giving meaning to social behavior, thereby creating society. *Feminist theory* is the analysis of women and men in society and is intended to improve women's lives. *Postmodernism* argues that constructed cultural products are the realities of our postindustrial society.

Key Terms

conflict theory 20	organic metaphor 17	social structure 06
debunking 07	positivism 14	sociological imagination 05
diversity 11	postmodernism 21	sociology 04
empirical 07	social change 19	symbolic interaction theory 20
Enlightenment 13	social Darwinism 17	troubles 06
feminist theory 21	social facts 14	verstehen 17
functionalism 19	social institution 19	
issues 06	social interaction 19	

Online Resources

Sociology: The Essentials Companion Website

www.cengage.com/sociology/andersen
Visit your book companion website where you will find more resources to help you study and write your research papers. Resources include web links, learning objectives, Internet exercises, quizzing, and flash cards.

 is an easy-to-use online resource that helps you study in less time to get the grade you want NOW.

www.cengage.com/login
Need help studying? This site is your one-stop study shop. Take a Pre-Test and CengageNOW will generate a Personalized Study Plan based on your test results. The Study Plan will identify the topics you need to review and direct you to online resources to help you master those topics. You can then take a Post-Test to determine the concepts you have mastered and what you still need to work on.

Chapter two
CHAPTER TWO
Chapter

Culture and the Media

Defining Culture

The Elements of Culture

Cultural Diversity

The Mass Media and Popular Culture

Theoretical Perspectives on Culture and the Media

Cultural Change

Chapter Summary

[*In one contemporary* society known for its technological sophistication, people—especially the young—walk around with plugs in their ears. The plugs are connected to small wires that are themselves coated with a plastic film. These little plastic-covered wires are then connected to small devices made of metal, plastic, silicon, and other modern components, although most of the people who use them have no idea how they are made. When turned on, the device puts music into people's ears or, in some cases, shows pictures and movies on a screen about the size of a postage stamp. Some of the people who use these devices wouldn't even consider walking around without them; it is as if the device shields them from some of the other elements of their culture.

The same people who carry these devices around have other habits that, when seen from the perspective of someone unfamiliar with this culture, might seem peculiar and certainly highly ritualized. Apparently, when the young people in this society go away to school, most take a large number of various technological devices along with them. According to recent reports from these young people, many of them sleep with one of these devices turned on all night. It looks like a large box—some square, others relatively flat—and it projects pictures and sound when the user clicks buttons on another small device that, though detached from the bigger box, can be placed anywhere in the room. If you click the buttons on this portable device, the pictures and sound coming forth from the larger box will change possibly hundreds of times, revealing a huge

continued

David Young-Wolff/Photographer's Choice/Getty Images

two **chapter**

The Tchikrin people of the Brazilian rain forest paint elaborate and beautiful designs on their bodies that define the relationship of people to social groups. Are there ways that cultural practices in the United States also define social relationships?

© Joan Bamburger/Anthro Photos

assortment of images that seem to influence what people in this culture believe and, in many cases, how they behave. They say that in over 40 percent of the households in this culture, this device is turned on 24 hours a day (Gitlin 2002)!

The young people in this culture seem to get up every day and immediately go to another device where they do things with unusual names, such as to "IM" their friends (who, by the way, may be nowhere near them), pushing buttons with their thumbs on a small device with a tiny screen. Indeed, it seems that everything these young people do involves looking at some kind of screen, enough so that one of the authors of this book has labeled their generation "Screenagers."

Not everyone in this culture has access to all of these devices, although many want them. Indeed, having more of the devices seems to be a mark of one's social status, that is, how you are regarded in this culture. But very few people know where the devices are made, what they are made of, or how they work, even though the young often ridicule older people for not understanding how the devices work or why they are so important to them.[1]

From outside the culture, these practices seem strange, yet few within the culture think the behaviors associated with these devices are anything but perfectly ordinary. Most of the time, people do not spend much time thinking about the meaning of the behaviors associated with these devices unless, for some reason, they suddenly do not work.

You have surely guessed that the practices described here are taken from U.S. culture: iPods, MP3 players, text messaging, television/video viewing. These are such daily practices that they practically define modern American culture. Unless they are somehow interrupted, most people do not think much about their influence on society, on people's relationships, or on people's definitions of themselves.

When viewed from the outside, cultural habits that seem perfectly normal often seem strange. Take an example from a different culture. The Tchikrin people—a remote culture of the central Brazilian rain forest—paint their bodies in elaborate designs. Painted bodies communicate to others the relationship of the person to his or her body, to society, and to the spiritual world. The designs and colors symbolize the balance the Tchikrin people think exists between biological powers and the integration of people into the social group. The Tchikrin also associate hair with sexual powers; lovers get a special thrill from using their teeth to pluck an eyebrow or eyelash from their partner's face (Turner 1969). To the Tchikrin people, these practices are no more unusual or exotic than the daily habits we practice in the United States.

To study culture, to analyze it and measure its significance in society, we must separate ourselves from judgments such as "strange" or "normal." We must see a culture as it is seen by insiders, but we cannot be completely taken in by that view. We should know the culture as insiders and understand it as outsiders.

[1] This introduction is inspired by a classic article on the "Naci-rema"—*American*, backwards—by Horace Miner (1956). But it is also written based on the essays students at the University of Delaware have written based on the media blackout exercise described on page 44. Students have written that, without access to their usual media devices, they felt they "had no personality!" and that the period of the blackout was "the worst forty-eight hours of my life!"

Defining Culture

Culture is the complex system of meaning and behavior that defines the way of life for a given group or society. It includes beliefs, values, knowledge, art, morals, laws, customs, habits, language, and dress, among other things. Culture includes ways of thinking as well as patterns of behavior. Observing culture involves studying what people think, how they interact, and the objects they use.

In any society, culture defines what is perceived as beautiful and ugly, right and wrong, good and bad. Culture helps hold society together, giving people a sense of belonging, instructing them on how to behave, and telling them what to think in particular situations. Culture gives meaning to society.

Culture is both material and nonmaterial. **Material culture** consists of the objects created in a given society its buildings, art, tools, toys, print and broadcast media, and other tangible objects, such as those discussed in the chapter opener. In the popular mind, material artifacts constitute culture because they can be collected in museums or archives and analyzed for what they represent. These objects are significant because of the meaning they are given. A temple, for example, is not merely a building, nor is it only a place of worship. Its form and presentation signify the religious meaning system of the faithful.

Nonmaterial culture includes the norms, laws, customs, ideas, and beliefs of a group of people. Nonmaterial culture is less tangible than material culture, but it has a strong presence in social behavior. Examples of nonmaterial culture are numerous and found in the patterns of everyday life. In some cultures, people eat with utensils, in others, people do not. The eating utensils are part of material culture, but the belief about whether to use them is nonmaterial culture.

It is cultural patterns that make humans so interesting. Is it culture that distinguishes human beings from animals? Some animal species develop what we might call culture. Chimpanzees, for example, learn behavior through observing and imitating others, a point proved by observing the different eating practices among chimpanzees in the same species but raised in different groups (Whiten et al. 1999). Others have observed elephants picking up the dead bones of other elephants and fondling them, perhaps evidence of grieving behavior (Meredith 2003). Dolphins are known to have a complex auditory language. And most people think that their pets communicate with them. Apparently, humans are not unique in their ability to develop systems of communication. But some scientists generally conclude that animals lack the elaborate symbol-based cultures common in human societies. Perhaps, as even Charles Darwin wrote, "The difference in mind between man and the higher animals, great as it is, certainly is one of degree and not of kind" (Darwin, cited in Gould 1999).

Studying animal groups reminds us of the interplay between biology and culture. Human biology sets limits and provides certain capacities for human life

Zave Smith/Photolibrary

and the development of culture. Similarly, the environment in which humans live establishes the possibilities and limitations for human society. Nutrition, for instance, is greatly influenced by environment, thereby affecting human body height and weight. Not everyone can swim like Michael Phelps or lob a tennis ball like Venus and Serena Williams, but with training and conditioning, people can enhance their physical abilities. Biological limits exist, but cultural factors have an enormous influence on the development of human life.

Characteristics of Culture

Across societies, there are common characteristics of culture, even when the particulars vary. These different characteristics are as follows.

1. *Culture is shared.* Culture would have no significance if people did not hold it in common. Culture is collectively experienced and collectively agreed upon. The shared nature of culture is what makes human society possible. The shared basis of culture may be difficult to see in complex societies where groups have different traditions, perspectives, and ways of thinking and behaving. In the United States, for example, different racial and ethnic groups have unique histories, languages, and beliefs—that is, different cultures. Even within these groups, there are diverse cultural traditions. Latinos, for example, comprise many groups with distinct origins and cultures. Still, there are features of Latino culture, such as the Spanish language and some values and traditions, that are shared. Latinos also share a culture that is shaped

Different culture traditions create rituals that mark the passage from one phase of life to another. Here, a young boy undertakes the Bar Mitzvah ceremony marking his passage to adulthood (upper left); a young Apache girl experiences four-day-long puberty rites following the first onset of menstruation, celebrating her newfound womanhood (top right); a Navjote ceremony, the most important event in a young child's life, noting acceptance into the Parsi community in India (bottom left); a young girl awaits her quinceañera ceremony in Chiapas, Mexico (bottom right).

by their common experiences as minorities in the United States. Similarly, African Americans have created a rich and distinct culture that is the result of their unique experience within the United States. What identifies African American culture are the practices and traditions that have evolved from both the U.S. experience and African and Caribbean traditions. Placed in another country, such as an African nation, African Americans would likely recognize elements of their culture, but they would also feel culturally distinct as Americans.

Within the United States, culture varies by age, race, region, gender, ethnicity, religion, class, and other social factors. A person growing up in the South is likely to develop different tastes, modes of speech, and cultural interests than a person raised in the West. Despite these differences, there is a common cultural basis to life in the United States. Certain symbols, language patterns, belief systems, and ways of thinking are distinctively American and form a common culture, even though great cultural diversity exists.

2. *Culture is learned.* Cultural beliefs and practices are usually so well learned that they seem perfectly natural, but they are learned none the less. How do people come to prefer some foods to others? How is musical taste acquired? Culture may be taught through direct instruction, such as a parent teaching a child how to use silverware or teachers instructing children in songs, myths, and other traditions in school.

Culture is also learned indirectly through observation and imitation. Think of how a person learns what it means to be a man or a woman. Although the "proper" roles for men and women may never be explicitly taught, one learns what is expected from observing others. A person becomes a member of a culture through both formal and informal transmission of culture. Until the culture is learned, the person will feel like an outsider. The process of learning culture is referred to by sociologists as socialization, discussed in Chapter 4.

3. *Culture is taken for granted.* Because culture is learned, members of a given society seldom question the culture of which they are a part, unless for some reason they become outsiders or establish some critical distance from the usual cultural expectations. People engage unthinkingly in hundreds of specifically cultural practices every day; culture makes these practices seem "normal." If you suddenly stopped participating in your culture and questioned each belief and every behavior, you would soon find yourself feeling detached and perhaps a little disoriented; you would also become increasingly ineffective at functioning within your group. Little wonder that tourists stand out so much in a foreign culture. They rarely have much knowledge of the culture they are visiting and, even when they are well informed, typically approach the society from their own cultural orientation.

Think, for example, of how you might feel if you were a Native American student in a predominantly White classroom. Many, though not all, Native American people are raised to be quiet and not outspoken. If students in a classroom are expected to assert themselves and state what is on their minds, a Native American student may feel awkward, as will others for whom these expectations are contrary to their cultural upbringing. If the professor is not aware of these cultural differences, he or she may penalize students who are quiet, resulting perhaps in a lower grade for the student from a different cultural background. Culture binds us together, but lack of communication across cultures can have negative consequences, as this example shows.

4. *Culture is symbolic.* The significance of culture lies in the meaning it holds for people. **Symbols** are things or behaviors to which people give meaning; the meaning is not inherent in a symbol but is bestowed by the meaning people give it. The U.S. flag, for example, is literally a decorated piece of cloth. Its cultural significance derives not from the cloth of which it is made but from its meaning as a symbol of freedom and democracy, as was witnessed by the widespread flying of the flags after the terrorist attacks on the United States on September 11, 2001.

That something has symbolic meaning does not make it any less important or influential than objective facts. Symbols are powerful expressions of human culture. Think of the Confederate flag. Those who object to the Confederate flag being displayed on public buildings see it as a symbol of racism and the legacy of slavery. Those who defend it see it as representing southern heritage, a symbol of group pride and regional loyalty. Similarly, the use of Native American mascots to name and represent sports teams is symbolic of the exploitation of Native Americans. Native American activists and their supporters see the use of Native American mascots as derogatory and extremely insulting, representing gross caricatures of Native American traditions. (Think of the Washington Redskins, the Cleveland Indians, or the Atlanta Braves' "tomahawk chop.") The protests that have developed over controversial symbols are indicative of the enormous influence of cultural symbols.

Debunking Society's Myths

Myth: The use of Native American names for school mascots is just for fun and is no big deal.

Sociological perspective: Language carries with it great meaning that reflects the perceived social value of diverse groups. Research finds that exposure to the trivial or degrading use of Native American images for such things as school mascots and sports teams actually lowers Native American children's sense of self-worth (Fryberg 2003).

> doing sociological research

Tattoos: Status Risk or Status Symbol?

Research Question: Not so long ago tattoos were considered a mark of social outcasts. They were associated with gang members, sailors, and juvenile delinquents. But now tattoos are in vogue—a symbol of who's trendy and hip. How did this happen that a once stigmatized activity associated with the working class became a statement of middle-class fashion?

Research Method: This is what sociologist Katherine Irwin wanted to know when she first noticed the increase in tattooing among the middle class. Irwin first encountered the culture of tattooing when she accompanied a friend getting a tattoo in a shop she calls Blue Mosque. She started hanging out in the shop and began a four-year study using participant observation in the shop along with interviews of people getting their first tattoos. Irwin also interviewed some of the parents of tattooees and potential tattooees.

Research Results: Irwin found that middle-class tattoo patrons were initially fearful that their desire for a tattoo would associate them with low-status groups, but they reconciled this by adopting attitudes that associated tattooing with middle-class values and norms. Thus, they defined tattooing as symbolic of independence, liberation, and freedom from social constraints. Many of the women defined tattooing as symbolizing toughness and strength—values they thought rejected more conventional ideals of femininity.

Some saw tattoos as a way of increasing their attachment to alternative social groups or to gain entrée into "fringe" social worlds. Although tattoos held different cultural meanings to different groups, people getting tattooed used various techniques (what Irwin calls "legitimation techniques") to counter the negative stereotypes associated with tattooing.

Conclusions and Implications: Irwin concludes that people try to align their behavior with legitimate cultural values and norms even when that behavior seemingly falls outside of prevailing standards.

Questions to Consider

1. Do you think of tattoos as fashionable or deviant? What do you think influences your judgment about this, and how might your judgment be different were you in a different culture, age group, or historical moment?

2. Are there fashion adornments that you associate with different social classes? What are they, and what kinds of judgment (positive and negative) do people make about them? Where do these judgments come from, and why are they associated with social class?

Source: Irwin, Katherine. 2001. "Legitimating the First Tattoo: Moral Passage through Informal Interaction." *Symbolic Interaction* 24 (March): 49–73.

NBAE/Getty Images

Tattooing, once considered a working class symbol, has now become a middle-class fashion statement. Some, known as collectors, use extensive tattooing as a symbol of identity.

Symbolic attachments can guide human behavior. For example, people stand when the national anthem is sung and may feel emotional from displays of the cross or the Star of David. Under some conditions, people organize mass movements to protest what they see as the defamation of important symbols, such as the burning of a flag or the burning of a cross. The significance of the symbolic value of culture can hardly be overestimated. Learning a culture means not just engaging in particular behaviors but also learning their symbolic meanings within the culture.

5. *Culture varies across time and place.* Culture develops as humans adapt to the physical and social environment around them. Culture is not fixed from one place to another. In the United States, for example, there is a strong cultural belief in scientific solutions to human problems; consequently,

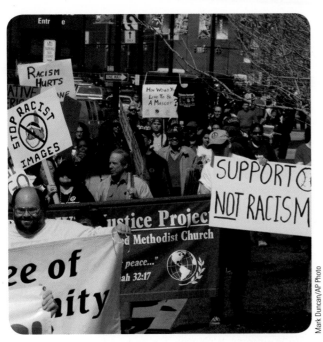

Mark Duncan/AP Photo

The use of American Indian stereotypes icons for school mascots is highly offensive to many and has been shown to lower the self-esteem of Native American children exposed to these negative images. Should similar stereotypes appear on team names that negatively stereotyped your ethnic group, how would you react?

many think that problems of food supply and environmental deterioration can be addressed by scientific breakthroughs, such as genetic engineering to create high-yield tomatoes or cloning cows to eliminate mad cow disease. In other cultural settings, different solutions may seem preferable. Indeed, some religions think of genetic engineering as trespassing on divine territory.

Culture also varies over time. As people encounter new situations, the culture that emerges is a mix of the past and present. Second generation immigrants to the United States are raised in the traditions of their culture of origin, and children of immigrants typically grow up with both the traditional cultural expectations of their parents' homeland and the cultural expectations of a new society. Adapting to the new society can create conflict between generations, especially if the older generation is intent on passing along their cultural traditions. The children may be more influenced by their peers and may choose to dress, speak, and behave in ways that are characteristic of their new society but unacceptable to their parents.

To sum up, culture is concrete because we can observe the cultural objects and practices that define human experience. Culture is abstract because it is a way of thinking, feeling, believing, and behaving. Culture links the past and the present because it is the knowledge that makes us part of human groups. Culture gives shape to human experience.

The Elements of Culture

Culture is multifaceted, consisting of both material and nonmaterial things, some parts of culture being abstract, others more concrete. The different elements of culture include language, norms, beliefs, and values (see Table 2.1).

Language

Language is a set of symbols and rules that, combined in a meaningful way, provides a complex communication system. The formation of culture among humans is made possible by language. Learning the language of a culture is essential to becoming part of a society, and it is one of the first things children learn. Indeed,

table2.1	**Elements of Culture**	
	Definition	**Examples**
Language	A set of symbols and rules that, put together in a meaningful way, provides a complex communication system	English; Spanish; hieroglyphics
Norms	The specific cultural expectations for how to behave in a given situation	Behavior involving use of personal space; manners
Folkways	General standards of behavior adhered to by a group	Cultural forms of dress; food habits
Mores	Strict norms that control moral and ethical behavior	Religious doctrines; formal law
Values	Abstract standards in a society or group that define ideal principles	Liberty; freedom
Beliefs	Shared ideas about what is true held collectively by people within a given culture	Belief in a higher being

Living in a multicultural society often juxtaposes diverse cultures, even in public places.

until children acquire at least a rudimentary command of language, they seem unable to acquire other social skills. Language is so important to human interaction that it is difficult to think of life without it; indeed, as one commentator on language has said, "Life is lived as a series of conversations" (Tannen 1990: 13).

Think about the experience of becoming part of a social group. When you enter a new society or a different social group, you have to learn its language to become a member of the group. This includes any special terms of reference used by the group. Lawyers, for example, have their own vocabulary and their own way of constructing sentences called, not always kindly, "legalese." Becoming a part of any social group—a friendship circle, fraternity or sorority, or any other group—involves learning the language they use. Those who do not share the language of a group cannot participate fully in its culture.

Language is fluid and dynamic and evolves in response to social change. Think, for example, of how the introduction of computers has affected the English language. People now talk about needing "downtime" and providing "input." Only a few years ago, had you said you were going to "text message" your friends, no one would have known what you were talking about. Text messaging has also introduced its own language: BFN (bye for now), LOL (laughing out loud), and GTG (got to go)—a new language shared among those in the text-messaging culture. There are now even online dictionaries listing and defining such "words" (see Table 2.2).

Does Language Shape Culture? Language is clearly a big part of culture. Edward Sapir (writing in the 1920s) and his student Benjamin Whorf (writing in the 1950s) thought that language was central in determining social thought. Their theory, the **Sapir–Whorf hypothesis,** asserts that language determines other aspects of culture because language provides the categories through which social reality is defined. Sapir and Whorf thought that language determines what people think because language forces people to perceive the world in certain terms (Sapir 1921; Whorf 1956).

If the Sapir–Whorf hypothesis is correct, then speakers of different languages have different perceptions of reality. Whorf used the example of the social meaning of time to illustrate cultural differences in how language shapes perceptions of reality. He noted that the Hopi Indians conceptualize time as a slowly turning cylinder, whereas English-speaking people conceive of time as running forward in one direction at a uniform pace. Linguistic constructions of time shape how the two different cultures think

table2.2		The Language of the Internet	
Language often changes as society and culture change. One example is messaging, some samples of which are provided below.			
AAS	alive and smiling	MTFBWU	May the force be with you.
B/F	boyfriend	OP	on phone
BM&Y	between me and you	POS	parent over shoulder
BTDT	been there, done that	SLAP	sounds like a plan
CRBT	crying really big tears	SNAG	sensitive New Age guy
FYEO	for your eyes only	SUP	what's up
GFI	go for it	TMB	text me back
HAGO	have a good one	UCMU	you crack me up
KIR	keeping it real	WDYK	what do you know
KOTL	kiss on the lips	WOMBAT	waste of money, brains, and time
KWIM	know what I mean?	WWJD	What would Jesus do?
LD	later dude	YRYOCC	You're running your own cuckoo clock.
		ZZ	bored

Source: "Text Messaging Abbreviations" 2007. **www.webopedea.com; www.netlingo.com**

about time and therefore how they think about reality. In Hopi culture, events are located not in specific moments of time but in "categories of being," as if everything is in a state of becoming, not fixed in a particular time and place (Carroll 1956). In contrast, the English language locates things in a definite time and place, placing great importance on verb tense, with things located precisely in the past, present, or future.

Recent critics do not think that language determines culture to the extent that Sapir and Whorf proposed. Language does not single-handedly dictate the perception of reality—but, no doubt, language has a strong influence on culture. Most scholars now see two-way causality between language and culture. Asking whether language determines culture or vice versa is like asking which came first, the chicken or the egg. Language and culture are inextricable, each shaping the other.

Consider again the example of time. Contemporary Americans think of the week as divided into two parts: *weekdays* and *weekends*, words that reflect how we think about time. When does a week end? Having language that defines the weekend encourages us to think about the weekend in specific ways. It is a time for rest, play, chores, and family. In this sense, language shapes how we think about the passage of time—we look forward to the weekend, we prepare ourselves for the workweek—but the language itself (the very concept of the weekend) stems from patterns in the culture—specifically, the work patterns of advanced capitalism. The capitalist work ethic makes it morally offensive to merely "pass the time"; instead, time is to be managed. Concepts of time in preindustrial, agricultural societies follow a different pattern. In agricultural societies, time and calendars are based on agricultural and seasonal patterns; the year proceeds according to this rhythm, not the arbitrary units of time of weeks and months. This shows how language and culture shape each other.

Social Inequality in Language. The language of any culture reflects the nature of that society. Thus, in a society where there is group inequality, language is likely to communicate assumptions and stereotypes about different social groups. What people say—including what people are called—reinforces patterns of inequality in society (Moore 1992). We see this in what different groups in the United States are called (see also the box on page 34, "Understanding Diversity: The Social Meaning of Language"). What someone is called can be significant because it imposes an identity on that person. This is why the names for various racial and ethnic groups have been so heavily debated. Thus, for years, many Native Americans objected to being called "Indian," because it was a term created about them by White conquerors. To emphasize their native roots in the Americas,

the term *Native American* was adopted. Now, though many prefer to be called by their actual origin, *Native American* and *American Indian* are also used interchangeably. Likewise, Asian Americans tend to be offended by being called "Oriental," an expression that stemmed from Western (that is, European and American) views of Asian nations.

Language reflects the social value placed on different groups, and it reflects power relationships, depending on who gets to name whom. Derogatory terms such as *redneck, white trash,* or *trailer park trash* stigmatize people based on regional identity and social class. This is also why it is so demeaning when derogatory terms are used to describe racial–ethnic groups. For example, throughout the period of Jim Crow segregation in the American South, Black men, regardless of their age, were routinely called "boy" by Whites. Calling a grown man a "boy" is an insult; it diminishes his status by defining him as childlike. Referring to a woman as a "girl" has the same effect. Why are young women, even well into their twenties, routinely referred to as "girls"? Just as does calling a man "boy," this diminishes women's status.

Debunking Society's Myths

Myth: Bilingual education discourages immigrant children from learning English and thus blocks their assimilation into American culture and reduces their chances for a good education.

Sociological perspective: Studies of students who are fluent bilinguals show that they outperform both English only students and students with limited bilingualism. Moreover, preserving the use of native languages can better meet the need for skilled bilingual workers in the labor market (Portes 2002).

Note, however, that terms such as *girl* and *boy* are pejorative only in the context of dominant and subordinate group relationships. African American women, as an example, often refer to each other as "girl" in informal conversation. The term *girl* used between those of similar status is not perceived as derogatory, but when used by someone in a position of dominance, such as when a male boss calls his secretary a "girl," it is demeaning. Likewise, terms such as *dyke, fag,* and *queer* are terms lesbians and gay men sometimes use without offense in referring to each other, even though the same terms are offensive to lesbians and gays when used about them by others. By reclaiming these terms as positive within their own culture, lesbians and gays build cohesiveness and solidarity (Due 1995). These examples show that power relationships between groups supply the social context for the connotations of language.

In sum, language can reproduce the inequalities that exist in society. At the same time, changing the

The Social Meaning of Language

Language reflects the assumptions of a culture. This can be seen and exemplified in several ways:

- Language affects people's perception of reality. *Example:* Researchers have found that using male pronouns, even when intended to be gender-neutral, produces male-centered imagery and ideas. Studies also find that when college students look at job descriptions written using masculine pronouns, they assume that women are not qualified for the job (Gastil 1990; Hamilton 1988; Switzer 1990).
- Language reflects the social and political status of different groups in society. *Example:* A term such as *woman doctor* suggests that men are the standard and women the exception. The term *working woman* (used to refer to women who are employed) also suggests that women who do not work for wages are not working. Ask yourself what the term *working man* connotes and how this differs from *working woman*.
- Groups may advocate changing language referring to them as a way of asserting a positive group identity. *Example:* Some advocates for the disabled challenge the term *handicapped*, arguing that it stigmatizes people who may have many abilities, even if they are physically distinctive. Also, though someone may have one disabling condition, she or he may otherwise be perfectly able.
- The implications of language emerge from specific historical and cultural contexts. *Example:* The naming of so-called races comes from the social and historical processes that define different groups as inferior or superior. Racial labels do not come just from physical, national, or cultural differences. The term *Caucasian,* for example, was coined in the seventeenth century when racist thinkers developed alleged scientific classification systems to rank different societal groups. Alfred Blumenbach used the label Caucasian to refer to people from the Caucasus of Russia whom he thought were more beautiful and intelligent than any group in the world.
- Language can distort actual group experience. *Example:* Terms used to describe different racial and ethnic groups homogenize experiences that may be unique. Thus, the terms *Hispanic* and *Latino* lump together Mexican Americans, island Puerto Ricans, U.S.-born Puerto Ricans, as well as people from Honduras, Panama, El Salvador, and other Central and South American countries. *Hispanic* and *Latino* point to the shared experience of those from

language that people use can, to some extent, alter social stereotypes and thereby change the way people think.

Norms

Social norms are another component of culture. **Norms** are the specific cultural expectations for how to behave in a given situation. Society without norms would be chaos; with norms in place, people know how to act, and social interactions are consistent, predictable, and learnable. There are norms governing every situation. Sometimes they are implicit; that is, they need not be spelled out for people to understand them. For example, when joining a line, there is an implicit norm that you should stand behind the last person, not barge in front of those ahead of you. Implicit norms may not be formal rules, but violation of these norms may nonetheless produce a harsh response. Implicit norms may be learned through specific instruction or by observation of the culture; they are part of a society's or group's customs. Norms are explicit when the rules governing behavior are written down or formally communicated. Typically, specific sanctions are imposed for violating explicit norms.

thinkingsociologically

Identify a *norm* that you commonly observe. Construct an experiment in which you, perhaps with the assistance of others, violate the norm. Record how others react and note the sanctions engaged through this norm violation exercise. *Note:* Be careful not to do anything that puts you in danger or causes serious problems for others.

In the early years of sociology, **William Graham Sumner** (1906) identified two types of norms: folkways and mores. **Folkways** are the general standards of

Latin cultures, but like the terms *Native American* and *American Indian*, they obscure the experiences of unique groups, such as the Sioux, Nanticoke, Cherokee, Yavapai, or Navajo.

- Language shapes people's perceptions of groups and events in society.
  Example: Following Hurricane Katrina in New Orleans, African American people taking food from abandoned stores were described as "looting" and White people as "finding food." Also, Native American victories during the nineteenth century are typically described as "massacres"; comparable victories by White settlers are described in heroic terms (Moore 1992).

- Terms used to define different groups change over time and can originate in movements to assert a positive identity.
  Example: In the 1960s, *Black American* replaced the term *Negro* because the civil rights and Black Power movements inspired Black pride and the importance of self-naming (Smith 1992). Earlier, *Negro* and *colored* were used to define African Americans. Currently, it is popular to refer to all so-called racial groups as "people of color." This phrase was derived from the phrase "women of color," created by feminist African American, Latina,[1] Asian American, and Native American women to emphasize their

common experiences. Some people find the use of "color" in this label offensive since it harkens back to the phrase "colored people," a phrase generally seen as paternalistic and racist because it was a label used by dominant groups to refer to African Americans prior to the civil rights movement. The phrase "women of color" now has a more positive meaning than the earlier term *colored women* because it is meant to recognize common experiences, not just label people because of their presumed skin color.

In this book, we have tried to be sensitive to the language used to describe different groups. We recognize that the language we use is fraught with cultural and political assumptions and that what seems acceptable now may be offensive later. Perhaps the best way to solve this problem is for different groups to learn as much as they can about one another, becoming more aware of the meaning and nuances of naming and language and more conscious of the racial assumptions embedded in the language. Greater sensitivity to the language used in describing different group experiences is an important step in promoting better intergroup relationships.

[1]*Latina* is the feminine form in Spanish and refers to women; *Latino*, to men.

behavior adhered to by a group. You might think of folkways as the ordinary customs of different group cultures. Men wearing pants and not skirts is an example of a cultural folkway. Other examples are the ways that people greet each other, decorate their homes, and prepare their food. Folkways may be loosely defined and loosely adhered to, but they nevertheless structure group customs and implicitly govern much social behavior.

Mores (pronounced "more-ays") are strict norms that control moral and ethical behavior. Mores provide strict codes of behavior, such as the injunctions, legal and religious, against killing others and committing adultery. Mores are often upheld through rules or **laws,** the written set of guidelines that define right and wrong in society. Basically, laws are formalized mores. Violating mores can bring serious repercussions. When any social norm is violated, the violator is typically punished. **Social sanctions** are

mechanisms of social control that enforce norms. The seriousness of a social sanction depends on how strictly the norm is held. The strictest norms in society are **taboos**—those behaviors that bring the most serious sanctions. Dressing in an unusual way that violates the folkways of dress may bring ridicule but is usually not seriously punished. In some cultures, the rules of dress are strictly interpreted, such as the requirement by Islamic fundamentalists that women who appear in public have their bodies cloaked and faces veiled. It would be considered a taboo for a woman in this culture to appear in public without being veiled. The sanctions for doing so can be as severe as whipping, branding, banishment, even death.

Sanctions can be positive or negative, that is, based on rewards or punishment. When children learn social norms, for example, correct behavior may elicit positive sanctions; the behavior is reinforced through praise, approval, or an explicit reward. Early on, for example, parents might praise children for learning to put on their own clothes; later, children might get an allowance if they keep their rooms clean. Bad behavior earns negative sanctions, such as getting spanked or grounded. In society, negative sanctions may be mild or severe, ranging from subtle mechanisms of control, such as ridicule, to overt forms of punishment, such as imprisonment, physical coercion, or death.

One way to study social norms is to observe what happens when they are violated. Once you become aware of how social situations are controlled by norms, you can see how easy it is to disrupt situations where adherence to the norms produces social order. **Ethnomethodology** is a theoretical approach in sociology based on the idea that you can discover the normal social order through disrupting it. As a technique of study, ethnomethodologists often deliberately disrupt social norms to see how people respond and try to reinstate social order (Garfinkel 1967).

In a famous series of ethnomethodological experiments, college students were asked to pretend they were boarders in their own homes for a period of fifteen minutes to one hour. They did not tell their families what they were doing. The students were instructed to be polite, circumspect, and impersonal; to use terms of formal address; and to speak only when spoken to. After the experiment, two of the participating students reported that their families treated the experiment as a joke; another's family thought the daughter was being extra nice because she wanted something. One family believed that the student was hiding some serious problem. In all the other cases, parents reacted with shock, bewilderment, and anger. Students were accused of being mean, nasty, impolite, and inconsiderate; the parents demanded explanations for their sons' and daughters' behavior (Garfinkel 1967). Through this experiment, the student researchers were able to see that even the informal norms governing behavior in one's home are carefully structured. By violating the norms of the household, the norms were revealed.

Ethnomethodological research teaches us that society proceeds on an "as if " basis. That is, society exists because people behave as if there were no other way to do so. Usually, people go along with what is expected of them. Culture is actually "enforced" through the social sanctions applied to those who violate social norms. Usually, specific sanctions are unnecessary because people have learned the normative expectations. When the norms are violated, their existence becomes apparent (see also Chapter 5).

Beliefs

As important as social norms are the beliefs of people in society. **Beliefs** are shared ideas held collectively by people within a given culture about what is true. Shared beliefs are part of what binds people together in society. Beliefs are also the basis for many norms and values of a given culture. In the United States, beliefs that are widely held and cherished are the belief in God and the belief in democracy.

Some beliefs are so strongly held that people find it difficult to cope with ideas or experiences that contradict them. Someone who devoutly believes in God may find atheism intolerable; those who believe in reincarnation may seem irrational to those who think life ends at death. Similarly, those who believe in magic may seem merely superstitious to those with a more scientific and rational view of the world.

Whatever beliefs people hold, they orient us to the world. They provide answers to otherwise imponderable questions about the meaning of life. Beliefs provide a meaning system around which culture is organized. Whether belief stems from religion, myth, folklore, or science, it shapes what people take to be possible and true. Although a given belief may be logically impossible, it nonetheless guides people through their lives.

Values

Deeply intertwined with beliefs are the values of a culture. **Values** are the abstract standards in a society or group that define ideal principles. Values define what is desirable and morally correct; thus, values determine what is considered right and wrong, beautiful and ugly, good and bad. Although values are abstract, they provide a general outline for behavior. Freedom, for example, is a value held to be important in U.S. culture, as is equality. Values are ideals forming the abstract standards for group behavior, but they are also ideals that may not be realized in every situation.

Values can be a basis for cultural cohesion, but they can also be a source of conflict. You may remember the conflicts that developed over the case of Terri Schiavo—a woman who collapsed from a chemical imbalance brought on by an eating disorder. She was left severely brain-damaged in a "permanent vegetative state" for over fifteen years. Her husband, Michael Schiavo, wanted to remove her feeding tube, thereby ending her life, something he argued she would have wanted. But her parents objected on religious grounds and took the case through numerous court decisions over several years and ultimately to the U.S. Supreme Court. By the time Terri Schiavo died in 2005 (after her feeding tube had been removed), the public had become embroiled in a huge national debate about differing values. There were those who believed she had the right to die in peace and with dignity, those who defined her death as murder, and others (in fact, the majority of Americans) who thought that the government simply had no business interfering in the private affairs of this family. The intensity of this issue, even the language by which people described Terri Schiavo, reflected different positions in a national conflict over competing values.

Cultural values can clash when groups have strongly held, but clashing, value systems. Values can be a source of cultural cohesion, but also of cultural conflict. What are some of the different values that are being debated in society?

Values guide the behavior of people in society; they also shape the social norms in a given culture. An example of the impact that values have on people's behavior comes from an American Indian society known as the Kwakiutl (pronounced "kwa-kee-YOO-tal"), a group from the coastal region of southern Alaska, Washington State, and British Columbia. The Kwakiutl developed a practice known as *potlatch,* in which wealthy chiefs would periodically pile up their possessions and give them away to their followers and rivals (Benedict 1934; Harris 1974; Wolcott 1996). The object of potlatch was to give away or destroy more of one's goods than did one's rivals. The potlatch reflected Kwakiutl values of reciprocity, the full use of food and goods, and the social status of the wealthiest chiefs in Kwakiutl society. (By the way, chiefs did not lose their status by giving away their goods because the goods were eventually returned in the course of other potlatches. They would even burn large piles of goods, knowing that others would soon replace their wealth through other potlatches.)

Compare this practice with the patterns of consumption in the United States. Imagine the CEOs of major corporations regularly gathering up their wealth and giving it away to their workers and rival CEOs! In the contemporary United States, *conspicuous consumption* (consuming for the sake of displaying one's wealth) celebrates values similar to those of the potlatch: High-status people demonstrate their position by accumulating more material possessions than those around them (Veblen 1899).

Together, norms, beliefs, and values guide the behavior of people in society. It is necessary to understand how they operate in a situation in order to understand why people behave as they do.

Cultural Diversity

It is rare for a society to be culturally uniform. As societies develop and become more complex, different cultural traditions appear. The more complex the society, the more likely its culture will be internally varied and diverse. The United States, for example, hosts enormous cultural diversity stemming from religious, ethnic, and racial differences, as well as regional, age, gender, and class differences. Currently, more than 12.5 percent of people in the United States are foreign born. In a single year, immigrants from more than 100 countries come to the United States (U.S. Census Bureau 2009a). Whereas earlier immigrants were predominantly from Europe, now Latin America and Asia are the greatest sources of new immigrants. One result is a large increase in the number of U.S. residents for whom English is the second language. Cultural diversity is clearly a characteristic of contemporary American society.

The richness of American culture stems from the many traditions that different groups have brought with them to this society, as well as from the cultural forms that have emerged through their experience within the United States. Jazz, for example, is one of the few musical forms indigenous to the United States. An indigenous art form refers to something that originated in a particular region or culture. However, jazz also has roots in the musical traditions of slave communities and African cultures. Since the birth of jazz, cultural greats such as Ella Fitzgerald, Count Basie, Duke Ellington, Billie Holiday, and numerous others have not only enriched the jazz tradition but have also influenced other forms of music, including rock and roll.

map2.1 *MAPPING AMERICA'S DIVERSITY*

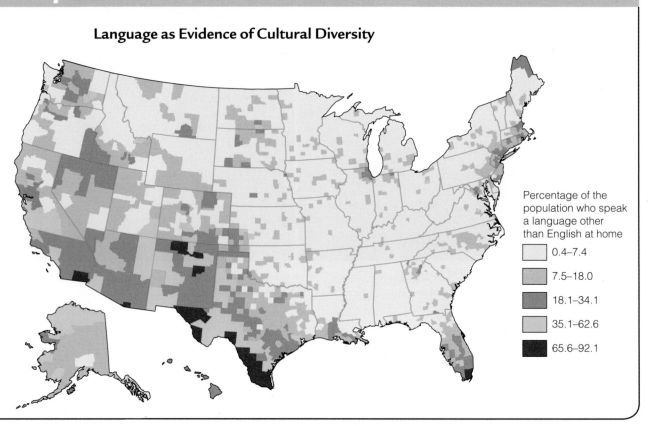

Language as Evidence of Cultural Diversity

Percentage of the population who speak a language other than English at home

- 0.4–7.4
- 7.5–18.0
- 18.1–34.1
- 35.1–62.6
- 65.6–92.1

With increased immigration and greater diversity in the U.S. population, evidence of cultural diversity can be seen in many homes—language being one type of evidence. This map shows the regional differences in the percentage of the population over age five who speak a language other than English at home. For the United States as a whole, 17.9 percent of the population—almost one-fifth—fit into this category. Eight percent of the population say they speak English less than very well. What implications does this have for the regions most affected? How might it influence relations between different generations within households?

Source: U.S. Census Bureau. 2009b. "American FactFinder." **www.census.gov**

Native American cultures have likewise enriched the culture of our society, as have the cultures that various immigrant groups have brought with them to the United States. With such great variety, how can the United States be called one culture? The culture of the United States, including its language, arts, food customs, religious practices, and dress, can be seen as the sum of the diverse cultures that constitute this society.

Dominant Culture

Two concepts from sociology help us understand the complexity of culture in a given society: dominant culture and subculture. The **dominant culture** is the culture of the most powerful group in a society. It is the cultural form that receives the most support from major institutions and that constitutes the major belief system. Although the dominant culture is not the only culture in a society, it is commonly believed to be "the" culture of a society despite the other cultures

present. Social institutions in the society perpetuate the dominant culture and give it a degree of legitimacy that is not shared by other cultures. Quite often, the dominant culture is the standard by which other cultures in the society are judged.

A dominant culture need not be the culture of the majority of people; rather, it is simply the culture of that group in society with enough power to define the cultural framework. As an example, think of a college or university that has a strong system of fraternities and sororities. On campus, the number of students belonging to fraternities and sororities is probably a numerical minority of the total student body, but the cultural system established by the Greeks may dominate campus life nonetheless. In a society as complex as the United States, it is hard to isolate a single dominant culture, although there is a widely acknowledged "American" culture that is considered to be the dominant one. Stemming from middle-class values, habits, and economic resources, this culture is strongly influenced by instruments of culture such as television, the fashion industry, and Anglo-European traditions and includes diverse elements such as fast food, Christmas shopping, and professional sports. It is also a culture that emphasizes achievement and individual effort—a cultural tradition that we will later see has a tremendous impact on how many in the United States view inequality (see Chapter 8).

Subcultures

Subcultures are the cultures of groups whose values and norms of behavior differ to some degree from those of the dominant culture. Members of subcultures tend to interact frequently with one another and share a common worldview. They may be identifiable by their appearance (style of clothing or adornments) or perhaps by language, dialect, or other cultural markers. You can view subcultures along a continuum of how well they are integrated into the dominant culture. Subcultures typically share some elements of the dominant culture and coexist within it, although some subcultures may be quite separated from the dominant one. This separation occurs because they are either unwilling or unable to assimilate into the dominant culture, that is, share its values, norms, and beliefs (Dowd and Dowd 2003).

Rap and hip-hop music first emerged as a subculture where young African Americans developed their own style of dress and music to articulate their resistance to the dominant White culture. Now, rap and hip-hop have been incorporated into mainstream youth culture.

Indeed, they are now global phenomena, as cultural industries have turned hip-hop and rap into a profitable commodity. Even so, rap still expresses an oppositional identity for Black and White youth and other groups who feel marginalized by the dominant culture (Rose 1994).

Some subcultures retreat from the dominant culture, as do the Amish, some religious cults, and some communal groups. In these cases, the subculture is actually a separate community that lives as independently from the dominant culture as possible. Other subcultures may coexist with the dominant society, and members of the subculture may participate in both the subculture and the dominant culture.

Subcultures also develop when new groups enter a society. Puerto Rican immigration to the U.S. mainland, for example, has generated distinct Puerto Rican subcultures within many urban areas. Although Puerto Ricans also partake in the dominant culture, their unique heritage is part of their subcultural experience. Parts of this culture are now entering the dominant culture. Salsa music, now heard on mainstream radio stations, was created in the late 1960s by Puerto Rican musicians who were expressing the contours of their working-class culture (Sanchez 1999; Boggs 1992). The themes in salsa reflect the experience of barrio people and mix the musical traditions of other Latin music, including rumba, mambo, and cha-cha. As with other subcultures, the boundaries between the dominant culture and the subculture are permeable, resulting in cultural change as new groups enter society.

Amy Pancetta/AP Photo

The Amish people form a subculture in the United States, although preserving their traditional way of life can be a challenge in the context of contemporary society.

Countercultures

Countercultures are subcultures created as a reaction against the values of the dominant culture. Members of the counterculture reject the dominant cultural values, often for political or moral reasons, and develop cultural practices that explicitly defy the norms and values of the dominant group. Nonconformity to the dominant culture is often the hallmark of a counterculture. Youth groups often form countercultures. Why? In part, they do so to resist the culture of older generations, thereby asserting their independence and identity. Some also argue that young people establish countercultures because they have so little power in society that they have to construct their own cultures to have some sort of status, or social standing, at least among their peers (Milner 2004). Thus, countercultures among youth, like other countercultures, usually have a unique way of dress, their own special language, perhaps even different values and rituals.

Some countercultures directly challenge the dominant society. The white supremacist movement is an example. People affiliated with this movement have an extreme worldview, one that is in direct opposition to dominant values. White supremacist groups have developed a shared worldview, one based on extreme hostility to racial minorities, gays, lesbians, and feminists. Because of their self-contained culture—one focused on hate—they can be very dangerous (Ferber 1998; Stern 1996).

Countercultures may also develop in situations where there is political repression and some groups are forced "underground." Under a dictatorship, for example, some groups may be forbidden to practice their religion or speak their own language. In Spain under the dictator Francisco Franco, people were forbidden to speak Catalan—the language of the region around Barcelona. When Franco died in 1975 and Spain became

The styles and practices of subcultures can be a source of innovation in society. Hip-hop, for example, once a subculture associated only with Black urban youth, has now influenced so-called mainstream style.

more democratic, the Catalan language flourished—both in public speaking and in the press.

Ethnocentrism and Cultural Relativism

Because culture tends to be taken for granted, it can be difficult for people within a culture to see their culture as anything but "the way things are." It can thus be difficult to view other cultures without making judgments based on one's own cultural views. **Ethnocentrism** is the habit of seeing things only from the point of view of one's own group. Judging one culture by the standards of another culture is ethnocentric. An ethnocentric perspective prevents you from understanding the world as it is experienced by others, and it can lead to narrow-minded conclusions about the worth of diverse cultures.

Any group can be ethnocentric. Also, ethnocentrism can be extreme or subtle—as in the example of social groups who think their way of life is better than that of any other group. Is there such a ranking among groups in your community? Fraternities and sororities often build group rituals around such claims, youth groups see their way of life as superior to adults, and urbanites may think their cultural habits are more sophisticated than those of groups labeled "country hicks." Ethnocentrism is a powerful force because it combines a strong sense of group solidarity with the idea of group superiority.

Ethnocentrism can build group solidarity, but it also discourages intergroup understanding. Understanding ethnocentrism is critical to understanding some of the major conflicts that are shaping current history. Taken to extremes, ethnocentrism can lead to overt political conflict, war, terrorism, even *genocide,* which is the mass killing of people based on their membership in a particular group. Understanding ethnocentrism can help explain the belief of groups such as al Qaeda that terrorism is justified as a religious *jihad* (defined as a religious struggle to defend Islamic faith). You might wonder how someone could believe so much in the righteousness of their religious faith that they would murder people. Ethnocentrism is a key part of the answer. Understanding ethnocentrism does not excuse such behavior, but it helps you understand how such murderous behavior can occur, though it would be overly simple to explain current political conflicts only in terms of ethnocentrism.

Ethnocentrism can also help you to understand the view that many nations now have of the United States (see Figure 2.1)—a fact that people within the United States have difficulty understanding because we hold ethnocentric views of our own culture, as if it is superior to all others. Many other nations do not see U.S. culture in the positive light that U.S. citizens might expect. As this figure shows, cultural values in the Islamic world can clash with those of the West and are part of the complexity of U.S. relations with those cultures.

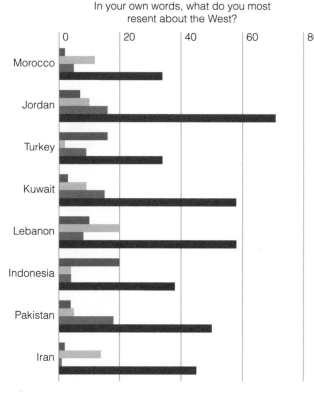

In your own words, what do you most resent about the West?

■ Arrogance, think they are the best/most powerful, advanced, civilized

■ High crime rate, drugs, violence, alcoholism, corruption

■ Negative attitude toward Arabs/Muslims

■ Social aspects (too free, low morals, free sex, impolite to elders, dislocation of family relations)

FIGURE 2.1 *Islamic Views of the West*

Data: Burkholder, Richard. 2003. "Iraq and the West: How Wide Is the Morality Gap?" *The Gallup Poll*, Princeton, NJ. **www.gallup.com**

Contrasting with ethnocentrism is cultural relativism. **Cultural relativism** is the idea that something can be understood and judged only in relation to the cultural context in which it appears. This does not make every cultural practice morally acceptable, but it suggests that without knowing the cultural context, it is impossible to understand why people behave as they do. For example, in the United States, burying or cremating the dead is the cultural practice. It may be difficult for someone from this culture to understand that in parts of Tibet, with a ruggedly cold climate and the inability to dig the soil, the dead are cut into pieces and left for vultures to eat. Although this would be repulsive (and illegal) in the United States, within Tibetan culture, this practice is understandable.

Understanding cultural relativism gives insight into some controversies, such as the international debate about the practice of clitoridectomy—a form of genital mutilation. In a clitoridectomy (sometimes called female circumcision), all or part of a young

woman's clitoris is removed, usually not by medical personnel, often in very unsanitary conditions, and without any pain killers. Sometimes, the lips of the vagina may be sewn together. Human rights and feminist organizations have documented this practice in some countries on the African continent, in some Middle Eastern nations, and in some parts of Southeast Asia; estimates are that around two million girls per year, worldwide, are at risk. This practice is most frequent in cultures where women's virginity is highly prized and where marriage dowries depend on some accepted proof of virginity (Toubia and Izett 1998).

From the point of view of Western cultures, clitoridectomy is genital mutilation and an example of violence against women. Many have called for international intervention to eliminate the practice, but there is also a debate about whether disgust at this practice should be balanced by a reluctance to impose Western cultural values on other societies. Should cultures have the right of self-determination or should cultural practices that maim people be treated as violations of human rights? This controversy is unresolved. The point is to see that understanding a cultural practice requires knowing the cultural values on which it is based.

The Globalization of Culture

The infusion of Western culture throughout the world seems to be accelerating as the commercialized culture of the United States is marketed worldwide. One can go to quite distant places in the world and see familiar elements of U.S. culture, whether it is McDonald's in Hong Kong, The Gap in South Africa, or Disney products in western Europe. From films to fast food, the United States dominates international mass culture, largely through the influence of capitalist markets, as conflict theorists would argue. The diffusion of a single culture throughout the world is referred to as **global culture.** Despite the enormous diversity of cultures worldwide, fashion, food, entertainment, and other cultural values are increasingly dominated by U.S. markets, thereby creating a more homogenous world culture. Global culture is increasingly marked by capitalist interests, squeezing out the more diverse folk cultures that have been common throughout the world (Rieff 1993).

Does increasing globalization of culture change traditional cultural values? Some worry that globalization imposes Western values on non-Western cultures, thus eroding long-held cultural traditions. But global economic change can also introduce more tolerant values to cultures that might have had a narrower worldview previously. As globalization occurs, *both* economic changes *and* traditional cultural values shape the emerging national culture of different societies (Inglehart and Baker 2000).

The conflict between traditional and more commercial values is now being played out in world affairs, with some arguing that the conflicts we see in international relations are partially rooted in a struggle between the values of a consumer-based, capitalist Western culture and the traditional values of local communities. Benjamin Barber (1995) expresses this as the struggle between "McWorld" and "Jihad"—a struggle he interprets as the tension between global commerce and parochial values. As some people resist the influence of market-driven values, movements to reclaim or maintain ethnic and cultural identity can intensify. Thus, you can witness a proliferation of culturally based movements, including strong feelings of nationalism, such as among extremist groups in the Middle East.

The Mass Media and Popular Culture

Increasingly, culture in the United States—as well as in many parts of the world—is dominated and shaped by the media. Indeed, the culture of the United States is so infused by the media that, when people think of U.S. culture, they are likely thinking of something connected to the media—television, film, and so forth. The term **mass media** refers to the channels of communication that are available to wide segments of the population—the print, film, and electronic media (radio and television), as well as the Internet.

The mass media have extraordinary power to shape culture, including what people believe and the information available to them. If you doubt this, observe how much the mass media affect your everyday life. Friends may talk about last night's episode of a particular show or laugh about the antics of their favorite sitcom character. Your way of dressing, talking, and even thinking has likely been shaped by the media. Now, even relationships are found and maintained via electronic networks such as Facebook—networks that are part of the mass media.

You can find the mass media everywhere—in airports, elevators, classrooms, bars and restaurants, and hospital waiting rooms. Enter an elevator in a major hotel and you might find CNN or The Weather Channel on twenty-four hours a day. You may even be born to the sounds and images of television, since they are turned on in many hospital delivery rooms. Television is now so ever-present in our lives that 42 percent of all U.S. households are now called "constant television households"—that is, those households where television is on most of the time (Gitlin 2002). For many families, TV is the "babysitter." Ninety-eight percent of all homes in the United States have at least one television. The average person consumes some form of media sixty-nine hours per week—more time than they likely spend in school or at work; thirty-three of these hours are spent watching television. On average, people spend only about two hours per week reading books (U.S. Census

Bureau 2009a). In fact, television dominates leisure time for most Americans; one-quarter of all Americans say watching TV is their favorite way to spend an evening, compared to 9 percent who would rather read and 1 percent whose favorite evening activity is listening to music (Saad 2002).

Television is a powerful transmitter of culture, but it also portrays a very homogeneous view of culture because in seeking the widest possible audience, networks and sponsors find the most common ground and take few risks. The mass media (and television, in particular) also play a huge role in shaping people's perception and awareness of social issues. For example, even though crime has actually decreased, the amount of time spent reporting crime in the media has increased. Sociologists have found that people's fear of crime is directly related to the time they spend watching television or listening to the radio (Angotti 1997; Chiricos et al. 1997). The media shape our definition of social problems by determining the range of opinion or information that is defined as legitimate and by deciding which experts will be called on to elaborate an issue (Gitlin 2002).

Perhaps you think that, despite this information, the mass media have little effect on you—no matter how much you might enjoy it. But try to do without it—even for a brief period of time—by trying the experiment in the "See for Yourself" feature on page 44. Simply getting away from all of the forms of media that permeate daily life may be extremely difficult to do!

The Organization of Mass Media

Mass media are not only a pervasive part of daily life, they are big business. On average, consumers spend $1000 per year on media consumption, half of that for television. That may not seem like much until you realize that the motion picture and video industry is a $73 billion industry; TV, $38 billion; telecommunications (including cell phones, Internet access, cable, and so forth), over $550 billion!

Thus, mass media are organized via powerful economic interests. And, increasingly, the media are owned by a small number of companies—companies that form huge media monopolies. This means that a few very powerful groups—media conglomerates—are the major producers and distributors of culture. A single corporation can control a huge share of television, radio, newspapers, music, publishing, film, and the Internet, as shown in Figure 2.2. As the production of popular culture becomes concentrated in the hands of just a few, there may be less diversity in the content.

The organization of the mass media as a system of economic interests means that there is enormous power in the hands of a few to shape the culture of the whole society. Sociologists refer to the concentration of cultural power as **cultural hegemony** (pronounced "heh-JEM-o-nee"), defined as the pervasive and excessive influence of one culture throughout society. Cultural hegemony means that people may conform to

Hearst Corporation owns:

17 magazines: including *Cosmo, Cosmo Girl, Marie Claire, O* (The Oprah Magazine), *Good Housekeeping, Redbook, Seventeen, Popular Mechanics,* and others), 12 newspapers, 27 local TV stations, several cable networks, including The History Channel, part of ESPN, Lifetime, A&E, Hearst Books, King Features TV syndicate; holds investments in Netscape, XM Satellite radio, drugstore.com, and more . . .

News Corporation owns:

World's largest newspaper publisher; Fox News; myspace.com (purchased in 2005 for $580 million); William Morrow and Avon Books; 35 local radio stations; 25 international newspapers; New York Post; Broadcasting rights for NFL in Asia; Harper Collins Publishing; TV Guide; 20th Century Fox Film; Los Angeles Dodgers (sold in 2004); Los Angeles Lakers; Los Angeles Kings; Staples Center (LA); and more . . .

Time Warner owns:

HBO; CNN; AOL Instant Messenger; Warner Brothers; Warner Brothers Studios; WB Television Network; Hanna-Barbera Cartoons; Castle Rock Entertainment; Warner Home Video; Turner Network Television; TBS Superstation; Netscape; Amazon.com (partial owner); Warner Brothers International Theaters (owns/operates multiplex theatres in over 12 countries); over 50 magazines (including *Time; Fortune; People; Life; Sports Illustrated; Money; In Style; Cooking Light; This Old House; Field and Stream; Popular Science; Snowboard Life; Yachting Magazine; Travel and Leisure;* and others); Atlanta Braves; and more . . .

Viacom owns:

MTV; BET (Black Entertainment Television); Nickelodeon; Paramount Pictures; Dreamworks; CBS (split in 2006); Simon and Schuster Publishers; College TV Network; Blockbuster Video; 39 television stations; 185 local radio stations; VH1; Comedy Central; Outdoor Systems billboards; Madison Square Garden (sold in 1995 for $1 billion); TV Land; Pearson Publishing (publisher of elementary, high school, and college textbooks); Rumored to be considering purchasing facebook.com for $2 billion; and more . . .

Disney owns:

Disney Pictures; Miramax Films; Touchstone Pictures; Buena Vista Home Entertainment; Pixar; ABC television network; 10 local television stations; ESPN network; ABC Family Television; Disney Channel; 68 local radio stations; 15 magazines; Buena Vista Music Group; 5 Disneylands; Disney stores; Disney Cruise Line; Disney Shopping Inc.; and more . . .

FIGURE 2.2 *Cultural Monopolies*

Note: Because corporate mergers are so common, these data were current as of 2007.

cultural patterns and interests that benefit powerful elites, even without those elites overtly forcing people into conformity. Thus, on the one hand, while there seems to be enormous choice in what media forms people consume, the cultural messages are largely homogenous (meaning "sameness"). Cultural hegemony produces a homogeneous mass culture, even when it appears that there is vast choice. Thus, cultural monopolies are a means through which powerful groups gain the assent of those they rule. The concept of cultural hegemony implies that culture is highly politicized, even if it does not appear so. Through cultural hegemony, those who control cultural institutions can also control people's political awareness because they create cultural beliefs that make the rule of those in power seem inevitable and right. As a result, political resistance to the dominant culture is blunted (Gramsci 1971). We explore this idea further in the section below on sociological theories of culture.

The Media and Popular Culture

Because the mass media pervade the whole society, the media influence such things as popular styles, language, and value systems. Together, these create **popular culture,** meaning the beliefs, practices, and objects that are part of everyday traditions. In a society so dominated by the mass media, popular culture, including such things as music and films, mass-marketed books and magazines, large-circulation newspapers, and Internet websites, are mass-produced. Popular culture is distinct from elite culture, which

The power of the mass media can be staggering, as evidenced by the popularity of American Idol.

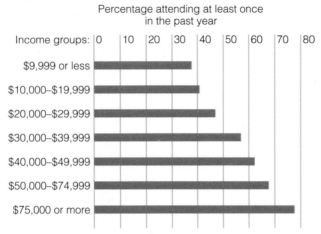

FIGURE 2.3 *Who Goes to the Movies?*

Source: U.S. Census Bureau. 2009a. *Statistical Abstract of the United States 2009*. Washington, DC: U.S. Department of Commerce, p. 747.

is shared by only a select few but is highly valued. Unlike elite culture (sometimes referred to as "high culture"), popular culture is mass-consumed and has enormous significance in the formation of public attitudes and values. Popular culture is also supported by patterns of mass consumption, as the many objects associated with popular culture are promoted and sold to a consuming public.

The distinction between popular and elite culture means that culture is consumed in different ways by various segments of the population. This, too, is affected by patterns of social class, race, and gender in the society. While popular culture may be widely available and be relatively cheap for the consumer, some groups derive their cultural experiences from expensive theater shows or opera performances where tickets may cost hundreds of dollars. Meanwhile, millions of "ordinary" citizens get their primary cultural experience from television, movie rentals, and increasingly, the Internet. Even something as seemingly common as Internet usage reflects patterns of social class differences in society, as you can see in Figure 2.4. This inequality has led many to advocate for free wireless service in some cities, as a way of trying to make Internet access more democratic.

Race, Gender, and Class in the Media

Many sociologists argue that the mass media can promote narrow definitions of who people are and what they can be. What is considered beautiful, for example? Is there a universal idea of beauty, or do the media promote different ideals for different groups? The mass media have a huge impact on how we see beauty and who or what is defined as beautiful; moreover, these beauty ideals change over time. Aging is defined as not beautiful, youth is. Light skin is promoted as more beautiful than dark skin, regardless of race, although being tan is seen as more beautiful than being pale. In African American women's magazines, the models typified as most beautiful are generally those with the clearly Anglo features of light skin, blue eyes, and straight or wavy hair. European facial features are also pervasive in the images of Asian and Latino women appearing in U.S. magazines. The point is that the media communicate that only certain forms of beauty are culturally valued. These ideals are not somehow "natural." They are constructed by those who control cultural and economic institutions (Craig 2002; Gimlin 2002).

You can learn a lot about how the media shape popular culture by looking carefully and systematically at the images produced and disseminated via the media. Content analyses of the media (a research

SEE *FOR YOURSELF*

TWO DAYS WITHOUT THE MEDIA

Suppose that you lived for a few days without use of the mass media that permeate our lives. How would this affect you? In an intriguing experiment, Charles Gallagher (a sociologist at La University) has developed a research project for students in which he asks them to stage a media blackout in their lives for just forty-eight hours. You can try this LaSalle yourself.

Begin by keeping a written log for forty-eight hours of exactly how much time you spend with some form of media. Include all time spent watching television, on the Internet, reading books and magazines, listening to music, viewing films, even using cell phones—any

activity that can be construed as part of the media monopoly on people's time.

Next, eliminate all use of the media, except for that required for work, school, and emergencies for a forty-eight-hour period, keeping a journal as you go of what happens, what you are thinking, what others say, and how people interact with you. *Warning: If you try the media blackout, be sure to have some plan in place for having your family and/or friends contact you in case of an emergency!* When one of the authors of this book (Andersen) had her students do this experiment, they complained even before starting that they wouldn't be able to do it! But they had to try. What happened?

First, Andersen's students had help: the week of the assignment came during a hurricane on the East Coast when many were without power for several

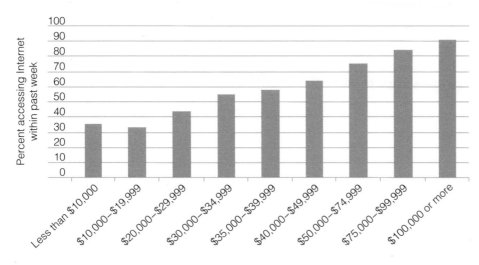

FIGURE 2.4 *Social Class and the Internet*

Source: Mediamark Research and Intelligence. 2007. New York.

method discussed in the following chapter) show distinct patterns of how race and gender and class are depicted in various media forms. On prime-time television, men are still a large majority of the characters shown. Over the years, there has been an increase in the extent to which women are depicted in professional jobs, but such images usually depict professional women as young (suggesting that career success comes early), thin, and beautiful women (Signorielli and Bacue 1999). In music videos, women are more present, but typically wearing sexy and skimpy clothing and more often the object of another's gaze than is true for their male counterparts; Black women especially are represented in sexualized ways (Emerson 2002; Signorielli et al. 1994; Collins 2004).

Even though African Americans and Hispanics watch more television than White people do, they are a smaller proportion of TV characters, generally confined to a narrow variety of character types, depicted in stereotypical ways. Latinos are often stereotyped

as criminals, passionate lovers, or comic figures. African American men are most often seen as athletes and sports commentators, criminals, or entertainers. Women who work as football sports commentators are typically on the sidelines, reporting not so much on the play of the game, as the human interest or injury reports—suggesting that women's role in sports is limited to that of the nurturer. It is difficult to find a single show where Asians are the principal characters—usually they are depicted in silent roles as domestics or behind-the-scenes characters. Native Americans make occasional appearances, where they usually are depicted as mystics or warriors. Jewish women are generally invisible on popular TV programming, except when they are ridiculed in stereotypical roles. Arab Americans are likewise stereotyped, depicted as terrorists, rich oil magnates, or in the case of women, as perpetually veiled and secluded (Read 2003; Mandel 2001).

In a good example of how race and gender stereotypes merge in the mass media, one analyst carefully

days. This did not deter the students from thinking they just *had to have* their DVD players, music, TV, and cell phones! Many of the students said they could not stand being without access to the media—even for a few hours. Most could not go the full two days without using the media.

Most reported that they felt isolated during the media exercise, not just from information, but mostly from other people. They were excluded from conversations with friends about what happened on a given television episode or about film characters or movie stars profiled in magazines and from playing computer games. One even wrote that without the media she felt that she had no personality! Without their connection to the media, students were alienated, isolated, and detached, although most also reported that they studied more without the distrac-

tion of the media. A most interesting finding was that several reported that they were much more reflective during this time and had more meaningful conversations with friends.

After trying this experiment, think about the enormous influence that the mass media have in shaping everyday life, including your self-concept and your relationship with other people. What does this exercise teach you about *cultural hegemony*? The role of the mass media in shaping society? How would each of the following theoretical frameworks explain what happened during your media blackout: functionalism, conflict theory, feminist theory, symbolic interaction?

Source: Personal correspondence, Charles Gallagher, LaSalle University.

analyzed multiple episodes of the popular show, *The Bachelor* (Dubrofsky 2006). Supposedly, the women all have an equal chance at being selected as the bachelor's mate, the basic concept of the show asserting heterosexual relationships as the most appropriate romance. But, as Dubrofsky shows, women of color are never chosen as the bachelor's mate; they are, in fact, eliminated early from the competition. Equally revealing, Dubrofsky shows how the show's set suggests a harem-like quality—multiple women available to one man, women lounging around on plush furniture, assembled to resemble a stereotypical harem—with plush, overstuffed cushions, lush gardens, and often Middle-Eastern tapestries on the walls, thereby producing stereotypes about the supposed sexual excess and availability of Middle Eastern women. Research documents numerous examples of stereotyped portrayals in the media—stereotypes you will see for yourself if you step outside of the taken-for-granted views with which you ordinarily observe the media.

thinkingsociologically

Watch a particular kind of television show (situation comedy, sports broadcast, children's cartoon, or news program, for example) and make careful written notes on the depiction of different groups in this show. How often are women and men or boys and girls shown?

How are they depicted? You could also observe the portrayal of Asian Americans, Native Americans, African Americans, or Latinos. What do your observations tell you about the cultural ideals that are communicated through *popular culture*?

Class stereotypes abound in the media and popular culture as well, with working-class men typically portrayed as being ineffectual, even buffoonish (Butsch 1992; Dines and Humez 2002). A recent study of TV talk shows also demonstrates how such shows exploit working-class people. The researcher, Laura Grindstaff, spent six months working on two popular talk shows, carefully doing participant observation and interviewing staff and guests. She found that guests had to enact stereotypes of their groups to get airtime. She argues that, although these shows give ordinary people a place to air their problems and be heard, they exploit the working class, making a spectacle of their troubles (Grindstaff 2002; Press 2002).

Recently, there has been increased representation of gays and lesbians in the media, after years of being virtually invisible or only the subject of ridicule. As advertisers have sought to expand their commercial markets, they are showing more gay and lesbian characters on television. This makes gays and lesbians more visible, although critics point out that they are still cast in narrow and stereotypical terms, showing little about real life for gays and lesbians.

Nonetheless, cultural visibility for any group is important because it validates people and can influence the public's acceptance of and generate support for equal rights protection (Gamson 1998a).

Television is not the only form of popular culture that influences public consciousness, class, gender, and race. Music, film, books, and other industries play a significant role in molding public consciousness. What images are produced by these cultural forms? You can look for yourself. Try to buy a birthday card that contains neither an age or gender stereotype, or watch TV or a movie and see how different gender and race groups are portrayed. You will likely find that women are depicted as trying to get the attention of men; African Americans are more likely than Whites to be seen singing and dancing.

Do these images matter? Studies find that exposure to traditional sexualized imagery in music videos has a negative effect on college students' attitudes, for example, holding more adversarial attitudes about sexual relationships (Kalof 1999). Other studies find that even when viewers see media images as unrealistic, they think that others find the images important and will evaluate them accordingly; this has been found especially true for young White girls who think boys will judge them by how well they match the media ideal (Milkie 1999). Although people do not just passively internalize media images and do distinguish between fantasy and reality (Hollander 2002; Currie 1997), such images form cultural ideals that have a huge impact on people's behavior, values, and self-image.

Theoretical Perspectives on Culture and the Media

Sociologists study culture and the media in a variety of ways, asking a variety of questions about the relationship of culture to other social institutions and the role of culture in modern life. One important question for sociologists studying the mass media is whether these images have any effect on those who see them. Do the media create popular values or reflect them? The **reflection hypothesis** contends that the mass media reflect the values of the general population (Tuchman 1979). The media try to appeal to the most broad-based audience, so they aim for the middle ground in depicting images and ideas. Maximizing popular appeal is central to television program development; media organizations spend huge amounts on market research to uncover what people think and believe and what they will like. Characters are then created with whom people will identify. Interestingly, the images in the media with which we identify are distorted versions of reality. Real people seldom live like the characters on television, although part of the appeal of these shows is how they build upon, but then mystify, the actual experiences of people.

The reflection hypothesis assumes that images and values portrayed in the media reflect the values existing in the public, but the reverse can also be true—that is, the ideals portrayed in the media also influence the values of those who see them. As an example, social scientists have studied the stereotyped images commonly found in children's programming. Among their findings, they have shown that the children who watch the most TV hold the most stereotypic gender attitudes (Signorielli 1991). Although there is not a simple and direct relationship between the content of mass media images and what people think of themselves, clearly these mass-produced images can have a significant impact on who we are and what we think.

Culture and Group Solidarity

Many sociologists have studied particular forms of culture and have provided detailed analyses of the content of cultural artifacts, such as images in certain television programs or genres of popular music. Other sociologists take a broader view by analyzing the relationship of culture to other forms of social organization. Beginning with some of the classical sociological theorists (see Chapter 1), sociologists have studied the relationship of culture to other social institutions. Max Weber looked at the impact of culture on the formation of social and economic institutions. In his classic analysis of the Protestant work ethic and capitalism, Weber argued that the Protestant faith rested on cultural beliefs that were highly compatible with the development of modern capitalism. By promoting a strong work ethic and a need to display material success as a sign of religious salvation, the Protestant work ethic indirectly but effectively promoted the interests of an emerging capitalist economy. (We revisit this issue in Chapter 13.) In other words, culture influences other social institutions.

Many sociologists have also examined how culture integrates members into society and social groups. Functionalist theorists, for example, believe that norms and values create social bonds that attach people to society. Culture therefore provides coherence and stability in society. Robert Putnam examines this idea in his book *Bowling Alone* (2000), in which he argues that there has been a decline in *civic engagement*—defined as participation in voluntary organizations, religious activities, and other forms of public life—in recent years. As people become less engaged in such activities, there is a decline in the shared values and norms of the society so that social disorder results. Sociologists are debating the extent to which there has been such a decline in public life, but from a functionalist perspective, the point is that participation in a common culture is an important social bond—one that unites society (Etzioni et al. 2001).

Classical theoretical analyses of culture have placed special emphasis on nonmaterial culture—the values, norms, and belief systems of society. Sociologists who use this perspective emphasize the integrative function of culture, that is, its ability to give people a sense of belonging in an otherwise complex social system (Smelser 1992a). In the broadest sense, they see culture as a major integrative force in society, providing societies with a sense of collective identity and commonly shared worldviews.

Culture, Power, and Social Conflict

Whereas the emphasis on shared values and group solidarity drives one sociological analysis of culture, conflicting values drives another. Conflict theorists (see Chapter 1) have analyzed culture as a source of power in society. You can find numerous examples throughout human history where conflict between different cultures has actually shaped the course of world affairs. One such example comes from the Middle East

table2.3   *Theoretical Perspectives on Culture*

	According to:		
Functionalism	**Conflict Theory**	**Symbolic Interaction**	**New Cultural Studies**
Culture . . .			
Integrates people into groups	Serves the interests of powerful groups	Creates group identity from diverse cultural meanings	Is ephemeral, unpredictable, and constantly changing
Provides coherence and stability in society	Can be a source of political resistance	Changes as people produce new cultural meanings	Is a material manifestation of a consumer-oriented society
Creates norms and values that integrate people in society	Is increasingly controlled by economic monopolies	Is socially constructed through the activities of social groups	Is best understood by analyzing its artifacts—books, films, television images

and the situation for the Kurdish people. The Kurds are an ethnic group (see Chapter 10) who speak their own language and inhabit an area in the Middle East that includes parts of Iraq, Iran, Turkey, and Syria, though they mostly live in northern Iraq. Most are Sunni Muslims, and they have experienced years of political and economic repression and, under Saddam Hussein, mass murder. Attempting to eliminate Kurds altogether, Saddam Hussein ordered the execution of over 180,000 people in Kurdish villages, often using chemical and biological weapons (O'Leary 2002). This and other examples of so-called ethnic cleansing show how cultural conflict can be driven by intense group hatred and powerful forms of domination.

Conflict theorists see contemporary culture as produced within institutions that are based on inequality and capitalist principles. As a result, the cultural values and products that are produced and sold promote the economic and political interests of the few—those who own or benefit from these cultural industries. As we have seen, this is especially evident in the study of the mass media and popular culture, marketed to the masses by entities with a vast economic stake in distributing their products. Conflict theorists conclude that the cultural products most likely to be produced are consistent with the values, needs, and interests of the most powerful groups in society. The evening news, for example, typically is sponsored by major financial institutions, and oil companies. Conflict theorists then ask how this commercial sponsorship influences the content of the news. If the news were sponsored by labor unions, would conflicts between management and workers always be defined as "labor troubles," or might newscasters refer instead to "capitalist troubles"?

Conflict theorists see culture as increasingly controlled by economic monopolies. Whether it is books, music, films, news, or other cultural forms, monopolies in the communications industry (where culture is increasingly located) have a strong interest in protecting the status quo. As media conglomerates swallow up smaller companies and drive out smaller, less-efficient competitors, the control that economic monopolies have over the production and distribution of culture becomes enormous. Mega-communications companies then influence everything—from the movies and television shows you see to the books you read in school.

However, culture can also be a source of political resistance and social change. Reclaiming an indigenous culture that had been denied or repressed is one way that groups mobilize to assert their independence. An example from within the United States is the *repatriation movement* among American Indians who have argued for the return of both cultural artifacts and human remains held in museum collections. Many American Indians believe that, despite the public good that is derived from studying such remains and objects, cultural independence and spiritual respect outweigh such scientific arguments (Thornton 2001). Other social movements, such as the gay and lesbian movement, have also used cultural performance as a means of political and social protest. Cross-dressing, drag shows, and other forms of "gender play" can be seen as cultural performances that challenge homophobia and traditional sexual and gender roles (Rupp and Taylor 2003).

A final point of focus for sociologists studying culture from a conflict perspective lies in the concept of cultural capital. **Cultural capital** refers to the cultural resources that are deemed worthy (such as knowledge of elite culture) and that give advantages to groups possessing such capital. This idea has been most developed by the French sociologist Pierre Bourdieu (1984), who sees the appropriation of culture as one way that groups maintain their social status.

Bourdieu argues that members of the dominant class have distinctive lifestyles that mark their status in society. Their ability to display this cultural lifestyle signals their importance to others; that is, they possess cultural capital. From this point of view, culture has a role in reproducing inequality among groups. Those with cultural capital use it to improve their social and economic position in society. Sociologists have found a significant relationship, for example, between cultural capital and grades in school; that is, those from the more well-to-do classes (those with more cultural capital) are able to parlay their knowledge into higher grades, thereby reproducing their social position by being more competitive in school admissions and, eventually, in the labor market (Hill 2001; Treiman 2001).

Symbolic Interaction and the Study of Culture

Especially productive when applied to the study of culture has been *symbolic interaction theory*—a perspective that analyzes behavior in terms of the meaning people give it. (See Chapter 1.) The concept of culture is central to this orientation. Symbolic interaction emphasizes the interpretive basis of social behavior, and culture provides the interpretive framework through which behavior is understood.

Symbolic interaction also emphasizes that culture, like all other forms of social behavior, is socially constructed; that is, culture is produced through social relationships and in social groups, such as the media organizations that produce and distribute culture. People do not just passively submit to cultural norms. Rather, they actively make, interpret, and respond to the culture around them. Culture is not one-dimensional; it contains diverse elements and provides people with a wide range of choices from which to select how they will behave (Swidler 1986). Culture, in fact, represents the creative dimension of human life.

In recent years, a new interdisciplinary field known as *cultural studies* has emerged that builds on the insights of the symbolic interaction perspective in sociology. Sociologists who work in cultural studies are often critical of classical sociological approaches to studying culture, arguing that the classical approach has overemphasized nonmaterial

[*what would* *they say now?*

Classical Theorists on Hip-Hop!

Suppose some of the classical theorists of sociology were asked to comment on the popularity of hip-hop? What might they say?

Emile Durkheim: I notice that young people can name hip-hop musicians whom others in the society do not recognize. This commonly happens as different generations tend to grow up within a shared music culture. Whether it's hip-hop, country, or pop, music cultures bind groups together by creating a sense of shared and collective identity. For young people, this makes them feel like part of a generation instead of being completely alienated from an otherwise adult-dominated culture.

Karl Marx: It is interesting that White youth are now the major consumers of hip-hop. Hip-hop originated from young, Black youth who are disadvantaged by the economic system of society. Now capitalism has appropriated this creative work and turned it into a highly profitable commodity that benefits dominant groups who control the music industry. As this has happened, the critical perspective originated by young, Black urban men has been supplanted by race and gender stereotypes that support the interests of the powerful.

Max Weber: Emile and Karl, you just see it one way. It's not that you are wrong, but you have to take a multi-dimensional view. Yes, hip-hop is an economic and a cultural phenomenon, but it is also linked to power in society. Haven't you noticed how political candidates try to use popular music to appeal to different political constituencies? Don't be surprised to find hip-hop artists performing at political conventions! That's what I find so intriguing: Hip-hop is an economic, cultural, and political phenomenon.

W.E.B. Du Bois: I've said that Black people have a "double consciousness"—one where they always have to see themselves through the eyes of a world that devalues them—American and "Black" at the same time. But, concurrently, the "twoness" that Black people experience generates wonderful cultural forms such as hip-hop that reflect the unique spirit of African Americans. I once wrote that "there is no true American music but the wild sweet melodies of the Negro slave" (DuBois 1903: 14), but I wish I had lived to see this new spirited and soulful form of musical expression!

culture, that is, ideas, beliefs, values, and norms. The new scholars of cultural studies find that material culture has increasing importance in modern society (Crane 1994; Walters 1999). This includes cultural forms that are recorded through print, film, artifacts, or the electronic media. Postmodernist theory has greatly influenced new cultural studies (see Chapter 1). *Postmodernism* is based on the idea that society is not an objective thing; rather, it is found in the words and images that people use to represent behavior and ideas. Given this orientation, postmodernism often analyzes common images and cultural products found in everyday life.

Classical theorists have tended to study the unifying features of culture; cultural studies researchers tend to see culture as more fragmented and unpredictable. To them, culture is a series of images—images that can be interpreted in multiple ways, depending on the viewpoint of the observer. From the perspective of new cultural studies theorists, the ephemeral and rapidly changing quality of contemporary cultural forms is reflective of the highly technological and consumer-based culture on which the modern economy rests.

Modern culture, for example, is increasingly dominated by the ever-changing, but ever-present, images that the media bombard us with in everyday life. The fascination that cultural studies theorists have for these images is partially founded in illusions that such a dynamic and rapidly changing culture produces.

Cultural Change

In one sense, culture is a conservative force in society; it tends to be based on tradition and is passed on through generations, conserving and regenerating the values and beliefs of society. Culture is also increasingly based on institutions that have an economic interest in maintaining the status quo. People are also often resistant to cultural change because familiar ways and established patterns of doing things are hard to give up. But in other ways, culture is completely taken for granted, and it may be hard to imagine a society different from that which is familiar.

Imagine, for example, the United States without fast food. Can you do so? Probably not. Fast food is so much a part of contemporary culture that it is hard to imagine life without it. Consider these facts about fast-food culture:

- The average person in the United States consumes three hamburgers and four orders of French fries per week.
- People in the United States spend more money on fast food than on movies, books, magazines, newspapers, videos, music, computers, and higher education combined.
- One in eight workers has at some point been employed by McDonald's.
- McDonald's is the largest private operator of playgrounds in the United States and the single largest purchaser of beef, pork, and potatoes.
- Ninety-six percent of American schoolchildren can identify Ronald McDonald—only exceeded by the number who can identify Santa Claus (Schlosser 2001).

Eric Schlosser, who has written about the permeation of society by fast-food culture, has written that "a nation's diet can be more revealing than its art or literature" (2001: 3). He relates the growth of the fast-food industry to other fundamental changes in American society, including the vast numbers of women entering the paid labor market, the development of an automobile culture, the increased reliance on low-wage service jobs, the decline of family farming, and the growth of agribusiness. One result is a cultural emphasis on uniformity, not to mention increased fat and calories in people's diets.

This example shows how cultures can change over time, sometimes in ways that are hardly visible to us unless we take a longer-range view or, as sociologists would do, question that which surrounds us. Culture is a dynamic, not static, force in society, and it develops as people respond to various changes in their physical and social environments.

Culture Lag

Sometimes cultures adjust slowly to changing cultural conditions, and the result can be **culture lag** (Ogburn 1922). Some parts of culture may change more rapidly than others; thus, one aspect of culture may "lag" behind another. Rapid technological change is often attended by culture lag because some elements of the culture do not keep pace with technological innovation. In today's world, we have the technological ability to develop efficient, less-polluting rapid transit, but changing people's transportation habits is difficult.

When culture changes rapidly or someone is suddenly thrust into a new cultural situation, the result can be **culture shock,** the feeling of disorientation when one encounters a new or rapidly changed cultural situation. Even moving from one cultural environment to another within one's own society can make a person feel out of place. The greater the difference between cultural settings, the greater the culture shock. Some people displaced from New Orleans following Hurricane Katrina experienced culture shock in the host communities where they relocated. Accustomed to the food, customs, and environment in their New Orleans homes, many evacuees were relocated to remote, mostly White, rural parts of the country. On top of their disorientation from the trauma of the storm itself, living in these new environments could be very disorienting (Wilkerson 2005).

Sources of Cultural Change

There are several causes of cultural change, including (1) a change in the societal conditions, (2) cultural diffusion, (3) innovation, and (4) the imposition of cultural change by an outside agency. Let us examine each.

1. *Cultures change in response to changed conditions in the society.* Economic changes, population changes, and other social transformations all influence the development of culture. A change in the makeup of a society's population may be enough by itself to cause a cultural transformation. The high rate of immigration in recent years has brought many cultural changes to the United States. Some cities, such as Miami and Los Angeles, have a Latin feel because of the large Hispanic population. But even outside urban areas, cultural change from immigration is apparent. Markets selling Asian, Mexican, and Middle Eastern foods are increasingly common; school districts include students who speak a huge variety of languages; popular music bears the imprint of different world cultures. This is not the first time U.S. culture has changed because of immigration. Many national traditions stem from the patterns of immigration that marked the earlier part of the twentieth century—think of St. Patrick's Day parades, Italian markets, and Chinatowns.

2. *Cultures change through cultural diffusion.* **Cultural diffusion** is the transmission of cultural elements from one society or cultural group to another. In our world of instantaneous communication, cultural diffusion is swift and widespread. This is evident in the degree to which worldwide cultures have been Westernized. Cultural diffusion also occurs when subcultural influences enter the dominant group. Dominant cultures are regularly enriched by minority cultures. An example is the influence of Black and Latino music on other musical forms. Rap music, for example, emerged within inner-city African American neighborhoods, describing and analyzing in its own form the economic and political

conditions of the urban ghetto. Now, rap music is listened to by White as well as Black audiences and is part of youth culture in general. Cultural diffusion is one thing that drives cultural evolution, especially in a society such as ours that is lush with diversity.

3. *Cultures change as the result of innovation, including inventions and technological developments.* Cultural innovations can create dramatic changes in society. Think, for example, of how the invention of trolleys, subways, and automobiles changed the character of cities. People no longer walked to work; instead, cities expanded outward to include suburbs. Furthermore, the invention of the elevator let cities expand not just out, but up.

Now the development of computer technology infiltrates every dimension of life. It is hard to overestimate the effect of innovation on contemporary cultural change. Technological innovation is so rapid and dynamic that one generation can barely maintain competence with the hardware of the next. The smallest laptop or handheld computer today weighs hardly more than a few ounces, and its capabilities rival that of computers that filled entire buildings only twenty years ago. Downloading music was not even imaginable just a few years ago; now it is a common practice.

What are some of the social changes that technology change is creating? People can now work and be miles—even nations—away from their places of employment. Families can communicate from multiple sites; children can be paged; grandparents can receive live photos of a family event; criminals are tracked via cellular technology; music can be stolen without even going in to a music store. Conveniences multiply with the growth of such technology, but so do the invasions of privacy and, perhaps, identity theft.

[a sociological eye on the media

The Blogging Culture

With the widespread use of the Internet, a new form of culture has emerged: the blog. A *blog* (short for "web log") is a chronological display of entries that people make on the Internet on topics that can be about anything, but most often involve politics or popular culture. A blog is an active diary of sorts in which multiple people participate, either as those who make the entries ("bloggers") or those who simply read them.

This new cultural phenomenon raises interesting questions for sociological research. Studies of blogs to date find, for example, that women are a small proportion of bloggers—only 10 percent of the bloggers on the most widely used political sites. Some use blogs as support systems—for example, a gay person in a very traditional and isolated community may participate in a blog that provides a national community of support. One study in China found that many women are using blogs to subvert traditional concepts of womanhood.

Blogs are also increasingly important in political organizing—not just in mainstream campaigns but in grassroots movements. The presidential campaign of President Barack Obama is a case in point. The extensive and innovative use of blogs and the Internet in the Obama campaign has been widely heralded as contributing to his support—especially among young, Internet-savvy voters.

The use of blogs is a good example of how technological innovation can create new forms of culture. Unlike traditional communities, blogging communities can cross vast geographic distances that connect people who might not ever meet face to face. Just as town meetings might have created a sense of community in the past, cyberspace communities can now create new "imagined communities." Some suggest that blogs can actually create a more democratic society by directly engaging more people in political discussion and activity.

Sources: Dolan, Jill. 2006. "Blogging on Queer Connections in the Arts and the Five Lesbian Brothers." *GLQ* 12: 491–506; Harp, Dustin, and Mark Tremayne. 2006. "The Gendered Blogosphere: Examining Inequality Using Network and Feminist Theory." *Journalism & Mass Communication Quarterly* 83: 247–264; Lawson-Borders, Gracie, and Rita Kirk. 2005. "Blogs in Campaign Communication." *American Behavioral Scientist* 49: 548–559; Macdougall, Robert. 2005. "Identity, Electronic Ethos, and Blogs: A Technologic Analysis of Symbolic Exchange on the New News Medium." *American Behavioral Scientist* 49: 575–599; Schaffer, Kay, and Song Xianlin. 2007. "Unruly Spaces: Gender, Women's Writing and Indigenous Feminism in China." *Journal of Gender Studies* 16 (1): 17–30; Perlmutter, David. 2008. *Blogwars: The New Political Battleground*. New York: Oxford University Press.

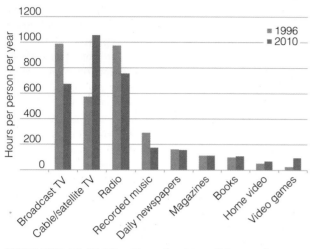

FIGURE 2.5 *The Changing Use of the Media*

Data: U.S. Census Bureau. 2009a. *The 2009 Statistical Abstract*. Washington, DC: U.S. Department of Commerce, p. 695.

Notes: Does not include Internet-related use of traditional media.

Examples include music downloaded from a computer, reading downloaded to e-books, listening to radio via the Internet, reading web-based newspapers, and so forth.

In such a rapidly changing technological world, it is hard to imagine what will be common in just a few years.

4. *Cultural change can be imposed.* Change can occur when a powerful group takes over a society and imposes a new culture. The dominating group may arise internally, as in a political revolution, or it may appear from outside, perhaps as an invasion. When an external group takes over the society of a "native," or indigenous, group—as White settlers did with Native American societies—they typically impose their own culture while prohibiting the indigenous group from expressing its original cultural ways. Manipulating the culture of a group is a way of exerting social control. Many have argued that public education in the United States, which developed during a period of mass immigration, was designed to force White, northern European, middle-class values onto a diverse immigrant population that was perceived to be potentially unruly and politically disruptive. Likewise, the schools run by the Bureau of Indian Affairs have been used to impose dominant group values on Native American children (Snipp 1996).

Resistance to political oppression often takes the form of a cultural movement that asserts or revives the culture of an oppressed group; thus, cultural expression can be a form of political protest. Identification with a common culture can be the basis for group solidarity, as found in the example of the "Black is beautiful" movement in the 1970s that encouraged Black Americans to celebrate their African heritage with Afro hairstyles, African dress, and African awareness. Cultural solidarity has also been encouraged among Latinos through La Raza Unida (meaning "the race," or "the people, united"). Cultural change can promote social change, just as social change can transform culture.

Chapter Summary

- ### What is culture?

  Culture is the complex and elaborate system of meaning and behavior that defines the way of life for a group or society. It is shared, learned, taken for granted, symbolic, and emergent and varies from one society to another.

- ### How do sociologists define norms, beliefs, and values?

  Norms are rules of social behavior that guide every situation and may be formal or informal. When norms are violated, *social sanctions* are applied. *Beliefs* are strongly shared ideas about the nature of social reality. *Values* are the abstract concepts in a society that define the worth of different things and ideas.

- ### What is the significance of diversity in human cultures?

  As societies develop and become more complex, cultural diversity can appear. The United States is highly diverse culturally, with many of its traditions influenced by immigrant cultures and the cultures of African Americans, Latinos, and Native Americans. The *dominant culture* is the culture of the most powerful group in society. *Subcultures* are groups whose *values* and cultural patterns depart significantly from the *dominant culture.*

- ***What is the sociological significance of the mass media and popular culture?***

  Elements of popular culture, such as the *mass media,* have an enormous influence on groups' beliefs and values, including images associated with racism and sexism. *Popular culture* includes the beliefs, practices, and objects of everyday traditions.

- ***What do different sociological theories reveal about culture?***

  Sociological theory provides different perspectives on the significance of culture. *Functionalist theory* emphasizes the influence of values, norms, and beliefs on the whole society. *Conflict theorists* see culture as influenced by economic interests and power relations in society. *Symbolic interactionists* emphasize that culture is socially constructed. This has influenced new cultural studies, which interpret culture as a series of images that can be analyzed from the viewpoint of different observers.

- ***How do cultures change?***

  There are several sources of cultural change, including change in societal conditions, *cultural diffusion,* innovation, and the imposition of change by dominant groups. As cultures change, *culture lag* can result, meaning that sometimes cultural adjustments are out of synchrony with each other. Persons who experience new cultural situations may experience *culture shock.*

Key Terms

beliefs 36
counterculture 40
cultural capital 48
cultural diffusion 50
cultural hegemony 42
cultural relativism 41
culture 27
culture lag 50
culture shock 50
dominant culture 38

ethnocentrism 40
ethnomethodology 36
folkways 34
global culture 41
language 31
laws 35
mass media 42
material culture 27
mores 35
nonmaterial culture 27

norms 34
popular culture 43
reflection hypothesis 46
Sapir–Whorf hypothesis 32
social sanctions 35
subculture 39
symbols 29
taboo 35
values 36

Online Resources

Sociology: The Essentials Companion Website

www.cengage.com/sociology/andersen
Visit your book companion website where you will find more resources to help you study and write your research papers. Resources include web links, learning objectives, Internet exercises, quizzing, and flash cards.

 is an easy-to-use online resource that helps you study in less time to get the grade you want NOW.

www.cengage.com/login
Need help studying? This site is your one-stop study shop. Take a Pre-Test and Cengage NOW will generate a Personalized Study Plan based on your test results. The Study Plan will identify the topics you need to review and direct you to online resources to help you master those topics. You can then take a Post-Test to determine the concepts you have mastered and what you still need to work on.

Duneier used a method of sociological research called *participant observation*. In this chapter we examine this method plus other methods of sociological research. Each method is different from the others, but they all share a common goal: a deeper understanding of how society operates.

The Research Process

Sociological research is the tool sociologists use to answer questions. There are various methods that sociologists use to do research, all of which involve rigorous observation and careful analysis.

As we saw in the chapter opener, sociologist Mitch Duneier (1999) examined several questions about a group of homeless people by living with them. Duneier was engaged in what is called **participant observation**—a sociological research technique in which the researcher actually becomes simultaneously both participant in and observer of that which she or he studies. In another example of participant observation, sociologist Peter Moskos (2008), as research for his doctoral dissertation, actually went through a police academy and spent two years as a beat policeman in a major American city, thus subjecting himself to both the rigid discipline of the police force as well as the dangers of the street in this role.

There are other kinds of sociological research that sociologists do as well. Some approaches are more structured and focused than participant observation, such as survey research. Other methods may involve the use of official records or interviews. The different approaches used reflect the different questions asked in the first place. Other methods may require statistical analysis of a large set of quantitative information. Either way, the chosen research method must be appropriate to the sociological question being asked. (In the Doing Sociological Research boxes throughout this book, we explore different research projects that sociologists have done, showing what question they started with, how they did their research, and what they found.)

However it is done, research is an engaging and demanding process. It requires skill, careful observation, and the ability to think logically about the things that spark your sociological curiosity.

Sociology and the Scientific Method

Sociological research derives from what is called the *scientific method,* originally defined and elaborated by the British philosopher **Sir Francis Bacon** (1561–1626). The **scientific method** involves several steps in a research process, including observation, hypothesis testing, analysis of data, and generalization. Since its beginnings, sociology has attempted to adhere to the scientific method. To the

degree that it has succeeded, sociology is a science; yet, there is also an art to developing sociological knowledge. Sociology aspires to be both scientific and humanistic, but sociological research varies in how strictly it adheres to the scientific method. Some sociologists test hypotheses (discussed later); others use more open-ended methods, such as in Duneier's study of homeless men or in Moskos's study of police officers on the beat.

Science is empirical, meaning it is based on careful and systematic observation, not just on conjecture. Although some sociological studies are highly *quantitative* and statistically sophisticated, others are *qualitatively* based, that is, based on more interpretive observations, not statistical analysis. Both quantitative and qualitative studies are empirical. Sociological studies may be based on surveys, observations, and many other forms of analysis, but they always depend on an empirical underpinning.

Sociological knowledge is not the same as philosophy or personal belief. Philosophy, theology, and personal experience can deliver insights into human behavior, but at the heart of the scientific method is the notion that a theory must be testable. This requirement distinguishes science from purely humanistic pursuits such as theology and literature.

One wellspring of sociological insight is **deductive reasoning.** When a sociologist uses deductive reasoning, he or she creates a specific research question about a focused point that is based on a more general or universal principle (see Figure 3.1). Here is an example of deductive reasoning: One might reason that because Catholic doctrine forbids abortion, Catholics would then

FIGURE 3.1 *The Research Process*

Research can begin by asking a research question derived from general theory or earlier studies, but it can also begin with an observation or even from the conclusion of prior research. One's research question is the basis for a research design and the subsequent collection of data. As this figure shows, the steps in the research process flow logically from what is being asked.

be less likely than other religious groups to support abortion rights. This notion is "deduced" from a general principle (Catholic doctrine). You could test this notion (the research question) via a survey. As it turns out, the testing of this hypothesis shows that it is incorrect: Surveys show that Catholics are more likely to support abortion rights than are other religious groups.

Inductive reasoning—another source of sociological insight—reverses this logic: that is, it arrives at general conclusions from specific observations. For example, if you observe that most of the demonstrators protesting abortion in front of a family planning clinic are evangelical Christians, you might infer that strongly held religious beliefs are important in determining human behavior. Again, referring to Figure 3.1, inductive reasoning would begin with one's observations. Either way—deductively or inductively—you are engaged in research.

Research Design

When sociologists do research, they engage in a process of discovery. They organize their research questions and procedures systematically—their research site being the social world. Through research, sociologists organize their observations and interpret them.

Developing a Research Question. Sociological research is an organized practice that can be described in a series of steps (see Figure 3.1). The first step in sociological research is to develop a research question. One source of research questions is past research. For any number of reasons, the sociologist might disagree with a research finding and decide to carry out further research or develop a detailed criticism of previous research. A research question can also begin from an observation that you make in everyday life, such as wondering about the lives of homeless people.

Developing a sociological research question typically involves reviewing existing studies on the subject, such as past research reports or articles. This process is often called a *literature review*. Digital technology has vastly simplified the task of reviewing past studies, that is, the "literature." Researchers who once had to burrow through paper indexes and card catalogs to find material relevant to their studies can now scan much larger swaths of material in far less time using online databases. The catalogs of most major libraries in the world are accessible on the Internet, as are specialized indexes, discussion groups, and other research tools developed to assist sociological researchers.

Increasingly, many journals that report new sociological research are now available online in full-text format. You must be careful using the Internet for research, however. How do you know when something found on the Web is valid or true? Much of what is found on the Web is invalid, that is, unsubstantiated by accurate research or empirical study. Pay attention, for example, to what person or group has

The research process involves several operations that can be performed on the computer, such as entering data in numerical form and writing the findings in a research report.

posted the website. Is it a political organization? An organization promoting a cause? A person expressing an opinion? See the box "A Sociological Eye on the Media: Research and the Media" on page 60 for some guidelines about interpreting what you see on the Web and in the media.

When you review prior research, you may wonder if the same results would be found if the study were repeated, perhaps examining a different group or studying the phenomenon at a different time. Research that is repeated exactly, but on a different group of people or in a different time or place, is called a **replication study.** Suppose earlier research found that women managers have fewer opportunities for promotion than do men. You might want to know if this still holds true. You would then replicate the original study, probably using a different group of women and men managers, but asking the same questions that were asked earlier. A replication study can tell you what changes have occurred since the original study and may also refine the results of the earlier work. Research findings should be reproducible; if research is sound, other researchers who repeat a study should get the same results unless, of course, some change has occurred in the interim.

Sociological research questions can also come from casual observation of human behavior. Perhaps you have observed the seating patterns in your college dining hall at lunch and wondered why people sit with the same group day after day. Does the answer point to similarity on the basis of race, gender, age, or perhaps political views? Answering this question would be an example of inductive reasoning: going from a specific observation (such as seating patterns at lunch) to a generalization (a theory about the effects of race and gender).

Creating a Research Design. A research **design** is the overall logic and strategy underlying a research project. Sociologists engaged in research may distribute questionnaires, interview people, or make direct observations in a social setting or laboratory. They might analyze cultural artifacts, such as magazines, newspapers, television shows, or other media. Some do research using historical records. Others base their work on the analysis of social policy. All these are forms of sociological observation. Research design consists of choosing the observational technique best suited to a particular research question.

thinkingsociologically

If you wanted to conduct research that would examine the relationship between student alcohol use and family background, what measures, or *indicators*, would you use to get at the two variables: alcohol use and family background? How might you *design* your study?

Suppose you wanted to study the career goals of student athletes. In reviewing earlier studies, perhaps you found research discussing how athletics is related to academic achievement (Messner 1992; Schacht 1996). You might also have read an article in your student newspaper reporting that the rate of graduation for women college athletes is much higher than the rate for men athletes and wondered if women athletes are better students than men athletes. In other words, are athletic participation, academic achievement, and gender interrelated, and if so, how?

Your research design would lay out a plan for investigating these questions. Which athletes would

Some research is done by analyzing the content of various cultural artifacts. Content analysis is one tool of sociological research.

you study? How will you study them? To begin, you will need to get sound data on the graduation rates of the groups you are studying to verify that your assumption of better graduation rates among women athletes is actually true. Perhaps, you think, the differences between men and women are not so great when the men and women play the same sports. Or perhaps the differences depend on other factors, such as what kind of financial support they get or whether coaches encourage academic success. To observe the influence of coaches, you might observe interactions between coaches and student athletes, recording what coaches say about class work. As you proceed, you would probably refine your research design and even your research question. Do coaches encourage different traits in men and women athletes? To answer this question, you have to build into your research design a comparison of coaches interacting with men and with women. Perhaps you even want to compare female and male coaches and how they interact with women and men. *The details of your research design flow from the specific questions you ask.*

Quantitative Versus Qualitative Research. The research design often involves deciding whether the research will be qualitative or quantitative or perhaps some combination of both. **Quantitative research** is that which uses numerical analysis. In essence, this approach reduces the data into numbers, for example, the percentage of teenage mothers in California. **Qualitative research** is somewhat less structured than quantitative research, yet still focuses on a central research question. Qualitative

Surveys and polls are a tool for sociological research. Here, presidential candidate Barack Obama is the winner of an election-night "exit poll" in the state of South Carolina. Exit polls gather the information from respondents as soon as they leave the polling place.

research allows for more interpretation and nuance in what people say and do and thus can provide an in-depth look at a particular social behavior. Both forms of research are useful, and both are used extensively in sociology.

Some research designs involve the testing of **hypotheses.** A hypothesis (pronounced "hy-POTH-i-sis") is a prediction or a hunch, a tentative assumption that one intends to test. If you have a research design that calls for the investigation of a very specific hunch, you might formulate a hypothesis. Hypotheses are often formulated as if–then statements. For example:

> *Hypothesis:* If a person's parents are racially prejudiced, then that person will, on average, be more prejudiced than a person whose parents are relatively free of prejudice.

This is merely a hypothesis or expectation, not a demonstration of fact. Having phrased a hypothesis, the sociologist must then determine if it is true or false. To test the preceding example, one might take a large sample of people and determine their prejudice level by interviews or some other mechanism. One would then determine the prejudice level of their parents. According to the hypothesis, one would expect to find more prejudiced children among prejudiced parents and more nonprejudiced children among nonprejudiced parents. If this association is found, the hypothesis is supported. If it is not found, then the hypothesis would be rejected.

Not all sociological research follows the model of hypothesis testing, but all research does include a plan for how **data** will be gathered. (Note that *data* is the plural form; one says, "data are used ...," not "data is used ....") Data can be qualitative or quantitative; either way, they are still data. Sociologists often try to convert their observations into a quantitative form (see the "Statistics in Sociology" box on page 64).

Sociologists frequently design research to test the influence of one variable on another. A **variable** is a characteristic that can have more than one value or score. A variable can be relatively straightforward, such as age or income, or a variable may be more abstract, such as social class or degree of prejudice. In much sociological research, variables are analyzed to understand how they influence each other. With proper measurement techniques and a good research design, the relationships between different variables can be discerned. In the example of student athletes given above, the variables you use would likely be student graduation rates, gender, and perhaps the sport played. In the hypothesis about race prejudice, parental prejudice and their child's prejudice would be the two variables you would study.

An **independent variable** (see Figure 3.2) is one that the researcher wants to test as the presumed cause of something else. The **dependent variable** is one on which there is a presumed effect. That is, if *X* is the independent variable, then *X* leads to *Y,* the dependent variable. In the previous example of the hypothesis, the amount of prejudice of the parent is the independent variable, and the amount of prejudice of the child is the dependent variable. In some sociological research, *intervening variables* are also studied—variables that fall between the independent and dependent variables.

Sociological research proceeds through the study of concepts. A **concept** is any abstract characteristic or attribute that can potentially be measured. Social class and social power are concepts. These are not things that can be seen directly, although they are key concepts in the field of sociology. When sociologists want to study concepts, they must develop ways of "seeing" them.

Variables are sometimes used to show more abstract concepts that cannot be directly measured, such as the concept of social class. In such cases the variables are **indicators**—something that points to or reflects an abstract concept. An indicator is a way of "seeing" a concept. An example is shown in Map 3.1 (page 62) on the United Nations' human development index. Here, the human development index is composed of several *indicators,* including life expectancy and educational attainment, combined to show levels of well-being. "Level of well-being" is the *concept.*

The **validity** of a measurement (an indicator) is the degree to which it accurately measures or reflects a concept. To ensure the validity of their findings, researchers usually use more than one indicator for a particular concept. If two or more chosen measures of a concept give similar results, it is likely that the measurements are giving an accurate—that is, valid—depiction of the concept. For example, using a person's occupation, years of formal education, and annual earnings—namely, using three indicators of her or his social class—would likely be more valid than using only one indicator.

Sociologists also must be concerned with the **reliability** of their research results. A measurement is reliable if repeating the measurement under the same circumstances gives the same result. If a person is given a test daily and every day the test gives different results, then the reliability of the test is poor. One way to ensure that sociological measurements are reliable is to use measures that have proved sound in past studies. Another technique is to have

FIGURE 3.2 *The Analysis of Variables*

Sociological research seeks to find out whether some independent variable (X) affects an intervening variable (Z), which in turn affects a dependent variable (Y).

a variety of people gather the data to make certain the results are not skewed by the tester's appearance, personality, and so forth. The researcher must be sensitive to all factors that affect the reliability of a study.

Sometimes sociologists want to gather data that would almost certainly be unreliable if the subjects (the people in the study) knew they were being studied. Knowing that they are being studied might cause people to change their behavior, a phenomenon in research known as the *Hawthorne effect*—an effect first discovered while observing work groups at a Western Electric plant in Hawthorne, Illinois. The work groups increased their productivity right after they were observed by the researchers—an effect not noticed at first by the researchers themselves.

a sociological eye on the media

Research and the Media

On any given day, if you watch the news, read a newspaper, or search the Web, you are likely to learn about various new research studies purporting some new finding. How do you know if the research results reported in the media are accurate?

Most people are not likely to check the details of the study or have the research skills to verify the study's claims. But one benefit of learning the basic concepts and tools of sociological research is to be able to critically assess the research frequently reported in the media. The following questions will help.

1. **What are the major variables in the study? Are the researchers claiming a causal connection between two or more variables?** For example, the press reported that one way parents can reduce the chances of their children becoming sexually active at an early age is to quit smoking (O'Neil 2002). The researcher who conducted this study actually claimed there was no direct link between parental smoking and teen sex, although she did find a correlation between parents' risky behaviors—smoking, heavy drinking, and not using seat belts—and children's sexual activity. She argued that parents who engage in unsafe activities provide a model for their children's own risky behavior (Wilder and Watt 2002).

   Just because there is a link, or "correlation," between two variables does not necessarily mean one caused the other. Seeing parental behavior as a model for what children do is hardly the same thing as seeing parents' smoking as the cause of early sexual activity!

2. **How have researchers defined and measured the major topics of their study?** For example, if someone claims that 10 percent of all people are gay, how is "being gay" defined? Does it mean having had only one such experience over one's entire lifetime or does it mean actually having a gay identity? The difference matters because one definition will likely inflate the number reported. Sometimes you must look up the original study, which may be online, to learn how things are defined or how they are measured. Ask yourself if the same conclusions would be reached had the researchers used different definitions and measurements.

3. **Is the research based on a truly random scientific sample, or is it biased?** You might have to go to the original source of the study to learn this, but often the sample will be reported in the press (even if in nonscientific language). For example, a study widely reported in the media had headlines exclaiming: "Study Links Working Mothers to Slower Learning" (Lewin 2002). But, if you read even the news report closely, you will learn that this study included only White, non-Hispanic families. Black and Hispanic children were dropped from some of the published results because there were too few cases in the sample to make meaningful statistical comparisons (Brooks-Gunn et al. 2002). Another study by the same research team found that there were no significant effects of mother's employment on children's intellectual development among African American or Hispanic children (Waldfogel et al. 2002). The point is not that the study is invalid, but that its results have more limited implications than the headlines suggest.

4. **Is there false generalization in the media report?** Often a study has more limited claims in the scientific version than what is reported in the media. Using the example just given about the connection between maternal employment and children's learning, it would be a big mistake to generalize from the study's results to all children and families. Remember that some groups were

An example of this effect would be a professor who wants to measure student attentiveness by observing how many notes are taken during class. Students who know they are being scrutinized will magically become more diligent!

Gathering Data. After research design comes data collection. During this stage of the research process, the researcher interviews people, observes behaviors, or collects facts that throw light on the research question. When sociologists gather original material, the product is known as *primary data.* Examples include the answers to questionnaires or notes made while observing group behavior. Sociologists often rely on *secondary data,* namely data that have already been gathered and organized by some other party. This can include national opinion polls, census data, national crime statistics, or data from an earlier study made available by the original researcher. Secondary data may also come from official sources, such as university records, city or county records, national health statistics, or historical records.

not included. Even within the study, the researchers found less effect due to mothers' employment in female-headed households than among married couples. You cannot generalize the findings reported in the media to all families.

5. **Can the study be replicated?** Unless there is full disclosure of the research methodology (that is, how the study was conducted), this will not be possible. But you can ask yourself how the study was conducted, whether the procedures used were reasonable and logical, and whether the researchers made good decisions in constructing their research question and research design.

6. **Who sponsored the study and do they have a vested interest in the study's results?** Find out if a group or organization with a particular vested interest in the outcome sponsors the research. For example, would you give as much validity to a study of environmental pollution that was funded and secretly conducted by a chemical company as you would a study on the same topic conducted by independent scientists who openly report their research methods and results? Research sponsored by interested parties does not necessarily negate research findings, but it can raise questions about the researchers' objectivity and the standards of inquiry they used.

7. **Who benefits from the study's conclusions?** Although this question does not necessarily challenge the study's findings, it can help you think about whom the findings are likely to help.

8. **What assumptions did the researchers have to make to ask the question they did?** For example, if you started from the assumption that poverty is not the individual's fault, but is the result of how society is structured, would you study the values of the poor or perhaps the values of policymakers? When research studies explore matters where social values influence people's opinions, it is especially important to identify the assumptions made by certain questions.

9. **What are the implications of the study's claims?** Thinking through the policy implications of a given result can often help you see things in a new light, particularly given how the media tend to sensationalize much of what is reported.

Consider the study of maternal employment and children's intellectual development examined in question 3 above. If you take the media headlines at face value, you might leap to the conclusion that working mothers hurt their children's intellectual development, and you might then think it would be best if mothers quit their jobs and stayed at home. But is this a reasonable implication of this study? Does the study not have just as many implications for day-care policies as it does for encouraging stay-at-home mothers? Especially when reported research studies involve politically charged topics (such as issues of "family values"), it is important to ask questions that explore various implications of social policies.

10. **Do these questions mean you should never believe anything you hear in the media?** Of course not. Thinking critically about research does not mean being negative or cynical about everything you hear or read. The point is not to reject all media claims out of hand, but instead to be able to evaluate good versus bad research. All research has limitations. Learning the basic tools of research, even if you never conduct research yourself or pursue a career where you would use such skills, can make you a better-informed citizen and prevent you from being duped by claims that are neither scientifically nor sociologically valid.

map**3.1** *VIEWING SOCIETY IN GLOBAL PERSPECTIVE*

The Human Development Index

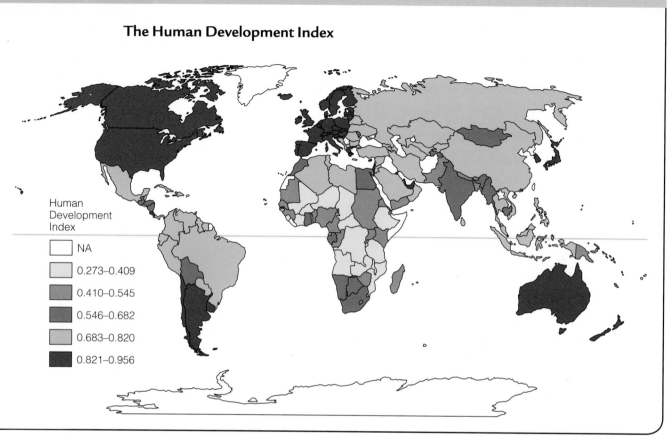

Human
Development
Index

☐	NA
	0.273–0.409
	0.410–0.545
	0.546–0.682
	0.683–0.820
	0.821–0.956

The human development index is a series of indicators developed by the United Nations and used to show the differing levels of well-being in nations around the world. The index is calculated using a number of indicators, including life expectancy, educational attainment, and standard of living. (Are these reasonable indicators of well-being? What else might you use?)

Data: United Nations. 2007. *Human Development Annual Report 2007.* **www.undp.org**

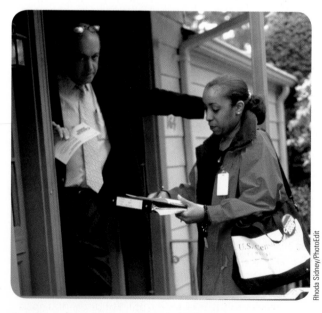

Rhoda Sidney/PhotoEdit

A census taker interviews a man in his home.

When gathering data, often the groups that sociologists want to study are so large or so dispersed that research on the whole group is impossible. To construct a picture of the entire group, sociologists take data from a subset of the group and extrapolate to get a picture of the whole. A **sample** is any subset of people (or groups or categories) of a population. A **population** is a relatively large collection of people (or groups or categories) that a researcher studies and about which generalizations are made. Suppose a sociologist wants to study the students at your school. All the students together constitute the population being studied. A survey could be done that reached every student, but conducting a detailed interview with every student would be highly impractical. If the sociologist wants the sort of information that can be gathered only during a personal interview, she would study only a portion, or sample, of all the students at your school.

How is it possible to draw accurate conclusions about a population by studying only part of it?

The secret lies in making sure that the sample is *representative* of the population as a whole. The sample should have the same mix of people as the larger population and in the same proportions. If the sample is representative, then the researcher can *generalize* what she finds from the sample to the entire population. For example, if she interviews a sample of 100 students and finds that 10 percent of them are in favor of a tuition increase, and if the sample is representative of the population, then she can conclude that about 10 percent of *all* the students at your school are in favor of a tuition increase. Note that a sample of 5 or 6 students would probably result in generalizations of poor quality, because the sample is not large enough to be representative. A *biased* (nonrepresentative) sample can lead to grossly inaccurate conclusions.

The best way to ensure a representative sample is to make certain that the sample population is selected randomly. A scientific **random sample** gives everyone in the population an equal chance of being selected. Quite often, striking and controversial research findings prove to be distorted by inadequate sampling. The man-on-the-street survey, much favored by radio and TV news reports, and certain other media as well, is the least-scientific type of sample and the least representative.

Analyzing the Data. After the data have been collected, whether primary or secondary data, they must be analyzed. **Data analysis** is the process by which sociologists organize collected data to discover the patterns and uniformities that the data reveal. The analysis may be statistical or qualitative. When the data analysis is completed, conclusions and generalizations can be made.

Data analysis is labor intensive, but it is also an exciting phase of research. Here is where research discoveries are made. Sometimes while pursuing one question, a researcher will stumble across an unexpected finding, referred to by researchers as **serendipity.** A serendipitous finding is something that emerges from a study that was not anticipated, perhaps the discovery of an association between two variables that the researcher was not looking for or some pattern of behavior that was outside the scope of the research design. Such findings can be minor sidelines to the researcher's major conclusions or, in some cases, lead to major new discoveries. They are part of the excitement of doing sociological research.

Reaching Conclusions and Reporting Results. The final stage in research is developing conclusions, relating findings to sociological theory and past research, and reporting the findings. An important question researchers will ask at this stage is whether their findings can be generalized. **Generalization** is the ability to draw conclusions from specific data and to apply them to a broader population.

Researchers ask, do my results apply only to those people who were studied, or do they also apply to the world beyond? Assuming that the results have wide application, the researcher can then ask if the findings refine or refute existing theories and whether the research has direct application to social issues. Using the earlier example of the relationship between parent and offspring prejudice, if you found that racially prejudiced people did tend to have racially prejudiced parents (thus supporting your hypothesis), then you might report these results in a paper or research report. You might also ask, what kinds of programs for reducing prejudice do the results of your study suggest?

The Tools of Sociological Research

There are several tools or techniques sociologists use to gather data. Among the most widely used are survey research, participant observation, controlled experiments, content analysis, historical research, and evaluation research.

The Survey: Polls, Questionnaires, and Interviews

Whether in the form of a questionnaire, interview, or telephone poll, surveys are among the most commonly used tools of sociological research. Questionnaires are typically distributed to large groups of people. The *return rate* is the percentage of questionnaires returned out of all those distributed. A low return rate introduces possible bias because the small number of responses may not be representative of the whole group.

Like questionnaires, interviews provide a structured way to ask people questions. They may be conducted face to face, by phone, even by electronic mail (email). Interview questions may be open-ended or closed-ended, though the open-ended form is particularly accommodating if respondents wish to elaborate.

Typically, a survey questionnaire will solicit data about the respondent, such as income, occupation or employment status (employed or unemployed), years of formal education, yearly income, age, race, and gender, coupled with additional questions that throw light upon a particular research question. For *closed-ended* questions, people must reply from a list of possible answers, like a multiple choice test. For *open-ended* questions, the respondent is allowed to elaborate on her or his answer. Closed-ended questions are generally analyzed quantitatively, and open-ended questions are generally analyzed qualitatively.

Participant observation combines subjective knowledge gained through personal involvement and objective knowledge acquired by disciplined recording of what one has seen. The subjective component supplies a dimension of information that is lacking in survey data.

Debunking Society's Myths

Myth: People who are just hanging out together and relaxing don't care much about social differences between them.

Sociological research: Even casual groups have organized social hierarchies. That is, they make distinctions within the group that give some people higher status than others. This has been shown in participant observation studies such as Duneier's (1999) study of the homeless on New York City's lower East Side; in Anderson's (1976) study of the people just hanging out in "Jelly's Bar"; in Anderson's (1999) "Code of the Street" study, which showed a rigid hierarchy among those engaged in street crime; and in the study of "Pam's Place" (Gimlin 1996, 2002), which showed the classlike distinctions hair dressers and their customers made between each other.

The men in this bar, as shown by participant observation as in Anderson's (1976) classic study of "Jelly's Bar" in A Place on the Corner, *reveal status differences among themselves that they create, such as (in descending status order) "regulars," "hoodlums," and "winos."*

Street Corner Society (1943), a classic work by sociologist William Foote Whyte, documents one of the first qualitative participant observation studies ever done. Whyte studied the Cornerville Gang, a group of Italian American men whose territory was a street corner in Boston in the late 1930s and early 1940s. Although not Italian, Whyte learned to speak the language, lived with an Italian family, and then infiltrated the gang by befriending the gang's leader, whose pseudonym was "Doc." Doc was the *informant* for Whyte, a person with whom the participant observer works closely in order to learn about the group. For the duration of the study, Doc was the only gang member who knew that Whyte was doing research on his gang. This represents what is called *covert participant observation*—in which the members of the group being studied do not know that they are being researched. (If the group is told that they are being studied and that they are the research subjects, then it is called *overt participant observation.*)

Most social scientists of the 1940s and 1950s thought gangs were socially disorganized, random deviant groups, but Whyte's study showed otherwise— as have participant observation studies since then, notably those of Anderson (1976, 1990, 1999). He found that the Cornerville Gang, and by implication other urban street corner gangs as well, was a highly organized mini-society with its own social hierarchy (social stratification), morals, practices, and punishments (sanctions) for deviating from the norms of the gang.

There are a few built-in weaknesses to participant observation as a research technique. We already mentioned that it is very time-consuming. Participant observers have to cull data from vast amounts of notes. Such studies usually focus on fairly small groups, posing problems of generalization. Participant observation can also pose real physical dangers to the researcher, such as being "found out" or "outed" if one is studying a street gang (Sanchez-Jankowski 1991). Observers may also lose their objectivity by becoming too much a part of what they study. If this happens—the observer becomes so much a part of the group that she or he is no longer observer but participant—it is called "going native" and is seen as one of the disadvantages of participant observation research. These limitations aside, participant observation has been the source of some of the most arresting and valuable studies in sociology (see "A Cop in the Hood" on page 70.).

Controlled Experiments

Controlled experiments are highly focused ways of collecting data and are especially useful for determining a pattern of cause and effect. To conduct a controlled experiment, two groups are created, an *experimental group,* which is exposed to the factor or variable one is examining, and the *control group,* which is not. In a controlled experiment, external influences are either eliminated or equalized, that is, held constant, between the experimental and the control group. This is necessary in order to establish cause and effect.

Suppose you wanted to study whether violent television programming causes aggressive behavior in children. You could conduct a controlled experiment to investigate this question. The behavior of children would be the dependent variable; the independent variable is whether or not the children are exposed to violent programming. To investigate your question, you would expose an experimental group of children (under monitored conditions) to a movie containing lots of violence (ultimate fighting, for example, or gunfighting). The control group would watch a movie that is free of violence. Aggressiveness in the children (the dependent variable) would be measured twice: a *pretest* measurement made before the movies are shown and a *posttest* measurement made afterward. You would take pretest and posttest measures on both the control and the experimental groups. Studies of this sort actually find that the children who watched the violent movie are indeed more violent and aggressive afterward than those who watched a movie containing no violence (Taylor et al. 2009; Worchel et al. 2000; Bushman 1998).

Among its advantages, a controlled experiment can establish causation, and it can zero in on a single independent variable. On the downside, controlled experiments can be artificial. They are for the most part done in a contrived laboratory setting (unless it is a *field experiment*), and they tend to eliminate many real-life effects. Analysis of controlled experiments includes making judgments about how much the artificial setting has affected the results (see Table 3.1).

Content Analysis

Researchers can learn a vast amount about a society by analyzing *cultural artifacts* such as newspapers, magazines, TV programs, or popular music. **Content analysis** is a way of measuring by examining the cultural artifacts of what people write, say, see, and hear. The researcher studies not people but the

Media violence tends to desensitize children to the effects of violence, including engendering less sympathy for victims of violence (Baumeister and Bushman 2008; Huesmann et al. 2003; Cantor 2000). Many also think that violent video games (another form of media) may be a cause of school shootings. Perhaps there is some link here; it is too simplistic to see a direct causal connection between viewing violence and actually engaging in it. For one thing, such an argument ignores the broader social context of violent behavior (including such things as the availability of guns, family characteristics, youth alienation from school, to name a few [Steinhamer 2007]).

communications the people produce as a way of creating a picture of their society.

Content analysis is frequently used to measure cultural change and to study different aspects of culture (Lamont 1992). Sociologists also use content analysis as an indirect way to determine how social groups are perceived—they might examine, for example, how Asian Americans are depicted in television dramas or how women are depicted in advertisements.

Children's books have been the subject of many content analyses. In acknowledgment of their impact on the development of youngsters, a team of sociologists compared images of Black Americans in children's books from the 1930s to the present (Pescosolido et al. 1997). They obtained three important findings: First, they found a declining representation of African Americans from the 1930s through the 1950s, with practically no representation from 1950 through 1964. Beginning in 1964, an increase in representation lasted until the mid-1970s, when the appearance of African American characters leveled off. Second, they found that the symbolic images of African Americans did change significantly over time. In the 1960s—a period of much racial unrest—African Americans were mostly portrayed in "safe," distant images, such as in secondary and nearly invisible occupational roles. Third, they found few portrayals of Black adults in intimate, egalitarian, or interracial relationships. Recent research on stereotyping generally confirms these three findings (Baumeister and Bushman 2008: 419–421).

Content analysis has the advantage of being *unobtrusive,* or "nonreactive." The research has no effect on the person being studied because the cultural artifact has already been produced. Content analysis is limited in what it can study, however, because it is based on mass communication—either visual, oral, or written. It cannot tell us what people really think about these images or whether they affect people's behavior. Other methods of research, such as interviewing or participant observation, would be used to answer these questions.

Historical Research

Historical research examines sociological themes over time. It is commonly done in historical archives, such as official records, church records, town archives, private diaries, or oral histories. The sources of this sort of material are critical to its quality and applicability. Oral histories have been especially illuminating, most dramatically in revealing the unknown histories of groups that have been ignored or misrepresented in other historical accounts. For example, when developing an account of the spirituality of Native Americans, one would be misguided to rely solely on the records left by Christian missionaries or U.S. Army officials. These records would give a useful picture of how Whites perceived Native American religion, but they would be a very poor source for discovering how Native Americans understood their own spirituality.

In a similar vein, the writings of a slave owner can deliver fascinating insights into slavery, but a slave owner's diary will certainly present a different picture of slavery as a social institution than will the written or oral histories of former slaves themselves.

Handled properly, comparative and historical research is rich with the ability to capture long-term social changes and is the perfect tool for sociologists who want to ground their studies in historical or comparative perspectives.

Evaluation Research

Evaluation research assesses the effect of policies and programs on people in society. If the research is intended to produce policy recommendations, then it is called *policy research*.

Suppose you want to know if an educational program is actually improving student performance. You could design a study that measured the academic performance of two groups of students, one that participates in the program and one that does not (a "control" group). If the academic performance of students in the program is better than that of those not in the program, and if the groups are alike in other ways (they are often matched to accomplish this), you would conclude that the program was effective. If the academic performance of the students in the program ended up being the same (or even worse) as those not in the program, then you would conclude that the program was not effective. If you use this research to recommend social policy, you would be doing policy research.

Research Ethics: Is Sociology Value Free?

The topics dealt with by sociology are often controversial. People have strong opinions about social questions, and in some cases, the settings for sociological work are highly politicized. Imagine spending time in an urban precinct house to do research on police brutality or doing research on acquired immune deficiency syndrome (AIDS) and sex education in a

table 3.1 Comparison of Six Research Techniques

Technique (Tool)	Qualitative Analysis or Quantitative Analysis	Advantages	Disadvantages
The survey (polls, questionnaires, interviews)	Usually quantitative, often qualitative	Permits the study of a large number of variables; results can be generalized to a larger population if sampling is accurate	Difficult to focus in great depth on a few variables; difficult to measure subtle nuances in people's attitudes
Participant observation	Usually qualitative	Studies actual behavior in its home setting; affords great depth of inquiry	Is very time-consuming; difficult to generalize beyond the research setting
Controlled experiment	Usually quantitative	Focuses on only two or three variables; able to study cause and effect	Difficult or impossible to measure large number of variables; may have an artificial quality
Content analysis	Can be either qualitative or quantitative	A way of measuring culture	Limited by studying only cultural products or artifacts (music, TV programs, stories, other), rather than people's actual attitudes
Historical research	Usually qualitative	Saves time and expense in data collection; takes differences over time into account	Data often reflect biases of the original researcher and reflect cultural norms that were in effect when the data were collected
Evaluation research	Can be either qualitative or quantitative	Evaluates the actual outcomes of a program or strategy; often direct policy application	Limited in the number of variables that can be measured; maintaining objectivity is problematic if research is done or commissioned by administrators of the program being evaluated

conservative public school system. Under these conditions, can sociology be scientifically objective? How do researchers balance their own political and moral commitments against the need to be objective and open-minded? Sociological knowledge has an intimate connection to political values and social views. Often the very purpose of sociological research is to gather data as a step in creating social policy. Can sociology be value free? Should it be?

This is an important question without a simple answer. Most sociologists do not claim to be value free, but they do try as best they can to produce objective research. It must be acknowledged that researchers make choices throughout their research that can influence their results. The problems sociologists choose to study, the people they decide to observe, the research design they select, and the type of media they use to distribute their research can all be influenced by the personal values of the researcher.

Sociological research often raises ethical questions. In fact, ethical considerations of one sort or another exist with any type of research. In a survey, the person being questioned is often not told the purpose of the survey or who is funding the study. Is it ethical to conceal this type of information?

In controlled experiments, *deception* is often employed, as in the now-famous studies by Stanley Milgram, to be reviewed later in Chapter 6, where people were led to believe that they were causing harm to another, when in fact they were not. Researchers often reveal the true purpose of an experiment only after it is completed. This is called **debriefing.** The deception is therefore temporary. But does that lessen the potential ethical violation? Does it lessen any potential damage to the self-concept of the subject or respondent?

One of the clearest ethical violations in all of the history of science has come to be known as the *Tuskegee Syphilis Study*. The study was conducted at the Tuskegee Institute in Macon County, Alabama, a historically Black college. For this study, begun in 1932 by the government's United States Health Service, a sample of about 400 Black males who were infected with the sexually transmitted disease syphilis (this was the "experimental" group) were allowed to go untreated medically for over forty years. Another 200 Black males who had not contracted syphilis were used as a control group. The purpose of the study was to examine the effects of "untreated syphilis in the male negro." The study was not unlike similar "studies" carried out against Jews by Hitler's Nazi regime in Germany at the same time—just before and during World War II in the 1930s and early 1940s. Untreated syphilis causes blindness, mental retardation, and death, and this is how many of the untreated men fared over the forty-plus-year period. In the 1950s, penicillin was discovered as an effective treatment for infectious diseases, including syphilis, and was widely available. Nonetheless, the scientists conducting the study decided *not* to give penicillin to the

Corbis/Sygma

Here a man with syphilis is being examined in order to determine the "progress" of syphilis in his case. This unfortunate man was an experimental subject in the U.S. government's infamous Tuskegee Syphilis Study, *one of the clearest ethical violations in all the history of science.*

infected men in the study on the grounds that it would "interfere" with the study of the physical and mental harm caused by untreated syphilis. The U.S. government itself authorized the study to be continued until the early 1970s—that is, until quite recently. By the mid-1970s pressure from the public and the press caused the federal government to terminate the study, but by then it was too late to save approximately 100 men who had already died of the ravages of untreated syphilis, plus many others who were forced to live with major mental and physical damage.

Following the ethical horrors of studies such as the Tuskegee study, the American Sociological Association (ASA) has since developed a professional code of ethics (see the ASA website for the full code of ethics). The federal government also has many regulations about the protection of human subjects. Ethical researchers adhere to these guidelines and must ensure that research subjects are not subjected to physical, mental, or legal harm. Research subjects must also be informed of the rights and responsibilities of both researcher and subject. Sociologists, like other scientists, also should not involve people in research without what is called **informed consent**—that is, getting agreement beforehand to participate from the respondents or subjects after the purposes of the study are explained to them. There may be exceptions to the need for informed consent, such as when observing people in public places. Sociologists also take measures to avoid identifying their respondents and to assure confidentiality through the use of pseudonyms or by not using names at all and by assigning random ID (code) numbers to all respondents during data analysis.

>doing sociological research

A Cop in the Hood: Participant Observation

An excellent example of participant observation research of the overt type is sociologist Peter Moskos's eighteen months as a bone fide police officer in Baltimore, Maryland. For his doctoral dissertation research, Moskos, who is White, underwent the standard six months of training in the police academy and was then assigned to Baltimore's Eastern District, a heavily African American and depressed ghetto with a heavy drug trade. He got to know and trust the other officers with whom he worked, and he became familiar with the social life of the homeless individuals, drug dealers, and neighborhood residents in East Baltimore. He lived minute by minute and day by day with the ever-present extreme dangers of police work, carried a Glock semi-automatic pistol with a seventeen-shot clip (which he never had to fire but had to "show" on occasion), and discovered that "danger creates a bond" among police officers. He wrote his field notes each day after work—numbering overall 350 typed, single-spaced pages. His study ranks with other classic participant observation studies in sociology, such as Whyte's *Street Corner Society* (1943), Anderson's *A Place on the Corner* (1976) and *Streetwise* (1990), and Duneier's *Sidewalk* (1999).

Moskos's study is important because, among other things, it dispels a number of myths that the public has about police officers and police work. For example, many think that summoning the police by calling 911 will get a quick solution to the problem—whether it be a drug deal taking place, an incident of domestic violence, or gunfire on the street. While the police are indeed generally quick to respond, in reality the drug deal or the domestic violence reconvenes immediately after the police leave the scene. Moskos even concludes that, unfortunately, 911 is "a joke."

Many assume that if a suspected drug dealer is standing close to a vial of cocaine in the street, the observing police officer will report that he or she "saw" the dealer throw the vial into the street. Moskos found, however, that this was rarely the case: The vast majority of officers over the vast majority of such incidents only reported "seeing" the dealer toss the vial if they indeed *saw* the dealer do so and were able to verify this act by another officer witnessing it. A veteran officer warned Moskos that "if you don't see him drop it, then just kick it or crush it."

Also of importance is Moskos's discovery of certain elements of social structure characterizing street drug trade. For example, virtually each and every illicit drug transaction on the street corner involves five social roles in addition to the person who actually purchases a drug or drugs: lookouts (who watch for police cars, the lowest-status role in the street transaction); steerers (who "hawk" or advertise their drug to passersby); money-men (who collect the money paid for the drug); slingers (who actually give the drugs to the purchaser); and gunmen (who stand ready in the shadows in case they feel needed). Engaging in such roles serves the function of limiting the legal liability of each individual in the event of arrests.

Such insights into the social structure and culture of street activity (in this case, street drug trade) ranks Moskos's work with other participant observation studies that reveal the structures and cultures of, for example, those of Whyte, Anderson, and Duneier.

Source: Moskos, Peter. 2008. *Cop in the Hood: My Year Policing Baltimore's Eastern District.* Princeton, NJ: Princeton University Press.

Chapter Summary

- ### What is sociological research?
  Sociological research is used by sociologists to answer questions and in many cases test hypotheses. The research method one uses depends upon the question that is asked.

- ### Is sociological research scientific?
  Sociological research is derived from the scientific method, meaning that it relies on empirical observation and, sometimes, the testing of *hypotheses*. The research process involves several steps: developing a

research question, designing the research, collecting data, analyzing data, and developing conclusions. Different research designs are appropriate to different research questions, but sociologists have to be concerned with the *validity,* the *reliability,* and the *generalization* of their results. Applying one's results obtained from a sample to a broader population is an example of generalization.

• *What is the difference between qualitative research and quantitative research?*

Qualitative research is research that is relatively unstructured, does not rely heavily upon statistics, and is closely focused on a question being asked. *Quantitative research* is research that uses statistical methods. Both kinds of research are used in sociology.

• *What are some of the statistical concepts in sociology?*

Through research, sociologists are able to make statements of *probability,* or likelihood. Sociologists use *percentages* and *rates.* The *mean* is the same as an average. The *median* represents the midpoint in an array of values or scores. The *mode* is the most common value or score. *Correlation* and *cross-tabulation* are statistical procedures that allow sociologists to see how two (or more) different variables are associated. There have been

instances of misuse of statistics in the behavioral and social sciences, including sociology, and these have resulted in incorrect conclusions.

• *What different tools of research do sociologists use?*

The most common tools of sociological research are surveys and interviews, participant observation, controlled experiments, content analysis, comparative and historical research, and evaluation research. Each method has its own strengths and weaknesses. You can better generalize from surveys, for example, than participant observation, but participant observation is better for capturing subtle nuances and depth in social behavior. Table 3.1 gives a comparison of these six research techniques.

• *Can sociology be value free?*

Although no research in any field can always be value free, sociological research nonetheless strives for objectivity while recognizing that the values of the researcher may have some influence on the work. One of the worst cases of ethical violation in scientific research was the Tuskegee Syphilis Study. There are ethical considerations in doing sociological research, such as whether one should collect data without letting research subjects (people) know they are being observed.

Key Terms

concept 59
content analysis 67
controlled experiment 66
correlation 64
cross-tabulation 64
data 59
data analysis 63
debriefing 69
deductive reasoning 56
dependent variable 59
evaluation research 68
generalization 63

hypothesis 59
independent variable 59
informed consent 69
indicator 59
inductive reasoning 57
mean 64
median 64
mode 64
participant observation 56
percentage 64
population 62
qualitative research 58

quantitative research 58
random sample 63
rate 64
reliability 59
replication study 57
research design 58
sample 62
scientific method 56
serendipity 63
spurious correlation 64
validity 59
variable 59

Online Resources

Sociology: The Essentials **Companion Website**

www.cengage.com/sociology/andersen
Visit your book companion website where you will find more resources to help you study and write your research papers. Resources include web links, learning objectives, Internet exercises, quizzing, and flash cards.

 is an easy-to-use online resource that helps you study in less time to get the grade you want NOW.

www.cengage.com/login
Need help studying? This site is your one-stop study shop. Take a Pre-Test and Cengage NOW will generate a Personalized Study Plan based on your test results. The Study Plan will identify the topics you need to review and direct you to online resources to help you master those topics. You can then take a Post-Test to determine the concepts you have mastered and what you still need to work on.

4

Socialization and the Life Course

The Socialization Process

Agents of Socialization

Theories of Socialization

Growing Up in a Diverse Society

Aging and the Life Course

Resocialization

Chapter Summary

[**During the summer** of 2000, scientists working on the human genome project announced that they had deciphered the human genetic code. By mapping the complex structure of DNA (deoxyribonucleic acid) on high-speed computers, scientists identified the 3.12 billion chemical base pairs in human DNA and put them in proper sequence, unlocking the genetic code of human life. Scientists likened this to assembling "the book of life," that is, having the knowledge to make and maintain human beings. The stated purpose of the human genome project is to see how genetics influences the development of disease, but it raises numerous ethical questions about human cloning and the possibility of creating human life in the laboratory. Is our genetic constitution what makes us human? Suppose you created a human being in the laboratory but left that creature without social contact. Would the "person" be human?

continued

© Jim Erickson/Corbis

four CHAPTER four CHAPTER FO

Rare cases of **feral children,** who have been raised in the absence of human contact, provide some clues as to what happens during human development when a person has little or no social contact. One such case, discovered in 1970, involved a young girl given the pseudonym of Genie. When her blind mother appeared in the Los Angeles County welfare office seeking assistance for herself, case workers first thought the girl was six years old. In fact, she was thirteen, although she weighed only 59 pounds and was 4 feet, 6 inches tall. She was small and withered, unable to stand up straight, incontinent, and severely malnourished. Her eyes did not focus, and she had two nearly complete sets of teeth. A strange ring of calluses circled her buttocks. She could not talk.

As the case unfolded, it was discovered that the girl had been kept in nearly total isolation for most of her life. The first scientific report about Genie stated:

> In the house Genie was confined to a small bedroom, harnessed to an infant's potty seat. Genie's father sewed the harness himself; unclad except for the harness, Genie was left to sit on that chair. Unable to move anything except her fingers and hands, feet and toes, Genie was left to sit, tied-up, hour after hour, often into the night, day after day, month after month, year after year (Curtiss 1977: 5).

At night, she was restrained in a handmade sleeping bag that held her arms stationary and placed in a crib. If she made a sound, her father beat her. She was given no toys and was allowed to play only with two old raincoats and her father's censored version of *TV Guide*. (He had deleted everything "suggestive," such as pictures of women in bathing suits.) Genie's mother, timid and blind, was also victimized by her husband. Shortly after the mother sought help after years of abuse, the father committed suicide.

Genie was studied intensively by scientists interested in language acquisition and the social effects of extreme confinement. They hoped that her development would throw some light on the question of nature versus nurture; that is, are people relatively more the product of their genes, or of their social training? After intense language instruction and psychological treat-

ment, Genie developed some verbal ability and, after a year, showed progress in her mental and physical development. Yet the years of isolation and severe abuse had taken their toll, and she was never able to catch up to her age peers. Genie was placed in a home for mentally disabled adults (Rymer 1993).

This rare case of a feral child sheds some light on the consequences of life without social contact. Knowing the sequence of the human genome may raise the specter of making human beings in the laboratory, but without society, what would humans be like? Genes may confer skin and bone and brain, but only by learning the values, norms, and roles that culture bestows on people do we become social beings—literally, *human* beings. Sociologists refer to this process as *socialization*—the subject of this chapter.

The Socialization Process

Socialization is the process through which people learn the expectations of society. **Roles** are the expected behavior associated with a given status in society. When you occupy a given social role, you tend to take on the expectations of others. For example, when you enter a new group of friends, you probably observe their behavior, their language, their dress, perhaps even their opinions of others, and often modify your own behavior accordingly. Before you know it, you are a member of the group, perhaps socializing others into the same set of expectations. We explore roles further

Formal and informal learning, through schools and other socialization agents, are important elements of the socialization process. Here hearing impaired children are learning sign language.

My Childhood (Bong Hwan Kim)

Childhood is a time when children learn their gender, as well as their racial and ethnic identity. This excerpt from an interview with Bong Hwan Kim, a Korean American man, is a reflection on growing up and learning both Korean and American culture.

I came to the United States in 1962, when I was three or four years old. My father had come before us to get a Ph.D. in chemistry. He had planned to return to Korea afterward, but it was hard for him to support three children in Korea while he was studying in the United States, and he wasn't happy alone, so he brought the family over. . . . The Bergenfeld, New Jersey, community where I grew up was a blue-collar town of about 40,000 people, mostly Irish and Italian Americans. I lived a schizophrenic existence. I had one life in the family, where I felt warmth, closeness, love, and protection, and another life outside—school, friends, television, the feeling that I was on my own. I accepted that my parents would not be able to help me much.

I can remember clearly my first childhood memory about difference. I had been in this country for maybe a year. It was the first day of kindergarten, and I was very excited about having lunch at school. All morning I could think only of the lunch that was waiting for me in my desk. My mother had made kimpahp [rice balls rolled up in dried seaweed] and wrapped it all up in aluminum foil. I was eagerly looking forward to that special treat. I could hardly wait. When the lunch bell rang I happily took out my foil-wrapped kimpahp. But all the other kids pointed and gawked. "What is that? How could you eat that?" they shrieked. I don't remember whether I ate my lunch or not, but I told my mother I would only bring tuna or peanut butter sandwiches for lunch after that.

I have always liked Korean food, but I had to like it secretly, at home. There are things you don't show to your non-Korean friends. At various times when I was growing up, I felt ashamed of the food in the refrigerator, but only when friends would come over and wonder what it was. They'd see a jar of garlic and say, "You don't eat that stuff, do you?" I would say, "I don't eat it, but my parents do; they do a lot of weird stuff like that."

. . . As a child you are sensitive; you don't want to be different. You want to be like the other kids. They made fun of my face. They called me "flat face." When I got older, they called me "Chink" or "Jap" or said "Remember Pearl Harbor." In all cases, it made me feel terrible. I would get angry and get into fights. Even in high school, even the guys I hung around with on a regular basis, would say, "You're just a Chink" when they got angry. Later, they would say they didn't mean it, but that was not much consolation. When you are angry, your true perceptions and emotions come out. The rest is a facade.

They used to say, "We consider you to be just like us. You don't seem Korean." That would give rise to such mixed feelings in me. I wanted to believe that I was no different from my White classmates. It was painful to be reminded that I was different, which people did when they wanted to put me in my place, as if I should be grateful to them for allowing me to be their friend.

I wanted to be as American as possible—playing football, dating cheerleaders. I drank a lot and tried to be cool. I had convinced myself that I was "American," whatever that meant, all the while knowing underneath that I'd have to reconcile myself to try to figure out where I would fit in a society that never sanctioned that identity as a public possibility. Part of growing up in America meant denying my cultural and ethnic identity. . . . When I got to college, I experienced an identity crisis. . . . I decided to go to Korea, hoping to find something to make me feel more whole. Being in Korea somehow gave me a sense of freedom I had never really felt in America. It also made me love my parents even more. I could imagine where they came from and what they experienced. I began to understand and appreciate their sacrifice and love and what parental support means. Visiting Korea didn't provide answers about the meaning of life, but it gave me a sense of comfort and belonging, the feeling that there was somewhere in this world that validated that part of me that I knew was real but few others outside my immediate family ever recognized.

Source: Edited by Karin Aguilar-San Juan. Copyright © 1994. *The State of Asian America.* Boston, MA: South End Press. Reprinted with permission of South End Press.

in Chapter 5, but it is important to know here that roles are learned through the socialization process.

Through socialization, people absorb their culture—customs, habits, laws, practices, and means of expression. Socialization is the basis for **identity:** how one defines oneself. Identity is both personal and social. To a great extent, it is bestowed by others, because we come to see ourselves as others see us. Socialization also establishes **personality,** defined as a person's relatively consistent pattern of behavior, feelings, predispositions, and beliefs.

The socialization experience differs for individuals, depending on factors such as age, race, gender, and class, as well as more subtle factors such as introversion-extroversion and other such aspects of personality. Women and men encounter different socialization patterns as they grow up because each gender brings with it different social expectations (see Chapter 11). Likewise, growing up Jewish, Asian, Latino, or African American involves different socialization experiences, as the box "Understanding Diversity: My Childhood (Bong Hwan Kim)" shows. In the example, a Korean American man reflects on the cultural habits he learned growing up in two cultures, Korean and American. His comments reveal the strain felt when socialization involves competing expectations—as when his schoolmates comment, "We consider you to be just like us—you don't seem Korean." Such strain can be particularly acute when a person grows up within different, even overlapping, cultures.

Through socialization, people internalize cultural expectations, then pass these expectations on to others. **Internalization** occurs when behaviors and assumptions are learned so thoroughly that people no longer question them, but simply accept them as correct. Through socialization, one internalizes the expectations of society. The lessons that are internalized can have a powerful influence on attitudes and behavior. For example, someone socialized to believe that homosexuality is morally repugnant is unlikely to be tolerant of gays and lesbians. If such a person, say a man, experiences erotic feelings about another man, he is likely to have deep inner conflicts about his identity. Similarly, someone socialized to believe that racism is morally repugnant is likely to be more accepting of different races. However, people can change the cultural expectations they learn. New experiences can undermine narrow cultural expectations. Attending college often has a liberalizing effect, supplanting old expectations with new ones generated by exposure to the diversity of college life.

The Nature–Nurture Controversy

Examining the socialization process helps reveal the degree to which our lives are *socially constructed,* meaning that the organization of society and the life outcomes of people within it are the result of social definitions and processes. Is it "nature" (what is natural) or is it "nurture" (what is social)—or both—that makes us human? This question has been the basis for debate for many years.

From a sociological perspective, what a person becomes results more from social experiences than from *innate* (inborn or natural) traits, although innate traits have some influence, as we saw in Chapter 2 on culture. For example, a person may be born with a great capacity for knowledge, but without a good education that person is unlikely to achieve his or her full potential and may not be recognized as intellectually gifted.

From a sociological perspective, nature provides a certain stage for what is possible, but society provides the full drama of what we become. The expression *tabula rasa,* for example, means a human is born as a "blank slate." But our values and social attitudes are not inborn; they emerge through the interactions we have with others and our social position in society. Such factors as your family environment, how people of your social group are treated, and the historic influences of the time all shape how we are nurtured by society or not.

Perhaps the best way to understand the nature–nurture controversy is not that one or the other fully controls who we become, but that life involves a complex interplay, or *interaction*, between natural and social influences on human beings. The emphasis in sociology, however, is to see the social realities of our lives as far more important in shaping human experience (Freese 2008; Freese et al. 2003; Ridley 2003; Guo and Stearns 2002).

Socialization as Social Control

Sociologist Peter Berger pointed out that not only do people live in society but society also lives in people (Berger 1963). Socialization is, therefore, a mode of social control. **Social control** is the process by which groups and individuals within those groups are brought into conformity with dominant social expectations. Sometimes the individual rebels and attempts to resist this conformity, but because people generally conform to cultural expectations, socialization gives society a certain degree of predictability. Patterns are established that become the basis for social order.

To understand how socialization is a form of social control, imagine the individual in society as surrounded by a series of concentric circles (see Figure 4.1). Each circle is a layer of social controls, ranging from the most subtle, such as the expectations of others, to the most overt, such as physical coercion and violence. Coercion and violence are usually not necessary to extract conformity because learned beliefs and the expectations of others are enough to keep people in line. These socializing forces can be subtle, because even when a person disagrees with others, he or she can feel pressure to conform and may experience stress and discomfort in choosing not to conform. People

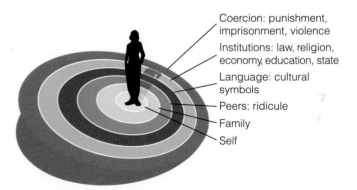

Coercion: punishment, imprisonment, violence

Institutions: law, religion, economy, education, state

Language: cultural symbols

Peers: ridicule

Family

Self

FIGURE 4.1 *Socialization as Social Control*

Though we are all individuals, the process of socialization also keeps us in line with society's expectations. This may occur subtly through peer pressure or, in some circumstances, through coercion and/or violence.

learn through a lifetime of experience that deviating from the expectations of others invites peer pressure, ridicule, and other social judgments that remind one of what is expected.

Conformity and Individuality

Saying that people conform to social expectations does not eliminate individuality. We are all unique to some degree. Our uniqueness arises from different experiences, different patterns of socialization, the choices we make, and the imperfect ways we learn our roles; furthermore, people resist some of society's expectations. Sociologists warn against seeing human beings as totally passive creatures because people interact with their environment in creative ways. Yet, most people conform, although to differing degrees. Socialization is profoundly significant, but this does not mean that people are robots. Instead, socialization emphasizes the adaptations people make as they learn to live in society.

Some people conform too much, for which they pay a price. Socialization into men's roles can encourage aggression and a zeal for risk-taking. Men have a lower life expectancy and higher rate of accidental death than do women, probably because of the risky behaviors associated with men's roles, that is, simply "being a man" (National Center for Health Statistics 2007; Kimmel and Messner 2004). Women's gender roles carry their own risks. Striving excessively to meet the beauty ideals of the dominant culture can result in feelings of low self-worth and may encourage harmful behaviors, such as smoking or severely restricting eating to keep one's weight down. It is not that being a man or woman is inherently bad for your health, but conforming to gender roles to an extreme can compromise your physical and mental health. (A convincing case in point is the affliction called *anorexia*, a pathological condition resulting from overly severe dieting, discussed more fully in Chapter 14.)

The Consequences of Socialization

Socialization is a lifelong process with consequences that affect how we behave toward others and what we think of ourselves. First, socialization *establishes self-concepts*. **Self-concept** is how we think of ourselves as the result of the socialization experiences we have over a lifetime. Socialization is also influenced by various social factors, as shown in Figure 4.2, which describes how students' self-concepts are shaped by gender.

Second, *socialization creates the capacity for role-taking*, or, put another way, for the ability to see oneself through the perspective of another. Socialization

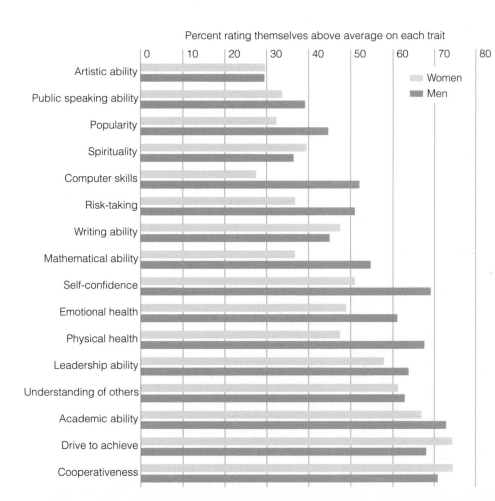

FIGURE 4.2 *Student Self-Concepts: The Difference Gender Makes* *

*Based on national sample of first-year college students, Fall 2003.

Source: Sax, L. J., et al. 2003. *The American Freshman: National Norms for Fall 2003.* Higher Education Research Institute. Los Angeles, CA: University of California, Los Angeles. Used by permission.

is fundamentally reflective; that is, it involves self-conscious human beings seeing and reacting to the expectations of others. The capacity for reflection and the development of identity are ongoing. As we encounter new situations in life, such as going away to college or getting a new job, we are able to see what is expected and to adapt to the situation accordingly. Of course, not all people do so successfully. This can become the basis for social deviance (explored in Chapter 7) or for many common problems in social and psychological adjustment.

Third, *socialization creates the tendency for people to act in socially acceptable ways.* Through socialization, people learn the normative expectations attached to social situations and the expectations of society in general. As a result, socialization creates some predictability in human behavior and brings some order to what might otherwise be social chaos.

Finally, *socialization makes people bearers of culture.* Socialization is the process by which people learn and internalize the attitudes, beliefs, and behaviors of their culture. At the same time, socialization is a two-way process—that is, a person is not only the recipient of culture but also is the creator of culture, passing cultural expectations on to others. The main product of socialization, then, is society itself.

Agents of Socialization

Socialization agents are people, or sources, or structures that pass on social expectations. Everyone is a socializing agent because social expectations are communicated in countless ways and in every interaction people have, whether intentionally or not. When people are simply doing what they consider "normal,"

they are communicating social expectations to others. When you dress a particular way, you may not feel you are telling others they must dress that way. Yet, when everyone in the same environment dresses similarly, some expectation about appropriate dress is clearly being conveyed. People feel pressure to become what society expects of them even though the pressure may be subtle and unrecognized.

thinkingsociologically

Think about the first week that you were attending college. What expectations were communicated to you and by whom? Who were the most significant *socialization agents* during this period? Which expectations were communicated formally and which informally? If you were analyzing this experience sociologically, what would be some of the most important concepts to help you understand how one "becomes a college student"?

Socialization does not occur simply between individual persons; it occurs in the context of social institutions. Recall from Chapter 1 that institutions are established patterns of social behavior that persist over time. Institutions are a level of society above individuals. Many social institutions shape the process of socialization, including, as we will see, the family, the media, peers, religion, sports, and schools.

The Family

For most people, the family is the first source of socialization. Through families, children are introduced to the expectations of society. Children learn to see themselves through their parents' eyes. Thus, how parents define and treat a child is crucial to the development of the child's sense of self.

An interesting example of the subtlety in familial socialization comes from a study comparing how U.S. and Japanese mothers talk to their children (Fernald and Morikawa 1993). Observers watched mothers from both cultures speak to their infant children, ranging in age from six to nineteen months. Both Japanese and U.S. mothers simplified and repeated words for their children, but strong cultural differences were evident in their context. U.S. mothers focused on naming objects for their babies: "Is that a *car*?" "Kiss the *doggy.*" Japanese mothers were more likely to use their verbal interactions with children as an opportunity to practice social routines such as "give me" and "thank you." The behavior of the Japanese mothers implied that the name of the object was less important than the polite exchange. Japanese mothers also were more likely to use sounds to represent the objects, such as "oink-oink" for a pig, or "vroom-vroom" for a vehicle. U.S. mothers were more likely to use the actual names for objects. The researchers interpreted these interactions as reflecting the beliefs and practices of each culture. Japanese mothers used objects as part of a ritual of social exchange, emphasizing polite routines,

Dress at work is a form of socialization.

whereas U.S. mothers focused on labeling things. In each case, the child receives a message about what is most significant in the culture.

What children learn in families is certainly not uniform. Even though families pass on the expectations of a given culture, within that culture families may be highly diverse, as we will see in Chapter 13. Some families may emphasize educational achievement; some may be more permissive, whereas others emphasize strict obedience and discipline. Even within families, children may experience different expectations based on gender or birth order (being first, second, or third born). Researchers have found, for example, that fathers' and mothers' support for gender equity increases when they have only daughters (Warner and Steel 1999). Living in a family experiencing the strain of social problems such as alcoholism, unemployment, domestic violence, or teen pregnancy also affects how children are socialized. The specific effects of different family structures and processes are the basis for ongoing and extensive sociological research.

As important as the family is in socializing the young, it is not the only socialization agent. As children grow up, they encounter other socializing influences, sometimes in ways that might contradict family expectations. Parents who want to socialize their children in less gender-stereotyped ways might be frustrated by the influence of the media, which promotes highly gender-typed toys and activities to boys and girls. These multiple influences on the socialization process create a reflection of society in us.

The Media

As we saw in Chapter 2, the mass media increasingly are important agents of socialization. Television alone has a huge impact on what we are socialized to believe and become. Add to that the print messages received in

books, comics, newspapers, and now blogs and the Internet, plus images from film, music, video games, and radio, and you begin to see the enormous influence the media have on the values we form, our images of society, our desires for ourselves, and our relationships with others. These images are powerful throughout our lifetimes, but many worry that their effect during childhood may be particularly deleterious.

The high degree of violence in the media has led to the development of a

Dima Gavrysh/AP Photo

The media have the power to transform people into superstars so that they are recognizable to people worldwide.

rating system for televised programming. There is no doubt that violence is extensive in the media. Analysts estimate that by age eighteen, the average child will have witnessed at least 18,000 simulated murders on television (Wilson et al. 2002). Moreover, violence in children's programming is frequently shown as humorous, or with no serious consequences (National Television Violence Study 1997).

Media violence also tends to desensitize children to the effects of violence, including engendering less sympathy for victims of violence (Baumeister and Bushman 2008; Huesmann et al. 2003; Cantor 2000). Many also think that violent video games (another form of media) may be a cause of school shootings. Perhaps there is some link here, but; it is too simplistic to see a direct causal connection between viewing violence and actually engaging in it. For one thing, such an argument ignores the broader social context of violent behavior (including such things as the availability of guns, family characteristics, youth alienation from school, to name a few (Steinhamer 2007).

Still, children tend to imitate the aggressive behavior they see in the media. Violence in the media is not solely to blame for violent behavior in society, however. Children do not watch television in a vacuum; they live in families where they learn different values and attitudes about violent behavior, and they observe the society around them, not just the images they see in fictional representations. Most likely, children are influenced not only by the images of televised and filmed violence but also by the social context in which they live. The images of violence in the media in some ways only reflect the violence in society.

The media expose us to numerous images that shape our definitions of ourselves and the world around us. What we think of as beautiful, sexy, politically acceptable, and materially necessary is strongly influenced by the media. If every week, as you read a newsmagazine, someone shows you the new car that will give you status and distinction, or if every weekend, as we watch televised sports, someone tells us that to have fun we should drink the right beer, it is little wonder that we begin to think that our self-worth can be measured by the car we drive and that parties are perceived as better when everyone is drunk. The values represented in the media, whether they are about violence, racist and sexist stereotypes, or any number of other social images, have a great effect on what we think and who we come to be.

Peers

Peers are those with whom you interact on equal terms, such as friends, fellow students, and coworkers. Among peers, there are no formally defined superior and subordinate roles, although status distinctions commonly arise in peer group interactions. Without peer approval, most people find it hard to feel socially accepted.

Peers are an important agent of socialization. Young girls and boys learn society's images of what they are supposed to be through the socialization process.

Support from peers and family is an important source of strong self-esteem. Organized peer groups such as the Special Olympics can also foster a desire for achievement and enhance one's sense of self-worth.

Peers are enormously important in the socialization process. Peer cultures for young people often take the form of *cliques*—friendship circles where members identify with each other and hold a sense of common identity. You probably had cliques in your high school and may even be able to name them. Did your school have "jocks," "preps," "goths," "nerds," "freaks," "stoners," and so forth? Sociologists studying cliques have found that they are formed based on a sense of exclusive membership, like the in-groups and out-groups we will examine in Chapter 6. Cliques are cohesive but also have an internal hierarchy, with certain group leaders having more power and status than other members. Interaction techniques, like making fun of people, produce group boundaries, defining who's in and who's out. The influence of peers is strong in childhood and adolescence, but it also persists into adulthood.

As agents of socialization, peers are important sources of social approval, disapproval, and support. This is one reason groups without peers of similar status are often at a disadvantage in various settings, such as women in male-dominated professions or minority students on predominantly White campuses. Being a "token" or an "only," as it has come to be called, places unique stresses on those in settings with relatively few peers from whom to draw support (Thoits 2009). This is one reason those who are minorities in a dominant group context often form same-sex or same-race groups for support, social activities, and the sharing of information about how to succeed in their environment (Tatum 1997).

Religion

Religion is another powerful agent of socialization, and religious instruction contributes greatly to the identities children construct for themselves. Children tend to develop the same religious beliefs as their parents; switching to a religious faith different from the one in which a person is raised is rare (Hadaway and Marler 1993). Even those who renounce the religion of their youth are deeply affected by the attitudes, images, and beliefs instilled by early religious training. Very often those who disavow religion return to their original faith at some point in their life, especially if they have strong ties to their family of origin and if they form families of their own (Wuthnow 1998; Wilson 1994).

Religious socialization shapes the beliefs that people develop. An example comes from studies of people who believe in creationism. **Creationism** is a set of beliefs that largely reject the theory of human biological evolution and instead argue that human beings as now exist were created by a central force or God. Those who believe this have generally been taught to believe so over a long period; that is, they have been specifically socialized to believe in the creationist view of the world's origin and to reject scientific explanations, such as the theory of evolution. Sociological research further finds that socialization into creationist beliefs is more likely to be effective among people who grow up in small-town environments where they are less exposed to other influences. Those who believe in creationism are also likely to have mothers who have filled the traditional homemaker's role (Eckberg 1992). This shows the influence that social context has on the religious socialization people experience.

Religious socialization influences a large number of beliefs that guide adults in how they organize their lives, including beliefs about moral development and behavior, the roles of men and women, and sexuality, to name a few. One's religious beliefs strongly influence belief about gender roles within the family, including men's engagement in housework and the odds that wives will be employed outside the home (Ellison and Bartkowski 2002; Becker and Hofmeister 2001; Scott 2000). Religious socialization also influences beliefs about sexuality, including the likelihood of tolerance for gay and lesbian sexuality (Reynolds 2003; Sherkat 2002). Religion can even influence child-rearing practices. Thus, sociologists have found that conservative Protestants are more likely to use strict discipline in raising children, but they are also more likely to hug and praise their children than are parents with less conservative religious views.

Sports

Most people perhaps think of sports as something that is just for fun and relaxation—or perhaps to provide opportunities for college scholarships and athletic careers—but sports are also an agent of socialization. Through sports, men and women learn concepts of self that stay with them in their later lives.

Sports are also where many ideas about gender differences are formed and reinforced (Messner 2002; Dworkin and Messner 1999). For men, success or failure as an athlete can be a major part of a man's identity. Even for men who have not been athletes, knowing about and participating in sports is an important source of men's gender socialization. Men learn that being competitive in sports is considered a part of manhood. Indeed, the attitude that "sports builds character" runs deep in the culture. Sports are supposed to pass on values such as competitiveness, the work ethic, fair play, and a winning attitude. Sports are considered to be where one learns to be a man.

Debunking Society's Myths

Myth: Many people feel that sports are for the most part played just for the fun of it.

Sociological perspective: Although sports are a form of entertainment, playing sports is also a source for socialization into roles, such as gender roles.

Michael Messner's research on men and sports reveals the extent to which sports shape masculine identity. Messner interviewed thirty former athletes: Latino, Black, and White men from poor, working-class, and middle-class backgrounds. All of them spoke of the extraordinary influence of sports on them as they grew up. Not only are sports a major source of gender socialization, but many working-class, African American, and Latino men often see sports as their only possibility for a good career, even though the number of men who succeed in athletic careers is a minuscule percentage of those who hold such hopes.

Messner's research shows that, for most men, playing or watching sports is often the context for developing relationships with fathers, even when the father is absent or emotionally distant in other areas of life. Older brothers and other male relatives also socialize young men into sports. For many of the men in Messner's study, the athletic accomplishments of other family members created uncomfortable pressure to perform and compete, although on the whole, they recalled their early sporting years with positive emotions. It was through sports relationships with male peers, more than anyone else, however, that the men's identity was shaped. As boys, the men could form "safe" bonds with other men; still, through sports activity, men learned homophobic attitudes (that is, fear and hatred of homosexuals) and rarely developed intimate, emotional relationships with each other (Messner 2002, 1992).

Sports historically have been less significant in the formation of women's identity, although this has changed, largely as the result of Title IX. Title IX opened more opportunities in athletics to girls and women by legally defining the exclusion of women from school sports as sex discrimination. Women who participate in sports typically develop a strong sense of bodily competence—something usually denied to them by the prevailing, unattainable cultural images of women's bodies. Sports also give women a strong sense of self-confidence and encourage them to seek challenges, take risks, and set goals (Blinde et al. 1994, 1993).

Still, athletic prowess, highly esteemed in men, is not tied to cultural images of womanliness. Quite the contrary, women who excel at sports are sometimes stereotyped as lesbians, or "butches," and may be ridiculed for not being womanly enough. These stereotypes reinforce traditional gender roles for women, as do media images of women athletes that emphasize family images and the personality of women athletes (Cavalier 2003; Blinde and Taub 1992a, 1992b). Research in the sociology of sports shows how activities as ordinary as shooting baskets on a city lot, playing on the soccer team for one's high school, or playing touch football on a Saturday afternoon can convey powerful cultural messages about our identity and our place in the world. Sports are a good example of the power of socialization in our everyday lives.

Schools

Once young people enter kindergarten (or, even earlier, day care), another process of socialization begins. At home, parents are the overwhelmingly dominant source of socialization cues. In school, teachers and other students are the source of expectations that encourage children to think and behave in particular ways. The expectations encountered in schools vary for different groups of students. These differences are shaped by a number of factors, including teachers' expectations for different groups and the resources that different parents can bring to bear on the educational process. The parents of children attending elite, private schools, for example, often have more influence on school policies and classroom activities than do parents in low-income communities. In any context, studying socialization in the schools is an excellent way to see the influence of gender, class, and race in shaping the socialization process.

Debunking Society's Myths

Myth: Schools are primarily places where young people learn skills and other knowledge.

Sociological perspective: There is a hidden curriculum in schools where students learn expectations associated with race, class, and gender relations in society as influenced by the socialization process.

For example, research finds that teachers respond differently to boys and girls in school. Boys receive more attention from teachers than do girls. Even when teachers respond negatively to boys who are misbehaving, they are paying more attention to the boys (American Association of University Women 1998; Sadker and Sadker 1994). Social class stereotypes also affect teachers' interactions with students. Teachers are likely to perceive working-class children and poor children as less bright and less motivated than middle-class children; teachers are also more likely to define working-class students as troublemakers (Oakes et al. 2000; Bowditch 1993). These negative appraisals are *self-fulfilling prophecies,* meaning that the expectations they create often become the cause of actual behavior in the children; thus, they affect the odds of success for children. (We will return to a discussion of self-fulfilling prophecies in Chapter 14.)

Boys also receive more attention in the curriculum than girls. The characters in texts are more frequently boys; the accomplishments of boys are more likely portrayed in classroom materials; and boys and men are more typically depicted as active players in history, society, and culture (American Association of University Women 1998; Sadker and Sadker 1994). This is called the *hidden curriculum* in the schools—the informal and often subtle messages about social roles that are conveyed through classroom interaction and classroom materials—roles that are clearly linked to gender, race, and class.

In schools, boys and girls are quite often segregated into different groups, with significant sociological consequences. Differences between boys and girls become exaggerated when they are defined as distinct groups (Thorne 1993). Seating boys and girls in separate groups or sorting them into separate play groups heightens gender differences and greatly increases the significance of gender in the children's interactions with each other. Equally important is that gender becomes less relevant in the interactions between boys and girls when they are grouped together in common working groups, although gender does not disappear altogether as an influence. Barrie Thorne, who has observed gender interaction in schools, concludes from her observations that gender has a "fluid" character and that gender relations between boys and girls can be improved through conscious changes that discourage gender separation.

While in school, young people acquire identities and learn patterns of behavior that are congruent with the needs of other social institutions. Sociologists using conflict theory to understand schools would say that U.S. schools reflect the needs of a capitalist society. School is typically the place where children are first exposed to a hierarchical, bureaucratic environment. Not only does school teach them the skills of reading, writing, and other subject areas, but it is also where children are trained to respect authority, be punctual, and follow rules—thereby preparing them for their future lives as workers in organizations that value these traits. Schools emphasize conformity to societal needs, although not everyone internalizes these lessons to the same degree (Bowles and Gintis 1976; Lever 1978). Research has found, for example, that working-class schoolchildren form subcultures in school that resist the dominant culture.

thinkingsociologically

Visit a local day-care center, preschool, or elementary school and observe children at play. Record the activities they are involved in, and note what both girls and boys are doing. Do you observe any differences between boys' and girls' play? What do your observations tell you about *socialization* patterns for boys and girls?

Theories of Socialization

Knowing that people become socialized does not explain how it happens. Different theoretical perspectives explain socialization, including psychoanalytic theory, social learning theory, and symbolic interaction theory. Each perspective, including functionalism as well as conflict theory, carries a unique set of assumptions about socialization and its effect on the development of the self (see Table 4.1).

Psychoanalytic Theory

Psychoanalytic theory originates in the work of Sigmund Freud (1856–1939). Perhaps Freud's greatest contribution was the idea that the unconscious mind shapes human behavior. Freud is also known for developing the technique of *psychoanalysis* to help discover the causes of psychological problems in the recesses of troubled patients' minds.

Psychoanalytic theory depicts the human psyche in three parts: the id, the superego, and the ego. The **id** consists of deep drives and impulses. Freud was particularly absorbed by the sexual component of the id, which he considered an especially forceful denizen of the unconscious mind. The **superego** is the dimension of the self that represents the standards of society. The superego incorporates or internalizes acquired values and norms—in short, culture. According to Freud, an ordered society requires that people repress the wild impulses generated by the id.

table 4.1	Theories of Socialization				
	Psychoanalytic Theory	**Social Learning Theories**	**Functional Theory**	**Conflict Theory**	**Symbolic Interaction Theory**
How each theory views:					
Individual learning process	The unconscious mind shapes behavior.	People respond to social stimuli in their environment.	People internalize the role expectations that are present in society.	Individual and group aspirations are shaped by the opportunities available to different groups.	Children learn through taking the role of significant others.
Formation of self	The self (ego) emerges from tension between the id and the superego.	Identity is created through the interaction of mental and social worlds.	Internalizing the values of society reinforces social consensus.	Group consciousness is formed in the context of a system of inequality.	Identity emerges as the creative self interacts with the social expectations of others.
Influence of society	Societal expectations are represented by the superego.	Young children learn the principles that shape the external world.	Society relies upon conformity to maintain stability and social equilibrium.	Social control agents exert pressure to conform.	Expectations of others form the social context for learning social roles.

Consequently, the id is in permanent conflict with the superego. The superego represents what Freud saw as the inherent repressiveness of society. People cope with the tension between social expectations (the superego) and their impulses (the id) by developing defense mechanisms, typically repression, avoidance, or denial (Freud 1960/1923, 1961/1930, 1965/1901). Suppose someone has a great desire for the wrong person or even for another person's property. This person might refuse to admit this (repression); or, acknowledging the impulse, might avoid the opportunity for temptation (avoidance); or might indulge in misconduct, believing it was not misconduct (denial).

The third component of the self in Freud's theory, the **ego,** is the seat of reason and common sense. The ego plays a balancing act between the id and the superego, adapting the desires of the id to the social expectations of the superego. In psychoanalytic theory, the conflict between the id and the superego occurs in the subconscious mind, yet it shapes human behavior. We get a glimpse of the unconscious mind in dreams and in occasional slips of the tongue—the famous "Freudian slip" that is believed to reveal an underlying state of mind. For example, someone might intend to say, "There were six people at the party," but instead says, "There were sex people at the party"!

Some sociologists have criticized Freud's work for not being generalizable because he worked with only a small and unrepresentative group of clients. Still, psychoanalytic theory is an influential analysis of human personality. We often speak of what motivates people, as if motives were internal, unconscious states of mind that direct human behavior.

The psychoanalytic perspective interprets human identity as relatively fixed at an early age in a process greatly influenced by one's family. Nancy Chodorow, a sociologist, uses psychoanalytic theory to explain how gendered personalities develop. She argues that infants are strongly attached to their primary caregiver in our society, typically the mother. As they grow older, they try to separate themselves from their parents, both physically and emotionally, becoming freestanding individuals. However, early attachments to the primary caregiver persist. At the same time, children identify with their same-sex parent, meaning that boys and girls separate themselves from their parents differently. Because girls identify with the mother, they are less able to detach themselves from their primary caregiver. This is the basis, according to Chodorow, for why—even in adulthood—women tend to have personalities based on attachment and an orientation toward others and men have personalities based on greater detachment (Chodorow 1978).

Social Learning Theory

Whereas psychoanalytic theory places great importance on the *internal* unconscious processes of the human mind, **social learning theory** considers the formation of identity to be a learned response to *external* social stimuli (Bandura and Walters 1963). Social learning theory emphasizes the societal context of socialization. Identity is regarded not as the product of the unconscious but as the result of modeling oneself in response to the expectations of others. According to social learning theory, behaviors and attitudes develop in response to reinforcement and encouragement from those around us.

Early models of social learning theory regarded learning rather simplistically in terms of stimulus and response. People were seen as passive creatures who merely responded to stimuli in their environment. This mechanistic view of social learning was transformed by the work of the Swiss psychologist **Jean Piaget** (1896–1980), who believed that learning was crucial to socialization but that imagination also had a critical role. He argued that the human mind organizes experience into mental categories, or configurations, which he called *schema*. Schema are then modified and developed as social experiences accumulate. Schema might be compared to a person's understanding of the rules of a game. Humans do not simply respond to stimulus but actively absorb experience and figure out what they are seeing to construct a picture (a schema) of the world.

Piaget proposed that children go through distinct stages of cognitive development as they learn the basic rules of reasoning. They must master the skills at each level before they go on to the next (Piaget 1926). In the initial *sensorimotor stage,* children experience the world only through their senses—touch, taste, sight, smell, and sound. Next comes the *preoperational stage,* in which children begin to use language and other symbols. Children in the preoperational stage cannot think in abstract terms, but they do gain an appreciation of meanings that go beyond their immediate senses. They also begin to see things as others might see them. Third, the *concrete operational stage* is when children learn logical principles regarding the concrete world. This stage prepares them for more abstract forms of reasoning. In the *formal operational stage,* children are able to think abstractly and imagine alternatives to the reality in which they live.

Building on Piaget's model of stages of development, psychologist Lawrence Kohlberg (1969) developed a theory of what he called moral development. Kohlberg interpreted the process of developing moral reasoning as occurring in several stages grouped into three levels: the *preconventional stage,* the *conventional stage,* and the *postconventional stage.* In the first, young children judge right and wrong in simple terms of obedience and punishment, based on their own needs and feelings. In adolescence (the conventional stage), Kohlberg argued, young people develop moral judgment in terms of cultural norms, particularly social acceptance and following authority. In the final stage of moral development, the postconventional stage, people are able to consider abstract ethical questions, thereby showing maturity in their moral reasoning. In Kohlberg's original research, men, he argued, reached a "higher" level of moral development than women because women remained more concerned with feelings and social opinions (a lower phase), whereas men were concerned with authority.

Kohlberg's work was later criticized by social psychologist Carol Gilligan. Gilligan (1982) found that women conceptualize morality in different terms than men. Instead of judging women by a standard set by men's experience, Gilligan showed that women's moral judgments were more contextual than those of men. In other words, when faced with a moral dilemma, women were more likely to consider the different relationships in the social context that would be affected by any decision instead of making moral judgments according to abstract principles. Gilligan's research made an important point, not just about the importance of including women in studies of human development but also about being careful not to assume that social learning follows a universal course for all groups.

Functionalism and Conflict Theory

Sociologists use a variety of theoretical perspectives to understand the socialization process, including those just described. They can also draw from the major theoretical frameworks we have introduced to understand socialization. From the vantage point of functionalist theory, socialization integrates people into society because it is the mechanism through which they internalize social roles and the values of society. This reinforces social consensus because it encourages at least some degree of conformity. Thus, socialization is one way that society maintains its stability.

Conflict theorists would see this differently. Because of the emphasis in conflict theory on the role of power and coercion in society, in thinking about socialization conflict theorists would be most interested in how group identity is shaped by patterns of inequality in society. A person's or group's identity always emerges in a context, and if that context is one marked by different opportunities for different groups, then one's identity will be shaped by that fact. This may help you understand why, for example, women are more likely to choose college majors in areas of study that have traditionally been associated with women's work opportunities (that is, in the so-called helping professions and in the arts and humanities and less frequently

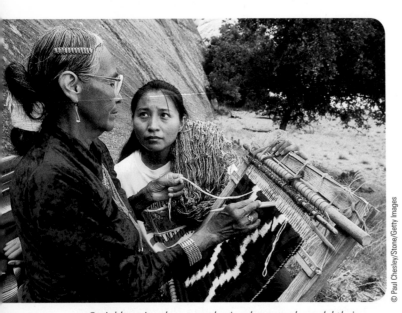

Social learning theory emphasizes how people model their behaviors and attitudes on those of others.

© Paul Chesley/Stone/Getty Images

in math and sciences). Furthermore, though social control agents pressure people to conform, people also resist oppression. Thus, the identities of people oppressed in society often include some form of resistance to oppression. This can help you understand why members of racial groups who identify with their own group, not the dominant White group, tend to have higher self-esteem (that is, a stronger valuing of self). In other words, resisting the expectations of a dominant group (such as being subservient or internalizing a feeling of inferiority) can actually heighten one's perceived self-worth.

Symbolic Interaction Theory

Recall that *symbolic interaction* theory centers on the idea that human actions are based on the meanings people attribute to behavior; these meanings emerge through social interaction (Blumer 1969). Symbolic interaction has been especially important in developing an understanding of socialization. People learn identities and values through socialization. For example, learning to become a good student, then, means taking on the characteristics associated with that role. Because roles are socially defined, they are not real, like objects or things, but are real because of the meanings people give them.

For symbolic interactionists, meaning is constantly reconstructed as people act within their social environments. The **self** is what we imagine we are; it is not only an interior bundle of drives, instincts, and motives. Because of the importance attributed to reflection in symbolic interaction theory, symbolic interactionists use the term *self,* rather than the term *personality,* to refer to a person's identity. Symbolic interaction theory emphasizes that human beings make conscious and meaningful adaptations to their social environment. From a symbolic interactionist perspective, identity is not something that is unconscious and hidden from view, but is socially bestowed and socially sustained (Berger 1963).

Two theorists have greatly influenced the development of symbolic interactionist theory in sociology. **Charles Horton Cooley** (1864–1929) and **George Herbert Mead** (1863–1931) were both sociologists at the University of Chicago in the early 1900s (see Chapter 1). Cooley and Mead saw the self developing in response to the expectations and judgments of others in their social environment.

Charles Horton Cooley postulated the **looking glass self** to explain how a person's conception of self arises through considering our relationships to others (Cooley 1902, 1967/1909). The development of the looking glass self emerges from (1) how we think we appear to others; (2) how we think others judge us; and (3) how the first two make us feel—proud, embarrassed, or other feelings. The looking glass self involves perception and effect, the perception of how others see us and the effect of others' judgment on us (see Figure 4.3).

How others see us is fundamental to the idea of the looking glass self. In seeing ourselves as others do, we respond to the expectations others have of us. This means that the formation of the self is fundamentally a social process—one based in the interaction people have with each other, as well as the human capacity for self-examination. One unique feature of human life is the ability to see ourselves through others' eyes. People can imagine themselves in relationship to others and develop a definition of themselves accordingly. From a symbolic interactionist perspective, the

FIGURE 4.3 *The Looking Glass Self*

The looking glass self refers to the process by which we attempt to see ourselves as others see us.

Drawing conceptualized by Norman Andersen.

reflective process is key to the development of the self. If you grow up with others who think you are smart and sharp-witted, chances are you will develop this definition of yourself. If others see you as dull-witted and withdrawn, chances are good that you will see yourself this way. George Herbert Mead agreed with Cooley that children are socialized by responding to others' attitudes toward them. According to Mead, social roles are the basis of all social interaction.

Taking the role of the other is the process of putting oneself into the point of view of another. To Mead, role-taking is a source of self-awareness. As people take on new roles, their awareness of self changes. According to Mead, identity emerges from the roles one plays. He explained this process in detail by examining childhood socialization, which he saw as occurring in three stages: the imitation stage, the play stage, and the game stage (Mead 1934). In each phase of development, the child becomes more proficient at taking the role of the other. In the first stage, the **imitation stage,** children merely copy the behavior of those around them. Role-taking in this phase is nonexistent because the child simply mimics the behavior of those in the surrounding environment without much understanding of the social meaning of the behavior. Although the child in the imitation stage has little understanding of the behavior being copied, he or she is learning to become a social being. For example, think of young children who simply mimic the behavior of people around them (such as pretending to read a book, but doing so with the book upside down).

In the second stage, the **play stage,** children begin to take on the roles of significant people in their environment, not just imitating but incorporating their relationship to the other. Especially meaningful is when children take on the role of **significant others,** those with whom they have a close affiliation. A child pretending to be his mother may talk to himself as the mother would. The child begins to develop self-awareness, seeing himself or herself as others do.

In the third stage of socialization, the **game stage,** the child becomes capable of taking on multiple roles at the same time. These roles are organized in a complex system that gives the child a more general or comprehensive view of the self. In this stage, the child begins to comprehend the system of social relationships in which he or she is located. The child not only sees himself or herself from the perspective of a significant other, but also understands how people are related to each other and how others are related to him or her. This is the phase where children internalize (incorporate into the self) an abstract understanding of how society sees them.

Mead compared the lessons of the game stage to a baseball game. In baseball, all roles together make the game. The pitcher does not just throw the ball past the batter as if they were the only two people on the field; rather, each player has a specific role, and each role intersects with the others. The network of social roles and the division of labor in the baseball game is a social system, like the social systems children must learn as they develop a concept of themselves in society.

In the game stage, children learn more than just the roles of significant others in their environment. They also acquire a concept of the **generalized other**—the abstract composite of social roles and social expectations. In the generalized other, they have an example of community values and general social expectations that adds to their understanding of self; however, children do not all learn the same generalized other. Depending on one's social position (that is, race, class, gender, region, or religion), one learns a particular set of social and cultural expectations.

If the self is socially constructed through the expectations of others, how do people become individuals? Mead answered this by saying that the self has two dimensions: the "I" and the "me." The "I" is the unique part of individual personality, the active, creative, self-defining part. The "me" is the passive, conforming self, the part that reacts to others. In each person, there is a balance between the I and the me, similar to the tension Freud proposed between the id and the superego. Mead differed from Freud, however, in his judgment about when identity is formed. Freud felt that identity was fixed in childhood and henceforth driven by internal, not external, forces. In Mead's version, social identity is always in flux, constantly emerging (or "becoming") and dependent on social situations. Over time, identity stabilizes as one learns to respond consistently to common situations.

Social expectations associated with given roles change as people redefine situations and as social and historical conditions change; thus, the social expectations learned through the socialization process are not permanently fixed. For example, as more women enter the paid labor force and as men take on additional responsibilities in the home, the expectations associated with motherhood and fatherhood are changing. Men now experience some of the role conflicts that women have faced in balancing work and family. As the roles of mother and father are redefined, children are learning new socialization patterns; however, traditional gender expectations maintain a remarkable grip. Despite many changes in family life and organization, young girls are still socialized for motherhood and young boys are still socialized for greater independence and autonomy.

Growing Up in a Diverse Society

Understanding the institutional context of socialization is important for understanding how socialization affects different groups in society. Socialization makes us members of our society. It instills in us the values of the culture and brings society into our self-definition, our perceptions of others, and our understanding of the world around us. Socialization is not, however, a uniform process, as the different examples developed in this chapter show. In a society as complex and diverse as the United States, no two people will have exactly the same experiences. We can find similarities between us, often across vast social and cultural differences, but variation in social contexts creates vastly different socialization experiences.

Furthermore, current changes in the U.S. population are creating new multiracial and multicultural environments in which young people grow up. Schools, as an example, are in many places being transformed by the large number of immigrant groups entering the school system. In such places, children come into contact with other children from a variety of different groups. This creates a new context in which children form their social values and learn their social identities (see the box "Doing Sociological Research: Children's Understanding of Race").

One task of the sociological imagination is to examine the influence of different contexts on socialization. Where you grow up; how your family is structured; what resources you have at your disposal; your racial-ethnic identity, gender, and nationality—all shape the socialization experience. Socialization experiences for all groups are shaped by many factors that intermingle and intersect to form the context for socialization.

One way that this has been demonstrated is in research by sociologist Annette Lareau (2003). Over an extended period of time, Lareau and her research assistants carefully observed White and Black families from middle-class, working-class, and poor back-

The family serves as a major agent of socialization, especially of the young.

© Kate Powers/Getty Images/Taxi

grounds. The researchers spent many hours in the homes of the families studied, including following the children and parents as they went about their daily routines. Based on these detailed observations, they observed important class differences in how families—both Black and White—socialize their children.

The middle-class children were highly programmed in their activities, their lives filled with various organized activities—music lessons, sports, school groups, and so forth. In contrast, the working-class and poor children, regardless of race, were less structured in their activities, and economic constraints were a constant theme in their daily lives. But the pace of life for working-class and poor children was slower and more relaxed. These children had more unstructured play time, whereas middle-class children's lives were a constant barrage of highly structured activities with intense time demands. Lareau argues that middle-class families engaged in *concerted cultivation* of childhood, meaning they made "deliberate and sustained effort to stimulate children's development and to cultivate their cognitive and social skills" (Lareau 2003: 238). Working-class and poor children experienced more "natural growth," that is, childhood experiences that allow them to develop in a less-structured environment with more time for creative play.

As a result, middle-class children tend to learn a more individualized self-concept and a sense of entitlement, but the price is an overly programmed daily life. Working-class and poor children experience obvious costs in that they have more financial constraints, but even more fundamentally, Lareau argues, they are left unable to negotiate their way through various social institutions as effectively as the middle-class. In a sense, these childhood socialization patterns are also reshaping the class system in which children will likely find themselves as adults. Middle-class children are being prepared, even if inadvertently, for lives with a sense of privilege and entitlement; working-class children, for responding to the directives of others. In this way, patterns of socialization occurring because of social class origins are training children to take their place in the class system that will likely mark their adult lives. Thus, social class is an important—although often invisible—force shaping the socialization of young people.

Aging and the Life Course

Socialization begins the moment a person is born. As soon as the sex of a child is known (which now can be even before birth), parents, grandparents, brothers, and sisters greet the infant with different expectations, depending on whether it is a boy or a girl. Socialization does not come to an end as we reach adulthood; rather, it continues through our lifetime. As we enter new situations, and even as we interact in familiar ones, we learn new roles and undergo changes in identity.

Sociologists use the term **life course** perspective to describe and analyze the connection between people's personal attributes, the roles they occupy, the life events they experience, and the social and historical aspects of these events (Stoller and Gibson 2000). The life course perspective underscores the point made by C. Wright Mills (introduced in Chapter 1) that personal biographies are linked to specific social–historical periods. Thus, different generations are strongly influenced by large-scale events (such as war, immigration, economic prosperity, or depression, for example).

The phases of the life course are familiar: childhood, youth and adolescence, adulthood, and old age. These phases of the life course bind different generations and define some of life's most significant events, such as birth, marriage, retirement, and death.

Childhood

During childhood, socialization establishes one's initial identity and values. In this period, the family is an extremely influential source of socialization, but experiences in school, peer relationships, sports, religion, and the media also have a profound effect. Children acquire knowledge of their culture through countless subtle cues that provide them with an understanding of what it means to live in society.

Socializing cues begin as early as infancy, when parents and others begin to describe their children based on their perceptions. Frequently, these perceptions are derived from the cultural expectations parents have

for children. Parents of girls may describe their babies as "sweet" and "cuddly," whereas boys are described as "strong" and "alert." Even though when they are tiny infants it is difficult to physically identify baby boys and girls, in this culture parents dress even their tiny infants in colors and styles that typically distinguish one gender from the other.

The lessons of childhood socialization come in myriad ways, some more subtle than others. In an example of how gender influences childhood socialization, researchers observed mothers and fathers who were walking young children through public places. Although the parents may not have been aware of it, both mothers and fathers were more protective toward girl toddlers than boy toddlers. Parents were more likely to let boy toddlers walk alone, but held girls' hands, carried them, or kept them in strollers (Mitchell et al. 1992).

Much socialization in early childhood takes place through play and games. Games that encourage competition help instill the value of competitiveness throughout someone's life. Likewise, play with other children and games that are challenging give children important intellectual, social, and interpersonal skills. Extensive research has been done on how children's play and games influence their identities as boys and girls (Campenni 1999). Generally, the research finds that boys' play tends to be rougher, more aggressive, and involve more specific rules. Boys are also more likely to be involved in group play, and girls engage in more conversational play (Moller et al. 1992). Sociologists have concluded that the games children play significantly influence their development into adults.

Another enormous influence on childhood socialization is what children observe of the adult world. Children are keen observers, and what they perceive will influence their self-concept and how they relate to others. This is vividly illustrated by research revealing how many adult child abusers were themselves victims of child abuse (Fattah 1994). Children become socialized by observing the roles of those around them and internalizing the values, beliefs, and expectations of their culture.

Adolescence

Only recently has adolescence been thought of as a separate phase in the life cycle. Until the early twentieth century, children moved directly from childhood roles to adult roles. It was only when formal education was extended to all classes that adolescence emerged as a particular phase in life when young people are regarded as no longer children, but not yet adults. There are no clear boundaries to adolescence, although it generally lasts from junior high school until the time one takes on adult roles by getting a job, marrying, and so forth. Adolescence can include the period through high school and extend right up through college graduation.

Erik Erikson (1980), the noted psychologist, stated that the central task of adolescence is the

>doing sociological research

Children's Understanding of Race

In a racially stratified society, people learn concepts about race that shape their interactions with others. Sociologists Debra Van Ausdale and Joe Feagin wanted to know how children understand racial and ethnic concepts and how this influences their interaction with other children.

Research Questions: How do children learn about race? Prior to Van Ausdale and Feagin's study, most knowledge about children's understandings of race came from experimental studies in a laboratory or from psychological tests and interviews with children. Van Ausdale and Feagin wanted to study children in a natural setting so they observed children in school, systematically observing children's interactions with one another.

Research Method: They observed three-, four-, and five-year-olds in an urban preschool. Twenty-four of the children were White; nineteen, Asian; four, Black; three, biracial; three, Middle Eastern; two, Latino; and three classified as "other." The children's racial designations were provided by their parents.

The researchers observed in the school five days a week for eleven months and saw one to three episodes involving significant racial or ethnic matters every day.

Research Results: The researchers found that young children use racial and ethnic concepts to exclude other children from play. Sometimes language is the ethnic marker; other times, skin color. The children showed an awareness of negative racial attitudes, even though they were attending a school that prided itself on limiting children's exposure to prejudice and discrimination and used a multicultural curriculum to teach students to value racial and ethnic diversity. At times, the children also used racial–ethnic understandings to include others—teaching other students about

racial–ethnic identities. Race and ethnicity were also the basis for children's concepts of themselves and others. As an example, one four-year-old White child insisted that her classmate was Indian because she wore her long, dark hair in a braid. When the classmate explained that she was not Indian, the young girl remarked that maybe her mother was Indian.

Conclusions and Implications: Throughout the research, the children showed how significant racial–ethnic concepts were in their interactions with others. Race and ethnicity are powerful identifiers of self and others. Despite the importance of race in the children's interactions, Van Ausdale and Feagin also noted a strong tendency for the adults they observed to *deny* that race and ethnicity were significant to the children. The implication is that *while adults tend to deny the reality of race in their everyday lives*, observing the interaction of children helps to instruct adults about the relevance of race and how racial awareness develops.

© Caroline Penn/Corbis

Through the socialization process, young children learn the values of their culture. These values shape their relationships with other people.

Source: Van Ausdale, Debra, and Joe R. Feagin. 2000. *The First R: How Children Learn Race and Racism.* Lanham, MD: Rowman and Littlefield.

formation of a consistent identity. Adolescents are trying to become independent of their families, but they have not yet moved into adult roles. Conflict and confusion can arise as the adolescent swings between childhood and adult maturity. Some argue that adolescence is a period of delayed maturity. Although society expects adolescents to behave like adults, they

are denied many privileges associated with adult life. Until age eighteen, they cannot vote or marry without permission, and sexual activity is condemned. In addition, until age twenty-one they cannot legally drink alcohol. The tensions of adolescence have been blamed for numerous social problems, such as drug and alcohol abuse, youth violence, and the school dropout rate.

The issues that young people face are a good barometer of social change across generations. Today's young people face an uncertain world where adult roles are less predictable than in the past. Marriage later in life, high divorce rates, frequent technological change, and economic recession all create a confusing environment for young people (Csikzentmihalyi and Schneider 2000). Studies of adolescents find that, in this context, young people understand the need for flexibility, specialization, and, likely, frequent job change. Although the media stereotype adolescents as slackers, most teens are willing to work hard, do not engage in criminal or violent activity, and have high expectations for an education that will lead to a good job. Many, however, find that their expectations are out of alignment with the opportunities that are actually available, particularly during periods of economic downturn, such as the recession of 2008–2009.

Patterns of adolescent socialization vary significantly by race and social class. National surveys find some intriguing class and race differences in how young people think about work and play in their lives. In general, the most economically privileged young people see their activities as more like play than work, whereas those less well off are more likely to define their activities as work. Likewise, White youth (boys especially) are more likely than other groups to see their lives as playful. The researchers interpret these findings to mean that being economically privileged allows you to think of your work as if it were play. Being in a less advantaged position, on the other hand, makes you see the world as more "worklike." This is supported by further findings that young people from less advantaged backgrounds spend more time in activities they define as purposeless (Schneider and Stevenson 1999).

Adulthood

Socialization does not end when one becomes an adult. Building on the identity formed in childhood and adolescence, adult socialization is the process of learning new roles and expectations in adult life. More so than at earlier stages in life, **adult socialization** involves learning behaviors and attitudes appropriate to specific situations and roles.

map4.1 VIEWING SOCIETY IN GLOBAL PERSPECTIVE

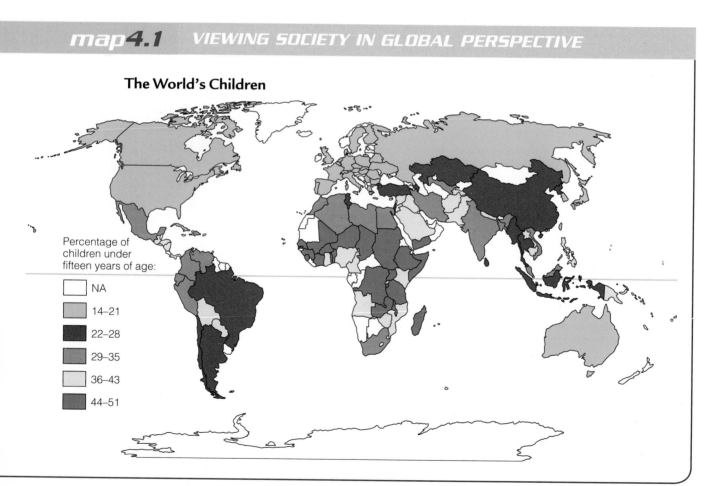

The World's Children

Percentage of children under fifteen years of age:

- NA
- 14–21
- 22–28
- 29–35
- 36–43
- 44–51

Throughout the world, the proportion of children as a percentage of the population of a given country tends to be higher in those countries that are most economically disadvantaged and most overpopulated. In such countries, children are also more likely to die young but are needed to contribute to the labor that families do. What consequences do you think the proportion of children in a given society has for the society as a whole?

Data: From the U.S. Census Bureau. 2008. *Statistical Abstract of the United States 2007.* Washington, DC: U.S. Government Printing Office

Youths entering college, to take an example from young adulthood, are newly independent and have new responsibilities. In college, one acquires not just an education but also a new identity. Those who enter college directly from high school may encounter conflicts with their family over their newfound status. Older students who work and attend college may experience difficulties (defined as *role conflict*; see Chapter 5) trying to meet dual responsibilities, especially if their family is not supportive. Meeting multiple and conflicting demands may require the returning student to develop different expectations about how much she can accomplish or to establish different priorities about what she will attempt. These changes reflect a new stage in her socialization (Settersten and Lovegreen 1998).

Adult life is peppered with events that may require the adult to adapt to new roles. Marriage, a new career, starting a family, entering the military, getting a divorce, or dealing with a death in the family all transform an individual's previous social identity. In today's world, these transitions through the life course are not as orderly as they were in the past. Where there was once a sequential and predictable trajectory of schooling, work, and family roles through one's twenties and thirties, that is no longer the case. Younger generations now experience diverse patterns in the sequencing of work, schooling, and family formation—even returning home—than was true in the past. These changes complicate the life course, and people have to make different adaptations to these changing roles (Cooksey and Rindfuss 2001; Rindfuss et al. 1999).

Becoming an adult is also taking longer than before. People stay in school longer, are marrying later, and postponing childbearing (see Table 4.2). With these social changes, people have to be inventive in the roles they occupy because some of the old expectations no longer apply.

Another part of learning a new role is **anticipatory socialization,** the learning of expectations associated with a role a person expects to enter in the future. One might rehearse the expectations associated with being a professor by working as a teaching assistant, taking a class in preparation for becoming a father, or attending a summer program to prepare for entering college. Anticipatory socialization allows a person to foresee the expectations associated with a new role and to learn what is expected in that role in advance.

In the transition from an old role to a new one, individuals often vacillate between their old and new identities as they adjust to fresh settings and expectations. An interesting example is *coming out,* the process of identifying oneself as gay or lesbian. This can be either a public coming out or a private acknowledgment of sexual orientation. The process can take years, and generally means coming out to a few people, at first selective family members or friends who are likely to have the most positive reaction. Coming out is rarely a single event, but occurs in stages on the way to developing a new identity—one that is not only a new sexual identity but also a new sense of self (Due 1995).

Age and Aging

Passage through adulthood involves many transitions. In our society, one of the most difficult transitions is the passage to old age. We are taught to fear aging in this society, and many people spend a lot of time and money trying to keep looking young. Unlike many other societies, ours does not revere the elderly, but instead devalues them, making the aging process even more difficult.

It is easy to think that aging is just a natural fact. Despite desperate attempts to hide gray hair, eliminate wrinkles, and reduce middle-aged bulge, aging is inevitable. The skin creases and sags, the hair thins, metabolism slows, and bones become less dense and more brittle by losing bone mass. Although aging is a physical process, the social dimensions of aging are just as important, if not more so, in determining the aging process. Just think about how some people appear to age much more rapidly than others. Some sixty-year-olds look only forty, and some forty-year-olds look sixty.

table4.2	*Showing the Transition to Adulthood*	
	1980	**2007**
Percentage aged 20–21 in school	31.9%	33.0%
Median age at first marriage		
Women	22.0 yrs	25.8 yrs
Men	24.7 yrs	27.5 yrs
Fertility rate, women aged 15–19	68.4 (per 1000 women)	60.0 (per 1000 women)
Percentage aged 16–19 in labor force		
Women	9.6%	4.9%
Men	8.1%	4.3%

Source: U.S. Census Bureau. 2008. *Statistical Abstract of the United States 2007.* Washington, DC: U.S. Government Printing Office.

The stresses of life that accompany age can change a person in many ways, as is evident in these "before" and "after" photographs of former President George W. Bush.

These differences result from combinations of biological and social factors, such as genetics, eating and exercise habits, stress, smoking habits, pollution in the physical environment, and many other factors. The social dimensions of aging are what interest sociologists.

Although the physiology of aging proceeds according to biological processes, what it *means* to grow older is a social phenomenon. **Age stereotypes** are preconceived judgments about what different age groups are like. Stereotypes abound for both old and young people. Young people, especially teenagers, are stereotyped as irresponsible, addicted to loud music, lazy ("slackers"), sloppy, and so on; the elderly are stereotyped as forgetful, set in their ways, mentally dim, and unproductive. Though like any stereotype, these stereotypes are largely myths, they are widely believed. Age stereotypes also differ for different groups. Older women are stereotyped as having lost their sexual appeal, contrary to the stereotype of older men as handsome or "dashing" and desirable. Gender is, in fact, one of the most significant factors in age stereotypes. Women may even be viewed as becoming old sooner than men because people describe women as old a decade sooner than they do men (Stoller and Gibson 2000)!

Age stereotypes are also reinforced through popular culture. Advertisements depict women as needing creams and lotions to hide "the telltale signs of aging." Men are admonished to cover the patches of gray hair that appear or to use other products to prevent baldness. Entire industries are constructed on the fear

of aging that popular culture promotes. Facelifts, tummy tucks, and vitamin advertisements all claim to "reverse the process of aging," even though the aging process is a fact of life.

Age Prejudice and Discrimination Age **prejudice** refers to a negative attitude about an age group that is generalized to all people in that group. Prejudice against the elderly is prominent. The elderly are often thought of as childlike and thus incapable of adult responsibility. Prejudice relegates people to a perceived lower status in society and stems from the stereotypes associated with different age groups.

Age discrimination is the different and unequal treatment of people based solely on their age. Whereas age prejudice is an attitude, age discrimination involves actual behavior. As an example, people may talk "baby talk" to the elderly. This reinforces the stereotype of the elderly as childlike and incompetent. Some forms of age discrimination are illegal. The Age Discrimination Employment Act, first passed in 1967 but amended several times since, protects people from age discrimination in employment. It states that age discrimination is a violation of the individual's civil rights. An employer can neither hire nor fire someone based solely on age, nor segregate or classify workers based on age. Age discrimination cases have become one of the most frequently filed cases through the Equal Employment Opportunity Commission (EEOC), the federal agency set up to monitor violations of civil rights in employment.

Ageism is a term sociologists use to describe the institutionalized practice of age prejudice and discrimination. More than a single attitude or an explicit act of discrimination, ageism is structured into the institutional fabric of society. Like racism and sexism, ageism encompasses both prejudice and discrimination, but it is also manifested in the structure of institutions. As such, it does not have to be intentional or overt to affect how age groups are treated. Ageism in society means that, regardless of laws that prohibit age discrimination, a person's age is a significant predictor of his or her life chances. Resources are distributed in society in ways that advantage some age groups and disadvantage others; cultural belief systems devalue the elderly; society's systems of care are often inadequate to meet people's needs as they grow old—these are the manifestations of ageism, a persistent and institutionalized feature of society.

Age Stratification. Most societies produce age hierarchies—systems in which some age groups have more power and better life chances than others. **Age stratification** refers to the hierarchical ranking of different age groups in society. Age stratification exists because processes in society ensure that people of different ages differ in their access to society's rewards, power, and privileges. As we will see, in the United States and elsewhere, age is a major source of inequality (see Figure 4.4).

Age is an *ascribed status*; that is, age is determined by when you were born. Different from other ascribed statuses, which remain relatively constant over the duration of a person's life, age changes steadily throughout your life. Still, you remain part of a particular generation—something sociologists call an **age cohort**—an aggregate group of people born during the same period.

People in the same age cohort share the same historical experiences—wars, technological developments, and economic fluctuations—although they might do so in different ways, depending on other life factors. Living through the Great Depression, for example, shaped an entire generation's attitudes and behaviors, as did growing up in the 1960s, as will being a member of the contemporary youth generation. Depending on how the major economic recession of 2009 shakes out, it too may significantly shape the current youth generation. Recall from Chapter 1 that C. Wright Mills saw the task of the sociological imagination as analyzing the relationship between biography and history. Understanding the experiences of different age cohorts is one way you can do this. People who live through the same historic period experience a similar impact of that period in their personal lives. The troubles and triumphs they experience and the societal issues they face are rooted in the commonality established by their age cohort. The shared historical experiences of age cohorts result in discernible generational patterns in social attitudes and similarity of life chances (Stoller and Gibson 2000).

Different generations must grapple with and respond to different social contexts. Someone born just after World War II would, upon graduation from high school or college, enter a labor market where there was widespread availability of jobs and, for many, expanding opportunities. Now young people face a labor market where entry-level jobs in secure corporate environments are rare and where many are trapped in low-level jobs with little opportunity for advancement. Many young people worry, as a result, about whether they will be able to achieve even the same degree of economic status as their parents—the first time this has happened in U.S. history. Understanding how society shapes the experiences of different generations is what sociologists mean by saying that age is a structural feature of society (see Figure 4.5).

The age structure of society shapes people's opportunities and is the basis for cultural understandings of age itself. In the very poor regions of northeast Brazil, for example, mothers show little attachment to those who are born small and weak; if they die, there is little ceremony, and their graves remain unmarked. Because sick infants are believed to be angels who fly to heaven, mother's tears are believed to dampen their wings, risking their flight. Anthropologists interpret this as the mothers' reaction to their impoverishment: They cannot invest attention or emotion in the lives of children who are unlikely to live (Scheper-Hughes 1992; Peoples and Bailey 2003).

There is also great variation in how the old are treated. In many societies, older people are given enormous respect. There may be traditions to honor the elders, and they may be given authority over decisions in society, as they are perceived as most wise. On the other hand, among some cultures, adults who can no longer contribute to the society because of old

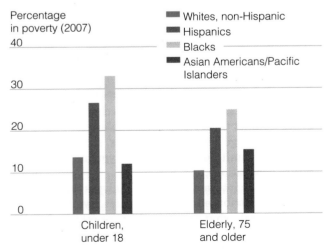

FIGURE 4.4 *Poverty by Age and Race*

Source: DeNavas-Walt, Carmen, Bernadette D. Proctor, and Cheryl Hill Lee. *Income, Poverty, and Health Insurance Coverage in the United States: 2008.* Washington, DC: U.S. Census Bureau. **www.census.gov**

Every culture has important rites of passage that mark the transition from one phase in the life course to another. Here different cultural traditions distinguish the rites of passage associated with marriage: a traditional Nigerian wedding (upper left); a young American couple (upper right); a Shinto (Japanese) bride taking a marital pledge by drinking sake (lower left); and a newlywed orthodox Christian couple in Macedonia (lower right).

the individual's worthiness, preparedness, or community acceptance.

Sociologists have noted that in the United States there is no standard and formalized rite of passage marking the transition from childhood to adulthood. As a consequence, the period of adolescence is attended by ambivalence and uncertainty. As adolescents hover between adult and child status, they may not have the clear sense of identity that a rite of passage can provide. However, although there is no universal ceremony in our culture by which young people are noted as moving from child to adult, some subcultures do mark the occasion. Among the wealthy, the debutante's coming out celebration is a traditional introduction of a young woman to adult society. Latinos may celebrate the *quinceañera* (fifteenth birthday) of young girls. A tradition of the Catholic Church, this rite recognizes the girl's coming of age, while also keeping faith with an ethnic heritage. Dressed in white, she is introduced by her parents to the larger community. Formerly associated mostly with working-class families and other Latinos, the quinceañera has also become popular among affluent Mexican Americans, who may

match New York debutante society by spending as much as $50,000–$100,000 on the event.

Resocialization

Most transitions people experience in their lifetimes involve continuity with the former self as it undergoes gradual redefinition. Sometimes, however, adults are forced to undergo a radical shift of identity. **Resocialization** is the process by which existing social roles are radically altered or replaced (Fein 1988). Resocialization is especially likely when people enter institutional settings where the institution claims enormous control over the individual. Examples include the military, prisons, monastic orders, and some cults (see also Chapter 6 for a discussion of total institutions). When military recruits enter boot camp, they are stripped of personal belongings, their heads are shaved, and they are issued identical uniforms. Although military recruits do not discard their former identities, the changes brought about by becoming a soldier can be dramatic and are meant to make the military one's *primary group*, not one's family, friends, or personal history. The military represents an extreme form of resocialization in which individuals are expected to subordinate their identity to that of the group. In such organizations, individuals are interchangeable, and group consensus (meaning, in the military, unanimous, unquestioned subordination to higher ranks) is an essential component of group cohesion and effectiveness. Military personnel are expected to act as soldiers, not as individuals. Understanding the importance of resocialization on entry to the military helps us understand such practices as "the rat line" at VMI (Virginia Military Institute) where members of the senior class taunt and harass new recruits.

Resocialization often occurs when people enter hierarchical organizations that require them to respond to authority on principle, not out of individual loyalty. The resocialization process promotes group solidarity and generates a feeling of belonging. Participants in these settings are expected to honor the symbols and objectives of the organization; disloyalty is seen as a threat to the entire group. In a convent, for example, nuns are expected to subordinate their own identity to the calling they have taken on, a calling that requires obedience both to God and to an abbess.

Hazings are good examples of rites of passage that often accompany induction into a group. The hazing scene here (hair pulling) is relatively mild compared to many high school hazing incidents.

thinking sociologically

Find a group of adults (young or old) who have just entered a new stage of life (getting a new or first job, getting married, becoming a grandparent, retiring, entering a nursing home, and so forth), and ask them to describe this new experience. Ask questions such as what others expect of them in this new role, how these expectations are communicated to them, what changes they see in their own behavior, and what expectations they have of their new situation. What do your observations tell you about *adult socialization*?

Resocialization may involve degrading initiates physically and psychologically with the aim of breaking down or redefining their old identity. They may be given menial and humiliating tasks and be expected to act in a subservient manner. Social control in such a setting may be exerted by peer ridicule or actual punishment. Fraternities and sororities offer an interesting everyday example of this pattern of resocialization. Intense resocialization rituals, whether in jailhouses, barracks, convents, or sorority and fraternity houses, serve the same purpose: imposing some sort of ordeal to cement the seriousness and permanence of new roles and expectations.

The Process of Conversion

Resocialization also occurs during what people popularly think of as conversion. A conversion is a far-reaching transformation of identity, often related to religious or political beliefs. People usually think of conversion in the context of cults, but it happens in other settings as well.

John Walker Lindh was a U.S. citizen when the United States entered the Iraq war. He joined the Taliban in Afghanistan and was later charged with conspiring to kill Americans abroad and supporting terrorist organizations. Lindh is an example of an *extreme conversion*. He was raised Catholic in an affluent family, but he converted to Islam as a teenager, changing not just his ideas, but also his dress. Neighbors described him as being transformed from "a boy who wore blue jeans and T-shirts to an imposing figure in flowing Muslim garb" (Robertson and Burke 2001). As a young man, he traveled to Yemen and Pakistan to study language and the Koran and was introduced there to the Taliban. News sources have since named him the "American Taliban."

As when people join religious cults, this is extreme conversion, but conversion happens in less extreme situations, too. People may convert to a different religion, thereby undergoing resocialization by changing beliefs and religious practices. Or someone may become strongly influenced by the beliefs of a social movement and abruptly or gradually change beliefs—even identity—as a result.

The Brainwashing Debate

Extreme examples of resocialization are seen in the phenomenon popularly called "brainwashing." In the popular view of brainwashing, converts have their previous identities totally stripped; the transformation is seen as so complete that only deprogramming can restore the former self. Potential candidates of brainwashing include people who enter religious cults, prisoners of war, and hostages. Sociologists have examined brainwashing to illustrate the process of resocialization.

As the result of their research, sociologists have cautioned against using the word *brainwashing* when referring to this form of conversion. The term implies that humans are mere puppets or passive victims whose free will can be taken away during these conversions (Robbins 1988). In religious cults, however, converts do not necessarily drop their former identity.

Sociological research has found that the people most susceptible to cult influence are the most suggestible, primarily young adults who are socially isolated, drifting, and having difficulty performing in other areas (such as in their jobs or in school). Such people may choose to affiliate with cults voluntarily. Despite the widespread belief that people have to be deprogrammed to be freed from the influence of cults, many people are able to leave on their own (Robbins 1988). So-called brainwashing is simply a manifestation of the *social influence people experience through interaction with others*. Even in cult settings, socialization is an interactive process, not just a transfer of group expectations to passive victims.

Forcible confinement and physical torture can be instruments of extreme resocialization. Under severe captivity and deprivation, a captured person may come to identify with the captor; this is known as the **Stockholm Syndrome.** In such instances, the captured person has become dependent on the captor. On release, the captive frequently needs debriefing, or deprogramming. Prisoners of war and hostages may not lose free will altogether, but they do lose freedom of movement and association, which makes prisoners intensely dependent on their captors and therefore vulnerable to the captor's influence.

The Stockholm Syndrome can help explain why some battered women do not leave their abusing spouses or boyfriends. Dependent on their abuser both financially and emotionally, battered women often develop identities that keep them attached to men who abuse them. In these cases, outsiders often think the women should leave instantly, whereas the women themselves may find leaving difficult, even in the most abusive situations.

Chapter Summary

- ### What is socialization, and why is it significant for society?
  Socialization is the process by which human beings learn the social expectations of society. Socialization creates the expectations that are the basis for people's attitudes and behaviors. Through socialization, people conform to social expectations, although people still express themselves as individuals.

- ### What are the agents of socialization?
  Socialization agents are those who pass on social expectations. They include the family, the media, peers, sports, religious institutions, and schools, among others. The family is usually the first source of socialization. The media also influence people's values and behaviors. Peers are an important source of individual identity; without peer approval, most people find it hard to be socially accepted. Schools also pass on expectations that are influenced by gender, race, and other social characteristics of people and groups.

- ### What theoretical perspectives do sociologists use to explain socialization?
  Psychoanalytic theory sees the self as driven by unconscious drives and forces that interact with the expectations of society. *Social learning theory* sees identity as a learned response to social stimuli such as role models. *Functionalism* interprets socialization as key to social stability because socialization establishes shared roles and values. *Conflict theory* interprets socialization in the context of inequality and power relations. *Symbolic interaction theory* sees people as constructing the self as they interact with the environment and give meaning to their experience. Charles Horton Cooley described this process as the *looking glass self*. Another sociologist, George Herbert Mead, described childhood socialization as occurring in three stages: imitation, play, and games.

- ***Does socialization mean that everyone grows up the same?***

  Socialization is not a uniform process. Growing up in different environments and in such a diverse society means that different people and different groups are exposed to different expectations. Factors such as family structure, social class, regional differences, and many others influence how one is socialized.

- ***Does socialization end during childhood?***

  Socialization continues through a lifetime, although childhood is an especially significant time for the formation of *identity*. Adolescence is also a period when peer cultures have an enormous influence on the formation of people's self-concepts. *Adult socialization* involves the learning of specific expectations associated with new roles.

- ***What are the social dimensions of the aging process?***

  Although aging is a physiological process, its significance stems from social meanings attached to aging. *Age prejudice* and *age discrimination* result in the devaluation of older people. *Age stratification*—referring to the inequality that occurs among different age groups—is the result.

- ***What does resocialization mean?***

  Resocialization is the process by which existing social roles are radically altered or replaced. It can take place in an organization that maintains strict social control and demands that the individual conform to the needs of the group or organization.

Key Terms

adult socialization 90	generalized other 86	role 74
age cohort 93	id 82	self 85
age discrimination 92	identity 75	self-concept 77
age prejudice 92	imitation stage 86	significant others 86
age stereotype 92	internalization 76	social control 76
age stratification 93	life course 88	social learning theory 83
ageism 93	looking glass self 85	socialization 74
anticipatory socialization 91	peers 79	socialization agents 78
creationism 80	personality 75	Stockholm Syndrome 98
disengagement theory 94	play stage 86	superego 82
ego 83	psychoanalytic theory 82	taking the role of the other 86
feral children 74	resocialization 96	
game stage 86	rite of passage 95	

Online Resources

Sociology: The Essentials Companion Website

www.cengage.com/sociology/andersen
Visit your book companion website where you will find more resources to help you study and write your research papers. Resources include web links, learning objectives, Internet exercises, quizzing, and flash cards.

 is an easy-to-use online resource that helps you study in less time to get the grade you want NOW.

www.cengage.com/login
Need help studying? This site is your one-stop study shop. Take a Pre-Test and Cengage NOW will generate a Personalized Study Plan based on your test results. The Study Plan will identify the topics you need to review and direct you to online resources to help you master those topics. You can then take a Post-Test to determine the concepts you have mastered and what you still need to work on.

5

Chapter five
CHAPTER FIVE
Chapte

Social Interaction and Social Structure

What Is Society?

What Holds Society Together?

Types of Societies

Social Interaction and Society

Theories About Analyzing Social Interaction

Interaction in Cyberspace

Chapter Summary

[*Picture a college* classroom on your campus. Students sit, and some are taking notes; others, listening; a few, perhaps, sleeping. The class period ends and students stand, gathering their books, backpacks, bags, and other gear. As they stand, many whip out their cell phones, place them to their ears, and quickly push buttons that connect them to a friend. As the students exit the room, many are engaged in *social interaction*—chatting with their friends: some by phone, others by text messaging, some by talking face-to-face. Few, if any, of them realize that their behavior is at that moment influenced by *society*—a society whose influence extends into their immediate social relationships, even when the contours of that society—its *social structure*—are likely invisible to them.

These same students might plug a music player into their ears as they move on to their next class, possibly tuning in to the latest sounds while tuning out the sounds of the environment around them. Some will return to their residences and perhaps text message friends, download some music, or connect with "friends" on Facebook or MySpace. Some might watch a video or podcast on a small, handheld device. Surrounding all of this behavior are social changes

continued

Darrin Klimek/Digital Vision/Getty Images

five Chapter five CHAPTER FIV

that are taking place in society, including changes in technology, in global communication, and in how people now interact with each other. How we make sense of these changes requires an understanding of the connection between society and social interaction. In this way, a sociological perspective can help you see the relationship between individuals and the larger society of which they are a part.

What Is Society?

In Chapter 2, we studied culture as one force that holds society together. *Culture* is the general way of life, including norms, customs, beliefs, and language. Human **society** is a system of social interaction that includes both culture and social organization. Within a society, members have a common culture, even though there may also be great diversity within it. Members of a society think of themselves as distinct from other societies, maintain ties of social interaction, and have a high degree of interdependence. The interaction they have, whether based on harmony or conflict, is one element of society. Within society, **social interaction** is behavior between two or more people that is given meaning by them. Social interaction is how people relate to each other and form a social bond.

Social interaction is the foundation of society, but society is more than a collection of individual social actions. Emile Durkheim, the classical sociological theorist, described society as *sui generis*—a Latin phrase meaning "a thing in itself, of its own particular kind." To sociologists, seeing society *sui generis* means that society is more than just the sum of its parts. Durkheim saw society as an organism, something comprising different parts that work together to create a unique whole. Just as a human body is not just a collection of organs but is alive as a whole organism with relationships between its organs, society is not only a simple collection of individuals, groups, or institutions but is a whole entity that consists of all these elements and their interrelationships.

Durkheim's point—central to sociological analysis—is that society is much more than the sum of the individuals in it. Society takes on a life of its own. It is patterned on humans and their interactions, but it is something that

RayArt/Graphics/Alamy

The introduction of new technologies is transforming how people interact with each other.

endures and takes on shape and structure beyond the immediacy of any given group of people. This is a basic idea that guides sociological thinking.

You can think of it this way: Imagine how a photographer views a landscape. The landscape is not just the sum of its individual parts—mountains, pastures, trees, or clouds—although each part contributes to the whole. The power and beauty of the landscape is that all its parts *relate* to each other, some in harmony and some in contrast, to create a panoramic view. The photographer who tries to capture this landscape will likely use a wide-angle lens. This method of photography captures the breadth and comprehensive scope of what the photographer sees. Similarly, sociologists try to picture society as a whole, not only by seeing its individual parts but also recognizing the relatedness of these parts and their vast complexity.

Macro- and Microanalysis

Sociologists use different lenses to see the different parts of society. Some views are more macroscopic—that is, sociologists try to comprehend the whole of society, how it is organized, and how it changes. This is called **macroanalysis,** a sociological approach that takes the broadest view of society by studying large patterns of social interaction that are vast, complex, and highly differentiated. You might do this by looking at a whole society or comparing different total societies to each other. For example, as we opened this chapter, you saw that large-scale changes in technology influence even the most immediate social interaction that we have with other people. Thus, whereas only a few years ago it would not have been imaginable to create a network of friends in cyberspace, today it is a common practice, especially for young people.

Other views are more microscopic—that is, the focus is on the smallest, most immediately visible parts of social life, such as specific people interacting with each other. This is called **microanalysis.** In this approach, sociologists study patterns of social interactions that are relatively small, less complex, and less differentiated—the microlevel of society. Again, thinking of how this chapter opened, you might want to study how people engage in texting each other on a one-to-one basis. How are they similar or different, on the basis of age, or gender, or social class or race? For example, do people text (that is, interact) with each other within racial groups more than between racial groups? Observing this would be an example of microanalysis. Thus a sociologist who studies social interaction via texting or on the Internet would be engaging in microanalysis but might interpret what is found in the context of macrolevel processes (such as race relations in society). Just as a photographer might use a wide-angle lens to photograph a landscape or a telephoto lens for a closer view, sociologists use both macro- and microanalyses to reveal different dimensions of society.

In this chapter, we continue our study of sociology by starting with the macrolevel of social life (by studying total social structures), then continuing through the microlevel (by studying groups and face-to-face interaction). The idea is to help you see how large-scale dimensions of society shape even the most immediate forms of social interaction.

Sociologists use the term **social organization** to describe the order established in social groups at any level. Specifically, social organization brings regularity and predictability to human behavior; social organization is present at every level of interaction, from the whole society to the smallest groups.

Social Institutions

Societies are identified by their cultural characteristics and the social institutions that compose each society. A **social institution** (or simply an institution) is an established and organized system of social behavior with a recognized purpose. The term refers to the broad systems that organize specific functions in society. Unlike individual behavior, social institutions cannot be directly observed, but their impact and structure can still be seen. For example, the family is an institution that provides for the care of the young and the transmission of culture. Religion is an institution that organizes sacred beliefs. Education is the institution through which people learn the skills needed to live in the society.

The concept of the social institution is important to sociological thinking. You can think of social institutions as the enduring consequences of social behavior, but what fascinates sociologists is how social institutions take on a life of their own. For example, you were likely born in a hospital, which itself is part of the health care institution. The simple act of birth, which you might think of as an individual experience, is shaped by the structure of this social institution. Thus, you were likely delivered by a doctor, accompanied by nurses and, perhaps, a midwife—each of whom exists in a specific social relationship to the health care institution. Each of these persons is in an *institutional* role. Moreover, the practices surrounding your birth were also shaped by this social institution. Thus, you might be initially removed from your mother and examined by a doctor, which is very different from the institutional practices in other societies.

The major institutions in society include the family, education, work and the economy, the political institution (or state), religion, and health care, as well as the mass media, organized sports, and the military. These are all complex structures that exist to meet certain needs that are necessary for society to exist. *Functionalist theorists* have traditionally identified these needs (functions) as follows (Aberle et al. 1950; Parsons 1951a; Levy 1949):

1. *The socialization of new members of the society.* This is primarily accomplished by the family, but involves others institutions as well, such as education.

2. *The production and distribution of goods and services.* The economy is generally the institution that performs this set of tasks, but this may also involve the family as an institution—especially in societies where production takes place within households.

3. *Replacement of society's members.* All societies must have a means of replacing members who die, move or migrate away, or otherwise leave the society. Families are typically organized to do this.

4. *The maintenance of stability and existence.* Certain institutions within a society (such as the government, the police force, and the military) contribute toward the stability and continuance of the society.

5. *Providing the members with an ultimate sense of purpose.* Societies accomplish this task by creating national anthems, for instance, and by encouraging patriotism in addition to providing basic values and moral codes through institutions such as religion, the family, and education.

Birth, though a natural process, occurs within social institutions—institutions that vary in different societies, depending on the social organization of society. Here you see how birth in the United States, which is mainly defined as a medical event, contrasts with a health assistant attending a birth in rural Mexico.

In contrast to functional list theory, conflict theory further notes that because conflict is inherent in most societies, the social institutions of society do not provide for all its members equally. Some members are provided for better than others, thus demonstrating that institutions affect people by granting more power to some social groups than to others. Using the example of the health care institution given above, some groups have considerably less power within the institution than do others. Thus, nurses are generally subordinate to doctors and doctors to hospital administrators. And beyond these specific actors within the health care institutions, different social groups in society have more or less power within social institutions. Thus, racial and ethnic minorities in general have poorer access to health care than others; the poor have less access, as do those of lower social class status. (For more information, see Chapter 14 on health care.)

Social Structure

Sociologists use the term **social structure** to refer to the organized pattern of social relationships and social institutions that together compose society. Social structures are not immediately visible to the untrained observer; nevertheless, they are present, and they affect all dimensions of human experience in society. Social structural analysis is a way of looking at society in which the sociologist analyzes the patterns in social life that reflect and produce social behavior.

Social class distinctions are an example of a social structure. Class shapes the access that different groups have to the resources of society, and it shapes many interactions people have with each other. People may form cliques with those who share similar class standing, or they may identify with certain values associated with a given class. Class then forms a social structure—one that shapes and guides human behavior at all levels, no matter how overtly visible or invisible this structure is to someone at a given time.

The philosopher Marilyn Frye aptly describes the concept of social structure in her writing. Using the metaphor of a birdcage, she writes that if you look closely at only one wire in a cage, you cannot see the other wires. You might then wonder why the bird within does not fly away. Only when you step back and see the whole cage instead of a single wire do you understand why the bird does not escape. Frye writes:

> It is perfectly obvious that the bird is surrounded by a network of systematically related barriers, no one of which would be the least hindrance to its flight, but all of which, by their relations to each other, are as confining as the solid walls of a dungeon.... One can study the elements of ... structure with great care and some good will without seeing or being able to understand that one is looking at a cage and that there are people there who are caged, whose motion and mobility are restricted, whose lives are shaped and reduced. (Frye 1983: 4–5)

Just as a birdcage is a network of wires, society is a network of social structures, both micro and macro.

thinkingsociologically

Using Marilyn Frye's analogy of the birdcage, think of a time when you believed your choices were constrained by *social structure*. When you applied to college, for example, could you go anywhere you wanted? What social structural conditions guided your ultimate selection of schools to attend?

What Holds Society Together?

What holds societies together? We ask this question throughout this chapter. This central question in sociology was first addressed by Emile Durkheim, the French sociologist writing in the late 1800s and early 1900s. He argued that people in society had a **collective consciousness,** defined as the body of beliefs common to a community or society that give people a sense of belonging and a feeling of moral obligation to its demands and values. According to Durkheim, collective consciousness gives groups social solidarity because members of a group feel they are part of one society.

Where does the collective consciousness come from? Durkheim argued that it stems from people's participation in common activities, such as work, family, education, and religion—in short, society's institutions.

Mechanical and Organic Solidarity

According to Durkheim, there are two types of social solidarity: mechanical and organic. **Mechanical solidarity** arises when individuals play similar—rather than different—roles within the society. Individuals in societies marked by mechanical solidarity share the same values and hold the same things sacred. This particular kind of cohesiveness is weakened when a society becomes more complex. Contemporary examples of mechanical solidarity are rare because most societies of the world have been absorbed in the global trend for greater complexity and interrelatedness. Native American groups before European conquest were bound together by mechanical solidarity; indeed, many Native American groups are now trying to regain the mechanical solidarity on which their cultural heritage rests, but they are finding that the superimposition of White institutions on Native American life interferes with the adoption of traditional ways of thinking and being, which prevents mechanical solidarity from gaining its original strength.

In contrast, **organic solidarity** occurs when people play a great variety of roles, and unity is based on role differentiation, not similarity. The United States and other industrial societies are built on organic solidarity, and each is cohesive because of the differentiation within each. Roles are no longer necessarily similar, but they are necessarily interlinked—the performance of multiple roles is necessary for the execution of society's complex and integrated functions.

Durkheim described this state as the **division of labor,** defined as the relatedness of *different* tasks that develop in complex societies. The labor force within the contemporary U.S. economy, for example, is divided according to the kinds of work people do. Within any division of labor, tasks become distinct from one another, but they are still woven into a whole.

The division of labor is a central concept in sociology because it represents how the different pieces of society fit together. The division of labor in most contemporary societies is often marked by distinctions such as age, gender, race, and class. In other words, if you look at who does what in society, you will see that women and men tend to do different things; this is the gender division of labor. Similarly, old and young to some extent do different things; this is a division of labor by age. This is crosscut by the racial division of labor, the pattern whereby those in different racial–ethnic groups tend to do different work—or are often forced to do different work—in society. At the same time, the division of labor is also marked by class distinctions, with some groups providing work that is highly valued and rewarded and others doing work that is devalued and poorly rewarded. As you will see throughout this book, gender, race, and class intersect and overlap in the division of labor in society.

Durkheim's thinking about the origins of social cohesion can bring light to contemporary discussions over "family values." Some want to promote traditional family values as the moral standards of society. Is such a thing necessarily good, or even possible? The United States is an increasingly diverse society, and family life differs among different groups. It is unlikely that a single set of family values can be the basis for social solidarity.

Gemeinschaft and Gesellschaft

Different societies are held together by different forms of solidarity. Some societies are characterized by what sociologists call **gemeinschaft,** a German word that means "community"; others are characterized as **gesellschaft,** which literally means "society" (Tönnies 1963/1887). Each involves a type of solidarity or cohesiveness. Those societies that are *gemeinschafts* (communities) are characterized by a sense of "we" feeling, a moderate division of labor, strong personal ties, strong family relationships, and a sense of personal loyalty. The sense of solidarity between members of the gemeinschaft society arises from personal ties; small, relatively simple social institutions;

and a collective sense of loyalty to the whole society. People tend to be well integrated into the whole, and social cohesion comes from deeply shared values and beliefs (often, sacred values). Social control need not be imposed externally because control comes from the internal sense of belonging that members share. You might think of a small community church as an example.

In contrast, in societies marked by *gesellschaft,* an increasing importance is placed on the secondary relationships people have—that is, less intimate and more instrumental relationships such as work roles instead of family or community roles. Gesellschaft is characterized by less prominence of personal ties, a somewhat diminished role of the nuclear family, and a lessened sense of personal loyalty to the total society. The solidarity and cohesion remain, and it can be very cohesive, but the cohesion comes from an elaborated *division of labor* (thus, *organic* solidarity), greater flexibility in social roles, and the instrumental ties that people have to one another.

Social solidarity under gesellschaft is weaker than in the gemeinschaft society, however. Gesellschaft is more likely than gemeinschaft to be torn by class conflict because class distinctions are less prominent, though still present, in the gemeinschaft. Racial–ethnic conflict is more likely within gesellschaft societies because the gemeinschaft tends to be ethically and racially very homogeneous; often it is characterized by only one racial or ethnic group. This means that conflict between gemeinschaft societies, such as ethnically based wars, can be very high, because both groups have a strong internal sense of group identity that may be intolerant of others (for example, Palestinians versus Israelis, or Shiite Muslims versus Sunni Muslims in Iraq).

In sum, complexity and differentiation are what make the gesellschaft cohesive, whereas similarity and unity cohere the gemeinschaft society. In a single society, such as the United States, you can conceptualize the whole society as gesellschaft, with some internal groups marked by gemeinschaft. Our national motto seems to embody this idea: *E pluribus unum* (Unity within diversity), although clearly this idealistic motto has only been partly realized.

Types of Societies

In addition to comparing how different societies are bound together, sociologists are interested in how social organization evolves in different societies. Simple things such as the size of a society can also shape its social organization, as do the different roles that men and women engage in as they produce goods, care for the old and young, and pass on societal traditions. Societies also differ according to their resource base—whether they are predominantly agricultural or industrial, for example, and whether they are sparsely or densely populated.

Thousands of years ago, societies were small, sparsely populated, and technologically limited. In the competition for scarce resources, larger and more technologically advanced societies dominated smaller ones. Today, we have arrived at a global society with highly evolved degrees of social differentiation and inequality, notably along class, gender, racial, and ethnic lines (Nolan and Lenski 2005).

Sociologists distinguish six types of societies based on the complexity of their social structure, the amount of overall cultural accumulation, and the level of their technology. They are *foraging, pastoral, horticultural, agricultural* (these four are called *preindustrial* societies), then *industrial* and *postindustrial* societies (see Table 5.1). Each type of society can still be found on Earth, though all but the most isolated societies are rapidly moving toward the industrial and postindustrial stages of development.

These different societies vary in the basis for their organization and the complexity of their division of labor. Some, such as foraging societies, are subsistence economies, where men and women hunt and gather food but accumulate very little. Others, such as pastoral societies and horticultural societies, develop a more elaborate division of labor as the social roles that are needed for raising livestock and farming become more numerous. With the development of agricultural societies, production becomes more large scale and strong patterns of social differentiation develop, sometimes taking the form of a caste system or even slavery.

The key driving force that distinguishes these different societies from each other is the development of technology. All societies use technology to help fill human needs, and the form of technology differs for the different types of society.

Preindustrial Societies

A **preindustrial society** is one that directly uses, modifies, and/or tills the land as a major means of survival. There are four kinds of preindustrial societies, listed here by degree of technological development: foraging (or hunting–gathering) societies, pastoral societies, horticultural societies, and agricultural societies (see Table 5.1).

In *foraging (hunting–gathering) societies,* the technology enables the hunting of animals and gathering of vegetation. The technology does not permit the refrigeration or processing of food, hence these individuals must search continuously for plants and game. Because hunting and gathering are activities that require large amounts of land, most foraging societies are nomadic; that is, they constantly travel as they deplete the plant supply or follow the migrations of animals. The central institution is the family, which serves as the means of distributing food, training children, and protecting its members. There is usually role differentiation on the basis of gender, although the specific form of the gender division of labor varies in different societies. They occasionally wage war with other clans or similar societies, and spears and bows and arrows are the weapons used. An example of a foraging society are the Pygmies of Central Africa.

In *pastoral societies,* technology is based on the domestication of animals. Such societies tend to develop in desert areas that are too arid to provide rich vegetation. The pastoral society is nomadic,

Different types of societies produce different kinds of social relationships. Some may involve more direct and personal relationships (called gemeinschafts), *whereas others produce more fragmented and impersonal relationships (called* gesellschafts).

table5.1		Types of Societies		
		Economic Base	**Social Organization**	**Examples**
Preindustrial Societies	*Foraging Societies*	Economic sustenance dependent on hunting and foraging	Gender is important basis for social organization, although division of labor is not rigid; little accumulation of wealth	Pygmies of Central Africa
	Pastoral Societies	Nomadic societies, with substantial dependence on domesticated animals for economic production	Complex social system with an elite upper class and greater gender role differentiation than in foraging societies	Bedouins of Africa and Middle East
	Horticultural Societies	Society marked by relatively permanent settlement and production of domesticated crops	Accumulation of wealth and elaboration of the division of labor, with different occupational roles (farmers, traders, craftspeople, and so on)	Aztecs of Mexico; Inca empire of Peru
	Agricultural Societies	Livelihood dependent on elaborate and large-scale patterns of agriculture and increased use of technology in agricultural production	Caste system develops that differentiates the elite and agricultural laborers; may include system of slavery	American South, pre–Civil War
Industrial Societies		Economic system based on the development of elaborate machinery and a factory system; economy based on cash and wages	Highly differentiated labor force with a complex division of labor and large formal organizations	Nineteenth- and most of twentieth-century United States and Western Europe
Postindustrial Societies		Information-based societies in which technology plays a vital role in social organization	Education increasingly important to the division of labor	Contemporary United States, Japan, and others

necessitated by the endless search for fresh grazing grounds for the herds of their domesticated animals. The animals are used as sources of hard work that enable the creation of a material surplus. Unlike a foraging society, this surplus frees some individuals from the tasks of hunting and gathering and allows them to create crafts, make pottery, cut hair, build tents, and apply tattoos. The surplus generates a more complex and differentiated social system with an elite or upper class and more role differentiation on the basis of gender. The nomadic Bedouins of Africa and the Middle East are pastoral societies.

In *horticultural societies,* hand tools are used to cultivate the land, such as the hoe and the digging stick. The individuals in horticultural societies practice ancestor worship and conceive of a deity or deities (God or gods) as a creator. This distinguishes them from foraging societies that generally employ the notion of numerous spirits to explain the unknowable. Horticultural societies recultivate the land each year and tend to establish relatively permanent settlements and villages. Role differentiation is extensive, resulting in different and interdependent occupational roles such as farmer, trader, and craftsperson. The Aztecs of Mexico and the Incas of Peru represent examples of horticultural societies.

The *agricultural society* is exemplified by the pre–Civil War American South, a society of slavery. Such societies have a large and complex economic system that is based on large-scale farming. Such societies rely on technologies such as use of the wheel and use of metals. Farms tend to be considerably larger than the cultivated land in horticultural societies. Large and permanent settlements characterize agricultural societies, which also exhibit dramatic social inequalities. A rigid caste system develops, separating the peasants, or slaves, from the controlling elite caste, which is then freed from manual work allowing time for art, literature, and philosophy, activities of which they can then claim the lower castes are incapable.

Industrial Societies

An *industrial society* is one that uses machines and other advanced technologies to produce and distribute goods and services. The Industrial Revolution began over 250 years ago when the steam engine was invented in England, delivering previously unattainable amounts of mechanical power for the performance of work. Steam engines powered locomotives, factories, and dynamos and transformed societies as the Industrial Revolution spread. The growth of science led to advances in farming techniques such as crop rotation, harvesting, and ginning cotton, as well as industrial-scale projects such as dams for generating hydroelectric power. Joining these advances were developments in medicine, new techniques to prolong and improve life, and the emergence of birth control to limit population growth.

Unlike agricultural societies, industrial societies rely on a highly differentiated labor force and the intensive use of capital and technology. Large formal organizations are common. The task of holding society together, falling on institutions such as religion in preindustrial societies, now falls more on the institutions that have a high division of labor, such as the economy and work, government, politics, and large bureaucracies.

Within industrial societies, the forms of gender inequality that we see in contemporary U.S. society tend to develop. With the advent of industrialization, societies move to a cash-based economy, with labor performed in factories and mills paid on a wage basis and household labor remaining unpaid. This introduced what is known as the *family-wage economy,* in which families become dependent on wages to support themselves, but work within the family (housework, child care, and other forms of household work) is unpaid and therefore increasingly devalued (Tilly and Scott 1978). In addition, even though women (and young children) worked in factories and mills from the first inception of industrialization, the family-wage economy is based on the idea that men are the primary breadwinners. A system of inequality in men's and women's wages was introduced—an economic system that even today continues to produce a wage gap between men and women.

Industrial societies tend to be highly productive economically, with a large working class of industrial laborers. People become increasingly urbanized as they move from farmlands to urban centers or other areas where factories are located. Immigration is common in industrial societies, particularly because industries are forming where there is a high demand for more, cheap labor.

Industrialization has brought many benefits to U.S. society—a highly productive and efficient economic system, expansion of international markets, extraordinary availability of consumer products, and for many, a good working wage. Industrialization has, at the same time, also produced some of the most serious social problems that our nation faces: industrial pollution, an overdependence on consumer goods, wage inequality and job dislocation for millions,

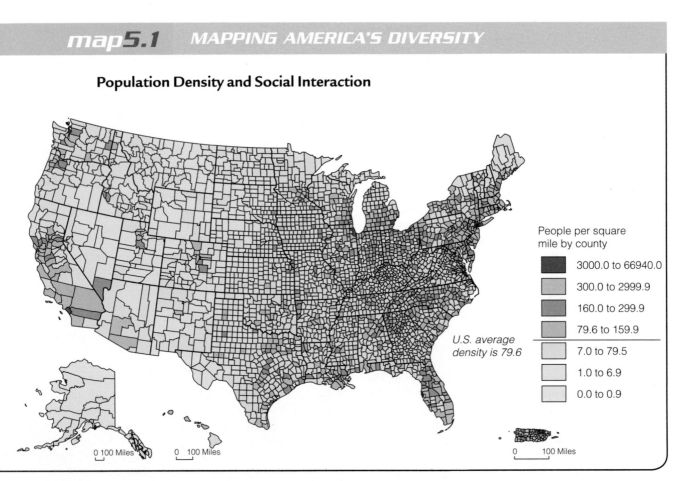

map 5.1 **MAPPING AMERICA'S DIVERSITY**

Population Density and Social Interaction

U.S. average density is 79.6

People per square mile by county

- 3000.0 to 66940.0
- 300.0 to 2999.9
- 160.0 to 299.9
- 79.6 to 159.9
- 7.0 to 79.5
- 1.0 to 6.9
- 0.0 to 0.9

As this map shows, population density (measured as the number of people per square mile) varies enormously in different regions and areas of the country. In what ways do you think the density of a given area might affect people's social interaction?

Data: U.S. Census Bureau. 2006. **www.census.gov/population**

and problems of crime and crowding in urban areas (see Map 5.1 on population density). The recent economic recession/depression that began in 2008 has been significantly responsible for many of these ills—including an exceptionally high rate of unemployment. Understanding the process of industrialization and its accompanying process of urbanization is a major avenue for sociological research and is explored further in Chapter 16.

Postindustrial Societies

In the contemporary era, a new type of society is emerging. Whereas most twentieth-century societies can be characterized in terms of their generation of material goods, **postindustrial society** depends economically on the production and distribution of services, information, and knowledge. Postindustrial societies are information-based societies in which technology plays a vital role in the social organization. The United States is fast becoming a postindustrial society, and Japan may be even further along. Many of the workers provide services such as administration, education, legal services, scientific research, and banking, or they engage in the development, management, and distribution of information, particularly in the areas of computer use and design. Central to the economy of the postindustrial society are the highly advanced technologies of computers, robotics, genetic engineering, and laser technology. Multinational corporations globally link the economies of postindustrial societies.

The transition to a postindustrial society has a strong influence on the character of social institutions. Educational institutions acquire paramount importance in the postindustrial society, and science takes an especially prominent place. For some, the transition to a postindustrial society means more discretionary income for leisure activities—tourism, entertainment, and relaxation industries (spas, massage centers, and exercise) become more prominent—at least for people in certain classes and in the absence of severe economic recession, which has recently plagued not only the United States but Japan, Germany, and other technologically advanced countries as well. The transition to postindustrialism can mean permanent joblessness for many or the need to hold down more than one job simply to make ends meet. Workers without highly technical skills may not fit in such a society, and millions have already found themselves out of work.

Social Interaction and Society

You can see by now that society is an entity that exists above and beyond individuals. Also, different societies are marked by different forms of *social organization*.

Although societies differ, emerge, and change, they are also highly predictable. Your society shapes virtually every aspect of your life from the structure of its social institutions to the more immediate ways that you interact with people. It is to that level—the microlevel of society—that we now turn.

Groups

At the microlevel, society is made up of many different social groups. At any given moment, each of us is a member of many groups simultaneously, and we are subject to their influence: family, friendship groups, athletic teams, work groups, racial and ethnic groups, and so on. Groups impinge on every aspect of our lives and are a major determinant of our attitudes and values regarding everything from personal issues such as sexual attitudes and family values to major social issues such as the death penalty and physician-assisted suicide.

To sociologists a **group** is a collection of individuals who

- interact and communicate with each other;
- share goals and norms; and,
- have a subjective awareness of themselves as "we," that is, as a distinct social unit.

To be a group, the social unit in question must possess all three of these characteristics. We will examine the nature and behavior of groups in greater detail in Chapter 6.

In sociological terms, not all collections of people are groups. People may be lumped together into *social categories* based on one or more shared characteristics, such as teenagers (an age category), truck drivers (an occupational category), and even those who have lost their life savings and pensions as a result of criminal Ponzi investment schemes, such as occurred in the fall of 2008 with those who unknowingly invested money with the now-infamous Bernard Madoff (more about him in Chapter 7).

Social categories can become social groups, depending on the amount of "we" feeling the group has. Only when there is this sense of common identity, as defined in the characteristics of groups above, is a collection of people an actual group. For example, all people nationwide watching TV programs at 8 o'clock Wednesday evening form a distinct social unit, an *audience*. But they are not a group because they do not interact with one another, nor do they possess an awareness of themselves as "we." However, if many of the same viewers were to come together for a large meeting, such as fans of the long-running TV series *Star Trek* (the fans call themselves "Trekkies") coming together for a convention where they could interact and develop a "we" feeling, then they would constitute a group.

We now know that people do not need to be face-to-face in order to constitute a group. Online communities, for example, are people who interact with each other regularly, share a common identity, and think of themselves as being a distinct social unit. On the Internet community Facebook, for example, you may have a group of "friends," some of whom you know personally and others whom you only know online. But these *friends*, as they are known on Facebook, make up a social group that might interact on a regular, indeed, daily basis—possibly even across great distances.

Groups also need not be small or "close-up" and personal. *Formal organizations* are highly structured social groupings that form to pursue a set of goals. Bureaucracies, such as business corporations or municipal governments or associations such as the Parent-Teacher Association (PTA), are examples of formal organizations. A deeper analysis of bureaucracies and formal organizations appears in Chapter 6.

Status

Within groups, people occupy different statuses. **Status** is an established position in a social structure that carries with it a degree of prestige (that is, social value). A status is a rank in society. For example, the position "Vice President of the United States" is a status, one that carries relatively high prestige as well as a complex set of expectations. "High school teacher" is another status; it carries less prestige than "Vice President of the United States," but more prestige than, say, "cabdriver." Statuses occur within institutions. "High school teacher" is a status within the education institution. Other statuses in the same institution are "student," "principal," and "school superintendent."

Typically, a person occupies many statuses simultaneously. The combination of statuses composes a **status set,** which is the complete set of statuses occupied by a person at a given time (Merton 1968). A person may occupy different statuses in different institutions. Simultaneously, a person may be a bank president (in the economic institution), voter (in the political institution), church member (in the religious institution), and treasurer of the PTA (in the education institution). Each status may be associated with a different level of prestige.

Sometimes the multiple statuses of an individual conflict with one another. **Status inconsistency** exists where the statuses occupied by a person bring with them significantly different amounts of prestige and thus differing expectations. For example, someone trained as a lawyer, but working as a cabdriver, experiences status inconsistency. Some recent immigrants from Vietnam and Korea have experienced status inconsistency. Many refugees who had been in high-status occupations in their home country, such as teachers, doctors, and lawyers, could find work in the United States only as grocers or technicians—jobs of relatively lower status than the jobs they left behind. A relatively large body of research in sociology has demonstrated that status inconsistency can

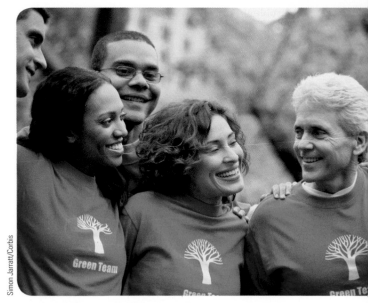

Social groups are organized around different kinds of relationships, but involve a "we" feeling.

lead to stress and depression (Thoits 2009; Taylor et al. 2009; Blalock 1991; Taylor and Hornung 1979; Hornung 1977).

Achieved statuses are those attained by virtue of individual effort. Most occupational statuses—police officer, pharmacist, or boatbuilder—are achieved statuses. In contrast, **ascribed statuses** are those occupied from the moment a person is born. Your biological sex is an ascribed status. Yet, even ascribed statuses are not exempt from the process of social construction. For most individuals, race is an ascribed status fixed at birth, although an individual with one light-skinned African American parent and one White parent may appear to be White and may go through life as a White person. This is called *passing*, although this term is used considerably less often now than it was several years ago. Ascribed status may not be rigidly defined, as for individuals who are *biracial* or *multiracial* (see also Chapter 10.) Finally, ascribed statuses can arise through means beyond an individual's control, such as severe disability or chronic illness.

Some seemingly ascribed statuses, such as gender, can become achieved statuses. Gender, typically thought of as fixed at birth, is a social construct. You can be born female or male, but becoming a woman or a man is the result of social behaviors associated with your ascribed status. In other words, gender is also achieved. People who cross-dress, have a sex change, or develop some characteristics associated with the other sex are good examples of how gender is achieved, but you do not have to see these exceptional behaviors to observe that. People "do" gender in everyday life. They put on appearances and behaviors that are associated with their presumed gender (West and Zimmerman 1987; West and Fenstermaker 1995; Andersen 2009).

If you doubt this, ask yourself what you did today to "achieve" your gender status. Did you dress a certain way? Wear "manly" cologne or deodorant? Splash on a "feminine" fragrance? These behaviors—all performed at the microlevel—reflect the macrolevel of your gender status.

Debunking Society's Myths

Myth: Gender is an *ascribed status* where one's gender identity is established at birth.

Sociological perspective: Although one's biological sex identity is an ascribed status, gender is a social construct and thus is also an *achieved status*—that is, accomplished through routine, everyday behavior, including patterns of dress, speech, touch, and other social behaviors. Sex is not the same as gender. (Andersen 2009).

The line between achieved and ascribed status can be hard to draw. Social class, for example, is determined by occupation, education, and annual income—all of which are achieved statuses—yet one's job, education, and income are known to correlate strongly with the social class of one's parents. Hence, one's social class status is at least partly—though not perfectly—determined at birth. It is an achieved status that includes an inseparable component of ascribed status as well.

Although people occupy many statuses at one time, it is usually the case that one status is dominant, called the **master status,** overriding all other features of the person's identity. The master status may be imposed by others, or a person may define his or her own master status. A woman judge, for example, may carry the master status "woman" in the eyes of many. She is seen not just as a judge, but as a woman judge, thus making gender a master status (Webster and Hysom 1998). A master status can completely supplant all other statuses in someone's status set. Being in a wheelchair is another example of a master status. People may see this, at least at first, as the most important, or salient, part of identity, ignoring other statuses that define someone as a person.

thinkingsociologically

Make a list of terms that describe who you are. Which of these are *ascribed statuses* and which are *achieved statuses*? What do you think your *master status* is in the eyes of others? Does one's *master status* depend on who is defining you? What does this tell you about the significance of social judgments in determining who you are?

Roles

A **role** is the behavior others expect from a person associated with a particular status. Statuses are occupied;

roles are acted or "played." The status of police officer carries with it many expectations; this is the role of police officer. Police officers are expected to uphold the law, pursue suspected criminals, assist victims of crimes, fill out forms for reports, and so on. Usually, people behave in their roles as others expect them to, but not always. When a police officer commits a crime, such as physically brutalizing someone, he or she has violated the role expectations. Role expectations may vary according to the role of the observer—whether the person observing the police officer is a member of a minority group, for example.

As we saw in Chapter 4, social learning theory predicts that we learn attitudes and behaviors in response to the positive reinforcement and encouragement received from those around us. This is important in the formation of our own identity in society. "I am Linda, the skater," or "I am John, the guitarist." These identities are often obtained through **role modeling,** a process by which we imitate the behavior of another person we admire who is in a particular role. A ten-year-old girl or boy who greatly admires the teenage expert skateboarder next door will attempt, through role modeling, to closely imitate the tricks that neighbor performs on the skateboard. As a result, the formation of the child's self-identity is significantly influenced.

A person may occupy several statuses and roles at one time. A person's **role set** includes all the roles occupied by the person at a given time. Thus, a person may be not only Linda the skater but also Linda the student, the daughter, and the lover. Roles may clash with each other, a situation called **role conflict,**

In role modeling, a person imitates the behavior of an admired other.

wherein two or more roles are associated with contradictory expectations. Notice that in Figure 5.1 some of the roles diagrammed for this college student may conflict with others. Can you speculate about which might and which might not?

In U.S. society, some of the most common forms of role conflict arise from the dual responsibilities of job and family. The parental role demands extensive time and commitment and so does the job role. Time given to one role is time taken away from the other. Although the norms pertaining to working women and men are rapidly changing, it is still true that women are more often expected to uphold traditional role expectations associated with their gender role and are more likely held responsible for minding the family when job and family conflict. The sociologist Arlie Hochschild captured the predicament of today's women when she described the "second shift": a working mother spends time and energy all day on the job, only to come home to the "second shift" of family and home responsibilities. These are sometimes delegated to a husband or boyfriend, who encounters less well-formed role expectations that he will take on those responsibilities and is therefore more likely to leave the jobs undone (Hochschild 1997).

Hochschild has found that some companies have instituted "family friendly" policies, designed to reduce the conflicts generated by the "second shift." Ironically, however, in her study she found that few workers take advantage of programs such as more flexible hours, paid maternity leave, and job sharing—except for the on-site child care that actually allowed parents to work more!

Hochschild's studies point to the conflict between two social roles: family roles and work roles. Her

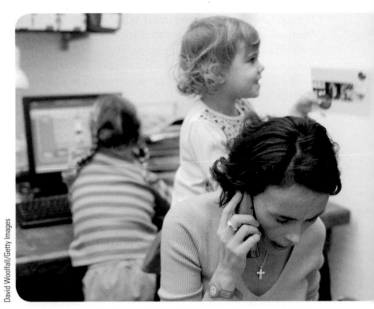

David Woolfall/Getty Images

Changes in the roles for women who are mothers can create role strain.

research is also illustrative of a different sociological concept: **role strain,** a condition wherein a single role brings conflicting expectations. Different from role conflict, which involves tensions *between* two roles, role strain involves conflicts within a single role. In Hochschild's study, the work role has not only the expectations traditionally associated with work, but also the expectation that one "love" one's work and be as devoted to it as to one's family. The result is role strain. The role of student often involves role strain. For example, students are expected to be independent thinkers, yet they feel—quite correctly—that they are often required to simply repeat on an exam what a professor tells them. The tension between the two competing expectations is an example of role strain.

Everyday Social Interaction

You can also see the influence of society in everyday behavior, including such basics as how you talk, patterns of touch, and who you are attracted to. Although you might think of such things as "just coming naturally," they are deeply patterned by society. The cultural context of social interaction really matters in our understanding of what given behaviors mean. An action is defined as positive or negative by the cultural context because social behavior is that to which people give meaning. An action that is positive in one culture can be negative in another. For example, shaking the right hand in greeting is a positive action in the United States, but the same action in East India or certain Arab countries might be an insult. Social and cultural context matter. A kiss on the lips is a positive act in most cultures, yet if you were kissed on the lips

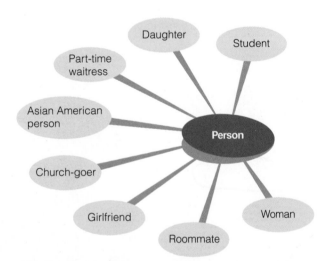

FIGURE 5.1 *Roles in a College Student's Role Set*

Identify the different roles that you occupy and draw a similar diagram of your own role set. Then identify which roles are consistent with each other and which might produce role conflict *and* role strain. *Are there institutional reasons why these conflicts and strains occur in your case?*

by a stranger, you would probably consider it a negative act, perhaps even repulsive.

Verbal and Nonverbal Communication.

We saw in the culture chapter (Chapter 2) how patterns of social interaction are embedded in the language we use, and language is deeply influenced by culture and society. Furthermore, communication is not just what you say, but also how you say it and to whom. You can see the influence of society on *how* people speak, especially in different contexts. Under some circumstances, a pause in speaking may communicate emphasis; for others, it may indicate uncertainty. Cultural differences across society make this obvious. Thus, during interactions between Japanese businessmen, long periods of silence often occur. Unlike U.S. citizens, who are experts in "small talk" and who try at all costs to avoid periods of silence in conversation, Japanese people do not need to talk all the time and regard periods of silence as desirable opportunities for collecting their thoughts (Worchel et al. 2000; Fukuda 1994). American businesspeople in their first meetings with Japanese executives often think, erroneously, that these silent interludes mean the Japanese are responding negatively to a presentation. Even though some find the Japanese mode of conversation highly uncomfortable, getting used to it is a key tool in successful negotiations. The fate of a deal may depend on a glance, an exhalation, or a smile.

Nonverbal communication is also a form of social interaction and can be seen in various social patterns. A surprisingly large portion of our everyday communication with others is nonverbal, although we are generally only conscious of a small fraction of the nonverbal "conversations" in which we take part. Consider all the nonverbal signals exchanged in a casual chat: body position, head nods, eye contact, facial expressions, touching, and so on. Studies of nonverbal communication, like those of verbal communication, show that it is much influenced by social forces, including the relationships between diverse groups of people. The meanings of nonverbal communications depend heavily on race, ethnicity, social class, and gender, as we shall see.

For example, patterns of touch (called **tactile communication**) are strongly influenced by gender. Parents vary their touching behavior depending on whether the child is a boy or a girl. Boys tend to be touched more roughly; girls, more tenderly and protectively. Such patterns continue into adulthood, where women touch each other more often in everyday conversation than do men. Women are on the average more likely to touch and hug as an expression of emotional support, whereas men touch and hug more often to assert power or to express sexual interest (Worchel et al. 2000; Wood 1994). Clearly, there are also instances where women touch to express sexual interest and/or dominance, but research shows that,

in general, for women touch is a supportive activity. For men, touch is often a dominance-asserting activity, except in athletic contexts where hugging and patting among men is a supportive activity (Worchel et al. 2000; Wood 1994; Tannen 1990).

In observing patterns of touch, you can see where social status influences the meaning of nonverbal behaviors. Professors, male or female, may pat a man or woman student on the back as a gesture of approval; students will rarely do this to a professor. Male professors touch students more often than do female professors, showing the additional effect of gender. Because such patterns of touching reflect power relationships between women and men, they can also be offensive and may even involve sexual harassment (see Chapters 11 and 15).

You can also see the social meaning of interaction by observing how people use personal space. **Proxemic communication** refers to the amount of space between interacting individuals. Although people are generally unaware of how they use personal space, usually the more friendly people feel toward each other, the closer they will stand. In casual conversation, friends stand closer to each other than do strangers. People who are sexually attracted to each other stand especially close, whether the sexual attraction is gay, lesbian, or heterosexual. According to anthropologist E. T. Hall (1966; Hall and Hall 1987), we all carry around us a *proxemic bubble* that represents our personal, three-dimensional space. When people we do not know enter our proxemic bubble, we feel threatened and may take evasive action. Friends stand close; enemies tend to avoid interaction and keep far apart. According to Hall's theory, we attempt to exclude from our private space those whom we do not know or do not like, even though we may not be fully aware that we are doing so.

Patterns of touch reflect differences in the power that is part of many social relationships.

SEE *FOR YOURSELF*

RIDING IN ELEVATORS

1. Try a simple experiment. Ride in an elevator and closely observe the behavior of everyone in the elevator with you. Write down in a notebook such things as how far away people stand from each other. What do they look at? Do they tend to stand in the corners? Do they converse with strangers or the people they are with? If so, what do they talk about?

2. Now return to the same elevator and do something that breaks the usual norms of elevator behavior, such as standing too close to someone. How did people react? How did you feel?

How does this experiment show how social norms are maintained through informal norms of social control?

The proxemic bubbles of different ethnic groups on average have different sizes. Hispanic people tend to stand much closer to each other than do White, middle-class Americans; their proxemic bubble is, on average, smaller. Similarly, African Americans also tend to stand close to each other while conversing. Interaction distance is quite large between White, middle-class, British males—their average interaction distances can be as much as several feet.

Proxemic interactions also differ between men and women (Taylor et al. 2009; Romain 1999; Tannen 1990). Women of the same race and culture tend to stand closer to each other in casual conversation than do men of the same race and culture. When a Middle Eastern man (who has a relatively small proxemic bubble) engages in conversation with a White, middle-class, U.S. man (who has a larger proxemic bubble), the Middle Eastern man tends to move toward the White American, who tends to back away. You can observe the negotiations of proxemic space at cocktail parties or any other setting that involves casual social interaction.

In a society as diverse as the United States, understanding how diversity shapes social interaction is an essential part of understanding human behavior. Ignorance of the meanings that gestures have in a society can get you in trouble. For example, some Mexicans and Mexican Americans may display the right hand held up, palm inward, all fingers extended, as an obscene gesture meaning "screw you many times over." This provocative gesture has no meaning at all in Anglo (White) society. Likewise, people who grow up in urban environments learn to avoid eye contact on the streets. Staring at someone for only two or three seconds can be interpreted as a hostile act, if done man to man (Anderson 1999, 1990).

If a woman maintains mutual eye contact with a male stranger for more than two or three seconds, she may be assumed by the man to be sexually interested in him. In contrast, during sustained conversation with acquaintances, women maintain mutual eye contact longer than do men (Romain 1999; Wood 1994).

Interpersonal Attraction

We have already asked, "What holds society together?" This was asked at the macroanalysis level—that is, the level of society. But what holds relationships together—or, for that matter, makes them fall apart? You will not be surprised to learn that formation of relationships has a strong social structural component—that is, it is patterned by social forces and can to a great extent be predicted.

Humans have a powerful desire to be with other human beings; in other words, they have a strong need for *affiliation*. We tend to spend about 75 percent of our time with other people when doing all sorts of activities—eating, watching television, studying, doing hobbies, working, and so on (Cassidy and Shaver 1999). People who lack all forms of human contact are very rare in the general population, and their isolation is usually rooted in psychotic or schizophrenic disorders. Extreme social isolation at an early age causes severe disruption of mental, emotional, and language development, as we saw in Chapter 4.

The affiliation tendency has been likened to **imprinting,** a phenomenon seen in newborn or newly hatched animals who attach themselves to the first living creature they encounter, even if it is of another species (Lorenz 1966). Studies of geese and squirrels show that once the young animal attaches itself to a human experimenter, the process is irreversible. The young animal prefers the company of the human to the company of its own species. A degree of imprinting may be discernible in human infant attachment, but researchers note that in infants the process is more complex, more changeable, and more influenced by social factors (Brown 1986).

Somewhat similar to affiliation is interpersonal attraction, a nonspecific positive response toward another person. Attraction occurs in ordinary day-to-day interaction and varies from mild attraction (such as thinking your grocer is a nice person) all the way to deep feelings of love. According to one view, attractions fall on a continuum ranging from hate to strong dislike to mild dislike to mild liking to strong liking to love. Another view is that attraction and love are two different continua, able to exist separately. In this view, you can actually like someone a whole lot, but not be in love. Conversely, you can feel passionate love for someone, including strong sexual feelings and intense emotion, yet not really "like" the person. Have you ever been in love with someone you did not particularly like?

Can attraction be scientifically predicted? Can you identify with whom you are most likely to fall in love?

Konrad Lorenz, the animal behaviorist, shows that adult Graylag geese *that have imprinted on him the moment they were hatched will follow him anywhere, as though he were their mother goose (Tweed Roosevelt, personal communication)!*

Romantic love is idealized in this society as something that "just happens," but research shows that interpersonal attraction follows predictable patterns.

The surprising answer to these questions is a loud, although somewhat qualified, yes. Most of us have been raised to believe that love is impossible to measure and certainly impossible to predict scientifically. We think of love, especially romantic love, as quick and mysterious—a lightning bolt. Couples report falling in love at first sight, thinking that they were "meant to be together" (McCollum 2002). Countless novels and stories support this view, but extensive research in sociology and social psychology suggests otherwise: In a probabilistic sense, love can be predicted beyond the level of pure chance. Let us take a look at some of these intriguing findings.

A strong determinant of your attraction to others is simply whether you live near them, work next to them, or have frequent contact with them. You are more likely to form friendships with people from your own city than with people from a thousand miles away. One classic study even showed that you are more likely to be attracted to someone on your floor, your residence hall, or your apartment building than to someone even two floors down or two streets over (Festinger et al. 1950). Such is the effect of proximity in the formation of human friendships.

Now, though the general principle still holds, many people form relationships without being in close proximity, such as in online dating. Studies of Internet dating show that, even in this cyberworld, social norms still apply. Studies of Internet dating find, for example, that unlike other dating behavior, on the Internet there is pressure to disclose more secrets about oneself in a shorter period of time (Lawson and Leck 2006).

Our attraction to another person is also greatly affected simply by how frequently we see that person or even his or her photograph. When watching a movie, have you ever noticed that the central character seems more attractive at the end of the movie than at the beginning? This is particularly true if you already find the person very attractive when the movie begins. Have you ever noticed that the fabulous-looking person sitting next to you in class looks better every day? You may be experiencing *mere exposure effect:* The more you see someone in person—or even in a photograph—the more you like him or her. In studies where people are repeatedly shown photographs of the same face, the more often a person sees a particular face, the more he or she likes that person (Moreland and Beach 1992; Zajonc 1968).

There are two qualifications to the effect. First, overexposure can result when a photograph is seen too often. The viewer becomes saturated and ceases to like the pictured person more with each exposure. Second, the initial response of the viewer can determine how much liking will increase. If someone starts out liking a particular person, seeing that person frequently will increase the liking for that person; however, if one starts out disliking the pictured person, the amount of dislike tends to remain about the same, regardless of how often one sees the person (Taylor et al. 2009).

We hear that "beauty is only skin deep." Apparently, that is deep enough. To a surprisingly large degree, the attractions we feel toward people of either gender are based on our perception of their physical attractiveness. A vast amount of research over the years has consistently shown the importance of attractiveness in human interactions: Adults react more leniently to the bad behavior of an attractive child than to the same behavior of an unattractive child (Berscheid and

Reis 1998; Dion 1972). Teachers evaluate cute children of either gender as "smarter" than unattractive children with identical academic records (Worchel et al. 2000; Clifford and Walster 1973). In studies of mock jury trials, attractive defendants, male or female, receive lighter jury-recommended sentences on average than unattractive defendants convicted of the same crime (Gilbert et al. 1998; Sigall and Ostrove 1975).

Of course, standards of attractiveness vary between cultures and between subcultures within the same society. What is highly attractive in one culture may be repulsive in another. In the United States, there is a maxim that you can never be too thin—a major cause of eating disorders such as *anorexia* and *bulimia*, especially among White women (Hesse-Biber 2007). The maxim is oppressive for women in U.S. society, yet it is clearly highly culturally relative, even within U.S. culture. Among certain African Americans, chubbiness in women is considered attractive. Such women are called "healthy" and "phatt" (not "fat"), which means the same as "stacked" or curvaceous. Similar cultural norms often apply in certain U.S. Hispanic populations. The skinny woman is not considered attractive. Nonetheless, studies show that anorexia and bulimia are now increasing among women of color, showing how cultural norms can change—even though Black women, in general, are more satisfied with their body image than White women (Lovejoy 2001; Fitzgibbon and Stolley 2000; see also Chapter 14).

Studies of dating patterns among college students show that the more attractive one is, the more likely one will be asked on a date. This applies to gay and lesbian dating as well as to heterosexual dating (Berscheid and Reiss 1998; Speed and Gangestad 1997). However, one very important exception can be added to this finding: Physical attractiveness predicts only the early stages of a relationship. When one measures relationships that last a while, other factors come into play, principally religion, political attitudes, social class background, educational aspirations, and race. Perceived physical attractiveness may predict who is attracted to whom initially, but other variables are better predictors of how long a relationship will last (Berscheid and Reis 1998).

So, do "opposites attract"? Not according to the research. We have all heard that people are attracted to their "opposite" in personality, social status, background, and other characteristics. Many of us grow up believing this to be true. However, if the research tells us one thing about interpersonal attraction, it is that with few exceptions we are attracted to people who are similar or even identical to us in socioeconomic status, race, ethnicity, religion, perceived personality traits, and general attitudes and opinions (Taylor et al. 2009; Brehm et al. 2002). "Dominant" people tend to be attracted to other dominant people, not to "submissive" people. Couples tend to have similar opinions about political issues of great importance to them, such as attitudes about abortion, crime, animal rights, and

urban violence. Overall, couples tend to exhibit strong cultural or subcultural similarity, not difference.

There are exceptions, of course. We sometimes fall in love with the *exotic*—the culturally or socially different. Novels and movies return endlessly to the story of the rich young woman who falls in love with a rough-and-ready biker, but such a pairing is by far the exception and not the rule. That rich young woman is far more likely to fall in love with a rich young man. When it comes to long-term relationships, including both friends and lovers (whether heterosexual, gay, or lesbian), humans vastly prefer a great degree of similarity, even though, if asked, they might deny it. In fact, the less similar a heterosexual relationship is with respect to race, social class, age, and educational aspirations (how far in school the person wants to go), the quicker the relationship is likely to break up (Silverthorne and Quinsey 2000; Worchel et al. 2000; Berscheid and Reis 1998; Stover and Hope 1993; Hill et al. 1976).

Debunking Society's Myths

Myth: Love is purely an emotional experience that you cannot predict or control.

Sociological perspective: Whom you fall in love with can be predicted beyond chance by such factors as proximity, how often you see the person, how attractive you perceive the person to be, and whether you are similar (not different) to her or him in social class, race–ethnicity, religion, age, educational aspirations, and general attitudes, including political attitudes and beliefs.

Most young romantic relationships, regrettably, come to an end. On campus, relationships tend to break up most often during gaps in the school calendar, such as winter and spring break. Summers are especially brutal on relationships formed during the academic year. Breakups are seldom mutual. Almost always, only one member of the pair wants to break off the relationship, whereas the other wants to keep it going. The sad truth means that the next time someone tells you that their breakup last week was "mutual," you know they are probably lying or deceiving themselves (Taylor et al. 2009).

Theories About Analyzing Social Interaction

Groups, statuses, and roles form a web of social interaction. The interaction people have with one another is a basic element of society. Sociologists have developed different ways of understanding social interaction. Functionalist theory, discussed in Chapter 1, is one such concept. Here we detail four others: the social construction of reality, ethnomethodology, impression management, and social exchange. The first three theories come directly from the symbolic interaction perspective.

The Social Construction of Reality

What holds society together? This is a basic question for sociologists, one that, as we have seen, has long guided sociological thinking. Sociologists note that society cannot hold together without something that is shared—a shared social reality.

Some sociological theorists have argued convincingly that there is little actual reality beyond that produced by the process of social interaction itself. This is the principle of the *social construction of reality*, the idea that our perception of what is real is determined by the subjective meaning that we attribute to an experience, a principle central to symbolic interaction theory (Berger and Luckmann 1967; Blumer 1969). Hence, there is no objective "reality" in itself. Things do not have their own intrinsic meaning. We subjectively impose meaning on things.

Children do this routinely—impose inherent meaning on things. Upon seeing a marble roll off a table, the child attributes causation (meaning) to the marble: The marble rolled off the table "because it wanted to." Such perceptions carry into adulthood: The man walking down the street who accidentally walks smack into a telephone pole, at first thought glares at the pole, as though the pole somehow caused the accident! He inadvertently attributes causation and meaning to an inanimate object—the telephone pole (Heider 1958).

Considerable evidence exists that people do just that; they force meaning on something when doing so allows them to see or perceive what they want to perceive—even if that perception seems to someone else to be contrary to actual fact. They then come to believe that what they perceived is indeed "fact." A classic and convincing study of this is Hastorf and Cantril's (1954) study of Princeton and Dartmouth students who watched a film of a game of basketball between the two schools. Both sets of students watched the same film. The students were instructed to watch carefully for rule infractions by each team. The results were that the Princeton students reported twice as many rule infractions involving the Dartmouth team as the Dartmouth students saw. The Dartmouth students saw about twice as many rule infractions by Princeton as the Princeton students saw! Remember that they all saw exactly the same game—the same "facts." We see the "facts" we want to see, as a result of the social construction of reality. Subsequent research has strongly supported the Hastorf and Cantril findings (Taylor et al. 2009; Ross 1977; Jones and Nisbett 1972).

As we saw in Chapter 1, our perceptions of reality are determined by what is called the *definition of the situation:* We observe the context in which we find ourselves and then adjust our attitudes and perceptions accordingly. Sociological theorist W. I. Thomas embodies this idea in his well-known dictum that *situations defined as real are real in their consequences* (Thomas 1966/1931). The Princeton and Dartmouth students saw different "realities" depending on what college they were attending, and the consequences (the perceived rule infractions) were very real to them.

The definition of the situation is a principle that can also affect a "factual" event such as whether an emergency room patient is perceived to be dead by the doctors. In his research in the emergency room of a hospital, Sudnow (1967) found that patients who arrived at the emergency room with no discernible heartbeat or breathing were treated differently by the attending physician depending on the patient's age. A person in his or her early twenties or younger was not immediately pronounced "dead on arrival" (DOA). Instead, the physicians spent a lot of time listening for and testing for a heartbeat, stimulating the heart, examining the patient's eyes, giving oxygen, and administering other stimulation to revive the patient. If the doctor obtained no lifelike responses, the patient was pronounced dead. Older patients, however, were on the average less likely to receive such extensive procedures. The older person was examined less thoroughly and often was pronounced dead on the spot with only a stethoscopic examination of the heart. In such instances, how the physicians defined the situation—how they socially constructed reality—was indeed real in its consequence for the patient!

Understanding the social construction of reality helps one see many aspects of society in a new light. Race and gender are significant influences on social experience because people believe them to be so. Indeed, society is constructed based on certain assumptions about the significance of race and gender. These assumptions have guided the formation of social institutions, including what work people do, how families are organized, and how power is exercised.

Ethnomethodology

Our interactions are guided by rules that we follow. Sometimes these rules are non-obvious and subtle. These rules are the *norms* of social interaction. Again, what holds society together? Society cannot hold together without norms, but what rules do we follow? How do we know what these rules or norms are? An approach in sociology called *ethnomethodology* is a clever technique for finding out.

Ethnomethodology (Garfinkel 1967), after *ethno* for "people" and *methodology* for "mode of study," is a clever technique for studying human interaction by deliberately disrupting social norms and observing how individuals attempt to restore normalcy. The idea is that to study such norms, one must first break them, because the subsequent behavior of the people involved will reveal just what the norms were in the first place.

Ethnomethodology is based on the premise that human interaction takes place within a consensus, and interaction is not possible without this consensus. The consensus is part of what holds society together. According to Garfinkel, this consensus will be revealed by people's *background expectancies,* namely, the norms for behavior that they carry with them into situations of interaction. It is presumed that these expectancies are to a great degree shared, and thus studying norms by deliberately violating them will reveal the norms that most people bring with them into interaction. The ethnomethodologist argues that you cannot simply walk up to someone and ask what norms the person has and uses, because most people will not be able to articulate them. We are not wholly conscious of what norms we use even though they are shared. Ethnomethodology is designed to "uncover" those norms.

Ethnomethodologists often use ingenious procedures for uncovering those norms by thinking up clever ways to interrupt "normal" interaction. William Gamson, a sociology professor, had one of his students go into a grocery store where jelly beans, normally priced at that time at 49 cents per pound, were on sale for 35 cents. The student engaged the saleswoman in conversation about the various candies and then asked for a pound of jelly beans. The saleswoman then wrapped them and asked for 35 cents. The rest of the conversation went like this:

> *Student:* Oh, only 35 cents for all those nice jelly beans? There are so many of them. I think I will pay $1 for them.
>
> *Saleswoman:* Yes, there are a lot, and today they are on sale for only 35 cents.
>
> *Student:* I know they are on sale, but I want to pay $1 for them. I just love jelly beans, and they are worth a lot to me.

> *Saleswoman:* Well, uh, no, you see, they are selling for 35 cents today, and you wanted a pound, and they are 35 cents a pound.
>
> *Student:* (voice rising) I am perfectly capable of seeing that they are on sale at 35 cents a pound. That has nothing to do with it. It is just that I personally feel that they are worth more, and I want to pay more for them.
>
> *Saleswoman:* (becoming quite angry) What is the matter with you? Are you crazy or something? Everything in this store is priced more than what it is worth. Those jelly beans probably cost the store only a nickel. Now do you want them or should I put them back?

At this point the student became quite embarrassed, paid the 35 cents, and hurriedly left (Gamson and Modigliani 1974).

The point here is that the saleswoman approached the situation with a presumed consensus, a consensus that becomes revealed by its deliberate violation by the student. The puzzled saleswoman took measures to attempt to normalize the interaction, even to *force* it to be normal.

Impression Management and Dramaturgy

Another way of analyzing social interaction is to study impression management, a term coined by symbolic interaction theorist Erving Goffman (1959). **Impression management** is a process by which people control how others perceive them. A student handing in a term paper late may wish to give the instructor the impression that it was not the student's fault but was because of uncontrollable circumstances ("my computer hard drive crashed," "my dog ate the last hard copy," and so on). The impression that one wishes to "give off" (to use Goffman's phrase) is that "I am usually a very diligent person, but today—just today—I have been betrayed by circumstances."

table 5.2	Theories of Social Interaction				
	The Social Construction of Reality	**Ethnomethodology**	**Dramaturgy**	**Social Exchange Theory**	**Game Theory**
Interprets society as:	Organized around the subjective meaning that people give to social behavior	Held together through the consensus that people share around social norms; you can discover these norms by violating them	A stage on which actors play their social roles and give impression to those in their "audience"	A series of interactions that are based on estimates of rewards and punishments	A system in which people strategize "winning" and "losing" in their interactions with each other
Analyzes social interaction as:	Based on the meaning people give to actions in society	A series of encounters in which people manage their impressions in front of others	Enactment of social roles played before a social audience	A rational balancing act involving perceived costs and benefits of a given behavior	Calculated risks to balance rewards and punishments

Impression management can be seen as a type of con game. We willfully attempt to manipulate others' impression of us. Goffman regarded everyday interaction as a series of attempts to con the other. In fact, trying in various ways to con the other is, according to Goffman, at the very center of much social interaction and social organization in society: Social interaction is just a big con game!

Perhaps this cynical view is not true of all social interaction, but we do present different "selves" to others in different settings. The settings are, in effect, different stages on which we act as we relate to others. For this reason, Goffman's theory is sometimes called the *dramaturgy model* of social interaction, a way of analyzing interaction that assumes the participants are actors on a stage in the drama of everyday social life. People present different faces (give off different impressions) on different stages (in different situations or different roles) with different others. To your mother, you may present yourself as the dutiful, obedient daughter, which may not be how you present yourself to a friend. Perhaps you think acting like a diligent student makes you seem a jerk, so you hide from your friends that you are really interested in a class or enjoy your homework. Analyzing impression management reveals that we try to con the other into perceiving us as we want to be perceived. The box "Doing Sociological Research: Doing Hair, Doing Class" shows how impression management can be involved in many settings, including the everyday world of the hair salon.

A clever study by Albas and Albas (1988) demonstrates just how pervasive impression management is in social interaction. The Albases studied how college students interacted with one another when the instructor returned graded papers during class. Some students got good grades ("aces"), others got poor grades ("bombers"), but both employed a variety of devices (cons) to maintain or give off a favorable impression. For example, the aces wanted to show off their grades, but they did not want to appear to be braggarts, so they casually or "accidentally" let others see their papers. In contrast, the bombers hid or covered their papers to hide their poor grades, said they "didn't care" what they got, or simply lied about their grades.

One thing that Goffman's theory makes clear is that social interaction is a very perilous undertaking. Have you ever been embarrassed? Of course you have; we all have. Think of a really big embarrassment that you experienced. Goffman defines embarrassment as a spontaneous reaction to a sudden or transitory challenge to our identity: We attempt to restore a prior perception of our "self" by others. Perhaps you were giving a talk before a class and then suddenly forgot the rest of the talk. Or perhaps you recently bent over and split your pants. Or perhaps you are a man and barged accidentally into a women's bathroom. All these actions will result in embarrassment, causing you to "lose face."

Impression management is the display of how you want others to define you. It can be especially obvious when you feel strong attachments to a particular group or identity.

Rich Abrahamson/The Fort Collins Coloradoan/AP Photo

You will then attempt to *restore face* ("save face"), that is, eliminate the conditions causing the embarrassment. You thus will attempt to con others into perceiving you as they might have before the embarrassing incident. One way to do this is to shift blame from the self to some other, for example, claiming in the first example that the teacher did not give you time to adequately memorize the talk; or in the second example, claiming that you will never buy that particular, obviously inferior brand of pants again; or in the third example, claiming that the sign saying "Women's Room" was not clearly visible. All these represent deliberate manipulations (cons) to save face on your part—to restore the other's prior perception of you.

Social Exchange and Game Theory

Another way of analyzing social interaction is through the social exchange model. The *social exchange model* of social interaction holds that our interactions are determined by the rewards or punishments that we receive from others (Homans 1974; Blau 1986; Cook and Gerbasi 2006). A fundamental principle of exchange theory is that an interaction that elicits approval from another (a type of reward) is more likely to be repeated than an interaction that incites disapproval (a type of punishment). According to the exchange principle, one can predict whether a given interaction is likely to be repeated or continued by calculating the degree of reward or punishment inspired by the interaction. If the reward for an interaction exceeds the punishment, then a potential for *social profit* exists and the interaction is likely to occur or continue.

Rewards can take many forms. They can include tangible gains such as gifts, recognition, and money, or subtle everyday rewards such as smiles, nods, and

>doing sociological research

Doing Hair, Doing Class

Research Question: Sociologist Debra Gimlin was curious about a common site for social interaction—hair salons. She noticed that the interaction that occurs in hair salons is often marked by differences in the social class status of clients and stylists. Her research question was, How do women attempt to cultivate the cultural ideals of beauty, and in particular, how is this achieved through the interaction between hair stylists and their clients?

Research Method: She did her research by spending more than 200 hours observing social interaction in a hair salon. She watched the interaction between clients and stylists and conducted interviews with the owner, the staff, and twenty women customers. During the course of her fieldwork, she recorded her observations of the conversations and interaction in the salon, frequently asking questions of patrons and staff. The patrons were mostly middle- and upper-middle class; the stylists, working

class. All the stylists were White, as were most of the clients.

Research Results: "Beauty work" as Gimlin calls it, involves the stylist bridging the gap between those who seek beauty and those who define it; her (or his) role is to be the expert in beauty culture, bringing the latest fashion and technique to clients. Beauticians are also expected to engage in some "emotion work"—that is, they are expected to nurture clients and be interested in their lives; often they are put in the position of sacrificing their professional expertise to meet clients' wishes.

According to Gimlin, since stylists typically have lower class status than their clients, this introduces an element into the relationship that stylists negotiate carefully in their routine social interaction. Hairdressers emphasize their special knowledge of beauty and taste as a way of reducing the status

---

pats on the back. Similarly, punishments come in many varieties, from extremes such as public humiliation, beating, banishment, or execution, to gestures as subtle as a raised eyebrow or a frown. For example, if you ask someone out for a date and the person says yes, you have gained a reward, and you are likely to repeat the interaction. You are likely to ask the person out again, or to ask someone else out. If you ask someone out, and he or she glares at you and says, "No way!", then you have elicited a punishment that will probably cause you to shy away from repeating this type of interaction with that person.

Social exchange theory has grown partly out of **game theory,** a mathematic and economic theory that predicts that human interaction has the characteristics of a "game," namely, strategies, winners and losers, rewards and punishments, and profits and costs (Nash 1951; Dixit and Sneath 1997; Kuhn and Nasar 2002). Simply asking someone out for a date indeed has a gamelike aspect to it, and you will probably use some kind of strategy to "win" (have the other agree to go out with you) and "get rewarded" (have a pleasant or fun time) at minimal "cost" to you (you don't want to spend a large amount of money on the date or you do not want to get into an unpleasant argument on the date). The interesting thing about game theory is

that it sees human interaction as just that: a game. Impression management theory also contains a gamelike element in its hypothesis that human interaction is a big con game. The mathematician John Nash is one of the inventors of game theory and was featured in the movie, *A Beautiful Mind.*

Interaction in Cyberspace

When people interact and communicate with one another by means of personal computers—through email, chat rooms, computer bulletin boards, virtual communities such as LinkedIn and the like, and other computer-to-computer interactions—they are engaging in **cyberspace interaction** (or virtual interaction).

The character of cyberspace interaction is changing as new technologies emerge. Not long ago, nonverbal interaction was absent in cyberspace as people could not "see" what others were like. But with the introduction of video-based cyberspace, such as YouTube and photos on Facebook and MySpace, people can now display still and moving images of themselves. These differences between themselves and their clients. They also try to nullify the existing class hierarchy by conceiving an alternative hierarchy, not one based on education, income, or occupation but only on the ability to style hair competently. Thus, stylists describe clients as perhaps "having a ton of money," but unable to do their hair or know what looks best on them. Stylists become confidantes with clients, who often tell them highly personal information about their lives. Appearing to create personal relationships with their clients, even though they never see them outside the salon, also reduces status differences.

Conclusions and Implications: Gimlin concludes that beauty ideals are shaped in this society by an awareness of social location and cultural distinctions. As she says, "Beauty is . . . one tool women use as they make claims to particular social statuses" (1996: 525).

Questions to Consider

The next time you get your hair cut, you might observe the social interaction around you and ask how class, gender, and race shape interaction in the salon or barbershop that you use. Try to get someone in class to collaborate with you so that you can compare observations in different salon settings. In doing so, you will be studying how gender, race, and class shape social interaction in everyday life.

1. Would you expect the same dynamic in a salon where men are the stylists?

2. Do Gimlin's findings hold in settings where the customers and stylists are not White or where they are all working class?

3. In your opinion, would Gimlin's findings hold in an African American *men's* barbershop?

Source: Gimlin, Debra. 1996. "Pamela's Place: Power and Negotiation in the Hair Salon." *Gender & Society* 10 (October): 505–526.

images provide new opportunities, as we noted above, for what sociologists would call the presentation of self and impression management. Sometimes this comes with embarrassing consequences. The young college student who displays a seminude photo of herself, projecting a sexual presentation of self, may be horrified if one of her parents or a potential employer visits her Facebook site!

Cyberspace interaction is becoming increasingly common among all age, gender, and race groups, although clear patterns are also present in who is engaged in this form of social interaction and how people use it (see Table 5.3 on page 124; Hargittai 2008). Women, for example, still lag behind men in use of the Internet, but young women and Black women use the Internet more than their male peers. Older women lag behind older men in their use. Men are also, in general, more intense users, meaning that they log on more often, spend more time online, and are more likely to use broadband.

Gender differences also prevail in how women and men use the Internet. Women are more likely to use email to write to friends and family, share news, plan events, and forward jokes. And women are more likely to report that email nurtures their relationships. Men, on the other hand, use the Internet more

to transact business, and they look for a wider array of information than women do. Men are also more likely to use the Internet for hobbies, including such things as sports fantasy leagues, downloading music, and listening to radio (Pew Internet and American Life Project 2009).

More than 75 percent of those aged 18–24 now use online social networking sites, with women among the most frequent users of such sites. The popularity of these sites is also growing rapidly among youth. Again, gender patterns emerge in how young people interact online. Boys are more likely than girls to say they use these sites to make new friends and to flirt; girls use it more to stay in touch with friends they already have (Pew Internet and American Life Project 2009).

It is too early to know the implications of these cyberspace interactions. Some think it will make social life more alienating, with people developing weaker social skills and less ability for successful face-to-face interaction. Some studies have noted that people can develop extremely close and in-depth relationships as a result of their interaction in cyberspace (Hargittai 2008). The Internet also creates more opportunity for people to misrepresent themselves or even create completely false—or even stolen—identities.

Social Interaction in an Age of Technology

Just as the invention of new technologies during the Industrial Revolution changed how people worked and lived, so is the Cyberspace Revolution transforming social interaction in society. Only a few years ago, forms of interaction and communication that are now common would have been seen as science fiction. What are some of the changes in social interaction—and, thus, in society—that have come as the result of technological changes?

© Bob Daemmrich/The Image Works

How have new technologies affected the interaction in social groups?

If you look, you will frequently see groups of friends, often young people, gathered together, but with everyone talking to someone else on a cell phone. Although new technologies allow people to stay closely in touch, do they also intrude on the intimacy that one might otherwise share in face-to-face interaction?

Matt Sayles/AP Photo

Is there a "digital divide" in new technologies—that is, inequalities of usage across different groups?

Many people have noted that, even with the increasing presence of computer technology, there is still a large racial and class divide in who has access to these various devices. White adults are far more likely, for example, to have Internet access than are Black adults, as are groups in higher income brackets. It remains to be seen whether this will hold up over time, since younger people under age 30 are far more likely to have Internet access than older adults.

Joanne Carole/AP Photo

How do new technologies change parts of the culture?

New forms of technology also have introduced new language into society—a form of cultural innovation. If you were asked to "TMB," would you "KWIM"? (If not, see the table in Chapter 2 on page 32.)

Kevork Djansezian/AP Photo

Can new technologies create new social problems?

The new forms of social interaction that computer technologies have created also raise new social issues. For example, what social issues arise from the intrusions on privacy that these new technologies can involve?

Do new technologies have any negative effects on social interaction?

Using new technologies is commonplace. The ability to communicate instantly, including around the globe, is a tremendous convenience and can draw distant groups closer together, but do they also produce more alienating social relationships?

© Jeff Greenberg/PhotoEdit

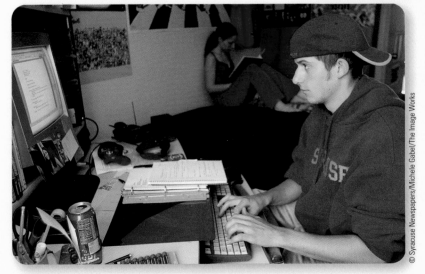

© Syracuse Newspapers/Michele Gabel/The Image Works

What new work skills do new technologies require and how do these affect different social groups?

The introduction of new technologies relies on new skills that people learn to be able to work (and play) effectively, but what new forms of inequality might be produced as the result?

table5.3	*Teens Online: Who is Most Likely to Create an Online Profile?*		
	Percentage in Each Group Who Create Online Profiles		**Percentage in Each Group Who Create Online Profiles**
Sex		**Age by Sex**	
Boys	51%	Boys aged 15–17	57%
Girls	58	Girls aged 15–17	70
Age		**Household Income**	
12–14	45	Less than $50,000	55
15–17	64	$50,000 or more	56
Age by Sex		**Race–Ethnicity**	
Boys aged 12–14	46	White non-Hispanic	53
Girls aged 12–14	44	Non-White	58

Source: Pew Internet and American Life Project: *Teens and Parents Survey.* 2009. **www.pewinternet.org.** Reprinted by permission.

But studies find that computer-mediated interactions also follow some of the same patterns that are found in face-to-face interaction. People still "manage" identities in front of a presumed audience; they project images of self to others that are consistent with the identity they have created for themselves, and they form social networks that become the source for evolving identities, just as people do in traditional forms of social interaction (Brignall and Van Valey 2005).

In this respect, cyberspace interaction is the application of Goffman's principle of *impression management.* The person can put forward a totally different and wholly created self, or identity. One can "give off," in Goffman's terms, any impression one wishes and, at the same time, know that one's true self is protected by anonymity. This gives the individual quite a large and free range of roles and identities from which to choose. As predicted by symbolic interaction theory, of which Goffman's is one variety, *the reality of the situation grows out of the interaction process itself.* This is a central point of symbolic interaction theory and is central to sociological analysis generally: Interaction creates reality.

Cyberspace interaction has thus resulted in new forms of social interaction in society—in fact, a new social order containing both deviants and conformists. These new forms of social interaction have their own rules and norms, their own language, their own sets of beliefs, and practices or rituals—in short, all the elements of culture, as defined in Chapter 2. For sociologists, cyberspace also provides an intriguing new venue in which to study the connection between society and social interaction.

Chapter Summary

- **What is society?**

  Society is a system of social interaction that includes both culture and *social organization.* Society includes *social institutions,* or established organized social behavior, and exists for a recognized purpose; *social structure* is the patterned relationships within a society.

- **What holds society together?**

  According to theorist Emile Durkheim, society with all its complex social organization and culture, is held together, depending on overall type, by *mechanical solidarity* (based on individual similarity) and *organic solidarity* (based on a *division of labor* among dissimilar individuals).

Two other forms of social organization also contribute to the cohesion of a society: *gemeinschaft* ("community," characterized by cohesion based on friendships and loyalties) and *gesellschaft* ("society," characterized by cohesion based on complexity and differentiation).

• **What are the types of societies?**
Societies across the globe vary in type, as determined mainly by the complexity of their social structures, their division of labor, and their technologies. From least to most complex, they are *foraging, pastoral, horticultural, agricultural* (these four constitute *preindustrial* societies), *industrial,* and *postindustrial* societies.

• **What are the forms of social interaction in society?**
All forms of social interaction in society are shaped by the structure of its social institutions. A *group* is a collection of individuals who interact and communicate with each other, share goals and norms, and have a subjective awareness of themselves as a distinct social unit. *Status* is a hierarchical position in a structure; a *role* is the expected behavior associated with a particular status. A *role* is the behavior others expect from a person associated with a particular status. Patterns of social interaction influence nonverbal interaction as well as patterns of attraction and affiliation.

• **What theories are there about social interaction?**
Social interaction takes place in society within the context of social structure and social institutions. Social interaction is analyzed in several ways, including the *social construction of reality* (we impose meaning and reality on our interactions with others); *ethnomethodology* (deliberate interruption of interaction to observe how a return to "normal" interaction is accomplished); *impression management* (a person "gives off" a particular impression to "con" the other and achieve certain goals, as in *cyberspace interaction*); and *social exchange* and *game theory* (one engages in gamelike reward and punishment interactions to achieve one's goals).

• **How is technology changing social interaction?**
Increasingly, people engage with each other through *cyberspace interaction*. Social norms develop in cyberspace as they do in face-to-face interaction, but a person in cyberspace can also manipulate the impression that he or she gives off, thus creating a new "virtual" self.

Key Terms

achieved status   110
ascribed status   110
collective consciousness   104
cyberspace interaction   120
division of labor   105
ethnomethodology   117
game theory   120
gemeinschaft   105
gesellschaft   105
group   109
impression management   118
imprinting   114

macroanalysis   102
master status   111
mechanical solidarity   104
microanalysis   102
nonverbal communication   113
organic solidarity   105
postindustrial society   109
preindustrial society   106
proxemic communication   113
role   111
role conflict   111
role modeling   111

role set   111
role strain   112
social institution   103
social interaction   102
social organization   103
social structure   104
society   102
status   110
status inconsistency   110
status set   110
tactile communication   113

Online Resources

Sociology: The Essentials Companion Website

www.cengage.com/sociology/andersen
Visit your book companion website where you will find more resources to help you study and write your research papers. Resources include web links, learning objectives, Internet exercises, quizzing, and flash cards.

 is an easy-to-use online resource that helps you study in less time to get the grade you want NOW.

www.cengage.com/login
Need help studying? This site is your one-stop study shop. Take a Pre-Test and Cengage NOW will generate a Personalized Study Plan based on your test results. The Study Plan will identify the topics you need to review and direct you to online resources to help you master those topics. You can then take a Post-Test to determine the concepts you have mastered and what you still need to work on.

6

Groups and Organizations

Types of Groups

Social Influence in Groups

Formal Organizations and Bureaucracies

Diversity: Race, Gender, and Class in Organizations

Functional, Conflict, and Symbolic Interaction: Theoretical Perspectives

Chapter Summary

Scott Olson/Getty Images/Reportage

[*Twelve citizens sit* together in an elevated enclosure, like a choir loft, and silently watch a drama unfold before them, day after day. They are respectfully addressed by highly paid professionals: lawyers, judges, expert witnesses. Their job, ruling on the innocence or guilt of a defendant or the settlement of a legal claim, was once the prerogative only of kings. Their decision may mean freedom or incarceration, fortune or penury, even life or death.

Juries have been the focus of much research, in part because they fill such a vital role in our society and in part because within this curiously artificial, yet intimate, group of random strangers can be found a wealth of interesting sociological phenomena. Jury verdicts and jury deliberations show the same inescapable influences of status, race, and gender that affect the rest of society (Vidmar and Hans 2007; Hans 2006). They also show the powerful effects of group pressure on the individual. Juries are, in some ways, society in miniature.

During jury selection, in a process called the *voir dire*, lawyers on both sides are entitled to eliminate any potential jurors and no explanation is required. Many lawyers who have great faith in their ability to judge jurors consider jury selection to be the most important part of a trial. By choosing jurors, they are choosing the verdict. Consider some of the folk wisdom clung to by trial lawyers in the past, with varying degrees of accuracy: Farmers believe in strict responsibility, whereas waiters

continued

and bartenders are forgiving; avoid the clergy; select married women. A guideline for Dallas, Texas, prosecutors advised against selecting "Yankees . . . unless they appear to have common sense" (Guinther 1988: 54). High-powered legal teams now make room for a new breed of legal specialist—the *trial consultant,* trained in sociological and psychological techniques—who contributes nothing but juror analysis as part of the jury selection process.

These analyses go beyond simply identifying the bias of a given juror. Juries are *groups,* and groups behave differently from individuals. Understanding group behavior is critical to predicting the performance of a jury. For instance, it is possible to make an educated prediction about who in a jury will become the most influential. Researchers have found that people with high status in society do the most talking in jury deliberations, and other jurors consider them to be the most helpful in reaching a verdict (Hans 2006; Hans and Martinez 1994; Berger and Zelditch 1985).

Factions or subgroupings form during jury deliberations, and if jury analysts expect a difficult decision, they can attempt to influence how factionalized juries will resolve their disputes based on sociological and psychological data about small group decision making (Saks and Marti 1997). For instance, jurors are much less likely to defect from large factions than from small ones. The larger the faction, the less willing a juror will be to defy the weight of group opinion. As we shall see

below, this is a *group size effect*: an effect of sheer numbers in the group independent of the effects of their own personalities.

What does this say about the state of justice in our legal system, when guilt or innocence depends not only on the legal facts but also on sociological aspects of the jury? Like society as a whole, and like organizations and bureaucracies, groups are subject to social influences. Whether a relatively small group, such as a jury, or a large organization, such as a major corporation, people and groups are influenced by sociological forces.

Types of Groups

Each of us is a member of many groups simultaneously. We have relationships in groups with family, friends, team members, and professional colleagues. Within these groups are gradations in relationships: We are generally closer to our siblings (our sisters and brothers) than to our cousins; we are intimate with some friends, merely sociable with others. If we count all our group associations, ranging from the powerful associations that define our daily lives to the thinnest connections with little feeling (other pet lovers, other company employees), we will uncover connections to literally hundreds of groups.

What is a group? Recall from Chapter 5, a **group** is two or more individuals who interact, share goals and norms, and have a subjective awareness as "we." To be considered a group, a social unit must have all three characteristics. The hazy boundaries of the definition are necessary. Consider two superficially similar examples: The individuals in a line waiting to board a train are unlikely to have a sense of themselves as one group. A line of prisoners chained together and waiting to board a bus to the penitentiary is more likely to have a stronger sense of common feeling.

As you remember from the previous chapter, certain gatherings are not groups in the strict sense, but may be *social categories* (for example, teenagers, truck drivers) or *audiences* (everyone watching a movie). The importance of defining a group is not to perfectly decide if a social unit is a group—an unnecessary endeavor—but to help us understand the behavior of people in society. As we inspect groups, we can identify characteristics that reliably predict trends in the behavior of the group and even the behavior of individuals in the group.

The study of groups has application at all levels of society, from the attraction between people who fall in love to the characteristics that make some corporations drastically outperform their competitors—or that lead them into bankruptcy. The aggregation of individuals into groups has a transforming power, and sociologists understand the social forces that make these transformations possible.

Michael Kelly/Getty Images

This jury of twelve persons is voting on something, perhaps on whether to acquit a person accused of a serious crime. The social pressures in a jury are extremely strong, making the lone "holdout" person very unlikely.

In this chapter, we move from the *microlevel* of analysis (the analysis of groups and face-to-face social influence) to the relatively more *macrolevel* of analysis (the analysis of formal organizations and bureaucracies).

Dyads and Triads: Group Size Effects

Even the smallest groups are of acute sociological interest. A **dyad** is a group consisting of exactly two people. A **triad** consists of three people. This seemingly minor distinction, first scrutinized by the German sociologist **Georg Simmel** (1858–1918), can have critical consequences for group behavior (Simmel 1902). Simmel was interested in discovering the effects of size on groups, and he found that the mere difference between two and three people spawned entirely different group dynamics (the behavior of a group over time).

Imagine two people standing in line for lunch. First one talks, then the other, then the first again. The interaction proceeds in this way for several minutes. Now a third person enters the interaction. The character of the interaction suddenly changes: At any given moment, two people are interacting more with each other than either is with the third. When the third person wins the attention of the other two, a new dyad is formed, supplanting the previous pairing. The group, a triad, then consists of a dyad (the pair that is interacting) plus an *isolate*.

Triadic segregation is what Simmel called the tendency for triads to segregate into a pair and an isolate (a single person). A triad tends to segregate into a *coalition* of the dyad against the isolate. The isolate then has the option of initiating a coalition with either member of the dyad. This choice is a type of social advantage, leading Simmel to coin the principle of *tertius gaudens*, a Latin term meaning "the third one gains." Simmel's reasoning has led to numerous contemporary studies of coalition formation in groups (for example, Konishi and Ray 2003; Caplow 1968; and many others).

For example, interactions in a triad often end up as "two against one." You may have noticed this principle of coalition formation in your own conversations. Perhaps two friends want to go to a movie you do not want to see. You appeal to one of them to go instead to a minor league baseball game. She wavers and comes over to your point of view. Now you have formed a coalition of two against one. The friend who wants to go to the movies is now the isolate. He may recover lost social ground by trying to form a new coalition by suggesting a new alternative (going bowling or to a different movie). This flip-flop interaction may continue for some time, demonstrating another observation by Simmel: A triad is a decidedly unstable social grouping, whereas dyads are relatively stable. The minor distinction between dyads and triads is one person but has important consequences because it changes the character of the interaction within the group. Simmel is known as the discoverer of **group size effects**—the effects of group number on group behavior *independent of the personality characteristics of the members themselves.*

Primary and Secondary Groups

Charles Horton Cooley (1864–1929), a famous sociologist of the Chicago School of sociology, introduced the concept of the **primary group**, defined as a group consisting of intimate, face-to-face interaction and relatively long-lasting relationships. Cooley had in mind the family and the early peer group. In his original formulation, *primary* was used in the sense of "first," the intimate group of the formative years (Cooley 1967/1909). The insight that there was an important distinction between intimate groups and other groups proved extremely fruitful. Cooley's somewhat narrow concept of family and childhood peers has been elaborated upon over the years to include a variety of intimate relations as examples of primary groups.

Primary groups have a powerful influence on an individual's personality or self-identity. The effect of family on an individual can hardly be overstated. The weight of peer pressure on schoolchildren is particularly notorious. Street gangs are a primary group, and their influence on the individual is significant; in fact, gang members frequently think of themselves as a family. Inmates in prison very frequently become members of a gang—primary groups based mainly upon race–ethnicity—as a matter of their own personal survival. The intense camaraderie formed among Marine Corps units in boot camp and in war, such as the war in Iraq, is another classic example of primary group formation and the resulting intense effect on the individual and upon her or his survival.

In contrast to primary groups are **secondary groups,** those that are larger in membership, less intimate, and less long-lasting. Secondary groups tend to be less significant in the emotional lives of people. Secondary groups include all the students at a college or university, all the people in your neighborhood, and all the people in a bureaucracy or corporation.

One of the best examples of the primary group is that consisting of parent and child.

>doing sociological research

Sharing the Journey

Modern society is often characterized as remote, alienating, and without much feeling of community or belonging to a group. This image of society has been carefully studied by sociologist Robert Wuthnow, who noticed that in the United States people are increasingly looking to small groups as a place where they can find emotional and spiritual support and where they find meaning and commitment, despite the image of society as an increasingly impersonal force.

Research Question: Wuthnow began his research by noting that, even with the individualistic culture of U.S. society, small groups play a major role in this society. He saw the increasing tendency of people to join recovery groups, reading groups, spiritual groups, and myriad other support groups. Wuthnow began his research by asking some specific questions, including, "What motivates people to join support groups?" "How do these groups function?" and "What do

members like most and least about such groups?" His broadest question, however, was to wonder how the wider society is influenced by the proliferation of small support groups.

Research Methods: To answer these questions, a large research team of fifteen scholars designed a study that included both a quantitative and a qualitative dimension. They distributed a survey to a representative sample of more than 1000 people in the United States. Supplementing the survey were interviews with more than 100 support group members, group leaders, and clergy. The researchers chose twelve groups for extensive study; researchers spent six months to three years tracing the history of these groups, meeting with members and attending group sessions.

Research Results: Based on this research, Wuthnow concludes that the small group movement is

SEE *FOR YOURSELF*

ANALYZING GROUPS

Identify a *group* of which you are a part. How does one become a member of this group? Who gets included and who gets excluded? Does the group share any unique language or other cultural characteristics (such as dress, jargon, or other group identifiers)? Does anyone ever leave the group, and if so, why?

1. Would you describe this group mainly as a *primary* or a *secondary group*? Why?

2. Now think about this group from the perspective of functionalist theory, conflict theory, and symbolic interaction theory. What are the *manifest* and *latent functions* of the group? Is there a hierarchy within the group? Is there competition between group members? What social meanings do members of the group share?

Secondary groups occasionally take on the characteristics of primary groups. The process can be accelerated in situations of high stress or crisis. For example, when a neighborhood meets with catastrophe, such as a flood, people who may be only acquaintances often come to depend on each other and in the process become more intimate. The secondary group of neighbors becomes, for a time, a primary group. This is precisely what happened in the otherwise highly impersonal neighborhoods in New York City near ground zero after the September 11, 2001, terrorist attack on the World Trade Center: Thousands of people pitched in to help, and as a result, many primary groups formed.

The same phenomenon occurred immediately after US Airways Flight 1549, on January 15, 2009—guided by the superior skills of pilot Chesley B. "Sully" Sullenberger—was forced to "land" in the cold waters of New York City's Hudson River, after both of the aircraft's jet engines had been blown out by a bird strike. Because of the tremendous skill of Sullenberger, the plane floated, and not a single death occurred. Immediately thereafter, the passengers, a secondary group before the landing, became intimately acquainted with each other—a quickly formed primary group. They even vowed to have annual reunions, further evidence of a secondary group's transformation into a primary group.

Primary and secondary groups serve different needs. Primary groups give people intimacy, companionship, and emotional support. These human fundamentally altering U.S. society. Forty percent of all Americans belong to some kind of small support group. As the result of people's participation in these groups, social values of community and spirituality are undergoing major transformation. People say they are seeking community when they join small groups whether the group is a recovery group, a religious group, a civic association, or some other small group. People turn to these small groups for emotional support more than for physical or monetary support.

Conclusions and Implications: Wuthnow argues that large-scale participation in small groups has arisen in a social context in which the traditional support structures in U.S. society, such as the family, no longer provide the sense of belonging and social integration that they provided in the past. Geographic mobility, mass society, and the erosion of local ties all contribute to this trend.

People still seek a sense of community, but they create it in groups that also allow them to maintain their individuality. In voluntary small groups, you are free to leave the group if it no longer meets your needs.

Wuthnow also concludes that these groups represent a quest for spirituality in a society when, for many, traditional religious values have declined. As a consequence, support groups are redefining what is sacred. They also replace explicit religious tenets imposed from the outside with internal norms that are implicit and devised by individual groups. At the same time, these groups reflect the pluralism and diversity that characterize society. In the end, they buffer the trend toward disintegration and isolation that people often feel in mass societies.

Source: Wuthnow, Robert. 1994. *Sharing the Journey: Support Groups and America's New Quest for Community.* New York: Free Press.

---

desires are termed **expressive needs** (also called socioemotional needs). Family and friends share and amplify your good fortune, rescue you when you misbehave, and cheer you up when life looks grim. Primary groups are a major influence on social life and an important source of social control. They are also a dominant influence on your likes and dislikes, preferences in clothing, political views, religious attitudes, and other characteristics. Many studies have shown the overwhelming influence of family and friendship groups on religious and political affiliation, as shown in the box "Doing Sociological Research: Sharing the Journey" (Wuthnow 1994).

Secondary groups serve **instrumental needs** (also called task-oriented needs). Athletic teams form to have fun and win games. Political groups form to raise funds and influence the government. Corporations form to make profits, and employees join corporations to earn a living. Needless to say, intimacies can develop in the act of fulfilling instrumental needs, and primary groups may also devote themselves to meeting instrumental needs. The true distinction between primary and secondary groups is in how intimate the group members feel about one another and how dependent they are on the group for sustenance and identity. Both primary and secondary groups are indispensable elements of life in society.

Reference Groups

Primary and secondary groups are groups to which members belong. Both are called *membership groups.*

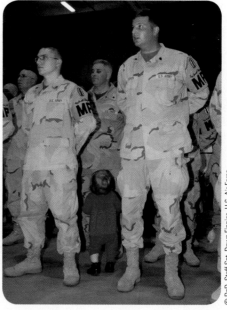

Identification with a reference group has a significant influence on one's identity.

© DoD, Staff Sgt. Dawn Finniss, U.S. Air Force

In contrast, **reference groups** are those to which you may or may not belong but use as a standard for evaluating your values, attitudes, and behaviors (Merton and Rossi 1950). Reference groups are generalized versions of role models. They are not "groups" in the sense that the individual interacts within (or in) them. Do you pattern your behavior on that of sports stars, musicians, military officers, or business executives? If so, those models are reference groups for you.

Imitation of reference groups can have both positive and negative effects. Members of a Little League baseball team may revere major league baseball players and attempt to imitate laudable behaviors such as tenacity and sportsmanship. But young baseball fans are also liable to be exposed to tantrums, fights, and tobacco chewing and spitting. This illustrates that the influence of a reference group can be both positive and negative.

Research has shown that identification with reference groups can strongly influence self-evaluation and self-esteem. Before the push for school desegregation began, it was thought that all-Black schools contributed to negative self-evaluation among Black students. Desegregation was expected to raise the self-esteem of Black children (Clark and Clark 1947). The original Clark and Clark study suggested that Black children in all-Black schools preferred playing with White dolls over Black dolls. This behavior was presumed to demonstrate a negative self-evaluation and self-rejection on the part of Black children.

In some cases desegregation did raise the self-esteem of Black children, but later research has also found that identification with a positive reference group was more important than desegregation. When racial or ethnic groups were consistently presented in a positive way, as in later multicultural educational programs designed to increase pride in Black culture, the self-esteem of the children was greater than that of Black children in integrated programs with no multicultural component. The same has been found for Latino children enrolled in Latino cultural awareness programs. Plainly, the representation of racial and ethnic groups in a society can have a striking positive effect on children acquiring their lifetime set of group affiliations (Harris 2006; Zhou and Bankston 2000; Baumeister and Bushman 2008; Steele 1996, 1992; Steele and Aronson 1995; Banks 1976).

In-Groups and Out-Groups

When groups have a sense of themselves as "us," they will also have a complementary sense of other groups as "them." The distinction is commonly characterized as *in-groups* versus *out-groups*. The concept was originally elaborated by the early sociological theorist **W. I. Thomas** (Thomas 1931). College fraternities and sororities certainly exemplify "in" versus "out." So do families. So do gangs—especially so. The same can be true of the members of your high school class, your sports team, your racial group, your gender, and your social class. Members of the upper classes in the United States occasionally refer to one another as PLUs "people like us" (Graham 1999; Frazier 1957).

Attribution theory is the principle that we all make inferences about the personalities of others, such as concluding what the other is "really like." These attributions depend on whether you are in the in-group or the out-group. Thomas F. Pettigrew has summarized the research on attribution theory, showing that individuals commonly generate a significantly distorted perception of the motives and capabilities of other people's acts based on whether that person is an in-group or out-group member (Baumeister and Bushman 2008; Pettigrew 1992; Gilbert and Malone 1995). Pettigrew describes the misperception as **attribution error,** meaning errors made in attributing causes for people's behavior to their membership in a particular group, such as a racial group. Attribution error has several dimensions, all tending to favor the in-group over the out-group. *All else being assumed equal, we tend to perceive people in our in-group positively and those in out-groups negatively, regardless of their actual personal characteristics:*

1. When onlookers observe improper behavior by an out-group member, onlookers are likely to attribute the deviance to the disposition (the personality) of the wrongdoer. Disposition refers to the perceived "true nature" or "inherent nature" of the person, often considered to be genetically determined. *Example:* A White person sees a Hispanic person carrying a knife and, without further information, attributes this behavior to the presumed "inherent tendency" for Hispanics to be violent. The same would be true if a Hispanic person, without

People often counteract the impersonality of society by forming informal friendship groups, as here, or joining a more formal small support group.

additional information, assumed that all Whites have an "inherent tendency" to be racist.

2. When the *same* behavior is exhibited by an in-group member, the perception is commonly held that the act is due to the *situation* of the wrongdoer, not to the in-group member's inherent disposition or personality. *Example:* A White person sees another White person carrying a knife and concludes, without further information, that the weapon must be carried for protection in a dangerous area.

3. If an out-group member is seen to perform in some laudable way, the behavior is often attributed to a variety of special circumstances, and the out-group member is seen as "the exception."

4. An in-group member who performs in the same laudable way is given credit for a worthy personality disposition.

Typical attribution errors include misperceptions between racial groups and between men and women. If a White police officer shoots a Black or Latino, a White individual, given no additional information, is likely to assume that the victim instigated the shooting and thus "deserved" to be shot. On the other hand, a Black person is more likely to assume that the police officer fired unnecessarily, perhaps because the officer is dispositionally assumed to be a racist (Taylor et al. 2009; Kluegel and Bobo 1993; Bobo and Kluegel 1991).

A related phenomenon has been seen in men's perceptions of women coworkers. Meticulous behavior in a man is perceived positively and is seen by other men as "thorough"; in a woman, the *exact same* behavior is perceived negatively and is considered "picky." Behavior applauded in a man as "aggressive" is condemned in a woman exhibiting the same behavior as "pushy" or "bitchy" (Baumeister and Bushman 2008; Uleman et al. 1996; Wood 1994).

Social Networks

As already noted, no individual is a member of only one group. Social life is far richer than that. A **social network** is a set of links between individuals, between groups, or between other social units, such as bureaucratic organizations or even entire nations (Aldrich and Ruef 2006; Centeno and Hargittai 2003; Mizruchi 1992). One could say that any given person belongs simultaneously to several networks (Wasserman and Faust 1994). Your group of friends, or all the people on an electronic mailing list to which you subscribe, or all of your Facebook subscribers are social networks.

The network of people you are closest to, rather than those merely linked to you in some impersonal way, is probably most important to you. Numerous research studies indicate that people get jobs via their personal networks more often than through formal job listings, want ads, or placement agencies (Ruef et al. 2003; Petersen et al. 2000; Granovetter 1995, 1974). This is especially true for high-paying, prestigious

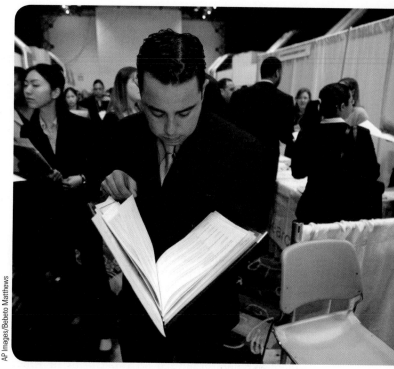

This job candidate does a last-minute check of his resume just before being interviewed by a company representative who contacted the job candidate through a social network. Like many others, he was let go from his previous job.

jobs. Getting a job is more often a matter of who you know than what you know. Who you know, and whom they know in turn, is a social network that may have a marked effect on your life and career.

Networks form with all the spontaneity of other forms of human interaction (Mintz and Schwartz 1985; Wasserman and Faust 1994; Knoke 1992). Networks evolve, such as social ties within neighborhoods, professional contacts, and associations formed in fraternal, religious, occupational, and volunteer groups. Networks to which you are only *weakly* tied (you may know only one person in your neighborhood) provide you with access to that entire network, hence the sociological principle that there is "strength in weak ties" (Granovetter 1973; Petersen et al. 2000; Montgomery 1992).

Networks based on race, class, and gender form with particular readiness. This has been especially true of job networks. The person who leads you to a job is likely to have a similar social background. Recent research indicates that the "old boy network"—any network of White, male corporate executives—is less important than it used to be, although it is certainly not by any means gone. The diminished importance of the old boy network is because of the increasing prominence of women and minorities in business organizations. In fact, among African American and Latino individuals, one's family can provide network contacts that can lead to jobs and upward mobility

(Dominguez and Watkins 2003). Still, as we will see later in this chapter, women and minorities are considerably underrepresented in corporate life, especially in high-status jobs (Padavic and Reskin 2002; Green et al. 1999; Collins-Lowry 1997; Gerson 1993). Some recent research shows that despite recent gains on the part of minorities, relative to Whites, Blacks and Latinos are still disproportionately harmed by lack of network contacts (Smith 2007).

Networks can reach around the world, but how big is the world? How many of us, when we discover someone we just met is a friend of a friend, have remarked, "My, it's a small world"? Research into what has come to be known as the *small world problem* has shown that networks make the world a lot smaller than you might think.

Original small world researchers Travers and Milgram wanted to test whether a document could be routed via the U.S. postal system to a complete stranger more than 1000 miles away using only a chain of acquaintances (Travers and Milgram 1969; Lin 1989; Kochen 1989; Watts and Strogatz 1998; Watts 1999). If so, how many steps would be required? The researchers organized an experiment in which approximately 300 senders were all charged with getting a document to one receiver, a complete stranger. (Remember that all this was well before the advent of the desktop computer in the 1980s.) The receiver was a male, Boston stockbroker. The senders were one group of Nebraskans and one group of Bostonians chosen completely at random. Every sender in the study was given the receiver's name, address, occupation, alma mater, year of graduation, wife's maiden name, and hometown. They were asked to send the document directly to the stockbroker if they knew him on a first-name basis. Otherwise, they were asked to send the folder to a friend, relative, or acquaintance known on a first-name basis who might be more likely than the sender to know the stockbroker.

How many intermediaries do you think it took, on average, for the document to get through? (Most people estimate from twenty to hundreds.) *The average number of intermediate contacts was only 6.2!* However, only about one-third of the documents actually arrived at the target. This was quite impressive, considering that the senders did not know the target person—hence the current expression that any given person in the country is on average only about "six degrees of separation" from any other person. In this sense the world is indeed "small."

This original small world research has recently been criticized on two grounds: First, only one-third of the documents actually reached the target person. The 6.2 average intermediaries applied only to these completed chains. Thus two-thirds of the initial documents never reached the target person. For these persons, the world was certainly not "small." Second, the sending chains tended to closely follow occupational, social class, and ethnic lines, just as general network

theory would predict (Wasserman and Faust 1994). Thus, the world may indeed be "small," but only for people in your immediate social network (Ruef et al. 2003; Kleinfeld 2002; Watts 1999).

A study of Black national leaders by Taylor and associates shows that Black leaders form a very closely knit network, one considerably more closely knit than longer-established White leadership networks (Taylor 1992b; Domhoff 2002; Jackson 2000; Jackson et al. 1995, 1994; Alba and Moore 1982; Moore 1979; Kadushin 1974; Mills 1956). The world is indeed quite "small" for America's Black leadership. Included in the study were Black members of Congress, mayors, business executives, military officers (generals and full colonels), religious leaders, civil rights leaders, media personalities, entertainment and sports figures, and others. The study found that when considering only direct personal acquaintances—not indirect links involving intermediaries—one-fifth of the entire *national* Black leadership network know each other directly as a friend or close acquaintance. The Black leadership network is considerably more closely connected than White leadership networks. The Black network had greater *density*. Add only one intermediary, the friend of a friend, and the study estimated that almost three-quarters of the entire Black leadership network are included. Therefore, any given Black leader can generally get in touch with three-quarters of all other Black leaders in the country either by knowing them personally (a "friend") or via only one common acquaintance (a "friend of a friend"). That's pretty amazing when one realizes that the study is considering the population of Black leaders in the entire country.

Social Influence in Groups

The groups in which we participate exert tremendous influence on us. We often fail to appreciate how powerful these influences are. For example, who decides what you should wear? Do you decide for yourself each morning, or is the decision already made for you by fashion designers, role models, and your peers? Consider how closely your hair length, hair styling, and choice of jewelry have been influenced by your peers. Did you invent your baggy pants, your dreadlocks, or your blue blazer? People who label themselves as nonconformists often conform rigidly to the dress code and other norms of their in-group. This was true of the Beatniks in the 1950s, the hippies of the 1960s and early 1970s, the punk rockers of the 1970s and 1980s, and the grunge kids and goths of the 1990s.

After the rebelliousness of youth has faded, the influences of our youth extend to adulthood. The choices of political party among adults (Republican, Democratic, or Independent) correlate strongly with the party of one's parents, again demonstrating the power of the primary group. Seven out of ten people vote with the political party of their parents, even though these same people insist that they think for themselves when voting (Worchel et al. 2000; Jennings and Niemi 1974). Furthermore, most people share the religious affiliation of their parents, although they will insist that they chose their own religion, free of any influence by either parent.

We all like to think we stand on our own two feet, immune to a phenomenon as superficial as group pressure. The conviction that one is impervious to social influence results in what social psychologist Philip Zimbardo calls the *not-me syndrome:* When confronted with a description of group behavior that is disappointingly conforming and not individualistic, most individuals counter that some people may conform to social pressure, "but not me" (Zimbardo et al. 1977; Taylor et al. 2009). But sociological experiments often reveal a dramatic gulf between what people think they will do and what they actually do. The conformity study by Solomon Asch discussed next is a case in point.

The Asch Conformity Experiment

We learned in the previous section that social influences are evidently quite strong. Are they strong enough to make us disbelieve our own senses? Are they strong enough to make us misperceive what is objective, actual fact? In a classic piece of work known as the Asch conformity experiment, Solomon Asch showed that even simple objective facts cannot withstand the distorting pressure of group influence (Asch 1955, 1951).

Examine the two illustrations in Figure 6.1. Which line on the right is more nearly equal in length to the line on the left (line S)? Line B, obviously. Could anyone fail to answer correctly?

In fact, Solomon Asch discovered that social pressure of a rather gentle sort was sufficient to cause

an astonishing rise in the number of wrong answers. Asch lined up five students at a table and asked which line in the illustration on the right is the same length as the line on the left. Unknown to the fifth student, the first four were *confederates*—collaborators with the experimenter who only pretended to be participants. For several rounds, the confederates gave correct answers to Asch's tests. The fifth student also answered correctly, suspecting nothing. Then on subsequent trials the first student gave a wrong answer. The second student gave the same wrong answer. Third, wrong. Fourth, wrong. Then came the fifth student's turn.

In Asch's experiment, fully *one-third* of all students in the fifth position gave the same wrong answer as the confederates at least half the time. Forty percent gave "some" wrong answers. Only one-fourth of the students consistently gave correct answers in defiance of the invisible pressure to conform.

Line length is not a vague or ambiguous stimulus. It is clear and objective. Wrong answers from one-third of all subjects is a very high proportion. The subjects fidgeted and stammered while doing it, but they did it nonetheless. Those who did not yield to group pressure showed even more stress and discomfort than those who yielded to the (apparent) opinion of the group.

Would you have gone along with the group? Perhaps, perhaps not. Sociological insight grows when we acknowledge the fact that fully one-third of all participants will yield to the group. The Asch experiment has been repeated many times over the years, with students and nonstudents, old and young, in groups of different sizes, and in different settings (Worchel et al. 2000; Cialdini 1993). The results remain essentially the same. One-third to one-half of the participants make a judgment contrary to fact, yet in conformity with the group. Finally, the Asch findings have consistently revealed a *group size effect*: The greater the number of individuals (confederates) giving an incorrect answer (from five up to fifteen confederates), the greater the number of subjects per group giving an incorrect answer.

The Milgram Obedience Studies

What are the limits of social pressure? In terms of moral and psychological issues, judging the length of a line is a small matter. What happens if an authority figure demands obedience—a type of conformity—even if the task is something the test subject (the person) finds morally wrong and reprehensible? A chilling answer emerged from the now famous Milgram Obedience Studies done from 1960 through 1973 by Stanley Milgram (Milgram 1974).

In this study, a naive research subject entered a laboratory-like room and was told that an experiment on learning was to be conducted. The subject was to act as a "teacher," presenting a series of test questions to another person, the "learner." Whenever

FIGURE 6.1 *Lines from Asch Experiment*

Source: Asch, Solomon, 1956. "Opinion and Social Pressure." *Scientific American* 19 (July): 31–36.

the learner gave a wrong answer, the teacher would administer an electric shock.

The test was relatively easy. The teacher read pairs of words to the learner, such as

blue	**box**
nice	**house**
wild	**duck**

The teacher then tested the learner by reading a multiple-choice answer, such as

blue   sky   ink   box   lamp

The learner had to recall which term completed the pair of terms given originally, in this case, "blue box."

If the learner answered incorrectly, the teacher was to press a switch on the shock machine, a formidable-looking device that emitted an ominous hum when activated (see Figure 6.2). For each successive wrong answer, the teacher was to increase the intensity of the shock by 15 volts.

The machine bore labels clearly visible to the teacher: Slight Shock, Moderate Shock, Strong Shock, Very Strong Shock, Intense Shock, Extreme Intensity Shock, Danger: Severe Shock, and lastly, XXX at 450 volts. As the voltage rose, the learner responded with squirming, groans, then screams.

(a)

(b)

(c)

(d)

FIGURE 6.2 Milgram's Setup

These photographs show how intimidating and authoritative the Milgram experiment must have been. The first picture (a) shows the formidable-looking shock generator. The second (b) shows the role player, who pretends to be getting the electric shock, being hooked up. The third (c) shows an experimental subject (seated) and the experimenter (in lab coat, standing). The fourth picture (d) shows a subject terminating the experiment prematurely, that is, before giving the maximum shock level (voltage) of 450 volts. A large majority (65 percent) of subjects did not do this and actually went all the way to the maximum shock level.

Source: Milgram, Stanley. 1974. *Obedience to Authority: An Experimental View.* New York: Harper & Row, p. 26.

The experiment was rigged. The learner was a confederate. No shocks were actually delivered. The true purpose of the experiment was to see if any "teacher" would go all the way to 450 volts. If the subject (teacher) tried to quit, the experimenter responded with a sequence of prods:

"Please continue."

"The experiment requires that you continue."

"It is absolutely essential that you continue."

"You have no other choice, you must go on."

In the first experiment, fully 65 percent of the volunteer subjects ("teachers") went *all the way* to 450 volts on the shock machine!

Milgram himself was astonished. Before carrying out the experiment, he had asked a variety of psychologists, sociologists, psychiatrists, and philosophers to guess how many subjects would actually go all the way to 450 volts. The opinion of these consultants was that only one-tenth of 1 percent (one in one thousand) would actually do it.

What would you have done? Remember the "not-me" syndrome. Think about the experimenter saying, "You have no other choice, you must go on." Most people claim they would refuse to continue as the voltage escalated. The importance of this experiment derives in part from how starkly it highlights the difference between what people *think* they will do and what they *actually* do.

Milgram devised a series of additional experiments in which he varied the conditions to find out what would cause subjects *not* to go all the way to 450 volts. He moved the experiment from an impressive university laboratory to a dingy basement to counteract some of the tendency for people to defer to a scientist conducting a scientific study. One learner was then instructed to complain of a heart condition. Still, well over half of the subjects delivered the maximum shock level. Speculating that women might be more humane than men (all prior experiments used only male subjects), Milgram did the experiment again using only women subjects. The results? Exactly the same. Social class background made no difference. Racial and ethnic differences had no detectable effect on compliance rate.

At the time that the Milgram experiments were conceived, the world was watching the trial in Jerusalem of World War II Nazi Adolf Eichmann. Millions of Jews, Gypsies, homosexuals, and communists were murdered between 1939 and 1945 by the Nazi party, led by Adolf Hitler. As head of the Gestapo's "Jewish section," Eichmann oversaw the deportation of Jews to concentration camps and the mass executions that followed. Eichmann disappeared after the war, was abducted in Argentina by Israeli agents in 1961, and transported to Israel, where he was tried and ultimately hanged for crimes against humanity.

The world wanted to see what sort of monster could have committed the crimes of the Holocaust, but a jarring picture of Eichmann emerged. He was slight and mild mannered, not the raging ghoul that everyone expected. He insisted that although he had indeed been a chief administrator in an organization whose product was mass murder, he was guilty only of doing what he was told to do by his superiors. He did not hate Jews, he said. In fact, he had a Jewish half-cousin whom he hid and protected. He claimed, "I was just following orders."

How different was Adolph Eichmann from the rest of us? The political theorist Hannah Arendt dared to suggest in her book *Eichman in Jerusalem* (1963) that evil on a giant scale is banal. It is not the work of monsters, but an accident of civilization. Arendt argued that to find the villain, we need only look into ourselves.

The Iraqi Prisoners at Abu Ghraib: Research Predicts Reality?

We have just learned that ordinary people will do horrible things to other humans simply because of the influence of the group, because of an authority figure, or because of a combination of both. This has been the lesson of the Asch studies and the Milgram studies. Recent events in the world have once again shown vividly and clearly how accurate such sociological and psychological experiments are in the prediction of actual human behavior.

In the spring of 2004, it was revealed that American soldiers who were military police guards at a prison in Iraq (the prison was named Abu Ghraib) had engaged in severe torture of Iraqi prisoners of war. The torture included sexual abuse of the prisoners—having male prisoners simulate sex with other male prisoners, positioning their mouths next to the genitals of another male prisoner, being forced to masturbate in view of others, and other such acts. Still other acts of torture involved physical abuse such as beatings, stomping on the fingers of prisoners (thus fracturing them), and a large number of other physical acts of torture, including bludgeoning, some allegedly resulting in deaths of prisoners. Such tortures are clearly outlawed by the Geneva Conventions and by clearly stated U.S. principles of war. Both male and female guards participated in these acts of torture, and although most of the Iraqi victims were male, some were female.

The guards later claimed that they were simply following orders, either orders directly given or indirectly assumed. At the time, President George W. Bush and then-Secretary of Defense Donald H. Rumsfeld both claimed that the acts of torture were merely the acts of a "corrupt few" and that the vast majority of American soldiers would never engage in such horrible acts. Since then, it has come to light via CIA memoranda that certain kinds of torture were indeed formal U.S. policy.

Now consider what we know from research. The Milgram studies strongly suggest that many ordinary

soldiers who were not at all "corrupt," at least not more than average, would indeed engage in these acts of torture, particularly if they believed that they were under orders to do so, *or* if they believed that they would not be punished in any way if they did. The American soldiers must bear a significant portion of the responsibility for their own behavior. *Nonetheless, the causes of the soldiers' behaviors lie not in their personalities (their "natures") but in the social structure and group pressures of the situation.*

The soldiers (guards) in the Abu Ghraib prison may not have received *direct* orders to torture prisoners. But they did so nonetheless. A now classic study of a simulated prison by Haney, Banks, and Zimbardo (1973) shows this effect quite clearly. In this study, Stanford University students were told by an experimenter to enter a dungeon-like basement. Half were told to pretend to be guards (to *role-play* being a guard) and half were told that they were prisoners (to role-play being a prisoner). Which students were told what was *randomly determined*.

After two or three days, the guards, completely on their own, began to act very sadistically and brutally toward the prisoners—having them strip naked, simulate sex, act subservient, and so on. Interestingly, the prisoners for the most part did just what the guards wanted them to do, no matter how unpleasant the requested act! The experiment was so scary that the researchers terminated the experiment after six days—more than one week early.

Remember that this study was conducted in 1973—thirty-one years *before* Abu Ghraib. Yet, this simulated prison study predicted quite precisely how both "guards" and "prisoners" would act in a prison situation. Group influence effects uncovered by the Asch as well as the Milgram studies ruled in both the simulated prison of 1973 as well as the only too real Iraq prison of 2004.

Debunking Society's Myths

Myth: People in groups are just individuals who make up their own minds about how to think and behave.

Sociological perspective: The Asch, Milgram, and simulated prison experiments conclusively show that people are profoundly influenced by group pressure, often causing them to make up their minds contrary to objective fact and even to deliberately cause harm to another person.

Groupthink

Wealth, power, and experience are apparently not enough to save us from social influences. **Groupthink,** as described by I. L. Janis, is the tendency for group members to reach a consensus opinion, even if that decision is downright stupid (Janis 1982).

Janis reasoned that because major government policies are often the result of group decisions, it would be fruitful to analyze group dynamics that operate at the highest level of government—for instance, in the Office of the President of the United States. The president makes decisions based upon group discussions with his advisers. The president is human and thus susceptible to group influence. To what extent have past presidents and their advisers been influenced by group decision making instead of just the facts?

Janis investigated five ill-fated decisions, all the products of group deliberation:

- The decision of the Naval High Command in 1941 not to prepare for the attack on Pearl Harbor by Japan, which occurred anyway.
- President Harry Truman's decision to send U.S. troops to North Korea in 1950.
- President John F. Kennedy's attempt to overthrow Cuba by launching the ill-fated invasion at the Bay of Pigs in 1961.
- President Lyndon B. Johnson's decision in 1967 to increase the number of U.S. troops in Vietnam.
- The fateful decision by President Richard M. Nixon's advisers in 1972 to break into the Democratic Party headquarters at the Watergate apartment complex, launching the famed Watergate affair.

All the preceding were group decisions, and all were absolute fiascoes. For example, the Bay of Pigs invasion was a major humiliation for the United States—a covert outing so ill conceived it is hard to imagine how it survived discussion by a group of foreign policy experts. Fifteen hundred Cuban exiles trained by the CIA to parachute into heavily armed Cuba landed in an impassibly dense 80-mile swamp far from their planned drop zone, with inadequate weapons and incorrect maps. A sea landing was demolished by well-prepared, warned defenders.

Janis discovered a common pattern of misguided thinking in his investigations of presidential decisions. He surmised that outbreaks of groupthink had several things in common:

1. *An illusion of invulnerability.* "With such a brilliant team, and such a nation, how could any plan fail?" thought those in the group.
2. *A falsely negative impression of those who are antagonists to the group's plans.* Fidel Castro was perceived to be clownish, and Cuban troops were supposed to be patsies. In truth, the defenders at the Bay of Pigs were actually highly trained commandoes. Castro remained in power for many decades after the failed invasion.
3. *Discouragement of dissenting opinion.* As groupthink takes hold, dissent is equated with disloyalty. This can discourage dissenters from voicing their objections.

4. *An illusion of unanimity.* In the aftermath, many victims of groupthink recall their reservations, but at the moment of decision there is a prevailing sense that the entire group is in complete agreement.

We now might ask if groupthink influenced the torture of Iraq prisoners at the Abu Ghraib prison. (For that matter, we might ask if groupthink influenced the Bush administration's decision to enter the war in Iraq in the first place.) The actions of the military guards there were, it seems at least in part, directly or indirectly the result of high-level group decisions among presidential advisers. Groupthink is not inevitable when a team gathers to make a decision, but it is common and appears in all sorts of groups, from student discussion groups to the highest councils of power (Flowers 1977; McCauley 1989; Aldag and Fuller 1993; Kelley et al. 1999; Paulus et al. 2001).

Debunking Society's Myths

Myth: A group of experts brought together in a small group will solve a problem according to their collective expertise.

Sociological perspective: Groupthink can lead even the most qualified people to make disastrous decisions because people in groups in the United States tend to seek consensus at all costs.

Risky Shift

The term *groupthink* is commonly associated with group decision making with consequences that are not merely unexpected but disastrous. Another group phenomenon, **risky shift** (also called *polarization shift*), may help explain why the products of groupthink are frequently calamities. Have you ever found yourself in a group engaged in a high-risk activity that you would not do alone? When you created mischief as a child, were you not usually part of a group? If so, you were probably in the thrall of risky shift—the general tendency for groups to be more risky than individuals taken singly.

Risky shift was first observed by James Stoner (1961). Stoner gave study participants descriptions of a situation involving risk, such as one in which persons seeking a job must choose between job security and a potentially lucrative but risky advancement. The participants were then asked to decide how much risk the person should take. Before performing his study, Stoner believed that individuals in a group would take less risk than individuals alone, but he found the opposite: After his groups had engaged in open discussions, they favored greater risk than they would have before discussion.

thinkingsociologically

Think of a time when you engaged in some risky behavior. What group were you part of, and how did the group influence your behavior? How does this illustrate the concept of *risky shift*? Is there more risky shift with more people in the group? If so, this would illustrate a *group size effect*.

Stoner's research has stimulated literally hundreds of studies using males and females, different nationalities, different tasks, and other variables (Pruitt 1971; Blaskovitch 1973; Johnson et al. 1977; Hong 1978; Worchel et al. 2000; Taylor et al. 2009). The results are complex. Most, but not all, group discussion leads to greater risk-taking. In subcultures that value caution above daring, as in some work groups of Japanese and Chinese firms, group decisions are less risky after discussion than before. The shift can occur in either direction, driven by the influence of group discussion, but there is generally some kind of shift, in one direction or the other, rather than no shift at all (Kerr 1992).

What causes risky shifts? The most convincing explanation is that deindividuation occurs. **Deindividuation** is the sense that one's self has merged with a group. In terms of risk-taking, one feels that responsibility (and possibly blame) is borne not only by oneself but also by the group. This seems to have happened among the American prison guards who tortured prisoners at Abu Ghraib prison: Each

guard could convince himself or herself that responsibility, hence blame, was to be borne by the group as a whole. The greater the number of people in a group, the greater the tendency toward deindividuation. In other words, deindividuation is a *group size effect*. As groups get larger, trends in risk-taking are amplified.

Formal Organizations and Bureaucracies

Groups, as we have seen, are capable of greatly influencing individuals. The study of groups and their effects on the individual represent an example of *microanalysis*, to use a concept introduced in Chapter 5. In contrast, the study of formal organizations and bureaucracies, a subject to which we now turn, represents an example of *macroanalysis*. The focus on groups drew our attention to the relatively small and less complex, whereas the focus on organizations draws our attention to the relatively large and structurally more complex.

A **formal organization** is a large secondary group, highly organized to accomplish a complex task or tasks and to achieve goals efficiently. Many of us belong to various formal organizations: work organizations, schools, and political parties, to name a few. Organizations are formed to accomplish particular tasks and are characterized by their relatively large size, compared with a small group such as a family or a friendship circle. Often organizations consist of an array of other organizations. The federal government is a huge organization comprising numerous other organizations, most of which are also vast. Each organization within the federal government is also designed to accomplish specific tasks, be it collecting your taxes, educating the nation's children, or regulating the nation's transportation system and national parks.

Organizations develop cultures and routine practices. The culture of an organization may be reflected in certain symbols, values, and rituals. Some organizations develop their own language and styles of dress. The norms can be subtle, such as men being expected to wear long-sleeve shirts and ties or women being expected to wear stockings, even on hot summer days. It does not take explicit rules to regulate this behavior; comments from coworkers or bosses may be enough to enforce such organizational norms. Some work organizations have instituted a practice called "casual day" or "dress down day" one day per week, usually Friday, when workers can dress less formally.

Organizations tend to be persistent, although they are also responsive to the broader social environment where they are located (DiMaggio and Powell 1991). Organizations are frequently under pressure to respond to changes in the society by incorporating

© Sutcliffe/Sipa Press

Streaking, or running nude in a public place—relatively popular among college students in the 1970s and early 1980s and still popular on some campuses—is more common as a group activity than as a strictly individual one. This illustrates how the group can provide the persons in it with deindividuation, or merging of self with group. This allows the individual to feel less responsibility or blame for his or her actions, thus convincing herself or himself that the group must share the blame.

new practices and beliefs into their structure. Business corporations, as an example, have had to respond to increasing global competition; they do so by expanding into new international markets, developing a globally focused workforce, and trimming costs by *downsizing,* that is, by eliminating workers and various layers of management. Another recent response to increased global competition is *outsourcing*—having manufacturing tasks ordinarily performed by the home company (such as the manufacture of athletic shoes or soccer balls) performed instead by foreign workers.

Organizations can be tools for innovation, depending on the organization's values and purpose. Rape crisis centers are examples of organizations that originally emerged from the women's movement because of the perceived need for services for rape victims. Rape crisis centers have, in many cases, changed how police departments and hospital emergency personnel respond to rape victims. By advocating changes in rape law and services for rape victims, rape crisis centers have generated change in other organizations as well (Schmitt and Martin 1999; Fried 1994).

Types of Organizations

Sociologists Blau and Scott (1974) and Etzioni (1975) classify formal organizations into three categories distinguished by their types of membership affiliation: normative, coercive, and utilitarian.

Normative Organizations. People join **normative organizations** to pursue goals that they consider worthwhile. They obtain personal satisfaction, but no monetary reward for membership in such an organization. In many instances, people join the normative organization for the social prestige that it offers. Many are service and charitable organizations and are often called *voluntary organizations.* They include organizations such as the PTA, Kiwanis clubs, political parties, religious organizations, the National Association for the Advancement of Colored People (NAACP), B'nai B'rith, La Raza, and other similar voluntary organizations that are concerned with specific issues. Civic and charitable organizations, such as the League of Women Voters, and political organizations, such as the National Women's Political Caucus, reflect the fact that for decades women have been excluded from traditionally all-male voluntary organizations and political networks, such as the Kiwanis and Lions clubs—organizations that are now less influential than they were two or three decades ago. Like other service and charitable organizations, these groups have been created to meet particular needs, ones that members see as not being served by other organizations.

Gender, class, race, and ethnicity all play a role in who joins what voluntary organization. Social class is reflected in the fact that many people do not join certain organizations simply because they cannot afford to join. Membership in a professional organization, as

one example, can cost hundreds of dollars each year. Those who feel disenfranchised, however, may join grassroots organizations—voluntary organizations that spring from specific local needs that people think are unmet. Tenants may form an organization to protest rent increases or lack of services, or a new political party may emerge from people's sense of alienation from existing party organizations. African Americans, Latinos, and Native Americans have formed many of their own voluntary organizations in part because of their historical exclusion from traditional White voluntary organizations—which are now vibrant, ongoing organizations in their own right (such as the African American organizations Delta Sigma Theta and Alpha Kappa Alpha sororities and the fraternities Alpha Phi Alpha, Kappa Alpha Psi, and Omega Psi Phi; see Giddings 1994).

The NAACP, founded in 1909 by W.E.B. Du Bois (recall from Chapter 1), and the National Urban League are two other large national organizations that have historically fought racial oppression on the legal and urban fronts, respectively. La Raza Unida, a Latino organization devoted to civic activities as well as combating racial–ethnic oppression, has a large membership, with Latinas holding major offices. In fact, such voluntary organizations dedicated to the causes of people of color have in recent years had more women in leadership positions than have many standard, White organizations. Similarly, Native American voluntary organizations have boasted increasing numbers of women in leadership positions (Feagin and Feagin 1993; Snipp 1996, 1989).

Coercive Organizations. **Coercive organizations** are characterized by membership that is largely involuntary. Prisons are an example of organizations that people are coerced to "join" by virtue of punishment for their crime. Similarly, mental hospitals are coercive organizations: People are placed in them, often involuntarily, for some form of psychiatric treatment. In many respects, prisons and mental hospitals are similar in their treatment of inmates or patients. They both have strong security measures such as guards, locked and barred windows, and high walls (Goffman 1961; Rosenhan 1973). Sexual harassment and sexual victimization are common in both prisons and mental hospitals (Andersen 2003; Chesney-Lind 1992).

The sociologist Erving Goffman has described coercive organizations as total institutions. A **total institution** is an organization that is cut off from the rest of society and one in which resident individuals are subject to strict social control (Goffman 1961). Total institutions include two populations: the "inmates" and the staff. Within total institutions, the staff exercises complete power over inmates, for example, nurses over mental patients and guards over prisoners. The staff administers all the affairs of everyday life, including basic human functions such as eating and sleeping. Rigid routines are characteristic of total institutions, thus explaining the common complaint by those in hospitals that they cannot sleep because nurses repeatedly enter their rooms at night, regardless of whether the patient needs medication or treatment. However, the problem of such rigid routines has eased somewhat in some institutions.

Utilitarian Organizations. The third type of organization named is **utilitarian**. These are large organizations, either for-profit or nonprofit, that individuals join for specific purposes, such as monetary reward. Large business organizations that generate profits (in the case of for-profit organizations) and salaries and wages for the organization's employees (as with either for-profit or nonprofit organizations) are utilitarian organizations. Examples of large, for-profit organizations include Microsoft, Amazon.com, and Procter & Gamble. Examples of large nonprofit organizations that pay salaries to employees are colleges and, universities, churches, and organizations such as the National Collegiate Athletic Association (NCAA).

Bureaucracy

As formal organizations develop, many become a **bureaucracy,** a type of formal organization characterized by an authority hierarchy, a clear division of labor, explicit rules, and impersonality. Bureaucracies are notorious for their unwieldy size and complexity as well as their reputation for being remote and cumbersome organizations that are highly impersonal and machinelike in their operation. The federal government is a good example of a cumbersome bureaucracy that many believe is ineffective because of its sheer size. Numerous other formal organizations have developed into huge bureaucracies: IBM, Disney, many universities, hospitals, state motor vehicle registration systems, and some law firms. Other formal organizations, such as Enron and WorldCom, quickly developed into large bureaucracies, then subsequently collapsed under fraudulent accounting procedures. More recently, large national banks such as Merrill Lynch, now in bankruptcy, and several others have fallen under close scrutiny by Congress for questionable practices such as exorbitant salaries and excess bonuses for top management. Automakers such as General Motors, now in bankruptcy, have been criticized on similar grounds.

The early sociological theorist **Max Weber** (1947/1925) analyzed the classic characteristics of the bureaucracy. These characteristics represent what he called the *ideal type bureaucracy*—a model rarely seen in reality but that defines the principal characteristics of a social form. The characteristics of bureaucracies described as an ideal type are:

1. **High degree of division of labor and specialization.** The notion of the specialist embodies this criterion. Bureaucracies ideally employ specialists in the various positions and occupations, and these specialists are responsible for a specific set of duties. Sociologist Charles Perrow (2007, 1994, 1986) notes that many modern bureaucracies have hierarchical authority structures and an elaborate division of labor.

2. **Hierarchy of authority.** In bureaucracies, positions are arranged in a hierarchy so that each is under the supervision of a higher position. Such hierarchies are often represented in an *organization chart*, a diagram in the shape of a pyramid that shows the relative rank of each position plus the lines of authority between each. These lines of authority are often called the "chain of command," and they show not only who has authority, but also who is responsible to whom and how many positions are responsible to a given position.

3. **Rules and regulations.** All the activities in a bureaucracy are governed by a set of detailed rules and procedures. These rules are designed, ideally, to cover almost every possible situation and problem that might arise, including hiring, firing, salary scales, and rules for sick pay and absences.

4. **Impersonal relationships.** Social interaction in the (ideal) bureaucracy is supposed to be guided by *instrumental* criteria, such as the organization's rules, rather than by *expressive needs*, such as personal attractions or likes and dislikes. The ideal is that the objective application of rules will minimize matters such as personal favoritism—giving someone a promotion simply because you like him or her or firing someone because you do not like him or her. Of course, as we will see, sociologists have pointed out that bureaucracy has "another face"—the *informal* social interaction that keeps the bureaucracy working and often involves interpersonal friendships and social ties, typically among people taken for granted in these organizations, such as the support staff.

5. **Career ladders.** Candidates for the various positions in the bureaucracy are supposed to be selected on the basis of specific criteria, such as education, experience, and standardized examinations. The idea is that advancement through the organization becomes a career for the individual. Some organizations, such as some universities and some law firms, have a policy of **tenure**—a guarantee of continued employment until one's retirement from the organization.

6. **Efficiency.** Bureaucracies are designed to coordinate the activities of many people in pursuit of organizational goals. Ideally, all activities have been designed to maximize this efficiency. The whole system is intended to keep social–emotional relations and interactions at a minimum and instrumental interaction at a maximum.

Bureaucracy's Other Face

All the characteristics of Weber's "ideal type" are general defining characteristics. Rarely do actual bureaucracies meet this exact description. A bureaucracy has, in addition to the ideal characteristics of structure, an *informal structure*. This includes social interactions, even network connections, in bureaucratic settings that ignore, change, or otherwise bypass the formal structure and rules of the organization. This informal structure often develops among those who are taken for granted in organizations, such as secretaries and administrative assistants—who are most often women. Sociologist Charles Page (1946) coined the phrase *bureaucracy's other face* to describe this condition.

This other face is informal culture. It has evolved over time as a reaction to the formality and impersonality of the bureaucracy. Thus, administrative assistants and secretaries will sometimes "bend the rules a bit" when asked to do something more quickly than usual for a boss they like and bend the rules in another direction for a boss they do not like by slowing down or otherwise sabotaging the boss's work. Researchers have noted, for example, that secretaries and assistants have more authority than their job titles and salaries suggest. As a way around the cumbersome formal communication channels within the organization, the informal network, or "grapevine," often works better, faster, and sometimes even more accurately than the formal channels. As with any culture, the informal culture in the bureaucracy has its own norms or rules. One is not supposed to "stab friends in the back," such as by "ratting on" them to a boss or spreading a rumor about them that is intended to get them fired. Yet, just as with any norms, there is deviation from the norms, and "back-stabbing" and "ratting" does happen.

Bureaucracy's other face can also be seen in the workplace subcultures that develop, even in the largest bureaucracies. Some sociologists interpret the subcultures that develop within bureaucracies as people's attempts to humanize an otherwise impersonal organization. Keeping photographs of family and loved ones in the office, placing personal decorations on one's desk (if you are allowed), and organizing office parties are some ways people resist the impersonal culture of bureaucracies. Of course, this informal culture can also become exclusionary, increasing the isolation that some workers feel at work. Gay and lesbian workers may feel left out when other workers gossip about people's heterosexual dates; minority workers may be excluded from the casual conversations in the workplace that connect nonminority people to one another.

The informal norms that develop within the modern day bureaucracy often cause worker productivity to go up or down, depending on the norms and how they are informally enforced. The classic 1930s Hawthorne Studies, so named because they were carried out at the Western Electric telephone plant in Hawthorne, Illinois (Roethlisberger and Dickson 1939), discovered that small groups of workers developed their own ideas—their own norms—about how much work they should produce each day. If someone produced too many completed tasks in a day, he would make the rest of the workers "look bad" and run the risk of having the organization raise its expectations of how much work the group might be expected to produce. Because of this, anyone producing too much was informally labeled a "rate buster," and that person was punished by some act, such as punches on the shoulder (called "binging") or by group ridicule (called "razzing"). By the same token, one could be accused of producing too little, in which case he was labeled a "chiseler" and punished in the same way by either binging or razzing. This informal culture of bureaucracy's other face continues today in a manner similar to the culture initially discovered in the early Hawthorne studies (Ritzer 2007; Perrow 2007, 1986).

Problems of Bureaucracies

In contemporary times, problems have developed that grow out of the nature of the complex bureaucracy. Two problem areas already discussed are the occurrence of risky shift in work groups and the development of groupthink. Additional problems include a tendency to *ritualism* and the potential for *alienation* on the part of those within the organization.

Ritualism. Rigid adherence to rules can produce a slavish following of them, regardless of whether it accomplishes the purpose for which the rule was originally designed. The rules become ends in themselves rather than means to an end: This is ritualism.

Two now classic examples of the consequences of *organizational ritualism* have come to haunt us: the explosion of the space shuttle *Challenger* on January 28, 1986, and, to our horror, the breakup of yet another space shuttle, the *Columbia,* on February 1, 2003. People in the United States became bound together at the moment of the *Challenger* accident. Many remember where they were and exactly what they were doing when they heard about the tragedy. The failure of the essential O-ring gaskets on the solid fuel booster rockets of the *Challenger* shuttle caused the catastrophic explosion. It was revealed later that the O-rings became brittle at below-freezing temperatures, as was the temperature at the launch pad the evening before the *Challenger* lifted off.

Why did the managers and engineers at NASA (National Aeronautics and Space Administration) allow the shuttle to lift off given these prior conditions? The managers had all the information about the O-rings before the launch. Furthermore, engineers had warned them against the danger. In a detailed analysis of the decision to launch, sociologist Diane Vaughan (1996) uncovered both risky shift and organizational ritualism within the organization. The NASA insiders, confronted with signals of danger, proceeded as if nothing was wrong when they were repeatedly faced with the evidence that something was

The horror of the explosion of the space shuttle Challenger *in 1986 is seen in the faces of the observers here. All seven astronauts died in the explosion (top right). Sociologist Diane Vaughan (1996) attributes the disaster to an ill-formed launch decision in the bureaucracy of NASA based on group interaction phenomena such as risky shift, ritualism, groupthink, and the normalization of deviance. Tragedy struck again in February 2003, when the space shuttle* Columbia *broke up upon reentry into the atmosphere, killing all seven of the astronauts on board (bottom right).*

indeed *very* wrong. They in effect *normalized* their own behavior so that their actions became acceptable to them, representing nothing out of the ordinary. This is an example of organizational ritualism, as well as what Vaughan calls the "normalization of deviance."

Unfortunately, history repeated itself on February 1, 2003, when the space shuttle *Columbia,* upon its return from space, broke up in a fiery descent into the atmosphere above Texas, killing all who were aboard. The evidence shows that a piece of hard insulating foam separated from an external fuel tank during launch and struck the shuttle's left wing, damaging it and dislodging its heat-resistant tiles that are necessary for reentry. The absence of these tiles caused a burn-up upon reentry into the atmosphere. With eerie similarity to the earlier 1986 *Challenger* accident, a recent research report concludes that a "flawed institutional culture" and—citing sociologist Diane Vaughan—a normalization of deviance accompanying a gradual erosion of safety margins were among the causes of the *Columbia* accident (Schwartz and Wald 2003).

No single individual was at fault in either accident. The story is not one of evil but rather of the ritualism of organizational life in one of the most powerful bureaucracies in the United States. It is a story of rigid group conformity within an organizational setting and of how deviant behavior is redefined, that is, socially constructed, to be perceived as normal. Organizational procedures in this case, or rituals, dominate so much that the means toward goals become the goals themselves. A critical decrease in overall safety and a dramatic increase in risk are the results. Vaughan's analysis is a powerful warning about the hidden hazards of bureaucracy in a technological age.

Alienation. The stresses on rules and procedures within bureaucracies can result in a decrease in the overall cohesion of the organization. This often psychologically separates a person from the organization and its goals. This state of *alienation* results in increased turnover, tardiness, absenteeism, and overall dissatisfaction with the organization.

Alienation can be widespread in organizations where workers have little control over what they do or where workers themselves are treated like machines employed on an assembly line, doing the same repetitive action for an entire work shift. Alienation is not restricted to manual labor, however. In organizations where workers are isolated from others, where they are expected only to implement rules, or where they think they have little chance of advancement, alienation can be common. As we will see, some organizations have developed new patterns of work to try to minimize worker alienation and thus enhance their productivity.

The McDonaldization of Society

Sometimes the problems and peculiarities of bureaucracy can have effects on the total society. This has been the case with what George Ritzer (2007) has called the **McDonaldization,** a term coined from the well-known fast-food chain. In fact, each month 90 percent of the children in the United States between ages 3 and 9 visit McDonald's! Ritzer noticed that the principles that characterize fast-food organizations are increasingly dominating more aspects of U.S. society, indeed, of societies around the world. McDonaldization refers to the increasing and ubiquitous presence of the fast-food *model* in most organizations that shape daily life. Work, travel, leisure, shopping, health care, politics, and even education, have all become subject to McDonaldization. Each industry is based on a principle of high and efficient productivity, which translates into a highly rational social organization, with workers employed at low pay but with customers experiencing ease, convenience, and familiarity.

Ritzer argues that McDonald's has been such a successful model of business organization that other industries have adopted the same organizational characteristics, so much so that their nicknames associate them with the McDonald's chain: McPaper for *USA Today,* McChild for child-care chains like KinderCare,

and McDoctor for the drive-in clinics that deal quickly and efficiently with minor health and dental problems. Finally, the efficiency and predictability characteristic of the various Starbuck coffee shops represent what Ritzer calls "Starbuckization"—a type of McDonaldization.

Based in part upon Max Weber's concept of the ideal bureaucracy mentioned earlier, Ritzer identifies four dimensions of the McDonaldization process: efficiency, calculability, predictability, and control:

1. **Efficiency** means that things move from start to finish in a streamlined path. Steps in the production of a hamburger are regulated so that each hamburger is made exactly the same way—hardly characteristic of a home-cooked meal. Business can be even more efficient if the customer does the work once done by an employee. In fast-food restaurants, the claim that you can "have it your way" really means that you assemble your own sandwich or salad.

2. **Calculability** means there is an emphasis on the quantitative aspects of products sold: size, cost, and the time it takes to get the product. At McDonald's, branch managers must account for the number of cubic inches of ketchup used per day; likewise, ice cream scoopers in chain stores measure out predetermined and exact amounts of ice cream. Workers are monitored for how long it takes them to complete a transaction; every bit of food and drink is closely monitored by computer, and everything has to be accounted for.

3. **Predictability** is the assurance that products will be exactly the same, no matter when or where they are purchased. Eat an Egg McMuffin in New York, and it will likely taste just the same as an Egg McMuffin in Los Angeles or Paris! Ditto for a Decaf Tall Cappuccino from Starbucks: It will taste the same in whatever town or city you are drinking it.

4. **Control** is the primary organizational principle that lies behind McDonaldization. Behavior of the customers and workers is reduced to a series of machinelike actions. Ultimately, efficient technologies replace much of the work that humans once performed.

McDonaldization—and Starbuckization—clearly brings many benefits. There is a greater availability of goods and services to a wide proportion of the population; instantaneous service and convenience to a public with less free time; predictability and familiarity in the goods bought and sold; and standardization of pricing and uniform quality of goods sold, to name a few benefits. However, this increasingly rational system of goods and services also spawns irrationalities. For example, the majority of workers at McDonald's lack full-time employment, have no worker benefits, have no control over their workplace, and quit on average after only four or five months. Ritzer argues

Evidence of the "McDonaldization of society" can be seen everywhere, perhaps including on your own campus. Shopping malls, food courts, sports stadiums, even cruise ships reflect this trend toward standardization.

Tony Dejak/AP Photo

that, as we become more dependent on the familiar and that which is taken for granted, there is the danger of dehumanization. People lose their creativity, and there is little concern with the quality of goods and services, thereby disrupting something fundamentally human—the capacity for error, surprise, and imagination.

Diversity: Race, Gender, and Class in Organizations

The hierarchical structuring of positions within organizations results in the concentration of power and influence with a few individuals at the top. Since organizations tend to reflect patterns within the broader society, this hierarchy, like that of society, is marked by inequality in race, gender, and class relations. Although the concentration of power in organizations is incompatible with the principles of a democratic society, discrimination against women and minorities still occurs (Perrow 2007, 1994). There have been widespread disparities in the promotion rates for White and Black workers, with Whites more likely to be promoted and promoted more quickly—a pattern repeated in many work organizations (Perrow 2002; Eichenwald 1996; McGuire and Reskin 1993; Collins 1989).

Traditionally, within organizations the most powerful positions are held by White men of upper social class status. Women and minorities, on average, occupy lower positions in the organization. Although very small numbers of minorities and women do get promoted, there is typically a "glass ceiling" effect, meaning that women and minorities may be promoted but only up to a point. The glass ceiling acts as a barrier to the promotion of women and minorities into higher ranks of management, as discussed further in Chapters 10 and 11. Over the last five years, promotions of minorities and women in the organization have become slightly more common, but the overall picture is still one of discrimination and a glass ceiling.

What then are the barriers that prevent more inclusiveness in the higher ranks of organizations? Sociological research finds that organizations are sensitive to the climate in which they operate. The more egalitarian the environment in which a firm operates, the more equitable is its treatment of women and minorities.

Yet studies find that patterns of race and gender discrimination persist throughout organizations even to this day, even when formal barriers to advancement have been removed. Many minorities are now equal to Whites in education, particularly in organizational jobs that require advanced graduate degrees such as the master of business administration (MBA). Still, White men in organizations are somewhat more likely to receive promotions than African American, Hispanic, and Native American workers *with the same education* (Smith 2007: DeWitt 1995; Zwerling and Silver 1992). The same thing often happens to both White and minority women in organizations: Women are less likely to receive promotions than a White male with the same education, and sometimes even less education. Studies consistently find that women are held to higher promotion standards than men; for men, the longer they are in a position in an organization, the more likely they will be promoted, but the same is not true for women. Studies find that women change jobs more frequently within organizations than do men, but these tend to be lateral moves. For men, job changes are more likely to mark a jump from

Few organizational boards and executive committees contain minorities and women: when present, they are often tokens.

© Bananastock/Jupiter Images

a lower level to a higher level in the organization, thus constituting a promotion.

Things work the same way with respect to people being discharged or fired. Studies show quite clearly that Black federal employees (men and women) were more than *twice as likely* to be dismissed as their White counterparts (DeWitt 1995; Zwerling and Silver 1992). This disparity occurred regardless of education and regardless of occupational category, pay level, type of federal agency, age, performance rating, seniority, or attendance record. The main reasons cited in the studies for such disparities were lack of network contacts with the old boy network and racial bias within the organization (Smith 2007). These studies were particularly well designed because they took into account quite a few factors besides race–ethnicity that are often given as reasons for not promoting minorities. The studies showed quite conclusively that with all factors taken into account, it is still race that is an important reason for the way people are treated in organizations today. Even though there has been some improvement in recent years, these studies strongly suggest that racism still thrives in the bureaucracy. Will things change as the result of having a Black man in the highest office of the United States? Some would say yes, because President Obama is an inspiring role model, but sociological processes occur beyond the influence of any one individual, so change requires more than the influence of one person. What do you think?

Debunking Society's Myths

Myth: Programs designed to enhance the number of women and minorities in organizational leadership are no longer needed because discriminatory barriers have been removed.

Sociological perspective: Research right up to the present continues to find significant differences in the promotion rates for women and minorities in most organizational settings. Even with the removal of formal discriminatory barriers, organizational practices persist that block the mobility of these workers.

A classic study by Rosabeth Moss Kanter (1977) shows how the structure of organizations leads to obstacles in the advancement of groups that are tokens—rarely represented—in the organizational environment. Kanter demonstrated how the hierarchical structure of the bureaucracy negatively affects both minorities and women who are underrepresented in the organization. In such cases, they represent *token* minorities or women; they feel put "out front" and under the all-too-watchful eyes of their superiors as well as coequals. As a result—as research since Kanter's has shown—they often suffer severe stress

(Smith 2007; Jackson 2000; Jackson et al. 1995, 1994; Yoder 1991; Spangler et al. 1978). Coworkers often accuse women and minorities of getting a position simply because they are women, minorities, or both. It is a fairly widespread phenomenon in universities and colleges that not only minorities, but women as well, are often accused of being admitted simply because of their gender and/or their race, *even in instances where the person has had superior admissions qualifications*. This is stressful for the person and shows that tokenism can have very negative consequences.

Social class, in addition to race and gender, plays a part in determining people's place within formal organizations. Middle- and upper-class employees in organizations make higher salaries and wages and are more likely to get promoted than are people of lower social class status, even for individuals who are of the same race or ethnicity. This even holds for people coming from families of lower social class status who are as well educated as their middle- and upper-class coworkers. Thus, their lower salaries and lack of promotion cannot necessarily be attributed to a lack of education. In this respect, their treatment in the bureaucracy only perpetuates rather than lessens the negative effects of the social class system in the United States.

The social class stratification system in the United States produces major differences in the opportunities and life chances of individuals, and the bureaucracy simply carries these differences forward. Class stereotypes also influence hiring practices in organizations. Personnel officers look for people with "certain demeanors," a code phrase for those who convey middle-class or upper-middle-class standards of dress, language, manners, and so on, which some people may be unable to afford or may not possess.

Patterns of race, class, and gender inequality in organizations persist at the same time that many organizations have become more aware of the need for a more diverse workforce. Responding to the simple fact of more diversity within the working-age population, organizations have developed human relations experts who work within organizations to enhance sensitivity to diversity. Such diversity training has become commonplace in many large organizations.

Functional, Conflict, and Symbolic Interaction: Theoretical Perspectives

All three major sociological perspectives—functional, conflict, and symbolic interaction—are exhibited in the analysis of formal organizations and bureaucracies (see Table 6.1). The functional perspective, based in this case on the early writing of Max Weber, argues

table 6.1	Theoretical Perspective on Organizations		
	Functionalist Theory	**Conflict Theory**	**Symbolic Interaction Theory**
Central Focus	Positive functions (such as efficiency) contribute to unity and stability of the organization.	Hierarchical nature of bureaucracy encourages conflict between superior and subordinate, men and women, and people of different racial or class backgrounds.	Stresses the role of self in the bureaucracy and how the self develops and changes.
Relationship of Individual to the Organization	Individuals, like parts of a machine, are only partly relevant to the operation of the organization.	Individuals subordinated to systems of power and experience stress and alienation as a result.	Interaction between superiors and subordinates forms the structure of the organization.
Criticism	Hierarchy can result in dysfunctions such as ritualism and alienation.	De-emphasizes the positive ways that organizations work.	Tends to downplay overall social organization.

that certain functions, called *eufunctions* (that is, positive functions), characterize bureaucracies and contribute to their overall unity. The bureaucracy exists to accomplish these eufunctions, such as efficiency, control, impersonal relations, and a chance for the individual to develop a career within the organization. As we have seen, however, bureaucracies develop the "other face" (informal interaction and culture, as opposed to formal or bureaucratic interaction and culture) as well as the problems of ritualism and alienation of the person from the organization. These latter problems are called *dysfunctions* (negative functions), which have the consequence of contributing to the disunity, lack of harmony, and less efficiency in the bureaucracy.

The conflict perspective argues that the hierarchical or stratified nature of the bureaucracy in effect encourages rather than inhibits conflict among the individuals within it. These conflicts are between superior and subordinate, as well as between racial and ethnic groups, men and women, and people of different social class backgrounds, hampering the smooth and efficient running of the bureaucracy.

Consider the symbolic interaction perspective as underlying two management theories, those of Argyris (1990) and of Ouchi (1981). Symbolic interaction stresses the role of the self in any group and especially how the self develops as a product of social interaction. Argyris's theory advocates increased involvement of the self within the organization as a way of "actualizing" the self. This helps reduce the disconnection between individual and organization as well as other organizational problems and dysfunctions. Ouchi's theory argues that increased interaction between superior and subordinate, based on the Japanese organization model of executives "walking around" and interacting more on a primary group basis, will reduce organizational dysfunction.

Chapter Summary

• ***What are the types of groups?***

Groups are a fact of human existence and permeate virtually every facet of our lives. Group size is important, as is the otherwise simple distinction between dyads and triads. *Primary groups* form the basic building blocks of social interaction in society. *Reference groups* play a major role in forming our attitudes and life goals, as do our relationships with in-groups and out-groups. *Social networks* partly determine things such as who we know and the kinds of jobs we get. Networks based on race–ethnicity, social class, and other social factors are extremely closely connected—very dense.

- **How strong is social influence?**

  The social influence groups exert on us is tremendous, as seen by the Asch conformity experiments. The Milgram experiments demonstrated that the interpersonal influence of an authority figure can cause an individual to act against his or her deep convictions. The torture and abuse of Iraqi prisoners of war by American soldiers as prison guards serves as testimony to the powerful effects of both social influence and authority structures.

- **What is the importance of groupthink and risky shift?**

  Groupthink can be so pervasive that it adversely affects group decision making. *Risky shift* similarly often compels individuals to reach decisions that are at odds with their better judgment.

- **What are the types of formal organizations and bureaucracies, and what are some of their problems?**

  There are several types of *formal organizations,* such as *normative, coercive,* or *utilitarian.* Weber typified *bureaucracies* as organizations with an efficient division of labor, an authority hierarchy, rules, impersonal relationships, and career ladders. Bureaucratic rigidities often result in organizational problems such as ritualism and resulting "normalization of deviance," which may have been significantly responsible for the

space shuttle *Challenger* explosion in 1986 and the space shuttle *Columbia* breakup in 2003. The **McDonaldization** of society has resulted in greater efficiency, calculability, and control in many industries, probably at the expense of some individual creativity.

- **What are the problems of diversity in organizations?**

  Formal organizations perpetuate society's inequalities on the basis of race–ethnicity, gender, and social class. Even today, Blacks, Hispanics, and Native Americans are less likely to get promoted, and more likely to get fired, than Whites of comparable education and other qualifications. Women experience similar effects of inequality, especially negative effects of tokenism, such as stress and lowered self-esteem. Finally, persons of less than middle-class origins make less money and are less likely to get promoted than a middle-class person of comparable education.

- **What do the functional, conflict, and symbolic interaction theories say about organizations?**

  Functional, conflict, and symbolic interaction theories highlight and clarify the analysis of organizations by specifying both organizational functions and dysfunctions (*functional theory*); by analyzing the consequences of hierarchical, gender, race, and social class conflict in organizations (*conflict theory*); and, finally, by studying the importance of social interaction and integration of the self into the organization (*symbolic interaction theory*).

Key Terms

attribution error 132	group 128	risky shift 138
attribution theory 132	group size effect 129	secondary group 129
bureaucracy 141	groupthink 137	social network 133
coercive organization 140	instrumental needs 131	tenure 141
deindividuation 139	McDonaldization 144	total institution 140
dyad 129	normative organization 140	triad 129
expressive needs 131	primary group 129	triadic segregation 129
formal organization 139	reference group 132	utilitarian organization 141

Online Resources

Sociology: The Essentials Companion Website

www.cengage.com/sociology/andersen

Visit your book companion website where you will find more resources to help you study and write your research papers. Resources include web links, learning objectives, Internet exercises, quizzing, and flash cards.

 is an easy-to-use online resource that helps you study in less time to get the grade you want NOW.

www.cengage.com/login

Need help studying? This site is your one-stop study shop. Take a Pre-Test and Cengage NOW will generate a Personalized Study Plan based on your test results. The Study Plan will identify the topics you need to review and direct you to online resources to help you master those topics. You can then take a Post-Test to determine the concepts you have mastered and what you still need to work on.

Chapter seven
CHAPTER SEVEN

Deviance and Crime

Defining Deviance

Sociological Theories of Deviance

Forms of Deviance

Crime and Criminal Justice

Terrorism as International Crime: A Global Perspective

Chapter Summary

[*In the early* 1970s, an airplane carrying forty members of an amateur rugby team crashed in the Andes Mountains in South America. The twenty-seven survivors were marooned at 12,000 feet in freezing weather and deep snow. There was no food except for a small amount of chocolate and some wine. A few days after the crash, the group heard on a small transistor radio that the search for them had been called off.

Scattered in the snow were the frozen bodies of dead passengers. Preserved by the freezing weather, these bodies became, after a time, sources of food. At first, the survivors were repulsed by the idea of eating human flesh, but as the days wore on, they agonized over the decision about whether to eat the dead crash victims, eventually concluding that they had to eat if they were to live.

In the beginning, only a few ate the human meat, but soon the others began to eat too. The group experimented with preparations as they tried different parts of the body. They developed elaborate rules about how, what, and whom they would eat. Some could not bring themselves to cut the meat from the human body, but would slice it once someone else had cut off large chunks. They all refused to eat certain parts—the lungs, skin, head, and genitals.

After two months, the group sent out an expedition of three survivors to find help. The group was rescued, and the world learned of their ordeal. Their cannibalism (the eating of other human beings) was generally accepted as something they had to do to survive. Although people might have been repulsed by the story, the survivors' behavior was understood as a necessary

continued

Nic Cleave Photograph / Alamy

adaptation to their life-threatening circumstances. The survivors also maintained a sense of themselves as good people even though what they did profoundly violated ordinary standards of socially acceptable behavior in most cultures in the world (Read 1974; Miller 1991; Henslin 1993).

Was the behavior of the Andes crash survivors socially deviant? Were the people made crazy by their experience, or was this a normal response to extreme circumstances?

Compare the Andes crash to another case of human cannibalism. In 1991, in Milwaukee, Wisconsin, Jeffrey Dahmer pled guilty to charges of murdering at least fifteen men in his home. Dahmer lured the men—eight of them African American, two White, and one a fourteen-year-old Laotian (Asian) boy—to his apartment, where he murdered and dismembered them, then cooked and ate some of their body parts. For those he considered most handsome, he boiled the flesh from their heads so that he could save and admire their skulls. Dahmer was seen as a total social deviant, someone who violated every principle of human decency. Even hardened criminals were disgusted by Dahmer. In fact, he was killed in prison by another inmate in 1994.

Why was Dahmer's behavior considered so deviant when that of the Andes survivors was not? The answer can be found by looking at the situation in which these behaviors occurred. For the Andes survivors, eating human flesh was essential for survival. For Dahmer, however, it was murder. From a sociological perspective, the deviance of cannibalism resides *not just in the act itself but also in the social context in which it occurs*. The exact same behavior—eating other human beings—is considered reprehensible in one context and acceptable in another. That is the essence of the sociological explanation: The nature of deviance is not only in the personality of the deviant person, nor is it inherently in the deviant act itself, but instead it is a significant part and product of the social structure.

Defining Deviance

Sociologists define **deviance** as *behavior that is recognized as violating expected rules and norms*. Deviance is more than simple nonconformity; it is behavior that departs significantly from social expectations. In the sociological perspective on deviance, there are four main identifying characteristics:

- Deviance emerges in a social context, not just the behavior of individuals; sociologists see deviance in terms of group processes and judgments.

- Not all behaviors are judged similarly by all groups; what is deviant to one group may be normative (not deviant) to another.

- Established rules and norms are socially created, not just morally decided or individually imposed.

- Deviance lies not just in behavior itself but in the social responses of groups to behavior by others.

Sociological Perspectives on Deviance

Strange, unconventional, or nonconformist behavior is often understandable in its sociological context. Consider suicide. Are people who commit suicide mentally disturbed, or might their behavior be explained by social factors? Are there conditions under which suicide is acceptable behavior—for example, someone who commits suicide in the face of a terminal illness compared to a despondent person who jumps from a window?

Sociologists distinguish two types of deviance: formal and informal. *Formal deviance* is behavior that breaks laws or official rules. Crime is an example. There are formal sanctions against formal deviance, such as imprisonment and fines. *Informal deviance* is behavior that violates customary norms (Schur 1984). Although such deviance may not be specified in law, it is judged to be deviant by those who uphold the society's norms.

The study of deviance can be divided into the study of why people violate laws or norms and the study of how society reacts. *Labeling theory* is discussed in detail later, but it recognizes that deviance is not just in the breaking of norms or rules but it includes how people react to those behaviors. Social groups actually *create* deviance "by making the rules whose infraction constitutes deviance, and by applying those rules to particular people and labeling them as outsiders" (Erikson 1966; Becker 1963 : 9).

The Context of Deviance. Even the most unconventional behavior can be understood if we know the context in which it occurs. Behavior that is deviant in one circumstance may be normal in another, or behavior may be ruled deviant only when performed by certain people. For example, people who break gender stereotypes may be judged as deviant even though their behavior is considered normal for the other sex. Heterosexual men and women who kiss in public are the image of romance; lesbians and gay men who even dare to hold hands in public are often seen as flaunting their sexual orientation.

The definition of deviance can also vary over time. Acquaintance rape (also called "date rape"), for example, was not considered social deviance until fairly recently. Women have been presumed to mean yes when they said no, and men were expected to "seduce" women through aggressive sexual behavior. Even now, women who are raped by someone they know may not think of it as rape.

If they do, they may find that prosecuting the offender is difficult because others do not think of it as rape, especially under certain circumstances, such as the woman being drunk. What is in fact rape may not be seen as such by everyone.

The sociologist Emile Durkheim argued that one reason acts of deviance are publicly punished is that the social order is threatened by deviance. Judging those behaviors as deviant and punishing them confirms general social standards. Therein lies the value of widely publicized trials, public executions, or the historical practice of displaying a wrongdoer in the stocks, which held one's feet fast, or the pillory, which held the hands and head. Passersby were permitted to hurl both stones and large rocks at those persons so immobilized. The punishment affirms the collective beliefs of the society, reinforces social order, and inhibits future deviant behavior, especially as defined by those with the power to judge others.

Debunking Society's Myths

Myth: Deviance is bad for society because it disrupts normal life.

Sociological perspective: Deviance tends to stabilize society. By defining some forms of behavior as deviant, people are affirming the social norms of groups. In this sense, society actually to some extent *creates* deviance.

© RAWA/WorldPicture News

This widely distributed photo of a woman being executed by the Taliban in Afghanistan illustrates the extreme sanctions that can be brought against those defined as deviant by a powerful group. In this case, the photo also mobilized world condemnation of the Taliban regime for its treatment of women.

Durkheim argued that societies actually *need* deviance to know what presumably normal behavior is. In this sense, Durkheim considered deviance "functional" for society (Durkheim 1951/1897; Erikson 1966). You could observe Durkheim's point in the aftermath of the terrorist attacks of September 11, 2001. Horrified by the sight of hijacked planes flying into the World Trade Center and the Pentagon and crashing in a Pennsylvania field, U.S. citizens responded through publicly demonstrating strong patriotism. Durkheim would interpret these terrorist acts as deviance producing strong social solidarity. This was one of Durkheim's most important insights: *Deviance produces social solidarity. Instead of breaking society up, deviance produces a pulling together, or social solidarity.*

The Influence of Social Movements. The perception of deviance may also be influenced by social movements, which are networks of groups that organize to support or resist changes in society (see Chapter 16). With a change in the social climate, formerly acceptable behaviors may be newly defined as deviant. Smoking, for instance, was once considered glamorous, sexy, and "cool." Now, smokers are widely scorned as polluters and, despite strong lobbying by the tobacco industry, regulations against smoking have proliferated.

Whereas in 1987 only 17 percent of the public thought that smoking should be banned in restaurants, by 2008 over half (60 percent) thought so (Gallup Organization 2008). The increase in public disapproval of smoking results as much from social and political movements as it does from the known health risks. The success of the antismoking movement has come from the mobilization of constituencies able to articulate to the public that smoking is dangerous. Note that the key element here is the ability of people to mobilize—not just the evidence of risk. In other words, there has to be a social response for deviance to be defined as such; scientific evidence of harm in and of itself is not enough.

The Social Construction of Deviance. Perhaps because it violates social conventions or because it sometimes involves unusual behavior, deviance captures the public imagination. Commonly, however, the public understands deviance as the result of individualistic or personality factors. Many people see deviants as crazy, threatening, "sick," or in some other ways inferior, but sociologists see deviance as influenced by society—the same social processes and institutions that shape all social behavior.

Deviance, for example, is not necessarily irrational or "sick" and may be a positive and rational adaptation to a situation. Think of the Andes survivors discussed in this chapter's opener. Was their action (eating human flesh) irrational, or was it an inventive and rational response to a dreadful situation? To use another example, are gangs the result of the irrational

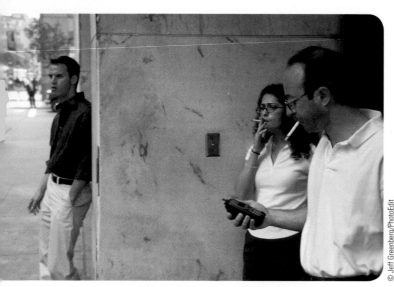

© Jeff Greenberg/PhotoEdit

Once considered "cool," smokers are now considered to be deviants, scorned as polluters, and often banished to outside office buildings, as here.

behavior of maladjusted youth, or are they rational responses to social situations?

Sociological studies of gangs in the United States shed light on this question. The family situations of gang members are often problematic, although girls in gangs tend to be more isolated from their families than are boys in gangs (Fleisher 2000; Esbensen-Finn et al. 1999). Given the class, race, and gender inequality faced by minority youth, many turn to gangs for the social support they lack elsewhere (Walker-Barnes and Mason 2001; Moore and Hagedorn 1996). For example, some poor, young Puerto Rican girls live in relatively confined social environments with little opportunity for educational or occupational advancement. Their community expects them to be "good girls" and to remain close to their families. Joining a gang is one way to reject these restrictive roles (Messerschmidt 1997; Campbell 1987). Are these young women irrational or just doing the best they can to adapt to their situation? Sociologists interpret their behavior as an understandable adaptation to conditions of poverty, racism, and sexism.

Also, in some subcultures or situations, deviant behavior is encouraged and praised. Have you ever been egged on by friends to do something that you thought was deviant, or have you done something you knew was wrong? Many argue that the reason so many college students drink excessively is that the student subculture encourages them to do so—even though students know it is harmful. Similarly, the juvenile delinquent regarded by school authorities as defiant and obnoxious is rewarded and praised by peers for the very behaviors that school authorities loathe. Much deviant behavior occurs, or escalates, because of the social support received from others.

Some behavior patterns defined as deviant are also surprisingly similar to so-called normal behavior.

Is a heroin addict who buys drugs with whatever money he can find so different from a business executive who spends a large proportion of his discretionary income on alcohol? Each may establish a daily pattern that facilitates drug use; each may select friends based on shared interests in drinking or taking drugs; and each may become so physically, emotionally, and socially dependent on their "fix" that life seems unimaginable without it. Which of the two is more likely to be considered deviant?

The point is that deviance is both created and defined within a social context. It is not just weird, pathological, or irrational behavior. Sociologists who study deviance understand it in the context of social relationships and society. They define deviance in terms of existing social norms and the social judgments people make about one another. Indeed, deviant behavior can sometimes be indicative of changes that are taking place in the cultural folkways. Thus, whereas only a few years ago body piercing and tattooing were associated with gangs and disrespectable people, now it is considered fashionable among young, middle-class people.

In sum, a sociological perspective on deviance asks: Why is deviance more common in some groups than others? Why are some more likely to be labeled deviant than others, even if they engage in the exact same behavior? How is deviance related to patterns of inequality in society? Sociologists do not ignore individual psychology but integrate it into an explanation of deviance that focuses on the social conditions surrounding the behavior, going beyond explanations of deviance that root it in the individual personality.

The Medicalization of Deviance

Commonly, people will say that someone who commits a very deviant act is "sick." This common explanation is what sociologists call the **medicalization of deviance** (Conrad and Schneider 1992). Medicalizing deviance attributes deviant behavior to a "sick" state of mind, where the solution is to "cure" the deviant through therapy or other psychological treatment.

An example is found in alcoholism. There is some evidence that there may be a genetic basis to alcoholism, and certainly alcoholism must be understood at least in part in medical terms, but viewing alcoholism *solely* from a medical perspective ignores the social causes that influence the development and persistence of this behavior. Practitioners know that medical treatment alone does not solve the problem. The social relationships, social conditions, and social habits of alcoholics must be altered, or the behavior is likely to recur.

thinkingsociologically

Ask some of your friends to explain why rape occurs. What evidence of the *medicalization of deviance* exists in your friends' answers?

Sociologists criticize the medicalization of deviance for ignoring the effects of social structures on the development of deviant behavior. From a sociological perspective, deviance originates in society, not just in individuals. Changing the incidence of deviant behavior requires changes in society in addition to changes in individuals. Deviance, to most sociologists, is not a pathological state but an *adaptation to the social structures* in which people live. Factors such as family background, social class, racial inequality, and the social structure of gender relations in society produce deviance, and these factors must be considered in order to explain it.

Sociological Theories of Deviance

Sociologists have drawn on several major theoretical traditions to explain deviant behavior, including functionalism, conflict theory, and symbolic interaction theory.

Functionalist Theories of Deviance

Recall that functionalism is a theoretical perspective that interprets all parts of society, even those that may seem dysfunctional, as instead contributing to the stability of the whole. At first glance, deviance seems to be dysfunctional for society. Functionalist theorists argue otherwise (see Table 7.1). They contend that deviance is functional *because it creates social cohesion*. Branding certain behaviors as deviant provides contrast with behaviors that are considered normal, giving people a heightened sense of social order. Norms are meaningless unless there is deviance from them; thus, deviance is necessary to clarify what society's norms are. Group coherence then comes from sharing a common definition of legitimate, as well as deviant, behavior. The collective identity of the group is affirmed when those who are defined as deviant are ridiculed or condemned by group members (Erikson 1966).

To give an example, think about how many people define gay men as deviant. Although lesbians and gay men have rejected this label, labeling homosexuality as deviant is one way of affirming the presumed normality of heterosexual behavior. Labeling someone else an outsider is, in other words, a way of affirming one's "insider" identity.

Durkheim: The Study of Suicide. The functionalist perspective on deviance stems originally from the work of **Emile Durkheim**. Recall that one of Durkheim's central concerns was how society maintains its coherence (or social order). Durkheim saw deviance as functional for society because it produces solidarity among society's members. He developed his analysis of deviance in large part through his analysis of suicide. Through this work, he discovered a number of important sociological points. First, he criticized the usual psychological interpretations of why people commit suicide, turning instead to sociological explanations with data to back them up. Second, he emphasized the role of social structure in producing deviance. Third, he pointed to the importance of people's social attachments to society in understanding deviance. Finally, he elaborated the functionalist view that deviance provides the basis for social cohesion. His studies of suicide illustrate these points.

Durkheim was the first to argue that the causes of suicide were to be found in social factors, not individual personalities. Observing that the rate of suicide in a society varied with time and place, Durkheim looked for causes linked to these factors other than emotional stress. Durkheim argued that suicide rates are affected by the different social contexts in which they emerge. He looked at the degree to which people feel integrated into the structure of society and their social surroundings as social factors producing suicide.

Durkheim analyzed three types of suicide: anomic suicide, altruistic suicide, and egoistic suicide. **Anomie,** as defined by Durkheim, is the condition that exists when social regulations in a society break down: The controlling influences of society are no longer effective, and people exist in a state of relative normlessness.

table7.1 *Sociological Theories of Deviance*

Functionalist Theory	Symbolic Interaction Theory	Conflict Theory
Deviance creates social cohesion.	Deviance is a learned behavior, reinforced through group membership.	Dominant classes control the definition of and sanctions attached to deviance.
Deviance results from structural strains in society.	Deviance results from the process of social labeling, regardless of the actual commission of deviance.	Deviance results from social inequality in society.
Deviance occurs when people's attachment to social bonds is diminished.	Those with the power to assign deviant labels themselves produce deviance.	Elite deviance and corporate deviance go largely unrecognized and unpunished.

The term *anomie* refers not to an individual's state of mind, but instead to social conditions.

Anomic suicide occurs when the disintegrating forces in the society make individuals feel lost or alone. Teenage suicide is often cited as an example of anomic suicide. Studies of college campuses, for example, trace the cause of campus suicides to feelings of depression and hopelessness (Langhinrichsen-Rohling et al. 1998). These feelings are, however, more likely to arise in certain sociological contexts. Thus, suicide is more likely committed by those who have been sexually abused as children or by those whose parents are alcoholics (Thakkar et al. 2000; Bryant and Range 1997).

Altruistic suicide occurs when there is excessive regulation of individuals by social forces. An example is someone who commits suicide for the sake of a religious or political cause. For example, after hijackers took control of four airplanes—crashing two into the World Trade Center in New York, one into the Pentagon, and, through the intervention of passengers, one in a Pennsylvania field—many wondered how anyone could do such a thing, killing themselves in the process. Although sociology certainly does not excuse such behavior, it can help explain it. Terrorists and *suicide bombers* are so regulated by their extreme beliefs that they are willing to die to kill as many people as possible to achieve their goals. As Durkheim argued, altruistic suicide results when individuals are excessively dominated by the expectations of their social group. People who commit altruistic suicide subordinate themselves to collective expectations, even when death is the result.

Egoistic suicide occurs when people feel totally detached from society. This helps explain the high rate of suicide among the elderly in the United States. People over seventy-five years of age have one of the highest rates of suicide, presumably because the elderly lose many of their functional ties to society (National Center for Health Statistics 2008). Ordinarily, people are integrated into society by work roles, ties to family and community, and other social bonds. When these bonds are weakened through retirement or loss of family and friends, the likelihood of egoistic suicide increases. Suicide is also more likely to occur among people who are not well integrated into social networks (Berkman et al. 2000). Thus, it should not be surprising that women have lower suicide rates than men (National Center for Health Statistics 2008). Sociologists explain this fact as a result of men being less embedded in social relationships of care and responsibility than women (Watt and Sharp 2001).

Durkheim's major point is that suicide is a social, not just an individual, phenomenon. Recall from Chapter 1 that Durkheim sees sociology as the discovery of the social forces that influence human behavior. One recent study shows that these social forces that affect behavior such as suicides among youths are *multilevel*: The higher the degree of integration of the

Strong ties among the Navajo produce social integration, resulting in the fact that the Navajo have one of the lowest suicide rates of any group in the United States, and also lowest among other Native American tribal groups.

individual into structural "levels," such as the family, the peer group, religion, the neighborhood, and the school, then the lower the risk of suicide (Mainon and Kuhl 2008). As individualistic as suicide might seem, Durkheim uncovered the influence of social structure even here. In fact, suicide also varies considerably by state (see Map 7.1)—another structural "level."

Durkheim's theory of suicide can help you understand such horrific acts as the Virginia Tech student massacre in 2007. Seung-Hui Cho, a college student, armed with two semiautomatic pistols, shot and killed thirty-two people at Virginia Tech University, wounded fourteen others, and then killed himself, bringing the total killed to thirty-three. This act shocked the nation, and the media instantly dubbed it the largest mass rampage killing in U.S. history. There are social–structural elements that are common to both the Virginia Tech rampage and other "school shootings," such as the rampage killings at Columbine High School in Littleton, Colorado, in 1999. Both acts were committed by individuals who could be characterized as extremely socially isolated and utterly outside a network of peers. All three perpetrators (Cho in the case of Virginia Tech and Dylan Klebold and Eric Harris in the case of Columbine High School) were social isolates, and all three committed suicide immediately after their carnage. In Durkheim's sense, all three instances represented examples of egoistic suicide, given the attributes of social isolation, lack of integration into society, troubled individual histories, and a desire to "make their mark" in history by killing the largest number of individuals possible in a single attack (Newman 2006).

Merton: Structural Strain Theory. The functionalist perspective on deviance has been further elaborated by the sociologist **Robert Merton**

map7.1   *MAPPING AMERICA'S DIVERSITY*

Suicide Rates by State

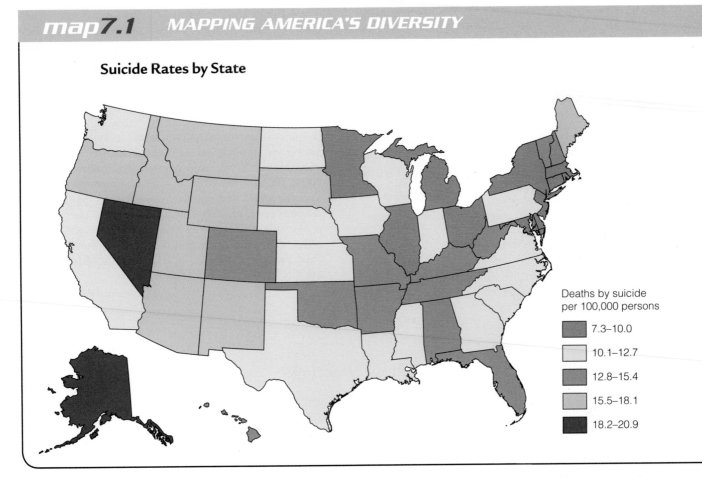

Deaths by suicide
per 100,000 persons

7.3–10.0

10.1–12.7

12.8–15.4

15.5–18.1

18.2–20.9

Many factors can influence the suicide rate in different contexts. As discussed in the text, suicides can be caused by a multiple of structural and cultural factors, and sometimes these factors may be differently distributed by state or region. What are some of the social facts about the different states and regions that might affect the different rates of suicide you see in this map? What, in particular, might you guess about such social facts characterizing the states with the highest suicide rates?
Data: U.S. Census Bureau, 2008. *The 2007 Statistical Abstract: National Data Book.* Washington, DC: U.S. Government Printing Office.

(1910–2003). Merton's **structural strain theory** traces the origins of deviance to the tensions caused by the gap between cultural goals and the means people have available to achieve those goals. Merton noted that societies are characterized by both culture and social structure. Culture establishes goals for people in society; social structure provides, or fails to provide, the means for people to achieve those goals. In a well-integrated society, according to Merton, people use accepted means to achieve the goals society establishes. In other words, the goals and means of the society are in balance. When the means are out of balance with the goals, deviance is likely to occur. According to Merton, this imbalance, or disjunction, between cultural goals and structurally available means can actually compel the individual into deviant behavior (Merton 1968).

To explain further, a collective goal in U.S. society is the achievement of economic success. The legitimate means to achieve such success are education and jobs, but not all groups have equal access to those means. The result is structural strain that produces deviance.

According to Merton, lower-class individuals are most likely to experience these strains because they internalize the same goals and values as the rest of society but have blocked opportunities for success. Structural strain theory therefore helps explain the high correlation that exists between unemployment and crime.

Figure 7.1 illustrates how strain between cultural goals and structurally available means can produce deviance. *Conformity* is likely to occur when the goals are accepted and the means for attaining the goals are made available to the individual by the social structure. If this does not occur, then cultural–structural strain exists, and at least one of four possible forms of deviance is likely to result: innovative deviance, ritualistic deviance, retreatism deviance, or rebellion.

Consider the case of female prostitution: The prostitute has accepted the cultural values of the dominant society—obtaining economic success and material wealth. Yet if she is poor, then the structural means to attain these goals are less available to her and turning to prostitution may result.

	Cultural goals accepted?	Institutionalized means toward goal available?
Conformity	Yes	Yes
Innovative deviance	Yes	No
Ritualistic deviance	No	Yes
Retreatism deviance	No	No
Rebellion	No (old goals) Yes (new goals)	No (old means) Yes (new means)

FIGURE 7.1 *Merton's Structural Strain Theory*

Other forms of deviance also represent strain between goals and means. *Retreatism deviance* becomes likely when neither the goals nor the means are available. Examples of retreatism are the severe alcoholic or the homeless person or the hermit. *Ritualistic deviance* is illustrated in the case of some eating disorders among college women, such as *bulimia* (purging oneself after eating). The cultural goal of extreme thinness is perceived as unattainable, even though the means for trying to attain it are plentiful, for example, good eating habits and proper diet methods (Sharp 2000). Finally, *rebellion* as a form of deviance is likely to occur when new goals are substituted for more traditional ones, and also new means are undertaken to replace older ones, as by force or armed combat. Many right-wing extremist groups, such as the American Nazi Party, "skinheads," and the Ku Klux Klan (KKK), are examples of this type of deviance.

what would they say now?

Classical Theorists Reflect on School Shootings

Sociological theory can help explain the underlying causes of current events. Were some of the classical theorists to be brought back to life, what might they say about school shootings?

Emile Durkheim: As with suicide, people tend to think of most acts of deviance in individualistic terms, thus missing some important sociological dimensions even to horrific individual crimes. Shooters who go on a rampage are often alienated people, living outside a strong support system of peers. Often they are people who have been identified as "deviant" by peers because of their lack of integration into peer networks. The public mourning that follows school shootings, though certainly marked by great individual and collective sadness, can also be seen as a process of social cohesion that unites communities, even in their great sorrow.

Max Weber: School shootings involve the interplay of cultural factors with social structural realities. As a culture, U.S. society has become quite bureaucratic and this can be quite dehumanizing; how else could a person kill so many other human beings, unless the culture has made him numb to his actions? But, at the same time, there is an economic ethic in this society—one that tells people they are most valued when they are successful. Young people facing such an interplay of culture and the economic structure may feel so devalued that they can be led by these multiple social forces into behaviors that would otherwise be unimaginable.

Karl Marx: Guns, and violence more generally, are highly profitable businesses in this society. Just look at the sales figures for things like Grand Theft Auto—the video game where people spend money to shoot and rape people for fun. Why would we not expect this to shape actual behavior? Ultimately, a capitalist society has to treat people like things because what really drives the society is a system of profit—here people are mere cogs in the capitalist system. Only by knowing this can you understand how people could go on such angry rampages as has happened in school shootings!

W.E.B. Du Bois: It is no accident that most of the mass school shootings have been committed by young, White men. I think they are, in part, angry about their class and race status in society. In the contemporary culture of White youth, "Black" is cool, White is not. And note that some of the targets of school shootings were "preppies," suggesting that class anger is part of the context of these terrible events. Of course, gun violence is also a major social problem in African American, low-income communities, but it takes a different form than we have seen in mass school shootings and rarely generates the kind of sympathy and empathy that has followed mass school shootings.

Social Control Theory. Taking functionalist theory in another direction, Travis Hirschi has developed social control theory to explain deviance. **Social control theory,** a type of functionalist theory, suggests that deviance occurs when a person's (or group's) attachment to social bonds is weakened (Hirschi 1969; Gottfredson and Hirschi 1995, 1990). According to this view, people internalize social norms because of their attachments to others. People care what others think of them and therefore conform to social expectations because they accept what people expect. You can see here that social control theory, like the functionalist framework from which it stems, assumes the importance of the socialization process in producing conformity to social rules. When that conformity is broken, deviance occurs.

Social control theory assumes there is a common value system within society, and breaking allegiance to that value system is the source of social deviance. This theory focuses on how deviants are attached (or not) to common value systems and what situations break people's commitment to these values. Social control theory suggests that most people probably feel some impulse toward deviance at times but that the attachment to social norms prevents them from actually participating in deviant behavior. Sociologists find that juveniles whose parents exercise little control over violent behavior and who learn violence from aggressive peers are most likely to engage in violent crimes (Heimer 1997), as was the case with the two teenagers who killed twelve students and a teacher at Columbine High School (Newman 2006).

Functionalism: Strengths and Weaknesses. Functionalism emphasizes that social structure, not just individual motivation, produces deviance. Functionalists argue that social conditions exert pressure on individuals to behave in conforming or nonconforming ways. Types of deviance are linked to one's place in the social structure; thus, a poor person blocked from economic opportunities may use armed robbery to achieve economic goals, whereas a stockbroker may use insider trading to achieve the same. Functionalists acknowledge that people choose whether to behave in a deviant manner but believe that they make their choice from among socially prestructured options. The emphasis in functionalist theory is on social structure, not individual action. In this sense, functionalist theory is highly sociological.

Functionalists also point out that what appears to be dysfunctional behavior may actually be functional for the society. An example is the fact that most people consider prostitution to be dysfunctional behavior; from the point of view of an individual, that is true: It demeans the women who engage in it, puts them at physical risk, and subjects them to sexual exploitation. From the view of functionalist theory, however, prostitution supports and maintains a social system that links women's gender roles with sexuality, associates sex with commercial activity, and defines women as passive sexual objects and men as sexual aggressors. In other words, what appears to be deviant may actually serve various purposes for society.

Critics of the functionalist perspective argue that it does not explain how norms of deviance are first established. Despite its analysis of the ramifications of deviant behavior for society as a whole, functionalism does little to explain why some behaviors are defined as normative and others as illegitimate. Who determines social norms and on whom such judgments are most likely to be imposed are questions seldom asked by anyone using a functionalist perspective. Functionalists see deviance as having stabilizing consequences in society, but they tend to overlook the injustices that labeling someone deviant can produce. Others would say that the functionalist perspective too easily assumes that deviance has a positive role in society; thus, functionalists rarely consider the differential effects that the administration of justice has on different groups. The tendency in functionalist theory to assume that the system works for the good of the whole too easily ignores the inequities in society and how these inequities are reflected in patterns of deviance. These issues are left for sociologists who work from the perspectives of conflict theory and symbolic interaction.

Conflict Theories of Deviance

Recall that conflict theory emphasizes the unequal distribution of power and resources in society. It links the study of deviance to social inequality. Based on the work of Karl Marx (1818–1883; see Chapter 1), conflict theory sees a dominant class as controlling the resources of society and using its power to create the institutional rules and belief systems that support its power. Like functionalist theory, conflict theory is a *macrostructural* approach; that is, both theories look at the structure of society as a whole in developing explanations of deviant behavior.

Because some groups of people have access to fewer resources in capitalist society, they are forced into crime to sustain themselves. Conflict theory posits that the economic organization of capitalist societies produces deviance and crime. The high rate of crime among the poorest groups, especially economic crimes such as theft, robbery, prostitution, and drug selling, are a result of the economic status of these groups. Rather than emphasizing values and conformity as a source of deviance as do functional analyses, conflict theorists see crime in terms of power relationships and economic inequality (Grant and Martínez 1997).

The upper classes, conflict theorists point out, can also better hide crimes they commit because affluent groups have the resources to mask their deviance and crime. As a result, a working-class man who beats his wife is more likely to be arrested and prosecuted than an upper-class man who engages in the same behavior. In addition, those with greater resources can afford to buy their way out of trouble by paying bail, hiring expensive attorneys, or even resorting to bribes.

Corporate crime is crime committed within the legitimate context of doing business. Conflict theorists expand our view of crime and deviance by revealing the significance of such crimes. They argue that appropriating profit based on exploitation of the poor and working class is inherent in the structure of capitalist society. **Elite deviance** refers to the wrongdoing of wealthy and powerful individuals and organizations (Simon 2007). Elite deviance includes what early conflict theorists called *white-collar crime* (Sutherland 1940; Sutherland and Cressey 1978). Elite deviance includes tax evasion; illegal campaign contributions; corporate scandals, such as fraudulent accounting practices that endanger or deceive the public but profit the corporation or individuals within it; and even government actions that abuse the public trust. Several examples of elite deviance are covered in detail later in this chapter.

The ruling groups in society develop numerous mechanisms to protect their interests according to conflict theorists who argue that law, for example, is created by elites to protect the interests of the dominant class. Thus law, supposedly neutral and fair in its form and implementation, works in the interest of the most well-to-do (Weisburd et al. 2001, 1991; Spitzer 1975). Another way that conflict theorists see dominant groups as using their power is through the excessive regulation of populations that are a potential threat to affluent interests. Periodically sweeping the homeless off city streets, especially when there is a major political event or other elite event occurring, is a good example.

Conflict theory emphasizes the significance of social control in managing deviance and crime. **Social control** is the process by which groups and individuals within those groups are brought into conformity with dominant social expectations. Social control, as we saw in Chapter 4, can take place simply through socialization, but dominant groups can also control the behavior of others through marking them as deviant. An example is the historic persecution of witches during the Middle Ages in Europe and during the early colonial period in America (Ben-Yehuda 1986; Erikson 1966). Witches often were women who were healers and midwives—those whose views were at odds with the authority of the exclusively patriarchal hierarchy of the church, then the ruling institution. "Witch hunt" is a term still used today to refer to the aggressive pursuit of those who dissent from prevailing political and social norms.

One implication of conflict theory, especially when linked with labeling theory, is that the power to define deviance confers an important degree of social control. **Social control agents** are those who regulate and administer the response to deviance, such as the police and mental health workers. Members of powerless groups may be defined as deviant for even the slightest infraction against social norms, whereas others may be free to behave in deviant ways without consequence. Oppressed groups may actually engage in more deviant behavior, but it is also true that they have a greater likelihood of being labeled deviant and incarcerated or institutionalized, whether or not they have actually committed an offense. This is evidence of the power wielded by social control agents.

When powerful groups hold stereotypes about other groups, the less powerful people are frequently assigned deviant labels. As a consequence, the least powerful groups in society are subject most often to social control. You can see this in the patterns of arrest data. Poor people are more likely to be considered criminals and therefore more likely to be arrested, convicted, and imprisoned than middle- and upper-class people. The same is true of Latinos, Native Americans, and African Americans. Sociologists point out that this does not necessarily mean that these groups are somehow more criminally prone; rather, they take it as evidence of the differential treatment of these groups by the criminal justice system.

Conflict Theory: Strengths and Weaknesses.

The strength of conflict theory is its insight into the significance of power relationships in the definition, identification, and handling of deviance. It links the commission, perception, and treatment of crime to inequality in society and offers a powerful analysis of how the injustices of society produce crime and result in different systems of justice

From the point of view of conflict theory, social control agents play a significant role in defining and causing deviant behavior, as in this case of cocaine use.

© David Young-Wolff/PhotoEdit

for disadvantaged and privileged groups. Not without its weaknesses, however, critics point out that laws protect most people, not just the affluent, as conflict theorists argue.

In addition, although conflict theory offers a powerful analysis of the origins of crime, it is less effective in explaining other forms of deviance. For example, how would conflict theorists explain the routine deviance of middle-class adolescents? They might point out that much of middle-class deviance is driven by consumer marketing. Profits are made from the accoutrements of deviance—rings in pierced eyebrows, "gangsta" rap music, and so on—but economic interests alone cannot explain all the deviance observed in society. As Durkheim argued, deviance is functional for the whole of society, not just those with a major stake in the economic system.

Symbolic Interaction Theories of Deviance

Whereas functionalist and conflict theories are *macrosociological* theories, certain *microsociological* theories of deviance look directly at the interactions people have with one another as the origin of social deviance. *Symbolic interaction theory* holds that people behave as they do because of the meanings people attribute to situations (see Chapter 1). This perspective emphasizes the meanings surrounding deviance, as well as how people respond to those meanings. Symbolic interaction emphasizes that deviance originates in the interaction between different groups and is defined by society's reaction to certain behaviors. Symbolic interactionist theories of deviance originate in the perspective of the Chicago School of sociology.

W. I. Thomas and the Chicago School. **W. I. Thomas** (1863–1947), one of the early sociologists from the University of Chicago, was among the first to develop a sociological perspective on social deviance. Thomas explained deviance as *a normal response to the social conditions in which people find themselves.* Thomas was one of the first to argue that delinquency was caused by the social disorganization brought on by slum life and urban industrialism; he saw deviance as a problem of social conditions, not individual character.

Differential Association Theory. Thomas's work laid the foundation for a classic theory of deviance: differential association theory. **Differential association theory,** a type of symbolic interaction theory, interprets deviance, including criminal behavior, as behavior one learns through interaction with others (Sutherland 1940; Sutherland and Cressey 1978). Edwin Sutherland argued that becoming a criminal or a juvenile delinquent is a matter of learning criminal ways within the primary groups to which one belongs. To Sutherland, people become criminals when they are more strongly socialized to break the law than to

obey it. Differential association theory emphasizes the interaction people have with their peers and others in their environment. Those who "differentially associate" with delinquents, deviants, or criminals learn to value deviance. The greater the frequency, duration, and intensity of their immersion in deviant environments, the more likely it is that they will become deviant.

Consider the career path of con artists and hustlers. Hustlers seldom work alone. Like any skilled worker, they have to learn the "tricks of the trade." A new recruit becomes part of a network of other hustlers who teach the recruit the norms of the deviant culture (Prus and Sharper 1991). Crime also tends to run in families. This does not necessarily mean that crime is passed on in genes from parent to child. It means that youths raised in deviant families are more likely socialized to become deviant themselves (Miller 1986). Differential association theory offers a compelling explanation for how deviance is culturally transmitted—that is, people pass on deviant expectations through the social groups in which they interact, of which the family is but one.

Critics of differential association theory have argued that this perspective tends to blame deviance on the values of particular groups. Differential association has been used, for instance, to explain the higher rate of crime among the poor and working class, arguing that this higher rate of crime occurs because they do not share the values of the middle class. Such an explanation, critics say, is class-biased, because it overlooks the deviance that occurs in the middle-class culture and among elites. Disadvantaged groups may share the values of the middle class but cannot necessarily achieve them through legitimate means (a point, you will remember, made by Merton's structural strain theory). Still, differential association theory offers a good explanation of why deviant activity may be more common in some groups than others, and it emphasizes the significant role that peers play in encouraging deviant behavior.

Labeling Theory. **Labeling theory,** a branch of symbolic interaction theory, interprets the responses of others as the most significant factor in understanding how deviant behavior is both created and sustained (Becker 1963). Labeling theory stems from the work of W. I. Thomas, who it will be recalled wrote, "If men define situations as real, they are real in their consequences" (Thomas and Thomas 1928: 572). A *label* is the assignment or attachment of a deviant identity to a person by others, including by agents of social institutions; therefore, people's reactions, not the action itself, produce deviance as a result of the labeling process. Once applied, the deviant label is difficult to shed.

Linked with conflict theory, labeling theory shows how those with the power to label an act or a person deviant and to impose sanctions—such as police, court officials, school authorities, experts, teachers, and official agents of social institutions—wield great power in determining societal understandings of deviance. When they apply the "deviant" label, it sticks. Furthermore, because deviants are handled through bureaucratic organizations, bureaucratic workers "process" people according to rules and procedures, seldom questioning the basis for those rules or willing or able to challenge them (Cicourel 1968; Kitsuse and Cicourel 1963; Margolin 1992; Montada and Lerner 1998).

Once the label is applied, it is difficult for the deviant to recover a nondeviant identity. To give an example, once a social worker or psychiatrist labels a client mentally ill, that person will be treated as mentally ill, regardless of his or her actual mental state. Pleas by the accused that he or she is mentally sound are typically taken as further evidence of mental illness! Once labeled, a person may have great difficulty changing his or her classification; the label itself has consequences.

A person need not have actually engaged in deviant behavior to be labeled deviant; yet, once applied, the label has social consequences. Labeling theory helps explain why convicts released from prison have such high rates of *recidivism* (return to criminal activities). Convicted criminals are formally and publicly labeled wrongdoers. They are treated with suspicion ever afterward and have great difficulty finding legitimate employment: The label "ex-con" defines their future options.

It is in fact exceedingly difficult for an ex-con to find employment after release from prison, even more so if the person is male and Black or Hispanic. In a clever study, Pager (2007) had pretrained role-players pose as ex-cons looking for a job. These role-players went into the job market and were interviewed for various jobs; all of them used the same preset script during the interview. The idea of the study was to see how many of them would be invited back for another interview. The results were staggering: A very small percentage of the male Black and Hispanic role-players were invited back for interviews, yet a large percentage of the male White role-players were. Blacks who were *not* ex-cons were *less* likely to be invited back for the job interview than were Whites who *were* ex-cons! So the effect of race alone exceeded the effect of incarceration alone. These upsetting differences could not be attributed to differences in interaction displayed during the interview, because everyone used the exact same prepared script.

Researchers Bruce Western (2006) and Jeffrey Reiman (2007) note that the prison system in the United States is in effect designed to *train and socialize* prisoners into a career of secondary deviance and to tell the public that crime is a threat primarily from the poor (see the box, "Understanding Diversity: The Rich Get Richer and the Poor Get Prison," on page 171). Reiman sees that the goal of the prison system is not to reduce crime but to impress upon the public that crime is inevitable and that it originates only from the lower classes. Prisons accomplish this, even if unintentionally, by demeaning prisoners, not training them in marketable skills, and stigmatizing them as different from "decent citizens." As a consequence, the person will never be able to pay his or her debt to society, and the prison system has created the very behavior it intended to eliminate.

Labeling theory suggests that deviance refers not just to something one does but to something one becomes. **Deviant identity** is the definition a person has of himself or herself as a deviant. Most often, deviant identities emerge over time (Lemert 1972; Simon 2007). A drug addict, for example, may not think of herself as a junkie until she realizes she no longer has non-using friends. The formation of a deviant identity, like other identities, involves a process of social transformation in which a new self-image and new public definition of a person emerges. This is a process that involves how people view the deviant and how the deviant views himself or herself. Studies of tattoo "collectors" (that is, those who are very heavily tattooed) find, for example, that if collectors first learn to interpret tattooing as a desirable thing, they then begin to feel connected to a subculture of other collectors and eventually come to see their tattoos as part of themselves (Irwin 2001; Vail 1999; Montada and Lerner 1998; Scheff 1984).

thinkingsociologically

Perform an experiment by doing something deviant for a period, such as carrying around a teddy bear doll and treating it as a live baby or standing in the street and looking into the air, as though you are looking at something up there. Make a record of how others respond to you, and then ask yourself how *labeling theory* is important to the study of *deviance*. Then take your experiment a step further and ask yourself how people's reactions to you might have differed had you been of another race or gender. You might want to structure this question into your experiment by teaming up with a classmate of another race or gender. You could then compare responses to the same behavior by both of you. *A note of caution:* Do not do anything illegal or dangerous; even the most seemingly harmless acts of deviance can generate strong (and sometimes hostile) reactions, so be careful in planning your experiment!

Deviant Careers. In the ordinary context of work, a career is the sequence of movements a person makes through different positions in an occupational system (Becker 1963). A **deviant career**—a direct outgrowth of the labeling process—is the sequence of movements people make through a particular subculture of deviance. Deviant careers can be studied sociologically, like any other career. Within deviant careers, people are socialized into new "occupational" roles and encouraged, both materially and psychologically, to engage in deviant behavior. The concept of a deviant career emphasizes that there is a progression through deviance: Deviants are recruited, given or denied rewards, and promoted or demoted. As with legitimate careers, deviant careers involve an evolution in the person's identity, values, and commitment over time. Deviants, like other careerists, may have to demonstrate their commitment to the career to their superiors, perhaps by passing certain tests of their mettle, such as when a gang expects new members to commit a crime, perhaps even shoot someone.

Within deviant careers, there may be rites of passage that bring increased social status among peers. Punishments administered by the authorities may even become badges of honor within a deviant community. Similarly, labeling a teenager "bad" for behavior that others think is immoral may actually encourage the behavior to continue because the juvenile may take this as a sign of success as a deviant.

Deviant Communities. The preceding discussion continues to indicate an important sociological point: Deviant behavior is not just the behavior of maladjusted individuals; it often takes place within a group context and involves group response. Some groups are actually organized around particular forms of social deviance; these are called **deviant communities** (Becker 1963; Erikson 1966; Blumer 1969; Mizruchi 1983).

Like subcultures and countercultures, deviant communities maintain their own values, norms, and rewards for deviant behavior. Joining a deviant community closes one off from conventional society and tends to solidify deviant careers because the deviant individual receives rewards and status from the in-group. Disapproval from the out-group may only enhance one's status within. Deviant communities also create a worldview that solidifies the deviant identity of their members. They may develop symbolic systems such as emblems, forms of dress, publications, and other symbols that promote their identity as a deviant group. Gangs wear their "colors," prostitutes have their own vocabulary of *tricks* and *johns;* skinheads have their insignia and music. All are examples of deviant communities. Ironically, subcultural norms and values reinforce the deviant label both inside and outside the deviant group, thereby reinforcing the deviant behavior.

Some deviant communities are organized specifically to provide support to those in presumed deviant categories. Groups such as Alcoholics Anonymous, Weight Watchers, and various twelve-step programs help those identified as deviant overcome their deviant behavior. These groups, which can be quite effective, accomplish their mission by encouraging members to accept their deviant identity as the first step to recovery.

A Problem with Official Statistics. Because labeling theorists see deviance as produced by those with

Some deviance develops in deviant communities, such as the Neo-Nazis/"skin-heads" shown marching here. Such right-wing extremist groups have remained relatively constant in numbers for the last several years.

the power to assign labels, they question the value of official statistics as indicators of the true extent of deviance. *Reported rates of deviant behavior are themselves the product of socially determined behavior, specifically the behavior of identifying what is deviant.* Official rates of deviance are produced by people in the social system who define, classify, and record certain behaviors as deviant and others as legitimate. Labeling theorists are more likely to ask how behavior becomes labeled deviant than they are to ask what motivates people to become deviant (Best 2001; Kitsuse and Cicourel 1963).

For example, in the aftermath of the terrorist attacks on the World Trade Center, officials debated whether to count the deaths of thousands as murder or as a separate category of terrorism. The decision would change the official rate of deviance by inflating or deflating the reported crime rate of murder in New York City in that year. In the end, these deaths were not counted in the murder rate. Labeling theorists think that official rates of deviance do not necessarily reflect the actual commission of crimes or deviant acts; instead, the official rates reflect social judgments.

In another example, official rape rates are underestimates of the actual extent of rape, largely due to victims' reluctance to report. Also, some rapes are less likely to be "counted" as rape by police, such as if the victim is a prostitute, was drunk at the time of the assault, or had a previous relationship with the assailant. Moreover, rapes resulting in death are classified as homicides and therefore do not appear in the official statistics on rape. Discrimination by the police can also influence official statistics; thus, higher arrest rates for African American men tell us as much about police behavior as it does the behavior of African American men (Babbie 2007; DeFleur 1975). Given such problems, any official statistics must be interpreted with caution.

Labeling Theory: Strengths and Weaknesses. The strength of labeling theory is its recognition that the judgments people make about presumably deviant behavior have powerful social effects. Labeling theory does not, however, explain why deviance occurs in the first place. It may illuminate the consequences of a young man's violent behavior, but it does not explain the actual origins of the behavior. Put bluntly, it does not explain why some people become deviant and others do not.

Forms of Deviance

Although there are many forms of deviance, the sociology of deviant behavior has focused heavily on subjects such as mental illness, social stigmas, and crime. As we review each, you will also see how the different sociological theories about deviance contribute to understanding each subject. In addition, you will see how the social context of race, class, and gender relationships shape these different forms of deviance.

Mental Illness

Sociological explanations of mental illness look to the social systems in which mental illness is defined, identified, and treated, even though it is typical for many to think of mental illness only in psychological terms. This has several implications for understanding mental illness. Functionalist theory suggests, for example, that by recognizing mental illness, society also upholds normative values about more conforming behavior. Symbolic interaction theory tells us that mentally ill people are not necessarily "sick," but rather are the victims of societal reactions to their behavior. Some go so far as to say there is no such thing as mental illness, only people's reactions to unusual behavior. From this point of view, people learn faulty self-images and then are cast into the role of patient when they are treated

This man's extreme spiked hairdo shows deviant behavior in society but rigid conformity to the norms of the peer group or deviant community.

© Keith Erskine/Alamy

by therapists. Once someone is labeled as a "patient," he or she is forced into the "sick" role, as expected by those who reinforce it, and it becomes difficult to get out of the role (Szasz 1974).

Labeling theory, combined with conflict theory, suggests that those people with the fewest resources are most likely to be labeled mentally ill. Women, racial minorities, and the poor all suffer higher rates of reported mental illness and more serious disorders than do groups of higher social and economic status. Furthermore, research over the years has consistently shown that middle- and upper-class persons are more likely to receive some type of psychotherapy for their illness. Poorer individuals and minorities are more likely to receive only physical rehabilitation and medication, with no accompanying psychotherapy (Hollingshead and Redlich 1958; Simon 2007).

Debunking Society's Myths

Myth: Mental illness is an abnormality best studied exclusively by psychologists and physicians.

Sociological perspective: Mental illness follows patterns associated with race, class, and gender relations in society and is subject to a significant labeling effect. Those who study and treat mental illness benefit from combining a sociological perspective with both medical and psychological knowledge.

Sociologists give two explanations for the correlation between social status and mental illness. On the one hand, the stresses of being in a low-income group, being a racial minority, or being a woman in a sexist society all contribute to higher rates of mental illness; the harsher social environment is a threat to mental health. On the other hand, the same behavior that is labeled mentally ill for some groups may be tolerated and not so labeled in others. For example, behavior considered crazy in a homeless woman (who is likely to be seen as "deranged") may be seen as merely eccentric or charming when exhibited by a rich person.

Social Stigmas

A **stigma** is an attribute that is socially devalued and discredited. Some stigmas result in people being labeled deviant. The experiences of people who are disabled, disfigured, or in some other way stigmatized are studied in much the same way as other forms of social deviance. Like other deviants, people with stigmas are stereotyped and defined only in terms of their presumed deviance.

Think, for example, of how disabled people are treated in society. Their disability can become a

master status (Chapter 5), a characteristic of a person that overrides all other features of the person's identity (Goffman 1963). Physical disability can become a master status when other people see the disability as the defining feature of the person; a person with a disability becomes "that blind woman" or "that paralyzed guy." People with a particular stigma are often seen to be all alike. This may explain why stigmatized individuals of high visibility are often expected to represent the whole group.

People who suddenly become disabled often have the alarming experience of their new master status rapidly erasing their former identity. They may be treated and seen differently by people they know. A master status may also prevent people from seeing other parts of a person. A person with a disability may be assumed to have no meaningful sex life, even if the disability is unrelated to sexual ability or desire. Sociologists have argued that the negative judgments made about people with stigmas tend to confirm the "usualness" of others (Goffman 1963: 3). For example, when welfare recipients are stigmatized as lazy and undeserving of social support, others are indirectly promoted as industrious and deserving.

Stigmatized individuals are measured against a presumed norm and may be labeled, stereotyped, and discriminated against. In Goffman's words, people with stigmas are perceived to have a "spoiled identity." Seen by others as deficient or inferior, they are caught in a role imposed by the stigma.

Sometimes, people with stigmas bond with others, perhaps even strangers. This can involve an acknowledgment of "kinship" or affiliation that can be as subtle as an understanding look, a greeting that makes a connection between two people, or a favor extended to a stranger who the person sees as sharing the presumed stigma. Public exchanges are common between various groups that share certain forms of disadvantage, such as people with disabilities, lesbians and gays, or members of other minority groups.

Substance Abuse: Drugs and Alcohol

As with mental illness and stigmas, sociologists study the social factors that influence drug and alcohol use. Who uses what and why? How are users defined by others? These questions guide sociological research on substance abuse.

One of the first things to ask when thinking about drugs and alcohol is why using one substance is considered deviant and stigmatizing and using another is not. How do such definitions of deviance change over time?

For example, alcohol is a legal drug. Whether one is labeled an alcoholic depends in large part on the social context in which one drinks, not solely on the amount of alcohol consumed. For years, the businessman's lunch where executives drank two or three martinis was viewed as normative. Drinking wine from a bottle in a brown bag on the street corner is considered highly deviant; having martinis in a posh bar is seen as cool—even though one martini contains considerably more alcoholic content than a glass of wine.

Sociological understandings challenge views of drug and alcohol use as stemming solely from inherent individual propensities that lead to substance abuse. Patterns of use vary by factors such as age, gender, and race (see Figure 7.2). Age is one significant predictor of illegal drug use. Young people are on average more likely to use marijuana and cocaine and binge drink than are people who are somewhat older, though there are, of course, exceptions.

Crime and Criminal Justice

The concept of deviance in sociology is a broad one, encompassing many forms of behavior—both legal and illegal, ordinary and unusual. **Crime** is one form of deviance, specifically behavior that violates

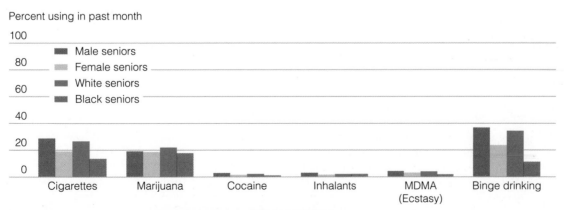

FIGURE 7.2 *Use of Selected Substances by High School Seniors*

Source: National Center for Health Statistics. 2008. *Health United States 2008.* Hyattsville, MD: U.S. Department of Health and Human Services.

particular criminal laws. Not all deviance is crime. Deviance becomes crime when it is designated by the institutions of society as violating a law or laws. *Deviance* is behavior that is recognized as violating rules and norms of society. Those rules may be formal laws, in which case the deviant behavior would be called *crime,* or informal customs or habits, in which case the deviant behavior would not be called crime.

Criminology is the study of crime from a scientific perspective. Criminologists include social scientists such as sociologists who stress the societal causes and treatment of crime. All the theoretical perspectives on deviance that we examined earlier contribute to our understanding of crime (see Table 7.2). According to the functionalist perspective, crime may be *necessary* to hold society together—a profound hypothesis. By singling out criminals as socially deviant, others are defined as good. The nightly reporting of crime on television is a demonstration of this sociological function of crime. Conflict theory suggests that disadvantaged groups are more likely to become criminal; it also sees the well-to-do as better able to hide their crimes and less likely to be punished. Symbolic interaction helps us understand how people learn to become criminals or come to be accused of criminality, even when they may be innocent. Each perspective traces criminal behavior to social conditions rather than only to the intrinsic tendencies or personalities of individuals.

Measuring Crime: How Much Is There?

Is crime increasing in the United States? One would certainly think so from watching the media. Images of violent crime abound and give the impression that crime is a constant threat and is on the rise. Data on crime actually show that violent crime peaked in 1990, but *decreased* through the 1990s and leveled off a bit through 2005–2007 (see Figure 7.3). Assault and robbery, in particular, decreased quite significantly through the 1990s. Murder and rape remained more constant, though they too have shown some decline since the 1990s.

Data about crime come from the Federal Bureau of Investigation (FBI) based on reports from police departments across the nation. The data are distributed annually in the *Uniform Crime Reports* and are the basis for official reports about the extent of crime and its rise and fall over time. *These data show that although media coverage of crime—especially as reported in TV news—has remained high and about the same, the officially reported rate of crime has decreased.*

A second major source of crime data is the *National Crime Victimization Surveys* published by the Bureau of Justice Statistics in the U.S. Department of Justice. These data are based on surveys in which national samples of people are periodically asked if they have been the victims of one or more criminal acts (see Figures 7.4 and 7.5 on pages 171–172). These surveys also show that violent crime, including rape, assault, robbery, and murder, declined by 15 percent over the 1990s.

Both of these sources of data—the *Uniform Crime Reports* and the *National Crime Victimization Surveys*—are subject to the problem of underreporting. About half to two-thirds of all crimes may not be reported to police, meaning that much crime never shows up in the official statistics. Certain serious crimes, such as rape, are significantly underreported, as we have already noted. Victims may be too upset to report a rape to the police, or they may believe that the police will not believe a rape has occurred. Equally significant, the victim may not want to undergo the continued emotional stress of an investigation and trial. Recall from earlier in this chapter that certain kinds of noncriminal deviance, such as suicide, are also underreported, particularly by upper-income families, because of embarrassment to the deceased person's family.

Another problem arises in the attempt to measure crime by means of official statistics. The FBI's *Uniform Crime Reports* stress what are called **index crimes,** which include the violent crimes of murder, manslaughter, rape, robbery, and aggravated assault, plus property crimes of burglary, larceny-theft, and motor

table 7.2 *Sociological Theories of Deviance*

Functionalist Theory	Symbolic Interaction Theory	Conflict Theory
Societies require a certain level of crime in order to clarify norms.	Crime is behavior that is learned through social interaction.	The lower the social class, the more the individual is *forced* into criminality.
Crime results from social structural strains (such as class inequality) within society.	Labeling criminals tends to reinforce rather than deter crime.	Inequalities in society by race, class, gender, and other forces tend to produce criminal activity.
Crime may be functional to society, thus difficult to eradicate.	Institutions with the power to label, such as prisons, actually produce rather than lessen crime.	Reducing social inequality in society is likely to reduce crime.

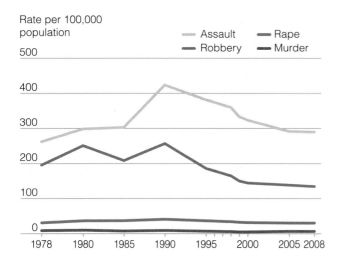

FIGURE 7.3 *Violent Crime in the United States*

Source: Federal Bureau of Investigation, 2009. *Uniform Crime Reports.* Washington, DC: U.S. Department of Justice. **http://www.fbi.gov**

vehicle theft. These crimes are committed mostly by individuals who are disproportionately Black, Hispanic, and poor. Statistics based on these offenses do not reflect the crimes that tend to be committed by middle-class and upper-class persons, such as tax violations, insider trading, fraudulent investment and accounting schemes, embezzlement, and other so-called elite crimes. The official statistics provide a relatively inflated picture for index crimes but an underreported picture of elite crimes, giving a biased picture of crime.

A final result is, unfortunately, that the public sees the stereotypic "criminal" as a lower-class person, most likely an African American or Latino male, not as a middle- or upper-class White person who has committed tax fraud. The official statistics give biased support to the stereotype. This in turn perpetuates the public belief that the "typical" criminal is lower class and minority instead of upper class and nonminority.

a sociological eye on the media

Images of Violent Crime

The media routinely drive home two points to the consumer: violent crime is always high and may be increasing over time and there is much random violence constantly around us. The media bombard us with stories of "wilding," in which bands of youths kill random victims. Many of us think road rage is extensive (which it is not) and completely random. The media vividly and routinely report such occurrences as pointless, random, and probably increasing.

The evidence shows that although violent crime in the United States increased during the 1970s and 1980s, it nonetheless began to decrease in 1990 and continues to decrease nationally through the present. For example, both robbery and physical assaults have declined dramatically since 1990 (see Figure 7.3). Yet according to research (Best 2007, 1999; Glassner 1999), the media have consistently given a picture that violent crime has increased during this same period and, furthermore, that the violence is completely unpatterned and random.

No doubt there are occasions when victims are indeed picked at random. But the statistical rule of randomness could not possibly explain what has come to be called *random violence,* a vision of patternless chaos that is advanced by the media. If randomness truly ruled, *then each of us would have an equal chance of being a victim*—and of being a criminal. This is assuredly not the case. The notion of random violence, and the notion that it is increasing, ignores virtually everything that criminologists, psychologists, sociologists, and extensive research studies know about crime: It is highly patterned and significantly predictable, beyond sheer chance, by taking into account the social structure, social class, location, race–ethnicity, gender, labeling, age, whom one's family members are, and other such variables and forces in society that affect both criminal and victim. The broad picture, then, is clearly not conveyed in the media. Some have speculated that the picture maintained in the media of increasing crime is simply a tool to increase viewer ratings. But criminal violence is not increasing, but decreasing, and it is not random, but highly patterned and even predictable.

Personal and Property Crimes. The *Uniform Crime Reports* are subject to the same biases in official statistics mentioned earlier, but they are the major source of information on patterns of crime and arrest, with crimes classified into four categories. **Personal crimes** are violent or nonviolent crimes directed against people. Included in this category are murder, aggravated assault, forcible rape, and robbery. As we see in Figure 7.3, aggravated assault is the most frequently reported personal crime.

Hate crimes refer to assaults and other malicious acts (including crimes against property) motivated by various forms of social bias, including that based on race, religion, sexual orientation, ethnic/national origin, or disability. This form of crime has been increasing in recent years, especially against gays and lesbians (Jenness and Broad 2002).

Property crimes involve theft of property without threat of bodily harm. These include burglary (breaking and entering), larceny (the unlawful taking of property, but without unlawful entry), auto theft, and arson. Property crimes are the most frequent criminal infractions.

Finally, **victimless crimes** violate laws but are not listed in the FBI's serious crime index. These include illicit activities, such as gambling, illegal drug use, and prostitution, in which there is no complainant. Nonetheless, there is clearly at least some degree of victimization in such crimes: Some researchers see in many instances prostitution as containing at least one victim—the prostitute herself. Enforcement of these crimes is typically not as rigorous as the enforcement of crimes against persons or property, although periodic crackdowns occur, such as the current policy of mandatory sentencing for drug violations.

Elite and White-Collar Crime. Sociologists use the term *white-collar crime* to refer to criminal activities by people of high social status who commit their crimes in the context of their occupation (Sutherland 1940; Sutherland and Cressey 1978). White-collar crime includes activities such as embezzlement (stealing funds from one's employer), involvement in illegal stock manipulations (insider trading), and a variety of violations of income tax law, including tax evasion. Also included are manipulations of accounting practices to make one's company appear profitable, thus artificially increasing the value of the company's stock.

Until very recently, white-collar crime seldom generated great concern in the public mind—far less than the concern directed at street crime. In terms of total dollars, however, white-collar crime is even more consequential for society. Scandals involving prominent white-collar criminals come to the public eye occasionally, such as the Ponzi scheme by white-collar criminal Bernard Madoff in 2009. Madoff ran a **Ponzi scheme**—*a con game* whereby a central person (Mr. Madoff) collects money from a large number of people, including friends and relatives, and then promises to invest their dollars with a high rate of interest for them. In Madoff's case, he promised a 10 percent rate of annual return, a very high rate, even while the U.S. economy was souring. Actually, the money was never invested at all, but was used to pay off earlier investors. This is the key principle of a Ponzi scheme. In the Madoff case, investors were led to believe that their money was under the competent control of Madoff himself, and because they trusted him, they never saw any records or stock certificates. For the scheme to work, Madoff had to convince new recruits that they could make a great deal of money if they just "left it to him." In the meantime Madoff siphoned off a portion of the collected funds for himself and lived lavishly.

Such schemes are often called *pyramid schemes*, with large numbers of recent investors at the bottom of the pyramid and smaller numbers of the older (original) investors at the top. Such illegal schemes are named for one Carlo Ponzi, an immigrant who invented and perfected it in the 1920s in the United States.

When arrested early in 2009, Madoff had processed over the years about $50 billion—that is $50 *billion*, not $50 million! Some (not all) of the early investors received annual interest, but the vast majority of all investors were simply told by Madoff that their money was safe with him. Madoff will spend the rest of his life in prison for his efforts. His scheme has been called the largest Ponzi scheme ever perpetrated in United States history. Madoff leaves in his crumbled wake hundreds of former friends and clients, several of whom are quite famous, such as New Jersey Senator Frank

Bernard Madoff, master-mind of a massive "Ponzi" investment scam that bilked over $50 billion from unsuspecting investors.

Brendan McDermid/Reuters/Landov

Lautenberg, Hall of Fame baseball pitcher Sandy Koufax, and actor John Malkovitch (*The New York Times* 2009). Many former clients lost everything—their life savings, retirement accounts, and other investments—so much so that many universities, charities, and other organizations have felt the weight of the pyramid's collapse.

Organized Crime

The structure of crime and criminal activity in the United States often takes on an organized, almost institutional character. This is crime in the form of mob activity and racketeering, known as organized crime. Also, there are crimes committed by bureaucracies, known as corporate crime. Both types of crime are so highly organized, complex, and sophisticated that they take on the nature of social institutions.

Organized crime is crime committed by structured groups typically involving the provision of illegal goods and services to others. Organized crime syndicates are typically stereotyped as the Mafia, but the term can refer to any group that exercises control over large illegal enterprises, such as the drug trade, illegal gambling, prostitution, weapons smuggling, or money laundering. These organized crime syndicates are often based on racial, ethnic, or family ties, with different groups dominating and replacing each other in different criminal "industries," at different periods in U.S. history.

A key concept in sociological studies of organized crime is that these industries are organized along the same lines as legitimate businesses; indeed, organized crime has taken on a corporate form (Best 2007; Carter 1999). There are likely senior partners who control the profits of the business, workers who manage and provide the labor for the business, and clients who buy the services that organized crime provides, such as prostitution and drug dealing. In-depth studies of the organized-crime underworld are difficult, owing to its secretive nature and dangers. As organized crime has moved into seemingly legitimate corporate organizations, it is even more difficult to trace, although some sociologists have penetrated underworld networks and provided fascinating accounts of how these crime networks are organized (Carter 1999).

Corporate Crime and Deviance: Doing Well, Doing Time

Corporations and even entire governments may engage in deviance—behavior that can be very costly to society. Sociologists estimate that the costs of corporate crime may be as high as $200 billion every year, dwarfing the take from street crime (roughly $15 billion), which most people imagine is the bulk of criminal activity. Tax cheaters in business alone probably skim $50 billion a year from the IRS, three times the value of street crime. Taken as a whole, the cost of corporate crime is almost 6000 times the amount taken in bank robberies in a given year and 11 times the total amount for all theft in a year (Reiman 2007).

Corporate crime and deviance is wrongdoing that occurs within the context of a formal organization or bureaucracy and is actually sanctioned by the norms and operating principles of the bureaucracy (Simon 2007).

This dinner gathering of HBO's long-running "Sopranos" gang illustrates both the hierarchy and group cohesion that characterizes organized crime.

This can occur within any kind of organization—corporate, educational, governmental, or religious. It exists once deviant behavior becomes institutionalized in the routine procedures of an organization. Sociological studies of corporate deviance show that this form of deviance is embedded in the ongoing and routine activities of organizations (Punch 1996; Lee and Ermann 1999). Individuals within the organization may participate in the deviant behavior with little awareness that their behavior is illegitimate. In fact, their actions are likely to be defined as in the best interests of the organization—business as usual. New members who enter the organization learn to comply with the organizational expectations or leave.

Instead of conceptualizing organizational deviance as merely the behavior of bad individuals, sociologists see it as the result of employees following rules and making decisions in more ordinary ways.

A recent example of corporate and accounting malfeasance involved the WorldCom Corporation, a telecommunications company of worldwide repute. The company engaged in a multimillion dollar accounting fraud that disguised mounting losses from early in the year 2000 through the summer of 2002. By incorrectly reporting its operating expenses as though they were capital gains—again, employing illegal accounting—the company was able to inflate the value of its own stock even though its own finances were rapidly deteriorating (Eichenwald 2002; Costello 2004).

Race, Class, Gender, and Crime

Arrest data show a clear pattern of differential arrests along lines of race, gender, and class. To sociologists the central question posed by such data is whether this reflects actual differences in the extent of crime among different groups or whether this reflects differential treatment by the criminal justice system. The answer is "both" (D'Alissio and Stolzenberg 2003).

Certain groups are more likely to commit crime than others, because crime is distinctively linked to patterns of inequality in society. Unemployment, for example, is one correlate of crime, as is poverty. For example, there is a clear link between the likelihood of lethal violence and the socioeconomic conditions for Latinos in urban areas (Martinez 2002).

Sociologists also show that prosecution by the criminal justice system is significantly related to patterns of race, gender, and class inequality. We see this in the bias of official arrest statistics, in treatment by the police, in patterns of sentencing, and in studies of imprisonment.

Arrest statistics show a strong correlation between social class and crime, the poor being more likely than others to be arrested for crimes. Does this mean that the poor commit more crimes? To some extent, yes. Unemployment and poverty are related to crime (Best 2007; Reiman 2007; Scarpitti et al. 1997; Hagan 1993; Britt 1994; M. Smith et al. 1992). The reason is simple: Those who are economically deprived often see no alternative to crime, as Merton's structural strain theory would predict.

Moreover, law enforcement is concentrated in lower income and minority areas. People who are better off are further removed from police scrutiny and better able to hide their crimes. When and if white-collar criminals are prosecuted and convicted, with few exceptions they typically receive light sentences—often a fine or community service instead of imprisonment. Middle- and upper-income people may be perceived as being less in need of imprisonment because they likely have a job and high-status people to testify for their good character. And, white-collar crime is simply perceived as less threatening than crimes by the poor. Class also predicts who most likely will be victimized by crime, with those at the highest ends of the socioeconomic scale least likely to be victims of violent crime (see Figure 7.4).

Research shows that African Americans and Hispanics are more than twice as likely to be arrested for a crime than are Whites. Native Americans and Asians are exceptions, with both groups having relatively low rates of arrest for crime (Federal Bureau of Investigation 2006).

Bearing in mind the factors that affect the official rates of arrest and conviction—bias of official statistics, influence of powerful individuals, discrimination in patterns of arrest, differential policing—there remains evidence that the actual commission of crime varies by race. Why? Again, sociologists find a compelling explanation in social structural conditions. Racial minority groups are far more likely than Whites to be poor, unemployed, and living in single-parent families. These social facts are all predictors of a higher rate of crime. Note, too, as Figure 7.5 shows, that African Americans are generally more likely to be victimized by crime.

Generally, women commit fewer crimes than do men. Why? Although the number of women arrested for crime has increased slightly in recent years, the numbers are still small relative to men, except for a few crimes such as fraud, embezzlement, and prostitution. Some argue that women's lower crime participation reflects their socialization into less risk-taking roles; others say that women commit crimes that are extensions of their gender roles—this would explain why the largest number of arrests of women are for crimes such as shoplifting, credit card fraud, and passing bad checks.

Nonetheless, women's participation in crime has been increasing in recent years, the result of several factors. Women are now more likely to be employed in jobs that present opportunities for crimes, such as property theft, embezzlement, and fraud. Violent crime by women has also increased notably since the early 1980s, possibly because the images that women have of themselves are changing, making new behaviors possible. Most significant, crime by women is related to their continuing disadvantaged status in society. Just

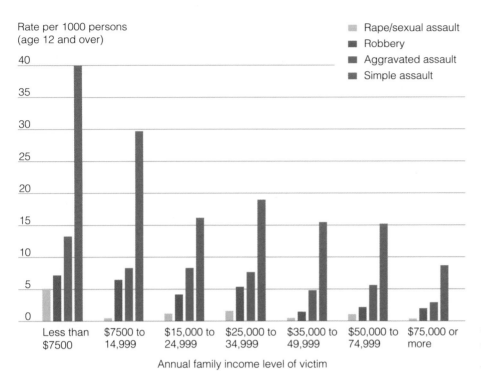

Rate per 1000 persons (age 12 and over)

Legend:
- Rape/sexual assault
- Robbery
- Aggravated assault
- Simple assault

Annual family income level of victim

FIGURE 7.4 Victimization in Crime: A Class Phenomenon

Source: U.S Bureau of Justice Statistics, 2008.

as crime is linked to socioeconomic status for men, so it is for women (Belknap 2001; Miller 1986).

Despite recent achievements, many women remain in disadvantaged, low-wage positions in the labor market. At the same time, changes in the social structure of families mean that more women are economically responsible for their children without the economic support of men. Disadvantaged women may turn to illegitimate means of support, a trend that may be exacerbated by reductions in welfare support.

[understanding diversity

The Rich Get Richer and the Poor Get Prison

Jeffrey H. Reiman (2007) notes that the prison system in the United States, instead of serving as a way to rehabilitate criminals, is in effect designed to train and socialize inmates into a career of crime. It is also designed in such a way as to assure the public that crime is a threat primarily from the poor and that it originates at the lower rungs of society. This observation has also been made by Bruce Western (Reiman 2007). Reiman and Western note that prisons contain elements that seem designed to accomplish this view. One can "construct" a prison that ends up looking like a U.S. prison.

First, continue to label as criminals those who engage in crimes that have no unwilling victim, such as prostitution or gambling. Second, give prosecutors and judges broad discretion to arrest, convict, and sentence based on appearance, dress, race, and apparent social class. Third, treat prisoners in a painful and demeaning manner, as one might treat children. Fourth, make certain that prisoners are not trained in a marketable skill that would be useful upon their release. And, finally, assure that prisoners will forever be labeled and stigmatized as different from "decent citizens," even after they have paid their debt to society. Once an ex-con, always an ex-con. One has thus socially constructed a U.S. prison, an institution that will continue to generate the very thing that it claims to eliminate.

Sources: Reiman, Jeffrey H. 2007. *The Rich Get Richer and the Poor Get Prison.* 8th edition. Boston, MA: Allyn and Bacon; Western, Bruce. 2007. *Punishment and Inequality in America.* New York: Russell Sage Foundation.

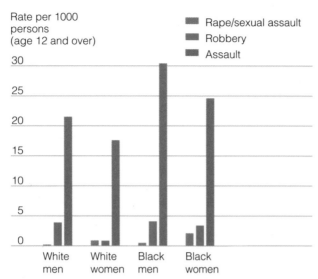

FIGURE 7.5 *Crime Victimization (by race and gender)*

Source: U.S. Bureau of Justice Statistics. 2008.

Women are less likely than men to be victimized by crime, although victimization by crime among women varies significantly by race and age. Black women are much more likely than White women to be victims of violent crime; young Black women are especially vulnerable. Divorced, separated, and single women are more likely than married women to be crime victims.

For all women, victimization by rape is probably the greatest fear. Although rape is the most underreported crime, until recently it has been one of the fastest growing—something criminologists explain as the result of a greater willingness to report *and* an actual increase in the extent of rape (Federal Bureau of Investigation 2006). More than 200,000 rapes (including attempted rapes) are reported to the police annually. Officials estimate that this is probably only about one in four of all rapes committed. Many women are reluctant to report rape because they fear the consequences of having the criminal justice system question them. Rape victims are least likely to report the assault when the assailant is someone known to them, even though a large number of rapes are committed by someone the victim knows.

A disturbingly frequent form of rape is *acquaintance rape*—rape committed by an acquaintance or someone the victim has just met. The extent of acquaintance rape is difficult to measure. The Bureau of Justice Statistics finds that 3 percent of college women experience rape or attempted rape in a given college year; 13 percent report being stalked (Fisher et al. 2000). Research finds that acquaintance rape is linked to men's acceptance of various rape myths (such as believing that a woman's "no" means "yes"), the use of alcohol, and the peer support that men receive in some all-male groups and organizations,

such as fraternities (Taylor et al. 2006; Ullman et al. 1999; Boeringer 1999; Belknap et al. 1999).

Sociologists have argued that the causes of rape lie in women's status in society—that women are treated as sexual objects for men's pleasure. The relationship between women's status and rape is also reflected in data revealing who is most likely to become a rape victim. African American women, Latinas, and poor women have the highest likelihood of being raped, as do women who are single, divorced, or separated. Young women are also more likely to be rape victims than older women (U.S. Bureau of Justice Statistics 2008). Sociologists interpret these patterns to mean that the most powerless women are also most subject to this form of violence.

The Criminal Justice System: Police, Courts, and the Law

Whether it is in the police station, the courts, or prison, the factors of race, class, and gender are highly influential in the administration of justice in this society. Those in the most disadvantaged groups are more likely to be defined and identified as deviant independently of their behavior and, having encountered these systems of authority, are more likely to be detained and arrested, found guilty, and punished.

Debunking Society's Myths

Myth: The criminal justice system treats all people according to the neutral principles of law.

Sociological perspective: Race, class, and gender continue to have an influential role in the administration of justice. For example, even when convicted of the same crime as Whites, African American and Latino male defendants with the *same* prior arrest record as Whites are more likely to be arrested, sentenced, and to be sentenced for longer terms than White defendants.

The Policing of Minorities. There is little question that minority communities are policed more heavily than White neighborhoods; moreover, policing in minority communities has a different effect from that in White, middle-class communities. To middle-class Whites, the presence of the police is generally reassuring, but for African Americans and Latinos an encounter with a police officer can be terrifying. Regardless of what they are doing at the time, minority people, men in particular, are perceived as a threat, especially if they are observed in communities where they "don't belong."

Racial profiling has recently come to the public's attention, although it is a practice that has a long history. Often referred to half in jest by African Americans as the offense of "DWB," or "driving while Black," **racial profiling** on the part of a police officer is the use of race alone as the criterion for deciding whether to stop and detain someone on suspicion of their having committed a crime. The police argue that racial profiling is justified because a high proportion of Blacks and Hispanics commit crimes; but, although the crime rate for Blacks and Hispanics is higher than that of Whites, race is a particularly bad basis for suspicion because the vast majority of Blacks and Hispanics, like the vast majority of Whites, do not commit any crime at all. As evidence of this, studies have found that *eight out of every ten* automobile searches carried out by state troopers on the New Jersey Turnpike over ten years were conducted on vehicles driven by Blacks and Hispanics; the vast majority of these searches turned up no evidence of contraband or crimes of any sort (Kocieniewski and Hanley 2000; Cole 1999).

Racial minorities are also more likely than the rest of the population to be victims of excessive use of force by the police, also called police brutality. Most cases of police brutality involve minority citizens, and there is usually no penalty for the officers involved. Increasing the number of minority police officers has some effect on how the police treat minorities. Simply increasing the number of African Americans in police departments does not, however, reduce crime dramatically, because it does not change the material conditions that create crime to begin with (Cashmore 1991).

Race and Sentencing. What happens once minority citizens are arrested for a crime? Bail is set higher for African Americans and Latinos than for Whites, and minorities have less success with plea bargains. Once on trial, minority defendants are found guilty more often than White defendants. At sentencing, African Americans and Hispanics are likely to get longer sentences than Whites, even when they have the same number of prior arrests and socioeconomic background as Whites. Young African American men, as well as Latino men, are sentenced more harshly than any other group, and once sentenced they are less likely to be released on probation (Western 2007; Steffensmeier and Demuth 2000; Mauer 1999; Steffensmeier et al. 1998; Chambliss and Taylor 1989; Bridges and Crutchfield 1988). In fact Blacks and Hispanics who have *already received* the death penalty are more likely to be executed, rather than being pardoned or having the execution postponed, than are Whites who have committed the same crime (Jacobs et al. 2007). Any number of factors influence judgments about sentencing, including race of the judge, severity of the crime, race of the victim, and the gender of the defendant, but throughout these studies, race is shown to consistently matter—and matter a lot.

Prisons: Rehabilitation or Mass Racialized Incarceration? Racial minorities account for *more than half* of the federal and state male prisoners in the United States. Blacks have the highest rates of imprisonment, followed by Hispanics, then Native Americans and Asians. (Native Americans and Asian Americans together are less than 1 percent of the total prison population.) Hispanics are the fastest growing minority group in prison (U.S. Bureau of Justice Statistics 2008). Native Americans, though a small proportion of the prison population, are still overrepresented in prisons. In theory, the criminal justice system is supposed to be unbiased, able to objectively weigh guilt and innocence. The reality is that the criminal justice system reflects the racial and class stratification and biases in society.

The United States and Russia have the highest rate of incarceration in the world (see Figure 7.6). In the United States, the rate of imprisonment has been rapidly growing (see Figure 7.7). By all signs, the population of state and federal prisons continues to grow, with the population in prisons exceeding the capacity of the facilities. The cost to the nation of keeping people behind bars is at least $160 billion (U.S. Bureau of Justice Statistics 2008).

Although it is certainly true that, as we have already noted, Blacks and Hispanics themselves commit proportionately more crime than Asians and Native American Indians, it is nonetheless *also* demonstrably true that the structure of the U.S. criminal justice system disproportionately *propels* Blacks and Hispanics into prison at a greater rate than same-aged Whites who have the same criminal record. This is because unemployment is much higher for Blacks and Hispanics. Also, federal and state officials have traditionally been more hostile to persons of color who run afoul of the criminal justice system than to Whites who do so. Finally, the "three strikes" law (giving mandatory long sentences if caught three times for the same drug offense) disproportionately affects minorities of color. The situation is so severe and there are so many minority persons now in prison that sociologist Bruce Western calls them a *new color caste* in U.S. society—in other words, a society unto itself (Western 2007).

The picture of incarceration in the United States seems contradictory. The overall violent crime rate has declined (see Figure 7.3); this would cause us to expect that the rate of admissions to prison would also decline. Yet, at the same time, the numbers of individuals in state and federal prisons have been increasing (Figure 7.7). Why is there such growth in the prison population when the crime rate has been declining? A major reason for the increasing number of individuals behind bars is the increased enforcement of drug offenses and the mandatory sentencing that has been introduced. Nearly one-quarter of those in state prisons are serving a drug sentence (U.S. Bureau of Justice Statistics 2008).

Women in prison face unique problems, in part because they are in a system designed for men and run mostly by men, which tends to ignore the particular needs of women. For example, 25 percent of the women entering prison are pregnant or have just given birth,

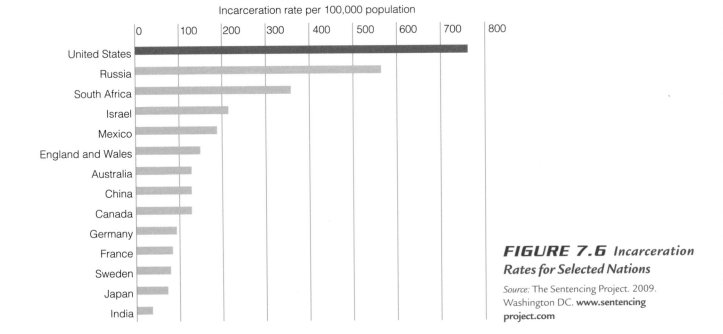

Incarceration rate per 100,000 population

FIGURE 7.6 Incarceration Rates for Selected Nations

Source: The Sentencing Project. 2009. Washington DC. **www.sentencing project.com**

but they often get no prenatal or obstetric care. Male prisoners are trained for such jobs as auto mechanics, whereas women are more likely to be trained in relatively lower-status jobs such as beauticians and launderers. The result is that few women offenders are rehabilitated by their experience in prison.

The United States, then, is putting offenders in prison at a record pace. Is crime being deterred? Are prisoners being rehabilitated? Or are they simply being *warehoused*—put on a shelf? If the deterrence argument were correct, we would expect that increasing the risk of imprisonment would lower the rate of crime. For example, we would expect drug use to decline as enforcement of drug laws increases. In the past few years, there has been a marked increase in drug law enforcement but not the expected decrease in drug use. Using drugs as an example, then, it appears that the threat of imprisonment does not deter crime (Mauer 1999).

There is also little evidence that the criminal justice system rehabilitates offenders. Using drugs as the example again, only about 20 percent of those

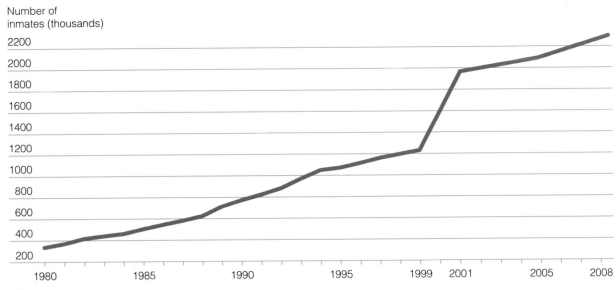

FIGURE 7.7 State and Federal Prison Population, 1980–2008

Source: U.S. Bureau of Justice Statistics. 2008. *"Prison and Jail Inmates."* **www.ojp.usdoj.gov/bjs/prisons.html**

imprisoned for drug offenses ever receive drug treatment. If the criminal justice system fails to reduce crime, what does it do? Some sociologists contend that the criminal justice system is not meant to reduce crime but has other *functions*, namely, to reinforce an image of crime as a threat from the poor and from racial groups.

Terrorism as International Crime: A Global Perspective

Crime now crosses international borders and has become global, as we see with terrorism.

The FBI includes *terrorism* in its definition of crime, seeing it as violent action to achieve political ends (White 2002). Thus, terrorism is a crime that violates both international and domestic laws. It is a crime that crosses national borders and to understand it requires a global perspective.

Terrorism is globally linked to other forms of international crime. A case in point was Osama bin Laden's role in the international opium trade. Afghanistan, where bin Laden's terrorist al Qaeda organization was headquartered, was the world's largest grower of opium-producing poppies. It is suspected that profits from the international drug trade helped finance the September 11 terrorist attacks. Therefore, a global perspective on crime involves recognizing the global basis of some international crime networks that cross national borders (Binns 2003).

Many nations have long experienced terrorism in the form of bombings, hijackings, suicide attacks, and other terrorist crimes. But the attacks of September 11 focused the world's attention on the problem of terrorism in new ways, including increased fears of **bioterrorism**—the form of terrorism involving the dispersion of chemical or biological substances intended to cause widespread disease and death.

Terrorists crashed two jets into the World Trade Center on September of 2001, causing the total collapse of both towers and killing over 3,000 people.

Another form of terrorism, and thus cause for international concern, is **cyberterrorism,** the use of the computer to commit one or more terrorist acts. Terrorists may use computers in a number of ways. Data-destroying computer viruses may be implanted electronically in an enemy's computer. Another use would be to employ "logic bombs" that lie dormant for years until they are electronically instructed to overwhelm a computer system. The use of the Internet to serve the needs of international terrorists has already become a reality (Jucha 2002).

Without understanding the political, economic, and social relations from which terrorist groups originate, terrorist acts seem like the crazed behavior of violent single individuals. Although sociologists in no way excuse such acts, they look to the social structure of conflicts from which terrorism emerges as the cause of such criminal and deviant behavior.

Chapter Summary

- ## What is the difference between deviance and crime?

  Deviance is behavior that violates norms and rules of society, and *crime* is a type of deviant behavior that violates the formal criminal law. *Criminology* is the study of crime from a scientific perspective.

- ## How do sociologists conceptualize and explain deviance and crime?

  Deviance is behavior that is recognized as violating expected rules and norms and that should be understood in the social context in which it occurs. Psychological explanations of deviance place the cause of deviance primarily within the individual. Sociologists emphasize the total social context in which deviance occurs. Sociologists see deviance more as the result of group and institutional, not individual, behavior.

• *What does sociological theory contribute to the study of deviance and crime?*

Functionalist theory sees both deviance and crime as functional for the society because it affirms what is acceptable by defining what is not. *Structural strain theory*, a type of functionalist theory, predicts that societal inequalities actually force and compel the individual into deviant and criminal behavior. *Conflict theory* explains deviance and crime as a consequence of unequal power relationships and inequality in society. *Symbolic interaction theory* explains deviance and crime as the result of meanings people give to various behaviors. *Differential association theory*, a type of symbolic interaction theory, interprets deviance as behavior learned through social interaction with other deviants. *Labeling theory*, also a type of symbolic interaction theory, argues that societal reactions to behavior produce deviance, with some groups having more power than others to assign deviant labels to people.

• *What are the major forms of deviance?*

Mental illness, *stigma*, and substance abuse are major forms of deviance studied by sociologists, although deviance comprises many different forms of behavior. Sociological explanations of mental illness focus on the social context in which mental illness develops and is treated. Social *stigmas* are attributes that are socially devalued. Substance abuse includes alcohol and drug abuse but is not limited to these two forms.

• *What are the connections between inequality, deviance, and crime?*

Sociological studies of crime analyze the various types of crimes, such as *elite crime, organized crime, corporate crime,* and *personal* and *property crimes.* Many types of crimes are underreported, such as rape and certain elite and corporate crimes. Sociologists study the conditions, including race, class, and gender inequality, that produce crime and shape how different groups are treated by the criminal justice system, such as showing group differences in sentencing.

• *How is crime related to race, class, and gender?*

In general, crime rates for a variety of crimes are higher among minorities than among Whites, among poorer persons than among middle- or upper-class persons, and among men than among women. Women, especially minority women, are more likely to be victimized by serious crimes such as rape or violence from a spouse or boyfriend.

• *How is globalization affecting the development of deviance and crime?*

International *terrorism* is a crime, and crime is thus global. Other global crimes of significance are *bioterrorism* and *cyberterrorism.* Osama bin Laden's al Qaeda organization, assumed to be the organization that destroyed the World Trade Center and killed over 3000 individuals, was centered in Afghanistan and was central to the international drug trade. Thus, crimes are clearly not just the acts of a crazed individual or small group of individuals, but the result of structural and cultural conditions.

Key Terms

altruistic suicide   156
anomic suicide   156
anomie   155
bioterrorism   175
crime   165
criminology   166
cyberterrorism   175
deviance   152
deviant career   162
deviant communities   163
deviant identity   162

differential association
   theory   161
egoistic suicide   156
elite deviance   160
hate crime   168
index crimes   166
labeling theory   161
master status   165
medicalization of deviance   154
organized crime   169
personal crimes   168

Ponzi scheme   168
property crimes   168
racial profiling   172
social control   160
social control agents   160
social control theory   159
stigma   164
structural strain theory   157
victimless crimes   168

Online Resources

Sociology: The Essentials Companion Website

www.cengage.com/sociology/andersen

Visit your book companion website where you will find more resources to help you study and write your research papers. Resources include web links, learning objectives, Internet exercises, quizzing, and flash cards.

 is an easy-to-use online resource that helps you study in less time to get the grade you want NOW.

www.cengage.com/login

Need help studying? This site is your one-stop study shop. Take a Pre-Test and Cengage NOW will generate a Personalized Study Plan based on your test results. The Study Plan will identify the topics you need to review and direct you to online resources to help you master those topics. You can then take a Post-Test to determine the concepts you have mastered and what you still need to work on.

8

Social Class and Social Stratification

Social Differentiation and Social Stratification

The Class Structure of the United States

Social Mobility: Myths and Realities

Why Is There Inequality?

Poverty

Chapter Summary

[*One afternoon in* a major U.S. city, two women go shopping. They are friends—wealthy, suburban women who shop for leisure. They meet in a gourmet restaurant and eat imported foods while discussing their children's private schools. They also talk about the volunteer work they do in the local hospital, but they don't worry too much about their own health care, at least not financially, because both are fully covered through ample health insurance policies. After lunch, they spend the afternoon in exquisite stores—some of them large, elegant department stores; others, intimate boutiques where the staff know them by name. When one of the women stops to use the bathroom in one store, she enters a beautifully furnished room with an upholstered chair, a marble sink with brass faucets, fresh flowers on a wooden pedestal, shining mirrors, an ample supply of hand towels, and jars of lotion and soaps. The toilet is in a private stall with solid doors. In the stall there is soft toilet paper and another small vase of flowers.

The same day, in a different part of town, another woman goes shopping. She lives on a marginal income earned as a stitcher in a textiles factory. Her daughter badly needs a new pair of shoes because she has outgrown last year's pair. The woman goes to a nearby discount store where she hopes to find a pair of shoes for under $15, but she dreads the experience. She knows her

continued

Beth A. Keiser/AP Photo

eight
chapter eight
CHAPTER EI

daughter would like other new things—a bathing suit for the summer, a pair of jeans, and a blouse. But this summer the daughter will have to wear hand-me-downs because medical bills over the winter have depleted the little money left after food and rent. For the mother, shopping is not recreation but a bitter chore reminding her of the things she is unable to get for her daughter.

While this woman is shopping, she, too, stops to use the bathroom. She enters a vast space with sinks and mirrors lined up on one side of the room and several stalls on the other. The tile floor is gritty and gray. The locks on the stall doors are missing or broken. Some of the overhead lights are burned out, so the room has dark shadows. In the stall, the toilet paper is coarse. When the woman washes her hands, she discovers there is no soap in the metal dispensers. The mirror before her is cracked. She exits quickly, feeling as though she is being watched.

Two scenarios, one society. The difference is the mark of a society built upon class inequality. The signs are all around you. Think about the clothing you wear. Are some labels worth more than others? Do others in your group see the same marks of distinction and status in clothing labels? Do some people you know never seem to wear the "right" labels? Whether it is clothing, bathrooms, schools, homes, or access to health care, the effect of class inequality is enormous, giving privileges and resources to some and leaving others struggling to get by.

Great inequality divides society. Nevertheless, most people think that in the United States equal opportunity exists for all. The tendency is to blame individuals for their own failure or attribute success to individual achievement. Many people think the poor are lazy and do not value work. At the same time, the rich are often admired for their supposed initiative, drive, and motivation. Neither is an accurate portrayal. There are many hard-working individuals who are poor, and most rich people have inherited their wealth rather than earned it themselves.

Observing and analyzing class inequality is fundamental to sociological study. What features of society cause different groups to have different opportunities? Why is there such an unequal allocation of society's resources? Sociologists respect individual achievements but have found the greatest cause for the disparities in material success is the organization of society. Instead of understanding inequality as the result of individual effort, sociologists thus study the social structural origins of inequality.

Social Differentiation and Social Stratification

All social groups and societies exhibit social differentiation. **Status,** as we have seen earlier, is a socially defined position in a group or society. **Social differentiation** is the process by which different statuses develop in any group, organization, or society. Think of a sports organization. The players, the owners, the managers, the fans, the cheerleaders, and the sponsors all have a different status within the organization. Together they constitute a whole social system, one that is marked by social differentiation.

thinkingsociologically

Take a shopping trip to different stores and observe the appearance of stores serving different economic groups. What kinds of bathrooms are there in stores catering to middle-class clients? The rich? The working class? The poor? Which ones allow the most privacy or provide the nicest amenities? What fixtures are in the display areas? Are they simply utilitarian with minimal ornamentation, or are they opulent displays of consumption? Take detailed notes of your observations, and write an analysis of what this tells you about *social class* in the United States.

Status differences can become organized into a hierarchical social system. Social stratification is a relatively fixed, hierarchical arrangement in society by which groups have different access to resources, power, and perceived social worth. **Social stratification** is a system of structured social inequality. Using sports as an example again, you can see that many of the players earn extremely high salaries, although most do not. Those who do are among the elite in this system of inequality. But it is the owners who control the resources of the teams and hold the most power in this system. Sponsors (including major corporations and media networks) are the economic engines on which this system of stratification rests; fans are merely observers who pay to watch the teams play, but the revenue they generate is essential for keeping this system intact. Altogether, sports are systems of stratification because the groups that constitute the organization are arranged in a hierarchy where some have more resources and power than others. Some provide resources; others take them. And, even within the field of sports, there are huge differences in which teams—and which sports—are among the elite.

All societies seem to have a system of social stratification, although they vary in the degree and complexity of stratification. Some societies stratify only along a single dimension, such as age, keeping the stratification system relatively simple. Most contemporary societies are more complex, with many factors interacting to create different social strata. In the United States,

table8.1	Inequality in the United States

Almost one in five (18 percent) children in the United States lives in poverty, including 35 percent of African American children, 29 percent of Hispanic children, 10 percent of White children, and 12 percent of Asian American children (DeNavas-Walt et al. 2008).

The rate of poverty among people in the United States has been steadily increasing since 2000 (DeNavas-Walt et al. 2008).

Among women heading their own households, 31 percent live below the poverty line (DeNavas-Walt et al. 2008).

One percent of the U.S. population controls 33 percent of the total wealth in the nation; the bottom 20 percent owe more than they own (Mishel et al. 2007).

When Leona Helmsley (the hotel financier) died, she left a $12 million inheritance to her dog, with the provision that the dog should eventually be buried in a mausoleum with Helmsley—a mausoleum that would be steam-cleaned once per week in perpetuity.

The average CEO of a major company has a salary of $13.1 million per year; workers earning the minimum wage make $10,712 per year if they work 40 hours a week for 52 weeks and hold only one job (Lavelle 2001).

social stratification is strongly influenced by class, which is in turn influenced by matters such as one's occupation, income, and education, along with race, gender, and other influences such as age, region of residence, ethnicity, and national origin (see Table 8.1).

Estate, Caste, and Class

Stratification systems can be broadly categorized into three types: estate systems, caste systems, and class systems. In an **estate system** of stratification, the ownership of property and the exercise of power is monopolized by an elite who have total control over societal resources. Historically, such societies were feudal systems where classes were differentiated into three basic groups—the nobles, the priesthood, and the commoners. Commoners included peasants (usually the largest class group), small merchants, artisans, domestic workers, and traders. The nobles controlled the land and the resources used to cultivate the land, as well as all the resources resulting from peasant labor.

Estate systems of stratification are most common in agricultural societies. Although such societies have been largely supplanted by industrialization, there are still examples of societies that have a small but powerful landholding class ruling over a population that works mainly in agricultural production. Unlike the feudal societies of the European Middle Ages, however, contemporary estate systems of stratification display the influence of international capitalism. The "noble class" comprises not knights who conquered lands in war, but international capitalists or local elites who control the labor of a vast and impoverished group of people, such as in some South American societies where landholding elites maintain a dictatorship over peasants who labor in agricultural fields.

In a **caste system,** one's place in the stratification system is an *ascribed status* (see Chapter 5), meaning it is a quality given to an individual by circumstances of birth. The hierarchy of classes is rigid in caste systems and is often preserved through formal law and cultural practices that prevent free association and

Social class differences make it seem as if some people are practically living in two different societies.

movement between classes. The system of *apartheid* in South Africa was a stark example of a caste system. Under apartheid, the travel, employment, associations, and place of residence of Black South Africans were severely restricted. Segregation was enforced using a pass system in which Black South Africans could not be in White areas unless for purposes of employment; those found without passes were arrested, often sent to prison without ever seeing their families again. Interracial marriage was illegal. Black South Africans were prohibited from voting; the system was one of total social control where anyone who protested was imprisoned. The apartheid system was overthrown in 1994 when Nelson Mandela, held prisoner for 27 years of his life, was elected president of the new nation of South Africa; a new national constitution guaranteeing equal rights to all was ratified in 1996.

In **class systems,** stratification exists, but a person's placement in the class system can change according to personal achievements; that is, class depends to some degree on *achieved status,* defined as status that is earned by the acquisition of resources and power, regardless of one's origins. Class systems are more open than caste systems because position does not depend strictly on birth, and classes are less rigidly defined than castes because the divisions are blurred by those who move between one class and the next.

Despite the potential for movement from one class to another, in the class system found in the United States, class placement still depends heavily on one's social background. Although ascription (the designation of ascribed status according to birth) is not the basis for social stratification in the United States, the class a person is born into has major consequences for that person's life. Patterns of inheritance; access to exclusive educational resources; the financial, political, and social influence of one's family; and similar factors all shape one's likelihood of achievement. Although there is no formal obstacle to movement through the class system, individual achievement is very much shaped by an individual's class of origin.

Defining Class

In common terms, *class* refers to style or sophistication. In sociological use, **social class** (or *class*) is the social structural position groups hold relative to the economic, social, political, and cultural resources of society. Class determines the access different people have to these resources and puts groups in different positions of privilege and disadvantage. Each class has members with similar opportunities who tend to share a common way of life. Class also includes a cultural component in that class shapes language, dress, mannerisms, taste, and other preferences. *Class is not just an attribute of individuals; it is a feature of society.*

The social theorist Max Weber described the consequences of stratification in terms of **life chances,** meaning the opportunities that people have in common by virtue of belonging to a particular class. Life chances include the opportunity for possessing goods, having an income, and having access to particular jobs. Life chances are also reflected in the quality of everyday life. Whether you dress in the latest style or wear another person's discarded clothes, have a vacation in an exclusive resort, take your family to the beach for a week, or have no vacation at all, these life chances are the result of being in a particular class.

Class is a structural phenomenon; it cannot be directly observed. Nonetheless, you can "see" class through various displays that people project, often unintentionally, about their class status. What clothing do you wear? Do some worn objects project higher class status than others? How about cars? What class status is displayed through the car you drive or, for that

Recovery from Hurricane Katrina was not easy for anyone, but the likelihood of recovery was shaped by race in in combination with class, evidenced in the progress in different neighborhoods, even several years after this massive disaster.

matter, whether you even have a car or use a bus to get to work? In these and myriad other ways, class is projected to others as a symbol of our presumed worth in society.

Social class can be observed in the everyday habits and presentations of self that people project. Common objects, such as clothing and cars, become symbols of one's class status. As such, they can be ranked not only in terms of their economic value, but also in terms of the status that various brands and labels carry. The interesting thing about social class is that a particular object may be quite ordinary, but with the right "label" it becomes a *status symbol* and thus becomes valuable. Take the example of Vera Bradley bags. These paisley bags are made of ordinary cotton with batting. Not long ago, such cloth was cheap and commonplace, associated with rural, working-class women. If such a bag were sewn and carried by a poor person living on a farm, the bag (and perhaps the person!) would be seen as ordinary, almost worthless. But, transformed by the right label (and some good marketing), Vera Bradley bags have become status symbols, selling for a high price (often a few hundred dollars—a price one would never pay for a simple cotton purse). Presumably, having such a bag denotes the status of the person carrying it. (See also the box "See for Yourself: Status Symbols in Everyday Life.")

Status symbols connote class status because of the meaning people attach to them, not necessarily their actual value.

is income; other common indicators are education, occupation, and place of residence. These indicators alone do not define class, but they are often accurate measures of the class standing of a person or group. We will see that these indicators tend to be linked. A good income, for example, makes it possible to afford a house in a prestigious neighborhood and an exclusive education for one's children. In the sociological study of class, indicators such as income and education have had enormous value in revealing the outlines and influences of the class system.

The Class Structure of the United States

The class structure of the United States is elaborate, arising from the interactions of old wealth, new wealth, intensive immigration, a culture of entrepreneurship and individualism, and in recent times, vigorous globalization. One can conceptualize the class system as a ladder, with different class groups arrayed up and down the rungs, each rung corresponding to a different level in the class system. Conceptualized this way, social class is the common position groups hold in a status hierarchy (Wright 1979; Lucal 1994); class is indicated by factors such as levels of income, occupational standing, and educational attainment. People are relatively high or low on the ladder depending on the resources they have and whether those resources are education, income, occupation, or any of the other factors known to influence people's placement (or ranking) in the stratification system. Indeed, an abundance of sociological research has stemmed from the concept of **status attainment,** the process by which people end up in a given position in the stratification

---

SEE *FOR YOURSELF*

STATUS SYMBOLS IN EVERYDAY LIFE

You can observe the everyday reality of social class by noting the status that different ordinary objects have within the context of a class system. Make a list of every car brand you can think of—or, if you prefer, every clothing label. Then rank your list with the highest status brand (or label) at the top of the list, going down to the lowest status. Then answer the following questions:

1. Where does the presumed value of this object come from? Does the value come from the actual cost of producing the object or something more subjective?

2. Do people make judgments about people wearing or driving the different brands you have noted? What judgments do they make? Why?

3. What consequences do you see (positive and negative) of the ranking you have observed? Who benefits from the ranking and who does not?

What does this exercise reveal about the influence of *status symbols* in society?

Because sociologists cannot isolate and measure social class directly, they use other indicators to serve as measures of class. A prominent indicator of class

Social Inequality

If you use your sociological imagination, you will begin to see the signs of social inequality in many places, including commonplace locations, such as shopping. Sociologist Christine Williams noted this in her research for the book, *Inside Toyland*, where she studied the influence of class, race, and gender in the different experiences of shopping in a "big box" toy store compared with shopping in a boutique toy store. Where do you see class inequality in your everyday life?

Thomas Barwick/Getty Images

Don Ryan/AP Photo

How does the experience of shopping vary by social class?

Of course, everybody loves a bargain, but social class inequality means that people will have very different experiences. For some, shopping is a cruel reminder of their class status; for others, it may be a form of leisure and a mark of their higher status. For the upper-class shopper, shopping is likely to be more personal and accommodating.

Rob Kim/Landov

Richard Vogel/AP Photo

How are status symbols related to the process of globalization?

Although globalization may not be immediately visible, many of the products by which people in one place accrue status are linked to the low status of workers in other nations. This is known as the global assembly line.

Michael Newman/PhotoEdit

Satcham/zefa/Corbis

How does social class shape the *division of labor* in the workplace?

Race and gender intersect with class to shape the division of labor. You can see this if you pay attention to who works where and in what capacity.

Andrew Holbrooke/Corbis News/Corbis

How wide is the class gap in U.S. society?

Differences in how people live are shaped by social class; class inequality has been growing in recent years, evidenced in part by differences in where and how people live.

Kareem Black/Getty Images

system. Status attainment research describes how factors such as class origins, educational level, and occupation produce class location.

This laddered model of class suggests that in the United States stratification is hierarchical but somewhat fluid. That is, the assumption is that people can move up and down different "rungs" of the ladder—or class system. In a relatively *open class system* such as the United States, people's achievements do matter, although the extent to which people rise rapidly and dramatically through the stratification system is less than the popular imagination envisions. Some people do begin from modest origins and amass great wealth and influence (celebrities such as Bill Gates, Oprah Winfrey, and millionaire athletes), but these are the exceptions, not the rule. Some people move down in the class system, but as we will see, most people remain relatively close to their class of origin. When people rise or fall in the class system, the distance they travel is usually relatively short, as we will see further in the section on social mobility.

The image of stratification as a laddered system, with different gradients of social standing, emphasizes that one's **socioeconomic status (SES)** is derived from certain factors. Income, occupational prestige, and education are the three measures of socioeconomic status that have been found to be most significant in determining people's placement in the stratification system.

Income is the amount of money a person receives in a given period. As we will see, income is distinct from **wealth,** which is the total value of what one owns, minus one's debts. The **median income** for a society is the midpoint of all household incomes. In other words, half of all households earn more than the median income; half earn less. In 2007, median household income in the United States was $50,233 (see Figure 8.1; DeNavas-Walt et al. 2008). Those bunched around the median income level are considered middle class, although sociologists debate which income brackets constitute middle-class standing because the range of what people think of as "middle class" is quite large. Nonetheless, income is a significant indicator of social class standing, although not the only one. Map 8.1 provides a visual image of regional differences in the distribution of income by showing median income in different parts of the country, organized by county.

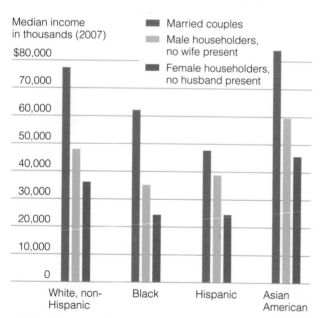

FIGURE 8.1 *Median Annual Income by Race and Household Status*

As illustrated in this graph, married couple households have the highest median income in all racial–ethnic groups; female-headed households, the least. The only households that reach median income status are White (non-Hispanic), African American, and Asian married households.

Source: U.S. Census Bureau. 2008. *Current Population Survey,* Table FINC-02. Age of Reference Person, by Total Money Income in 2007, Type of Family, Race and Hispanic Origin of Reference Person. **www.census.gov**

SEE *FOR YOURSELF*

INCOME DISTRIBUTION: SHOULD GRADES BE THE SAME?

Figure 8.2 shows the income distribution with in the United States. Imagine that grades in your class were distributed based on the same curve. Let's suppose that after students arrived in class and sat down, different groups received their grades based on where they were sitting in the room and in the same proprtion as the U.S. income distribution. Only students in the front receive As; the back, Ds and Fs. The middle of the room gets the Bs and Cs. Write a short essay answering the following questions based on this hypothetical scenario.

1. How many students would receive As, Bs, Cs, Ds, and Fs?

2. Would it be fair to distribute grades this way? Why or why not?

3. Which groups in the class might be more likely to support such a distribution? Who would think the system of grade distribution should be changed?

4. What might different groups do to preserve or change the system of grade distribution? What if you really needed an A, but got one of the Fs? What might you do?

5. Are there circumstances in actual life that are beyond the control of people and that shape the distribution of income?

6. How is social stratification maintained by the beliefs that people have about merit and fairness?

Adapted from: Brislen, William, and Clayton D. Peoples. 2005. "Using a Hypothetical Distribution of Grades to Introduce Social Stratification." *Teaching Sociology* 33 (January): 74–80.

Percent of population
in each income quintile

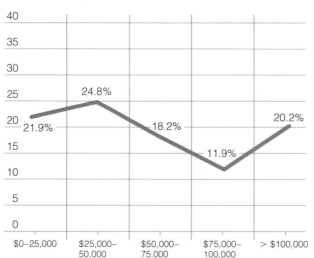

FIGURE 8.2 *Income Distribution in the United States*

This graph shows the percentage of the population that falls into each of the five income brackets.

Source: DeNavas-Walt, Carmen, Bernadette Proctor, and Jessica C. Smith. 2008. *Income, Poverty, and Health Insurance Coverage in the United States: 2007.* Washington, DC: U.S. Census Bureau, #P60-235.

Radius Images/PhotoLibrary

Ric Feld/AP Photo

"It's a dog's life," or so the saying goes. But even dogs have their experiences shaped by the realities of social class.

Occupational prestige is a second important indicator of socioeconomic status. **Prestige** is the value others assign to people and groups. **Occupational prestige** is the subjective evaluation people give to jobs. To determine occupational prestige, sociological researchers typically ask nationwide samples of adults to rank the general standing of a series of jobs. These subjective ratings provide information about how people perceive the worth of different occupations. People tend to rank professionals, such as physicians, professors, judges, and lawyers highly, with occupations such as electrician, insurance agent, and police officer falling in the middle. Occupations with low occupational prestige are maids, garbage collectors, and shoe-shiners (Nakao and Treas 2000; Davis and Smith 1984). These rankings do not reflect the worth of people within these positions but are indicative of the judgments people make about the worth of these jobs.

The final major indicator of socioeconomic status is **educational attainment,** typically measured as the total years of formal education. The more years of education attained, the more likely a person will have higher class status. The prestige attached to occupations is strongly tied to the amount of education the job requires; the more education people think is needed for a given occupation, the more occupational prestige people attribute to that job (Blau and Duncan 1967; MacKinnon and Langford 1994; Ollivier 2000).

Layers of Social Class

Taken together, income, occupation, and education are good indicators of people's class standing. How then do sociologists understand the array of classes in the United States? Most sociologists describe the class system in the United States as being divided into several classes: upper, upper middle, middle, lower middle, and lower class. The different classes are arrayed up and down, like a ladder, with those with the most money, education, and prestige on the top rungs and those with the least at the bottom.

In the United States, the *upper class* owns the major share of corporate and personal wealth; it includes those who have held wealth for generations as well as those who have recently become rich. Only a very small proportion of people actually constitute the upper class, but they control vast amounts of wealth and power in the United States. Those in this class are elites who exercise enormous control throughout society. Some wealthy individuals can wield as much power as entire nations (Friedman 1999).

Each year, the business magazine *Forbes* publishes a list of the 400 wealthiest families and individuals in the country. By 2007, you had to have at least $1.3 billion to be on the list! Bill Gates and Warren Buffet compete for being the richest person on this list; Gates has an estimated worth of $57 billion; Buffet, $50 billion, though some say now over $60 billion. Of all the wealth represented on the *Forbes 400* list, most is inherited, although there has been some increase in the number of people on the list with self-created wealth (Miller and Greenberg 2008). Still, the best predictor of future wealth is the family into which you are born.

Those in the upper class with newly acquired wealth are known as the *nouveau riche.* Luxury vehicles, high-priced real estate, and exclusive vacations may mark the lifestyle of the newly rich. Larry Ellison, who made his fortune as the founder of the software company Oracle, is the third wealthiest person in the United States; he has a megayacht that is 482

map8.1 MAPPING AMERICA'S DIVERSITY

Median Income in the United States

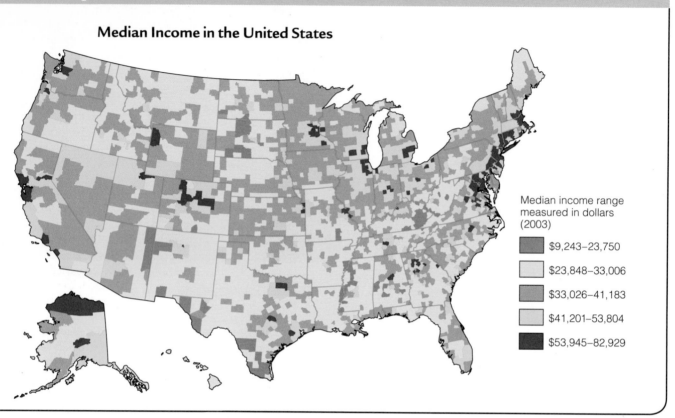

Median income range
measured in dollars
(2003)

- $9,243–23,750
- $23,848–33,006
- $33,026–41,183
- $41,201–53,804
- $53,945–82,929

If you look closely at this map, you will see that median income tends to be higher in more urban areas. Thus, in 2007, median income inside metropolitan areas was $51,831 and outside such areas, $40,615. What the map does not show, however, are differences within cities. Median income inside central cities is substantially lower ($44,205) than the median income within metropolitan areas out of the center city ($57,444)—that is, in suburban areas. Given this, what do you conclude about the significance of residence in the structure of the class system?

Data: U.S. Census Bureau. 2004. *American Fact Finder.* **www.census.gov.** U.S. Census Bureau. 2008. *Statistical Abstract of the United States 2008.* Washington, DC: U.S. Department of Commerce.

feet long, five stories high, with 82 rooms inside. The megayacht also includes an indoor swimming pool, a cinema, a space for a private submarine, and a basketball court that doubles as a helicopter launch pad.

The *upper middle-class* includes those with high incomes and high social prestige. They tend to be well-educated professionals or business executives. Their earnings can be quite high indeed, even millions of

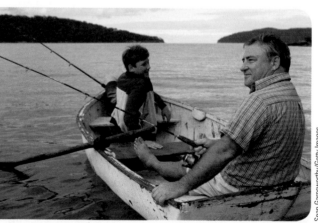

Social class influences many things, including the leisure time people experience. Few can even imagine having something like the yacht pictured on the left, owned by Larry Ellison, founder of Oracle.

dollars a year. It is difficult to estimate exactly how many people fall into this group because of the difficulty of drawing lines between the upper, upper middle, and middle classes. Indeed, the upper middle-class is often thought of as "middle class" because their lifestyle sets the standard to which many aspire, but this lifestyle is actually unattainable by most. A large home full of top-quality furniture and modern appliances, two or three relatively new cars, vacations every year (perhaps a vacation home), high-quality college education for one's children, and a fashionable wardrobe are simply beyond the means of a majority of people in the United States.

The *middle-class* is hard to define in part because being "middle class" is more than just economic position. A very large portion of Americans identify themselves as middle class even though they vary widely in lifestyle and in resources at their disposal. But the idea that the United States is an open class system leads many to think that the majority have a middle-class lifestyle; thus, the middle class becomes the ubiquitous norm even though many who call themselves middle class have a tenuous hold on this class position.

The *lower middle-class* includes workers in the skilled trades and low-income bureaucratic workers, many of whom may actually think they are middle class. Also known as the *working class,* this class includes blue-collar workers (those in skilled trades who do manual labor) and many service workers, such as secretaries, hair stylists, food servers, police, and firefighters. A medium to low income, education, and occupational prestige define the lower middle-class relative to the class groups above it. The term *lower* in this class designation refers to the relative position of the group in the stratification system, but it has a pejorative sound to many people, especially to people who are members of this class, many of whom think of themselves as middle class.

The *lower class* is composed primarily of displaced and poor. People in this class have little formal education and are often unemployed or working in minimum-wage jobs. People of color and women make up a disproportionate part of this class. The poor include the *working poor*—those who work at least 27 hours a week but whose wages fall below the federal poverty level. Six percent of all working people now live below the poverty line, a proportion that has generally increased over time. Among White workers, men and women are nearly equivalent in the likelihood of being among the working poor, but among African American workers, Black women are almost twice as likely as Black men to be among the working poor (U.S. Bureau of Labor Statistics 2008).

The concept of the **urban underclass** has been added to the lower class (Wilson 1987). The underclass includes those who are likely to be permanently unemployed and without much means of economic support. The underclass has little or no opportunity for movement out of the worst poverty. Rejected from the economic system, those in the underclass may become dependent on public assistance or illegal activities. Structural transformations in the economy have left large groups of people, especially urban minorities, in these highly vulnerable positions. The growth of the urban underclass has exacerbated the problems of urban poverty and related social problems (Wilson 1996, 1987).

Class Conflict

Derived from the early work of Karl Marx, sociologists also analyze class according to the perspective of conflict theory. Conflict theory defines classes in terms of their structural relationship to other classes and their relationship to the economic system. The analysis of class from this sociological perspective interprets inequality as resulting from the unequal distribution of power and resources in society (see Chapter 1). Sociologists who work from a conflict perspective see the classes as facing off against each other, with elites exploiting and dominating others. The key idea in this model is that class is not simply a matter of what individuals possess in terms of income and prestige; instead, class is defined by the relationship of the classes to the larger system of economic production (Vanneman and Cannon 1987; Wright 1985).

From a conflict perspective, the middle class, or the *professional–managerial class,* includes managers, supervisors, and professionals. Members of this group have substantial control over other people, primarily through their authority to direct the work of others, impose and enforce regulations in the workplace, and determine dominant social values. Although, as Marx argued, the middle class is controlled by the ruling class, members of this class tend to identify with the interests of the elite. The professional–managerial class, however, is caught in a contradictory position between elites and the working class. Like elites, those in this class have some control over others, but like the working class, they have minimal control over the economic system (Wright 1979). Karl Marx argued that as capitalism progresses, more and more of those in the middle class drop into the working class as they are pushed out of managerial jobs into working-class jobs or as professional jobs become organized more along the lines of traditional working-class employment.

Has this happened? Not to the extent Marx predicted. He thought that ultimately there would be only two classes—the capitalist and the proletariat. To some extent, however, this is occurring. Classes have become more polarized, with the well-off accumulating even more resources and the middle class seeing their income as either flat or falling, measured in constant dollars (Mishel et al. 2007). Rising levels of debt among the middle class have contributed to this growing inequality. Many now have a fragile hold on being middle class: The loss of a job, a family emergency, such as the death of a working parent, divorce, disability, or a prolonged illness, can quickly leave middle- and working-class families in a precarious financial state. At the same time, high salaries for CEOs, tax loopholes that favor the rich, and sheer greed are concentrating more wealth in the hands of a few.

Members of the working class have little control over their own work lives; instead, they generally have to take orders from others. This concept of the working class departs from traditional blue-collar definitions of working-class jobs because it includes many so-called white-collar workers (secretaries, salespeople, and nurses), any group working under the rules imposed by managers. The middle class may exercise some autonomy at work, but the working class has little power to challenge decisions of those who supervise them, except insofar as they can organize collectively, as in unions, strikes, or other collective work actions.

Whether you see the class system as a ladder or as a system of conflict, you can see that the class structure in the United States is hierarchical. Class position gives different people access to jobs, income, education, power, and social status, all of which bestow further opportunities on some and deprive others of success. People sometimes move from one class to another (although this is not the norm), but the class structure is a system with boundaries built into it, generating class conflict. The middle and working classes shoulder much of the tax burden for social programs, producing resentment by these groups toward the poor. At the same time, corporate taxes have declined and tax loopholes for the rich have increased, an indication of the privilege that is perpetuated by the class

Labor unions, traditionally dominated by White men in the skilled trades, are not only more diverse, but also represent workers in occupations typically thought of as "white-collar" work.

system. Whatever features of the class system different sociologists study, they see class stratification as a dynamic process—one involving the interplay of access to resources, judgments about different groups, and the exercise of power by a few.

Growing Inequality and the Distribution of Wealth and Income

Income is the amount of money brought into a household from various sources (wages, investment income, dividends, and so on) during a given period. In recent years, income growth has been greatest for those at the top of the population (see Figure 8.3); for everyone else, income (controlling for the value of the dollar) has been relatively flat. This has contributed to growing inequality in the United States. But inequality becomes even more apparent when you consider both wealth and income.

Wealth is the monetary value of everything one actually owns. It is calculated by adding all financial assets (stocks, bonds, property, insurance, savings, value of investments, and so on) and subtracting debts, which gives a dollar amount that is one's **net worth.** Wealth allows you to accumulate assets over generations, giving advantages to subsequent generations that they might not have had on their own. Unlike income, wealth is cumulative—that is, its value tends to increase through investment; it can be passed on to the next generation, giving those who inherit wealth a considerable advantage in accumulating more resources.

To understand the significance of wealth compared to income in determining class location, imagine two college graduates graduating in the same year, from the same college, with the same major and same grade point average. Imagine further that upon graduation, both get jobs with the same salary in the same organization. Yet, in one case parents paid all the student's college expenses and gave her a car upon graduation. The other student worked while in school and graduated with substantial debt from student loans. This student's family has no money with which to help support the new worker. Who is better off? Same salary, same credentials, but wealth (even if modest) matters. It gives one person an advantage that will be played out many times over as the young worker buys a home, finances her own children's education, and possibly inherits additional assets.

Where is all the wealth? The wealthiest 1 percent own 33 percent of all net worth; the bottom 80 percent control only 16 percent. The top 10 percent also own 88 percent of all stock; the bottom 40 percent own less than 1 percent of total stock holdings. Moreover, there has been an increase in the concentration of wealth since the 1980s, making the United States one of the most "unequal" nations in the world (Mishel et al. 2007). The growth of wealth by a select few, though long a feature of the U.S. class system, has also reached historic levels. As just one example,

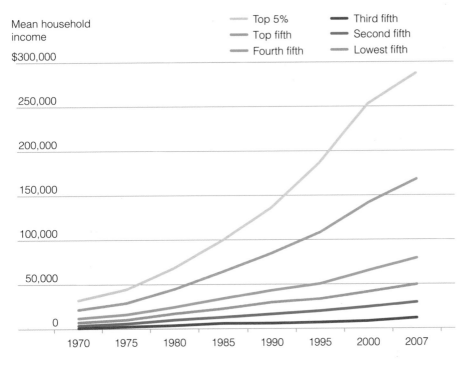

FIGURE 8.3 *Mean Household Income Received by Different Income Groups, 1970–2007*

As you can see in this graph, for most income groups in the United States, income (adjusted for rate of inflation) has remained relatively flat in recent years. The top groups, however, have seen huge increases in income, a trend resulting in growing inequality in the United States.

Source: U.S. Census Bureau. 2008. *Historical Income Tables Households, Table H-3.*
www.census.gov

John D. Rockefeller is typically heralded as one of the wealthiest men in U.S. history. But comparing Rockefeller with Bill Gates, controlling for the value of today's dollars, Gates has far surpassed Rockefeller's riches (Myerson 1998).

In contrast to the vast amount of wealth and income controlled by elites, a very large proportion of Americans have hardly any financial assets once debt is subtracted. Figure 8.4 shows the net worth of different income brackets in the United States, and you can see that most of the population has very low net worth. Moreover, 18 percent of the population have zero or negative net worth, usually because their debt exceeds their assets. The American Dream of owning a home, a new car, taking annual vacations, and sending one's children to good schools—not to mention saving for a comfortable retirement—is increasingly unattainable by many. When you see the amount of income and wealth controlled by a small segment of the population, a sobering picture of class inequality emerges.

Despite the prominence of rags-to-riches stories in American legend, most wealth in this society is inherited. A few individuals make their way into the elite class by virtue of their own success, but this is rare. The upper class is also overwhelmingly White and Protestant. The wealthy also exercise tremendous political power by funding lobbyists, exerting their social and personal influence on other elites, and contributing heavily to political campaigns. Studies of elites also find that they tend to be politically quite conservative (Burris 2000; Zweigenhaft and Domhoff 2006). They travel in exclusive social networks that tend to be open only to those in the upper class. They tend to intermarry, their children are likely to go to expensive schools, and they spend their leisure time in exclusive resorts.

Race also influences the pattern of wealth distribution in the United States; for every dollar of wealth held by White Americans, Black Americans have only 26 cents. Overall, White families have ten times the wealth of Black households. And, at all levels of income, occupation, and education, Black families have lower levels of wealth than similarly situated White families. Being able to draw on assets during times of economic stress means that families with some resources can better withstand difficult times

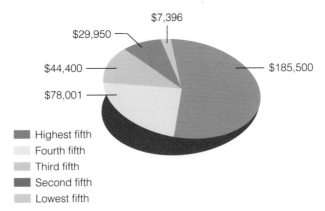

FIGURE 8.4 *Median Net Worth by Household Income*

Recall that one's net worth is the value of everything owned minus one's debt. You can see here the vast differences in wealth across different income groups. Remember that median is the midpoint; thus, in each of these groups, half of those in the group have more wealth; half, less.

Source: Orzechowski, Shawna, and Peter Sepielli. 2003. *Net Worth and Asset Ownership of Households: 1998 and 2000.* Washington, DC: U.S. Census Bureau, P70–88.

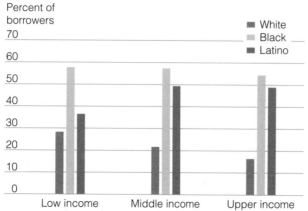

FIGURE 8.5 *Percentage of Borrowers with High-Cost Home Purchase Loans by Race and Income*

Source: ACORN. 2007. Foreclosure Exposure: A Study of Racial and Income Disparities in Home Mortgage Lending in 172 American Cities. Reprinted with permission. **www.acorn.org**

than those without assets. Even small assets, such as home ownership or a savings account, provide protection from crises such as increased rent, a health emergency, or unemployment. Because the effects of wealth are intergenerational—that is, they accumulate over time—just providing equality of opportunity in the present does not address the differences in class status that Black and White Americans experience (Oliver and Shapiro 2006; Mishel et al. 2007).

What explains the disparities in wealth by race? Wealth accumulates over time. Thus, government policies in the past have prevented Black Americans from being able to accumulate wealth. Discriminatory housing policies, bank lending policies, tax codes, and so forth have disadvantaged Black Americans, resulting in the differing assets Whites and Blacks in general hold now. Even though some of these discriminatory policies have ended, many continue. Either way, their effects persist, resulting in what sociologists Melvin Oliver and Thomas Shapiro call the *sedimentation of racial inequality.*

Understanding the significance of wealth in shaping life chances for different groups also challenges the view that all Hispanics have similar experiences and wealth. Cuban Americans and Spaniards are similar to Whites in their wealth holdings, whereas Mexicans, Puerto Ricans, Dominicans, and other Hispanic groups more closely resemble African Americans on the various indicators of wealth and social class. Likewise, one can better understand differences in class status among Asian American groups by carefully considering the importance not just of income, education, and occupation, but also patterns in the net assets of different groups (Oliver and Shapiro 2006).Without significant wealth holdings, families of any race are less able to transmit assets from previous generations to the next generation, one main support of social mobility.

Patterns of wealth contribute heavily to class inequality in the United States—inequality that is growing. Despite beliefs to the contrary, class divisions in the United States are real and becoming more pronounced. The elite are becoming better off, whereas many in the middle class, working class, and the poor seem to be struggling. Indeed, many middle-class people feel that their way of life is slipping away. Many factors have contributed to growing inequality in the United States, including the profound effects of national and global economic changes—changes that recently produced a so-called economic downturn. While some see the economic problems of the nation as stemming from individual greed on Wall Street, the problems are systemic, even though individual greed likely plays a role.

The structural problem is referred to as **economic restructuring,** referring to the decline of manufacturing jobs in the United States, the transformation of the economy by technological change, and the process of globalization. We examine economic restructuring more in Chapter 15 on the economy, but the point here is that these structural changes are having a profound effect on the life chances of people in different social classes. Many in the working class, for example, once largely employed in relatively stable manufacturing jobs with decent wages and good benefits, now likely work, if they work at all, in lower-wage jobs with fewer benefits, such as pensions or health care. Middle-class families may have amassed large sums of debt, sometimes to support a middle-class lifestyle, but also perhaps to pay for their children's education.

Economists have their explanations of why the nation has encountered such huge economic problems, but sociologists have their own way of thinking about this, namely, that economic problems are not purely economic: They are social, both in their origins and their consequences. Take home ownership as an example. Home ownership is one of the main components of the

The foreclosure crisis that unfolded in 2009 affected many, including whole communities, especially those of working class and minority residents.

>doing sociological research

The Fragile Middle Class

Research Question: The hallmark of the middle class in the United States is its presumed stability. Home ownership, a college education for children, and other accoutrements of middle-class status (nice cars, annual vacations, an array of consumer goods) are the symbols of middle-class prosperity. But the rising rate of bankruptcy among the middle class shows that the middle class is not as secure as it is presumed to be (see Figure 8.4). Personal bankruptcy has risen dramatically with now more than one million filings for bankruptcy per year. How can this be happening in such a prosperous society? This is the question examined by Teresa Sullivan, Elizabeth Warren, and Jay Lawrence Westbrook in their study of bankruptcy and debt among the middle class.

Research Method: They based their study on an analysis of official records of bankruptcy in five states, as well as on detailed questionnaires given to individuals who filed for bankruptcy.

Research Results: Their findings debunk the idea that bankruptcy is most common among poor people. Instead, they found bankruptcy is mostly a middle-class phenomenon representing a cross-section of those in this class (meaning that those who are bankrupt are matched on the demographic characteristics of race, age, and gender with others in the middle class). They also debunk the notion that bankruptcy is rising because it is so easy to file. Rather, they found many people in the middle class so overwhelmed with debt that they cannot possibly pay it off. Most often people file for bankruptcy as a result of job loss and lost wages. But divorce, medical problems, housing expenses, and credit card debt also drive many to bankruptcy court.

Conclusions and Implications: Sullivan and her colleagues explain the rise of bankruptcy as stemming from structural factors in society that fracture the stability of the middle class. The volatility of jobs under modern capitalism is one of the biggest factors, but add to this the "thin safety net"—no health insurance for many, but rising medical costs. Also, the American Dream of owning one's own home means many are "mortgage poor"—extended beyond their ability to keep up.

In addition, the United States is a credit-driven society. Credit cards are routinely mailed to people in the middle class, encouraging them to buy beyond their means. You can now buy virtually anything on credit: cars, clothes, doctor's bills, entertainment, groceries. You can even use one credit card to pay off other credit cards. Indeed, it is difficult to live in this society without credit cards. Increased debt is the result. Many are simply unable to keep up with compounding interest and penalty payments, and debt takes on a life of its own as consumers cannot keep up with even the interest payments on debt.

Sullivan, Warren, and Westbrook conclude that increases in debt and uncertainty of income combine to produce the fragility of the middle class. Their research shows that "even the most secure family may be only a job loss, a medical problem, or an out-of-control credit card away from financial catastrophe" (2000: 6).

Questions to Consider

1. Have you ever had a credit card? If so, how easy was it to get? Is it possible to get by without a credit card?

2. What evidence do you see in your community of the fragility or stability of different social class groups?

Source: T. A. Sullivan, E. Warren, and J. L. Westbrook, *The Fragile Middle Class: Americans in Debt.* Copyright © 2000 by Yale University Press.

American Dream. For most Americans, owning one's own home is the primary means of attaining economic security. It is also the key to other resources—good schools, cleaner neighborhoods, and an investment in the future. Likewise, losing your home is more than just a financial crisis—it reverberates through various aspects of your life.

Housing foreclosure is a trauma for anyone who experiences it, but foreclosure has hit some groups especially hard. Studies also find that one in ten African Americans will experience foreclosure, compared to one in twenty-five White Americans (Oliver and Shapiro 2008). Furthermore, entire communities have been affected by this crisis, with a strong correlation between foreclosure rates and the proportion of Black people living in a particular area (Bostic and Lee 2008). Why?

Some might argue that this occurred because individual people made bad decisions—buying homes beyond their means. Sociologists explain this as the result of institutional lending practices that target particular groups, making them more vulnerable to the economic forces that can shatter individual lives. One way institutions do this is through subprime loans—loans that carry interest rates higher than is conventional and that may also include adjustable interest rates that increase sharply over time. African Americans are 2.7 times more likely than White Americans to hold subprime mortgage loans; Latinos, 2.3 times more likely. Even more telling, these disparities exist even when the mortgage holders are at the same income level; upper-income African Americans and Latinos are *three times more likely* than White Americans in the same income bracket to hold subprime loans (ACORN 2007; Rivera at al. 2008). These facts reveal that there is a racial dimension to the foreclosure crisis that is not understood solely by individual economic decisions. Lenders may see African Americans as a greater credit risk, but they also know that the value of real estate is less in racially segregated neighborhoods. And, discriminatory practices in the housing market have been well documented (Oliver and Shapiro 2006; Squires 2007).

The sociological point is that economic problems have a sociological dimension and cannot be explained by individual decisions alone. Economic policies also have different effects for different groups—sometimes intended, sometimes not. Tax policies, for example, distribute economic benefits unevenly and according to specific sociological patterns. Corporations benefit the most from the tax structure, as corporate taxes have fallen in recent years; while most individual Americans are paying more in federal tax than ever before (an increase from 13 to 15 cents per dollar of income earned since 1990), corporate taxes since 1990 have fallen from 26 cents on the dollar to 20 cents, even though corporate profits were up 252 percent in that period (Johnston 2000). Tax policies also benefit those at the upper ends of the class system who can take advantage of numerous tax benefits and loopholes. And, as many in the public understood, the Congressional bailouts of recent years have had far more benefit for corporations than for the individuals suffering from the financial crisis.

Diverse Sources of Stratification

Class is only one basis for stratification in the United States. Factors such as age, ethnicity, and national origin have a tremendous influence on stratification. Race and gender are two primary influences in the stratification system in the United States. In fact, analyzing class without also analyzing race and gender can be misleading. Race, class, and gender, as we are seeing throughout this book, are overlapping systems of stratification that people experience simultaneously. A working-class Latina, for example, does not experience herself as working class at one moment, Hispanic at another moment, and a woman the next. At any given point in time, her position in

society is the result of her race, class, *and* gender status. In other words, class position is manifested differently, depending on one's race and gender, just as gender is experienced differently depending on one's race and class, and race is experienced differently depending on one's gender and class. Depending on one's circumstances, race, class, or gender may seem particularly salient at a given moment in a person's life. For example, a Black middle-class man stopped and interrogated by police when driving through a predominantly White middle-class neighborhood may at that moment feel his racial status as his single most outstanding characteristic, but at all times his race, class, and gender influence his life chances. As social categories, race, class, and gender shape all people's experience in this society, not just those who are disadvantaged (Andersen and Collins 2007).

Class also significantly differentiates group experience within given racial and gender groups. Latinos, for example, are broadly defined as those who trace their origins to regions originally colonized by Spain. The ancestors of this group include both White Spanish colonists and the natives who were enslaved on Spanish plantations. Today, some Latinos identify as White, others as Black, and others by their specific national and cultural origins. The very different histories of those categorized as Latino are matched by significant differences in class. Some may have been schooled in the most affluent settings; others may be virtually unschooled. Those of upper-class standing may have had little experience with prejudice or discrimination; others may have been highly segregated into barrios and treated with extraordinary prejudice. Latinos who live near each other geographically in the United States and who are the same age and share similar ancestry may have substantially different experiences based on their class standing (Massey 1993). Neither class, race, nor gender, taken alone, can be considered an adequate indicator of different group experiences.

The Race–Class Debate. The relationship between race and class is much debated among sociologists. The Black middle class goes all the way back to the small numbers of free Blacks in the eighteenth and nineteenth centuries (Frazier 1957), expanding in the twentieth century to include those who were able to obtain an education and become established in industry, business, or a profession. Although wages for Black middle-class and professional workers never matched those of Whites in the same jobs, within the Black community the Black middle class has had relatively high prestige. Many sociologists conclude that the class structure among African Americans has existed alongside the White class structure, separate and different.

In recent years, both the African American and Latino middle classes have expanded, primarily as the result of increased access to education and middle-class occupations for people of color (Higginbotham 2001; Pattillo-McCoy 1999). Although middle-class

Blacks and Latinos may have economic privileges that others in these groups do not have, their class standing does not make them immune to the negative effects of race. Asian Americans also have a significant middle class, but they have also been stereotyped as the most successful minority group because of their presumed educational achievement, hard work, and thrift. This stereotype is referred to as the *myth of the model minority* and includes the idea that a minority group must adopt alleged dominant group values to succeed. This myth about Asian Americans obscures the significant obstacles to success that Asian Americans encounter, and it ignores the hard work and educational achievements of other racial and ethnic groups. The idea that Asian Americans are the "model minority" also obscures the high rates of poverty among many Asian American groups (Lee 1996).

Despite recent successes, many in the Black middle class have a tenuous hold on this class status. The Black middle class remains as segregated from Whites as the Black poor, and continuing racial segregation in neighborhoods means that Black middle-class neighborhoods are typically closer to Black poor neighborhoods than the White middle-class neighborhoods are to White poor ones. This exposes many in the Black middle class to some of the same risks as those in poverty. This is not to say that the Black middle class has the same experience as the poor, but it challenges the view that the Black middle class "has it all" (Pattillo-McCoy 1999; Lacy 2007) Furthermore, Black Americans are still much more likely to be working class than middle class; they are also more likely to be working class than are Whites (Horton et al. 2000).

The Influence of Gender and Age. The effects of gender further complicate the analysis of class. In the past, women were thought to derive their class position from their husbands or fathers, but sociologists now challenge this assumption. Measured by their own income and occupation, the vast majority of women would likely be considered working class. The median income for women, even among those employed full time, is far below the national median income level. In 2007, when median income for men working year-round and full time was $45,113, the median income of women working year-round, full time was $35,102 (DeNavas-Walt et al. 2008). The vast majority of women work in low-prestige and low-wage occupations, even though women and men have comparable levels of educational attainment.

Age, too, is a significant source of stratification. The age group most likely to be poor are children, 18 percent of whom live in poverty in the United States. This represents a change from the recent past when the aged were the most likely to be poor (see Figure 8.6). Although many elderly people are now poor (10 percent of those 65 and over), far fewer in this age category are poor than was the case not many years ago (DeNavas-Walt et al. 2008). This shift reflects the greater affluence of the older segments of the population—a trend that is likely to continue as the current large cohort of middle-aged, middle-class Baby Boomers grow older.

Social Mobility: Myths and Realities

Popular legends extol the possibility of anyone becoming rich in the United States. The well-to-do are admired not just for their style of life but also for their supposed drive and diligence. The admiration for those who rise to the top makes it seem like anyone who is clever enough and works hard can become fabulously

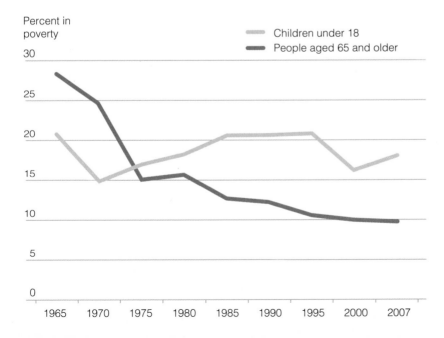

FIGURE 8.6 *Poverty Among the Old and the Young, 1965–2007*

Source: DeNavas-Walt, Carmen, Bernadette Proctor, and Jessica C. Smith. 2008. *Income, Poverty, and Health Insurance Coverage in the United States: 2007.* Washington, DC: U.S. Census Bureau, #P60-235.

rich. The assumption is that the United States class system is a **meritocracy**—that is, a system in which one's status is based on merit or accomplishments, not other social characteristics. As the word suggests, in a meritocracy people move up and down through the class system based on merit, not based on other characteristics. Is this the case in the United States?

Defining Social Mobility

Social mobility is a person's movement over time from one class to another. Social mobility can be up or down, although the American Dream emphasizes upward movement. Mobility can be either *intergenerational*, occurring between generations, as when a daughter rises above the class of her mother or father; or *intra*generational, occurring within a generation, as when a person's class status changes as the result of business success (or disaster).

Societies differ in the extent to which social mobility is permitted. Some societies are based on *closed class systems,* in which movement from one class to another is virtually impossible. In a caste system, for example, mobility is strictly limited by the circumstances of one's birth. At the other extreme are *open class systems,* in which placement in the class system is based on individual achievement, not ascription. In open class systems, there are relatively loose class boundaries, high rates of class mobility, and weak perceptions of class difference.

The Extent of Social Mobility

Does social mobility occur in the United States? Social mobility is much more limited than people believe. Success stories of social mobility do occur, but research finds that experiences of mobility over great distances are rare, certainly far less than believed. Most people remain in the same class as their parents. What mobility exists is typically short in distance, and some people actually drop to a lower status, referred to as *downward social mobility*. Research finds that rates of upward social mobility are highest among White men, followed by White women, then Black men, and finally, Black women (Mazunder 2008). Currently, half of all Americans say that they are not moving forward, and one-third say they have fallen further back (Acs and Zimmerman 2008). Evidence also suggests that mobility between generations may be becoming even more rigid than in the past (Sawhill and McClanahan 2006).

Social mobility is influenced most by factors that affect the whole society, not just by individual characteristics. Just being born in a particular generation can have a significant influence on one's life chances. The fears of today's young, middle-class people that they will be unable to achieve the lifestyles of their parents show the effect that being in a particular generation can have on one's life chances. In other words, when mobility occurs, it is usually because of societal changes that create or restrict opportunities, including such changes as economic cycles, changes in the occupational

structure, and demographic factors, such as the number of college graduates in the labor force (Beller and Hout 2006). But mobility in the United States is not impossible. Indeed, many have immigrated to this nation with the knowledge that their life chances are better here than in their countries of origin. And, the social mobility that does exist is greatly influenced by education. But, in sum, social mobility is much more limited than the American Dream of mobility suggests.

Class Consciousness

Because of the widespread belief that mobility is possible, people in the United States, compared to many other societies, tend not to be very conscious of class. **Class consciousness** is the perception that a class structure exists along with a feeling of shared identification with others in one's class—that is, those with whom you share life chances (Centers 1949). Notice that there are two dimensions to the definition of class consciousness: the idea that a class structure exists and one's class identification.

There has been a long-standing argument that Americans are not very class conscious because of the belief that upward mobility is possible and because of the belief in individualism that is part of the culture. Images of opulence also saturate popular culture, making it seem that such material comforts are available to anyone. The faith that upward mobility is possible ironically perpetuates inequality since, if people believe that everyone has the same chances of success, they are likely to think that whatever inequality exists must be fair or the result of individual success and failure.

Class inequality in any society is usually buttressed by ideas that support (or actively promote) inequality. Beliefs that people are biologically, culturally, or socially different can be used to justify the higher position of some groups. If people believe these ideas, the ideas provide legitimacy for the system. Karl Marx used the term **false consciousness** to describe the class consciousness of subordinate classes who had internalized the view of the dominant class. Marx argued that the ruling class controls subordinate classes by infiltrating their consciousness with belief systems that are consistent with the interests of the ruling class. If people accept these ideas, which justify inequality, they need not be overtly coerced into accepting the roles designated for them by the ruling class.

Class consciousness in the United States has been higher at some times than others. Now, 46 percent of the public identifies as working class, 46 percent as middle class, 5 percent as lower class, and 3 percent as upper class (National Opinion Research Center 2004). There have been times when class consciousness was higher, such as during the labor movement of the 1920s and 1930s. Then, working-class people had a very high degree of class consciousness and mobilized on behalf of workers' rights. But now the formation of

a sociological eye
on the media

Reproducing Class Stereotypes

The media have a substantial impact on how people view the social class system and different groups within it. Especially because people tend to live and associate with people in their own class, how they see others can be largely framed by the portrayal of different class groups in the media. Research has found this to be true and, in addition, has found that mass media have the power to shape public support for policies for public assistance.

To begin with, the media overrepresent the lifestyle of the most comfortable classes. It is the rare family that can afford the home decor and fashion depicted even in soap operas, ironically most likely watched by those in the working class. Media portrayals, such as those found on television talk shows as well as sports, tend to emphasize stories of upward mobility. When the working class is depicted, it tends to be shown as deviant, reinforcing class antagonism and giving viewers a sense of moral and "class superiority" (Gersch 1999).

Content analyses of the media also find that the poor are largely invisible in the media (Mantsios 2001). Those poor people who are depicted in television and magazines are more often portrayed as Black than is actually the case, leading people to overestimate the actual number of the Black poor. The elderly and working poor are rarely seen (Clawson and Trice 2000; Gilens 1996). Representations of welfare overemphasize themes of dependency, especially when the portrayal is of African Americans. Women are also more likely than men to be represented as dependent (Misra et al. 2003). And rarely are welfare activists shown as experts; rather, public officials are typically given the voice of authority (Ryan 1996). One result is that the media end up framing the "field of thinkable solutions to public problems" (Sotirovic 2000, 2001), but do so within a context that ignores the social structural context of social issues.

Further resources: See the film, *Class Dismissed: How TV Frames the Working Class*, Media Education Foundation. **www.mediaed.org**

Sources: Gersch, Beate. 1999. "Class in Daytime Talk Television." *Peace Review* 11 (June): 275–281; Sotirovic, Mira. 2001. "Media Use and Perceptions of Welfare." *Journal of Communication* 51 (December): 750–774; Sotirovic, Mira. 2000. "Effects of Media Use on Audience Framing and Support for Welfare." *Mass Communication & Society* 2-3 (Spring-Summer): 269–296; Bullock, Heather E., Karen Fraser, and Wendy R. Williams. 2001. "Media Images of the Poor." *The Journal of Social Issues* 57 (Summer): 229–246; Clawson, Rosalee A., and Rakuya Trice. 2000. "Poverty as We Know It: Media Portrayals of the Poor." *The Public Opinion Quarterly* 64 (Spring): 53–64; Gilens, Martin. 1996. "Race and Poverty in America: Public Misperceptions and the American News Media." *Public Opinion Quarterly* 60 (Winter): 515–541; Misra, Joy, Stephanie Moller, and Marina Karides. 2003. "Envisioning Dependency: Changing Media Depictions of Welfare in the 20th Century." *Social Problems* 50 (November): 482–504; Ryan, Charlotte. 1996. "Battered in the Media: Mainstream News Coverage of Welfare Reform." *Radical America* 26 (August): 29–41.

a relatively large middle class and a relatively high standard of living militates against class discontent. Racial and ethnic divisions also make strong alliances within various classes less stable. The growing inequality of today could result in a higher degree of class consciousness, but this has not yet developed into a significant class-based movement for change.

Why Is There Inequality?

Stratification occurs in all societies. Why? This question originates in classical sociology in the works of Karl Marx and Max Weber, theorists whose work continues to inform the analysis of class inequality today.

Karl Marx: Class and Capitalism

Karl Marx (1818–1883) provided a complex and profound analysis of the class system under capitalism—an analysis that, although more than 100 years old, continues to inform sociological analyses and has been the basis for major world change. Marx defined classes in relationship to *the means of production,* defined as the system by which goods are produced and distributed. In Marx's analysis, two primary classes exist under capitalism: the *capitalist class,* those who own the means of production, and the *working class* (or proletariat), those who sell their labor for wages. There are further divisions within these two classes: the *petty bourgeoisie,* small business owners and managers (those whom you might think of as middle class) who identify with the interests of the capitalist class but do not own the means of production, and the *lumpenproletariat,* those who have become unnecessary as workers and are then discarded. (Today, these would be the underclass, the homeless, and the permanently poor.)

Marx thought that with the development of capitalism, the capitalist and working class would become increasingly antagonistic (something he referred to as class struggle). As class conflicts became more intense, the two classes would become more polarized, with the petty bourgeoisie becoming deprived of their property and dropping into the working class. This analysis is still reflected in contemporary questions about whether the classes are becoming more polarized, with the rich getting richer and everyone else worse off, as we have seen.

In addition to the class struggle that Marx thought would characterize the advancement of capitalism, he also thought that capitalism was the basis for other social institutions. Capitalism is the *infrastructure* of society, with other institutions (such as law, education, the family, and so forth) reflecting capitalist interests. Thus, according to Marx, the law supports the interests of capitalists; the family promotes values that socialize people into appropriate work roles; and education reflects the interests of the capitalist class. Over time, capitalism increasingly penetrates society, as we can clearly see with the corporate mergers that characterize modern life and the predominance of capitalist values in society's institutions.

Why do people support such a system? Here is where ideology plays a role. **Ideology** refers to belief systems that support the status quo. According to Marx, the dominant ideas of a society are promoted by the ruling class. Through their control of the communications industries in modern society, the ruling class is able to produce ideas that buttress their interests.

Much of Marx's analysis boils down to the consequences of a system based on the pursuit of profit. If goods were exchanged at the cost of producing them, no profit would be produced. Capitalist owners want to sell commodities for more than their actual value—more than the cost of producing them, including materials and labor. Because workers contribute value to the system and capitalists extract value, Marx saw capitalist profit as the exploitation of labor. Marx believed that as profits became increasingly concentrated in the hands of a few capitalists, the working class would become increasingly dissatisfied. The basically exploitative character of capitalism, according to Marx, would ultimately lead to its destruction as workers organized to overthrow the rule of the capitalist class. *Class conflict* between workers and capitalists, he argued, was inescapable, with revolution being the inevitable result. Perhaps the class revolution that Marx predicted has not occurred, but the dynamics of capitalism that he analyzed are unfolding before us.

At the time Marx was writing, the middle class was small and consisted mostly of small business owners and managers. Marx saw the middle class as dependent on the capitalist class, but exploited by it, because the middle class did not own the means of production. He saw middle-class people as identifying with the interests of the capitalist class because of the similarity in their economic interests and their dependence on the capitalist system. Marx believed that the middle class failed to work in its own best interests because it falsely believed that it benefited from capitalist arrangements. Marx thought that in the long run the middle class would pay for their misplaced faith when profits became increasingly concentrated in the hands of a few and more and more of the middle class dropped into the working class. Because he did not foresee the emergence of the large and highly differentiated middle class we have today, not every part of Marx's theory has proved true. Still, his analysis provides a powerful portrayal of the forces of capitalism and the tendency for wealth to belong to a few, whereas the majority work only to make ends meet. He has also influenced the lives of billions of people under self-proclaimed Marxist systems that were created in an attempt, however unrealized, to overcome the pitfalls of capitalist society.

Max Weber: Class, Status, and Party

Max Weber (1864–1920) agreed with Marx that classes were formed around economic interests, and he agreed that material forces (that is, economic forces) have a powerful effect on people's lives. However, he disagreed with Marx that economic forces are the primary dimension of stratification. Weber saw three dimensions to stratification:

- *class* (the economic dimension);
- *status* (or prestige, the cultural and social dimension); and,
- *party* (or power, the political dimension).

Weber is thus responsible for a *multidimensional view* of social stratification because he analyzed the connections between economic, cultural, and political systems. Weber pointed out that, although the economic, social, and political dimensions of stratification are usually related, they are not always consistent. A person could be high on one or two dimensions, but low on another. A major drug dealer is an example: high wealth (economic dimension) and power (political dimension) but low prestige (social dimension), at least in the eyes of the mainstream society, even if not in other circles.

Weber defined *class* as the economic dimension of stratification—how much access to the material goods of society a group or individual has, as measured by income, property, and other financial assets. A family with an income of $100,000 per year clearly has more access to the resources of a society than a family living on an income of $40,000 per year. Weber understood that a class has common economic interests and that economic well-being was the basis for one's life chances. But, in addition, he thought that people were also stratified based on their status and power differences.

Status, to Weber, is the prestige dimension of stratification—the social judgment or recognition given to a person or group. Weber understood that class distinctions are linked to status distinctions—that is, those with the most economic resources tend to have the highest status in society, but not always. In a local community, for example, those with the most status may be those who have lived there the longest, even if newcomers arrive with more money. Although having power is typically related to also having high economic standing and high social status, this is not always the case, as you saw with the example of the drug dealer.

Finally, *party* (or what we would now call power) is the political dimension of stratification. It is the capacity to influence groups and individuals even in the face of opposition. Power is also reflected in the ability of a person or group to negotiate their way through social institutions. An unemployed Latino man wrongly accused of a crime, for instance, does not have much power to negotiate his way through the criminal justice system. By comparison, business executives accused of corporate crime can afford expensive lawyers and, thus, frequently go unpunished or, if they are found guilty, serve relatively light sentences in comparatively pleasant facilities. Again, Weber saw power as linked to economic standing, but he did not think that economic standing was always the determining cause of people's power.

Marx and Weber explain different features of stratification. Both understood the importance of the economic basis of stratification, and they knew the significance of class for determining the course of one's life. Marx saw people as acting primarily out of economic interests. Weber refined the sociological analyses of stratification to account for the subtleties that can be observed when you look beyond the sheer economic dimension to stratification, stratification being the result of economic, social, and political forces. Together, Marx and Weber provide compelling theoretical grounds for understanding the contemporary class structure.

Functionalism and Conflict Theory: The Continuing Debate

Marx and Weber were trying to understand why differences existed in the resources and power that different groups in society hold. The question persists of why there is inequality. Two major frameworks in sociological theory—functionalist and conflict theory—take quite different approaches to understanding inequality (see Table 8.2, p. 200).

The Functionalist Perspective on Inequality. Functionalist theory views society as a system of institutions organized to meet society's needs (see Chapter 1). The functionalist perspective emphasizes that the parts of society are in basic harmony with each other; society is characterized by cohesion, consensus, cooperation, stability, and persistence (Parsons 1951a; Merton 1957; Eitzen and Baca Zinn 2007). Different parts of the social system complement one another and are held together through social consensus and cooperation. To explain stratification, functionalists propose that the roles filled by the upper classes—such as governance, economic innovation, investment, and management—are essential for a cohesive and smoothly running society and hence are rewarded in proportion to their contribution to the social order (Davis and Moore 1945).

According to the functionalist perspective, social inequality serves an important purpose in society: It motivates people to fill the different positions in society that are needed for the survival of the whole. Functionalists think that some positions in society are more important than others and require the most talent and training. Rewards attached to those positions (such as higher income and prestige) ensure that people will make the sacrifices needed to acquire the training for functionally important positions (Davis and Moore 1945). Higher class status thus comes to those who acquire what is needed for success (such as education and job training). In other words, functionalist theorists see inequality as based on a reward system that motivates people to succeed.

The Conflict Perspective on Inequality. Conflict theory also sees society as a social system, but unlike functionalism, conflict theory interprets society as being held together through conflict and coercion. From a conflict-based perspective, society comprises competing interest groups, some with more power than others. Different groups struggle over societal resources and compete for social advantage. Conflict theorists argue that those who control society's resources also hold power over others. The powerful are also likely to act to reproduce their advantage and try to shape societal beliefs to make their privileges appear to be legitimate and fair. In sum, conflict theory emphasizes the friction in society rather than the coherence and sees society as dominated by elites.

From the perspective of conflict theory, derived largely from the work of Karl Marx, social stratification is based on class conflict and blocked opportunity. Conflict theorists see stratification as a system of domination and subordination in which those with the most resources exploit and control others. They also see the different classes as in conflict with each other, with the unequal distribution of rewards reflecting the class interests of the powerful, not the survival needs of the whole society (Eitzen and Baca Zinn 2007). According to the conflict perspective,

table 8.2   Functionalist and Conflict Theories of Stratification

Interprets	Functionalism	Conflict Theory
Inequality	Inequality serves an important purpose in society by motivating people to fill the different positions in society that are needed for the survival of the whole.	Inequality results from a system of domination and subordination where those with the most resources exploit and control others.
Class structure	Differentiation is essential for a cohesive society.	Different groups struggle over societal resources and compete for social advantage.
Reward system	Rewards are attached to certain positions (such as higher income and prestige) as a way to ensure that people will make the sacrifices needed to acquire the training for functionally important positions in society.	The more stratified a society, the less likely that society will benefit from the talents of all its citizens, because inequality prevents the talents of those at the bottom from being discovered and used.
Classes	Some positions in society are more functionally important than others and are rewarded because they require the greatest degree of talent and training.	Classes exist in conflict with each other as they vie for power and economic, social, and political resources.
Life chances	Those who work hardest and succeed have greater life chances.	The most vital jobs in society—those that sustain life and the quality of life—are usually the least rewarded.
Elites	The most talented are rewarded in proportion to their contribution to the social order.	The most powerful reproduce their advantage by distributing resources and controlling the dominant value system.
Class consciousness/ideology	Beliefs about success and failure confirm the status of those who succeed.	Elites shape societal beliefs to make their unequal privilege appear to be legitimate and fair.
Social mobility	Upward mobility is possible for those who acquire the necessary talents and tools for success (such as education and job training).	There is blocked mobility in the system because the working class and poor are denied the same opportunities as others.
Poverty	Poverty serves economic and social functions in society.	Poverty is inevitable because of the exploitation built into the system.
Social policy	Because the system is basically fair, social policies should only reward merit.	Because the system is basically unfair, social policies should support disadvantaged groups by redirecting society's resources for a more equitable distribution of income and wealth.

inequality provides elites with the power to distribute resources, make and enforce laws, and control value systems; elites use these powers in ways that reproduce inequality. Others in the class structure, especially the working class and the poor, experience blocked mobility.

Conflict theorists argue that the consequences of inequality are negative. From a conflict point of view, the more stratified a society, the less likely that society will benefit from the talents of its citizens; inequality limits the life chances of those at the bottom, preventing their talents from being discovered and used. To the waste of talent is added the restriction of human creativity and productivity.

The Debate Between Functionalist and Conflict Theories. Implicit in the argument of each perspective is criticism of the other perspective. Functionalism assumes that the most highly rewarded jobs are the most important for society, whereas conflict theorists argue that some of the most vital jobs in society—those that sustain life and the quality of life, such as farmers, mothers, trash collectors, and a wide range of other laborers—are usually the least rewarded. Conflict theorists also criticize functionalist theory for assuming that the most talented get the greatest rewards. They point out that systems of stratification tend to devalue the contributions of those left at the bottom and to underutilize the diverse talents of all people (Tumin 1953). In contrast, functionalist theorists contend that the conflict view of how economic interests shape social organization is too simplistic. Conflict theorists respond by arguing that functionalists hold too conservative a view of society and overstate the degree of consensus and stability that exists.

The debate between functionalist and conflict theorists raises fundamental questions about how people

[*what would* *they say now?*

Classical Theorists Reflect on Social Class and Sports

Suppose some of the classical theorists of sociology observed an American sports event? What might they say about social class and sports?

Emile Durkheim: Truthfully, my theories do not address social class that much because I am more interested in the cultural symbols and events that bind people together. So I notice that people wear a lot of sports symbols, like jerseys and hats, and they put a lot of these symbols on their cars! It certainly projects an identity to others that must make them feel part of some collective group. I can't help but wonder if some of these symbolic identities are connected to people's social class location. I mean, I don't see really rich people wearing NASCAR caps, but I do see them with various yacht club logos on their polo shirts and ties.

Max Weber: I think there is more than one dimension to this question of the connection between social class and sports. Everywhere I look I see class, power, and prestige all tangled up in sports. First, look at the class differences associated with different sports: Some seem to have much more prestige than others. But there is also a big connection to politics. Wow—during elections, there are politicians all over the place—at tailgate parties and hanging out in the expensive box seats. It just seems like sports are really embedded in the culture of this society, something especially suggested by the phrase an "All-American."

Karl Marx: I just see corporate profits and exploitation everywhere I look. My goodness, not only are the stadiums named for corporate sponsors, but I've even heard various plays labeled by a sponsor, like the "AT&T All-American Play of the Week!" And the money that is spent on the commercials! It makes me faint to think of how much is spent just to get the fans to buy a particular brand of beer! Social class is everywhere in sports, although the workers are often invisible. Sure, some of the athletes are very highly paid, but it's the working-class people holding this all together. Who do you think cleans up these stadiums after the fans leave, or cooks the food, and takes out all the trash? It's an amazing example of a capitalist social system!

W.E.B. Du Bois: There has been a strong tendency throughout U.S. history to use Black people as material resources that yield dividends for others. I am as thrilled as anyone else to see the incredible athletic accomplishments of my people, and it seems that sports are one place where the color line is at least blurred. But there are yet too many instances where racial prejudice rears its head. Black people will not be free in any context until the color line in sports, and elsewhere, is gone.

---

view inequality. Is it inevitable? How is inequality maintained? Do people basically accept it? This debate is not just academic. The assumptions made from each perspective frame public policy debates. Whether the topic is taxation, poverty, or homelessness, if people believe that anyone can get ahead by ability alone, they will tend to see the system of inequality as fair and accept the idea that there should be a differential reward system. Those who tend toward the conflict view of the stratification system are more likely to advocate programs that emphasize public responsibility for the well-being of all groups and to support programs and policies that result in more of the income and wealth of society going toward the needy.

Poverty

Many people in the United States were shocked when, following the devastation of Hurricane Katrina in New Orleans and the Gulf Coast, thousands of poor, mostly African American people were seen on national TV in horrific circumstances, struggling to stay alive and visibly poor. Katrina uncovered one of the faces of poverty in the United States. For many people it was surprising to see conditions that are normally associated with poor, underdeveloped nations right here in the United States in one of our major and beloved cities.

The truth is that, despite the relatively high average standard of living in the United States, poverty afflicts millions of people. And the particulars of poverty are deeply related to the social structures of class, race, and gender. In New Orleans, for example, when Katrina hit, 28 percent of the population was poor—twice the national rate. Two-thirds of families headed by women in New Orleans were poor, twice the national rate (U.S. Census Bureau 2004). Although a disaster like Katrina can hurt anyone— and did harm people of differing social and economic statuses—natural disasters tend to have a particularly hard impact on the most vulnerable social groups, such as the poor. This shows that disasters are not just natural phenomena; they also have sociological dimensions (Bobo and Dawson 2006; Hartman and Squires 2006).

The poverty that Katrina unmasked was not, of course, news to social scientists who have long analyzed the extent of poverty in the United States and how it affects society's problems. Poor health care, failures in the education system, and crime are all related to poverty. Who is poor, and why is there so much poverty in an otherwise relatively affluent society?

The federal government has established an official definition of poverty used to determine eligibility for government assistance and to measure the extent of poverty in the United States. The **poverty line** is the amount of money needed to support the basic needs of a household, as determined by government; below this line, one is considered officially poor. To determine the poverty line, the Social Security Administration takes a low-cost food budget (based on dietary information provided by the U.S. Department of Agriculture) and multiplies by a factor of three, assuming that a family spends approximately one-third of its budget on food. The resulting figure is the official poverty line, adjusted slightly each year for increases in the cost of living. In 2007, the official poverty line for a family of four was $21,203.

Although a cutoff point is necessary to administer antipoverty programs, this definition of poverty can be misleading. A person or family earning $1 above the cutoff point would not be officially categorized as poor.

thinkingsociologically

Using the current federal *poverty line* ($21,203 for a family of four) as a guide, develop a monthly budget that does not exceed this income level and that accounts for all of your family's needs. For purposes of this exercise, assume that you head a family of four, the figure on which this poverty threshold is based. Base your budget on the actual costs of such things in your locale (rent, food, transportation, utilities, clothing, and so forth). Don't forget to account for taxes (state, federal, and local), health care expenses, your children's education, car repairs, and so on. What does this exercise teach you about those who live below the poverty line?

Who Are the Poor?

There are now more than 37 million poor people in the United States, representing 12.5 percent of the population. After the 1950s, poverty declined in the United States; it increased sharply from about 1978 until the mid-1990s, declined, and has been on the rise again since 2000. Although the majority of the poor are White, disproportionately high rates of poverty are also found among Asian Americans, Native Americans, Black Americans, and Hispanics. One-third of Native Americans, 25 percent of African Americans, 22 percent of Hispanics, 10 percent of Asians and Pacific Islanders, and 8 percent of non-Hispanic Whites are poor (DeNavas-Walt et al. 2008). Among Hispanics, there are further differences among groups. Puerto Ricans—the Hispanic group with the lowest median income—have been most likely to suffer increased poverty, probably because of their concentration in the poorest segments of the labor market and their high unemployment rates (Tienda and Stier 1996; Hauan et al. 2000). Asian American poverty has also increased substantially in recent years, particularly among the most recent immigrant groups, including Laotians, Cambodians, Vietnamese, Chinese, and Korean immigrants; Filipino, Japanese, and Asian Indian families have lower rates of poverty (U.S. Census Bureau 2004; Lee 1994).

The vast majority of the poor have always been women and children, but the percentage of women and children considered to be poor has increased in recent years. The term **feminization of poverty** refers to the large proportion of the poor who are women and children. This trend results from several factors, including the dramatic growth of female-headed households, a decline in the proportion of the poor who are elderly (not matched by a decline in the poverty of women and children), and continuing wage inequality between women and men. The large number of poor women is associated with a commensurate large number of poor children. By 2007, 18 percent of all children (those under age 18) in the United States were poor, including 10 percent of non-Hispanic White children, 35 percent of Black children, 29 percent of Hispanic children, and 12 percent of Asian American children (DeNavas-Walt 2008).

One-third of all families headed by women are poor (see Figure 8.7). In recent years, wages for young workers have declined; since most unmarried mothers are quite young, there is a strong likelihood that their children will be poor. Because of the divorce rate and generally little child support provided by men, women are also increasingly likely to be without the contributing income of a spouse and for longer periods of their lives. Women are more likely than men to live with children and to be financially responsible for them. However, women without children also suffer a high poverty rate, compounded in recent times by the fact that women now live longer than before and are less likely to be married than in previous periods.

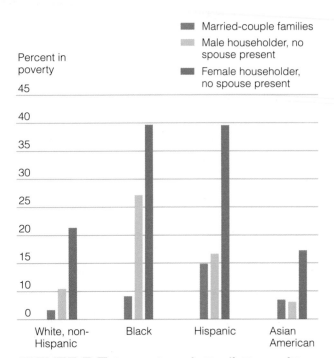

FIGURE 8.7 *Poverty Status by Family Type and Race*
Source: U.S. Census Bureau. 2008. *Historical Income Tables Households,* Table POV-02.

Debunking Society's Myths

Myth: Marriage is a good way to reduce women's dependence on welfare.

Reality: Although it is true that married-couple households are less likely to be poor than single-headed households, forcing women to marry encourages women's dependence on men and punishes women for being independent. Research indicates that poor women place a high value on marriage and want to be married, but also understand that men's unemployment and instability makes their ideal of marriage unattainable. In addition, large numbers of women receiving welfare have been victims of domestic violence (Scott et al. 2002; Edin and Kefalas 2005).

The poor are not a one-dimensional group. They are racially diverse, including Whites, Blacks, Hispanics, Asian Americans, and Native Americans. They are diverse in age, including not just children and young mothers, but men and women of all ages, and especially a substantial number of the elderly, many of whom live alone. They are also geographically diverse, to be found in areas east and west, south and north, urban and rural, although, as Map 8.2 shows, poverty rates are generally higher in the South and Southwest. One marked change in poverty is the growth of poverty in suburban areas to over 9 percent of all poverty. At the same time, half of the poor live inside central cities. But the focus on urban poverty should not cause you to lose sight of the extent of rural poverty as well.

Despite the idea that the poor "milk" the system, government supports for the poor are limited. Half of the poor receive no food stamps; 80 percent get no housing assistance; less than half receive Medicaid, which is the federal health care system for the poor (National Priorities Project 2007; U.S. Census Bureau 2008b). Budget reductions at the state and federal level also have an impact on poverty, especially among women who are more reliant than men on public sector jobs. All told, government support for poor women and their children is negligible, despite the fact that the system is reviled as overly generous and producing sloth and dependence.

Among the poor are the thousands of homeless. It is difficult to estimate the number of homeless people. Depending on how one defines and measures homelessness, the estimates vary widely. If you count the number of homeless on any given night, there may be about 444,000 to 842,000 homeless people (depending on the month measured), but measuring those experiencing homelessness over a period of one year, the estimates are about 3.5 million people (National Coalition for the Homeless 2008).

Whatever the actual numbers of homeless people, there has been an increase in homelessness over the past two decades, even without counting those displaced by Hurricane Katrina in 2005. Families are the fastest growing segment of the homeless—40 percent; children are 40 percent of the homeless.

map8.2 *MAPPING AMERICA'S DIVERSITY*

Poverty in the United States

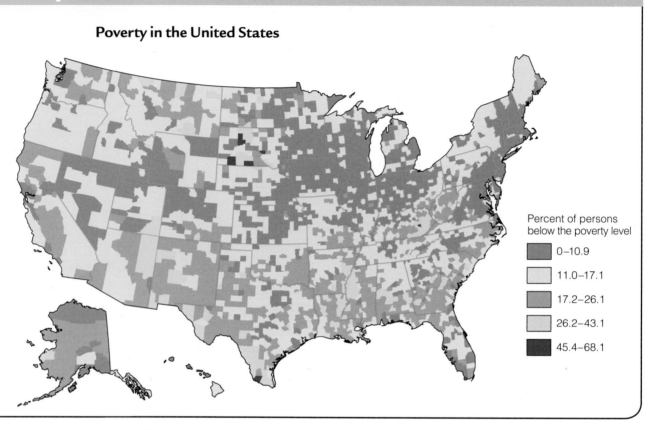

Percent of persons
below the poverty level

▓	0–10.9
�ढ	11.0–17.1
▒	17.2–26.1
░	26.2–43.1
■	45.4–68.1

This map shows regional differences in poverty rates (that is, the percentage of poor in different counties). As you can see, poverty is higher in the South (14.2 percent) than in the West (12 percent), Northeast (11.4 percent), and Midwest (11.1 percent). Various social factors explain different rates of poverty, including regional labor markets, the degree of urbanization, immigration patterns, and population composition, among other factors. What do you think the major causes of poverty are in your region?
Data: U.S. Census Bureau. 2004. *American Fact Finder.* **www.census.gov.**

Moreover, half of the women with children who are homeless have fled from domestic violence (National Coalition Against Domestic Violence 2001; Zorza 1991). Among homeless people, about 40 percent are African American; another 40 percent are White; 13 percent, Hispanic; 4 percent, Native American; and 2 percent, Asian (National Coalition for the Homeless 2006).

There are many reasons for homelessness. The great majority of the homeless are on the streets because of the lack of affordable housing and an increase in poverty, leaving many people with no choice but to live on the street. Add to that problems of inadequate health care, domestic violence, and addiction, and you begin to understand the factors that create homelessness. Some of the homeless (about 16 percent of single, homeless adults) are mentally ill; the movement to get mental patients out of institutional settings has left many without mental health resources that might help them (National Coalition for the Homeless 2008).

Causes of Poverty

Most agree that poverty is a serious social problem. There is far less agreement on what to do about it. Public debate about poverty hinges on disagreements about its underlying causes. Two points of view prevail: Some blame the poor for their own condition, and some look to social structural causes to explain poverty. The first view, popular with the public and many policy-makers, is that poverty is caused by the cultural habits of the poor. According to this point of view, behaviors such as crime, family breakdown, lack of ambition, and educational failure generate and sustain poverty, a syndrome to be treated by forcing the poor to fend for themselves. The second view is more sociological, one that understands poverty as rooted in the structure of society, not in the morals and behaviors of individuals.

Blaming the Victim: The Culture of Poverty. Blaming the poor for being poor stems from the myth that success requires only individual motivation and ability. Many in the United States adhere to this view and hence have a harsh opinion of the poor. This attitude is also reflected in U.S. public policy concerning poverty, which is rather ungenerous compared with other industrialized nations. Those who blame the poor for their own plight typically argue that poverty is the result of early childbearing, drug and alcohol abuse, refusal to enter the labor market, and crime. Such thinking puts the blame for poverty on individual choices, not on societal problems. In other words, it blames the victim, not the society, for social problems (Ryan 1971).

The **culture of poverty** argument attributes the major causes of poverty to the absence of work values and the irresponsibility of the poor. In this light, poverty is seen as a dependent way of life that is transferred, like other cultural values, from generation to generation. The culture of poverty argument has now been adapted by many policymakers to argue that the actual causes of poverty are found in the breakdown of major institutions, including the family, schools, and churches.

Is the culture of poverty argument true? To answer this question, we might ask: Is poverty transmitted across generations? Researchers have found only mixed support for this assumption. Many of those who are poor remain poor for only one or two years; only a small percentage of the poor are chronically poor. More often, poverty results from a household crisis, such as divorce, illness, unemployment, or parental death. People tend to cycle in and out of poverty. The public stereotype that poverty is passed through generations is thus not well supported by the facts.

A second question is, Do the poor want to work? The persistent public stereotype that they do not is central to the culture of poverty thesis. This attitude presumes that poverty is the fault of the poor, that if they would only change their values and adopt the American work ethic, then poverty would go away. What is the evidence for these claims?

Most of the able-bodied poor *do* work, even if only part-time. Moreover, as we saw above, the number of workers who constitute the *working poor* has increased. You can see why this is true when you calculate the income of someone working full-time for minimum wage. Someone working forty hours per week, fifty-two weeks per year, at minimum wage will have an income far below the poverty line. This is the major reason that many have organized a *living wage campaign,* intended to raise the federal minimum wage to provide workers with a decent standard of living.

Current policies that force those on welfare to work also tend to overlook how difficult it is for poor people to retain the jobs they get. Prior to welfare reform in the mid-1990s, poor women who went off welfare to take jobs often found they soon had to return to welfare because the wages they earned were not enough to support their families. Leaving welfare often means losing health benefits, yet incurring increased living expenses. The jobs that poor people find often do not lift them out of poverty. In sum, attributing poverty to the values of the poor is both unproven and a poor basis for public policy (Albelda and Tilly 1996; Catanzarite and Ortiz 1996).

Structural Causes of Poverty. From a sociological point of view, the underlying causes of poverty lie in the economic and social transformations taking place in the United States. Careful scholars do not attribute poverty to a single cause. There are many causes. Two of the most important are the *restructuring of the economy,* which has resulted in diminished earning power and increased unemployment, and *the status of women in the family and the labor market,* which has contributed to women being overrepresented among the poor. Add to these underlying conditions the federal policies in recent years that have *diminished social support for the poor* in the form of welfare, public housing, and job training. Given these reductions in federal support, it is little wonder that poverty is so widespread.

Debunking Society's Myths

Myth: Mothers on welfare have more children to increase the size of their welfare checks.

Sociological perspective: No causal relationship exists between the size of welfare benefits and the number of births by welfare recipients. The presence of "family cap" policies (those that prohibit welfare benefits with the birth of an additional child) are not associated with women's childbearing (Ryan et al. 2006).

The restructuring of the economy has caused the disappearance of manufacturing jobs, traditionally an avenue of job security and social mobility for many workers, especially African American and Latino workers (Baca Zinn and Eitzen 2007; Wilson 1996). The working class has been especially vulnerable to these changes. Economic decline in those sectors of the economy where men have historically received good pay and good benefits has meant that fewer men are the sole support for their families. Most families now need two incomes to achieve a middle-class way of life. The new jobs that are being created fall primarily in occupations that offer low wages and few benefits; they also tend to be filled by women, especially women of color, leaving women poor and men out of work (McCall 2001; Browne 1999; Andreasse 1997). Such jobs offer little chance to get out of poverty. New jobs are also typically located in neighborhoods far away from the poor, creating a mismatch between the employment opportunities and the residential base of the poor.

Declining wage rates caused by transformations taking place within the economy fall particularly hard on young people, women, and African Americans and Latinos, who are the groups most likely to be among the working poor. The high rate of poverty among women is also strongly related to women's status in the family and the labor market. Divorce is one cause of poverty, although minority women are more likely than White women to be poor even within marriage (Catanzarite and Ortiz 1996). Women's child-care responsibilities make working outside the home on marginal incomes difficult. Many women with children cannot manage to work outside the home, because it leaves them with no one to watch their children. More women now depend on their own earnings to support themselves, their children, and other dependents. Whereas unemployment has always been considered a major cause of poverty among men, for women low wages have a major role. Notice that the median income for all women ($27,337 in 2007) is not that far above the poverty line.

The persistence of poverty also increases tensions between different classes and racial groups. William Julius Wilson, one of the most noted analysts of poverty and racial inequality, has written, "The ultimate basis for current racial tension is the deleterious effect of basic structural changes in the modern American economy on Black and White lower-income groups, changes that include uneven economic growth, increasing technology and automation, industry relocation, and labor market segmentation" (1978: 154). Wilson's comments demonstrate the power of sociological thinking by convincingly placing the causes of both poverty and racism in their societal context, instead of the individualistic thinking that tends to blame the poor for their plight.

Welfare and Social Policy

Current welfare policy is covered by the 1996 Personal Responsibility and Work Reconciliation Act (PRWRA). This federal policy eliminated the longstanding welfare program titled Aid to Families with Dependent Children (AFDC), which was created in 1935 as part of the Social Security Act. Implemented during the Great Depression, AFDC was meant to assist poor mothers and their children. It acknowledged that some people are victimized by economic circumstances beyond their control and deserve assistance. For much of its lifetime, this law supported mostly White mothers and their children; not until the 1960s did welfare come to be identified with Black families.

The new welfare policy gives block grants to states to administer their own welfare programs through the program called **Temporary Assistance for Needy Families (TANF)**. TANF stipulates a lifetime limit of five years for people to receive aid and requires all welfare recipients to find work within two years—a policy known as *workfare*. Those who have not found work within two months of receiving welfare can be required to perform community service jobs for free.

In addition, welfare policy denies payments to unmarried teen parents under eighteen years of age unless they stay in school and live with an adult. It also requires unmarried mothers to identify the fathers of their children or risk losing their benefits (Hays 2003; Edin and Kefalas 2005). These broad guidelines are established at the federal level, but individual states can be more restrictive, as many have been. At the heart of welfare reform is the idea that public assistance creates dependence, discouraging people from seeking jobs. The very title of the new law, emphasizing personal responsibility and work, suggests that poverty is the fault of the poor. Low-income women, for example, are stereotyped as just wanting to have babies to increase the size of their welfare checks—even though research finds no support for this idea (Edin and Kefalas 2005).

Is welfare reform working? Many claim that welfare reform is working because, since passage of the new law, the welfare rolls have shrunk. But having

Is It True?*

	True	False
1. Income growth has been greatest for those in the middle class in recent years.		✓
2. The average American household has most of its wealth in the stock market.		✓
3. Social mobility is greater in the United States than in any other Western nation.		✓
4. The majority of welfare recipients are African American.		✓
5. Poor, teen mothers do not have the same values about marriage as middle-class people.		✓
6. Old people are the most likely to be poor.		✓
7. Poverty in U.S. suburbs is increasing.	✓	

*The answers can be found on page 207.

Is It True? (Answers)

1. FALSE. Income growth has been highest in the top 20 percent of income groups; most others have seen their incomes remain relatively flat (DeNavas-Walt et al. 2008).

2. FALSE. Eighty percent of all stock is owned by a small percentage of people. For most people, home ownership is the most common financial asset (Oliver and Shapiro 2006).

3. FALSE. The United States has lower rates of social mobility than Canada, Sweden, and Norway and ranks near the middle in comparison to other Western nations (Beller and Hout 2006).

4. FALSE. The majority of those receiving welfare (TANF) are White (U.S. Census Bureau 2008).

5. FALSE. Research finds that poor, teen mothers value marriage and want to be married, but associate marriage with economic security, which they do not think they can achieve (Edin and Kefalas 2005).

6. FALSE. Although those over 65 years of age used to be the most likely to be poor, poverty among the elderly has declined; the most likely to be poor are children (DeNavas-Walt et al. 2008).

7. TRUE. Although most of the poor live inside metropolitan areas, poverty in the suburbs has been increasing (De Navas-Walt et al. 2008).

---

fewer people on welfare does not mean poverty is reduced; in fact, as we have seen, poverty has actually increased since passage of welfare reform. Having fewer people on the rolls can simply mean that people are without a safety net.

Many studies also find that low-wage work does not lift former welfare recipients out of poverty (Hays 2003). Critics of the current policy also argue that forcing welfare recipients to work provides a cheap labor force for employers and potentially takes jobs from those already employed. In the first few years of welfare reform, the nation was also in the midst of an economic boom; jobs were thus more plentiful. But in an economic downturn, those who are on aid or in marginal jobs are vulnerable to economic distress, particularly given the time limits now placed on receiving public assistance (Albelda and Withorn 2002).

Research done to assess the impact of a changed welfare policy is relatively recent. Politicians brag that welfare rolls have shrunk, but reduction in the welfare rolls is a poor measure of the true impact of welfare reform because this would be true simply because people are denied benefits. And because welfare has been decentralized to the state level, studies of the impact of current law must be done on a state-by-state basis. Such studies are showing that those who have gone into workfare programs most often earn wages that keep them below the poverty line. Although some states report that family income has increased following welfare reform, the increases are slight. More people have been evicted because of falling behind on rent. Families also report an increase in other material hardships, such as phones and utilities being cut off. Marriage rates among former recipients have not changed, although more now live with nonmarital partners, most likely as a way of sharing expenses. The number of children living in families without either parent has also increased, probably because parents had to relocate to find work. In some states, the numbers of people neither working nor receiving aid also increased (Lewis et al. 2002; Acker et al. 2002; Bernstein 2002).

The public debate about welfare rages on, often in the absence of informed knowledge from sociological research and almost always without input from the subjects of the debate, the welfare recipients themselves. Although stigmatized as lazy and not wanting to work, those who have received welfare actually believe that it has negative consequences for them, but they say they have no other viable means of support. They typically have needed welfare when they could not find work or had small children and were without child care. Most were forced to leave their last job because of layoffs or firings or because the work was only temporary. Few left their jobs voluntarily.

Welfare recipients also say that the welfare system makes it hard to become self-supporting, because the wages one earns while on welfare are deducted from an already minimal subsistence. Furthermore, there is not enough affordable day care for mothers to leave home and get jobs. The biggest problem they face in their minds is lack of money. Contrary to the popular image of the conniving "welfare queen," welfare recipients want to be self-sufficient and provide for their families, but they face circumstances that make this very difficult to do. Indeed, studies of young, poor mothers find that they place a high value on marriage, but they do not think they or their boyfriends have the means to achieve the marriage ideals they cherish (Hays 2003; Edin and Kefalas 2005).

Another popular myth about welfare is that people use their welfare checks to buy things they do not need. But research finds that when former welfare recipients find work, their expenses actually go up. Although they may have increased income, their expenses (in the form of child care, clothing, transportation, lunch money, and so forth) increase, leaving them even less disposable income. Moreover, studies find that low-income mothers who buy "treats" for their children (name-brand shoes, a movie, candy, and so forth) do so because they want to be good mothers (Edin and Lein 1997).

Other beneficiaries of government programs have not experienced the same kind of stigma. Social

Security supports virtually all retired people, yet they are not stereotyped as dependent on federal aid, unable to maintain stable family relationships, or insufficiently self-motivated. Spending on welfare programs is also a pittance compared with the spending on other federal programs. Sociologists conclude that the so-called welfare trap is not a matter of learned dependency, but a pattern of behavior forced on the poor by the requirements of sheer economic survival (Hays 2003; Edin and Kefalas 2005).

Chapter Summary

• **What different kinds of stratification systems exist?**

Social stratification is a relatively fixed hierarchical arrangement in society by which groups have different access to resources, power, and perceived social worth. All societies have systems of stratification, although they vary in composition and complexity. *Estate systems* are those in which power and property are held by a single elite class; in *caste systems,* placement in the stratification is by birth, and in *class systems,* placement is determined by achievement.

• **How do sociologists define class?**

Class is the social structural position groups hold relative to the economic, social, political, and cultural resources of society. It is highly significant in determining one's *life chances.*

• **How is the class system structured in the United States?**

Social class can be seen as a hierarchy, like a ladder, where income, occupation, and education are indicators of class. *Status attainment* is the process by which people end up in a given position in this hierarchy. *Prestige* is the value assigned to people and groups by others within this hierarchy. Classes are also organized around common interests and exist in conflict with one another.

• **Is there social mobility in the United States?**

Social mobility is the movement between class positions. Education gives some boost to social mobility, but social mobility is more limited than people believe; most people end up in a class position very close to their class of origin. *Class consciousness* is both the perception that a class structure exists and the feeling of shared identification with others in one's class. The United States has not been a particularly class-conscious society because of the belief in upward mobility.

• **What analyses of social stratification do sociological theorists provide?**

Karl Marx saw class as primarily stemming from economic forces; Max Weber had a multidimensional view of stratification, involving economic, social, and political dimensions. Functionalists argue that social inequality motivates people to fill the different positions in society that are needed for the survival of the whole, claiming that the positions most important for society require the greatest degree of talent or training and are, thus, most rewarded. Conflict theorists see social stratification as based on class conflict and blocked opportunity, pointing out that those at the bottom of the stratification system are least rewarded because they are subordinated by dominant groups.

- *How do sociologists explain why there is poverty in the United States?*

  The *culture of poverty thesis* is the idea that poverty is the result of the cultural habits of the poor that are transmitted from generation to generation, but sociologists see poverty as caused by social structural conditions, including unemployment, gender inequality in the workplace, and the absence of support for child care for working parents.

- *What current policies address the problem of poverty?*

  Current welfare policy, adopted in 1996, provides support through individual states, but recipients are required to work after two years of support and have a lifetime limit of five years' support.

Key Terms

caste system   181
class consciousness   196
class system   182
culture of poverty   205
economic restructuring   192
educational attainment   187
estate system   181
false consciousness   196
feminization of poverty   202
ideology   198

income   196
life chances   182
median income   186
meritocracy   196
net worth   190
occupational prestige   187
poverty line   202
prestige   187
social class   182
social differentiation   180

social mobility   196
social stratification   180
socioeconomic status (SES)   186
status   180
status attainment   183
Temporary Assistance for Needy
   Families (TANF)   206
urban underclass   189
wealth   186

Online Resources

Sociology: The Essentials Companion Website

www.cengage.com/sociology/andersen
Visit your book companion website where you will find more resources to help you study and write your research papers. Resources include web links, learning objectives, Internet exercises, quizzing, and flash cards.

 is an easy-to-use online resource that helps you study in less time to get the grade you want NOW.

www.cengage.com/login
Need help studying? This site is your one-stop study shop. Take a Pre-Test and Cengage NOW will generate a Personalized Study Plan based on your test results. The Study Plan will identify the topics you need to review and direct you to online resources to help you master those topics. You can then take a Post-Test to determine the concepts you have mastered and what you still need to work on.

Chapter nine
CHAPTER NINE
Chapter

Global Stratification

Global Stratification

Theories of Global Stratification

Consequences of Global Stratification

World Poverty

Globalization and Social Change

Chapter Summary

"It takes a village to raise a child," the saying goes. But it also seems to take a world to make a shirt—or so it seems from looking at the global dimensions of the production and distribution of goods. Try this simple experiment: Look at the labels on your clothing. (If you do this in class, try to do so without embarrassing yourself and others!) What do you see? "Made in Indonesia," "Made in Vietnam," "Made in Malawi," all indicating the linkage of the United States to systems of production around the world. The popular brand Nike, as just one example, has not a single factory in the United States, although its founder and chief executive officer is one of the wealthiest people in America. Nike products are manufactured mostly in Southeast Asia.

Taking your experiment further, ask yourself: Who made your clothing? A young person trying to lift his or her family out of poverty? Might it have been a child? In many areas of the world, one in five children under age 15 works (International Labour Organization 2002). What countries benefit most from this system of production? Answering these questions reveals much about the interconnection among countries in the global stratification system, a system in which the status of the people in one country is intricately linked to the status of the people in others. Recall from Chapter 1 that C. Wright Mills identified the task of sociology as seeing the social forces that exist beyond the individual. This is particularly important when studying global stratification.

continued

© Paula Bronstein/Getty Images

nine
chapter nine

CHAPTER NI

The person in the United States (or western Europe or Japan) who thinks he or she is expressing individualism by wearing the latest style is actually part of a global system of inequality. The adornments available to that person result from a whole network of forces that produce affluence in some nations and poverty in others. Dominant in the system of global stratification are the United States and other wealthy nations. Those at the top of the global stratification system have enormous power over the fate of other nations. Although world conflict stems from many sources, including religious differences, cultural conflicts, and struggles over political philosophy, the inequality between rich and poor nations causes much hatred and resentment. One cannot help but wonder what would happen if the differences between the wealth of some nations and the poverty of others were smaller. In this chapter we examine the dynamics and effects of global stratification.

Global Stratification

In the world today, there are not only rich and poor people but also rich and poor countries. Some countries are well-off, some countries are doing so-so, and a growing number of countries are poor and getting poorer. There is, in other words, a system of **global stratification** in which the units we are considering are countries, much like a system of stratification within countries in which the units are individuals or families. Just as we can talk about the upper-class or lower-class individuals within a country, we can also talk of the equivalent upper-class or lower-class countries in this world system. One manifestation of global stratification is the great inequality in life chances that differentiates nations around the world. Simple measures of well-being, including life expectancy, infant mortality, and access to health services, reveal the consequences of a global system of inequality. And, the gap between the rich and poor is sometimes greater in nations where the average person is least well off. No longer can these nations be understood without considering the global system of stratification of which they are a part.

The effects of the global economy on inequality have become increasingly evident, as witnessed by public concerns about jobs being sent overseas. A coalition of unions, environmentalists, and other groups has also emerged to protest global trade policies that they believe threaten jobs and workers' rights in the United States, as well as contributing to environmental degradation. Such policies also encourage further McDonaldization (see Chapter 6). Thus, popular stores such as The Gap and Niketown often have been targets of political protests because they symbolize the expansion of global capitalism. Protestors see the growth of such stores as eroding local cultural values and spreading the values of unfettered consumerism around the globe. Protests over world trade policies also have emerged in a student-based movement against companies that manufacture apparel with college logos.

The relative affluence of the United States means that U.S. consumers have access to goods produced around the world. A simple thing, such as a child's toy, can represent this global system. For many young girls in the United States, Barbie™ is the ideal of fashion and romance. Young girls may have not just one Barbie, but several, each with a specific role and costume. Cheaply bought in the United States, but produced overseas, Barbie is manufactured by those probably not much older than the young girls who play with her and who would need all of their monthly pay to buy just one of the dolls that many U.S. girls collect by the dozens (Press 1996: 12).

The manufacturing of toys and clothing are examples of the global stratification that links the United States and other parts of the world. When companies export jobs, workers in the United States lose them. For example, note that in 1973, more than 56,000 U.S. workers were employed in toy factories. Now, even though the market has become glutted with the latest popular items, only 27,000 U.S. workers are employed in toy factories. The companies that make the toys amass profits, but U.S. workers lose jobs and then blame foreign workers for taking them. Major toy manufacturers have adopted policies that prohibit the exploitation of child labor, but clearly it is more profitable for companies to take their labor overseas where workers are paid far less and get fewer benefits than many workers in the United States.

Rich and Poor

One dimension of stratification between countries is wealth. Enormous differences exist between the wealth of the countries at the top of the global stratification system and the wealth of the countries at the bottom. Although there are different ways to measure the wealth of nations, one of the most common is to use the per capita **gross national income (GNI).** The GNI measures the total output of goods and services produced by residents of a country each year plus the income from nonresident sources, divided by the size of the population. This does not truly reflect what individuals or families receive in wages or pay; it is simply each person's annual share of their country's income, should, in theory, the proceeds be shared equally. But you can use this measure to get a picture of global stratification (see Map 9.1).

Per capita GNI is reliable only in countries that are based on a cash economy. It does not measure informal exchanges or bartering in which resources are exchanged without money changing hands. These noncash transactions are not included in the GNI

Global stratification often means that consumption in the more affluent nations is dependent on cheap labor in other less affluent nations.

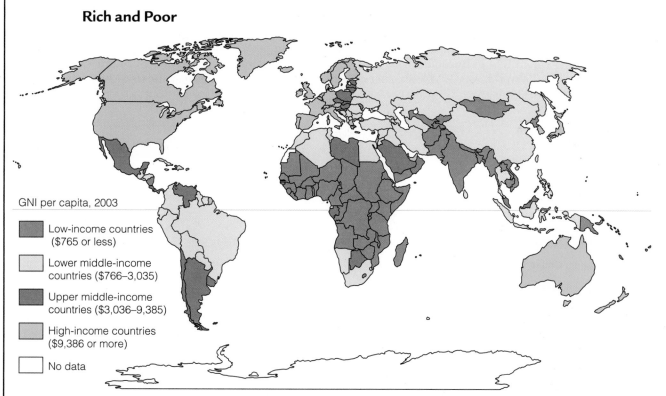

map 9.1 VIEWING SOCIETY IN GLOBAL PERSPECTIVE

Rich and Poor

GNI per capita, 2003

- Low-income countries ($765 or less)
- Lower middle-income countries ($766–3,035)
- Upper middle-income countries ($3,036–9,385)
- High-income countries ($9,386 or more)
- No data

Most nations are linked in a world system that produces wealth for some and poverty for others. The GNI (gross national income), depicted here on a per capita basis for most nations in the world, is an indicator of the wealth and poverty of nations.

Source: World Bank Atlas 2007. © 2007 by World Bank. Reprinted by permission. **www.worldbank.org**

Many common products marketed in the United States are produced in a global economy.

United States is 419 times wealthier than an average citizen of Burundi. In Afghanistan, the GNI was estimated to be $875 or less (World Bank 2009), indicating the relative affluence of people in the United States.

Which are the wealthiest nations? Figure 9.1 lists the ten richest countries in the world measured by the annual per capita GNI in 2007. Luxembourg is the richest nation in the world on a per capita basis; the United States, seventh. Of course, Luxembourg has a tiny population compared with the United States. Note that most of the wealthy countries are in western Europe. They are mostly industrialized countries (or support such countries through such businesses as banking), and they are mostly urban. These countries represent the equivalent of the upper class—even though many people within them are poor.

Now consider the ten poorest countries in the world, also shown in Figure 9.1, and again using per capita GNI as the measure of wealth. Not every country reports reliable statistics on poverty, so there may be other extremely poor nations, but we do know that most of the world's poorest countries are in eastern or central Africa. These countries have not become industrialized, are largely rural, have high fertility rates, and still depend heavily on subsistence agriculture. These countries rank at the bottom of the global stratification system.

Clearly, many countries in the world are very poor, whereas other countries are rich. This does not mean that all people in rich countries are rich or that all people in poor countries are poor. But on average, people in poor countries are much worse off than people in rich countries. In many poor countries, the life of an average citizen is desperate; these countries also have the largest populations. In a world with a population

calculation, but they are more common in developing countries. As a result, measures of wealth based on the GNI, or other statistics that count cash transactions, are less reliable among the poorer countries and may underestimate the wealth of the countries at the lower end of the economic scale.

The per capita GNI of the United States, which is one of the wealthier nations in the world (though not the wealthiest on a per capita basis), was $46,040 in 2007. The per capita GNI in Burundi, one of the poorest countries in the world, was $110. Using per capita GNI as a measure of wealth, the average person in the

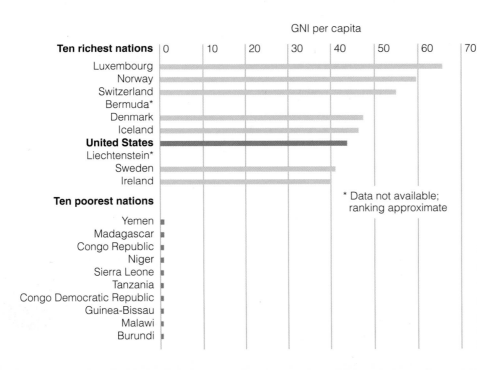

FIGURE 9.1 *The Rich and the Poor: A World View*

*GNI per capita, measured in U.S. dollars.

Source: World Bank Atlas 2002, p. 18-19. Copyright © by World Bank. Reprinted by permission. **www.worldbank.org**

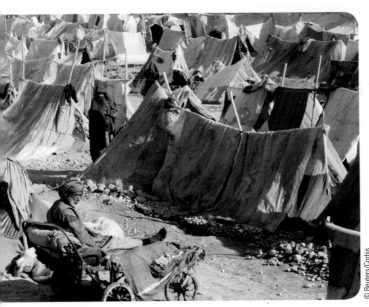

Global stratification means that enormous differences exist not only in the relative well-being of different countries but also within nations. Large numbers of people live in poverty such as in this refugee camp at Mazar-i-Sharif in Afghanistan—one of the poorest nations in the world.

© Reuters/Corbis

of nearly six billion, more than three billion—more than half the world's population—live in the poorest forty-five countries. Often poor nations are rich with natural resources but are exploited for such resources by more powerful nations. We will look more closely at the nature and causes of world poverty later in this chapter.

Global Networks of Power and Influence

Global stratification involves nations in a large and integrated network of both economic and political relationships. **Power**—meaning the ability of a country to exercise control over other countries or groups of countries—is a significant dimension of global stratification. Countries can exercise several kinds of power over other countries, including military, economic, and political power. The **core countries** have the most power in the world economic system. These countries control and profit the most from the world system, and thus they are the "core" of the world system. These include the powerful nations of Europe, the United States, Australia, and Japan.

Surrounding the core countries, both structurally and geographically, are the **semiperipheral countries** that are semi-industrialized and, to some degree, represent a kind of middle class (such as Spain, Turkey, and Mexico). They play a middle-man role, extracting profits from the poor countries and passing those profits on to the core countries. At the bottom of the world stratification system, in this

model, are the **peripheral countries.** These are the poor, largely agricultural countries of the world. Even though they are poor, they often have important natural resources that are exploited by the core countries. This exploitation, in turn, keeps them from developing and perpetuates their poverty. Often these nations are politically unstable, and, though they exercise little world power, political instability can cause a crisis for core nations that depend on their resources. Military intervention by the United States or European nations is often the result.

This categorizing system emphasizes the power of each country in the world economic system. Another way that these countries are sometimes labeled is as first-, second-, and third-world nations. This language grows out of the politics of the Cold War and reflects the political and economic dimensions of global stratification. **First-world countries** consist of the industrialized capitalist countries of the world, including the United States, New Zealand, Australia, Japan, and the countries of western Europe. They are industrialized and have a market-based economy and a democratically elected government. The **second-world countries** are socialist countries, which include the former Soviet Union, China, Cuba, North Korea, and, prior to the fall of the Berlin Wall, the eastern European nations. During the Cold War, these countries had a communist-based government and a state-managed economy, as some still do. Although less developed than the first-world countries, the second-world countries tried to provide citizens with services such as free education, health care, and low-cost housing, consistent with the principles of socialism, but poverty often prevented them from doing so. Moreover, their governments were not democratically elected. **Third-world countries** in this scheme are the countries that are poor, underdeveloped, largely rural, and with high levels of poverty. Many of the governments of the third-world countries are autocratic dictatorships (ruled by one person with absolute authority), though not all. Although these countries are generally poor, wealth is concentrated in the hands of only a few elite.

Because this system of categorization was based on the logic of the Cold War, it has changed. For instance, the oil-rich countries of the Middle East are not part of the first or second world, according to this scheme, but they also do not belong in the same category as the poor countries of Africa and Asia because they have considerably more wealth. The collapse of the Soviet Union and the change in the governments of eastern Europe has led to the transformation of almost all the second-world countries. Although some countries still have a communist-based government, many of them, including China, are moving toward a market economic system. Still, like the terms *core* and *periphery,* the terms *first-, second-,* and *third-world* are useful in denoting the power differences of nations within global stratification.

>doing sociological research

Servants of Globalization: Who Does the Domestic Work?

Research Question: International migration is becoming an increasingly common phenomenon. Women are one of the largest groups to experience migration, often leaving poor nations to become domestic workers in wealthier nations. What are these women's experiences in the context of global stratification? This is what Rhacel Salazar Parreñas wanted to know.

Research Method: Parreñas studied two communities of Filipina women, one in Los Angeles and one in Rome, Italy, conducting her research through extensive interviewing with Filipina domestic workers in these two locations. She supplemented the interviews with participant observation in church settings, after-work social gatherings, and in employers' homes. The interviews were conducted in English and Tagalog—sometimes a mixture of both.

Research Results: Parreñas found that Filipina domestics experienced many status inconsistencies. They were upwardly mobile in terms of their home country but were excluded from the middle-class Filipino communities in the communities where they lived. Thus, they experienced feelings of social exclusion in addition to being separated from their own families.

Conclusions and Implications: The women Parreñas studied are part of a new social form: *transnational families*—that is, families whose members live across the borders of nations. These Filipinas provide the labor for more affluent households while their own lives are disrupted by these new global forces. As global economic restructuring evolves, it may be that more and more families will take this form as they adapt to changing economic and social conditions.

Questions to Consider

1. Are there domestic workers in your community who provide child care and other household work for middle- and upper-class households? What is the race, ethnicity, nationality, and gender of these workers? What does this tell you about the division of labor in domestic work and its relationship to global stratification?

2. Why do you think domestic labor is so underpaid and undervalued? Are there social changes that might result in a reevaluation of the value of this work?

Source: Parreñas, Rhacel Salazar. 2001. *Servants of Globalization: Women. Migration and Domestic Work*. Stanford, CA: Stanford University Press.

SEE *FOR YOURSELF*

THE GLOBAL ECONOMY OF CLOTHING

Look at the labels in your clothes and note where your clothing was made. Where are the products bearing your college logos manufactured and sold? Who profits from the distribution of these goods? What does this tell you about the relationship of *core, semiperipheral,* and *peripheral* countries within world systems theory? What further information would reveal the connections between the country where you live and the countries where your clothing is made and distributed?

Race and Global Inequality

Along with class inequality, there is a racial component to world inequality, which can be seen in several ways. The rich core countries, those that dominate the world system, are largely European, with the exception of the United States and Japan. In Europe and the United States, the population is mostly White; in the poor countries of the world, mostly in Africa, Asia, or South America, the populations are largely people of color. On average, there are vast differences in life chances and lifestyle between the countries of the world with White populations and the countries of the world with Black populations.

Exploitation of the human and natural resources of regions populated by people of color has characterized the history of Western capitalism, with people of color being dominated by Western imperialism and colonialism. The inequities that have resulted are enormous. Patterns of malnutrition and hunger show these inequities. More than one billion people in the world suffer from malnutrition and hunger. The vast majority of these people are people of color—that is, those not of European descent (Uvin 1998).

How did this racial inequality come about? On the surface, global capitalism is not explicitly racist, as were earlier forms of industrial capitalism. Yet, in fact, it is the rapid expansion of the global capital system that has led to the increase in racial inequality between nations. In the new capitalist system, a new **international division of labor** has emerged that is not tied to place but can employ cheap labor anywhere. Cheap labor is usually found in non-Western countries. The exploitation of cheap labor has created a poor and dependent workforce that is mostly people of color. The profits accrue to the wealthy owners, who are mostly White, resulting in a racially divided world. Some have argued further that multinational corporations' exploitation of the poor peripheral nations has forced an exodus of unskilled workers from the impoverished nations to the rich nations. The flood of third-world refugees into the industrialized nations is thereby increasing racial tensions, fostering violence, and destroying worker solidarity (Sirvananadan 1995).

South Africa, the United States, and Brazil each developed different sets of racial categories. Although all three countries have many people of mixed descent, race is defined differently in each place. In South Africa, the particular history of Dutch and English colonialism led to strongly drawn racial categories that defined people in four separate categories: "White," "coloured" (including indigenous Khoi and San people, as well as people of mixed descent), "Black," and "Indian." Except for Black South Africans, who had no political representation under apartheid, there were three separate parliaments—one for each of the other groups. In the United States, given its history of slavery, the "one drop" rule was used, which defined anyone with any African heritage as Black, thus ruling out any category of mixed race.

Brazil is yet a different case. The Brazilian elite declared Brazil a racial democracy at the early stages of national development. Racial differences were thought not to matter. Yet, instead of creating an egalitarian society free of racism, Afro-Brazilians were still of lower social status and Euro-Brazilians remain at the highest social status, suggesting that color itself stratifies people—a sociological phenomenon sometimes referred to as "colorism" (Marx 1997; Frederickson 2003; Telles 2004).

Theories of Global Stratification

How did world inequality occur? Sociological explanations of world stratification generally fall into three camps: modernization theory, dependency theory, and world systems theory, each explained here (see Table 9.1).

Modernization Theory

Modernization theory views the economic development of countries as stemming from technological change. According to this theory, a country becomes more "modernized" by increased technological development, and this technological development is also dependent on other countries. Modernization theory was initially developed in the 1960s to explain why some countries had achieved economic development and why some had not (Rostow 1978).

Modernization theory sees economic development as a process by which traditional societies become more complex and differentiated. For economic development to occur, modernization theory predicts, countries must change their traditional attitudes, values, and institutions. Economic achievement is thought to

table 9.1 Theories of Global Stratification

	Modernization Theory	Dependency Theory	World Systems Theory
Economic Development	Arises from relinquishing traditional cultural values and embracing new technologies and market-driven attitudes and values	Exploits the least powerful nations to the benefit of wealthier nations that then control the political and economic systems of the exploited countries	Has resulted in a single economic system stemming from the development of a world market that links core, semiperipheral, and peripheral nations
Poverty	Results from adherence to traditional values and customs that prevent societies from competing in a modern global economy	Results from the dependence of low-income countries on wealthy nations	Is the result of core nations extracting labor and natural resources from peripheral nations
Social Change	Involves increasing complexity, differentiation, and efficiency	Is the result of neocolonialism and the expansion of international capitalism	Leads to an international division of labor that increasingly puts profit in the hands of a few while exploiting those in the poorest and least powerful nations

derive from attitudes and values that emphasize hard work, saving, efficiency, and enterprise. These values are said by the theory to be found in modern (developed) countries but are lacking in traditional societies. Modernization theory suggests that nations remain underdeveloped when traditional customs and culture discourage individual achievement and kin relations dominate.

As an outgrowth of functionalist theory, modernization theory derives some of its thinking from the work of Max Weber. In *The Protestant Ethic and the Spirit of Capitalism* (1958/1904), Weber saw the economic development that occurred in Europe during the Industrial Revolution as a result of the values and attitudes of Protestantism. The Industrial Revolution took place in England and northern Europe, Weber argued, because the people of this area were hardworking Protestants who valued achievement and believed that God helped those who helped themselves.

Modernization theory is similar to the argument of the culture of poverty, which sees people as poor because they have poor work habits, engage in poor time management, are not willing to defer gratification, and do not save or take advantage of educational opportunities (see Chapter 8). Countries are poor, in other words, because they have poor attitudes and poor institutions.

Modernization theory can partially explain why some countries have become successful. Japan is an example of a country that has made huge strides in economic development, in part because of a national work ethic (McCord and McCord 1986). But the work ethic alone does not explain Japan's success. In sum, modernization theory may partially explain the value context in which some countries become successful and others do not, but it is not a substitute for explanations that also look at the economic and political context of national development. It also rests on an arrogant perspective that the United States and other more economically developed nations have superior values compared to other nations. Critics point out that this perspective blames countries for being poor when other causes of their status in the world may be outside their control. Whether a country develops or remains poor may be the result of other countries exploiting the less powerful. Modernization theory does not sufficiently take into account the interplay and relationships between countries that can affect a country's economic or social condition.

Developing countries, modernization theory says, are better off if they let the natural forces of competition guide world development. Free markets, according to this perspective, will result in the best economic order. But, as critics argue, markets do not develop independently of government's influence. Governments can spur or hinder economic development, especially as they work with private companies to enact export strategies, restrict imports, or place embargoes on the products of nonfavored nations.

Dependency Theory

Although market-oriented theories may explain why some countries are successful, they do not explain why some countries remain in poverty or why some countries have not developed. It is necessary to look at issues outside the individual countries and to examine the connections between them. Drawing on the fact that many of the poorest nations are former colonies of European powers, another theory of world stratification focuses on the processes and results of European colonization and imperialism. This theory, called **dependency theory,** focuses on explaining the persistence of poverty in the world. It holds that the poverty of the low-income countries is a direct result of their political and economic dependence on the wealthy countries. Specifically, dependency theory argues that the poverty of many countries is a result of exploitation by powerful countries. This theory is derived from the work of Karl Marx, who foresaw that a capitalist world economy would create an exploited class of dependent countries, just as capitalism within countries had created an exploited class of workers.

Dependency theory begins by examining the historical development of this system of inequality. As the European countries began to industrialize in the 1600s, they needed raw materials for their factories, and they needed places to sell their products. To accomplish this, the European nations colonized much of the world, including most of Africa, Asia, and the Americas. **Colonialism** is a system by which Western nations became wealthy by taking raw materials from colonized societies and reaping profits from products finished in the homeland. Colonialism worked best for the industrial countries when the colonies were kept undeveloped to avoid competition with the home country. For example, India was a British colony from 1757 to 1947. During that time, Britain bought cheap cotton from India, made it into cloth in British mills, and then sold the cloth back to India, making large profits. Although India was able to make cotton into cloth at a much cheaper cost than Britain, and very fine cloth at that, Britain nonetheless did not allow India to develop its cotton industry. As long as India was dependent on Britain, Britain became wealthy and India remained poor.

Under colonialism, dependency was created by the direct political and military control of the poor countries by powerful developed countries. Most colonial powers were European countries, but other countries, particularly Japan and China, had colonies as well. Colonization came to an end soon after the Second World War, largely because of protests by colonized people and the resulting movement for independence. As a result, according to dependency theory, the powerful countries turned to other ways to control the poor countries and keep them dependent. The powerful countries still intervene directly in the affairs of the dependent nations by sending troops or, more often, by imposing economic or political restrictions and sanctions. But other methods, largely economic, have been developed to control the dependent poor countries, such as price controls, tariffs, and, especially, the control of credit. Indeed, the level of debt that some nations accrue is a major source of global inequality.

The rich industrialized nations, according to dependency theory, are able to set prices for raw materials produced by the poor countries at very low levels so that the poor countries are unable to accumulate enough profit to industrialize. As a result, the poor, dependent countries must borrow from the rich countries. However, debt creates only more dependence. Many poor countries are so deeply indebted to the major industrial countries that they must follow the economic edicts of the rich countries that loaned them the money, thus increasing their dependency. This form of international control has sometimes been called **neocolonialism,** a form of control of the poor countries by the rich countries but without direct political or military involvement.

Multinational corporations are companies that draw a large share of their profits from overseas investments and that conduct business across national borders. They play a role in keeping the dependent nations poor, dependency theory suggests. Although their executives and stockholders are from the industrialized countries, multinational corporations recognize no national boundaries and pursue business where they can best make a profit. Multinationals buy resources where they can get them cheapest, manufacture their products where production and labor costs are lowest, and sell their products where they can make the largest profits.

Many critics fault companies for perpetuating global inequality by taking advantage of cheap overseas labor to make large profits for U.S. stockholders.

Companies are, in fact, doing what they should be doing in a market system: trying to make a profit. Nonetheless, dependency theory views the practices of multinationals as responsible for maintaining poverty in the poor parts of the world.

One criticism of dependency theory is that many poor countries were never colonies, for example, Ethiopia. Some former colonies have also done well. Two of the greatest postwar success stories of economic development are Singapore and Hong Kong. Both of these countries were British colonies—Hong Kong until 1997—and clearly dependent on Britain, yet they have had successful economic development precisely because of their dependence on Britain. Other former colonies are also improving economically, such as India.

World Systems Theory

Modernization theory examines the factors internal to an individual country, and dependency theory looks to the relationship between countries or groups of countries. Another approach to global stratification is called **world systems theory.** Like the dependency theory, this theory begins with the premise that no nation in the world can be considered in isolation. Each country, no matter how remote, is tied in many ways to the other countries in the world. However, unlike dependency theory, world systems theory argues that there is a world economic system that must be understood as a single unit, not in terms of individual countries or groups of countries. This theoretical approach derives to some degree from the work of the dependency theorists and is most closely associated with the work of Immanuel Wallerstein in *The Modern World System* (1974) and *The Modern World System II* (1980). According to this theory, the level of economic development is explained by

The gap between the rich and poor worldwide can be staggering. At the same time that many struggle for mere survival, others enjoy the pleasantries of a gentrified lifestyle.

David Bacon/The Image Works

© Jorgen Schytte/Peter Arnold, Inc.

Guest workers programs provide immigrant labor in many nations of the world. Here Indian immigrant workers stage a hunger strike in Washington, DC to protest the treatment of immigrant workers helping to restore New Orleans in the aftermath of hurricane Katrina.

understanding each country's place and role in the world economic system.

This world system has been developing since the sixteenth century. The countries of the world are tied together in many ways, but of primary importance are the economic connections in the world markets of goods, capital, and labor. All countries sell their products and services on the world market and buy products and services from other countries. However, this is not a market of equal partners. Because of historical and strategic imbalances in this economic system, some countries are able to use their advantage to create and maintain wealth, whereas other countries that are at a disadvantage remain poor. This process has led to a global system of stratification in which the units are not people but countries.

World systems theory sees the world divided into three groups of interrelated nations: core or first-world countries, semiperiperal or second-world countries, and peripheral or third-world countries. This world economic system has resulted in a modern world in which some countries have obtained great wealth and other countries have remained poor. The core countries control and limit the economic development in the peripheral countries so as to keep the peripheral countries from developing and competing with them on the world market; thus the core countries can continue to purchase raw materials at a low price.

Although world systems theory was originally developed to explain the historical evolution of the world system, modern scholars now focus on the international division of labor and its consequences. This approach is an attempt to overcome some of the

shortcomings in world systems theory by focusing on the specific mechanism by which differential profits are attached to the production of goods and services in the world market. A tennis shoe made by Nike is designed in the United States; uses synthetic rubber made from petroleum from Saudi Arabia; is sewn in Indonesia; is transported on a ship registered in Singapore, which is run by a Korean management firm using Filipino sailors; and is finally marketed in Japan and the United States. At each of these stages, profits are taken, but at very different rates.

thinkingsociologically

What are the major industries in your community? Find out in what parts of the world they do business, including where their product is produced. How does the *international division of labor* affect jobs in your region?

World systems theorists call this global production process a **commodity chain,** the network of production and labor processes by which a product becomes a finished commodity. By following a commodity through its production cycle and seeing where the profits go at each link of the chain, one can identify which country is getting rich and which country is being exploited.

World systems theory helps explain the growing phenomenon of international migration. An international division of labor means that the need for cheap labor in some of the industrial and developing nations draws workers from poorer parts of the globe. International migration is also the result of refugees seeking asylum from war-torn parts of the world or from countries where political oppression, often against particular ethnic groups, forces some to leave. The

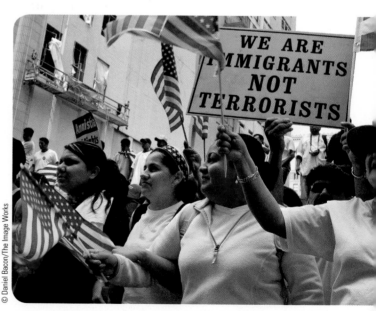

Public debates over immigration policy have mobilized many who point out that immigration has long been a part of our national heritage.

development of a world economy, however, is resulting in large changes in the composition of populations around the globe. **World cities,** that is, cities that are closely linked through the system of international commerce, have emerged. Within these cities, families and their surrounding communities often form *transnational communities,* communities that may be geographically distant but socially and politically close. Linked through various communication and transportation networks, transnational communities share information, resources, and strategies for coping with the problems of international migration.

International migration, sometimes legal, sometimes not, has radically changed the racial and ethnic composition of populations not only in the United States but also in many European and Asian nations (Rodriquez 1999; Light et al. 1998). Some of those who migrate internationally are professional workers, but many others remain in the lowest segments of the labor force where, although their work is critical to the world economy, they are treated with hostility and suspicion, discriminated against, and stereotyped as undeserving and threatening. In many nations, including the United States, this has led to numerous political tensions over immigration, even while the emergence of migrant groups in world cities is now a major feature of the urban landscape (White 1998).

There are a number of criticisms of world systems theory. Certainly, it is useful to see the world as an interconnected set of economic ties between countries and to understand that these ties often result in the exploitation of poor countries. For one, countries that were once at the center of this world system no longer occupy such a lofty position—England, for example. Peripheral countries can also improve their standard of living with investment by core countries, although the benefits do not accrue equally to groups within such nations. Low-wage factories may benefit managers, but not the working class. Even core countries can be hurt by the world system, such as when jobs move overseas. Who benefits from this world system is differentiated—in all countries—by one's placement, not just in the world class system but in the class system internal to each country within this global system. Also, low-wage sweatshops are found in all nations, not just the peripheral countries. Despite these criticisms, world systems theory has provided a powerful tool for understanding global inequality.

Consequences of Global Stratification

It is clear that some nations are wealthy and powerful and some are poor and powerless. What are the consequences of this world stratification system? Table 9.2 shows some of the basic indicators of national well-being for selected nations. You can see that there are considerable differences in the quality of life in these different places in the world.

Population

One of the biggest differences in rich and poor nations is population. The poorest countries comprise three billion people—over half the world's population (World Bank 2007). The poorest countries also have the highest birthrates and the highest death rates. The total *fertility rate,* how many live births a woman will have over her lifetime at current fertility rates, shows that in the poorest countries women on average have almost five children. Because of this high fertility rate, the populations of poor countries are growing faster than the populations of wealthy countries; these countries therefore also have a high proportion of young children.

In contrast, the richest countries have a total population of approximately one billion people—only 15 percent of the world's population. The populations of the richest countries are not growing nearly as fast

table9.2 *Quality of Life: A Comparative Perspective*

	Life Expectancy (years)	Infant Mortality (per 1000 births)	Adult Literacy (percent of population over 15)	Child Malnutrition (percent underweight)	Access to Safe Water (percent of population)
Afghanistan	45 yrs	165	29%	n/a	40%
Iran	71	32	77	n/a	94
Iraq	61	115	60	12%	81
Mexico	75.1	22.6	91	n/a	97
United States	77.9	6.9	95	1	100

n/a, not available.

Source: World Bank. 2007. **www.worldbank.org;** U.S. Census Bureau. 2006. *Statistical Abstract of the United States.* Washington, DC: U.S. Department of Commerce.

as the populations of the poorest countries. In the richest countries, women have about two children over their lifetime, and the populations of these countries are growing by only 1.2 percent. Many of the richest countries, including most of the countries of Europe, are actually experiencing population declines. With a low fertility rate, the rich countries have proportionately fewer children, but they also have proportionately more elderly, which can also be a burden on societal resources. Different from the poorest nations, the richest ones are largely urban.

Rapid population growth as a result of high fertility rates can make a large difference in the quality of life of the country. Countries with high birthrates are faced with the challenge of having too many children and not enough adults to provide for the younger generation. Public services, such as schools and hospitals, are strained in high-birthrate countries, especially because these countries are poor to begin with. However, very low birthrates, as many rich countries are now experiencing, can also lead to problems. In countries with low birthrates, there often are not enough young people to meet labor force needs, and workers must be imported from other countries.

Although the data clearly show that poor countries have large populations and high birthrates and rich countries have smaller populations and low birthrates, does this mean that the large population results in the low level of wealth of the country or that high fertility rates keep countries poor?

Scholars are divided on the relationship between the rate of population growth and economic development (Cassen 1994; Demeny 1991). Some theorize that rapid population growth and high birthrates lead to economic stagnation and that too many people keep a country from developing, thus miring the country in poverty (Ehrlich 1990). However, other researchers point out that some countries with very large populations have become developed (Coale 1986). After all, the United States has the third largest population in the world at 281 million people, yet it is one of the richest and most developed nations in the world. China and India, the two nations in the world with the largest populations, are also showing significant economic development. Scholars now believe that even though in some situations large population and high birthrates can impede economic development, in general fertility levels are affected by levels of industrialization, not the other way around. That is, as countries develop, their fertility levels decrease and their population growth levels off (Hirschman 1994; Watkins 1987).

Health and Environment

Significant differences are also evident in the basic health standards of countries, depending on where they are in the global stratification system. The high-income countries have lower childhood death rates, higher life expectancies, and fewer children born

underweight. People born today in wealthy countries can expect to live about seventy-seven years, and women outlive men by several years. Except for some isolated or poor areas of the rich countries, almost all people have access to clean water and acceptable sewer systems.

In the poorest countries the situation is completely different. Many children die within the first five years of life, people live considerably shorter lives, and fewer people have access to clean water and adequate sanitation. In the low-income countries, the problems of sanitation, clean water, childhood death rates, and life expectancies are all closely related. In many of the poor countries, drinking water is contaminated from poor or nonexistent sewage treatment. This contaminated water is then used to drink, to clean eating utensils, and to make baby formula. For adults, waterborne illnesses such as cholera and dysentery sometimes cause severe sickness but seldom result in death. However, children under age 5, and especially those under the age of 1, are highly susceptible to the illnesses carried in contaminated water. A common cause of childhood death in countries with low incomes is dehydration brought on by the diarrhea contracted by drinking contaminated water.

Degradation of the environment is a problem that affects all nations, which are linked in one vast environmental system. But global stratification also means that some nations suffer at the hands of others. Overdevelopment is resulting in deforestation. The depletion of this natural resource is most severe in South America, Africa, Mexico, and Southeast Asia (World Bank 2007). On the other hand, the overproduction of "greenhouse gas," emission of carbon dioxide from the burning of fossil fuels, is most severe in the United States, Canada, Australia, parts of western Europe and Russia, and, increasingly, China—places that use the most energy.

Although high-income countries have only 15 percent of the world population, together they use more than half of the world's energy. The United States alone uses one-quarter of the world's energy, though it holds only 4 percent of the world's population (see Figure 9.2). Safe water is also crucial; more than a

FIGURE 9.2 *Who Uses the World's Energy?*

Source: From World Development Indicators 2004, p. 114. Copyright © 2004 by World Bank. Reprinted by permission.

billion people do not have access to safe water. Moreover, water supplies are declining, a problem that will only be exacerbated by population growth and economic development. The World Bank has, in fact, warned that one-half of the world's population will face severe water shortage by the year 2025 (World Bank 2004). Clearly, global stratification has some irreversible environmental effects that are felt around the globe.

Education and Illiteracy

In the high-income nations of the world, education is almost universal, and the vast majority of people have attended school at least at some level. Literacy and school enrollment are now taken for granted in the high-income nations, although people in these wealthy nations who do not have a good education stand little chance of success. In the middle- and lower-income nations, the picture is quite different. Elementary school enrollment, virtually universal in wealthy nations, is less common in the middle-income nations and even less common in the poorest nations.

How do people survive who are not literate or educated? In much of the world, education takes place outside formal schooling. Just because many people in the poorer countries never go to school does not mean that they are ignorant or that they are uneducated. Most of the education in the world takes place in family settings, in religious congregations, or in other settings where elders teach the next generation the skills and knowledge they need to survive. This type of informal education often includes basic literacy and math skills that people in these poorer countries need for their daily lives.

The disadvantage of this informal and traditional education is that although it prepares people for their traditional lives, it often does not give them the skills and knowledge needed to operate in the modern world. In an increasingly technological world, this can perpetuate the underdeveloped status of some nations.

Gender Inequality

The position of a country in the world stratification system also affects gender relations within different countries. Poverty is usually felt more by women than by men. Although gender inequality has not been achieved in the industrialized countries, compared with women in other parts of the world women in the wealthier countries are much better off.

The United Nations is one of the organizations that carefully monitors the status of women globally. Their reports indicate mixed news with regard to women's status around the world. On the one hand, women's poverty has declined in some of the nations where it has been extreme, particularly in India, China, and some parts of Latin America. But in sub-Saharan Africa, women's poverty has increased. And though women's share of representation in governments has increased, they still hold only 16 percent of parliamentary seats worldwide. Also, although women have achieved near equity in levels of primary education, there are large gaps in the status of women and men in secondary and higher education—a fact that has huge implications for the work women do in a global economy that increasingly demands educational attainment (United Nations 2005a).

Perhaps most distressing is the global extent of violence against women. Violence takes many forms, including violence within the family, rape, sexual harassment, sex trafficking and prostitution, and state-based violence, among other things. The United Nations has concluded that "violence against women persists in every country in the world as a pervasive violation of human rights and a major impediment to achieving gender equality" (United Nations 2006a: 9). Several factors put women at risk of violence, ranging from individual level risk factors (such as a history of abuse as a child and substance abuse) to societal level factors, such as gender roles that entrench male dominance and societal norms that tolerate violence as a means of conflict resolution (see Table 9.3). Clearly, the inequalities that mark global stratification have particularly deleterious effects for the world's women.

War and Terrorism

The consequences of global stratification are also found in the international conflicts that bring war and an increased risk of terrorism. Although global inequality is certainly not the only cause of such problems, it contributes to the instability of world peace and the threat of terrorism. Global stratification generates inequities in the distribution of power between nations. Moreover, globalization has created a world-based capitalist class with unprecedented wealth and power. This is a class that now crosses national borders, thus some have defined it as a "transnational capitalist class" (Langman and Morris 2002). Coupled with the enormous poverty that exists, the visibility of this class and its association with Western values leads to resentment and conflict. Furthermore, attempts by wealthier nations to control access to the world's natural resources, such as oil, generate much political conflict. Thus, the same power and affluence that makes the United States a leader throughout the world makes it a target by those who resent its dominance.

In the Middle East, for example, oil production has created prosperity for some and exposed people in these nations to the values of Western culture. When people from different nations, such as those in the Middle East, study at U.S. universities and travel on business or vacations, they are exposed to Western values and patterns of consumption. As one commentator has noted, "Even those who have remained at home have not escaped exposure to Western culture.

table9.3	*Risk Factors for Violence Against Women: A Global Analysis*

The United Nations has studied the frequent use of violence against women in the world and identified the factors that put women at risk. These factors are found at various levels.

Individual Level:
- Frequent use of alcohol and drugs
- Membership in marginalized communities
- History of abuse as a child
- Witnessing marital violence in the home

Community Level:
- Women's isolation and lack of social support
- Community attitudes that tolerate and legitimate male violence
- High levels of social and economic inequality, including poverty

Family/Relationship Level:
- Male control of wealth
- Male control of decision making
- History of marital violence
- Significant disparities in economic, educational, or employment status

Societal Level:
- Gender roles that entrench male dominance and women's subordination
- Tolerance of violence as a means of conflict resolution
- Inadequate laws and policies to prevent and punish violence
- Limited awareness and sensitivity on the part of officials and social service providers

Source: United Nations. 2006a. *In-Depth Study on All Forms of Violence Against Women.* New York: United Nations.

In most of the countries of the modern Middle East western cultural influences are pervasive. They see western television programs, they watch western movies, they listen to western music, frequently wear western clothes and visit western web sites. Even western foods are locally available. McDonald's are now found in many of the major cities" (Bailey 2003: 341). Moreover, the sexual liberalism of Western nations and the relative equality of women also add

© Reuters/Damir Sagolj/Landov

Global conflicts, including war, can result from the inequality that global stratification produces..

to the volatile mix of nations clashing (Norris and Inglehart 2002).

As a result, some traditional leaders, including religious clerics, define Western culture as a source of degeneracy. Countries such as the United States, where consumerism is rampant, then become the target of those who see this as a threat to their traditional way of life (Ehrlich and Liu 2002). In this sense, global stratification and the dominance of Western culture are inseparable (Bailey 2003). Understood in this way, terrorism is not just a question of clashing religious values (although that is a contributing factor) but also stems from the global dominance of some nations over others. This is why those who commit atrocious acts, like flying jets into the World Trade Center towers, can believe themselves as fighting for a righteous cause.

Terrorism can be defined as premeditated, politically motivated violence perpetrated against noncombatant targets by persons or groups who use their action to try to achieve their political ends (White 2002). Terrorism can be executed through violence or threats of violence and can be executed through various means—suicide bombs, biochemical terror, cyberterror, or other means. Because terrorists operate outside the bounds of normative behavior, it is very difficult to prevent. Although rigid safeguards can be put in place, such safeguards also threaten the freedoms that are characteristics of open, democratic societies. The fact that terrorism is so difficult to stop contributes to the fear that it is intended to generate.

Inequality is also connected to the context in which terrorism emerges. A study of Al Qaeda terrorists finds that the leaders tend to come from middle-class backgrounds, though they often use those who are young, poorly educated, and economically disadvantaged to carry out suicide missions. Families of suicide bombers often receive large cash payments; at the same time they can feel they have served a sacred cause (Stern 2003). The fact that one-third of Iraqis now live in poverty—a change from having a thriving, largely middle-class economy in the 1970s and 1980s—helps explain the high rates of violence within Iraq now (United Nations News Center 2007). This also suggests that improving the lives of those who now feel collectively humiliated could provide some protection against terrorism.

World Poverty

One fact of global inequality is the growing presence and persistence of poverty in many parts of the world. There is poverty in the United States, but very few people in the United States live in the extreme levels of deprivation found in some of the poor countries of the world, as seen in Map 9.2.

In the United States, the poverty level is determined by the yearly income for a family of four that is considered necessary to maintain a suitable standard of living. Twelve and a half percent of Americans live in poverty (DeNavas-Walt et al. 2008). This definition of poverty in the United States identifies **relative poverty:** The households in poverty in the United States are poor compared with other Americans, but when one looks at other parts of the world, an income at the U.S. poverty line of $21,203 would make a family very well off.

The United Nations measures world poverty in two ways. **Absolute poverty** is the situation in which people live on less than $1 per day. **Extreme poverty** is defined as the situation in which people live on less than $275 a year, that is, on less than 75 cents a day. There are six hundred million people in the world who live at or below this extreme poverty level.

However, money does not tell the whole story because many people in the poor countries do not always deal in cash. In many countries, people survive by raising crops for personal consumption and by bartering or trading services for food or shelter. These activities do not show up in calculations of poverty levels that use amounts of money as the measure. As a result, the United Nations Development Program also defines what it calls the human poverty index.

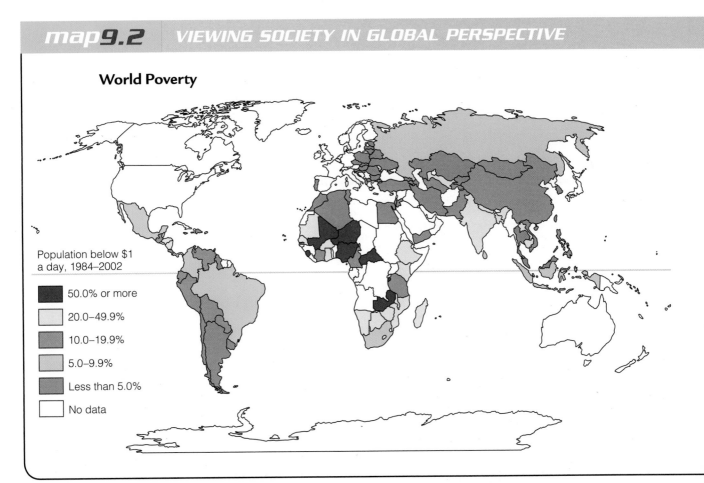

map 9.2   VIEWING SOCIETY IN GLOBAL PERSPECTIVE

World Poverty

Population below $1 a day, 1984–2002

- 50.0% or more
- 20.0–49.9%
- 10.0–19.9%
- 5.0–9.9%
- Less than 5.0%
- No data

Source: From World Bank Atlas 2007. Copyright © 2007 by World Bank. Reprinted by permission. **www.worldbank.org**

The **human poverty index** is a multidimensional measure of poverty, meant to indicate the degree of deprivation in four basic dimensions of human life: a long and healthy life, knowledge, economic well-being, and social inclusion. Different specific indicators of these different dimensions are used to measure poverty in the industrialized and developing countries because what constitutes these different dimensions of life can vary substantially in such different environments. In developing countries, the indicators are

- the percentage of people born not expected to live to age 40;
- the adult illiteracy rate;
- the proportion of people lacking access to health services and safe water; and,
- the percentage of children under age 5 who are moderately or severely underweight.

In industrialized countries, the human poverty index is measured by

- the proportion of people not expected to live to age 60;
- the adult functional illiteracy rate;
- the incidence of income poverty (because income is the largest source of economic provisioning in industrialized countries); and,
- long-term unemployment rates.

Figure 9.3 compares the human poverty index in select developing and industrialized nations (United Nations 2006c).

Who Are the World's Poor?

Using the United Nations' definition of absolute poverty (those whose level of consumption falls below $1 per day), one billion people, about one-fifth of the world's population, live in poverty. Another 1.5 billion live on $1–$2 per day, resulting in more than 40 percent of the world's population forming what the United Nations calls a *global underclass*. The good news is that the number of people living in poverty is declining, with most of the reduction attributable to progress in East Asia, particularly in the People's Republic of China. At the same time, poverty in other areas is increasing, including in eastern Europe and central Asia. Sub-Saharan Africa has the highest incidence of poverty of anywhere in the world, despite the rich natural resources of this region. Almost half of the population in this region live in poverty. As a result, infant mortality here is high, life expectancy is low, school enrollment is low, and there are very high death rates due to AIDS (United Nations 2005b).

The character of poverty differs around the globe. In Asia, the pressures of large population growth leave many without sustainable employment. And, as manufacturing has become less labor intensive with more mechanized production, the need for labor in certain industries has declined. Even though new technologies provide new job opportunities, they also create new forms of illiteracy because many people have neither the access nor the skills to use information technology. In sub-Saharan Africa, the poor live in marginal areas

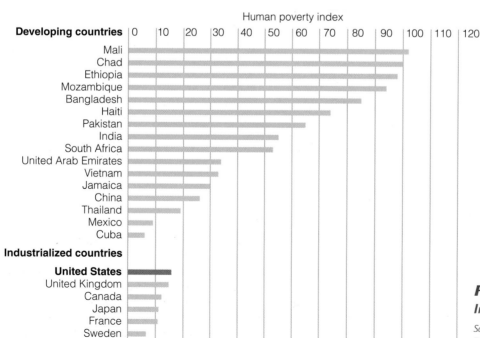

FIGURE 9.3 *Human Poverty Index*

Source: United Nations Human Development Report. 2006. **www.undp.org**

where poor soil, erosion, and continuous warfare have created extremely harsh conditions. Political instability and low levels of economic productivity contribute to the high rates of poverty. Solutions to world poverty in these different regions require sustainable economic development, as well as an understanding of the diverse regional factors that contribute to high levels of poverty.

Women and Children in Poverty

There is no country in the world in which women are treated as well as men. As with poverty in the United States, women bear a larger share of the burden of world poverty. Some have called this *double deprivation*—in many of the poor countries women suffer because of their gender and because they disproportionately carry the burden of poverty. For instance, in situations of extreme poverty, women have the burden of taking on much of the manual labor because in many cases the men have left to find work or food. The United Nations concludes that strengthening women's economic security through better work is essential for reducing world poverty (United Nations 2005a).

Because of their poverty, women tend to suffer greater health risks than men. Although women outlive men in most countries, the difference in life expectancy is *less* in the countries in poverty. This is explained by several factors. For one, fertility rates are higher in poor countries. Giving birth is a time of high risk for women, and women in poor countries with poor nutrition, poor maternal care, and the lack of trained birth attendants are at higher risk of dying during and after the birth.

High fertility rates are also related to the degree of women's empowerment in society—an often neglected aspect of the discussion between fertility and poverty. Societies where women's voices do not count for much tend to have high fertility rates as well as other social and economic hardships for women, including lack of education, job opportunities, and information about birth control. Empowering women through providing them with employment, education, property, and voting rights can have a strong impact on reducing the fertility rate (Sen 2000).

Women also suffer in some poor countries because of traditions and cultural norms. Most (though not all) of the poor countries are patriarchal, meaning that men control the household. As a result, in some situations of poverty the women eat after the men, and boys are fed before girls. In conditions of extreme poverty, baby boys may also be fed before baby girls because boys have higher status than girls. As a result, female infants have a lower rate of survival than male infants.

A distressing number of children in the world are also poor (see Figure 9.4). Children in poverty do not have the luxury of an education. Schools are usually few or nonexistent in poor areas of the world, and families are so poor that they cannot afford to send their children to school. Children from a very early age are required to help the family survive by working or performing domestic tasks such as fetching water. In extreme situations, children at a young age work as beggars, young boys and girls are sold to work in sweatshops, and young girls are sold into prostitution by their families. This may seem unusually cruel and harsh by Western standards, but it is difficult to imagine the horror of starvation and the desperation that many families in the world must feel that would force them to take such measures to survive. In poor countries, families feel they must have more children for their survival, yet having more children perpetuates the poverty. The United Nations estimates that there are 211 million children between age 5 and 14 in the paid labor force throughout the world. Most of the children, 127 million, are in Asia, and 48 million are in sub-Saharan Africa (International Labour Organization 2002). Many of these children work long hours in difficult conditions and enjoy few freedoms, making products (soccer balls, clothing, and toys, for example) for those who are much better off.

Another problem in the very poor areas of the world is homeless children (Mickelson 2000). In many situations, families are so poor that they can no longer care for their children, and the children must go without education out on their own, even at young ages. Many of these homeless children end up in the streets of the major cities of Asia and Latin America. In Latin America, it is estimated that there are thirteen million street children, some as young as six years old. Alone, they survive through a combination of begging, prostitution, drugs, and stealing. They sleep in alleys or in makeshift shelters. Their lives are harsh, brutal, and short.

Debunking Society's Myths

Myth: There are too many people in the world, and there is simply not enough food to go around.

Sociological perspective: Growing more food will not end hunger. If systems of distributing the world's food were more just, hunger could be reduced.

Poverty and Hunger

How can you live on less than $1 a day? The answer is that you cannot, or at least you cannot live very well. Malnutrition and hunger are growing problems because many of the people in poverty cannot find or afford food. The World Health Organization estimates that about eight hundred million people in the world are malnourished, which leads to disease (World Health Organization 2000).

Hunger results when there is not enough to eat to feed a designated area (such as a region or country). It may be that there is an inadequate supply of food or that households simply cannot afford to purchase enough food to feed themselves. Hunger stifles the mental and physical development of children and leads to disease and death. The trend in the world has been a reduction in the number of malnourished people, with the number of malnourished people expected to drop to under six hundred million by the year 2015. This is encouraging, but it is also short of the goal for reducing hunger that the World Health Organization has hoped for (World Health Organization 2000). Although the food supply is plentiful in the world, and is actually increasing faster than the population, the rate of malnutrition is still dangerously high.

Why are people hungry? Is there not enough food to feed all the people in the world? In fact, plenty of food is grown in the world. The world's production of wheat, rice, corn, and other grains is sufficient to adequately feed all the people in the world. Much grain grown in the United States is stored and not used. The problem is that the surplus food does not get to the truly needy. The people who are starving lack what they need for obtaining adequate food, such as arable land or a job that would pay a living wage. In many cases, in the past people grew food crops and were able to feed themselves, but today much of the best land has been taken over by agribusinesses that

[understanding diversity]

War, Childhood, and Poverty

"In 2003, surgeons were forced to amputate both of Ali Ismaeel Abbas's arms after an errant U.S. bomb slammed into his Baghdad home during the opening phase of the Iraq war. Pictures of the twelve-year-old, who lost his parents in the attack, soon appeared on TV screens and in newspapers around the world. Since then, Abbas, who was treated in Kuwait, has come to represent a grim reality: All too often the victims of war are innocent children" (McClelland 2003: 20).

In the past ten years alone, UNICEF estimates that over two million children have died in war, with even more injured, disabled, orphaned, or forced into refugee camps (Machel 1996). One estimate is that of all the victims of war, 90 percent are civilian—half of those, children (McClelland 2003).

In the aftermath of war children are also highly vulnerable to outbreaks of disease. In Iraq, following the war in 2003, many children died of diseases such as anemia and diarrhea—diseases that can be prevented. Children in Iraq were already living under extreme hardship under the regime of Saddam Hussein. Economic sanctions against Iraq during his regime also produced high infant mortality because of food shortages.

The United Nations has passed resolutions prohibiting the use of children under age 18 in combat. It has linked the threats to children from violence with high rates of poverty around the world. Although reducing poverty would not eliminate the threat of war, it would go a long way toward improving children's lives in war-torn regions.

Bassim Daham/AP Photo

War, though it may seem remote to some, affects millions both in the United States and in war-torn countries. Many of those most affected are children. Here relatives mourn the death of children killed during a U.S. raid in Tikrit, Iraq in 2006.

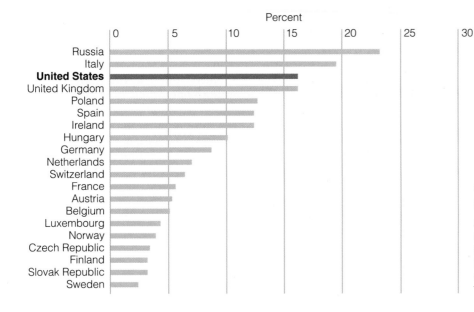

FIGURE 9.4 *Child Poverty in the Wealthier Nations*

Source: Velminckx, Koen, and Timothy Smeeding, eds. 2001. *Child Well-Being, Child Poverty, and Child Policy in Modern Nations.* Fig 8.4, p. 230. Tonawanda, NY: University of Toronto Press. Reprinted by permission of The Policy Press.

grow cash crops, such as tobacco or cotton, and subsistence farmers have been forced onto marginal lands on the flanks of the desert where conditions are difficult and crops often do not grow.

Causes of World Poverty

What causes world poverty, and why are so many people so desperately poor and starving? More to the point, why is poverty decreasing in some areas but increasing in others? We do know what does *not* cause poverty. Poverty is not necessarily caused by too rapid population growth, although high fertility rates and poverty are related. In fact, many of the world's most populous countries, India and China, for instance, have large segments of their population that are poor, but even with very large populations these countries have begun to reduce poverty levels. Poverty is not caused by people being lazy or uninterested in working. People in extreme poverty work tremendously hard just to survive, and they would work hard at a job if they had one. It is not that they are lazy; it is that there are no jobs for them.

Poverty is a result of a mix of causes. For one, the areas where poverty is increasing have a history of unstable governments or, in some cases, virtually no effective government to coordinate national development or plans that might alleviate extreme poverty and starvation. World relief agencies are reluctant to work in or send food to countries where the national governments cannot guarantee the safety of relief workers or the delivery of food and aid to where it should go. Food convoys may be hijacked or roads blocked by bandits or warlords.

In many countries with high proportions of poverty, the economies have collapsed and the governments have borrowed heavily to remain afloat. As a condition of these international loans, lenders, including the World Bank and the International Monetary Fund, have demanded harsh economic restructuring to increase capital markets and industrial efficiency. These economic reforms may make good sense for some and may lead these countries out of economic ruin over time, but in the short run, these imposed reforms have placed the poor in a precarious position because the reforms also called for drastically reduced government spending on human services.

Poverty is also caused by changes in the world economic system. Although poverty has been a long-term problem and has many causes, increases in poverty and starvation in Africa and Latin America can be attributed in part to the changes in world markets that favored Asia economically but put sub-Saharan Africa and Latin America at a disadvantage. As the price of products declined with more industrialization in places such as India, China, Indonesia, South Korea, Malaysia, and Thailand, commodity-producing nations in Africa and Latin America suffered. In Latin America, the poor have flooded to the cities, hoping to find work, whereas in Africa they did the opposite, fleeing to the countryside hoping to be able to grow subsistence crops. Governments often had to borrow to provide help to their citizens. Many governments collapsed or found themselves in such great debt that they were unable to help their own people. This has created massive amounts of poverty and starvation.

An often unrecognized cause of poverty is war. War disrupts the infrastructure of a society—its roads, utility systems, water, sanitation, even schools. For countries already struggling economically, this can be devastating. Food production may be disrupted and commerce can be threatened as it may be difficult, even impossible, to move goods in and out of a country. And the loss of life and major injury can mean that there are fewer productive citizens who can work, thus threatening family and community well-being (Pathways to Peace 2009). Moreover, the billions of dollars spent each year on military struggle rob societies of the resources that could be used to address humanitarian needs. Add to this the fact that wars are more likely to occur in nations that are already poor and you see the impact that war has on world poverty.

In sum, poverty has many causes. It is a major global problem that affects the billions who are living in poverty, but also affects all people in one way or another. In some areas, poverty rates are declining as some countries begin to improve their economic situation; however, in other areas of the world, poverty is increasing, and countries are sinking into financial, political, and social chaos.

Globalization and Social Change

Globalization is, in some ways, not a new thing. Nations have long been engaged through a global system of trade, travel, and tourism. But what is new about globalization is the extent to which it permeates daily life for people all over the world and the pace with which globalization is developing. New technologies now allow for extraordinarily fast transactions across tremendous distances, both linking people together in new ways and transferring goods, cultural symbols, and communication systems in ways that were unimaginable not that long ago (Eitzen and Baca Zinn 2006).

Globalization is thus ushering in social changes—some good, some not—that will continue to evolve in the years ahead. As we have seen, globalization has meant that many countries in the world are becoming better off, but many countries remain persistently poor, some very poor. Is the world getting better or worse? What will happen in the future?

There is some good news. In some areas of the world, particularly east Asia, but also in Latin America, many countries have shown rapid growth and are emerging as stronger nations. These countries are sometimes called the **newly industrializing countries (NICs),** and they include South Korea, Malaysia, Thailand, Taiwan, and Singapore. In these countries, the governments have invested in social and economic development, often with outside help from other nations and corporations. Because some of the NICs have large populations, their success demonstrates that economic development can occur in heavily populated countries. China, for example, has embarked on an aggressive policy of industrial growth, and India is also improving economically.

Yet for all the success stories that globalization has generated, many nations are not making it. These

include nations on all continents. In many cases, governments have collapsed or are corrupt, the economy is bankrupt, the standard of living is poor, and people are starving. In many areas of the world ethnic hatred has led to mass genocide and forced millions from their homes, creating huge numbers of refugees. In Darfur in the western region of Sudan, over 400,000 civilians have been murdered and millions have lost their homes, creating an international outcry demanding that Western governments intervene to stop the violence and provide massive humanitarian aid.

Globalization has also brought the expansion of the system of capitalism, including to nations once hostile to capitalist economics, such as China. This has opened new markets, increased global trade, but also expanded the reach of multinational corporations. The development of such world financial markets may bring prosperity and wealth to many nations and individuals, and it can allow some formerly poor countries to share in the world's wealth. But economic prosperity does not usually filter down to the people at the lower levels of society, and it can force nations into huge amounts of debt, thus allowing poverty and hunger to continue—or even worsen. Thus, while market economies create opportunities for some to become wealthy, both individuals and nations, many nations and individuals do not benefit from this global transformation.

Globalization is a strong force that will continue to shape the future of most nations. Some see globalization simply as the expansion of Western markets and culture into all parts of the world. Western civilization brings positive new values (including democracy and more equality for women), but it can also bring values that may not be seen as positive changes—such

In an increasingly global world, it is easier for people to become aware of international conflict and human rights abuses, thus generating social movements to advocate for more humane treatment.

as increased consumerism or a change in the nation's sexual mores. Globalization certainly brings new products to remote parts of the world (movies, clothing styles, and other commercial goods), but some see this as a form of imperialism—that is, the domination of Western nations. Resistance to Western globalization and imperialism produces some of the international problems now dominating United States and world history, as evidenced in the hostility felt by militant fundamentalist Islamic groups toward the United States.

Globalization has created great progress in the world—including trade, migration, the spread of diverse cultures, the dissemination and sharing of new knowledge, greater freedom for women, travel, and so forth. Moreover, globalization has not simply extended the values and knowledge of Western culture. Many of the things we now take for granted in our culture originated in non-Western cultures. For example, the decimal system—fundamental to modern math and science—originated in India between the second and sixth centuries and was soon further developed by Arab mathematicians. Western societies certainly get credit for the development of science and technology, but the credit is not theirs alone (Sen 2002).

It is no doubt true that globalization is contributing to the inequality between nations and to the exploitation of some nations and groups by others. Perhaps the solution is not in resisting globalization, but in working so that the benefits of the global economy, global technology, and knowledge reach parts of the world in less exploitative ways. As long as great disparities in standards of living, human rights and basic freedoms, environmental quality, and so forth persist, world conflict is likely to be the result.

Chapter Summary

- ### What is global stratification?

  Global stratification is a system of unequal distribution of resources and opportunities between countries. A particular country's position is determined by its relationship to other countries in the world. The countries in the global stratification system can be categorized according to their per capita gross national product, or wealth. The world's countries can also be categorized as *first-, second-,* or *third-world* countries, which describe their political affiliation and their level of development. The global stratification system can also be described according to the economic power countries have.

- ### How do systems of power affect different countries in the world?

  The countries of the world can be divided into three levels based on their power in the world economic system. The *core countries* are the countries that control and profit the most from the world system. *Semiperipheral countries* are semi-industrialized and play a middleman role, extracting profits from the poor countries and passing those profits on to the core countries. At the bottom of the world stratification system are the *peripheral countries,* which are poor and largely agricultural, but with important resources that are exploited by the core countries. Most of these nations are populated by people of color, perpetuating racism as part of the world system.

- ### What are the theories of global stratification?

  Modernization theory interprets the economic development of a country in terms of the internal attitudes and values. Modernization theory ignores that the development of a country may be due to its economic relationships with other more powerful countries. *Dependency theory* draws on the fact that many of the poorest nations are former colonies of European colonial powers that keep colonies poor and do not allow their industries to develop, thus creating dependency. *World systems theory* argues that no nation can be seen in isolation and that there is a world economic system that must be understood as a single unit.

- ### What are some of the consequences of global stratification?

  The poorest countries have more than half the world's population and have high birthrates, high mortality rates, poor health and sanitation, low rates of literacy and school attendance, and are largely rural. The richest countries have low birthrates, low mortality rates, better health and sanitation, high literacy rates, high school attendance, and largely urban populations. Although women in the wealthy countries are not completely equal to men, they suffer less inequality than do women in the poor countries.

• *How do we measure and understand world poverty?*

Relative poverty means being poor in comparison to others. *Absolute poverty* describes the situation where people do not have enough to survive, measured as having the equivalent of $1 per day. *Extreme poverty* is defined as the situation in which people live on less than 75 cents a day. The United Nations has developed a *human poverty index*—a multidimensional measure that accounts for life expectancy, knowledge,

economic well-being, and social inclusion. Poverty particularly affects women and children. Children in the very poor countries are forced to work at very early ages and do not have the opportunity for schooling. Street children are a growing problem in many cities of the world. Starvation is also a consequence of the global stratification system.

• *What is the future of global stratification?*

The future of global stratification is varied and depends on the country's position within the world economic system. Some countries, particularly those in east Asia—commonly referred to as *newly industrializing countries*—have shown rapid growth and emerged as developed countries. Many nations, though, are not making it. Governments collapse, countries suffer economic bankruptcy, the standard of living plummets, and people starve.

Key Terms

absolute poverty 225
colonialism 218
commodity chain 220
core countries 215
dependency theory 218
extreme poverty 225
first-world countries 215
global stratification 212
gross national income (GNI) 212

human poverty index 226
international division of labor 217
modernization theory 217
multinational corporation 219
neocolonialism 219
newly industrializing countries (NICs) 230
peripheral countries 215
power 215

relative poverty 225
second-world countries 215
semiperipheral countries 215
terrorism 224
third-world countries 215
world cities 221
world systems theory 219

Online Resources

Sociology: The Essentials Companion Website

www.cengage.com/sociology/andersen
Visit your book companion website where you will find more resources to help you study and write your research papers. Resources include web links, learning objectives, Internet exercises, quizzing, and flash cards.

 is an easy-to-use online resource that helps you study in less time to get the grade you want NOW.

www.cengage.com/login
Need help studying? This site is your one-stop study shop. Take a Pre-Test and CengageNOW will generate a Personalized Study Plan based on your test results. The Study Plan will identify the topics you need to review and direct you to online resources to help you master those topics. You can then take a Post-Test to determine the concepts you have mastered and what you still need to work on.

10

Chapter ten
CHAPTER TEN

Race and Ethnicity

Race and Ethnicity

Racial Stereotypes

Prejudice, Discrimination, and Racism

Theories of Prejudice and Racism

Diverse Groups, Diverse Histories

Attaining Racial and Ethnic Equality: The Challenge

Chapter Summary

[*You might expect* a society based on the values of freedom and equality, such as ours, not to be deeply afflicted by racial conflict, but think of the following situations:

- When Hurricane Katrina struck New Orleans and the Gulf Coast in 2006, hundreds of thousands of people were displaced and billions of dollars of property destroyed. Although the hurricane affected the lives (and deaths) of many, African Americans—many of them poor—were disproportionately killed or left homeless. Millions of Americans were shocked by the images of poor people desperate to survive but left without federal or state governmental help for a long time.

- In 2009, James von Brunn, an eighty-eight year-old self-proclaimed white supremacist, gunned down and killed a security guard at the U.S. Holocaust Memorial Museum in Washington, DC. James von Brunn was known by federal authorities to be affiliated with various hate groups. The shooter left an anti-Semitic letter in his car, parked outside the Museum, charging that "Obama was created by Jews." The guard who was shot and killed, Stephen T. Johns, a thirty-nine year old, African American guard who worked at the Museum. Von Brunn, shot and wounded by other security guards, was charged with murder.

- Derogatory labels, which reinforce prejudice and racism, have been applied throughout American history to persons of color, and despite minimal improvements in race relations, they persist only slightly diminished right up to the present: Blacks (niggers, coons, jigaboos, spades, sambos, jungle bunnies), Hispanics (spics, greasers, wetbacks, beaners), and Asians (slants, slopes, chinks, Japs, flips).

continued

Michael Newman/PhotoEdit

Not even White ethnic groups have been spared in recent history, as illustrated by derogatory labels applied to the Irish (micks, paddies), Germans (krauts), French (frogs), Italians (wops, ginnies, dagos), Poles (polacks), Hungarians (bohunks), and Jews (kikes, hebes, hymies, yids).

These incidents and labels are not pretty. They all have one thing in common—race-ethnicity. Along with gender and social class, race and ethnicity have fundamental importance in human social interaction and are an utterly integral part of social institutions.

Of course, the racial and ethnic groups do not always interact as enemies. Nor is interracial tension always obvious. It can be as subtle as a White person who simply does not initiate interactions with African Americans and Latinos or an elderly White man who almost imperceptibly leans backward at a cocktail party as a Japanese American man approaches him.

In everyday human interaction, race still matters and matters a lot. What is race, and what is ethnicity? Why does society treat racial and ethnic groups differently, and why is there social inequality—stratification—between these groups? How are these divisions and inequalities able to persist so stubbornly, and how extensive are they? These inequalities are so strong and persistent in American society that sociologists are not convinced by recent speculation in the popular press that the landslide election of President Barack Obama, the first African American U.S. president, will lead to the virtual disappearance of prejudice and racism. Such questions and issues fascinate sociologists who do research on racial and ethnic relations and stratification in our society.

Race and Ethnicity

Within sociology, the terms *ethnic, race, minority,* and *dominant group* have very specific meanings, different from the meanings these terms have in common usage. These concepts are important in developing a sociological perspective on race and ethnicity.

Ethnicity

An **ethnic group** is a social category of people who share a common culture, for example, a common language or dialect; a common nationality; a common religion; and common norms, practices, customs, and history. Ethnic groups have a consciousness of their common cultural bond. Italian Americans, Japanese Americans, Arab Americans, Polish Americans, Greek Americans, Mexican Americans, and Irish Americans are all examples of ethnic groups in the United States. Ethnic groups are

This band, playing for St. Patrick's Day, an Irish holiday, contains people of varied racial–ethnic backgrounds.

also found in other societies, such as the Pashtuns in Afghanistan or the Shiites and Sunnis in Iraq, whose ethnicity is based on religious differences.

An ethnic group does not exist only because of the common national or cultural origins of a group, however. Ethnic groups develop because of their unique historical and social experiences. These experiences become the basis for the group's *ethnic identity,* meaning the definition the group has of itself as sharing a common cultural bond. Prior to immigration to the United States, Italians, for example, did not necessarily think of themselves as a distinct group with common interests and experiences. Originating from different villages, cities, and regions of Italy, Italian immigrants identified themselves by their family background and community of origin. However, the process of immigration and the experiences Italian Americans faced as a group in the United States, including discrimination, created a new identity for the group (Waters and Levitt 2002; Waters 1990; Alba 1990).

The social and cultural basis of ethnicity allows ethnic groups to develop more or less intense ethnic identification at different points in time. Ethnic identification may grow stronger when groups face prejudice or hostility from other groups. Perceived or real threats and perceived competition from other groups may unite an ethnic group around common political and economic interests. Ethnic unity can develop voluntarily, or it may be involuntarily imposed when ethnic groups are excluded by more powerful groups from certain residential areas, occupations, or social clubs. These exclusionary practices strengthen ethnic identity.

Race

Like ethnicity, race is primarily, though not exclusively, a socially constructed category. A **race** is a

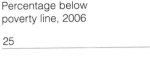
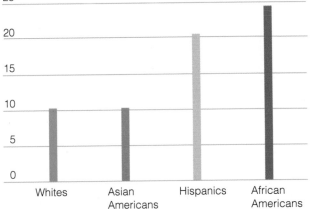

group treated as distinct in society based on certain characteristics, some of which are biological, that have been assigned social importance. Because of presumed biologically or culturally inferior characteristics (as defined by powerful groups in society), a race is often singled out for differential and unfair treatment. It is not the biological characteristics per se that define racial groups but *how groups have been treated historically and socially*.

Society assigns people to racial categories, such as Black, White, and so on, not because of science, logic, or fact, but because of opinion and social experience. In other words, how racial groups are defined is a *social* process. This is what is meant when one says that race is "socially constructed" (Higginbotham and Andersen 2006). Although the meaning of race begins with perceived biological differences between groups (such as differences in physical characteristics like skin color, lip form, and hair texture), on closer examination, the assumption that racial differences are purely biological breaks down.

Debunking Society's Myths

Myth: Racial differences are fixed, biological categories.

Sociological perspective: Race is a social concept, one in which certain physical or cultural characteristics take on social meanings that become the basis for racism and discrimination. The definition of race varies across cultures within a society, across different societies, and at different times in the history of a society.

The social categories used to presumably divide groups into races are not fixed, and they vary from society to society and at different times in the history of a given society (Morning 2005, 2007, 2008; Washington 2006). Within the United States, laws defining who is Black have historically varied from state to state. North Carolina and Tennessee law historically defined a person as Black if he or she had even one great-grandparent who was Black (thus being one-eighth Black). In other southern states, having any Black ancestry at all defined one as a Black person—the so-called one drop (that is, of Black blood) rule (Broyard 2007; Taylor 2006; Malcomson 2000). This "one drop" rule still applies to a great extent today in the United States.

This is even more complex when we consider the meaning of race in other countries. In Brazil, a light-skinned Black person could well be considered White, especially if the person is of high socioeconomic status; this demonstrates that one's race in Brazil is in part actually *defined* by one's social class. Thus, in parts of Brazil, it is often said that "money lightens" (*o dinheiro embranquece*). In this sense, a category such as social class can become *racialized*. In fact, in Brazil people are considered Black only if they are of African descent and have no discernible White ancestry at all. The vast majority of U.S. Blacks would not be considered Black in Brazil (Surratt and Inciardi 1998; Telles 2004). Although Brazil is often touted as being a utopia of race "mixing" and racial social equality, nonetheless, as sociologist Edward Telles notes, light-skinned Brazilians continue to be privileged and continue to hold a disproportionate share of the wealth and power (Telles 1994, 2004).

Racialization is a process whereby some social category, such as a social class or nationality, takes on what *society perceives* to be racial characteristics (Malcomson 2000; Harrison 2000; Omi and Winant 1994). The experiences of Jewish people provide a good example of what it means to say that race is a socially constructed category. Jews are more accurately called an *ethnic group* because of common religious and cultural heritage, but in Nazi Germany Hitler defined Jews as a "race." An ethnic group had thus become *racialized*. Jews were presumed to be biologically inferior to the group Hitler labeled the Aryans— white-skinned, blonde, tall, blue-eyed people. On the basis of this definition—which was supported through Nazi law, taught in Nazi schools, and enforced by the Nazi military—Jewish people were grossly mistreated. They were segregated, persecuted, and systematically murdered in what has come to be called the Holocaust during the Second World War.

Mixed-race people defy the biological categories that are typically used to define race. Is someone who is the child of an Asian mother and an African American father Asian or Black? Reflecting this issue, the U.S. Census's current practice is for a person to check several racial categories, rather than just one (Wright 1994; Waters 1990), although considerable controversy has arisen over this procedure (Harrison 2000). As Table 10.1 shows, the decennial U.S. census (taken every ten years) has dramatically changed its racial and ethnic classifications since 1890, reflecting the fact that society's thinking about racial and ethnic categorization has not remained constant through time (Rodriguez 2006; Harrison 2000; Mathews 1996; Lee 1993).

The biological characteristics that have been used to define different racial groups vary considerably both within and between groups. Many Asians, for example, are actually lighter skinned than many Europeans and White Americans and, regardless of their skin color, have been defined in racial terms as yellow. Some light-skinned African Americans are also lighter in skin color than some White Americans. Developing racial categories overlooks the fact that human groups defined as races are—biologically speaking—much more alike than they are different.

The biological differences that are presumed to define different racial groups are somewhat arbitrary. Why, for example, do we differentiate people based on skin color and not some other characteristic such as height or hair color? You might ask yourself how a society based on the presumed racial inferiority

Scott Olson/Getty Images

This is Barack Obama, the first African American ever to be elected U.S. president. His father is Black African (Kenyan) and his mother is White American. Why is his race African American?

very few are actually given this official status, and the criteria for tribal membership as well as definition as "Indian" or "Native American" have varied considerably throughout American history. Thus, as with African Americans, it has been the state or federal government, and not so much the racial or ethnic group *itself,* that has defined who is a member of the group and who is not!

Official recognition by the government matters. For example, only those groups officially defined as Indian tribes qualify for health, housing, and educational assistance from the Bureau of Indian Affairs (the BIA) or are allowed to manage the natural resources on Indian lands and maintain their own system of governance (Locklear 1999; Brown 1993; Snipp 1989).

This definition of race emphasizes that in addition to physical and cultural differences, race is created and maintained by the most powerful group (or groups) in society. This definition of race also incorporates presumed group differences in the context of social and historical experience. As a result, who is defined as a race is as much a political question as a biological one (Brodkin 2006). For example, although they probably did not think of themselves as a race, Irish Americans in the early twentieth century were defined by more powerful White groups as a "race" that was inferior to White people. At that time, Irish people were not considered by many even to be White (Ignatiev 1995)! In fact, a century ago the Irish were called "Negroes turned inside out," and Negroes (Black people) were called "smoked Irish" (Malcomson 2000).

The social construction of race has been elaborated on in an insightful perspective in sociology known as racial formation theory (Omi and Winant 1994; Brodkin 2006). **Racial formation** is the process by which a group comes to be defined as a race. This definition is supported through official social institutions such as the law and the schools. This concept emphasizes the importance of social institutions in producing and maintaining the meaning of race; it also connects the process of racial formation to the exploitation of so-called racial groups. A good example comes from African American history. During slavery, an African American was defined as being three-fifths of a person (equivalently, as "divested two-fifths the man") for the purposes of deciding how slaves would be counted for state representation in the new federal government and how they would be defined as property in order to be taxed as property. The process of defining slaves in this way served the purposes of White Americans, not slaves themselves, and it linked the definition of slaves as a race to the political and economic needs of the most powerful group in society (Higginbotham 1978).

The process of racial formation also explains how groups such as Asian Americans, American Indians, and Latinos have been defined as races, despite the different experiences and nationalities of the groups composing these three categories. Race, like ethnicity,

of red-haired people would compare to other racial inequalities in the United States. The likelihood is that if a powerful group defined another group as inferior because of some biological characteristics, and they used their power to create social institutions that treated this group unfairly, a system of racial inequality would result. In fact, *very few biological differences* exist between racial groups. Most of the variability in almost all biological characteristics, even blood type and various bodily chemicals, is *within* and not between racial groups (Rodriguez 2006; Malcomson 2000; Lewontin 1996).

Different groups use different criteria to define racial groups. To American Indians, being classified as an American Indian depends upon proving one's ancestry, but this proof varies considerably from tribe to tribe. Among some American Indians, one must be able to demonstrate 75 percent American Indian ancestry to be recognized as such; for other American Indians, demonstrating 50 percent American Indian ancestry is sufficient.

It also matters *who* defines racial group membership. The government makes tribes prove themselves as tribes through a complex set of federal regulations (called the "federal acknowledgment process");

table10.1	Comparison of U.S. Census Classifications, 1890–2000				
Census Date	**White**	**African American**	**Native American**	**Asian American**	**Other Categories**
1890	White	Black, Mulatto, Quadroon, Octoroon	Indian	Chinese Japanese	
1900	White	Black	Indian	Chinese Japanese	
1910	White	Black Mulatto	Indian	Chinese Japanese	Other
1990	White	Black or Negro	Indian (American) Eskimo Aleut	Chinese Japanese Filipino Korean Asian Indian Vietnamese	Hawaiian Guamanian Samoan Asian or Pacific Islander Other
2000[a, b]	White	Black or African American	American Indian Alaskan Native	Chinese Japanese Filipino Korean Asian Indian Vietnamese	Native Hawaiian Other Pacific Islander Other

[a]In 2000, for the first time ever, individuals could select more than one racial category. Only 2 percent actually did so.
[b]Hispanics are included under "Other."

Source: Lee, Sharon. 1993. "Racial Classification in the U.S. Census: 1890–1990." *Ethnic and Racial Studies* 16(1): 75–94; Rodriguez 2006; U.S. Census Bureau. 2003. "Racial and Ethnic Classification Used in Census 2000 and Beyond." Taylor & Francis Ltd.

lumps groups together that may have very different historical and cultural backgrounds, but once they are so labeled, the groups are perceived as a single entity. This reflects a more general principle in the social sciences called **out-group homogeneity effect,** where all members of any out-group are perceived to be similar or even identical to each other and differences among them are perceived to be minor or nonexistent. This has recently been the case in the United States with Middle Easterners: Lebanese, Syrians, Saudi Arabians, Iranians, Iraqis, Jordanians, Egyptians, Afghans, and many others are classified as one group and called Middle Easterners, or simply "Arabs."

Minority and Dominant Groups

Minorities are racial or ethnic groups, but not all racial or ethnic groups are always considered minorities. Irish Americans, for instance, are not now thought of as minorities, although they once certainly were in the early part of the twentieth century. A **minority group** is any distinct group in society that shares common group characteristics and is forced to occupy low status in society because of prejudice and discrimination. The group that assigns a racial or ethnic group to subordinate status in society is called the **dominant group.**

A group may be classified as a minority on the basis of ethnicity, race, sexual preference, age, or class status, for example. A minority group is not necessarily a numerical minority but is a group that holds low status in relation to other groups in society, regardless of the size of the group. In South Africa, Blacks outnumber Whites ten to one, but until Nelson Mandela's election as president and the dramatic change of the country's government in 1994, Blacks were an officially oppressed and politically excluded social minority under the infamous *apartheid* (pronounced "aparthate" or "apart-hite") system of government. In general, a racial or ethnic minority group has the following characteristics (Simpson and Yinger 1985):

1. The minority group possesses characteristics (such as race, ethnicity, sexual preference, age, or religion, and even gender) that are popularly regarded as different from those of the dominant group.

2. The minority group suffers prejudice and discrimination by the dominant group.

3. Membership in the group is frequently ascribed rather than achieved, although either form of status can be the basis for being identified as a minority.

4. Members of a minority group feel a strong sense of group solidarity. There is a "consciousness of kind" or "we" feeling. This bond grows from common cultural heritage and the shared experience of being a recipient of prejudice and discrimination.

Debunking Society's Myths

Myth: Minority groups are those with the least numerical representation in society.

Sociological perspective: A minority group is any group, regardless of size, that is singled out in society for unfair treatment and that generally occupies lower status in the society.

Racial Stereotypes

Racial and ethnic inequality is peculiarly resistant to change. Racial and ethnic inequality in society produces racial stereotypes, and these stereotypes become the lens through which members of different groups perceive one another.

Stereotypes and Salience

In everyday social interaction, people tend to *categorize* other people. Fortunately or unfortunately, we all do this. The most common bases for such categorizations are race, gender, and age. A person *immediately* identifies a stranger as Black, Asian, Hispanic, White, and so on; as a man or woman; and as a child, adult, or elderly person. Quick and ready categorizations help people process the huge amounts of information they receive about people with whom they have come into contact. People quickly assign others to a few categories, saving themselves the task of evaluating and remembering every little discernible detail about a person. People are taught from childhood to treat each person as a unique individual, but research over the years clearly shows that they do not. Instead, people routinely categorize others in some way or another. This allows thinking about people and processing information about them.

A **stereotype** is an oversimplified set of beliefs about members of a social group or social stratum. It is based on the tendency of humans to categorize a person based on a narrow range of perceived characteristics. Stereotypes are presumed, usually incorrectly, to describe the "typical" member of some social group.

Stereotypes based on race or ethnicity are called *racial–ethnic stereotypes.* Here are some common examples of racial–ethnic stereotypes: Asian Americans have been stereotyped as overly ambitious, sneaky, and clannish; African Americans often bear the stereotype of being loud, lazy, naturally musical, and so on. Hispanics are stereotyped as lazy, oversexed, and for Hispanic men, macho; Jews have been perceived as materialistic and unethical. Such stereotypes, presumed to describe the "typical" member of a group, are factually inaccurate for the vast majority of members of a group. *No group in U.S. history has escaped the process of categorization and stereotyping,*

not even White groups. For example, Italians have been stereotyped as overly emotional and prone to crime, the Irish as heavy drinkers and prone to politics, and so on for virtually any group in U.S. history.

The categorization of people into groups and the subsequent application of stereotypes is based on the **salience principle,** which states that we categorize people on the basis of what appears initially prominent and obvious—that is, salient—about them. Skin color is a salient characteristic; it is one of the first things that we notice about someone. Because skin color is so obvious, it becomes a basis for stereotyping. Gender and age are also salient characteristics of an individual and thus serve as notable bases for group stereotyping.

thinking sociologically

Observe several people on the street. What are the first things you notice about them (that is, what is *salient*)? Make a short list of these things. Do these lead you to *stereotype* these people? On what are your stereotypes based?

The choice of salient characteristics is culturally determined. In the United States, skin color, hair texture, nose form and size, and lip form and size have become salient characteristics, and these characteristics determine whether we perceive someone as "intelligent" or "stupid," as "attractive" or "unattractive," or even "trustworthy" or "untrustworthy," as shown in a clever experiment by Hunt (2005). We then use these features to categorize people in our minds on the basis of race. In other cultures, religion may be far more salient than skin color. In the Middle East, whether one is Muslim or Christian is far more important than skin color. Religion in the Middle East is a salient characteristic and takes considerable priority over race.

The Interplay Among Race, Gender, and Class Stereotypes

Alongside racial and ethnic stereotypes, gender and social class in addition to age are among the most prominent features by which people are categorized. In our society, there is a complex interplay among racial or ethnic, gender, and class stereotypes.

Among *gender stereotypes,* those based on a person's gender, the stereotypes about women are more likely to be negative than those about men. The "typical" woman has been traditionally stereotyped as subservient, overly emotional, and talkative, inept at math and science, and so on. Many of these are *cultural stereotypes.* They are conveyed and supported by the cultural media—music, TV, magazines, art, and literature—and also by one's family. Men, too, are painted in crude strokes, although usually not as negatively as women. Men in the media are stereotyped as macho, insensitive, and pigheaded and are portrayed in situation comedies as inept. Men are often depicted as wanting to have sex with as many women as possible in the shortest time available.

Social class stereotypes are based on assumptions about social class status. Upper-class people are stereotyped (by middle- and lower-class people) as snooty, aloof, condescending, and phony. Some of the stereotypes held about the middle class (by both the upper class and the lower class) are that they are overly ambitious, striving, and obsessed with "keeping up with the Joneses." Finally, stereotypes about lower-class people abound: They are perceived by the upper and middle classes as dirty, lazy, unmotivated, violent, and so on.

The principle of **stereotype interchangeability** holds that stereotypes, especially negative ones, are often interchangeable from one social class to another, from one racial or ethnic group to another, from a racial or ethnic group to a social class, or from a social class to a gender. Stereotype interchangeability is sometimes revealed through humor. Ethnic jokes often interchange different groups as the butt of the humor, stereotyping them as dumb and inept. Take the stereotype of African Americans as inherently lazy. This stereotype has also been applied in recent history to Hispanic, Polish, Irish, Italian, and other groups (illustrating interchangeability from one racial–ethnic group to another). It has even been applied to those people perceived as lower class (showing interchangeability from a racial–ethnic group to a social class). In fact, "laziness" is often used to explain *why* someone is lower class or poor.

Middle-class people are more likely to attribute the low status of a lower-class person to something *internal,* such as "inherent" laziness or lack of willpower (Morlan 2005; Krasnodemski 1996; Worchel et al. 2000). Lower-class people are more likely to attribute their status to discrimination or poor opportunities—that is, to an *external* societal factor (Morlan 2005; Krasnodemski 1996; Kluegel and Bobo 1993; Bobo and Kluegel 1991).

The same kinds of stereotypes have historically been applied to women. Many of the stereotypes applied to women in literature and the media—they are childlike, overly emotional, unreasonable, bad at mathematics, and so on—have also been applied to African Americans, lower-class people, the poor, and earlier in the twentieth century, Chinese Americans. This shows stereotype interchangeability among gender, racial groups, and social classes. A common theme is apparent: Whatever group occupies the lowest social status in society at a given time (whether racial or ethnic minorities, women, or lower-class people) is negatively stereotyped, and often the same negative stereotypes are used between and among these groups. The stereotype is then used as an "explanation" for the observed behavior of a stereotyped group's member to justify his or her lower status in society. This in turn subjects the stereotyped group to prejudice, discrimination, and racism, which will now be discussed.

Prejudice, Discrimination, and Racism

Many people use the terms *prejudice, discrimination,* and *racism* loosely, as if they were all the same thing. They are not. Typically, in common parlance, people also think of these terms as they apply to individuals, as if the major problems of race were the result of individual people's bad will or biased ideas, thus ignoring the social structural and institutional aspects of race in America. Sociologists use more refined concepts to understand race and ethnic relations, distinguishing carefully between prejudice, discrimination, and racism.

Prejudice

Prejudice is the evaluation of a social group and the individuals within it based on conceptions about the social group held despite facts that disprove them; the beliefs involve both prejudgment and misjudgment (Allport 1954; Jones 1997). Prejudices are usually defined as negative predispositions or as evaluations that are rarely positive. Thinking ill of people only because they are members of group X is prejudice.

A negative prejudice against someone not in one's own group is often accompanied by a positive prejudice in favor of someone who *is* in one's own group. Thus, the prejudiced person will have negative attitudes about a member of an *out-group* (any group other than one's own) and positive attitudes about someone simply because he or she is in one's *in-group* (any group a person considers to be one's own).

Most people disavow racial or ethnic prejudice, yet the vast majority of us carry around some prejudices, whether about racial–ethnic groups, men and women, old and young, upper class and lower class, or straight and gay. Virtually no one is free of prejudice—of both harboring it as well as being the recipient of it. Six decades of research have shown definitively that people who are more prejudiced are also more likely to stereotype and categorize others by race or ethnicity or by gender than those who are less prejudiced (Adorno et al. 1950; Jones 1997; Taylor et al. 2009).

Prejudice based on race or ethnicity is called *racial* or *ethnic prejudice*. If you are a Latino and dislike an Anglo only because he or she is White, then this constitutes prejudice: It is a negative judgment or prejudgment based on race and ethnicity and very little else. If the Latino individual attempts to justify these feelings by arguing that "all Whites have the same bad character," then the Latino is using a stereotype as justification for the prejudice. Note that prejudice can be held by any group against another group.

Prejudice is also revealed in the phenomenon of **ethnocentrism,** which was examined in Chapter 2 on culture. Ethnocentrism is the belief that one's group is superior to all other groups. The ethnocentric person feels that his or her own group is moral, just, and right and that an out-group—and thus any member of that out-group—is immoral, unjust, wrong, distrustful, or criminal. The ethnocentric individual uses his or her own in-group as the standard against which all other groups are compared.

Prejudice and Socialization. Where does racial–ethnic prejudice come from? How do moderately or highly prejudiced people end up that way? People are not born with stereotypes and prejudices. Research shows that these attitudes are learned and internalized through the socialization process, including both *primary socialization* (family, peers, teachers) as well as *secondary socialization* (such as the media). Children imitate the attitudes of their parents, peers, and teachers. If the parent complains about "Japs taking away jobs from Americans," then the child grows up thinking negatively about the Japanese, including Japanese Americans. Attitudes about race are formed early in childhood, at about age 3 or 4 (Allport 1954; Feagin 2000; Van Ausdale and Feagin 1996). There is a very close correlation between the racial and ethnic attitudes of parents and those of their children. The more ethnically or racially prejudiced the parent, the more ethnically or racially prejudiced the child will be. This is even true for individuals who *insist* that they can think for themselves and who think they are not influenced by their parents' prejudice (Taylor et al. 2009).

Major vehicles for the communication of racial–ethnic attitudes to both young and old are the media, especially television, magazines, newspapers, and books. For many decades, African Americans, Hispanics, Native Americans, and Asians were rarely presented in the media and then only in negatively stereotyped roles. The Chinese were shown in movies, magazines, and television in the 1950s as bucktoothed buffoons who ran shirt laundries. Japanese Americans were depicted as sneaky and untrustworthy. Hispanics were shown as either ruthless banditos or playful, happy-go-lucky people who took long siestas. American Indians were presented as either villains or subservient characters like the Lone Ranger's faithful sidekick, Tonto. Finally, there is the drearily familiar portrayal of the Black person as subservient, lazy, clowning, and bug-eyed, a stereotypical image that persisted from the nineteenth century all the way through the 1950s and early 1960s.

Discrimination

Discrimination is overt negative and unequal treatment of the members of some social group or stratum solely because of their membership in that group or stratum. Prejudice is an attitude; discrimination is overt behavior. *Racial–ethnic discrimination* is unequal treatment of a person on the basis of race or ethnicity.

Discrimination in housing has been a particular burden on minorities. Many studies have been able to reproduce the situation, showing that when two persons identical in nearly all respects (age, education, gender, social class, and other characteristics) present themselves as potential tenants for the same housing, if one is White and the other is a minority, the minority person will often be refused housing by a White landlord and the otherwise identical White applicant will not. A minority landlord who refuses housing to a White person while granting it to a minority person of similar social characteristics is also discriminating, but reverse discrimination of this sort is far less frequent and far less of a problem in society (Masey 2005; Feagin 2000; Feagin and Vera 1995; Feagin and Feagin 1993).

Prejudice is an attitude; discrimination is behavior. The discrimination affecting the nation's minorities takes a number of forms—for example, income discrimination and discrimination in housing. Discrimination in employment and promotion and discrimination in education (see Chapter 14) are two other forms of discrimination.

Although the median income of Black and Hispanic families has increased somewhat since 1950, the income gap between these two groups and Whites has remained virtually unchanged since 1967, as can be seen from Figure 10.1. Furthermore, per capita income since 1967 has grown at a faster rate for Whites.

Yet even these median income figures tell only part of the story. In addition to annual income, the *net worth* of White families has *consistently* grown faster than that of Black families (Shapiro and Oliver 1995, 2001, 2006). Net worth may well be a better indicator of economic inequality and annual income. Poverty among Blacks decreased from 1960 to 1970, but the poverty level has been somewhat steady since. The current poverty rate (the percent below the poverty level; see Figure 10.2) is highest for African Americans and Hispanics compared with Whites or Asians. In all these racial groups, children have the highest rate of poverty.

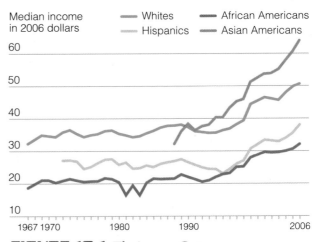

FIGURE 10.1 *The Income Gap*

Source: U.S. Census Bureau. 2007. "Current Population Reports." **www.census.gov**

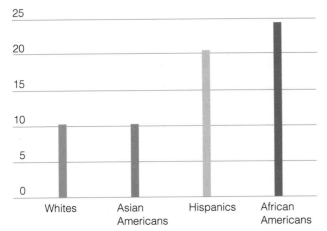

Percentage below
poverty line, 2006

FIGURE 10.2 *Poverty Among Racial Groups*

Source: U.S. Census Bureau. "Current Population Survey, March 2007." **www.census.gov**

Discrimination in housing is illegal under U.S. law. Nonetheless, for many years right on down through 2008, banks and mortgage companies often withheld mortgages from minorities based on "redlining," an illegal practice in which an entire minority neighborhood is designated as "no loan." (This situation may improve somewhat under the fiscal policies of the Obama administration, which as of this writing is holding banks more accountable.) Racial segregation may also be fostered by *gerrymandering,* the calculated redrawing of election districts, school districts, and similar political boundaries in order to maintain racial segregation. As a result, **residential segregation,** the spatial separation of racial and ethnic groups into different residential areas, called "American apartheid" by Massey and Denton (1993; Massey 2005; Frankenberg and Lee 2006), continues to be a reality in this country (see the box "Doing Sociological Research: American Apartheid" on page 255).

Racism

Racism includes both attitudes and behaviors. A negative attitude taken toward someone simply because he or she belongs to a racial or ethnic group is a prejudice, as has already been discussed. An attitude or prejudice is what you think and feel; a behavior is what you do. **Racism** is the *perception and treatment* of a racial or ethnic group, or member of that group, as intellectually, socially, and culturally inferior to one's own group. It is more than an attitude; it is institutionalized in society. Racism involves negative attitudes that are sometimes linked with negative behavior.

There are different **forms of racism.** Obvious, overt racism, such as physical assaults, from beatings to lynchings, has often been called **old-fashioned racism,** or *traditional racism* (or *Jim Crow racism*),

by researchers (for example, Bobo 2006; Hunter 2002; Bobo 1999). This form of racism has declined somewhat in our society since the 1950s, though it certainly has not disappeared (Parmalee 2001; Schumann et al. 1997). Instances of racially motivated lynchings still occur in the United States. Racism can also be subtle, covert, and nonobvious; this is known as **aversive racism,** another form of racism (Jones 1997; Kovel 1970). Consistently avoiding interaction with someone of another race or ethnicity is an example of aversive racism. This form of racism is quite common and has remained at roughly the same level for more than thirty years, with perhaps a slight increase (Bobo 2006; Schumann et al. 1997; Dovidio and Gaertner 1986; Katz et al. 1986; Kovel 1970). Even when overt forms of racism dissipate, aversive racism tends to persist. Because if it less visible than overt racism, people can believe racism is diminished when it is not.

After the Second World War and during the 1950s, a shift to **laissez-faire racism** occurred. This type of racism—also called *symbolic racism* by some (Taylor et al. 2009; Bobo 2006; Bobo and Smith 1998)—involves several elements:

1. The subtle but persistent negative stereotyping of minorities, particularly Black Americans, especially in the media.

2. A tendency to blame Blacks themselves for the gap between Blacks and Whites in socioeconomic standing, occupational achievement, and educational achievement.

3. Clear resistance to meaningful policy efforts (such as affirmative action, discussed later) designed to ameliorate America's racially oppressive social conditions and practices.

The last element is rooted in perceptions of threat to maintaining the status quo (Bobo 1999; Bobo and Smith 1998, Tarman and Sears 2005).

A close relative of laissez-faire racism is **color-blind racism**—so named because the individual affected by this type of racism prefers to ignore legitimate racial–ethnic, cultural, and other differences and insists that the race problems in America will go away if only race is ignored all together. Accompanying this belief is the opinion that race differences in America are merely an illusion and that race is not real. Simply refusing to perceive any differences at all between racial groups (thus, being color-blind) is in itself a form of racism (Gallagher 2006; Bonilla-Silva 2001). This will come as a surprise to many. These types of racism do not necessarily involve explicit or purposeful intent on the part of the nonminority individual to harm the minority person.

Color-blindness thus hides what is called **White privilege** behind a mask: It allows Whites to define themselves as politically and racially tolerant as they proclaim adherence to a belief system that does not see or judge individuals by "the color of their skin." They think of skin color as irrelevant. This view tends to ignore the structured racial dominance in society—White privilege—with its falsely assumed meritocracy and belief that racial barriers have been dismantled. Much of White America now sees a level playing field; yet a majority of Blacks continue to perceive it as still quite uneven. Persons of any racial background can now wear low-slung hip-hop clothing, listen to gangsta rap music, and root for their favorite majority-Black athletic team. This gives the false impression that racial barriers have fallen, but in fact White status (White privilege) remains (Gallagher 2006).

Institutional racism as a form of racism is the negative treatment and oppression of one racial or ethnic group *by society's existing institutions* based on the presumed inferiority of the oppressed group. It is a form of racism that exists at the level of social structure and is in Durkheim's sense *external* to the individual—thus institutional. Key to understanding institutional racism is seeing that dominant groups have the economic and political power to subjugate the minority group, *even if they do not have the explicit intent* of being prejudiced or discriminating against others. Power, or lack thereof, accrues to groups because of their position in social institutions, not just because of individual attitudes or behavior. The power that resides in society's institutions can be seen in such patterns as persistent economic inequality between racial groups, which is reflected in high unemployment among minorities, lower wages, lower net worth, and different patterns of job placement (Oliver and Shapiro 2006; Bobo 1999; Bonilla-Silva 1997).

Racial profiling, is an example of institutional racism in the criminal justice system. In addition African American and Hispanic persons are arrested considerably more often than Whites and Asians. In fact, an African American or Hispanic wrongdoer is more likely to be arrested than a White person who commits the *exact same crime,* even when the White person shares the same age, socioeconomic environment, and prior arrest record as the Black or Hispanic.

Institutional racism is also seen in educational institutions, such as when schools assign Blacks and Latinos to the lower cognitive ability tracks than Whites with the *same* test scores, as will be seen in Chapter 14. In these instances, racism is a characteristic of the institutions and not necessarily of the individuals within the institution. *This is why institutional racism can exist even without prejudice being the cause.*

Consider this: Even if every White person in the country lost all of his or her personal prejudices, and even if he or she stopped engaging in individual acts of discrimination, institutional racism would still persist for some time. Over the years, it has become so much a part of U.S. institutions (hence, the term *institutional racism*) that discrimination can occur even when no

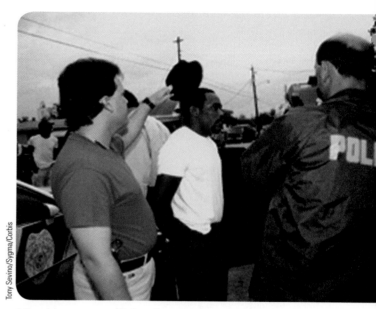

Minorities are more likely to be arrested than Whites for the same offense. Does this reflect institutional racism rather than any individual prejudice of the arresting police officer?

single person is deliberately causing it. Existing at the level of social structure instead of at the level of individual attitude *or* behavior, it is external to the individual personality and is thus a *social fact* of the sort sociological theorist Emile Durkheim observed (Chapter 1). It is for these reasons that current speculation that racism will disappear as the result of the election of America's first Black president, President Barack Obama, are decidedly premature (Bobo and Charles 2009).

Debunking Society's Myths

Myth: The primary cause of racial inequality in the United States is the persistence of *prejudice*.

Sociological perspective: Prejudice is one dimension of racial problems in the United States, but institutional racism can flourish even while prejudice is on the decline. Prejudice is an attribute of the individual, whereas institutional racism is an attribute of social structure.

Theories of Prejudice and Racism

Why do prejudice, discrimination, and racism exist? Two categories of theories have been advanced. The first category consists of psychological theories about prejudice. The second category consists of sociological theories of racism, including institutional racism as well as prejudice and discrimination.

Psychological Theories of Prejudice

Two traditional psychological theories of prejudice are the scapegoat theory and the theory of the authoritarian personality. **Scapegoat theory** argues that, historically, members of the dominant group in the United States have harbored various frustrations in their desire to achieve social and economic success (Feagin and Feagin 1993). As a result of this frustration, they vent their anger in the form of aggression. This aggression is directed toward some substitute that takes the place of the original perception of the frustration. Members of minority groups become these substitutes, that is, the scapegoats. The psychological principle that aggression often follows frustration (originally from the frustration–aggression hypothesis of Dollard et al. [1939]) is central to the scapegoat principle. For example, a White person who perceived that he or she was denied a job because "too many" Mexican immigrants were being permitted to enter the country would be using Mexican Americans as a scapegoat if he or she felt negatively (thus prejudiced) toward a specific Mexican American person, even if that person did not have the job in question and had nothing at all to do with the White person not getting the job.

The second theory, an older one, argues that individuals who possess an authoritarian personality are more likely to be prejudiced against minorities than are nonauthoritarian individuals. The **authoritarian personality** (after Adorno et al. [1950], who coined the term) is characterized by a tendency to rigidly categorize other people, as well as inclinations to submit to authority, strictly conform, be very intolerant of ambiguity, and be inclined toward superstition. The authoritarian person is more likely to stereotype or categorize another and thus readily places members of minority groups into convenient and oversimplified categories or stereotypes. There is some research that links strong authoritarianism with high religious orthodoxy and extreme varieties of political conservatism (Bobo and Kluegel 1991; Altemeyer 1988).

Sociological Theories of Prejudice and Racism

Current sociological theory focuses more on explaining the existence of racism, particularly institutional racism as well as other forms of racism, although speculation about the existence of prejudice is also a component. The three sociological theoretical perspectives considered throughout this text have bearing on the study of racism, discrimination, and prejudice: functionalist theory, symbolic interaction theory, and conflict theory.

Functionalist Theory. Functionalist theory argues that for race and ethnic relations to be functional and thus contribute to the harmonious conduct and stability of society, racial and ethnic minorities

The opening of the National Museum of the American Indian (part of the Smithsonian Institution in Washington, DC) was cause for celebration among diverse groups of Native Americans, as well as others.

must assimilate into that society. **Assimilation** is a process by which a minority becomes socially, economically, and culturally absorbed within the dominant society. The assimilation perspective assumes that to become fully fledged members of society, minority groups must adopt as much of the dominant society's culture as possible, particularly its language, mannerisms, and goals for success, and thus give up much of its own culture. Assimilationism stands in contrast to racial cultural **pluralism**—the maintenance and persistence of one's culture, language, mannerisms, practices, art, and so on.

Symbolic Interaction Theory. Symbolic interaction theory addresses two issues: first, the role of social interaction in reducing racial and ethnic hostility, and second, how race and ethnicity are socially constructed. Symbolic interactionism asks, What happens when two people of different racial or ethnic origins come into contact with each other, and how can such interracial or interethnic contact reduce hostility and conflict? **Contact theory,** which originated with the psychologist Gordon Allport (Allport 1954; Cook 1988), argues that interaction between Whites and minorities will reduce prejudice within both groups—but only if three conditions are met:

1. The contact must be between individuals of equal status; the parties must interact on equal ground. A Hispanic cleaning woman and the wealthier White woman who employs her may interact, but their interaction will not reduce prejudice. Instead, their interaction is more likely to perpetuate stereotypes and prejudices on the part of both.

2. The contact between equals must be sustained; short-term contact will not decrease prejudice. Bringing Whites together with Hispanics, Blacks, and Native Americans for short weekly meetings will not erase prejudice. Daily contact on the job between individuals *of equal job status* will tend to remove prejudice.

3. Social norms favoring equality must be agreed upon by the participants. Having African Americans and White skinheads interact on a TV talk show, such as the *Jerry Springer Show,* will probably not decrease prejudice; this interaction might well increase it.

Conflict Theory. The basic premise of conflict theory is that class-based conflict is an inherent and fundamental part of social interaction. To the extent that racial and ethnic conflict is tied to class conflict, conflict theorists argue that class inequality must be reduced to lessen racial and ethnic conflict in society.

The current "class versus race" controversy in sociology (reviewed in more detail later in this chapter and also in Chapter 8) concerns the question of whether class (namely, economic differences between races) or race is more important in explaining inequality and its consequences or whether they are of equal importance. Those focusing primarily upon class conflict, such as sociologist William Julius Wilson (1978, 1987, 1996), have argued that class and changes in the economic structure are sometimes more important than race in shaping the life chances of different groups. Wilson argues that being disadvantaged in the United States is more a matter of class, although he sees this clearly linked to race. Sociologists focusing primarily on the role of race (Bonilla-Silva and Baiocchi 2001; Feagin 2000; Bonilla-Silva 1997; Feagin and Feagin 1993; Willie 1979) argue the opposite: They say that race has been and is relatively more important than class—though class is still important—in

explaining and accounting for inequality and conflict in society and that directly addressing the question of race forthrightly is the only way to solve the country's race problems (see Table 10.2). Wilson has consistently argued, however, that group race differences are clearly causally related to class differences, and that, in addition, race has an effect independent of class.

A recent variety of the conflict perspective is the *intersection perspective.* This perspective refers to the interactive or combined effects of racism, classism (elitism), and gender in the oppression of people. Intersectional theory posits that any person is socially located in a position that involves race, class, and gender and, thus, looking at only one of them to explain their status is incomplete. This perspective notes that not only are the effects of gender and race intertwined, but also both are intertwined with the effects of class. Class, along with race and gender, are integral components of social structure, according to the intersection perspective (Andersen and Collins 2010; Collins 1990, 1998).

Diverse Groups, Diverse Histories

The different racial and ethnic groups in the United States have arrived at their current social condition through histories that are similar in some ways, yet quite different in other respects. Their histories are related because of a common experience of White supremacy, economic exploitation, and political disenfranchisement.

An historical perspective on each group follows, which will aid in understanding how prejudice, discrimination, and racism has operated throughout the history of U.S. society (see Map 10.1).

table10.2	*Comparing Sociological Theories of Race and Ethnicity*		
	Functionalism	**Conflict Theory**	**Symbolic Interaction**
The Racial Order	Has social stability when diverse racial and ethnic groups are assimilated into society	Is intricately intertwined with class stratification of racial and ethnic groups	Is based on social construction that assigns groups of people to diverse racial and ethnic categories
Minority Groups	Are assimilated into dominant culture as they adopt cultural practices and beliefs of the dominant group	Have life chances that result from the opportunities formed by the intersection of class, race, and gender	Form identity as the result of sociohistorical change
Social Change	Is a slow and gradual process as groups adapt to the social system	Is the result of organized social movements and other forms of resistance to oppression	Is dependent on the different forms of interaction that characterize intergroup relations

Jewish immigrants were questioned at Ellis Island, the point of entry to the United States for many early European immigrants.

Native Americans: The First of This Land

The exact size of the indigenous population in North America at the time of the Europeans' arrival with Columbus in 1492 has been estimated at anywhere from one million to ten million people. Native Americans were here tens of thousands of years before they were "discovered" by Europeans. Discovery quickly turned to conquest, and in the course of the next three centuries, the Europeans systematically drove the Native Americans from their lands, destroying their ways of life and crushing various tribal cultures. Native Americans were subjected to an onslaught of European diseases. Lacking immunity to these diseases, Native Americans suffered a population decline, considered by some to have been the steepest and most drastic of any people in the history of the world. Native American traditions have survived in many isolated places, but what is left is only a ghostly echo of the original 500 nations of North America (Nagel 1996; Thornton 1987; Snipp 1989).

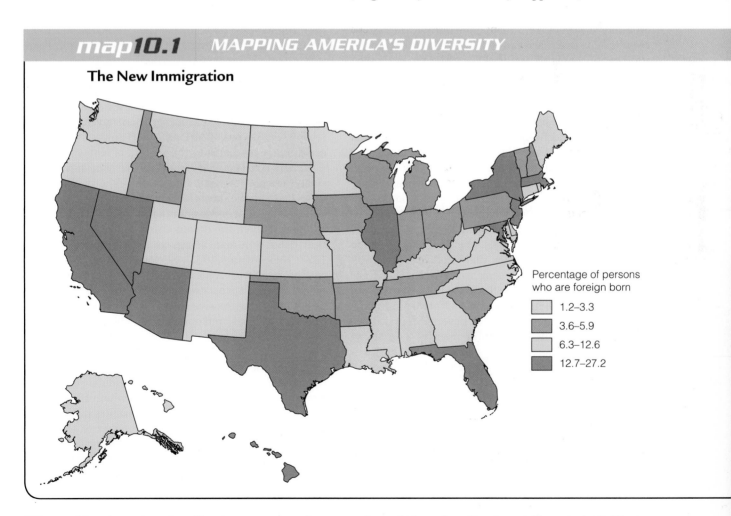

map 10.1 — MAPPING AMERICA'S DIVERSITY

The New Immigration

Percentage of persons who are foreign born

- 1.2–3.3
- 3.6–5.9
- 6.3–12.6
- 12.7–27.2

This map shows the total number of immigrants per state. Some states have a high number of immigrants (for example, California, Florida, New York), and other states have fewer immigrants (for example, Wyoming, South Dakota, and Vermont). Where does your state fall in the number of immigrants to the total population?

Data: U.S. Census Bureau. *2006 Community Survey. Using American Factfinder.* **www.census.gov**

At the time of European contact in the 1640s, there was great linguistic, religious, governmental, and economic heterogeneity among Native American tribes. Most historical accounts have underestimated the degree of cultural and social variety, however. Between the arrival of Columbus in the Caribbean in 1492 and the establishment of the first thirteen colonies in North America in the early 1600s, the ravages of disease and the encroachment of Europeans caused a considerable degree of social disorganization. Sketchy accounts of American Indian cultures by early colonists, fur traders, missionaries, and explorers underestimated the great social heterogeneity among the various American Indian tribal groups and the devastating effects of the European arrival on Indian society.

By 1800, the number of Native Americans had been reduced to a mere 600,000, and wars of extermination against the Indians were being conducted in earnest. Fifty years later, the population had fallen by another half. Indians were killed defending their land, or they died of hunger and disease when taking refuge in inhospitable country. In 1834, 4000 Cherokee died on a forced march from their homeland in Georgia to reservations in Arkansas and Oklahoma, a trip memorialized as the Trail of Tears. The Sioux were forced off their lands by the discovery of gold and the new push of European immigration. Their reservation was established in 1889 (quite recently in history), and they were designated as wards, subjecting them to capricious and humiliating governmental policies. The following year the U.S. Army mistook Sioux ceremonial dances for war dances and moved in to arrest the leaders. A standoff exploded into violence, during which federal troops killed 200 Sioux men, women, and children at the infamous Wounded Knee massacre.

Today about 55 percent of all Native Americans live on or near a reservation, which is land set aside by the U.S. government for their exclusive use. The other 45 percent live in or near urban areas (U.S. Census Bureau 2006; Snipp 1989). The reservation system has served the Indians poorly. Many Native Americans now live in conditions of abject poverty, deprivation, and alcoholism, and they suffer massive unemployment (currently more than 50 percent among males—extraordinarily high). They are at the lowest rung of the socioeconomic ladder, with the highest poverty rate. The first here in this land are now last in status, a painful irony of U.S. history.

African Americans

The development of slavery in the Americas is related to the development of world markets for sugar and tobacco. Slaves were imported from Africa to provide the labor for sugar and tobacco production and to enhance the profits of the slaveholders. It is estimated that somewhere between twenty and one hundred million Africans were transported under appalling conditions to the Americas—38 percent went to Brazil; 50 percent to the Caribbean; 6 percent to Dutch, Danish, and Swedish colonies; and only 6 percent to the United States (Genovese 1972; Jordan 1969).

Slavery evolved as a *caste system* in which one caste, the slaveholders, profited from the labor of another caste, the slaves. Central to the operation of slavery was the principle that human beings could be *chattel* (or property). As an economic institution, slavery was based on the belief that Whites are superior to other races, coupled with a belief in a patriarchal social order. The social distinctions maintained between Whites and Blacks were castelike, with rigid categorization and prohibitions, rather than merely classlike, which suggests more pliant social demarcations. Vestiges of this caste system remain in the United States to this day.

The slave system also involved the domination of men over women. In this combination of patriarchy and White supremacy, White males presided over their "property" of White women as well as their "property" of Black men and women. This in turn led to gender stratification among the slaves themselves, which reflected the White slaveholder's assumptions about the relative roles of men and women. Black women performed domestic labor for their masters and their own families. White men further exerted their authority in demanding sexual relations with Black women (Blassingame 1973; Raboteau 1978; White 1985; Davis 1981). The predominant attitude of Whites toward Blacks was paternalistic. Whites saw slaves as childlike and incapable of caring for themselves. The more recent stereotypes of African Americans as "childlike" are directly traceable to the system of slavery.

There exists a widespread belief that slaves passively accepted slavery. Scholarship shows this to be false. Instead, the slaves struggled to preserve both their culture and their sense of humanity and to resist, often by open conflict, the dehumanizing effects of a system that defined human beings as mere property (Myers 1998; Blassingame 1973). Slaves revolted against the conditions of enslavement in a variety of ways, from passive means such as work slowdowns and feigned illness to more aggressive means such as destruction of property, escapes, and outright rebellion.

After slavery was presumably ended by the Civil War (1861–1865) and the Emancipation Proclamation (1863), Black Americans continued to be exploited for their labor. In the South, the system of sharecropping emerged, an exploitative system in which Black families tilled the fields for White landowners in exchange for a share of the crop. With the onset of the First World War and the intensified industrialization of society came the Great Migration of Blacks from the South to the urban North. This massive movement, lasting from the late 1800s through the 1920s, significantly affected the status of Blacks in society because there was now a greater potential for collective action (Marks 1989).

In the early part of the century, the formation of Black ghettos had a dual effect. It victimized Black Americans with grim urban conditions and encouraged the development of Black resources, including volunteer organizations, settlement houses, social movements, political action groups, and artistic and cultural achievements. During the 1920s, Harlem in New York City became an important intellectual and artistic oasis for Black America—as did organizations and settlement houses at the time in other northern cities, such as famed Karamu House in Cleveland, Ohio.

The Harlem Renaissance gave the nation great literary figures, such as Langston Hughes, Jesse Fausett, Alain Locke, Arna Bontemps, Zora Neale Hurston, Wallace Thurman, and Nella Larsen (Bontemps 1972; Rampersad 1986, 1988; Marks and Edkins 1999). At the same time, many of America's greatest musicians, entertainers, and artists came to the fore, such as musicians Duke Ellington, Count Basie, Billie Holiday, Cab Calloway, and Louis Armstrong, and painter Hale Woodruff. The end of the 1920s and the stock market crash of 1929 brought everyone down a peg or two, Whites as well as Blacks, although in the words of Harlem Renaissance writer Langston Hughes, Black Americans at the time "had but a few pegs to fall" (Hughes 1967).

Latinos

Latino Americans include Chicanos and Chicanas (Mexican Americans), Puerto Ricans, Cubans, and other recent Latin American immigrants to the United States. They also include Latin Americans who have lived for generations in the United States; they are not immigrants but very early settlers from Spain and Portugal in the 1400s. The population of Latinos has grown considerably over the past few decades, with the largest increase among Mexican Americans. The terms *Hispanic* and *Latino* or *Latina* mask the great structural and cultural diversity among the various Hispanic groups. The use of such inclusive terms also tends to ignore important differences in their respective entries into U.S. society: Mexican Americans through military conquest of the Mexican War (1846–1848); Puerto Ricans through war with Spain in the Spanish–American War (1898); and Cubans as political refugees fleeing since 1959 from the Communist dictatorship of Fidel Castro, which was opposed by the U.S. government (Glenn 2002; Bean and Tienda 1987).

Mexican Americans. Before the Anglo (White) conquest, Mexican colonists had formed settlements and missions throughout the West and Southwest. In 1834, the U.S. government ordered the dismantling of these missions, bringing them under tight governmental control and creating a period known as the golden age of the ranchos. Land then became concentrated into the hands of a few wealthy Mexican ranchers, who had been given large land grants by the Mexican government. This economy created a class system *within* the Chicano community, consisting of the elite ranchers, mission farmers, and government administrators at the top; *mestizos,* who were small farmers and ranchers, as the middle class; a third

class of skilled workers; and a bottom class of manual laborers, who were mostly Indians (Mirandé 1985; Maldonado 1997).

With the Mexican–American War of 1846–1848, Chicanos lost claims to huge land areas that ultimately became Texas; New Mexico; and parts of Colorado, Arizona, Nevada, Utah, and California. White cattle ranchers and sheep ranchers enclosed giant tracts of land, thus cutting off many small ranchers, both Mexican and Anglo. Thus began a process of wholesale economic and social *exclusion* of Mexicans and Mexican Americans from U.S. society, much of which continues to this day, generation after generation (Telles and Ortiz 2008).

It was at this time that Mexicans, as well as early U.S. settlers of Mexican descent, became defined as an inferior race that did not deserve social, educational, or political equality. This is an example of the *racial formation process,* as noted earlier in this chapter (Omi and Winant 1994). Anglos believed that Mexicans were lazy, corrupt, and cowardly, which launched stereotypes that would further oppress Mexicans; these stereotypes were used to justify the lower status of Mexicans and Anglo control of the land that Mexicans were presumed to be incapable of managing (Telles and Ortiz 2008; Moore 1976). As has been noted several times in this chapter, stereotyping has been used in this society as a way of *falsely explaining and justifying* the lower social status of society's minorities.

During the twentieth century, advances in agricultural technology changed the organization of labor in the Southwest and West. Irrigation allowed year-round production of crops and a new need for cheap labor to work in the fields. Migrant workers from Mexico were exploited as a cheap source of labor. Migrant work was characterized by low earnings, poor housing conditions, poor health, and extensive use of child labor. The wide use of Mexican migrant workers as field workers, domestic servants, and other kinds of poorly paid work continues, particularly in the Southwest (Amott and Matthaei 1996). Barriers to equal educational opportunity for Mexican Americans, such as culturally and linguistically biased standardized tests, continue the process of exclusion of Mexican Americans (Telles and Ortiz 2008, Taylor 2009).

Puerto Ricans. The island of Puerto Rico was ceded to the United States by Spain in 1898. In 1917, the Jones Act extended U.S. citizenship to Puerto Ricans, although it was not until 1948 that Puerto Ricans were allowed to elect their own governor. In 1952, the United States established the Commonwealth of Puerto Rico, with its own constitution. Following the Second World War, the first elected governor launched a program known as Operation Bootstrap, which was designed to attract large U.S. corporations to the island of Puerto Rico by using tax breaks and other concessions. This program contributed to rapid

Activities such as this Puerto Rican Day Parade in New York City reflect pride in one's group culture and result in greater cohesiveness of the group.

overall growth in the Puerto Rican economy, although unemployment remained high and wages remained low. Seeking opportunity, unemployed farm workers began migrating to the United States. These migrants were interested in seasonal work, and thus a pattern of temporary migration characterized the Puerto Ricans' entrance into the United States (Amott and Matthaei 1996; Rodriguez 1989).

Unemployment in Puerto Rico became so severe that the U.S. government even went so far as to attempt a reduction in the population by some form of population control. Pharmaceutical companies experimented with Puerto Rican women in developing contraceptive pills, and the U.S. government actually encouraged the sterilization of Puerto Rican women. One source notes that by 1974 more than 37 percent of the women of reproductive age in Puerto Rico had been sterilized (Roberts 1997). More than one-third of these women have since indicated that they regret sterilization because they were not made aware at the time that the procedure was irreversible.

Cubans. Cuban migration to the United States is recent in comparison with the other Hispanic groups. The largest migration has occurred since the revolution led by Fidel Castro in 1959; between then and 1980, more than 800,000 Cubans—one-tenth of the entire island population—migrated to the United States. The U.S. government defined this as a political exodus, facilitating the early entrance and acceptance of these migrants. Many of the first migrants had been middle- and upper-class professionals and landowners under the prior dictatorship of Fulgencio Batista, but they had lost their land during the Castro revolution. In exile in the United States, some worked to overthrow Castro, often with the support of the federal government. Yet many other Cuban immigrants

were of modest means and, like other immigrant groups, came seeking freedom from political and social persecution and escape from poverty.

The most recent wave of Cuban immigration came in 1980, when the Cuban government, still under Castro, opened the port of Mariel to anyone who wanted to leave Cuba. In the five months following this action, 125,000 Cubans came to the United States—more than the combined total for the preceding eight years. The arrival of people from Mariel has produced debate and tension, particularly in Florida, a major center of Cuban migration. The Cuban government had previously labeled the people fleeing from Mariel as "undesirable"; some had been incarcerated in Cuba before leaving. They were actually not much different from previous refugees such as the "golden exiles," who were professional and high-status refugees (Portes and Rumbaut 1996). Because the refugees escaping from Mariel had been labeled as undesirables, and because they were forced to live in primitive camps for long periods after their arrival, they have been unable to achieve much social and economic mobility in the United States—thus ironically reinforcing the initial perception that they were "lazy" and "undesirable." In contrast, the earlier Cuban migrants, who were on average more educated and much more settled, have enjoyed a fair degree of success (Portes and Rumbaut 1996, 2001; Amott and Matthei 1996; Pedraza 1996a).

Asian Americans

Like Hispanic Americans, Asian Americans are from many different countries and diverse cultural backgrounds; they cannot be classified as the single cultural rubric Asians. Asian Americans include migrants from China, Japan, the Philippines, Korea, and Vietnam, as well as more recent immigrants from Cambodia and Laos.

Chinese. Attracted by the U.S. demand for labor, Chinese Americans began migrating to the United States during the mid-nineteenth century. In the early stages of this migration, the Chinese were tolerated because they provided cheap labor. They were initially seen as good, quiet citizens, but racial stereotypes turned hostile when the Chinese came to be seen as competing with White California gold miners for jobs. Thousands of Chinese laborers worked for the Central Pacific Railroad from 1865 to 1868. They were relegated to the most difficult and dangerous work, worked longer hours than the White laborers, and for a long time were paid considerably less than the White workers.

The Chinese were virtually expelled from railroad work near the turn of the twentieth century (in 1890–1900) and settled in rural areas throughout the western states. As a consequence, anti-Chinese sentiment and prejudice ran high in the West. This ethnic antagonism was largely the result of competition between the White and Chinese laborers for scarce jobs. In 1882, the federal government passed the Chinese Exclusion Act, which banned further immigration of unskilled Chinese laborers. Like African Americans, the Chinese and Chinese Americans were legally excluded from intermarriage with Whites (Takaki 1989). The passage of this openly racist act, which was preceded by extensive violence toward the Chinese, drove the Chinese populations from the rural areas into the urban areas of the West. It was during this period that several Chinatowns were established by those who had been forcibly uprooted and who found strength and comfort within enclaves of Chinese people and culture (Nee 1973).

Japanese. Japanese immigration to the United States took place mainly between 1890 and 1924, after which passage of the Japanese Immigration Act forbade further immigration. Most of these first-generation immigrants, called *Issei,* were employed in agriculture or in small Japanese businesses. Many Issei were from farming families and wished to acquire their own land, but in 1913 the Alien Land Law of California stipulated that Japanese aliens could lease land for only three years and that lands already owned or leased by them could not be bequeathed to heirs. The second generation of Japanese Americans, or *Nisei,* were born in the United States of Japanese-born parents. They became better educated than their parents, lost their Japanese accents, and in general became more "Americanized," that is, culturally assimilated. The third generation, called *Sansei,* became even better educated and assimilated, yet still met with prejudice and discrimination, particularly where Japanese Americans were present in the highest concentrations, as on the West Coast from Washington to Southern California (Takaki 1989; Glenn 1986).

The Japanese suffered the complete indignity of having their loyalty questioned when the federal government, thinking they would side with Japan after the Japanese attack on Pearl Harbor in December 1941, herded them into concentration camps. By executive order of President Franklin D. Roosevelt, much of the West Coast Japanese American population (many of them loyal second- and third-generation Americans) had their assets frozen and their real estate confiscated by the government. A media campaign immediately followed, labeling Japanese Americans "traitors" and "enemy aliens." Virtually all Japanese Americans in the United States had been removed from their homes by August of 1942, and some were forced to stay in relocation camps until as late as 1946. Relocation destroyed numerous Japanese families and ruined them financially (Glenn 1986; Kitano 1976; Takaki 1989).

In 1986, the U.S. Supreme Court allowed Japanese Americans the right to file suit for monetary reparations. In 1987, legislation was passed, awarding $20,000 to each person who had been relocated and offering an official apology from the U.S. government. One is motivated to contemplate how far this paltry sum and late apology could go in righting what many have argued was the "greatest mistake" the United States has ever made as a government.

Filipinos. The Philippine Islands in the Pacific Ocean fell under U.S. rule in 1899 as a result of the Spanish–American War, and for a while Filipinos could enter the United States freely. By 1934, the islands became a commonwealth of the United States, and immigration quotas were imposed on Filipinos. More than 200,000 Filipinos immigrated to the United States between 1966 and 1980, settling in major urban centers on the West and East Coasts. More than two-thirds of those arriving were professional workers; their high average levels of education and skill have eased their assimilation. By 1985, there were more than one million Filipinos in the United States. Within the next thirty years, demographers project that this population will become one of the largest group of Asian Americans in the United States. (Winnick 1990).

Koreans. Many Koreans entered the United States in the late 1960s after amendments to the immigration laws in 1965 raised the limit on immigration from the Eastern Hemisphere. The largest concentration of Koreans is in Los Angeles. As much as half of the adult Korean American population is college educated, an exceptionally high proportion. Many of the immigrants were successful professionals in Korea; upon arrival in the United States, though, they have been forced to take on menial jobs, thus experiencing downward social mobility and status inconsistency. This is especially true of those Koreans who migrated to the East Coast. However, nearly one in eight Koreans in the United States today owns a business; many own small greengrocer businesses. Many of these stores are located in predominantly African American communities and have become one among several sources of ongoing conflict between some African Americans and Koreans. This has fanned negative feeling and prejudice on both sides—among Koreans against African Americans and among African Americans against Koreans (Chen 1991).

Vietnamese. Among the more recent groups of Asians to enter the United States have been the South Vietnamese, who began arriving following the fall of South Vietnam to the Communist North Vietnamese at the end of the Vietnam War in 1975. These immigrants, many of them refugees who fled for their lives, numbered about 650,000 in the United States in 1975. About one-third of the refugees settled in California. Many faced prejudice and hostility, resulting in part from the same perception that has dogged many immigrant groups before them—that they were in competition for scarce jobs. A second wave of Vietnamese immigrants arrived after China attacked Vietnam in 1978. As many as 725,000 arrived in America, only to face discrimination in a variety of locations. Tensions became especially heated when the Vietnamese became a substantial competitive presence in the fishing and shrimping industries in the Gulf of Mexico on the Texas shore. Since that time, however, many communities have welcomed them, and most Vietnamese heads of households have become employed full time (Winnick 1990; Kim 1993).

Middle Easterners

Since the mid-1970s, immigrants from the Middle East have been arriving in the United States. They have come from countries such as Syria, Lebanon, Egypt, and Iran, and more recently, especially Iraq. Contrary to popular belief, the immigrants speak no single language and follow no singular religion and thus are ethnically diverse. Some are Catholic, some are Coptic Christian, and many are Muslim. Many are from working-class backgrounds, but many were professionals—teachers, engineers, scientists, and other such positions—in their homelands. About 65 percent of those residing in this country were born outside the United States; about half are college-educated (Kohut 2007). Like immigrant populations before them, Middle Easterners have formed their own ethnic enclaves in the cities and suburbs of this country as they pursue the often elusive American Dream (Abrahamson 2006).

After the terrorist attacks on the World Trade Center and the Pentagon on September 11, 2001, many male Middle Easterners of several nationalities became unjustly suspect in this country and were subjected to severe harassment; racially motivated physical attacks; and as already noted, out-and-out racial profiling, if only because they had dark skin and—as with some—wore a turban of some sort on their heads. Most of these individuals, of course, probably had no discernible connection at all with the terrorists. A recent survey shows that *most* Muslims in this country believe that the September 11, 2001, terrorist attacks were indeed the cause of the increased racial harassment and violence against them (Kohut 2007). Finally, the U.S. war with Iraq, begun in 2003, has not helped in easing tensions between White Americans and Middle Easterners generally because, as already noted, a majority of White Americans tend to lump Iraqis together with other Middle Easterners.

White Ethnic Groups

The story of White ethnic groups in the United States begins during the colonial period. White Anglo-Saxon Protestants (WASPs), who were originally immigrants from England and to some extent Scotland and Wales, settled in the New World (what is now North America). They were the first ethnic group to come into contact on a large scale with those people already here—namely, Native American Indians. WASPs came to dominate the newly emerging society earlier than any other White ethnic group.

In the late 1700s, the WASPs regarded the later immigrants from Germany and France as foreigners with odd languages, accents, and customs, and derogatory labels ("krauts," "frogs") were applied to them. Tension between the "old stock" and the "foreigners" continued through the Civil War era until around 1860, when the national origins of U.S. immigrants began to change (Handlin 1951). Of all racial and ethnic groups in the United States during that time and since, only WASPs do not think of themselves as a nationality. The WASPs came to think of themselves as the "original" Americans despite the prior presence of Native American Indians, who in turn the WASPs described and stereotyped as savages. As immigrants from northern, western, eastern, and southern Europe began to arrive, particularly during the mid- to late-nineteenth century, WASPs began to direct prejudice and discrimination against many of these newer groups. Long discriminated against by the male WASP establishment, women began to assert social and political power, challenging the power of male WASPs in the United States. Much of that WASP dominance remains, however, as is evident in their popular use of the terms "race" and "ethnicity" to describe virtually everyone but themselves (Andersen and Collins 2010).

There were two waves of migration of White ethnic groups in the mid- and late-nineteenth century. The first stretched from about 1850 through 1880 and included northern and western Europeans: English, Irish, Germans, French, and Scandinavians. The second wave of immigration occurred from 1890 to 1914 and included eastern and southern European populations: Italians, Greeks, Poles, Russians, and other eastern Europeans. The immigration of Jews to the United States extended for well over a century, but the majority of Jewish immigrants came to the United States during the period from 1880 to 1920.

The Irish arrived in large numbers in the mid-nineteenth century as a consequence of food shortages and massive starvation in Ireland. During the latter half of the nineteenth century and in the early twentieth century, the Irish in the United States were abused, attacked, and viciously stereotyped. It is instructive to remember that the Irish, particularly on the East Coast and especially in Boston, underwent a period of ethnic oppression of extraordinary magnitude. A frequently seen sign posted in Boston saloons of the day proclaimed "No dogs or Irish allowed." The sign was not intended as a joke. German immigrants were similarly stereotyped, as were the French and the Scandinavians. *It is easy to forget that virtually all immigrant groups have gone through times of oppression and prejudice,* although these periods were considerably longer for some groups than for others.

As illustrated by this photograph in Austin, Texas, immigrant neighborhoods often exemplify both residential segregation and community.

© Bob Daemmrich/PhotoEdit

As a rule, where the population density of an ethnic group in a town, city, or region was greatest, so too was the amount of prejudice, negative stereotyping, and discrimination to which that group was subjected.

More than 40 percent of the world's Jewish population lives in the United States, making it the largest community of Jews in the world. Most of the Jews in the United States arrived between 1880 and the First World War, originating from the eastern European countries of Russia, Poland, Lithuania, Hungary, and Romania. Jews from Germany arrived in two phases; the first wave came just prior to the arrival of those from eastern Europe, and the second came as a result of Hitler's ascension to power in Germany during the late 1930s. Because many German Jews were professionals who also spoke English, they assimilated more rapidly than those from the eastern European countries. Jews from both parts of Europe underwent lengthy periods of anti-Jewish prejudice, **anti-Semitism** (defined as the hatred of Jewish people), and discrimination, particularly on Manhattan's Lower East Side. Significant anti-Semitism still exists in the United States (Ferber 1999; Simpson and Yinger 1985; Essed 1991).

In 1924, the National Origins Quota Act was passed, one of the most discriminatory legal actions ever taken by the United States in the area of immigration. By this act, the first real establishment of *ethnic quotas* in the United States, immigrants were permitted to enter the country only in proportion to their numbers already existing in the United States. Thus, ethnic groups who were already here in relatively high proportions (English, Germans, French, Scandinavians, and others, mostly western and northern Europeans) were allowed to immigrate in greater numbers than were those from southern and eastern Europe, such as Italians, Poles, Greeks, and other eastern Europeans. Hence, the act discriminated against southern and eastern Europeans in favor of western and northern Europeans. It has been noted that the European groups who were discriminated against by the National Origins Quota Act tended to be those with darker skins, even though they were White and European.

Immigrants during this period were subject to literacy tests and even IQ tests given *in English* (Kamin 1974). The act barred anyone who was classified as a convict, lunatic, "idiot," or "imbecile" from immigration. On New York City's Ellis Island, non-English-speaking immigrants, many of them Jews, were given the 1916 version of the Stanford-Binet IQ test in English. Obviously, non-English-speaking persons taking this test were unlikely to score high. On the basis of this grossly biased test, governmental psychologist H. H. Goddard classified fully 83 percent of Jews, 80 percent of Hungarians, and 79 percent of Italians as "feebleminded." It did not dawn on Goddard or the U.S. government that the IQ test, in English, probably did not measure something called intelligence, as intended, but instead simply measured the immigrant's mastery of the English language (Kamin 1995; Gould 1981; Taylor 1980, 1992a, 2002).

Attaining Racial and Ethnic Equality: The Challenge

Race and ethnic relations in the United States have posed a major challenge for the nation, one that is becoming even more complex as the racial–ethnic population becomes more diverse. Even as the nation elected its first African American president, racial inequalities persist. Intergroup contact has been both negative and positive, obvious and subtle, tragic and helpful. How can the nation respond to its new diversity as well as to the issues faced by racial and ethnic minorities who have been present since the nation's founding? This question engages significant sociological thought and attention to the nation's record of social change with regard to race and ethnic groups.

The White Immigrants Made It: Why Can't They?

Many Americans believe that with enough hard work and loyalty to the dominant White culture of the country, any minority can make it and thus "assimilate" into American society. Quite a few older members of racial and ethnic minority groups still adhere to this belief. It is the often-heard argument that African Americans, Hispanics, and Native Americans need only to pull themselves up "by their own bootstraps" to become a success.

thinking sociologically

Write down your own racial–ethnic background and list one thing that people from this background have positively contributed to U.S. society or culture. Also list one experience (current or historical) in which people from your group have been victimized by society. Discuss how these two things illustrate the fact that racial–ethnic groups have both been *victimized* and have made *positive contributions* to this society. Share your comments with others: What does this reveal to you about the connections between different groups of people and their experiences as racial–ethnic groups in the United States?

This *assimilation perspective* dominated sociological thinking a generation ago and is still prominent in U.S. thought (Telles and Ortiz 2008; Alba and Nee 2003; Portes and Rumbaut 2001; Rumbaut 1996;

Glazer 1970). The assimilationist believes that to overcome adversity and oppression, the minority person need only imitate the dominant White culture as much as possible. In this sense, minorities must assimilate "into" White culture and White society. The general assumption is that with each new generation, assimilation becomes more and more likely. But one of the questions asked in this perspective is to what extent groups can maintain some of their distinct cultural values and still be incorporated into the society to which they have moved. One could argue, for example, that the Irish have been able to assimilate quite fully into American culture while still maintaining an ethnic identity—one that is particularly salient around St. Patrick's Day!

Many Asian American groups have followed this pattern and have thus been called by some the "model minority," but this label ignores the fact that Asians are still subject to considerable prejudice, discrimination, racism, and poverty (Woo 1998; Lee 1996; Takaki 1989).

There are problems with the assimilation model. First, it fails to consider the time that it takes certain groups to assimilate. Those from rural backgrounds (Native Americans, Hispanics, African Americans, White Appalachians, and some White ethnic immigrants) typically take much longer to assimilate than those from urban backgrounds.

Second, the histories of Black and White arrivals are very different, with lasting consequences. Whites came voluntarily; Blacks arrived in chains. Whites sought relatives in the New World; Blacks were sold and separated from close relatives. For these and other reasons, the experiences of African Americans and Whites as

Neighborhoods such as this one in Manhattan, New York, are indicative of residential segregation.

Jeff Greenberg/PhotoEdit

newcomers can hardly be compared, and their assimilation is unlikely to follow the same course.

Third, although White ethnic groups did indeed face prejudice and discrimination when they arrived in America, many entered at a time when the economy was growing rapidly and their labor was in high demand. Thus they were able to attain education and job skills. In contrast, by the time Blacks migrated to northern industrial areas from the rural South, Whites had already established firm control over labor and used this control to exclude Blacks from better-paying jobs and higher education.

Fourth, assimilation is more difficult for people of color because skin color is an especially *salient* characteristic, ascribed and relatively unchangeable. White ethnic group members can change their names, which many did, but people of color cannot easily change their skin color.

The assimilation model raises the question of whether it is possible for a society to maintain cultural pluralism, which is defined as different groups in society maintaining their distinctive cultures, while also coexisting peacefully with the dominant group. Some groups have explicitly practiced cultural pluralism: The Amish people of Lancaster County in Pennsylvania and of north central Ohio—who travel by horse and buggy; use no electricity; and run their own schools, banks, and stores—constitute a good example of a relatively complete degree of cultural pluralism. A somewhat lesser degree of cultural pluralism, but still present, is maintained by "Little Italy" neighborhoods in some U.S. cities and also by certain Black Muslim groups in the United States.

Segregation and the Urban Underclass

Segregation is the spatial and social separation of racial and ethnic groups. Minorities, who are often believed by the dominant group to be inferior, are compelled to

> doing sociological research

American Apartheid

The term *apartheid* (pronounced "apart-hate" or "apart-hite") was used to describe the society of South Africa prior to the election of Nelson Mandela in 1994. It refers to the rigid separation of the Black and White races. Sociological researchers Massey and Denton argue that the United States is now under a system of apartheid and that it is based on a very rigid residential segregation in the country.

Research Question: What is the current state of residential segregation? Massey and Denton note that the terms "segregation" and "residential segregation" practically disappeared from the American vocabulary in the late 1970s and early 1980s. These terms were spoken little by public officials, journalists, and even civil rights officials. This was because the ills of race relations in America were at the time attributed, though erroneously, to other causes such as a "culture of poverty" among minorities, or inadequate family structure among Blacks, or too much welfare for minority groups. The Fair Housing Act was passed in 1968, and the problem of segregation and discrimination in housing was declared solved. Yet nothing could be further from the truth.

Research Methods and Results: Researchers Massey and Denton amassed a large amount of data demonstrating that residential segregation not only has persisted in American society but that it has actually *increased* since the 1960s. Most Americans vaguely realize that urban America is still residentially segregated, but few appreciate the depth of Black and Hispanic segregation or the degree to which it is maintained by ongoing institutional arrangements and contemporary individual actions. Urban society is thus *hypersegregated,* or characterized by an extreme form of residential and educational segregation.

Conclusions and Implications: Massey and Denton find that most people think of racial segregation as a faded notion from the past, one that is decreasing over time. Today theoretical concepts such as the culture of poverty, institutional racism, and welfare are widely debated, yet rarely is residential segregation considered to be a major contributing cause of urban poverty and the underclass. Massey and Denton argue that their purpose is to redirect the focus of public debate back to race and racial segregation.

Source : Massey, Douglas S., and Nancy A. Denton. 1993. *American Apartheid: Segregation and the Making of the Underclass.* Cambridge, MA: Harvard University Press; Douglas S. Massey. 2005. *Strangers in a Strange Land: Humans in an Urbanizing World.* New York: Norton.

live separately under inferior conditions and are given lower-class educations, jobs, and protections under the law. Although desegregation has been mandated by law (thus eliminating *de jure segregation,* or legal segregation), *de facto segregation*—segregation in fact—still exists, particularly in housing and education.

Segregation has contributed to the creation of an **urban underclass,** a grouping of people, largely minorities and the poor, who live at the absolute bottom of the socioeconomic ladder in urban areas (Wilson 1987, 2009; Massey and Denton 1993). Indeed, the level of housing segregation is so high for some groups, especially poor African Americans and Latinos, that it has been termed **hypersegregation,** referring to a pattern of extreme segregation (Massey and Denton 1993; Massey 2005). Currently, the rate of segregation of Blacks and Hispanics in U.S. cities is actually increasing, thus allowing for less and less interaction between White and Black children and White and Hispanic children (Schmitt 2001; Massey and Denton 1993; see the box "Doing Sociological Research: American Apartheid"). In education, the extraordinary realization is that schools are also becoming more segregated, a phenomenon called *resegregation* since schools are now more segregated than they were even in the 1980s (Frankenberg and Lee 2002).

In a seminal study, W. J. Wilson (1987) attributes the causes of the urban underclass to economic and social structural deficits in society. He rejects the "culture of poverty" explanation, an earlier view that attributes the condition of minorities to their own presumed "cultural deficiencies," a view attributed to Lewis (1960, 1966) and to Moynihan's (1965) so-called Moynihan Report. The problems of the inner city, such as joblessness, crime, teen pregnancy, welfare dependency, and acquired immune deficiency syndrome (AIDS) are seen to arise from social class inequalities, that is, inequalities in the *structure of society*, and these inequalities have dire behavioral consequences at the individual level, in the form of drug abuse, violence, and lack of education (Wilson 1987, 1996, 2009; Sampson 1987). But despite these disadvantages, many individuals nonetheless manage to achieve upward occupational and economic mobility (Newman 1999). Wilson argues that the civil rights agendas need to be enlarged and that the major problem of the underclass, joblessness, needs to be addressed by fundamental changes in the economic institution, such as government-financed jobs and universal health care. Recent government policies came close to this recommendation.

The Relative Importance of Class and Race

The presence of an urban underclass—one composed mostly of African Americans, Latinos, and in some cities, poor Whites and Asian Americans—highlights the significant connection between race and social class. In some regards, class can be seen as having increasing importance in the lives of various groups, including African Americans. Thus, William Julius Wilson (1978, 1996, 2009) points out that there has been a simultaneous expansion of both the Black middle class as well as the Black urban underclass. The same pattern is true for Latinos. This does not mean that race is unimportant, but the influence of class is increasing, even though race still remains extremely important. In numerous studies, scholars find that race, in and of itself, still influences such things as income, wealth holdings, occupational prestige, place of residence, educational attainment, and numerous other measures of socioeconomic well-being (Oliver and Shapiro 2006; Brown et al. 2005; Patillo-McCoy 1999).

What is important is understanding the *intersecting* effects of race and class acting together (Andersen and Collins 2010). Racial–ethnic groups live in what has been called a *matrix of domination* (Collins 1990). That is, no single factor alone determines one's location in society. Rather, race—together with class, gender, age, even sexual orientation—place one in a system of social advantage and disadvantage. Understanding the *interrealtionship* among these social factors is critical to understanding any one of them, including race and ethnicity.

The Civil Rights Strategy

The history of racial and ethnic relations in the United States shows several strategies to achieve greater equality. Political mobilization, legal reform, and social policy have been the basis for much social change in race relations, but there are continuing questions about how best to achieve a greater degree of racial justice in this society. The major force behind most progressive social change in race relations was the civil rights movement. Marked by the strong moral and political commitment and courage of participants, the civil rights movement is probably the single most important source for change in race relations in the twentieth century. The civil rights movement eliminated the formal and legal basis of discrimination and awakened the national consciousness to the problems of race in America.

The civil rights movement was based on the passive resistance philosophy of Martin Luther King Jr., learned from the philosophy of *satyagraha* ("soul firmness and force") of the Indian Mahatma (meaning "leader") Mohandas Gandhi. This philosophy encouraged resistance to segregation through nonviolent techniques, such as sit-ins, marches, and appealing to human conscience in calls for brotherhood, justice, and equality. Although African Americans had worked for racial justice and civil rights long before this historic movement, the civil rights movement has brought greater civil rights under the law to many groups: women, disabled people, the aged, and gays and lesbians (Andersen 2007).

understanding diversity

Race and Hurricane Katrina

A brutally devastating hurricane, given the name Katrina, hit New Orleans, Louisiana, and other locations along the country's southern gulf coast, such as Biloxi, Mississippi, early in the fall of 2005. Katrina's winds, reaching at times 150 miles per hour, tore apart hundreds of houses, apartment buildings, hospitals, oil wells and derricks, and other structures. Massive flooding devastated New Orleans, with water reaching as high as 20 feet in some locations, stranding people, their pets, and farm animals. Because of the slowness of the federal government's response (it took the president of the United States one full week after the hurricane even to acknowledge the devastation as a national disaster and to visit New Orleans) and because of the extent of the flooding, over 1000 people died from drowning, from direct hits by flying debris, or from lack of medical attention.

Ben Sklar/AP Photo

The nation was stunned by images of human bodies floating down water-filled streets, the water itself slicked with oil and filled with sewage and other contaminants. In rest homes for the elderly, patients died as a result of lack of electricity and oxygen supplies needed for assisted breathing. In one case, over twenty patients were simply left alone, unattended for one full week by administrators and staff. Each and every patient died from neglect and lack of medical care. Many people died simply waiting to be evacuated from their communities or from temporary transfer locations—another effect of the slowness of the state and federal governments to act. The government's Federal Emergency Management Administration (FEMA) failed miserably in its intended role of organizing and coordinating responses to the devastation and getting people to safe locations quickly. FEMA's response was so anemic that its director was forced to resign within weeks.

The most negatively affected areas of New Orleans were those neighborhoods in the lowest-lying areas of the city, areas up to 20 feet or more below sea level. These neighborhoods were primarily poor and Black or Hispanic; many were *hypersegregated,* that is, almost entirely African American. Clearly, these neighborhoods had the highest potential for flood damage and the aftermath of contamination and disease. Many have argued that if the White and wealthy had been so concentrated in such neighborhoods, the response of the federal government and the president would probably have been much more rapid and perhaps more effective.

An additional manifestation of race and class bias was seen as a result of the evacuation of thousands to the New Orleans Superdome and the New Orleans Convention Center. Within days the Superdome and Convention Center became cauldrons of misery—no water, no food, unbearable heat, grossly inadequate facilities—including few and clogged toilets and a partially collapsed roof—and an utter lack of medical care for the injured, pregnant women, infants, and the elderly. The individuals so evacuated were mainly poor, and they were neglected and ignored for more than a week by the federal government, FEMA, and even the Louisiana state government.

In national polls taken after Hurricane Katrina, there was a racial divide: Three-quarters of African Americans thought that racism affected the poor response, but only one-third of White Americans thought so.

The major civil rights movement in the United States intensified shortly after the 1954 *Brown v. Board of Education* decision, the famous Supreme Court case that ruled that in education "separate but equal" was unconstitutional. In 1955, African American seamstress and NAACP secretary Rosa Parks made news in Montgomery, Alabama. By prior arrangement with the NAACP, Parks bravely refused to relinquish her seat in the "White only" section on a segregated bus when asked to do so by the White bus driver. At the time, the majority of Montgomery's bus riders were African American, and the action of Rosa Parks initiated the now-famous Montgomery bus boycott, led by the young Martin Luther King Jr. The boycott, which took place in many cities beyond Montgomery, was successful in desegregating the buses.

It got more African American bus drivers hired and catapulted Martin Luther King Jr. to the forefront of the civil rights movement.

Impetus was given to the civil rights movement and the boycott by the unspeakably brutal murder in 1954 of Emmett Till, a Black teenager from Chicago, who was killed in Mississippi merely for whistling at a White woman in a store. After he did so, a group of White men rousted Till from his bed at the home of a relative, beat him until he was dead and unrecognizable as a human being, tied a heavy cotton gin fan around his neck, and dumped him into the nearest river. Later, his mother, in Chicago, allowed a picture of his horribly misshapen head and body in his casket to be published (in *Jet* magazine) so that the public could contemplate the horror vested upon her son. No one was ever prosecuted for the Till murder.

The civil rights movement produced many episodes of both tragedy and heroism. In a landmark 1957 decision, President Dwight D. Eisenhower called out the national guard, after initial delay, to assist the entrance of nine Black students into Little Rock Central High School in Little Rock, Arkansas. Sit-ins followed throughout the South in which White and Black students perched at lunch counters until the Black students were served. Organized bus trips from North to South to promote civil rights, "freedom rides," forged on despite the murders of freedom riders Viola Liuzzo, a White Detroit housewife; Andrew Goodman and Michael Schwerner, two White students; and James Chaney, a Black student. The murders of civil rights workers—especially when they were White—galvanized public support for change.

Radical Social Change

While the civil rights movement developed throughout the late 1950s and 1960s, a more radical philosophy of change also developed, as more militant leaders grew increasingly disenchanted with the limits of the civil rights agenda (which was perceived as moving too slowly). The militant Black power movement, taking its name from the book *Black Power* (published in 1967 by political activist Stokely Carmichael, later Kwame Touré, and Columbia University political science professor Charles V. Hamilton), had a more radical critique of race relations in the United States and saw inequality as stemming not just from moral failures but from the institutional power that Whites had over Black Americans (Carmichael and Hamilton 1967).

Before breaking with the Black Muslims (the Black Nation of Islam in America) and his religious mentor, Elijah Muhammad, and prior to his assassination in 1965, Malcolm X advocated a form of pluralism, demanding separate business establishments, banks, churches, and schools for Black Americans. He echoed an earlier effort of the 1920s led by Marcus Garvey's back-to-Africa movement, the Universal Negro Improvement Association (UNIA).

The Black power movement of the late 1960s rejected assimilation and instead demanded pluralism in the form of self-determination and self-regulation of Black communities. Militant groups such as the Black Panther Party advocated fighting oppression with armed revolution. The U.S. government acted quickly, imprisoning members of the Black Panther Party and members of similar militant revolutionary groups and, in some cases, killing them outright (Brown 1992).

The Black power movement also influenced the development of other groups who were affected by the analysis of institutional racism that the Black power movement developed, as well as by the assertion of strong group identity that this movement encouraged. Groups such as La Raza Unida, a Chicano organization, encouraged "brown power," promoting solidarity and the use of Chicano power to achieve racial justice. Likewise, the American Indian Movement (AIM) used some of the same strategies and tactics that the Black power movement had encouraged, as have Puerto Rican, Asian American, and other racial protest groups. Elements of Black power strategy were also borrowed by the developing women's movement, and Black feminism was developed upon the realization that women, including women of color, shared in the oppressed status fostered by institutions that promoted racism (Collins 1990, 1998). Overall, the Black power movement dramatically altered the nature of political struggle and race and ethnic relations in the United States. It and the other movements it inspired changed the nation's consciousness about race and forced even academic scholars to develop a deeper understanding of how fundamental racism is to U.S. social institutions (Branch 1988, 1998, 2006; Morris 1984).

Affirmative Action

A continuing question from the dialogue between a civil rights strategy and more radical strategies for change is the debate between race-specific versus color-blind programs for change. *Color-blind policies* are those advocating that all groups be treated alike, with no barriers to opportunity posed by race, gender, or other group differences. Equal opportunity is the key concept in color-blind policies.

Race-specific policies are those that recognize the unique status of racial groups because of the long history of discrimination and the continuing influence of institutional racism. Those advocating such policies argue that color-blind strategies will not work because Whites and other racial–ethnic groups do not start from the same position. Even given equal opportunities, continuing disadvantage produces unequal results. The tension between these two strategies for change is a major source for many of the political debates surrounding race relations now.

Affirmative action, a heavily contested program for change, is a race-specific policy for reducing job and educational inequality that has had some limited success. Affirmative action means two things. First, it means recruiting minorities from a wide base in order

The "Tax" on Being a Minority in America: Give Yourself a True-False Test (An Illustration of White Privilege)

On the following test, give yourself one point for each statement that is true for you personally. When you are done, total up your points. The higher your score (the more points you have), the less "minority tax" you are paying in your own life.

1. My parents and grandparents were able to purchase a house in any neighborhood they could afford.

2. I can take a job in an organization with an affirmative action policy without people thinking I got my job because of my race.

3. My parents own their own home.

4. I can look at the mainstream media and see people who look like me represented in a wide variety of roles.

5. I can choose from many different student organizations on campus that reflect my interests.

6. I can go shopping most of the time pretty well assured that I will not be followed or harassed when I am in the store.

7. If my car breaks down on a deserted stretch of road, I can trust that the law enforcement officer who shows up will be helpful.

8. I have a wide choice of grooming products that I can buy in places convenient to campus and/or near where I live as a student.

9. I never think twice about calling the police when trouble occurs.

10. The schools I have attended teach about my race and heritage and present it in positive ways.

11. I can be pretty sure that if I go into a business or other organization (such as a university or college) to speak with the "person in charge" I will be facing a person of my race.

Your total points: _____

Your racial identity: _____

Your gender: _____

How would you describe your social class? _____

Now gather results from some of your classmates and see if their total points vary according to their own race, gender, and/or social class.

Source: Adapted from the *Discussion Guide for Race: The Power of an Illusion.* **www.pbs.org**

to ensure consideration of groups that have been traditionally overlooked, while not using rigid quotas based on race or ethnicity. Second, affirmative action means using admissions slots (in education) or set-aside contracts or jobs (in job hiring) to assure minority representation. The principal objection, heard from both sides of the racial line, is that applying either interpretation of affirmative action programs is, in effect, using a quota.

The Legal Defense Fund (LDF), established by the NAACP (National Association for the Advancement of Colored People), has argued forcefully and on legal grounds that the push for affirmative action on the basis of race must continue, though not in the form of rigid quotas. It has noted that in the tradition of its founding lawyer and the future member of the Supreme Court, Thurgood Marshall, the affirmative action policies of the LDF have helped to fundamentally change the composition of the nation's formerly segregated universities and colleges as well as to greatly expand the educational opportunities for African Americans and other minorities. It argues that standardized tests (such as the SATs) are of limited validity and do not adequately predict performance of minorities of color in college; they are thus not a legitimate basis on which to judge Black and White candidates against each other.

Data have shown that Blacks admitted to selective colleges and universities under affirmative action programs reveal high rates of social and economic success after graduation. For example, the percent-age of Blacks who were admitted to college under affirmative action programs and went on to graduate school and law school was *higher* than the percentage of Whites from the same schools who did so (Espenshade et al. 2005; Bowen and Bok 1998). This is clearly a benefit of affirmative action in the field of education.

The U.S. Supreme Court decided in 1978 (in *The University of California Regents vs. Bakke*) that race could be used as a criterion for admission to undergraduate, professional, and graduate schools or for job recruitment, as long as race is combined with other criteria and as long as rigid racial quotas are not used. Then twenty-five years later, in 2003, the U.S. Supreme Court decided two cases that modified its 1978 decision. In *Grutter v. Bollinger*, in a five-to-four decision, the high court decided that, as in the 1978 decision, race could indeed be used as a factor in admissions decisions for the University of Michigan School of Law as long as race was considered along with other factors and the decision to admit or not admit was made on a case-by-case basis. In the second case (*Gratz v. Bollinger*), a six-to-three decision, the Court threw out as unconstitutional any system of assigning favorable points to minority candidates seeking admission that would increase their chances for admission. This decision thus ruled out the use of any form of minority quotas, interpreting the point system as a type of quota, but it upheld as constitutional a system that considers race among many factors on an individual basis.

Chapter Summary

- ### How are race and ethnicity defined?

  In virtually every walk of life, race matters. A *race* is a social construction based loosely on physical criteria, whereas an *ethnic group* is a culturally distinct group. A group is *minority* or *dominant* not on the basis of their numbers in a society but on the basis of which group occupies lower average social status.

- ### What are stereotypes, and how are they important?

  Stereotyping and *stereotype interchangeability* reinforce racial and ethnic prejudices and thus cause them to persist in the maintenance of inequality in society. Racial and gender stereotypes have similar dynamics in society, and both racial and gender stereotypes receive ongoing support in the media. Stereotypes serve to justify and make legitimate the oppression of groups based on race, ethnicity, and gender. Stereotypes such as "lazy" support attributions made to the minority, which attempt to cast blame on the minority in question, thus removing blame from the social structure.

- ### What are the differences between prejudice, discrimination, and racism?

  Prejudice is an attitude usually involving negative prejudgment on the basis of race or ethnicity. *Discrimination* is overt behavior involving unequal treatment. *Racism* involves both attitude and behavior. Racism can take on several forms, such as traditional or *old-fashioned racism, aversive* or subtle racism, *laissez-faire racism, color-blind racism* (which masks *White privilege*), and institutional racism. *Institutional racism* is unequal treatment, carrying with it notions of cultural inferiority of a minority, which has become ingrained into the economic, political, and educational institutions of society.

- ### Do all minority groups have different histories, or are they similar?

  Historical experiences show that different groups have unique histories, although they are bound together by some similarities in the prejudice and discrimination they have experienced.

- ### What are the challenges in attaining racial and ethnic equality?

  Not all immigrant groups and minority groups *assimilate* at the same rate, and some groups (U.S. Black Muslims; the Amish) maintain cultural *pluralism.* An *urban underclass* remains entrenched in the United States, and cities remain *hypersegregated* on the basis of race and ethnicity.

- ### What are some of the approaches to attaining racial and ethnic equality?

  Approaches include Rev. Martin Luther King's nonviolent civil rights strategy, radical social change and movements such as the Black Power movement, La Raza Unida, and the American Indian Movement (AIM), all of which directly addressed *institutional racism. Affirmative action* policies, which are race-specific rather than race-blind programs, continue to be changed and modified through Supreme Court cases. Some policy researchers have argued that only deep-rooted economic changes in the social structure will alleviate persistent economic and racial stratification and inequality.

Key Terms

affirmative action   258
anti-Semitism   253
assimilation   245
authoritarian personality   245
aversive racism   243
color-blind racism   243
contact theory   245
discrimination   242
dominant group   239
ethnic group   236
ethnocentrism   241
forms of racism   243

hypersegregation   256
institutional racism   244
laissez-faire racism   243
minority group   239
old-fashioned racism   243
out-group homogeneity
   effect   239
pluralism   245
prejudice   241
race   236
racial formation   238
racial profiling   244

racialization   237
racism   243
residential segregation   243
salience principle   240
scapegoat theory   245
segregation   255
stereotype   240
stereotype
   interchangeability   241
urban underclass   256
White privilege   243

Online Resources

Sociology: The Essentials Companion Website

www.cengage.com/sociology/andersen
Visit your book companion website where you will find more resources to help you study and write your research papers. Resources include web links, learning objectives, Internet exercises, quizzing, and flash cards.

 is an easy-to-use online resource that helps you study in less time to get the grade you want NOW.

www.cengage.com/login
Need help studying? This site is your one-stop study shop. Take a Pre-Test and Cengage NOW will generate a Personalized Study Plan based on your test results. The Study Plan will identify the topics you need to review and direct you to online resources to help you master those topics. You can then take a Post-Test to determine the concepts you have mastered and what you still need to work on.

11

Chapter eleven
CHAPTER ELEVEN
Chapter

Gender

The Social Construction of Gender

Gender Socialization

Gender Stratification

Theories of Gender

Gender in Global Perspective

Gender and Social Change

Chapter Summary

[*Imagine suddenly becoming* a member of the other sex. What would you have to change? First, you would probably change your appearance—clothing, hairstyle, and any adornments you wear. You would also have to change some of your interpersonal behavior. Contrary to popular belief, men talk more than women, are louder, are more likely to interrupt, and are less likely to recognize others in conversation. Women are more likely to laugh, express hesitance, and be polite (Robinson and Smith-Lovin 2001; Anderson and Leaper 1998; Crawford 1995; Cameron 1998). Gender differences also appear in nonverbal communication. Women use less personal space, touch less in impersonal settings (but are touched more), and smile more, even when they are not necessarily happy (LaFrance 2002; Lombardo et al. 2001; Basow 1992). Researchers even find that men and women write email in a different style, women writing less opinionated email than men and using it to maintain rapport and intimacy (Sussman and Tyson 2000; Colley and Todd 2002). Finally, you might have to change many of your attitudes because men and women differ significantly on many, if not most, social and political issues (see Figure 11.1).

continued

HOANG DINH NAM/Getty Images

If you are a woman and became a man, perhaps the change would be worth it. You would probably see your income increase (especially if you became a White man). You would have more power in virtually every social setting. You would be far more likely to head a major corporation, run your own business, or be elected to a political office—again, assuming that you are White. Would it be worth it? As a man, you would be far more likely to die a violent death and would probably not live as long as a woman (Kung et al. 2008).

If you are a man who became a woman, your income would most likely drop significantly. More than forty-five years after passage of the Equal Pay Act in 1963, men still earn 23 percent more than women, even comparing those working year-round and full time (DeNavas-Walt et al. 2008). You would probably become resentful of a number of things because poll data indicate that women are more resentful than men about things such as the amount of money available for them to live on, the amount of help they get from their mates around the house, how men share child care, and how they look. Women also report being more fearful on the streets than men. However, women are more satisfied than men with their role as parents and with their friendships outside of marriage.

For both women and men, there are benefits, costs, and consequences stemming from the social

definitions associated with gender. As you imagined this experiment, you may have had difficulty trying to picture the essential change in your *biological* identity: But is this the most significant part of being a man or woman? Nature determines whether you are male or female but it is society that gives significance to this distinction. Sociologists see *gender* as a social fact because who we become as men and women is largely shaped by cultural and social expectations.

The Social Construction of Gender

From the moment of birth, gender expectations influence how boys and girls are treated. Now that it is possible to identify the sex of a child in the womb, gender expectations may begin even before birth. Parents and grandparents might select pink clothes and dolls for baby girls, sports clothing and brighter colors for boys. Even if they try to do otherwise, it will be difficult because baby products are so typed by gender. Much research shows how parents and others continue to treat children in stereotypical ways throughout their childhood. Girls may be expected to cuddle and be sweet, whereas boys are handled more roughly and given greater independence.

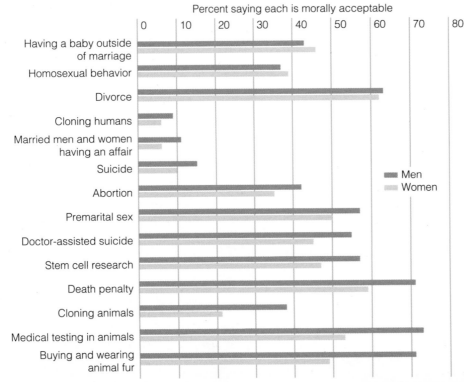

FIGURE 11.1 *The Gender Gap in Attitudes*

Data: Saad, Lydia. 2003. "Pondering 'Women's Issues,' Part II." *The Gallup Poll.* Princeton, NJ: The Gallup Organization. **www.gallup.com**

Defining Sex and Gender

Sociologists use the terms *sex* and *gender* to distinguish biological sex identity from learned gender roles. **Sex** refers to biological identity, being male or female. For sociologists, the more significant concept is **gender**—the socially learned expectations and behaviors associated with members of each sex. This distinction emphasizes that behavior associated with gender is culturally learned.

The cultural basis of gender is especially apparent when we look at other cultures. Across different cultures, the gender roles associated with masculinity and femininity vary considerably. In Western industrialized societies, people tend to think of masculinity and femininity in dichotomous terms, with men and women even defined as "opposite sexes." The views from other cultures challenge this assumption. Historically, the *berdaches* in Navajo society were anatomically normal men who were defined as a third gender considered to fall between male and female. Berdaches married other men who were not considered berdaches and were defined as ordinary men. Moreover, neither the berdaches nor the men they married were considered homosexuals, as they would be considered in many of today's Western cultures (Nanda 1998; Lorber 1994).

We need not look at striking departures from familiar gender categories to see how differently gender can be constructed in other cultures. For example, in China the Chinese Marriage Law of 1950 formally defines marriage as a relationship between equal companions who share responsibility for child care and the family. This is in sharp contrast to age-old gender roles in Chinese tradition. The roles did not change overnight in China. As in many societies backed by strong traditions, contemporary roles can coexist with more traditional ones.

There can also be substantial differences in the construction of gender across social classes or within subcultures

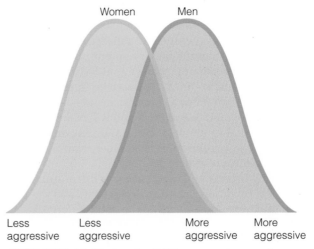

The hijras of India are a sexual minority group; these "men" (that is, born male) are considered to be a "third gender"— neither man nor woman. Hijras provide a good illustration of the socially constructed basis of sexuality.

sinopictures/Peter Arnold Inc.

in a given culture. Within the United States, as we will see, there is considerable variation in the experiences of gender among different racial and ethnic groups (Baca Zinn et al. 2005; Andersen and Collins 2010). In addition, even within a given culture, differences among people of a given gender can be greater than differences across gender (see Figure 11.2). Looking at gender sociologically quickly reveals the social and cultural dimensions of something often popularly defined as biologically fixed.

Sex Differences: Nature or Nurture?

Despite the known power of social expectations, many still believe that differences between men and women are biologically determined. Biology is, however, only one component in the differences between men and women. The important question in sociology is not whether biology or culture is more important in forming men and women, but how culture produces a person's gender identity.

Biological determinism refers to explanations that attribute complex social phenomena to physical characteristics. The argument that men are more aggressive because of hormonal differences (in particular, the presence of testosterone) is a biologically determinist argument. Although people popularly believe that testosterone causes aggressive behavior in men, studies find only a modest correlation between aggressive behavior and testosterone levels. Furthermore, changes in testosterone levels do not predict changes in men's aggression (such as by "chemical castration," the administration of drugs that eliminate the production or circulation of testosterone). What's more, there are minimal differences in the levels of sex hormones between girls and boys during early childhood, yet researchers find considerable

Women   Men

Less   Less   More   More
aggressive   aggressive   aggressive   aggressive

FIGURE 11.2 *Gender Differences: Aggression*

Even when men and women as a whole tend to differ on a given trait, within-gender differences can be just as great as across-gender differences. Some men, for example, are less aggressive than some women.

a sociological eye
on the media

Cultural Gatekeepers and the Construction of Femininity

Many have noted the distorted images of women that appear in the media. The common argument is that media images present an unrealistic image of women, which shapes women's self-concepts and limits their sense of possibilities for their appearance, their relationships, their careers, and so forth. Femininity is defined in the media by *cultural gatekeepers*—those who make decisions about what images to project. Cultural gatekeepers also have to respond to audience criticism. How they respond is an important part of the institutional process by which media images are sustained.

One sociologist, Melissa Milkie, wanted to explore how images of femininity are constructed in the media, particularly when producers encounter criticism from their audience. As readers of magazines, girls have protested many of the narrow and limiting images in the media, particularly those portraying girls' bodies.

Milkie interviewed ten top editors of leading girls' magazines to find out how they, as cultural gatekeepers, responded to the criticism from girls that images of girls in teen magazines do not reflect what "real girls" are like.

Milkie found that even the top editors think there are institutional limitations on what they can do to respond to girls' criticism. The editors who were very

sensitive to the criticisms they received either said there was not much they could do about it or they dismissed the girls' complaints as misguided. They would claim the image was beyond their control, either because of the artistic process, advertisers' needs, or the culture itself. Thus, despite their positions of power, editors believed they could not fully control the images that appear. They pointed to institutional constraints that, in effect, thwarted efforts for change. Some editors simply dismissed the criticisms as girls' misreading the intent or meaning of an image.

Either way, Milkie's research shows how the organizational complexity of media institutions limits how much change is possible in how images of femininity are constructed. Market forces, advertisers, the values of producers, and the values of the public all intertwine in shaping the decisions of cultural gatekeepers. Milkie also shows, however, that people are not passive about what they see in the media, suggesting that how people respond to those images is an important part of the effect of such images in society.

Source: Milkie, Melissa A. 2002. "Contested Images of Femininity: An Analysis of Cultural Gatekeepers' Struggles with the 'Real Girl' Critique." *Gender & Society* 16 (December): 839–859.

differences in the aggression exhibited by boys and girls as children (Fausto-Sterling 2000, 1992).

A person's sex identity is established at the moment of conception when the father's sperm provides either an X or a Y chromosome to the egg at fertilization. The mother contributes an X chromosome to the embryo. The combination of two X chromosomes makes a female, while the combination of an X and a Y makes a male. Under normal conditions, chemical events directed by genes on the sex-linked chromosomes lead to the formation of male or female genitalia.

Being **intersexed** (also known as **hermaphroditism**) is a condition caused by irregularities in the process of chromosome formation or fetal differentiation that produces persons with mixed biological sex characteristics. An intersexed infant may be born with ovaries or testes, but with ambiguous or mixed genitals. Or, an intersexed person may be a chromosomal male but have an incomplete penis and no urinary canal.

Case studies of intersexed persons reveal the extraordinary influence of social factors in shaping the person's identity (Preves 2003). Parents of intersexed

children are usually advised to have their child's genitals surgically assigned to either male or female and also to give the child a new name, a different hairstyle, and new clothes—all intended to provide the child with the social signals judged appropriate to a single gender identity. One physician who has worked on such cases gives the directive to parents that they "need to go home and do their job as child rearers with it very clear whether it's a boy or a girl" (Kessler 1990: 9).

Transgendered people are those who deviate from the binary (that is, male or female) system of gender; they include transsexuals (those whose sexual identity as a man or woman differs from their biological sex identity at birth), cross-dressers, and others who do not fit within the normative expectations of gender. Research on transgendered individuals shows that they experience enormous pressure to fit within the usual expectations. Likewise, those who change their sex as adults report enormous pressure, particularly during their transition period, because others expect them to be one sex or the other. Most find that whatever their biological sex, they are forced—from fear of rejection and the desire for self-preservation—to manage an identity that would fall into one category or another (Gagné and Tewksbury 1998; Stryker and Whittle 2006).

From a sociological perspective, biology alone does not determine gender identity. People must adjust to the expectations of others and the social understanding of what it means to be a man or a woman. A person may remain genetically one sex, while socially being the other—or perhaps something in between. In other words, there is not a fixed relationship between biological and social outcomes. If you only see men and women as biologically "natural" states, you miss some of the fascinating ways that gender is formed in society.

Physical differences between the sexes do, of course, exist. In addition to differences in anatomy, at birth boys tend to be slightly longer and weigh more than girls. As adults, men tend to have a lower resting heart rate, higher blood pressure, and higher muscle mass and muscle density. These physical differences contribute to the tendency for men to be physically stronger than women, but this can be altered, depending on level of physical activity. The public now routinely sees displays of women's athleticism and expects great performances from both men and women in world-class events such as the Olympics. Women can achieve a high degree of muscle mass and muscle density through bodybuilding and can win over men in activities that require high levels of endurance, such as the four women who have won the Iditarod—the Alaskan dog sled race considered to be one of the most grueling competitions in the world. In other words, until men and women really compete equally in activities from which women have historically been excluded, we may not know the real extent of physical differences between women and men.

Arguments based on biological determinism assume that differences between women and men are "natural" and, presumably, resistant to change. Like biological explanations of race differences, biological explanations of inequality between women and men tend to flourish during periods of rapid social change. They protect the *status quo* (existing social arrangements) by making it appear that the status of women or people of other races is "natural" and therefore should remain as it is. If social differences between women and men were biologically determined, we would find no variation in gender relations across cultures, but extensive differences are well documented. Moreover, even within the same culture, there can be vast *within-gender* differences. That is, the variation on a given trait, such as aggression or competitiveness, can be as great within a given gender group as the difference across genders. Thus, some women are more aggressive than some men, and some men are less competitive than some women (see Figure 11.2). We would not exist without our biological makeup, but we would not be who we are without society and culture.

In sum, sociologists emphasize the social basis for both sex and gender. Sociologists see people's sex identity and their gender identity as being influenced by society. Culture defines certain sexual behaviors as appropriate (or not) and establishes particular expectations for women and for men. This is learned through the process of *gender socialization*.

Gender Socialization

As we saw in Chapter 4, socialization is the process by which social expectations are taught and learned. Through **gender socialization,** men and women learn the expectations associated with their sex. The rules of gender extend to all aspects of society and daily life. Gender socialization affects the self-concepts of women and men, their social and political attitudes, their perceptions about other people, and their feelings about relationships with others. Although not everyone is perfectly socialized to conform to gender expectations, socialization is a powerful force directing the behavior of men and women in gender-typical ways.

Even people who set out to challenge traditional expectations often find themselves yielding to the powerful influence of socialization. Women who consciously reject traditional women's roles may still find themselves inclined to act as hostess or secretary in a group setting. Similarly, men may decide to accept equal responsibility for housework, yet they fail to notice when the refrigerator is empty or the child needs a bath—household needs they have been trained to let someone else notice (DeVault 1991). These expectations are so pervasive that it is also difficult to change them on an individual basis. If you doubt this, try buying clothing or toys for a young child without purchasing something that is gender-typed, or talk to parents who have tried to raise their children without conforming to gender stereotypes and see what they report about the influence of such things as children's peers and the media.

The Formation of Gender Identity

One result of gender socialization is the formation of **gender identity,** which is one's definition of oneself as a woman or man. Gender identity is basic to our self-concept and shapes our expectations for ourselves, our abilities and interests, and how we interact with others. Gender identity shapes not only how we think about ourselves and others but also influences numerous behaviors, including such things as the likelihood of drug and alcohol abuse, violent behavior, depression, or even how aggressive you are in driving (Andersen 2009).

One area in which gender identity has an especially strong effect is in how people feel about their appearance. Studies find strong effects of gender identity on body image. Concern with body image begins mostly during adolescence. Thus, studies of young children (that is, preschool age) find no gender differences in how boys and girls feel about their bodies (Hendy et al. 2001), but by early adolescence clear differences emerge. At this age, girls report comparing their bodies

Gender, Bodies, and Beauty

Sociologist Debra Gimlin argues that bodies are "the surface on which prevailing rules of a culture are written" (Gimlin 2002: 6). You see this in many ways, such as how people mark and adorn their bodies in different cultures. This is especially evident with regard to gender. Men and women alike may practice elaborate rituals to achieve particular ideals established by the culture—ideals that vary because of gender roles, as well as class and race relations.

© Mike Watson Images/SuperStock

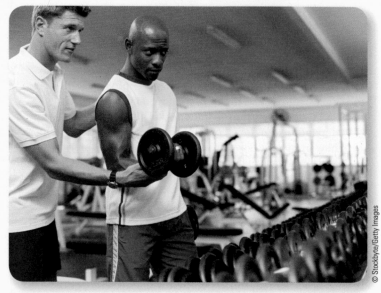

© Stockbyte/Getty Images

How does gender influence our concepts and practices about the body?

One of the gender differences some have noted is that men generally try to bulk themselves up, sometimes using steroids and other products to do so, while women generally strive to be thin. What are the dominant cultural images of beauty and good looks for women and men? Do they vary because of race?

© Markus Moellenberg/zefa/Corbis

Anorexia is particularly common among White, middle-class women, although the occurrence of anorexia is increasing among women of color. Do you see evidence of this on your campus?

Beauty ideals in a culture are strongly influenced by popular culture. Fashion models, for example, set some of the standards for women's and men's appearance, but they are often dangerously thin.

Stuart Ramson/AP Photo

Evan Hurd/Sygma/Corbis

Although it is now commonplace, popular fashion, including among young girls, tends to sexualize women's bodies.

© Digital Vision Ltd./SuperStock

Coneyl Jay/Photo Researchers, Inc.

Thinkstock Images/Jupiter Images

Cosmetic surgery is a rapidly growing, and highly profitable, industry. Some do it to try to eliminate signs of aging; others, for aesthetic purposes. The pervasive influence of western images means that some women in other nations pay for procedures intended to make them look more "western."

Many women—and some men—spend a great deal of money on procedures to alter their appearance. The use of Botox treatments, breast augmentation, liposuction, and other procedures are increasing, even among young women—and solely for cosmetic reasons. How do you explain this?

to others of their sex more often than boys do. By early adolescence, girls report lower *self-esteem* (that is, how well one thinks of oneself) than boys; they also report more negativity about their body image than do boys. This type of thinking among girls is related to lower self-esteem (Jones 2001; Polce-Lynch et al. 2001). Among college students, women also are more dissatisfied with their appearance than are men (Hoyt and Kogan 2001). These studies indicate that idealized images of women's bodies in the media, as well as peer pressures, have a huge impact on young girls' and women's gender identity and feelings about their appearance.

Sources of Gender Socialization

As with other forms of socialization, there are different agents of gender socialization: family, peers, children's play, schooling, religious training, mass media, and popular culture, to name a few. Gender socialization is reinforced whenever gender-linked behaviors receive approval or disapproval from these multiple influences.

Parents are one of the most important sources of gender socialization. Parents may discourage children from playing with toys that are identified with the other sex, especially when boys play with toys meant for girls. Research finds that parents are more tolerant of girls not conforming to gender roles than is true for boys (Kane 2006). Direct observations of fathers and mothers interacting with their young children also show that fathers are less likely to provide basic care, although both parents tend to share responsibility for discipline (LaFlamme et al. 2002).

Expectations about gender are changing, although researchers suggest that the cultural expectations about gender may have changed more than people's actual behavior. Thus, mothers and fathers now report that fathers should be equally involved in child rearing, but the reality is different. Mothers still spend more time in child-related activities and have more responsibility for children. Furthermore, the gap that mothers perceive between fathers' ideal and actual involvement in child rearing is a significant source of mothers' stress (Milkie et al. 2002).

Gender socialization patterns also vary within different racial–ethnic families. Latinas, as an example, have generally been thought to be more traditional in their gender roles, although this varies by generation and by the experiences of family members in the labor force. Within families, young women and men learn to formulate identities that stem from their gender, racial, and ethnic expectations.

Peers strongly influence gender socialization—sometimes more so than one's immediate family. Peer relationships shape children's patterns of social interaction. The play in which they engage—with peers and others—also shapes analytical skills and their values and attitudes. Studies find that boys and girls often organize their play in ways that reinforce not only gender but also race and age norms (Moore 2001). Peer relationships often reinforce the gender norms of the culture—norms that are typically even more strictly applied to boys than to girls.

Changes in gender roles have involved more men in parenting.

Thus, boys who engage in behavior that is associated with girls are likely to be ridiculed by friends—more so than are girls who play or act like boys (Sandnabba and Ahlberg 1999). In this way, homophobic attitudes, routinely expressed among peers, reinforce dominant attitudes about what it means to "be a man" (Pascoe 2007). While girls may be called "tomboys," boys who are called "sissies" are more harshly judged. Note, though, that beyond a certain age tomboy behavior among girls may result in the girl being labeled a "dyke."

Children's play is another source of gender socialization. You might think about what you played as a child and how this influenced your gender roles. Typically, though not in every case, boys are encouraged to play outside more; girls, inside. Boys' toys are more machine-like and frequently promote the development of militaristic values; they tend to encourage aggression, violence, and the stereotyping of enemies—values rarely associated with girls' toys. Children's books in schools also communicate gender expectations. Even with publishers' guidelines that discourage stereotyping, textbooks still depict men as aggressive, argumentative, and competitive. Men and boys are also more likely to be featured in children's books, although interestingly, systematic analysis of children's books shows that fathers are not very present and, when they are, are most often shown as ineffectual (Anderson and Hamilton 2005).

*thinking*sociologically

Visit a local toy store and try to purchase a toy for a young child that is not gender-typed. What could you buy? What could you not buy? What does your experiment teach you about gender role socialization? If you take a child with you, note what toys he or she wants. What does this tell you about the effectiveness of *gender socialization?*

Schools are particularly strong influences on gender socialization because of the amount of time children spend in them. Teachers often have different expectations for boys and girls. In school, boys get more attention, even if it is sometimes negative attention. When teachers of either sex respond more to boys, both positively and negatively, they heighten boys' sense of importance (American Association of University Women 1992, 1998; Sadker and Sadker 1994). Because of **Title IX** passed in 1972, schools are forbidden by law to discriminate based on gender. (See also the section on legislative change later in this chapter.) Yet gender inequality is nonetheless pervasive in the schools—reflected in course-taking patterns in the curriculum, teachers' interaction with students, and students' interactions with each other.

Religion is an often overlooked but significant source of gender socialization. The major Judeo-Christian religions in the United States place strong emphasis on gender differences, with explicit affirmation of the authority of men over women. In Orthodox Judaism, men offer a prayer to thank God for not having created them as a woman or a slave. The patriarchal language of most Western religions and, in some faiths, the exclusion of women from positions of religious leadership signifies the lesser status of women in religious institutions. Any religion, interpreted in a fundamentalist way, can be oppressive to women. Indeed, the most devout believers of any faith tend to hold the most traditional views of women's and men's roles. But the influence of religion on gender attitudes cannot be considered separately from other factors. For many, religious faith inspires a belief in egalitarian roles for women and men; both Christian and Islamic women have organized to resist fundamentalist and sexist practices (Gerami and Lehnerer 2001).

The *media* in its various forms (television, film, magazines, music, and so on) communicate strong—some would even say cartoonish—gender stereotypes. Despite some changes in recent years, television and films continue to depict highly stereotyped roles for women and men. Men on television heavily outnumber women, and women are underrepresented in the leading roles in film (Eschholz et al. 2002). Men are not only more visible but also more formidable, stereotyped in strong, independent roles. Women are more likely now to be portrayed as employed outside of the home and in professional jobs (Signorielli and Bacue 1999), but it is still more usual to see women depicted as sex objects. In fact, the sexualization of women is so extensive in the media that the American Psychological Association has concluded that there is "massive exposure to portrayals that sexualize women and girls and teach girls that women are sexual objects" (American Psychological Association 2007: 5).

SEE *FOR YOURSELF*

GENDER AND POPULAR CELEBRATIONS

Try to buy a friend a birthday card that does not stereotype women or men. Alternatively, try to buy a Father's Day or Mother's Day card and see if you can find one without gender stereotypes. How do the gender images in the cards you see overlap with stereotypes about aging, family, and images of beauty? After doing this experiment, ask yourself how products promoted in the media affect ideas about gender.

Even with social changes in gender roles, boys and girls tend to engage in play activities deemed appropriate for their gender.

Social scientists debate the extent to which people actually believe what they see on television, but research with children shows that they identify with the television characters they see. Both boys and girls rate the aggressive toys that they see on television commercials as highly desirable. They also judge them as more appropriate for boys' play, suggesting that something as seemingly innocent as a toy commercial reinforces attitudes about gender and violence (Klinger et al. 2001). Even with adults, researchers find that there is a link between viewing sexist images and having attitudes that support sexual aggression, antifeminism, and more traditional views of women (American Psychological Association 2007). It is easy to see how extensive such images are by just watching the media with a critical eye. Women in advertisements are routinely shown in poses that would shock people if the characters were male. Consider how often women are displayed in ads dropping their pants, skirts, or bathrobe, or are shown squirming on beds. How often are men shown in such poses? Men are sometimes displayed as sex objects in advertising, but not nearly as often as women. The demeanor of women in advertising—on the ground, in the background, or looking dreamily into space—makes them appear subordinate and available to men.

The Price of Conformity

A high degree of conformity to stereotypical gender expectations takes its toll on both men and women. One of the major ways to see this is in the very high rate of violence against women—both in the United States and worldwide. Too frequently, men's power in society is manifested in physical and emotional violence. Violence takes many forms, including rape, sexual abuse, intimate partner violence, stalking, genital mutilation, and honor killings. Around the world, the United Nations is working in various ways to reduce violence against women, including some initiatives to help men examine cultural assumptions about masculinity that promote violence—a topic we examine further in the next two chapters. For now, it is important to understand that violence against women stems from the attitudes of power and control that gender expectations produce and that can lead *some* men to engage in violent behavior—toward women, as well as toward other men.

Violence by men is only one of the harms to women stemming from dominant gender norms. Adhering to gender expectations of thinness for women and strength for men is related to a host of negative health behaviors, including eating disorders, smoking, and for men, steroid abuse. The dominant culture promotes a narrow image of beauty for women—one that leads many women, especially young women, to be disturbed about their body image. Striving to be thin, millions of women engage in constant dieting, fearing they are fat even when they are well within or below healthy weight standards. Many develop eating disorders by purging themselves of food or cycling through various fad diets—behaviors that can have serious health consequences.

Many young women develop a distorted image of themselves, thinking they are overweight when they may actually be dangerously thin. And, despite the known risks of smoking, increasing numbers of young women smoke not only because they think it "looks cool" but also because they think it will keep them thin. Eating disorders can be related to a woman having a history of sexual abuse, but they also come from the promotion of thinness as an ideal beauty standard for women—a standard that can put girls' and women's health in jeopardy (Hesse-Biber and Hagy 2007).

Men, too, pay the price of conformity if they too thoroughly internalize gender expectations that say they must be independent, self-reliant, and unemotional. Although many men are more likely now than in the past to express intimate feelings, gender socialization discourages intimacy among them, affecting the quality of men's friendships. While conformity to traditional gender roles denies women access to power, influence, achievement, and independence in the public world, it denies men the more nurturing and other-oriented worlds that women have customarily inhabited. Learning traditional gender roles can also produce physical daring and risk-taking that can result in early death or injury from accidents. The strong undercurrent of violence in today's culture of masculinity can in many ways be attributed to the learned gender roles that put men and women at risk.

Race, Gender, and Identity

Gender identity does not emerge apart from other social factors. Thus, race—in combination with gender (as well as social class)—means that men and women from different racial groups may have different expectations regarding gender roles. For example, when asked to rate desirable characteristics in men and women, White men and women are more likely than Latinos to select different traits for men and women, a fact that is counter to the stereotype of Hispanics as holding highly polarized views of manhood and womanhood. African Americans are the group most likely to find value in both sexes displaying traits such as being assertive, athletic, self-reliant, gentle, and eager to soothe hurt feelings (Harris 1994), although

Too much conformity to gender roles can be harmful to your health. Such is the case of anorexic women who starve themselves attempting to meet cultural standards of thinness.

>*doing* sociological research

Eating Disorders: Gender, Race, and the Body

Research Question: "A culture of thinness," "the tyranny of slenderness," "the beauty myth": These are terms used to describe the obsession with weight and body image that permeates the dominant culture, especially for girls and women. Just glance at the covers of popular magazines for women and girls and you will very likely find article after article promoting new diet gimmicks, each bundled with a promise that you will lose pounds in a few days if you only have the proper discipline or use the right products. Moreover, the models on the covers of such magazines are likely to be thin, often dangerously so because being too thin causes serious health problems. *Do these body ideals affect all women equally?*

Research Method: Meg Lovejoy wanted to know if the drive for thinness is unique to White women and how gendered images of the body might differ for African American and White women in the United States. Her research is based on reviewing the existing research literature on eating disorders, which has generally concluded that, compared with White women, Black women are less likely to develop eating disorders.

Research Results: Black women are less likely than White women to engage in excessive dieting and are less fearful of fat, although they are more likely to be obese and experience compulsive overeating. White women, on the other hand, tend to be very dissatisfied with their body size and overall appearance, with an increasing number engaging in obsessive dieting. Black and White women also tend to distort their own weight in opposite directions: White women are more likely to overestimate their own weight (that is, saying they are fat when they are not); Black women are more likely to underestimate their weight (saying they are average when they are overweight by medical standards). Why?

Conclusions and Implications: Lovejoy concludes that you cannot understand eating disorders without knowing the different stigmas attached to Black and White women in society. She suggests that Black women develop alternative standards for valuing their appearance as a way of resisting mainstream, Eurocentric standards. Black women who do so are then less susceptible to the controlling and damaging influence of the institutions that promote the ideal of thinness as feminine beauty. On the other hand, the vulnerability that Black women experience in society can foster mental health problems that are manifested in overeating. Eating disorders for Black women can also stem from the traumas that result from racism, especially when combined with sexism and other forms of oppression.

Lovejoy, along with others who have examined this issue, concludes that eating disorders must be understood in the context of social structures—gender, race, class, and ethnicity—that affect all women, although in different ways. The cultural meanings associated with bodies differ for different groups in society but are deeply linked to our concepts of ourselves and the basic behaviors—like eating—that we otherwise think of as "natural."

Questions to Consider

1. Pay attention to the music and visual images in popular culture and ask yourself what cultural messages are being sent to different race and gender groups? What messages are being conveyed about appropriate appearance? How do they affect people's body image and their self-esteem?

2. Lovejoy examines eating disorders in the context of gender, race, class, and ethnicity. What cultural meanings are broadcast with regard to age?

3. Is there a "culture of thinness" among your peers? If so, what impact do you think it has on people's self-concept? If not, are there other cultural meanings associated with weight among people in your social groups?

Source: Lovejoy, Meg. 2001. "Disturbances in the Social Body: Differences in Body Image and Eating Problems among African American and White Women." *Gender & Society* 15 (April): 239–261.

African American women tend to be more liberal in this regard than African American men (Hunter and Sellers 1998). And, Asian American women are more likely than Asian American men to value egalitarian roles for men and women (Chia et al. 1994).

African American women, like White women, are encouraged to become nurturing and other-oriented, but African American women are also socialized to become self-sufficient, aspire to an education, desire an occupation, and regard work as an expected part of a woman's role. This is most likely because African American mothers are more likely to have been employed and supporting themselves than has historically been true for White mothers; this encourages more self-sufficient attitudes on the part of daughters (Collins 1990).

Men's gender identity is also affected by race. Latino men, for example, bear the stereotype of *machismo*—exaggerated masculinity. Although machismo is associated with sexist behavior by men, within Latino culture it is also associated with honor, dignity, and respect (Mirandé 1979; Baca Zinn 1995). Researchers find that Latino families are rather egalitarian, with decision making frequently shared by men and women. To the extent that machismo exists, it is not just a cultural holdover from Latino societies, but can be how men defy their racial oppression.

Gender Socialization and Homophobia

Homophobia is the fear and hatred of homosexuals. Homophobia plays an important role in gender socialization because it encourages stricter conformity to traditional expectations, especially for men and young boys. Slurs directed against gays encourage boys to act more masculine as a way of affirming for their peers that they are not gay (Pascoe 2007). As a consequence, homophobia also discourages so-called feminine traits in men, such as caring, nurturing, empathy, emotion, and gentleness. Men who endorse the most traditional male roles also tend to be the most homophobic (Burgess 2001; Alden 2001; Basow and Johnson 2000). In this way, homophobia is one of the means by which socialization into expected gender roles takes place. The consequence is not only conformity to gender roles, but a learned hostility toward gays and lesbians.

SEE *FOR YOURSELF*

CHANGING YOUR GENDER

Try an experiment based on the example of changing *gender* that opens this chapter.

1. First, make a list of everything you think you would have to do to change your behavior if you were a member of a different gender. Separate the things in your list according to whether they are related to such factors as appearance, attitude, or behavior.

2. Second, for a period of twenty-four hours, try your best to change any of these things that you are willing to do. Keep a log that records how others react to you during this period and how the change makes you feel.

3. When your experiment is over, write a report on what your brief experiment tells you about "doing gender" (see page 282) and how gender identities are supported (or not) through social interaction.

Homophobia is a learned attitude, as are other forms of negative social judgments about particular groups. Homophobia is also deeply embedded in people's definitions of themselves as men and women. Boys are often raised to be manly by repressing so-called feminine characteristics in themselves. Being called a "fag" or a "sissy" is one of the peer sanctions that socializes a child to conform to particular gender roles. Similarly, pressures on adolescent girls to abandon tomboy behavior are a mechanism by which girls are taught to adopt the behaviors and characteristics associated with womanhood. Being labeled a lesbian may cause those with a strong attraction to women to repress this emotion and direct love only toward men. We can see, therefore, how homophobic ridicule, though it may be in the context of play and joking, has serious consequences for both heterosexual and homosexual men and women. Homophobia socializes most people into expected gender roles, and it produces numerous myths about gays and lesbians—examined in more detail in the following chapter.

The Institutional Basis of Gender

The process of gender socialization tells us a lot about how gender identities are formed, but gender is not just a matter of identity: *Gender is embedded in social institutions.* This means that institutions are patterned by gender, resulting in different experiences and opportunities for men and women. Sociologists analyze gender not just as interpersonal expectations but as characteristic of institutions. This is what is meant by the term *gendered institution.* This concept means that entire institutions are patterned by gender.

Gendered institutions are the total pattern of gender relations that structure social institutions, including the stereotypical expectations, interpersonal relationships, and the different placement of men and women that are found in institutions. Schools, for example, are not just places where children learn gender roles but are gendered institutions because they are founded on specific gender patterns. Seeing institutions as gendered reveals that gender is not just an attribute of individuals but is "present in the processes, practices, images and ideologies, and distributions of power in the various sectors of social life" (Acker 1992: 567).

As an example of the concept of gendered institution, think of what it is like to work as a woman in a work organization dominated by men. Women in this situation report that men's importance in the organization is communicated in subtle ways, whereas women are made to feel like outsiders. Important career connections may be made in the context of men's informal interactions with each other—both inside and outside the workplace. Women may be treated as tokens or may think that company policies are ineffective in helping them cope with the particular demands in their lives. These institutional patterns of gender affect men, too, particularly if they try to establish more balance between their personal and work lives. To say that work institutions are gendered institutions means that, taken together, there is a cumulative and systematic effect of gender throughout the institution.

Gender is not just a learned role; it is also part of social structure, just as class and race are structural dimensions of society. Notice that people do not think about the class system or racial inequality in terms of "class roles" or "race roles." It is obvious that race relations and class relations are far more than matters of interpersonal interaction. Race, class, and gender inequalities are experienced within interpersonal relationships, but they extend beyond relationships. Just as it would seem strange to think that race relations in the United States are controlled by race–role socialization, it is also wrong to think that gender relations are the result of gender socialization alone. Like race and class, gender is a system of privilege and inequality in which women are systematically disadvantaged relative to men. There are institutionalized power relations between women and men, and men and women have unequal access to social and economic resources.

Gender Stratification

Gender stratification refers to the hierarchical distribution of social and economic resources according to gender. Most societies have some form of gender stratification, although the specific form varies from country to country. Comparative research finds that women are more nearly equal in societies characterized by the following traits (Chafetz 1984):

- Women's work is central to the economy.
- Women have access to education.
- Ideological or religious support for gender inequality is not strong.
- Men make direct contributions to household responsibilities, such as housework and child care.
- Work is not highly segregated by sex.
- Women have access to formal power and authority in public decision making.

In Sweden, where there is a relatively high degree of gender equality, the participation of both men and women in the labor force and the household (including child care and housework) is promoted by government policies. Women also have a strong role in the political system, although women still earn less than men in Sweden. In many countries, women and girls have less access to education than men and boys, but that gap is closing. Still, in most countries, the illiteracy rate among women is much higher than among men (United Nations 2006b).

As the preceding list suggests, gender stratification is multidimensional. In some societies, women may be free in some areas of life but not in others. In Japan, for example, women tend to be well educated and participate in the labor force in large numbers. Within the family, however, Japanese women have fairly rigid gender roles. Yet the rate of violence against women in Japan (in the form of rape, prostitution, and pornography) is quite low in relation to other nations, even though women are widely employed as "sex workers" in hostess clubs, bars, and sex joints (Allison 1994). Patterns of gender inequality are most reflected in the wage differentials between women and men around the world, as Figure 11.3 shows.

Gender stratification can be extreme, as was the case in Afghanistan under the Taliban regime. The Taliban, an extremist militia group, seized power in Afghanistan in 1996 and stripped women and girls of basic human rights. Women were banished from the labor force, schools were closed to girls, and women who were enrolled in the universities were expelled. In Afghanistan, women were prohibited from leaving their homes unless accompanied by a close male relative. The windows of houses where women lived were painted black to keep women literally invisible to the public. This extreme segregation and exclusion of women from public life has been labeled **gender apartheid**. Gender apartheid is also evident in other nations, although not as extreme as it was under Taliban rule; in Saudi Arabia, women are not allowed to drive; in Kuwait, women were not allowed to vote until 2006.

Sexism and Patriarchy

Gender stratification is supported by beliefs that treat gender inequality as "natural." Sexism defines women as different from and inferior to men; it can be overt, but can also be subtle. Like racism, sexism makes gender roles seem natural when they are actually rooted in entrenched systems of power and privilege. In this sense, sexism is both a belief and is anchored in social institutions. For example, the idea that men should be paid more than women because they are the primary breadwinners is a sexist idea, but that idea is also embedded in the wage structure. Because of this institutional dimension to sexism, people no longer have to be individually sexist for there still to be consequences of sexism.

Like racism, sexism generates social myths that have no basis in fact but support the continuing advantage of dominant groups over subordinates. A case in point is the belief that women of color are being hired more often and promoted more rapidly than others. This belief misrepresents the facts. Women rarely take jobs away from men because most women of color work in gender- and race-segregated jobs. The truth is that women, especially women of color, are burdened by obstacles to job mobility that are not present for men, especially White men (Browne 1999; Padavic and Reskin 2002). The myth that women of color get all the jobs makes White men seem to be the victims of race and gender privilege. Although there may be occasional cases where a woman of color (or a man of color, for that matter) gets a job that a White man also applied for, usually gender and race privilege favor White men.

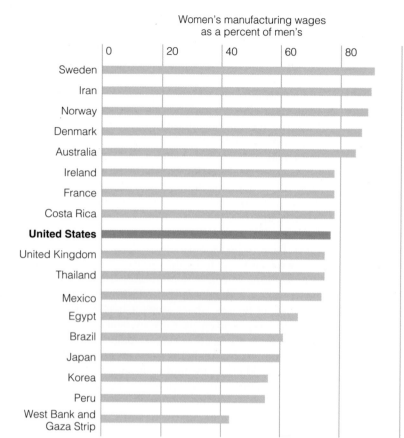

Women's manufacturing wages
as a percent of men's

| | 0 | 20 | 40 | 60 | 80 | 100 |

Sweden
Iran
Norway
Denmark
Australia
Ireland
France
Costa Rica
United States
United Kingdom
Thailand
Mexico
Egypt
Brazil
Japan
Korea
Peru
West Bank and
Gaza Strip

FIGURE 11.3 *The Wage Gap: An International Perspective*

Data: From United Nations. 2006b (August). United Nations Statistics Division: *Statistics and Indicators on Women and Men.* **www.unstats.un.org/ demographic/products/default.htm**

Sexism also works to devalue the work that women do—both in dollar terms and in more subjective perceptions. To give a historical example, in jobs that were once held by men, but became dominated by women, wages declined as women became more numerous. When this happens, the prestige of the occupation also tends to fall (Andersen 2009). You can also see this tendency in something labeled the "mommy tax," referring to the loss of income women experience if they reenter the labor market after staying home to raise children. According to the author who coined this term, a college-educated woman with one child will lose about one million dollars in lifetime earnings as the result of the "mommy tax" (Crittenden 2002).

Sexism emerges in societies structured by **patriarchy,** referring to a society or group in which men have power over women. It can be present in the private sector, such as in families where husbands have authority over their wives. But patriarchy also marks public institutions when men hold all or most of the powerful positions. Forms of patriarchy vary from society to society, but it is common throughout the world. In some societies, it is rigidly upheld in both the public and private spheres, and women may be formally excluded from voting, holding public office, or working outside the home. In societies like the contemporary United States, patriarchy may be somewhat diminished in the private sphere (at least in some households), but the public sphere continues to be based on patriarchal relations.

Matriarchy has traditionally been defined as a society or group in which women have power over men. Anthropologists have debated the extent to which such societies exist. New research finds that matriarchies do exist, though not in the form the customary definition implies. Based on her study of the Minangkabau matriarchal society in West Sumatra (in Indonesia), anthropologist Peggy Sanday argues that scholars have used a Western definition of power to define matriarchy that does not apply in non-Western societies. The Minangkabau define themselves as a matriarchy, meaning that women hold economic and social power. However, the Minangkabau are not ruled by women. The people believe that rule should be by consensus, including that of men and women. Thus, matriarchy exists but not as a mirror image of patriarchy (Sanday 2002).

In sum, gender stratification is an institutionalized system that rests on specific belief systems supporting the inequality of men and women. Although theoretically one could have a society stratified by gender where women hold power over men, that is not how gender stratification has evolved. In the next sections, different manifestations of gender stratification in the United States are examined, especially as it involves women's status relative to men.

Women's Worth: Still Unequal

Gender stratification is especially obvious in the persistent earnings gap between women and men (see Figure 11.4). Although the gap has closed somewhat since the 1960s when women earned 59 percent of what

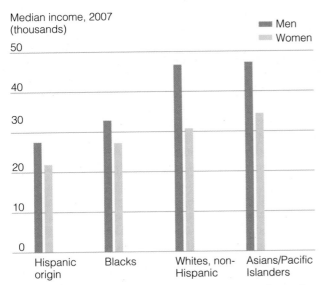

Median income, 2007 (thousands)

■ Men
■ Women

FIGURE 11.4 *Median Income by Race and Gender*

Data include only workers working a full-time job, 50 weeks per year or more; age 25 and older.

Source: U.S. Census Bureau. 2009c. *Detailed Income Tabulations from the Current Population Survey: Educational Attainment People 25 Years Old and Over, by Total Money Earnings in 2005, Work Experience in 2005, Age, Race, Hispanic Origin, and Sex,* Table PINC-03. Washington, DC: U.S. Department of Commerce. **www.census.gov**

percent of men. Since 1960, married women with children have nearly tripled their participation in the labor force. Two-thirds of mothers are now in the labor force, including more than half of mothers with infants. Current projections indicate that women's labor force participation will continue to rise and men's will decline slightly (Bureau of Labor Statistics 2009).

This pattern of women being in the labor market has long been true for women of color but now also characterizes the experience of White women; the labor force participation rates of White women and women of color have, in fact, converged. More women in all racial groups are also now the sole supporters of their families.

Why do women continue to earn less than men, even when laws prohibiting gender discrimination have been in place for more than forty-five years? The Equal Pay Act of 1963 was the first federal law to require that men and women should receive equal pay for equal work, an idea that is supported by the majority of Americans. But wage discrimination is rarely overt. Most employers do not even explicitly set out to pay women less than men. Despite good intentions and legislation, however, differences in men's and women's earnings persist. Research reveals four strong explanations for this: human capital theory, dual labor market theory, gender segregation, and overt discrimination.

Human Capital Theory. Human capital theory explains gender differences in wages as resulting from the individual characteristics that workers bring to jobs. Human capital theory assumes that the economic system is fair and competitive and that wage discrepancies reflect differences in the resources (or *human capital*) that individuals bring to their jobs. Factors such as age, prior experience, number of hours worked, marital status, and education are human capital variables. Human capital theory asserts that these characteristics will influence people's worth in the labor market. For example, higher job turnover

men earned, women who work year-round and full time still earn, on average, only 78 percent of what men earn. Women with bachelor's degrees earn the equivalent of men with associate's degrees (see Figure 11.5). In 2007, the median income for women working full time and year-round was $35,102; for men, it was $45,113 (DeNavas-Walt et al. 2008).

The income gap between women and men persists despite the increased participation of women in the labor force. The **labor force participation rate** is the percentage of those in a given category who are employed either part-time or full-time. Sixty percent of all women are in the paid labor force compared with 74

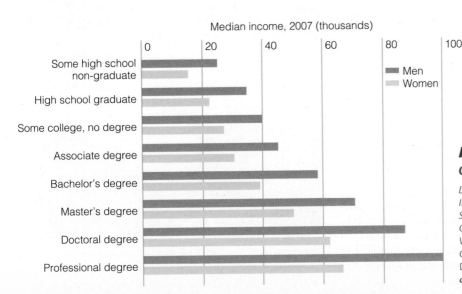

Median income, 2007 (thousands)

■ Men
■ Women

FIGURE 11.5 *Education, Gender, and Income*

Data: U.S. Census Bureau. 2009c. Detailed Income Tabulations from the Current Population Survey: Educational Attainment People 25 Years Old and Over, by Total Money Earnings in 2005, Work Experience in 2005, Age, Race, Hispanic Origin, and Sex, Table PINC-03. Washington, DC: U.S. Department of Commerce. www. census.gov rates or work records interrupted by child rearing and family responsibilities could negatively influence the earning power of women.

Much evidence supports the human capital explanation for the differences between men's and women's earnings because education, age, and experience do influence earnings. But, when you compare men and women who have the same level of education, previous experience, and number of hours worked per week, women still earn less than men (see Figure 11.5). Although human capital theory explains some of the difference between men's and women's earnings, it does not explain it all. Sociologists have looked to other factors to complete the explanation of wage inequality (Browne 1999).

The Dual Labor Market. A second explanation of discrepancies in men's and women's earnings is **dual labor market theory,** which contends that women and men earn different amounts because they tend to work in different segments of the labor market. The dual labor market reflects the devaluation of women's work because it is in low-wage jobs that women are most concentrated. Although it is hard to untangle cause and effect in the relationship between the devaluation of women's work and low wages in certain jobs, once such an earnings structure is established, it is difficult to change. As a result, although equal pay for equal work may hold in principle, it applies to relatively few people because most men and women are not engaged in equal work.

According to dual labor market theory, the labor market is organized in two different sectors: the *primary market* and the *secondary market*. In the primary labor market, jobs are relatively stable, wages are good, opportunities for advancement exist, fringe benefits are likely, and workers are afforded due process. Working for a major corporation in a management job is an example of this. Jobs in the primary labor market are usually in large organizations where there is greater stability, steady profits, benefits for workers, better wages, and a rational system of management. In contrast, the secondary labor market is characterized by high job turnover, low wages, short or nonexistent promotion ladders, few benefits, poor working conditions, arbitrary work rules, and capricious supervision. Many of the jobs students take—such as waiting tables, selling fast food, or cooking and serving fast food—fall into the secondary labor market. However, for students these jobs are usually short term.

Within the primary labor market, there are two tiers. The first consists of high-status professional and managerial jobs with potential for upward mobility, room for creativity and initiative, and more autonomy. The second tier comprises working-class jobs, including clerical work, skilled, and semiskilled blue-collar work. Women and minorities in the primary labor market tend to be in the second tier. Although these jobs may be more secure than jobs in the secondary

labor market, they are more vulnerable and do not have as much mobility, pay, prestige, or autonomy as jobs in the first tier of the primary labor market.

There is, in addition, an informal sector of the market where there is even greater wage inequality, no benefits, and little, if any, oversight of employment practices. Individuals may hire such workers as private service workers or under-the-table workers who perform a service for a fee (painting, babysitting, car repairs, and any number of services). Although there are no formal data on the informal sector because much of it tends to be in an underground economy, it is likely that women and minorities form a large segment of this market activity. White men in this sector are also disadvantaged because of the instability and lack of protection in this work.

Gender Segregation. Dual labor market theory explains wage inequality as a function of the structure of the labor market, not the individual characteristics of workers as suggested by human capital theory. Because of the dual labor market, men and women tend to work in different occupations and, when working in the same occupation, in different jobs. This is referred to as **gender segregation,** a pattern in which different groups of workers are separated into occupational categories based on gender. There is a direct association between the number of women in given occupational categories and the wages paid in those jobs. In other words, the greater the proportion of women in a given occupation, the lower the pay (Bureau of Labor Statistics 2009). Gender segregation is a specific form of **occupational segregation;** segregation in the labor market can also be based on factors such as race, class, age, or any combination thereof.

Despite several decades of legislation prohibiting discrimination against women in the workplace,

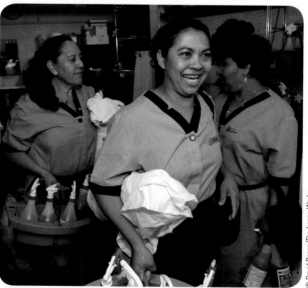

Women of color, such as these hotel maids, tend to be segregated in low-wage, service work.

Data show that occupations where women of color predominate also tend to have the lowest wages.

most women and men still work in gender-segregated occupations. That is, the majority of women work in occupations where most of the other workers are women, and the majority of men work mostly with men. Women also tend to be concentrated in a smaller range of occupations than men. To this day, more than half of all employed women work as clerical workers and sales clerks or in service occupations such as food-service workers, maids, health-service workers, hair-dressers, and child-care workers. Men are dispersed over a much broader array of occupations. Women make up 82 percent of elementary and middle school teachers, 97 percent of secretaries, 90 percent of bookkeepers, and 94 percent of child-care workers—stark evidence of the persistence of gender segregation in the labor force (Bureau of Labor Statistics 2009).

Gender segregation also occurs within occupations. Women usually work in different jobs from men, but when they work within the same occupation, they are segregated into particular fields or job types. For example, in sales work, women tend to do noncommissioned sales or to sell products that are of less value than those men sell. Or, among waiters and waitresses, women often work in lower-priced restaurants where they are likely to be tipped less than men (Hall 1993).

Overt Discrimination. A fourth explanation of the gender wage gap is discrimination. **Discrimination** refers to practices that single out some groups for different and unequal treatment. Despite the progress of recent years, overt discrimination continues to afflict women in the workplace. It is argued that men (especially White men), by virtue of being the dominant group in society, have an incentive to preserve their advantages in the labor market. They do so by establishing rules that distribute rewards unequally. Women pose a threat to traditional White

male privileges, and men may organize to preserve their own power and advantage (Reskin 1988).

Within institutions, dominant groups will use their position of power to perpetuate their advantage (Lieberson 1980). Historically, White men used labor unions to exclude women and racial minorities from well-paying, unionized jobs, usually in the blue-collar trades. A more contemporary example is seen in the efforts of some groups to dilute legislation that has been developed to assist women and racial–ethnic minorities. These efforts can be seen as an attempt to preserve group power.

Another example of overt discrimination is the harassment that women experience at work, including *sexual harassment* and other means of intimidation. Sociologists see such behaviors as ways for men to protect their advantages in the labor force. No wonder that women who enter traditionally male-dominated professions suffer the most sexual harassment; the reverse seldom occurs for men employed in jobs historically filled by women. Although men can be victims of sexual harassment, this is rare. Sexual harassment is a mechanism for preserving men's advantage in the labor force—a device that also buttresses the belief that women are sexual objects for the pleasure of men.

Each of these explanations—human capital theory, dual labor market theory, gender segregation, and overt discrimination—contributes to an understanding of the continuing differences in pay between women and men. Wage inequality between men and women is clearly the result of multiple factors that together operate to place women at a systematic disadvantage in the workplace.

The Devaluation of Women's Work

Across the labor market, women tend to be concentrated in those jobs that are the most devalued, causing some to wonder if the fact that the jobs are held by women leads to devaluation of the jobs. Why, for example, is pediatrics considered a less prestigious specialty than cardiology? Why are elementary school teachers (81 percent of whom are women) paid less than airplane mechanics (97 percent of whom are men)? The association of elementary school teaching with children and its identification as "women's work" lowers its prestige and economic value. Indeed, if measured by the wages attached to an occupation, child care is one of the least prestigious jobs in the nation—paying on average only $345 per week in 2006, which would come out to an income below the federal poverty line if you worked every week of the year (Bureau of Labor Statistics 2009).

Only a small proportion of women work in occupations traditionally thought to be men's jobs (such as the skilled trades). The representation of women in skilled blue-collar jobs has increased, but it is still a very small fraction (typically less than 3 percent) of those in skilled trades such as plumbers, electricians, and

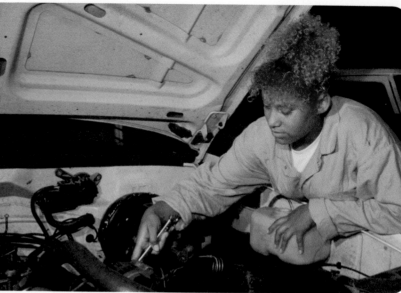

David Young-Wolff/PhotoEdit

Because gender segregation is so pervasive in the workplace, people are still surprised when they see women and men in non-traditional occupations.

carpenters (Bureau of Labor Statistics 2009). Likewise, very few men work in occupations historically considered to be women's work, such as nursing, elementary school teaching, and clerical work. Interestingly, men who work in occupations customarily thought of as women's work tend to be more upwardly mobile within these jobs than are women who enter fields traditionally reserved for men (Williams 1992; Budig 2002). Gender segregation in the labor market is so prevalent that most jobs can easily be categorized as men's work or women's work. Occupational segregation reinforces the belief that there are significant differences between the sexes. Think of the characteristics of a soldier. Do you imagine someone who is compassionate, gentle, and demure? Similarly, imagine a secretary. Is this someone who is aggressive, independent, and stalwart? The association of each characteristic with a particular gender makes the occupation itself a gendered occupation.

For all women, perceptions of gender-appropriate behavior influence the likelihood of success within institutions. Even something as simple as wearing makeup has been linked to women's success in professional jobs (Dellinger and Williams 1997). When men or women cross the boundaries established by occupational segregation, they are often considered to be gender deviants. They may be stereotyped as gay and have their "true gender identity" questioned. Men who are nurses may be stereotyped as effeminate or gay; women Marines may be stereotyped as "butch." Social practices like these serve to reassert traditional gender identities, perhaps softening the challenge to traditionally male-dominated institutions that women's entry challenges (Williams 1995).

As a result, many men and women in nontraditional occupations feel pressured to assert gender-appropriate behavior. Men in jobs historically defined as women's work may feel impelled to emphasize their masculinity, or if they are gay, they may feel even more pressure to keep their sexual orientation secret. Such social disguises can make them seem unfriendly and distant, characteristics that can have a negative effect on performance evaluations. Heterosexual women in male-dominated jobs may also feel obliged to squash suspicions that they are lesbians or are excessively mannish. And, studies have found that lesbian women are more likely to be open about their sexual identity at work when they work predominantly with women and have women as bosses (Schneider 1984).

Balancing Work and Family

As the participation of women in the labor force has increased, so have the demands of keeping up with work and home life. Although some changes are evident, women continue to hold primary responsibility for meeting the needs of families, as we will see in more detail in Chapter 13. Many men are now much more involved in housework and child care than has been true in the past, although most of this work still falls to women—a phenomenon that has been labeled "the second shift" (Hochschild 1989).

The social speedup that comes from increased hours of employment for both men and women (but especially women), coupled with the demands of maintaining a household, are a source of considerable stress (Jacobs and Gerson 2004). Women continue to provide most of the labor that keeps households

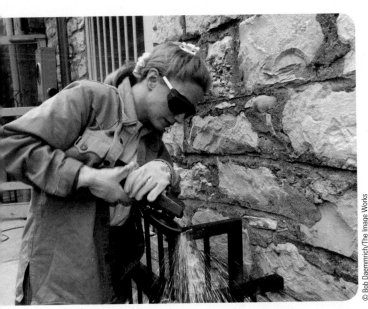

Although they are still a small minority, some women have entered jobs in the skilled trades that were traditionally held only by men.

© Bob Daemmrich/The Image Works

running—cleaning, cooking, running errands, driving children around, and managing household affairs. Although more men are engaged in housework and child care, a huge gender gap remains in the amount of such work done by women and men. Women are also much more likely to be providing care, not just for children, but also for their older parents. The strains produced by these demands have made the home seem more and more like work for many; a large number of women and men report that their days at both work and home are harried and that they find work to be the place where they find emotional gratification and social support. In this contest between home and work, simply finding time can be an enormous challenge (Hochschild 1997). It is not surprising then that women report stress as one of their greatest concerns (Newport 2000).

Theories of Gender

Why is there gender inequality? The answer to this question is important, not only because it makes us think about the experiences of women and men, but also because it guides attempts to address the persistence of gender injustice. The major theoretical frameworks in sociology provide some answers, but feminist scholars have also found that traditional perspectives in the discipline are inadequate to address the new issues that have emerged from feminist research.

The Frameworks of Sociology

The major frameworks of sociological theory—functionalism, conflict theory, and symbolic interaction—provide some answers to the question of why gender inequality exists, although, as we will see in the next section, feminist scholars have developed new theories to address women's lives directly. Functionalists, for example, have been criticized for interpreting gender as a fixed role in society. *Functionalist theory* purports that men fill instrumental roles in society whereas women fill expressive roles and presumes that this arrangement works to the benefit of society (see Chapter 1). Feminists object to such a characterization, arguing that this presumes sexist arrangements are functional for society. Limiting women's roles to expressive functions and men's to instrumental functions is dysfunctional according to feminists—both for men and women. Although few contemporary functionalist theorists would make such traditionalist arguments, functionalism does emphasize people's socialization into prescribed roles as the major impetus behind gender inequality. Thus, conditions such as wage inequality, a functionalist might argue, are the result of choices women make that may result in their inequality but that nonetheless involve functional adaptation to the competing demands of family and work roles.

Conflict theorists, in contrast, see women as disadvantaged by power inequities between women and men that are built into the social structure. This includes economic inequity, as well as women's disadvantages

Is It True?*

	True	False
1. Men are more aggressive than women.		
2. Parents have the most influence on children's gender identities.		
3. Most women hold feminist values.		
4. In all racial–ethnic groups, women earn less on average than men.		
5. The wage gap between women and men has closed since the 1970s largely as the result of women being more likely to enter the labor force.		
6. In terms of wages, middle-class women have most benefited from antidiscrimination policies.		

*The answers can be found on page 282.

Is It True? (Answers)

1. FALSE. Generalizations such as this ignore variation occurring within gender categories; moreover, "aggression" is a broad term that can have multiple meanings.

2. FALSE. There are numerous sources of gender socialization; even parents who try to raise their children not to conform too strictly to gender norms will find that peers, the media, schools, and other socialization agents all push people into the expected behaviors associated with gender.

3. TRUE. Although many women do not use the label "feminist" to define themselves, surveys show that the majority of women agree with basic feminist principles. Self-identification as a feminist is most likely among well-educated, urban women (McCabe 2005).

4. TRUE. However, the gap in median income is not as wide *within* some groups as it is in others. White women, for example, earn 66 percent of what White men earn, but Hispanic women earn 80 percent of what Hispanic men earn (because both have very low earnings on average). Black women earn 82 percent of Black men's earnings, and Asian women, 73 percent of men's earnings. And White and Asian American women, on average, earn *more* than Black and Hispanic men (U.S. Census Bureau 2009).

5. FALSE. The most significant reason for the decline in the wage gap between women and men is the decline in men's wages; a smaller portion of this closing gap is attributed to changes in women's wages (Mishel et al. 2008).

6. FALSE. Although all women do benefit from equal employment legislation, wage data indicate that the group whose wages have increased the most since the 1970s are women in the top 20 percent of earners. Middle- and working-class women have seen far lower gains, and poor women's wages have been relatively flat over this period of time (Mishel et al. 2008).

in political and social systems. Conflict theorists, for example, see wage inequality as produced from men's historic power to devalue women's work and to benefit as a group from the services that women's labor provides. At the same time, conflict theorists have been much more attuned to the interactions of race, class, and gender inequality because they see all forms of inequality as stemming from the differential access to resources that dominant groups in society have.

From an *ethnomethodological* perspective (see Chapters 2 and 5) feminist scholars have developed what is known as **doing gender** a theoretical perspective that interprets gender as something accomplished through the ongoing social interactions people have with one another (West and Zimmerman 1987; West and Fenstermaker 1995). Seen from this framework, people produce gender through the interaction they have with one another and through the interpretations they have of certain actions and appearances. In other words, gender is not something that is an attribute of different people, as functionalists suggest; rather, it is constantly made up and reproduced through social interaction. When you act like a man or act like a woman, you are confirming gender and reproducing the existing social order (Peralta 2002). From this point of view, gender relations would change if large numbers of people behaved differently. This is one reason the theory has been criticized by those with a more macrosociological point of view; they say it ignores the power differences and economic differences that exist based on gender, race, and class. In other words, it does not explain the structural basis of women's oppression (Collins et al. 1995).

Feminist Theory

Feminism is not a single way of thinking and acting; it fundamentally refers to advocating for a more just society for women. **Feminist theory** has emerged from the women's movement and refers to analyses that seek to understand the position of women in society for the explicit purpose of improving their position in society. Four major frameworks have developed in feminist theory: liberal feminism, socialist feminism, radical feminism, and multiracial feminism (see Table 11.1).

Liberal feminism emerged from a long tradition that began among British liberals in the nineteenth century. Liberal feminism argues that inequality for women originates in traditions of the past that pose barriers to women's advancement. It emphasizes individual rights and equal opportunity as the basis for social justice and reform. The framework of liberal feminism has been used to support many of the legal changes required to bring about greater equality for women in the United States. Liberal feminists contend that gender socialization contributes to women's inequality because it is through learned customs that inequality is perpetuated. In the interests of social change, liberal feminism advocates the removal of barriers to women's advancement and the development of policies that promote equal rights for women.

Socialist feminism is a more radical perspective that interprets the origins of women's oppression in the system of capitalism. Because women constitute a cheap supply of labor, they are exploited by capitalism in much the same way that the working class is exploited. In the view of socialist feminists, capitalism interacts with patriarchy to make women less powerful, both as women and as laborers. Socialist feminists are critical of liberal feminism for not addressing the fundamental inequalities built into capitalist and patriarchal systems. To these feminists, equality for women will come only when the economic and political system is changed.

Radical feminism interprets patriarchy as the primary cause of women's oppression. To radical

table 11.1	**Feminist Theory: Comparing Perspectives**			
	Liberal Feminism	**Socialist Feminism**	**Radical Feminism**	**Multiracial Feminism**
Gender Identity	Identity is learned through socialization.	Gender identity reflects the needs of a capitalist economic system.	Women's identification with men gives men power over women.	Women and men of color form an oppositional consciousness as a reaction against oppression.
Gender Inequality	Inequality is the result of formal barriers to equal opportunity.	Gender inequality stems from class relations.	Patriarchy is the basis for women's powerlessness.	Race, class, and gender intersect to form a matrix of domination.
Social Change	Change is accomplished through legal reform and attitudinal change.	Transformation of the gender division of labor accompanies change in the class division of labor.	Liberation comes as women organize on their own behalf.	Women of color become agents of feminist change through alliances with other groups.

feminists, the origins of women's oppression lie in men's control over women's bodies; thus, they see violence against women—in the form of rape, sexual harassment, domestic violence, and sexual abuse—as mechanisms that men use to assert their power in society. Radical feminists think that change cannot come about through the existing system because that system is controlled and dominated by men.

Most recently, **multiracial feminism** has developed new avenues of theory for guiding the study of race, class, and gender (Andersen and Collins 2010; Baca Zinn and Dill 1996; Collins 1990). Multiracial feminism evolves from studies pointing out that earlier forms of feminist thinking excluded from analysis women of color, which made it impossible for feminists to deliver theories that informed people about the experiences of all women. Multiracial feminism examines the interactive influence of gender, race, and class, showing how together they shape the experiences of all women and men (Baca Zinn and Dill 1996).

These perspectives provide unique ways of looking at the experiences of women and men in society. These theoretical orientations have been the bedrock on which feminists have built their programs of social and political change.

Gender in Global Perspective

Increasingly, the economic condition of women and men in the United States is also linked to the fortunes of people in other parts of the world. The growth of a global economy and the availability of a cheaper industrial labor force outside the United States mean that U.S. workers have become part of an international division of labor. U.S.–based multinational corporations looking around the world for less expensive labor often turn to the developing nations and find that the cheapest laborers are women or children. The

global division of labor is thus acquiring a gendered component, with women workers, usually from the poorest countries, providing a cheap supply of labor for manufacturing products that are distributed in the richer industrial nations.

Worldwide, women work as much or more than men. It is difficult to find a single place in the world where the workplace is not segregated by gender. On a worldwide scale, women also do most of the work associated with home, children, and the elderly. While women's paid labor has been increasing, their unpaid labor in virtually every part of the world exceeds that of men. The United Nations estimates that the value of women's unpaid work (both in the home and in the community) amounts to at least $11 trillion (www.un.org).

Despite these general trends, women's situations differ significantly from nation to nation. China is unusual in that there is far greater sharing of household responsibilities than in most other nations. In China, both women and men work long hours in paid employment, and women are encouraged to stay in the labor force when they have children. There are also extensive child-care facilities in China and a fifty-six-day paid maternity leave. Many work organizations have extended this paid leave to six months, although women can lose seniority rights when they are on maternity leave (something that is illegal in the United States).

In contrast, Japan has marked inequality in the domestic sphere. Women are far more likely to leave the labor force after marrying or following childbirth, and Japanese women's identities are more defined by their roles at home, although for many this is changing. Compared with China, Japanese women more closely resemble the pattern that exists in Britain and, to some extent, in the United States, although they are less involved in paid employment than in either of these countries. Ironically, when comparing China, Japan, and Britain, researchers have found that Chinese women are the most discontented with what they perceive as gender injustice, whereas Japanese and British women express greater satisfaction with more limited employment. This may seem surprising, given the greater gender equality of Chinese women with Chinese men. Sociologists explain it as the result of the gap Chinese women see between official ideologies of gender equality and their observations of continuing inequalities in promotions and other benefits of work (Xuewen et al. 1992).

Work is not the only measure by which the status of women throughout the world is inferior to that of men. Women are vastly underrepresented in national parliaments (or other forms of government) everywhere; in only nineteen countries of the entire world is women's representation in national parliaments above 25 percent. Worldwide, women hold only 18 percent of all parliamentary seats. Only twenty-eight nations have ever had a woman as head of state (www.ipu.org).

The United Nations has also concluded that violence against women and girls is a global epidemic and one of the most pervasive violations of human rights (UNICEF 2000; United Nations 2006b). Violence against women takes many forms, including rape, domestic violence, infanticide, incest, genital mutilation, and murder (including so-called honor killings, where a woman may be killed to uphold the honor of the family if she has been raped or accused of adultery). Although violence is pervasive, some specific groups of women are more vulnerable than others—namely, minority groups, refugees, women with disabilities, elderly women, poor and migrant women, and women living in countries where there is armed conflict. Statistics on the extent of violence against women are hard to report with accuracy, both because of the secrecy that surrounds many forms of violence and because of differences in how different nations might report violence. Nonetheless, the United Nations estimates that between 20 and 50 percent of women worldwide have experienced violence from an intimate partner or family member.

As we saw in Chapter 9, many factors put women at risk of violence, including cultural norms, women's economic and social dependence on men, and political practices that either provide inadequate legal protection or provide explicit support for women's subordination (as in the example of the Taliban given earlier).

Gender and Social Change

Few lives have not been touched by the transformations that have occurred in the wake of the feminist movement (see also Chapter 16). The women's movement has changed how women's issues are perceived in the public consciousness. It has opened work opportunities, generated laws that protect women's rights, and spawned organizations that lobby for public policies on behalf of women. Many young women and men now take for granted freedoms struggled for by earlier generations. These include access to birth control, equal opportunity legislation, and laws protecting

against sexual harassment, as well as an increased presence of women on corporate boards, increased athletic opportunities for women, more presence in political life, and greater access to child care, to name a few changes. These impressive changes occurred in a relatively short period. How have they affected public consciousness about gender?

Contemporary Attitudes

In recent years, public attitudes toward gender roles have changed noticeably, especially regarding beliefs about the ideal lifestyle. Only a small minority of people disapprove of women being employed while they have young children, and both women and men say it is not fair for men to be the sole decision maker in the household. Although the majority of women now want to combine work and family, they believe they will be discriminated against in the labor force if they do so (Greenhouse 2000).

People's beliefs about appropriate gender roles have evolved as women's and men's lives have changed. Still, nearly half of all men (45 percent) now think one parent should stay at home while the other is employed, compared to 38 percent of all women who think so. But, the majority of men (68 percent) and 50 percent of women say they would prefer to work outside the home (Moore 2005; Saad 2007; McComb 2001). In reality, in more than half (52 percent) of married couple families both husband and wife are employed (Bureau of Labor Statistics 2009).

Debunking Society's Myths

Myth: The men most likely to support equality for women are White, middle-class men with a good education.

Sociological perspective: Although it is true that younger men tend to be more egalitarian than older men, African American men are the most likely to support women's equal rights and the right of women to work outside the home. On most measures of feminist beliefs, African American men tend to be more liberal than White men (Hunter and Sellers 1998).

Men's support for women's roles in the family and at work, however, varies across different groups. Younger men and single men are more egalitarian than older, married men. Among college students, however, women hold more egalitarian views of women's roles than do men; although both become less traditional in their views during college, women change more than men (Bryant 2003). There are also racial and ethnic differences in how different groups view gender roles, with minority men usually being more supportive of egalitarian roles than White men. The mothers of minority men are more likely to have been employed, and these men have different educational and employment backgrounds than White men. Their attitudinal differences reflect the economic necessity that minority men attribute to women working (Wilkie 1993; Blee and Tickamyer 1995).

Old attitudes do not die easily, and changed attitudes do not necessarily mean changed behavior. Gender attitudes change as society changes. However, young people's expectations for being able to "have it all" can be unrealistic. As an example, sociologist Michele Hoffnung surveyed a random sample of college women in their senior year, surveying them again seven years later (in 2001). She found that, as seniors, most of the women wanted careers, marriage, and motherhood, with career development being their top priority in their twenties. But at the seven-year point, those who had become mothers had fewer advanced degrees and lower career status than the nonmothers; marriage was not related to career status (Hoffnung 2004).

Attitudes, however, are only part of the problem of persistent gender inequality. Social change requires more than changing individual attitudes; it also means changing social institutions.

Legislative Change

Much legislation prohibits overt discrimination against women. In addition to the Equal Pay Act of 1963, the Civil Rights Act of 1964, adopted as the result of political pressure from the civil rights movement, banned discrimination in voting and public accommodations and required fair employment practices. Specifically, Title VII of the Civil Rights Act of 1964 forbids discrimination in employment on the basis of race, color, national origin, religion, or sex.

The passage of the Civil Rights Act, and Title VII in particular, opened up new opportunities to women in employment and education. This was further supported by *Title IX*, adopted as part of the Educational Amendments of 1972, which forbids gender discrimination in any educational institution receiving federal

Young women are the group most likely to support feminist goals, although the feminist movement has support across generations.

funds. Title IX prohibits colleges and universities from receiving federal funds if they discriminate against women in any program, including athletics. Adoption of this bill radically altered the opportunities available to women students and laid the foundation for many of the coeducational programs that are now an ordinary part of college life. This law has been particularly effective in opening up athletics to women.

Passage of antidiscrimination policies does not, however, guarantee their universal implementation. Has equality been achieved? In college sports, men still outnumber women athletes by more than two to one, and there is still more scholarship support for male athletes than for women. Title IX allows institutions to spend more money on male athletes if they outnumber women athletes, but it also stipulates that the number of male and female athletes should be roughly proportional to their representation in the student body. Studies of student athletes show that although there has been improvement in support for women's athletics since the implementation of Title IX, there is still a long way to go toward equity in women's sports (Lederman 1992; Sigelman and Wahlbeck 1999). Title IX is being challenged by some who argue that it has reduced opportunities for men in sports. Proponents of maintaining strong enforcement of Title IX counter this, however, by noting that budget reductions in higher education, not Title IX per se, are responsible for any reduction in athletic opportunities for men. Furthermore, they point out that men still greatly predominate in school sports.

In the workplace, a strong legal framework for gender equity is in place, yet equity has not been achieved. The United States has never, as an example, approved the **Equal Rights Amendment,** which would provide a constitutional principle that "equality of rights under the law shall not be denied or abridged by the United States or by any state on the basis of sex" (www.equalrightsamendment.org).

Affirmative action, as we saw in Chapter 10, is a method for opening opportunities to women and minorities that specifically redresses past discrimination by taking positive measures to recruit and hire previously disadvantaged groups. Affirmative action has been especially effective in opening new opportunities for women. The national discussion of affirmative action, fueled in part by conservative resistance to these policies, is a good illustration of what many would call the limitations of liberal philosophy. Because affirmative action promotes gender- and race-specific actions to remedy the effects of past discrimination, its opponents have argued that it constitutes reverse discrimination. Those who want social policies that are gender- and race-blind (a classic liberal position, now articulated by conservatives, as well) find it difficult to support policies that are not. The problem is that gender and race inequalities continue even though these inequities may not always result from conscious discrimination. Proponents of affirmative action argue that as long as the structural conditions of gender and race inequality exist, there is still a need for race- and gender-conscious actions designed to address persistent injustices. This viewpoint was upheld by the U.S. Supreme Court in 2003 in the case of *Grutter v. Bollinger*.

Debunking Society's Myths

Myth: Because of affirmative action, Black women are taking a lot of jobs away from White men.

Sociological perspective: Sociological research finds no evidence of this claim. Quite the contrary, women of color work in gender- and race-segregated jobs and only rarely in occupations where they compete with White men in the labor market (Browne 1999; Padavic and Reskin 2002).

One solution to the problem of gender inequality is to have more women in positions of public power. Is increasing the representation of women in existing situations enough? Without reforming the sexism in the institutions, change will be limited and may generate benefits only for groups who are already privileged. Feminists advocate restructuring social institutions to meet the needs of all groups, not just those who already have enough power and privilege to make social institutions work for them. The successes of the

Fred Beckham/AP Photo

Women have made many strides in various institutions, including sports where Title IX has given them new opportunities and also career options. Pictured here is Ashleigh Braxton (University of Pittsburgh) being guarded by Carla Jacobs (University of Cincinnati) during the 2007 Big East Women's Basketball Championship.

women's movement demonstrate that change is possible, but change comes only when people are vigilant about their needs.

Chapter Summary

- ### *How do sociologists distinguish sex and gender?*

  Sociologists use the term *sex* to refer to biological identity and *gender* to refer to the socially learned expectations associated with members of each sex. *Biological determinism* refers to explanations that attribute complex social phenomena entirely to physical or natural characteristics.

- ### *How is gender identity learned?*

  Gender socialization is the process by which gender expectations are learned. One result of socialization is the formation of *gender identity*. Overly conforming to gender roles has a number of negative consequences for both women and men,

including eating disorders, violence, and poor self-concepts. *Homophobia* plays a role in gender socialization because it encourages strict conformity to gender expectations.

- ### *What is a gendered institution?*

  Gendered institutions are those where the entire institution is patterned by gender. Sociologists analyze gender both as a learned attribute and as an institutional structure.

- ### *What is gender stratification?*

  Gender stratification refers to the hierarchical distribution of social and economic resources according to gender. Most societies have some form of gender stratification, although they differ in the degree and kind. Gender stratification in the United States is obvious in the differences between men's and women's wages.

- **How do sociologists explain the continuing earnings gap between men and women?**

There are multiple ways to explain the pay gap. *Human capital theory* explains wage differences as the result of individual differences between workers. *Dual labor market theory* refers to the tendency for the labor market to be organized in two sectors: the primary and secondary markets. *Gender segregation* persists and results in differential pay and value attached to men's and women's work. *Overt discrimination* against women is another way that men protect their privilege in the labor market.

- **Are men increasing their efforts in housework and child care?**

Many men are now more engaged in housework and child care than was true in the past, although women still provide the vast majority of this labor. Balancing work and family has resulted in social speedup, making time a scarce resource for many women and men.

- **What is feminist theory?**

Different theoretical perspectives help explain the status of women in society. *Feminist theory* links this explanation to the desire for improving women's lives. The major theoretical perspectives in feminism are *liberal feminism, socialist feminism, radical feminism,* and *multiracial feminism.* Each emphasizes different aspects of women's place in society.

- **When seen in global perspective, what can be observed about gender?**

The economic condition of women and men in the United States is increasingly linked to the fortunes of people in other parts of the world. Women provide much of the cheap labor for products made around the world. Worldwide, women work as much or more than men, though they own little of the world's property and are underrepresented in positions of world leadership.

- **What are the major social changes that have affected women and men in recent years?**

Public attitudes about gender relations have changed dramatically in recent years. Women and men are now more egalitarian in their attitudes, although women still perceive high degrees of discrimination in the labor force. A legal framework is in place to protect against discrimination, but legal reform is not enough to create gender equity.

Key Terms

affirmative action   285
biological determinism   265
discrimination   279
doing gender   282
dual labor market theory   278
Equal Rights Amendment   285
feminism   282
feminist theory   282
gender   265
gender apartheid   275
gender identity   267

gender segregation   278
gender socialization   267
gender stratification   275
gendered institution   274
hermaphroditism   266
homophobia   274
human capital theory   277
intersexed   266
labor force participation
    rate   277
liberal feminism   282

matriarchy   276
multiracial feminism   283
occupational segregation   278
patriarchy   276
radical feminism   282
sex   265
socialist feminism   282
transgendered   266
Title IX   271

Online Resources

Sociology: The Essentials Companion Website

www.cengage.com/sociology/andersen
Visit your book companion website where you will find more resources to help you study and write your research papers. Resources include web links, learning objectives, Internet exercises, quizzing, and flash cards.

 is an easy-to-use online resource that helps you study in less time to get the grade you want NOW.

www.cengage.com/login
Need help studying? This site is your one-stop study shop. Take a Pre-Test and Cengage NOW will generate a Personalized Study Plan based on your test results. The Study Plan will identify the topics you need to review and direct you to online resources to help you master those topics. You can then take a Post-Test to determine the concepts you have mastered and what you still need to work on.

12

Chapter twelve
CHAPTER TWELVE

Sexuality

Sex and Culture

Contemporary Sexual Attitudes and Behavior

Sexuality and Sociological Theory

Sex, Diversity, and Inequality

Sex and Social Issues

Sex and Social Change

Chapter Summary

[*A visitor from* another planet might conclude that people in the United States are obsessed with sex. Young people watch videos where women gyrate in sexual movements. A stroll through a shopping mall reveals expensive shops selling delicate, skimpy women's lingerie. Popular magazines are filled with images of women in seductive poses trying to sell every product imaginable; even bumper stickers brag about sexual accomplishments. People dream about sex, form relationships based on sex, fight about sex, and spend money to have sex. On the one hand, the United States appears to be a very sexually open society; however, sexual oppression still exists. Gay men, lesbians, and bisexuals are viewed with prejudice and are discriminated against—that is, treated like minority groups, as defined in Chapter 10.

Sexuality, usually thought to be a most private matter, has taken on a public life by being at the center of some of our most heated public controversies. Should sex be confined to marriage between a man and a woman? Should young people be educated about birth control or just encouraged to abstain from sex? What should we do about teen pregnancy? Sex is clearly a subject that polarizes the public on a range of social issues. Thus, sexuality is seen as a private matter at the same time that it is one basis for social inequality and social controversy. For sociologists, studying sexuality reveals how deeply it is entrenched in social norms, values, and social structures. Using the sociological imagination, we can see that human sexuality, like other forms of social behavior, is shaped by society and culture.

© Jonathan Nourok/Getty Images/Stone

Sex and Culture

Sexual behavior would seem to be utterly natural. Pleasure and sometimes the desire to reproduce are reasons people have sex, but sexual relationships and identities develop within a social context. It is the social context that establishes what sexual relationships mean and how we define our sexual identities, as well as what social supports are given (or denied) to people based on their sexual identity. *Sexuality is socially defined and patterned.*

Sex: Is It Natural?

From a sociological point of view, little in human behavior is purely natural, as we have learned in previous chapters. Behavior that appears to be natural is the behavior accepted by cultural customs and sanctioned by social institutions. Sexuality creates intimacy between people. People engage in sex not just because it feels good, but also because it is an important part of our social identity. Void of a cultural context and the social meanings attributed to sexual behavior, people might not attribute the emotional commitments, psychological interpretations, spiritual meanings, and social significance to sexuality that it has in different human cultures.

Is there a biological basis to sexual identity? This question is debated in both popular and scientific literature. Gay and lesbian people often say their sexual orientation is natural, something they just are, not just something they choose. Two concepts are important in this discussion: *sexual orientation* and *sexual identity*.

Sexual orientation refers to the attraction that people feel for people of the same or different sex. The term *sexual orientation* implies something deeply rooted in a person. **Sexual identity** is the definition of oneself that is formed around one's sexual relationships. Sexual identity is learned in the context of our social relationships and the social structures in which we live. Although sometimes used interchangeably, sexual orientation and sexual identity are not the same thing, nor is one's sexual identity simply based on one's sexual practices. For example, a man may have sex with other men—perhaps even on a regular basis, but not have a sexual identity as gay. Or, you may have a sexual identity as heterosexual even in the absence of actual sexual relationships. Sexual identity, as gay, lesbian, heterosexual, or bisexual, emerges in a social context, as we will see below in discussing the social construction of sexual identity. The point here is to see that sexual identity is not necessarily simply based on biological or "natural" states.

It is important to make the distinction between sexual orientation and sexual identity to untangle the often heated discussion about a presumed biological basis to sexual orientation. Gay, lesbian, and bisexual people often say that they do not choose their sexual orientation and that it is something they just "are," as if it were a biological imperative. Part of the debate about this—heated at times—comes from rejecting the idea that being gay is a choice, as if people could change their sexual orientation at will. There are political reasons for rejecting the idea of homosexuality as a choice because, if it is something inherent in people,

Sexual relationships, although highly personal, are also shaped by society and culture.

then perhaps others will be more accepting of gay, lesbian, and bisexual people.

Perhaps there is some biological basis to sexual orientation, but the evidence is not yet there. Even if a biological influence exists, social experiences are far more significant in shaping sexual identity, even though they are rarely reported in the media with as much acclaim as alleged biological bases to human sexuality (Connell 1992; Lorber 1994; Brookey 2001). Whatever the origins of sexual orientation, there is no doubt that social influences are a very significant part of all people's sexual identity.

Sometimes there is a public claim that scientists have discovered a so-called "gay gene," which presumably directs the sexual orientation of gays and lesbians. Interestingly, there are never claims about a so-called "heterosexual gene" because the implicit assumption seems to be that heterosexuality is the natural state and gay or lesbian behavior is somewhat a mutant form. There is no such scientific evidence. The evidence usually cited to claim a genetic basis for homosexuality was based on a problematic study of gay brothers (identical twins) who were found to have similar DNA markers on one of their X chromosomes. On closer examination, this research was refuted when it was found that the shared DNA markers found in pairs of gay brothers were no more likely than would be expected by chance (Wickelgren 1999). Moreover, the original study did not control for the environment in which the brothers were raised. They grew up in the same family, so obviously they were raised in the same environment. A true scientific test of the hypothesis that homosexuality is genetically based would require a stricter standard of evidence. One could study identical twins raised together who were the offspring of gay parents to see if, controlling for the environments in which they were raised, both turned out to be gay. To date, there are no such studies of biological relatives raised apart (Hamer et al. 1993).

Even if there is some yet undiscovered basis for sexual orientation, there is extensive evidence of the social influences that shape people's sexual identities. Social and cultural environments play a huge part in creating sexual identities. This is what interests sociologists—how sexual identity is constructed through social relationships and in the context of social institutions.

The Social Basis of Sexuality

We can see the social and cultural basis of sexuality in numerous ways:

1. **Human sexual attitudes and behavior vary in different cultural contexts.** If sex were purely natural behavior, sexual behavior would also be uniform among all societies, but it is not. Sexual behaviors considered normal in one society might be seen as peculiar in another. Think about this: In some cultures women do not believe that orgasm exists, even though biologically it does. In the eighteenth century, European and American writers

advised men that masturbation robbed them of their physical powers and that instead they should apply their minds to the study of business. These cultural dictates encouraged men to conserve semen on the presumption that its release would lessen men's intelligence or cause insanity (Freedman and D'Emilio 1988; Schwartz and Rutter 1998).

Debunking Society's Myths

Myth: Over time in a society, sexual attitudes become more permissive.

Sociological perspective: All values and attitudes develop in specific social contexts; change is not always in a more permissive direction (Freedman and D'Emilio 1988).

2. **Sexual attitudes and behavior change over time.** Fluctuations in sexual attitudes are easy to document. For example, in 1977, one-third of teens though it was morally wrong to have sex outside of marriage; now 42 percent think premarital sex is wrong, suggesting that teen tolerance of casual sex has declined in recent years (Lyons 2004, 2002). But what teens say and what they do are not always the same. Whatever teens say, young people are having sex at an early age (see Figure 12.1), although research shows that, compared to 1995, teens (both men and women) are now less likely to have had sex before age 15 than was true in 1995. Yet, by age 17, seven in ten teens have had sexual intercourse. Perhaps surprisingly, the number of young men who have had sex before age 15 has declined more than has the number of women having sex before age 15 (Guttmacher Institute 2006b). Looking at all age groups, more people also have sex before marriage than in the past, and people have more sex partners over their lifetimes (Laumann et al. 1994).

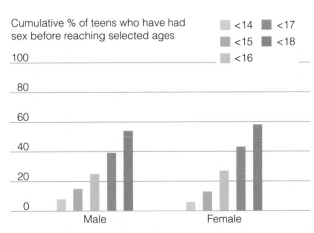

FIGURE 12.1 *Sex Among Teenagers*

Source: Guttmacher Institute. 2006b. "Facts on American Teens' Sexual and Reproductive Health." New York: Guttmacher Institute. **www.guttmacher.org**

3. **Sexual identity is learned.** Like other forms of social identity, sexual identity is acquired through socialization and ongoing relationships. Information about sexuality is transmitted culturally and becomes the basis for what we know of ourselves and others. Where did you first learn about sex? For that matter, what did you learn? For some, parents are the source of information about sex and sexual behavior. For many, peers have the strongest influence on sexual attitudes. Long before young people become sexually active, they learn **sexual scripts** that teach us what is appropriate sexual behavior for each gender (Schwartz and Rutter 1998). Children learn sexual scripts by playing roles—playing doctor as a way of exploring their bodies or hugging and kissing in a way that can mimic heterosexual relationships. Role playing teaches children social norms about sexuality. The roles learned in youth profoundly influence our sexual attitudes and behavior throughout life.

SEE *FOR YOURSELF*

CHILDREN'S MEDIA AND SEXUALITY

Watch two movies that are marketed to a specific population of young people (either children, teens, or young adults). Watch the movie very carefully and make a written list of any comments or behaviors you see that suggest *sexual scripts* for men and women. Make note of the assumptions in this script about heterosexual and homosexual behavior. What scripts did you see, and who is the intended audience? What do your observations suggest about how sexual scripts are promoted, overtly or not, through popular culture?

4. **Social institutions channel and direct human sexuality.** Social institutions, such as religion, education, or the family, define some forms of sexual expression as more legitimate than others. Debates about same-sex marriage illustrate this concept. Without being able to marry, gay and lesbian couples lose some of the privileges that married couples receive, such as employee benefits and the option to file joint tax returns. Social institutions, such as the media, also influence sexuality through the production and distribution of images that define cultural meanings attributed to sexuality.

5. **Sex is influenced by economic forces in society.** Sex sells. In the U.S. capitalist economy, sex appeal is used to hawk everything from cars and personal care products to stocks and bonds. In this sense, sex has become a commodity—something bought and sold in the marketplace of society. Sex also can be big business. By one estimate, Americans spend over $10 billion per year in the sex industry, including strip bars, peep shows, phone sex, sex acts, sex magazines, and pornography

rentals—and this is likely to be an underestimate (Barton 2006). Moreover, the business of sex comes as some people may actually be "bought and sold." Think of women who may feel that selling sexual services is their best option for earning a living wage. Sex workers are among some of the most exploited and misunderstood workers. And, as we will see later in this chapter, sex trafficking on a global basis is also more extensive than commonly thought.

6. **Public policies regulate sexual and reproductive behaviors.** In many ways, the government and other social policies intervene in people's sexual and reproductive decision making. Prohibiting federal spending on abortion, for example, eliminates reproductive choices for women who are dependent on state or federal aid. Government decisions about which reproductive technologies to endorse influence the choices of birth control technology available to men and to women. Government funding, or lack thereof, for sex education can influence how people understand sexual behavior. These facts challenge the idea that sexuality is a private matter and shows how social institutions can direct sexual behavior.

To summarize, human sexual behavior occurs within a cultural and social context. The culture defines certain sexual behaviors as appropriate or inappropriate. Like other forms of social identity, sexual identity is learned.

Contemporary Sexual Attitudes and Behavior

Sexual attitudes and behaviors in the United States are a mix of ideas and practices, both of which vary depending on different social factors that shape people's experiences. While on the one hand many support diverse sexual lifestyles and practices, the growth of a conservative movement has also shaped the sexual values and behaviors of others. These differences have also been at the heart of highly contested issues that have shaped U.S. politics in recent years, such as debates about same-sex marriage, sex education in the schools, and policies about reproductive and contraceptive health. This makes sociological research on sexuality all the more fascinating as public attitudes and behaviors shift with changes in the society itself.

Changing Sexual Values

Public opinion is now a mix of both liberal and conservative values about sexuality. Almost two-thirds of Americans (61 percent) think that sex before marriage is morally acceptable, but among Americans for whom religion is very important, only 38 percent think this is morally acceptable—but note that one-third still

a sociological eye on the media

Publicity Traps: Sex on TV Talk Shows

If you even casually peruse television talk shows, you are likely to see a wide array of sexual nonconformists—people who may be bisexual, transgendered, or in some way sexually nonconforming. Joshua Gamson, a sociologist who specializes in the study of media, sexuality, and social change, has studied television talk shows and used them to think about the cultural visibility of nonconforming sexual groups. Lesbians, gays, transgendered people, bisexuals, and others are usually marginalized in society—treated as invisible, seldom recognized, misunderstood, and usually suppressed in public space. But on television talk shows, the public can witness outrageous, boisterous, even wild guests. In fact, in an attempt to build larger audiences (because, after all, this is about profit), television talk shows have become less likely to feature calm, well-educated, distinguished guests. Instead, they often feature those who least conform to the dominant sexual value system. Gamson has asked, "What is the impact of the visibility of sexual nonconformists for gays and lesbians?"

Gamson sees no simple answer to this question. On the one hand, showcasing sexual nonconformity in a way that may seem freakish, foul-mouthed, and "abnormal" presents a distorted image of gay life and gives legitimacy to those who think that lesbian, gay, bisexual, transgender, and other diverse sexual life-styles are deviant. The presentation of people on "the fringe," as Gamson puts it, "makes social acceptability harder to gain by overemphasizing difference, often presented as frightening, pathological, pathetic, or silly" (1998b: 33).

But Gamson also sees another dimension to this question. He writes that although these images present a distorted image of gay life, the talk shows also make diverse sexual identities public, thereby having the added effect of making sexual nonconformity less shocking and, thus, in the long run, more acceptable. In addition, Gamson argues, these portrayals open up public space, where challenges to sexual conformity can transform the ordinarily fixed boundaries between gay and straight, normal and "queer."

The dilemma comes from the fact that these portrayals are generally not made by or on behalf of gay people. Yet becoming more visible in a public space such as the media can change people's understandings of who has a right to share public space.

Source: Gamson, Joshua. 1998. "Publicity Traps: Television Talk Shows and Lesbian, Gay, Bisexual, and Transgender Visibility." *Sexualities* 1: 11–41.

think so. But, if you put this question before an international audience, you might be surprised to find that Americans are more conservative on this matter than people in western Europe. In Germany, France, and the United Kingdom, a very large majority of people—in Germany (86 percent), France (88 percent), and the United Kingdom (80 percent)—say that they find this morally acceptable (Rheault and Mogahed 2008).

On gay rights, more Americans than ever before think that "homosexual relations between consenting adults should be legal" (55 percent compared to 43 percent in 1978). Eighty-one percent think "homosexuals should have equal rights in terms of job opportunities." Forty percent of Americans think marriage between same-sex couples should be recognized as legal, compared to only 27 percent who thought so in 1996 (Gallup Poll 2009; see also Figure 12.2). The public is clearly divided on the issue of same-sex marriage, and it is unclear what direction the law will take in the long run. In 2003, the U.S. Supreme Court ruled (*Lawrence v. Texas*) that private sexual relations are a constitutional liberty, a conclusion widely interpreted as a major victory for gay rights. The Massachusetts Supreme Court ruled in 2003 that same-sex marriages were constitutional in that state. Since then, other states have legally allowed same-sex marriage, although there have been various legal challenges to such rulings, making the legal future of same-sex marriage uncertain.

thinking sociologically

Keep a diary for one week and write down as many examples of *homophobia* and *heterosexism* as you observe in routine social behavior. What do your observations tell you about how *heterosexuality* is enforced?

Attitudes about sex vary significantly depending on various social characteristics. For example, men

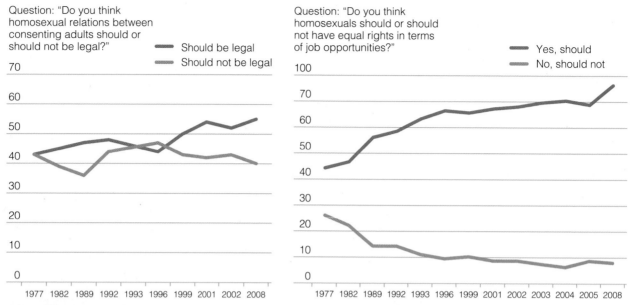

Question: "Do you think homosexual relations between consenting adults should or should not be legal?"

- Should be legal
- Should not be legal

Question: "Do you think homosexuals should or should not have equal rights in terms of job opportunities?"

- Yes, should
- No, should not

FIGURE 12.2 *Support for Gay Rights*

Data: Gallup Poll. 2009. "Homosexual Relations." Princeton, NJ: Gallup Poll. **www.gallup.com**

are more likely than women to think that homosexual relations are morally wrong. Sexual attitudes are also shaped by age. Younger people are more likely than older people to think that homosexuality is acceptable. These differences likely reflect not only the influence of age, but also historical influences on different generations. Religion also matters. Those who attend church weekly are far less likely to support gay rights compared to those who worship less often (Pelham and Crabtree 2009). Public opinion on matters about sexuality tap underlying value systems, thus generating public conflicts. In general, sexual liberalism is associated with greater education, youth, urban lifestyle, and political liberalism on other social issues.

Sexual Practices of the U.S. Public

Sexual practices are difficult to document. What we know about sexual behavior is typically drawn from surveys. Most of these surveys ask about sexual attitudes, not actual behavior. What people say they do may differ significantly from what they actually do.

As much as sex is in the news, national surveys of sexual practices are rare. Those that have been conducted tell us the following:

- Contrary to public opinion, teens are waiting longer to have sex than was true in the past, though by age 19 seven in ten teens have had sex (Guttmacher Institute 2006b).

- Having only one sex partner in one's lifetime is rare (Laumann et al. 1994).

- A significant number of people have extramarital sex (Laumann et al. 1994).

- A significant number of people are lesbian or gay (Laumann et al. 1994; Janus and Janus 1993).

- For those who are sexually active, sex is relatively frequent. Almost two-thirds of Americans report having sex a few times per month or two to three times per week (Laumann et al. 1994).

These facts, however, do not tell us much about the underlying social factors that produce sexual behavior. For that, we turn to sociological theory.

Sexuality and Sociological Theory

Should sex be restricted to marriage? Should prostitution be legal? How are sexual identities formed? These and other questions about sexuality are the subjects of sociological study. Sociological theory puts an analytical framework around the study of sexuality, examining its connection to social institutions and current social issues. How do the major sociological theories frame an understanding of sexuality?

Sex: Functional or Conflict-Based?

The three major sociological frameworks—functionalist theory, conflict theory, and symbolic interaction—take divergent paths in interpreting the social basis of human sexuality (see Table 12.1).

Functionalist theory, with its emphasis on the interrelatedness of different parts of society, tends to depict sexuality in terms of its contribution to the stability of social institutions. Norms that restrict sex to marriage encourage the formation of families. Similarly, beliefs that give legitimacy to heterosexual behavior but not homosexual behavior maintain a

>*doing* *sociological research*

Teens and Sex: Are Young People Becoming More Sexually Conservative?

Research Question: By the late 1990s, several national studies reported a decline in sexual activity among teens. The percentage of sexually active teens has dropped from the early 1990s, rates of teen pregnancy have fallen, teens are having fewer abortions, and the rate of sexually transmitted diseases among teens has declined. Does this herald a growth in sexual conservatism among young people and the success of policies encouraging sexual abstinence?

Research Method: Sociologists Barbara Risman and Pepper Schwartz based their research on a synthesis of all of the national studies on teen sexuality, as well as data from research organizations on the prevalence of teen sexuality.

Research Findings: Most of the change in teen sex activity is attributable to changes in behavior of boys, not girls. The number of high school boys who are virgins has increased. Girls' behavior has not changed significantly, except among African American girls, whose rates of sexual activity have declined, nearly matching those of White and Hispanic girls. Risman and Schwartz conclude that sexual behavior of boys is then becoming more like girls, the implication being that boys and girls are likely to begin their sexual lives within the context of romantic relationships.

Conclusions and Implications: While many declare that the changes in teen sexual behavior means a decline in the sexual revolution, Risman and Schwartz disagree. Certainly fear of AIDS, education about safe sex, and some growth in conservative values have contributed to changes in teen sexual norms. Risman and Schwartz show that numerous factors influence sexual behavior among teens, just as among adults. They suggest that sexuality is a normal part of adolescent social development and conclude that the sexual revolution—along with the revolution in gender norms—is generating more responsible, not more problematic, sexual behavior among young people.

Questions to Consider

1. Are people in your age group generally sexually conservative or sexually liberal? What factors influence young people's attitudes about sexuality?

2. Following from question 1, what evidence would you need in order to find out if young people in your community are more liberal than young people in the past? How would you design a study to investigate this question?

Source: Risman, Barbara, and Pepper Schwartz. 2002. "After the Sexual Revolution: Gender Politics in Teen Dating." *Contexts* 1 (Spring): 16–24.

---

particular form of social organization in which gender roles are easily differentiated and the nuclear family is defined as the dominant social norm. From this point of view, regulating sexual behavior is functional for society because it prevents the instability and conflict that more liberal sexual attitudes supposedly generate. Functionalists would also explain the call for a return to "family values" as producing the uniformity in values necessary for social order.

Conflict theorists see sexuality as part of the power relations and economic inequality in society. Power is the ability of one person or group to influence the behavior of another. Power relations in society influence the power that some sexual groups have over others and influence power within sexual relationships (discussed later in the chapter in "Sexual Politics"). Conflict theorists argue that sexual relations are linked to other forms of stratification, namely, race, class, and gender inequality. According

to this perspective, sexual violence such as rape or sexual harassment is the result of power imbalances, specifically between women and men.

At the same time, because conflict theorists see economic inequality as a major basis for social conflict, they tie the study of sexuality to economic institutions. They link the international sex trade to poverty, the status of women in society, and the economics of international development and tourism (Enloe 2001; Altman 2001). Still, conflict theorists do not see all sexual relations as oppressive. Sexuality is an expression of great social intimacy. In connecting sexuality and inequality, conflict theorists are developing a structural analysis of sexuality, not condemning sexual intimacy.

Because both functionalism and conflict theory are macrosociological theories (that is, they take a broad view of society, seeing sexuality in terms of the overall social organization of society), they do not

table 12.1　Theoretical Perspectives on Sexuality

Interprets:	Functionalism	Conflict Theory	Symbolic Interaction
Sexual norms	Sexual norms are functional for society because they encourage the formation of stable institutions.	Sexual norms are frequently contested by those who are subordinated by dominant groups.	Sexual norms emerge through social interaction and the construction of beliefs.
Sexual identity	Sexual identity is learned in the family; deviant identities contribute to social disorder.	Sexual identity is regulated by individuals and institutions that enforce compulsory heterosexuality.	Sexual identity is socially constructed as people learn the sexual scripts created in society. As queer theory suggests, multiple forms of sexual identity are possible and can be seen in how people cross the ordinarily assumed boundaries.
Changing sexual values	Regulating sexual values and norms is important for maintaining traditional and social stability; too much change results in social disorganization.	Social change comes through the activism of people who challenge dominant belief systems and practices.	Change in sexual value systems evolves as people construct new beliefs and practices over time. Also, sexual values can be changed through disrupting taken-for-granted categories of the dominant culture.

tell us much about the social construction of sexual identities. This is where the sociological framework of symbolic interaction is valuable.

Symbolic Interaction and the Social Construction of Sexual Identity

Symbolic interaction theory uses a **social construction perspective** to interpret sexual identity as learned, not inborn. To symbolic interactionists, culture and society shape sexual experiences. Patterns of social approval and social taboos make some forms of sexuality permissible and others not (Connell 1992; Lorber 1994).

The social construction of sexual identity is revealed by the coming out process. **Coming out**—the process of defining oneself as gay or lesbian—is a series of events and redefinitions in which a person comes to see herself or himself as having a gay identity. In coming out, a person consciously labels that identity either to himself or herself or to others (or both). This is usually not the result of a single homosexual experience. If it were, there would be far more self-identified homosexuals, because researchers find that a substantial portion of both men and women have some form of homosexual experience at some time in their lives.

The development of sexual identity is not necessarily a linear or unidirectional process, with persons moving predictably through a defined sequence of steps or phases. Although they may experience certain milestones in their identity development, some people experience periods of ambivalence about their identity and may switch back and forth between lesbian, heterosexual, and bisexual identity over time (Rust 1995, 1993). Some people may engage in lesbian or gay behavior but not adopt an identity as lesbian or gay. Certainly, many gays and lesbians never adopt a public definition of themselves as gay or lesbian, instead remaining "closeted" for long periods, if not for their entire lifetime.

One's sexual identity may also change. For example, a person who has always thought of himself or herself as a heterosexual may conclude at a later time that he or she is gay or lesbian. In more unusual cases, people may undergo a sex change operation, perhaps changing their sexual identity in the process. In the case of bisexuals, a person might adopt a dual sexual identity.

Although most people learn stable sexual identities, over the course of one's life, sexual identity evolves. Change is, in fact, a normal outcome of the process of identity formation. Changing social contexts (including dominant group attitudes, laws, and systems of social control), relationships with others, political movements, and even changes in the language used to describe different sexual identities all affect people's self-definition.

Queer theory is a perspective that has evolved from recognizing the socially constructed nature of sexual identity. Instead of seeing heterosexual or homosexual attraction as fixed in biology, queer theory interprets society as forcing these sexual boundaries, or dichotomies, on people. By challenging the "either/or" thinking that one is either gay or straight, queer theory challenges the idea that only one form of sexuality is normal and all other forms are deviant or wrong. As a result, queer theory has opened up fascinating new studies of gay, straight, bisexual, and transsexual identities and introduced the idea that sexual identity is a continuum of different possibilities for sexual expression and personal identity (Nardi and Schneider 1997; Seidman 2003; Stein and Plummer 1994).

Sexual Politics

Sexual politics refers to the link between sexuality and power, not just within individual relationships. The feminist movement first linked sexuality to the status of women in society, pointing to the possible exploitation of women within sexual relationships. Sexual politics also refers to the high rates of violence against women and sexual minorities and the privilege and power accorded to those presumed to be heterosexual.

The feminist and gay and lesbian liberation movements have put sexual politics at the center of the public's attention by challenging gender role stereotyping and sexual oppression (D'Emilio 1998). Among other things, this has profoundly changed public knowledge of gay and lesbian sexuality. Gay, lesbian, and feminist scholars have argued, and many now concur, that homosexuality is not the result of psychological deviance or personal maladjustment but is one of several alternatives for happy and intimate social relationships. The political mobilization of many lesbian women and gay men and the willingness of many to make their sexual identity public have also raised public awareness of the civil and personal rights of gays and lesbians. These changes make other changes in intimate relations possible.

The Influence of Race, Class, and Gender

Sexual behavior follows gendered patterns, as well as patterns established by race and class relations (Schwartz and Rutter 1998). Gender expectations emphasize passivity for women and assertiveness for men in sexual encounters. The "double standard" is the idea that different standards for sexual behavior

Gays, lesbians, and their allies have mobilized for social change, fostering pride and celebration as well as a reduction over time in homophobic attitudes.

Queer theory has also linked the study of sexuality to the study of gender, showing how transgressing (or violating) fixed gender categories can reconstruct the possibilities of how all people—men and women, gay, bisexual, or straight—construct their gender and sexual identity. Transgressing gender categories, such as by drag queens or cross-dressers, can show how sex and gender categories are usually constructed in dichotomous categories (that is, opposite or binary types). By violating these constructions, people are liberated from the social constraints that presumably fixed categories of identity create. Thus, queer theory emphasizes how performance and play with gender categories can be a political tool for deconstructing fixed sex and gender identities (Rupp and Taylor 2003).

Sexuality is, of course, also constructed in the context of social institutions. Social institutions tend to define heterosexuality as the only legitimate form of sexual identity and enforce it through social norms and sanctions, including peer pressure, socialization, law and other social policies, and, at the extreme, violence (Rich 1980).

Sex, Diversity, and Inequality

Patterns of sexuality reflect the social organization of society. When you understand this, you also see that sex is related to other social factors—such as race, class, and gender—and you see how sexuality is connected to social institutions and social change.

Having celebrities "come out" about being lesbian or gay has empowered others to be able to do so as well.

Classical Theorists Reflect on Sex and Popular Culture

Imagine theorists from another place in time observing sexuality in contemporary society. Perhaps they would be shocked, considering the norms of their time, but they would certainly use their sociological imagination. What would they say now?

Emile Durkheim: Remember that one of my ideas is that marking some behaviors as deviant is a way to identify what is considered "normal." Take the common phenomenon of boys, especially young boys, calling each other "faggots." Doing so is not so much saying the targeted person is gay. You can understand this is a way of asserting the dominant (and somewhat narrow) norms of what masculinity is presumed to be.

Max Weber: What I see is a popular culture where sexuality is rampant. Clothing styles, popular lingo, and styles of dance are all marked by overt displays of sexuality. I think these cultural displays go hand-in-hand with an economy that treats sexuality as part of the marketplace. That interplay between cultural ethics and the economic marketplace leads to social judgments about some sexual styles and identities being more highly valued than others.

Karl Marx: To me, it's all about exploitation, or should I say "sexploitation." Sexuality has become a commodity in modern society. It is used to sell things for the benefit of those who own the various industries where fashion, personal care products, and even sex itself, such as in the "sex business," profit some while exploiting others. My gosh, think of sex workers; they are among the most exploited, actually selling their bodies for the benefit of others.

W.E.B. Du Bois: I agree with Marx here, but add that exploitation is particularly harsh for Black people. This whole popular idea of so-called pimps and hos defines Black men as sexual predators and Black women as promiscuous. These are stereotypes that have a long history stemming from racism. And Black women are among the most sexually exploited. Their bodies are abused, they are portrayed in sexualized stereotypes in popular culture, and they are those who are most victimized by sexual violence. I think that Black women are beautiful and have had to endure a lot, so this all makes me very angry that they are used as sexualized symbols for the pleasure of others.

apply to men and women. Although this idea is weakening somewhat, men are still stereotyped as sexually overactive; women, less so. Women who openly violate this cultural double standard by being openly sexual are then cast in a negative light as "loose," as if the appropriate role for women is the opposite of "loose," say, "secured" or "caged." The double standard forces women into polarized roles as "good" girls or "bad" girls. The belief that women who are raped must have somehow brought it on themselves rests on the images of women as temptresses. Contrary to popular belief, men do not have a stronger sex drive than women. Men are, however, socialized more often to see sex in terms of performance and achievement, whereas women are more likely socialized to associate sex with intimacy and affection.

This double standard also applies in what is now known as the "hooking up" culture among young people. *Hooking up* is the term used to describe casual sexual relations ranging from kissing to sexual intercourse, relationships that occur without any particular commitment. Although the popular image of hooking up is that it is totally free and without constraint, in

fact there are very gendered norms within this behavior. Women who "hook up" too frequently or with too many different partners are likely to be judged "slutty." Men who do the same things are not judged in the same way. They may be seen as "players," but there is not the same shame or attribution of guilt that is targeted at women.

thinkingsociologically

A recent trend among young people on college campuses has been the creation of theme parties called "CEOs and corporate hos." Have such parties occurred on your campus? Why do you think they are so popular? What *sexual scripts* are being played out by such role playing, and how do these scripts involve gender, race, and class? You might also think about how such parties portray women as sexual objects.

Sexual politics are also integrally tied to race and class inequality in society. You can see this in the sexual stereotypes associated with race and class. Latinas are stereotyped as either "hot" or "virgins"; Latin men are stereotyped as "hot lovers"; African American men are stereotyped as overly virile; Asian American women are stereotyped as compliant and submissive, but passionate. Class relations also produce sexual stereotypes of women and men. Working-class and poor men may be stereotyped as dangerous, whereas working-class women may be disproportionately labeled "sluts."

You could see sexual politics play out during the public outcry when a young, unmarried woman on public assistance delivered octuplets, following in vitro fertilization. The young woman, Nadya Suleman (dubbed "Octomom" by the media), was already the mother of six children. She was completely scorned in the media for having poor judgment as a mother. At the very same time, a reality TV show, *Jon and Kate Plus Eight,* depicted a family with eight children as fun and lovable, not despicable and undeserving as was the case with Nadya Suleman. You can understand this seeming contradiction in a frame of sexual politics. Nadya Suleman was unmarried, an Iraqi immigrant, and also on public assistance—thus judged more harshly than the heterosexual, middle-class Gosselin family in *Jon and Kate Plus Eight.*

This example illustrates a point often made by sociologists: class, race, and gender hierarchies historically depict people of color and certain women as sexually promiscuous and uncontrollable (Nagel 2003). During slavery, for example, the sexual abuse of African American women was one way that slave owners expressed their ownership of African American people. Access to women slaves' sexuality was seen as a right of the slave owner. Under slavery, racist and sexist images of Black men and women were developed to justify the system of slavery. Black men were stereotyped as lustful beasts whose sexuality had to be controlled by the "superior" Whites. Black women were also depicted as sexual animals who were openly available to White men.

A Black man falsely accused of having had sex with a White woman could be murdered (that is, lynched) without penalty to his killers (Genovese 1972; Jordan 1968). Sexual abuse was also part of the White conquest of American Indians. Historical accounts show that the rape of Indian women by White conquerors was common (Freedman and D'Emilio 1988; Tuan 1984). These patterns also can be seen in the extensive rape of women that often accompanies war and military conquest. Most recently, this has been witnessed in Darfur, where the rape of women is used, not just as a horrid act of aggression against women, but also as a way to humiliate and control the Darfur people. Similar violence is occurring in the Democratic Republic of Congo where a civil war rages and thousands of women are raped each year as the result of the civil war.

In other contexts, poor women and women of color are the groups most vulnerable to sexual violence and exploitation. Becoming a prostitute, or otherwise working in the sex industry (as a topless dancer, striptease artist, pornographic actress, or other sex-based occupation), is often the last resort for women with limited options to support themselves. Women who sell sex also are condemned for their behavior more so than their male clients—further illustration of how gender stereotypes mix with race and class exploitation. Why, for example, are women, and not their male clients, arrested for prostitution? Although the Uniform Crime Statistics data do not report on the arrest rates of customers, prostitutes themselves claim that only about 10 percent of those arrested for prostitution are customers and that women of color are more likely to be arrested for prostitution than are White women, even though they are a smaller percentage of all prostitutes (Prostitutes Education Network 2009). Although these are not scientific data, they suggest the role that gender and race play in how laws against prostitution are enforced.

A Global Perspective on Sexuality

Cross-cultural studies of sexuality show that sexual norms, like other social norms, develop differently across cultures. Take sexual jealousy. Perhaps you think that seeing your sexual partner becoming sexually involved with another person would naturally evoke jealousy, no matter where it happened. Researchers have found this not to be true. In a study comparing patterns of sexual jealousy in seven different nations (Hungary, Ireland, Mexico, the Netherlands, the United States, Russia, and the former Yugoslavia), researchers found significant cross-national differences in the degree of jealousy when women and men saw their partners kissing, flirting, or being sexually involved with another person (Buunk and Hupka 1987).

Likewise, tolerance for gay and lesbian relationships varies significantly in different societies around the world. Germany has recently legalized gay and lesbian relationships, allowing them to register same-sex partnerships and have the same inheritance rights as heterosexual couples. The new law does not, however, give them the same tax advantages, nor can same-sex couples adopt children. Cross-cultural studies can make someone more sensitive to the varying cultural norms and expectations that apply to sexuality in different contexts. Different cultures simply view sexuality differently. In Islamic culture, for example, women and men are viewed as equally sexual, although women's sexuality is seen as potentially disruptive and needing regulation (Mernissi 1987).

Sex is also big business, and it is deeply tied to the world economic order. As the world has become more globally connected, an international sex trade has flourished—one that is linked to economic development, world poverty, tourism, and the subordinate status of women in many nations.

The *international sex trade,* sometimes also referred to as the "traffic in women" (Rubin 1975),

Sam Yeh/AFP Photo/Getty Images

The international trafficking of women for sex exploits women—and often children—and puts them at risk for disease and violence.

refers to the use of women worldwide as sex workers in an institutional context in which sex itself is a commodity. Sex is marketed in an international marketplace, and women as sex workers are used to promote tourism, cater to business and military men, and support a huge industry of nightclubs, massage parlors, and teahouses (Enloe 2001). **Sex trafficking** is a relatively new term, in which women, usually very young women, are forced by fraud or coercion into commercial sex acts. Sometimes identified as a form of slavery, sex trafficking can involve a system of debt and bondage, where young women (typically under age 18) are obligated to provide sexual services in exchange for alleged debt for the price of their housing, food, or other living expenses. Sometimes these young women are actually kidnapped; other times, they may simply be duped, initially lured into sex work by promises of marriage, large incomes, or the glamour of travel, but soon trapped in a cycle of debt and/or actual captivity (U.S. Department of Health and Human Services 2009).

Related to sex trafficking is **sexual tourism,** referring to the practice whereby people travel to particular parts of the world specifically to engage in commercial sexual activity. "Sex capitals" are places where prostitution openly flourishes, such as in Thailand and Amsterdam. Sex is an integral part of the world tourism industry. In Thailand, for example, men as tourists outnumber women by a ratio of three to one and, although certainly not all of them—or even a majority—are going to Thailand to explore sex tourism, there are men who go to Thailand, as well as other destinations, explicitly to buy sexual companionship. For example, hostess clubs in Tokyo cater to

corporate men. Although these clubs are not houses of prostitution, scholars who have studied them say that they are based on an environment in which women's sexuality is used as the basis for camaraderie among men (Allison 1994). Sex tourism is such a profitable enterprise that the International Labour Organization estimates that somewhere between 2 and 14 percent of the gross domestic product in Thailand, Indonesia, Malaysia, and the Philippines is derived from sex tourism. Much of this lucrative business involves the exploitation of children (U.S. Department of Justice 2009).

Sex tourism and sexual trafficking are now part of the global economy, contributing to the economic development of many nations and supported by the economic dominance of certain other nations. As with other businesses, the products of the sex industry may be produced in one region and distributed in others. (Think, for example, of the pornographic film industry centered in southern California, but distributed globally). The sex trade is also associated with world poverty; sociologists have found that the weaker the local economy, the more important the sex trade. The international sex trade is also implicated in problems such as the spread of AIDS worldwide, as well as the exploitation of women where women have limited economic opportunities (Altman 2001).

Understanding Gay and Lesbian Experience

Sociological understanding of sexual identity has developed largely through new studies of lesbian and gay experience. Long thought of only in terms of social deviance (see Chapter 7), gays and lesbians have been stereotyped in traditional social science. But the feminist and gay liberation movements have discouraged this approach, arguing that gay and lesbian experience is part of the broad spectrum of human sexuality.

The institutional context for sexuality within the United States, as well as other societies, is one in which homophobia permeates the culture. **Homophobia** is the fear and hatred of homosexuality. It is deeply embedded in people's definitions of themselves as men and women; it is manifested in prejudiced attitudes toward gays and lesbians, as well as overt hostility and violence against people suspected of being gay. Homophobia is a learned attitude, as are other forms of negative social judgments about particular groups.

Boys are often raised to be manly by repressing so-called feminine characteristics in themselves. Being called a "fag" or a "sissy" is one peer sanction that socializes a child to conform to particular gender roles. Similarly, verbal attacks on lesbians by using the term *butch* are a mechanism of social control because ridicule can be interpreted as encouraging social conformity to the presumed "normal" gender roles (Pascoe 2007). Homophobia produces many misunderstandings about gay people, such as the misunderstanding that

[understanding diversity]

Gay-Bashing and Gender Betrayal

Carmen Vázquez has related gay-bashing to the phenomenon she labels *gender betrayal*. She explains this by the following:

At the simplest level, looking or behaving like the stereotypical gay man or lesbian is reason enough to provoke a homophobic assault. Beneath the veneer of the effeminate gay male or the butch dyke, however, is a more basic trigger for homophobic violence. I call it gender betrayal.

The clearest expression I have heard of this sense of gender betrayal comes from Doug Barr, who was acquitted of murder in an incident of gay bashing in San Francisco that resulted in the death of John O'Connell, a gay man. Barr is currently serving a prison sentence for related assaults on the same night that O'Connell was killed. . . . When asked what he and his friends thought of gay men, he said, "We hate homosexuals. They degrade our manhood. We was brought up in a high school where guys are football players, mean and macho. Homosexuals are sissies who wear dresses. I'd rather be seen as a football player."

Doug Barr's perspective is one shared by many young men. I have made about 300 presentations to high school students in San Francisco, to boards of directors and staff of nonprofit organizations, and at conferences and workshops
on the topic of homophobia or "being lesbian and gay." Over and over again, I have asked, "Why do gay men and lesbians bother you?" The most popular response to the question is, "Because they act like girls," or "Because they think they're men." I have even been told, quite explicitly, "I don't care what they do in bed, but they shouldn't act like that."

They shouldn't act like that. Women who are not identified by the relationship to a man, who value their female friendship, who like and are knowledgeable about sports, or work as blue-collar laborers and wear what they wish are very likely to be "lesbian-baited" at some point in their lives. Men who are not pursuing sexual conquests of women at every available opportunity, who disdain sports, who choose to stay at home and be a house-husband, who are employed as hairdressers, designers, or housecleaners, or who dress in any way remotely resembling traditional female attire (an earring will do) are very likely to experience the taunts and sometimes the brutality of "fag bashing."

The straitjacket of gender roles suffocates many lesbians, gay men, and bisexuals, forcing them into closets without an exit and threatening our very existence when we tear the closet open.

Source: Vazquez, Carmen. 1992. "Appearances." (400 words) from Pp. 157–166 in *Homophobia: How We All Pay the Price*, edited by Warren J. Blumenfeld. Reprinted by permission of Beacon Press, Boston.

gays have a desire to seduce straight people. There is little evidence that this is true.

Fears that gay and lesbian parents will have negative effects on their children are also unsubstantiated by research. The ability of parents to form good relationships with their children is far more significant in children's social development than is their parents' sexual orientation (Stacey and Biblarz 2001).

Other myths about gay people are that they are mostly White men with large discretionary incomes who work primarily in artistic areas and personal service jobs (such as hairdressing). This stereotype prevents people from recognizing that gays and lesbians come from all racial–ethnic groups, may be working class or poor, and are employed in a wide range of occupations (Gluckman and Reed 1997). Some lesbians and gays are also elderly, though the stereotype defines gay people as primarily young or middle-aged (Smith 1983). These different misunderstandings reveal just a few of the many unfounded myths about gays and lesbians. Support for these attitudes comes from homophobia, not from empirical truth.

Heterosexism refers to the institutionalization of heterosexuality as the only socially legitimate sexual orientation. Heterosexism is rooted in the belief that heterosexual behavior is the only natural form of sexual expression and that homosexuality is a perversion of "normal" sexual identity. Heterosexism is reinforced through institutional mechanisms that project the idea that only heterosexuality is normal. Institutions also provide different benefits to people presumed to be heterosexual. Businesses and communities, for example, rarely recognize the legal rights of people in homosexual relationships, although this is changing. Within an institution, individual beliefs can reflect heterosexist assumptions. Thus, a person may be accepting of gay and lesbian people (that is, not be homophobic) but still benefit from heterosexual privileges. At the behavioral level, heterosexist practices can exclude lesbians and gays, such as when coworkers talk about dating the other sex, assuming that everyone is interested in a heterosexual partner. In the absence of institutional supports from the dominant culture, lesbians and gays have invented their own institutional support systems. Gay communities and gay rituals, such as gay pride marches, affirm gay and lesbian identities and provide a support system that is counter to the dominant heterosexual culture. Those who remain "in the closet" deny themselves this support system.

The absence of institutionalized roles for lesbians and gays affects the roles they adopt within relationships. Despite popular stereotypes, gay partners typically do not assume roles as the dominant or submissive sexual partner. They are more likely to adopt roles as equals. Gay couples and lesbian couples are also more likely than heterosexual couples to both be employed, another source of greater equality within the relationship. Researchers have also found that the quality of relationships among gay men is positively correlated with the social support the couple receives from others (Smith and Brown 1997; Metz et al. 1994; Harry 1979).

Lesbians and gays are a minority group in our society, denied equal rights and singled out for negative treatment in society. *Minority groups* are not necessarily numerical minorities; they are groups with similar characteristics (or at least perceived similar characteristics) who are treated with prejudice and discrimination (see Chapter 10). As a minority group, gays and lesbians have organized to advocate for their civil rights and to be recognized as socially legitimate citizens. Some organizations and municipalities have enacted civil rights protections on behalf of gays and lesbians, typically prohibiting discrimination in hiring. The Supreme Court ruled in 1996 (*Romer v. Evans*) that gays and lesbians cannot be denied equal protection under the law. This ruling overturned ordinances that had specifically denied civil rights protections to lesbians and gays (Greenhouse 1996). The implications of this case for domestic partner benefits, child custody and adoption by gay and lesbian parents, and gay marriage are not yet known, but will likely be determined in the courts in the years ahead.

Sex and Social Issues

In studying sexuality, sociologists tap into some highly contested social issues of the time. Birth control, reproductive technology, abortion, teen pregnancy, pornography, and sexual violence are all subjects of public concern and are important in the formation of social policy. Debates about these issues hinge in part on attitudes about sexuality and are shaped by race, class, and gender relations. These social issues can generate personal troubles that have their origins in the structure of society—recall the distinction made by C. Wright Mills between personal troubles and social issues (see Chapter 1).

Birth Control

The availability of birth control is now less debated than it was in the not too distant past, but this important reproductive technology has been strongly related to the status of women in society (Gordon 1977). Reproduction has been controlled by men; to this day, it is mostly men who define the laws and make the scientific decisions about what types of birth control will be available. Women are also most likely seen as being responsible for reproduction because it is a presumed part of their traditional role. At the same time, changes in birth control technology have also made it possible for women to change their roles in society, given that breaking the link between sex and reproduction has freed women from some traditional constraints.

The right to birth control is a recently won freedom. It was not until 1965 that the Supreme Court, in *Griswold v. Connecticut,* defined the use of birth control as a right, not a crime. This ruling originally applied only to married people; unmarried people were not extended the same right until 1972 in the Supreme Court decision *Eisenstadt v. Baird.* Today birth control is routinely available by prescription, but there is heated debate about whether access to birth control should be curtailed for the young—at a time when youths are experimenting with sex at younger ages and risking teenage pregnancy and AIDS. Some argue that increasing access to birth control will only encourage more sexual activity among the young.

Class and race relations also have had a role in shaping birth control policy. In the mid-nineteenth century, increased urbanization and industrialization ended the necessity for large families, especially in the middle class, because fewer laborers were needed to support the family. Early feminist activists such as Emma Goldman and Margaret Sanger also saw birth control as a way of freeing women from unwanted pregnancies and allowing them to work outside the home if they chose. As the birthrate fell among White upper- and middle-class families during this period,

Many gay and lesbian couples jubilantly took formal vows of marriage when Massachusetts legally recognized the right to gay unions.

Douglas McFadd/Getty Images

these classes feared that immigrants, the poor, and racial minorities would soon outnumber them.

The *eugenics* movement of the early twentieth century grew from the fear of domination by immigrant groups. **Eugenics** sought to apply scientific principles of genetic selection to "improve" the offspring of the human race. It was explicitly racist and class-based, calling for, among other things, the compulsory sterilization of those who eugenicists thought were unfit. Eugenicist arguments appeal to a public that fears the social problems that emerge from race and class inequality: Instead of attributing these problems (such as crime) to the structure of society, eugenicists blame the genetic composition of the least powerful groups in society.

In contemporary society, attitudes about the use of birth control have evolved and have also been very much shaped by greater awareness about the risks of HIV/AIDS infection. The proportion of women (aged 15–44) who use contraceptives has increased somewhat from 56 percent to 62 percent from 1995 to 2002. Among women users, the pill is the most frequent type of birth control used, followed by tubal ligation (sterilization), and then condoms. Similar data are not reported for men—itself a reflection of the belief that contraception is primarily a woman's responsibility. Among teenagers, where contraceptive use is important for preventing teen pregnancy, fewer than half of teen men who are having sex use condoms 100 percent of the time. White men are now less likely to use condoms the first time they have sex (compared to 1995), whereas Black and Hispanic men are now more likely to use condoms during their first intercourse (Guttmacher Organization 2006a, 2008).

New Reproductive Technologies

Practices such as surrogate mothering, in vitro fertilization, and new biotechnologies of gene splicing, cloning, and genetic engineering mean that reproduction is no longer inextricably linked to biological parents (Rifkin 1998). A child may be conceived through means other than sexual relations between one man and one woman. One woman may carry the child of another. Offspring may be planned through genetic engineering. A sheep can be cloned (that is, genetically duplicated). So can monkeys. Are humans next? With such developments, those who could not otherwise conceive children (infertile couples, single women, or lesbian couples) are now able to do so, thereby raising new questions. To whom are such new technologies available? Which groups are most likely to sell reproductive services, which groups to buy? What are the social implications of such changes?

There are no simple answers to such questions, but sociologists would point first to the class, race, and gender dimensions of these issues (Roberts 1997). Poor women, for example, are far more likely than middle-class or elite women to sell their eggs or offer their bodies as biological incubators. Groups that can afford new, costly methods of reproduction may do so at the expense of women whose economic need places them in the position of selling themselves for financial necessity.

Breakthroughs in reproductive technology raise especially difficult questions for makers of social policy. Developments in the technology of reproduction have ushered in new possibilities and freedoms but also raise questions for social policy. With new reproductive technologies, there is potential for a new eugenics movement. Sophisticated prenatal screenings make it possible to identify fetuses with presumed defects. Might society then try to weed out those perceived as undesirables—the disabled, certain racial groups, certain sexes? Will parents try to produce "designer children"? If boys and girls are differently valued, one sex may be more often aborted, a frequent practice in India and China—two of the most populous nations on Earth. Because of population pressures, state policy in China, for example, encourages families to have only one child. Because girls are less valued than boys, the aborting and selling of girls is common and has created a U.S. market for the adoption of Chinese baby girls.

There are no traditions to guide us on such questions. Public thinking about sexuality and reproduction will have to evolve. Although the concept of reproductive choice is important to most people, choice is conditioned by the constraints of race, class, and gender inequalities in society. Like other social phenomena, sexuality and reproduction are shaped by their social context.

Abortion

Abortion is one of the most seriously contested political issues. Of the U.S. public, 28 percent think abortion should be legal in any circumstances; 54 percent think it should be legal only under certain circumstances; and 17 percent think it should be illegal in all circumstances (see Figure 12.3). Public support for abortion has remained relatively constant over the last twenty years.

The right to abortion was first established in constitutional law by the *Roe v. Wade* decision in 1973. In *Roe v. Wade,* the Supreme Court ruled that at different points during a pregnancy, separate but legitimate rights collide—the right to privacy, the right of the state to protect maternal health, and the right of the state to protect developing life. To resolve this conflict of rights, the Supreme Court ruled that pregnancy

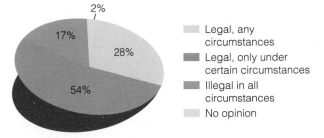

FIGURE 12.3 Attitudes Toward Abortion

Data: Gallup Poll. 2008. "Abortion." Princeton, NJ: The Gallup Organization. **www.gallup.com**

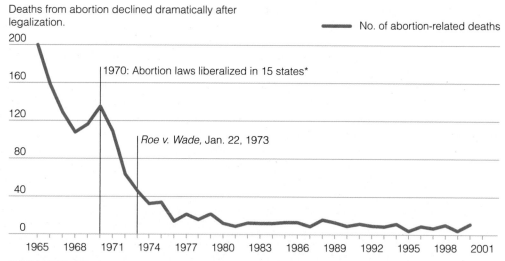

FIGURE 12.4 *Deaths from Abortion: Before and After Roe v. Wade*

*By the end of 1970, four states had repealed their antiabortion laws, and eleven states had reformed them.

Source: From Boonstra, H. D. et al., *Abortion in Women's Lives*, New York: Guttmacher Institute, 2006, Figure 2.2. Reprinted by permission of the Guttmacher Institute, **www.guttmacher.org**

occurred in trimesters. In the first, women's right to privacy without interference from the state prevails; in the second, the state's right to protect maternal health takes precedence; in the third, the state's right to protect developing life prevails. In the second trimester, the government cannot deny the right to abortion, but it can insist on reasonable standards of medical procedure. In the third, abortion may be performed only to save the life or health of the mother. More recently, the Supreme Court has allowed states to impose restrictions on abortion, but it has not, to date, overturned the legal framework of *Roe v. Wade.*

Data on abortion show that it occurs across social groups, although certain patterns do emerge. The abortion rate has declined since 1980 from a rate of 27.4 per 1000 women to 19.7 now (among women aged 15–44). And, as you can see in Figure 12.4, the number of deaths from illegal abortions plummeted in the years following the *Roe v. Wade* decision. Young women (aged 20–29) are the most likely group to get abortions, although the second most likely group is women under 20. Black women are four times more likely to have abortions than White women; poor women are four times more likely to have abortions than other women. You may be surprised to learn that the majority of women having abortions have already had at least one birth (U.S. Census Bureau 2009a; Boonstra et al. 2006).

The abortion issue provides a good illustration of how sexuality has entered the political realm. Abortion rights activists and antiabortion activists hold very different views about sexuality and the roles of women. Antiabortion activists tend to believe that giving women control over their fertility breaks up the stable relationships in traditional families. They tend to view sex as something that is sacred, and they are disturbed by changes that make sex less restrictive.

This belief has been fueled by the activism of the religious right, where strong passions against abortion have driven issues about sexual behavior directly into the political realm. Abortion rights activists, on the other hand, see women's control over reproduction as essential for women's independence. They also tend to see sex as an experience that develops intimacy and communication between people who love each other. The abortion debate can thus be interpreted as a struggle over the right to terminate a pregnancy as well as a battle over differing sexual values and a referendum on the nature of men's and women's relationships (Luker 1984).

Pornography and the Sexualization of Culture

Little social consensus has emerged about the acceptability and effects of pornography. Part of this debate is about defining what is obscene. The legal definition of obscenity is one that changes over time and in different political contexts. Public agitation over pornography has divided people into those who think it is solidly protected by the First Amendment, those who want it strictly controlled, those who think it should be banned for moral reasons, and those who think it must be banned because it harms women.

However pornography is legally defined, there is no question that it permeates contemporary culture. Once available only in more "underground" places like x-rated movie houses, pornography is now far more public than in the past. Hotel rooms have a huge array of pornographic films available on the television; pornographic spam appears regularly in people's email inboxes; casual references to pornography are made in popular shows on prime-time TV; images that once would have seemed highly pornographic are now commonly found on widely distributed magazines such as *Maxim*, *Cosmopolitan*, and others. Highly sexualized expressions and images are so widely seen throughout society that one commentator has said we are experiencing the *pornification of culture* (Levy 2005). Does this indicate that society has become more sexualized?

According to a major report from the American Psychological Association (APA), the answer is yes. The APA defines *sexualization* as including any one of the following conditions:

- People are judged based only on sexual appeal or behavior, to the exclusion of other characteristics;
- People are held to standards that equate physical attractiveness with being sexy;
- People are sexually objectified—meaning made into a "thing" for others' use;
- Sexuality is inappropriately imposed on a person (American Psychological Association 2007).

The APA report then details the specific consequences, especially for young girls, of a culture marked by sexualization. This report shows that overly sexualizing young girls harms them in psychological, physical, social, and academic ways. As examples, young girls may spend more time tending to their appearance than to their academic studies; they may engage in eating disorders to achieve an idealized, but unattainable image of beauty; or they may develop attitudes that put them at risk of sexual exploitation. Although the focus of this report is on young girls, one cannot help but wonder what effects the "pornification of culture" also has on young boys, as well as adult women and men.

Once confined to red-light districts, such as this one in Amsterdam, some argue that pornography and sexualized images have permeated even mainstream culture.

Guido Cozzi/Atlantide Phototravel/Corbis

Despite public concerns about pornography, most people believe that pornography should be protected by the constitutional guarantees of free speech and a free press. Yet people also believe that pornography dehumanizes women; women especially think so. Public controversy about pornography is not likely to go away because it taps so many different sexual values among the public.

Teen Pregnancy

Each year about 435,000 teenage girls (under age 19) have babies in the United States. The United States has the highest rate of teen pregnancy among developed nations, even though levels of teen sexual activity around the world are roughly comparable. Teen pregnancy has declined since 1990, a decline caused almost entirely from the increased use of birth control. Most teen pregnancies (82 percent) are unplanned, due largely to inconsistent use of birth control.

In recent years, teens have been encouraged to abstain from sex, causing researchers to wonder if abstinence accounts for the decline in teen pregnancy rates. Beginning in the early 1980s, the federal government encouraged abstinence policies, putting money behind the belief that encouraging chastity was the best way to reduce teen pregnancy. During the Bush administration (2000–2008), funding for abstinence programs tripled, spending $200 million annually on abstinence programs, especially in the public schools. Students were encouraged to take "virginity pledges," promising not to have sexual intercourse before marriage. Do abstinence pledges work?

In one of the most comprehensive and carefully controlled studies of abstinence, researchers compared a large sample of teen "virginity pledgers" and "nonpledgers" (*n* = 3440) who were matched on social attitudes such as religiosity and attitudes toward sex and birth control. The study compared the two groups over a five-year period. Results showed that over time there were *no differences* in the number of times they had sex, the age of first sex, and the practice of oral or anal sex. The main difference between the two groups was that pledgers were less likely to use birth control when they had sex. Also, five years after having taken an abstinence pledge, pledgers denied having done so. The researchers concluded that not only are abstinence pledges ineffective (supporting other research findings) but that taking the pledge makes pledgers less likely to protect themselves from pregnancy and disease when having sex (Rosenbaum 2009). Consistent with these findings are other studies that find that abstinence policies account for a very small portion of the decline in teen pregnancy—probably only about 10 percent of the difference (Santelli et al. 2007; Boonstra et al. 2006).

Although the rate of teen pregnancy has declined, so has the rate of marriage for teens who become pregnant. Thus, most babies born to teens will be raised by single mothers—a departure from the past when teen mothers often got married (U.S. Census Bureau 2009a). What concerns people about teen parents is that teens are more likely to be poor than other mothers, although sociologists have cautioned that this is because teen mothers are more likely poor *before* getting pregnant (Luker 1996). Teen parents are among the most vulnerable of all social groups.

Debunking Society's Myths

Myth: Providing sex education to teens only encourages them to become sexually active.

Sociological perspective: Comprehensive sex education actually delays the age of first intercourse; abstinence-only education has not been shown to be effective in delaying intercourse (Kirby 1997; Risman and Schwartz 2002).

Teenage pregnancy correlates strongly with poverty, lower educational attainment, joblessness, and health problems. Teen mothers have a greater incidence of problem pregnancies and are most likely to deliver low-birth-weight babies, a condition associated with myriad other health problems. Teen parents face chronic unemployment and are less likely to complete high school than those who delay childbearing. Many continue to live with their parents, although this is more likely among Black teens than among Whites.

Although teen mothers feel less pressure to marry now than in the past, if they raise their children alone, they suffer the economic consequences of raising children in female-headed households—the poorest of all income groups. Teen mothers report that they do not marry because they do not think the fathers are ready for marriage. Sometimes their families also counsel them against marrying

precipitously. These young women are often doubtful about men's ability to support them. They want men to be committed to them and their child, but they do not expect their hopes to be fulfilled (Farber 1990; Edin and Kefalas 2005). Research shows that low-income single mothers are distrustful of men, especially after an unplanned pregnancy. They think they will have greater control of their household if they remain unmarried. Many teen mothers also express fear of domestic violence as a reason for not marrying (Edin 2000). Both Black and White teen parents see marriage as an impediment to their career ambitions, although Black teen mothers are more likely than Whites to continue attending school (Farber 1990; Trent and Harlan 1990).

Why do so many teens become pregnant given the widespread availability of birth control? Teens typically delay the use of contraceptives until several months after they become sexually active. Teens who do not use contraceptives when they first have sex are twice as likely to get pregnant as those who use contraception (Guttmacher Institute 2006a). In recent years, the percentage of teens using birth control (especially condoms) has increased, although the pill is still the most widely used method (see Figure 12.5; Guttmacher Institute 2006a).

Sociologists have argued that the effective use of birth control requires a person to identify himself or herself as sexually active (Luker 1975). Teen sex, however, tends to be episodic. Teens who have sex on a couple of special occasions may not identify themselves as sexually active and may not feel obliged to take responsibility for birth control. Despite many teens initiating sex at an earlier age, social pressure continues to discourage them from defining themselves openly, or even privately, as sexually active.

Teen pregnancy is integrally linked to the gender expectations of men and women in society. Some teen

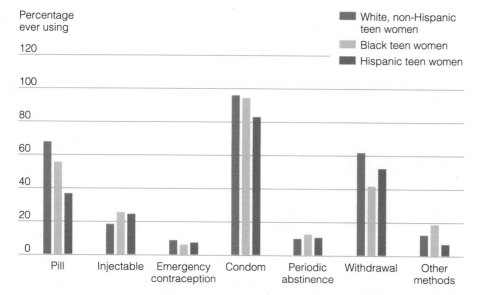

FIGURE 12.5 *Contraceptive Use Among Teen Women*

These data include teen (aged 15–19) women's contraceptive use, including only those who have had sexual intercourse. The data indicate contraceptive methods ever used, not necessarily those used regularly. What do you think explains the use of different methods by different groups? How do social factors influence the use of contraception among different groups?

Source: Abma, J.C. et al. 2002. *Teenagers in the United States: Sexual Activity, Contraceptive Use, and Childbearing, 2002.* Washington, DC: U.S. Department of Health and Human Services.

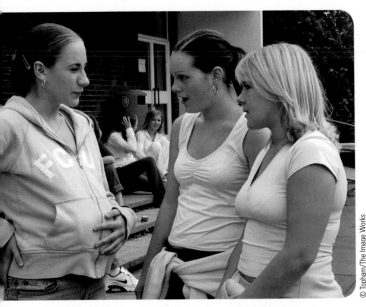

Despite public concerns about teen pregnancy, rates of teen pregnancy have actually declined in recent years.

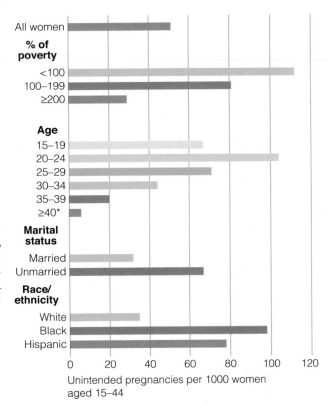

FIGURE 12.6 *Teens and Unintended Pregnancy*

*Denominator is women 40–44.

Source: From Boonstra, H. D. et al., *Abortion in Women's Lives*, New York: Guttmacher Institute, 2006, Figure 1.3. Reprinted by permission of the Guttmacher Institute, **www.guttmacher.org**

men consciously avoid birth control, thinking it takes away from their manhood. Teen women often romanticize motherhood, thinking that becoming a mother will give them social value they do not otherwise have. For teens in disadvantaged groups, motherhood confers a legitimate social identity on those otherwise devalued by society (Horowitz 1995). Although their hopes about motherhood are not realistic, they indicate how pessimistic the teenagers feel about their lives that are often marked by poverty, a lack of education, and few good job possibilities (Ladner 1986). For young women to romanticize motherhood is not surprising in a culture where motherhood is defined as a cultural ideal for women, but the ideal can seldom be realized when society gives mothers little institutional or economic support (see Figure 12.6).

Sexual Violence

Before the development of the feminist movement, sexual violence was largely hidden from public view. One great success of the women's movement has been to identify, study, and advocate better social policies to address the problems of rape, sexual harassment, domestic violence, incest, and other forms of sexual coercion. Sexual coercion is not just a matter of sexuality; it is also a form of power relations shaped by the social inequality between women and men. In the thirty years or so that these issues have been identified as serious social problems, volumes of research have been published on these different subjects, and numerous organizations and agencies have been established to serve victims of sexual abuse and to advocate reforms in social policy.

Rape and sexual violence were covered in Chapter 7 on deviance and crime, in keeping with the argument that these are forms of deviant and criminal behavior,

not expressions of human sexuality. Here we point out that various forms of sexual coercion (rape, domestic violence, and sexual harassment) can best be understood (and therefore changed) by understanding how social institutions shape human behavior and how social interactions are influenced by social factors such as gender, race, age, and class.

Take, for example, the phenomenon now known as *acquaintance rape* (sometimes also called *date rape*). Acquaintance rape is forced and unwanted sexual relations by someone who knows the victim (even if only a brief acquaintance). This kind of rape is common on college campuses, although it is also the most underreported form of rape. Researchers estimate, based on surveys, that 15 to 25 percent of college women will experience some form of acquaintance rape (Fisher et al. 2000; see also Chapter 6).

Studies show that although rape is an abuse of power, it is related to people's gender attitudes. Holding stereotypical attitudes about women is strongly related to adversarial sexual beliefs, accepting rape myths, and tolerating violence against women (American Psychological Association 2007).

Violence against women is more likely to occur in some contexts than others, especially in organizations that are set up around a definition of masculinity as competitive, where alcohol abuse occurs, and where

Public awareness of the problem of battered women has increased in recent years, although there is no evidence that battering itself has lessened.

women are defined as sexual prey. This is one explanation given for the high incidence of rape in some college fraternities (Martin and Hummer 1989; Armstrong et al. 2006; Stombler and Padavic 1997).

Research on violence against women also finds that among women, Black, Hispanic, and poor White women are more likely to be victimized by various forms of violence, including rape (U.S. Bureau of Justice Statistics 2007). African American and American Indian women and men report the highest incidence of intimate partner violence; Asian Americans and Pacific Islanders have the lowest incidence (Tjaden and Thoennes 2000). Studies also find that Black women are more aware of their vulnerability to rape than are White women and are more likely to organize themselves to resist rape collectively (Stombler and Padavic 1997).

In sum, sociological research on sexual violence shows how strongly sexual coercion is tied to the status of diverse groups of women in society. Rather than explaining sexual coercion as the result of maladjusted men or the behavior of victims, feminists have encouraged a view of sexual coercion that links it to an understanding of dominant beliefs about the sexual dominance of men and the sexual passivity of women. Researchers have shown that those holding the most traditional gender role stereotypes are most tolerant of rapists and least likely to give credibility to victims of rape (Marciniak 1998; Varelas and Foley 1998). Understanding sexual violence requires an understanding of the sociology of sexuality, gender, race, and class in society.

Sex and Social Change

As with other forms of social behavior, sexual behavior is not static. Sexual norms, beliefs, and practices emerge as society changes. As we saw in Chapter 11 on gender, some major changes affecting sexual relations come from changes in gender roles. But technological change, as well as the emphasis on consumerism in the United States, also affects sexuality. As you think about sex and social change, you might try to imagine what other social factors influence human sexual behavior.

The Sexual Revolution: Is It Over?

The **sexual revolution** refers to the widespread changes in men's and women's roles and a greater public acceptance of sexuality as a normal part of social development. Many changes associated with the sexual revolution have been changes in women's behaviors. Essentially, the sexual revolution has narrowed the differences in the sexual experiences of men and women. The feminist and the gay and lesbian movements have put the sexual revolution at the center of public attention by challenging gender role stereotyping and sexual oppression, profoundly changing our understanding of gay and lesbian sexuality. The sexual revolution has meant greater sexual freedom, especially for women, but it has not eliminated the influence of gender in sexual relationships.

Technology, Sex, and Cybersex

Technological change has also brought new possibilities for sexual freedom. One significant change is the widespread availability of the birth control pill. Sex is no longer necessarily linked with reproduction; new sexual norms associate sex with intimacy, emotional ties, and physical pleasure (Freedman and D'Emilio 1988). These sexual freedoms are not equally distributed among all groups, however. For women, sex is still more closely tied to reproduction than it is for men because women are still more likely to take the responsibility for birth control.

Contraceptives are not the only technology influencing sexual values and practices. Now the Internet has introduced new forms of sexual relations as many people seek sexual stimulation from pornographic websites or online sexual chat rooms. Cybersex, as sex via the Internet has come to be known, can transform sex from a personal, face-to-face encounter to a seemingly anonymous relationship with mutual online sex. This introduces new risks; for example, two-thirds of those visiting chat rooms are adults masquerading as children (Lamb 1998; Wysocki 1998; Cooper and Scherer 1999). The Internet has introduced new forms of deviance that are difficult to regulate.

Commercializing Sex

At the same time that there are new sexual freedoms for women, many worry that this will only increase their sexual objectification. Furthermore, sexuality is becoming more and more of a commodity in our highly consumer-based society. Thus, girls are being sexualized at younger ages, evidenced by the marketing of thongs to very young girls, the promotion of "sexy" dolls sold to young girls, and the highly sexualized content of media images that young boys and girls consume (Levy 2005). Often, these images are extremely violent and depict women as hypersexual

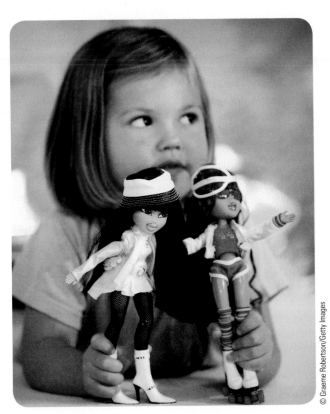

The popular Bratz dolls are being marketed to young girls, selling an image of women as sexual objects.

victims of men's aggression, such as the popular video game, "Grand Theft Auto."

Definitions of sexuality in the culture are heavily influenced by the advertising industry, which narrowly defines what is considered "sexy." Thin women, White women, and rich women are all depicted as more sexually appealing in the mainstream media. Images defining "sexy" also are explicitly heterosexual. The commercialization of sex uses women and, increasingly, men in demeaning ways. Thus, although the sexual revolution has removed sexuality from many of its traditional constraints, the inequalities of race, class, and gender still shape sexual relationships and values.

The combination of sexualization and commodification means that people become "made" into things for others' use. When people are held to narrow definitions of sexual attractiveness or are seen as valuable solely for their sexual appeal, you have social conditions that are ripe for exploitation and damage to people's sense of self-worth and value (American Psychological Association 2007). Thus, even with a seemingly increasingly "free" sexual society, sexuality is still nested in American culture within a system of power relations—power relations that, despite the sexual revolution, continue to influence how different groups are valued and defined.

Chapter Summary

- ***In what sense is sexuality, seemingly so personal an experience, a part of social structure?***
Sexual relationships develop within a social and cultural context. Sexuality is learned through socialization, is channeled and directed by social institutions, and reflects the race, class, and gender relations in society.

- ***What evidence is there of contemporary sexual attitudes and behavior?***
Contemporary sexual attitudes vary considerably by social factors such as age, gender, race, and religion. Sexual behavior has also changed in recent years, with mixed trends in both liberal and conservative sexual values. In general, attitudes on issues of premarital sex and gay and lesbian rights have become more liberal, though this depends on social characteristics such as age, gender, and degree of religiosity, among others.

- ***What does sociological theory have to say about sexual behavior?***
Functionalist theory depicts sexuality in terms of its contribution to the stability of social institutions. *Conflict theorists* see sexuality as part of the power relations and economic inequality in

society. *Symbolic interaction* focuses on the social construction of sexual identity.

• *What is meant by the social construction of sexual identity?*

Sociologists see *sexual identity,* whatever its form, as constructed through socialization experiences. They have moved away from studying gay and lesbian experience as deviant behavior. Debate continues about the underlying origins of homosexuality, although evidence for a biological basis of homosexuality is not as strong as the evidence for social causes. *Homophobia* is the fear and hatred of homosexuals. *Heterosexism* is the institutionalization of heterosexuality as the only socially legitimate sexual orientation.

• *How is sexuality related to contemporary social issues?*

Sexuality is related to some of the most difficult social problems—including birth control, abortion, reproductive technologies, teen pregnancy, pornography, and sexual violence. Such social problems can be understood by analyzing the sexual, gender, class, and racial politics of society.

• *How is sex related to social change?*

The *sexual revolution* refers to widespread changes in the roles of men and women and a greater acceptance of sexuality as a normal part of social development. The sexual revolution has been fueled by social movements, such as the feminist movement and the gay and lesbian rights movement. Technological changes, such as the development of the pill, have also created new sexual freedoms. Now, sexuality is influenced by the growth of cyberspace and its impact on personal and sexual interactions. At the same time, sex is treated as a commodity in this society; it is bought and sold and used to sell various products.

Key Terms

coming out   296
eugenics   303
heterosexism   301
homophobia   300
queer theory   296

sex tourism   300
sex trafficking   300
sexual identity   290
sexual orientation   290
sexual politics   297

sexual revolution   308
sexual scripts   292
social construction
   perspective   296

Online Resources

Sociology: The Essentials Companion Website

www.cengage.com/sociology/andersen
Visit your book companion website where you will find more resources to help you study and write your research papers. Resources include web links, learning objectives, Internet exercises, quizzing, and flash cards.

 is an easy-to-use online resource that helps you study in less time to get the grade you want NOW.

www.cengage.com/login
Need help studying? This site is your one-stop study shop. Take a Pre-Test and Cengage NOW will generate a Personalized Study Plan based on your test results. The Study Plan will identify the topics you need to review and direct you to online resources to help you master those topics. You can then take a Post-Test to determine the concepts you have mastered and what you still need to work on.

13

Chapter thirteen
CHAPTER THIRTEEN
Chapter

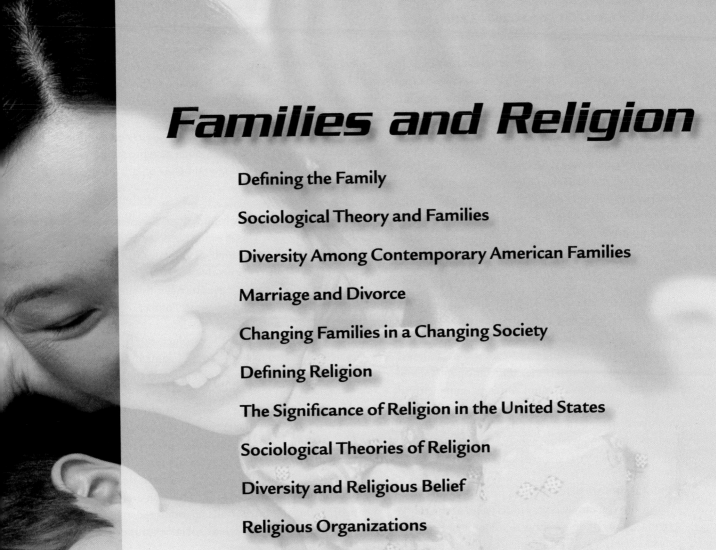

Families and Religion

Defining the Family

Sociological Theory and Families

Diversity Among Contemporary American Families

Marriage and Divorce

Changing Families in a Changing Society

Defining Religion

The Significance of Religion in the United States

Sociological Theories of Religion

Diversity and Religious Belief

Religious Organizations

Religion and Social Change

Chapter Summary

Tracy Kahn/Corbis

[***Suppose you were*** to ask a large group of people in the United States to describe their families. Many would describe divorced families. Some would describe single-parent families. Some would describe stepfamilies with new siblings and a new parent stemming from remarriage. Others would describe gay or lesbian households, perhaps with children present. Also included would be adoptive families and families with foster children. Others would describe the so-called traditional family with two parents living as husband and wife in the same residence as their biological children. Families have become so diverse that it is no longer possible to speak of "the family" as if it were a single thing.

continued

thirteen chapter thirteen

The traditional *family ideal*—a father employed as the breadwinner and a mother at home raising children—has long been the dominant cultural norm, communicated through a variety of sources, including the media, religion, and the law. Few families now conform to this ideal, and the number of families that ever did is probably fewer than generally imagined (Coontz 1992). Regardless of their form, families now face new challenges—possibly living on one income or managing family affairs when both parents are employed. Many families also feel that they are under siege by changes in society that are dramatically altering all family experiences.

Many view the changes taking place in families as positive. Women have new options and greater independence. Fathers are discovering that there can be great pleasure in domestic and child-care responsibilities. Change, however, also brings difficulties: balancing the demands of family and employment, coping with the interpersonal conflicts caused by changing expectations, and striving to make ends meet in families without sufficient financial resources. These changes bring new questions to the sociological study of families.

Family affairs are believed to be private, but as an institution, the family is very much part of the public agenda. Many people believe that "family breakdown" causes society's greatest problems. Public policies shape family life directly and indirectly, and family life is now being openly negotiated in political arenas, corporate boardrooms, and courtrooms, as well as in the bedrooms, kitchens, and "family" rooms of individual households.

As a social institution, the family is intertwined with other social institutions, such as religion. Religious values and customs tend to be learned first within families, and in the United States, religion influences beliefs about how families should be organized. Many major family events, such as weddings, christenings, and funerals, are observed with religious ceremonies; and many family behaviors, such as reproduction, marriage, divorce, and sexual behavior, are affected by religious values.

Religion thus has a profound effect on society and human behavior. This is easily observed in daily life outside the family. Church steeples dot the landscape. Invocations to a religious deity occur at the beginning of many public gatherings. The news frequently reports on events generated by religious conflict.

Religious beliefs have led to conflict, but religious beliefs have also been the soul of some of the most liberating social movements, including the civil rights movement and other human rights movements around the world. Some of life's sweetest moments are marked by religious celebration, and some of its most bitter conflicts persist because of unshakable religious conviction. Religion is both an integrative force in society and the basis for many of our most deeply rooted social conflicts. The family and religion are, for most people, the first institutions encountered in life. In this chapter we examine both as social institutions.

Defining the Family

The family is a *social institution*, that is, an established social system that emerges, changes, and persists over time. Institutions are "there"; we do not reinvent them every day, although people adapt in ways that make institutions constantly evolve, such as is the case with how families have changed over time.

Family diversity is the norm in American society, with no one type of family shaping people's experience.

We can define the **family** to refer to a primary group of people—usually related by ancestry, marriage, or adoption—who form a cooperative economic unit to care for offspring and each other and who are committed to maintaining the group over time (adapted from Lamanna and Riedmann 2008: 10).

Families are part of what are more broadly considered to be kinship systems. A **kinship system** is the pattern of relationships that define people's relationships to one another within a family. Kinship systems vary enormously across cultures and over time. In some societies, marriage is seen as a union of individuals; in others, marriage is seen as creating alliances between groups. In some kinship systems, marriages may be arranged, possibly even involving a broker whose job is to conduct the financial transactions and arrange the marriage ceremonies (Croll 1995). In still other kinship systems, maintaining multiple marriage partners may be the norm. Kinship systems can generally be categorized by the following features:

- how many marriage partners are permitted at one time;
- who is permitted to marry whom;
- how descent is determined;
- how property is passed on;
- where the family resides;
- how power is distributed.

Polygamy is the practice of men or women having multiple marriage partners. Polygamy usually involves one man having more than one wife, technically referred to as *polygyny; polyandry* is the practice of a woman having more than one husband, an extremely rare custom. Within the United States, polygamy is commonly associated with Mormons, even though polygamy is practiced now by only a few Mormon fundamentalists (who are estimated to be only about 2 percent of the state population of Utah); those who practice polygamy do so without official church sanction (Bachman and Esplin 1992; Driggs 1990; Brooke 1998).

Monogamy is the practice of a sexually exclusive marriage with one spouse at a time. It is the most common form of marriage in the United States and other Western industrialized nations. In the United States, monogamy is a cultural ideal that is prescribed through law and promoted through religious teachings. Lifelong monogamy is not always realized, however, as evidenced by the high rate of divorce and extramarital affairs. Many sociologists characterize modern marriage as *serial monogamy* in which individuals may, over a lifetime, have more than one marriage, but only one spouse at a time (Lamanna and Riedmann 2008).

In addition to defining appropriate marriage partners, kinship systems shape the distribution of property in society by determining descent. **Patrilineal kinship** systems trace descent through the father; **matrilineal kinship** systems, through the mother. **Bilateral kinship** (or bilineal kinship) traces descent through both. You can see the continuing influence of patrilineal kinship in contemporary society by noting how typical it is for children to assume the name of the father, not the mother. And, even though practices are evolving, it is still quite common for women to take men's names when they marry. **Matrilocal** and **patrilocal** are terms used to describe where a married couple resides. In some societies, married couples are expected to move into the husband's residence—or even the husband's family residence.

Kinship systems also determine whom one can marry. Even without specific laws, society establishes normative expectations about appropriate marriage partners. In general, people in the United States marry people with very similar social characteristics, such as class, race, religion, and educational backgrounds (Kalmijn 1991). Interracial marriage, although increasing, is still relatively infrequent (only about 4 percent of married couples; U.S. Census Bureau 2009a). Even though interracial marriages are not common, historically a tremendous amount of energy has been put into preventing them. Laws have prohibited marriage between various groups, including between Whites and African Americans and between Whites and Chinese, Japanese, Filipinos, Hawaiians, Hindus, and Native Americans. Not until 1967 were laws prohibiting interracial marriage declared unconstitutional by the U.S. Supreme Court (Kennedy 2003; Takaki 1989).

SEE *FOR YOURSELF*

ANALYZING FAMILY STRUCTURES

What is the kinship structure of your family? Make a list of all the people whom you consider to be part of your family. List them by name and then by your relationship (i.e., father, sister, partner, othermother, and so forth). Then using the concepts that describe *kinship systems*, identify the specific form of kinship that describes your family (nuclear, extended, matrilineal, patriarchal, and so forth). Are there characteristics of your family that suggest any need to revise these concepts? How does your family structure compare to that of other students in the class?

Extended Families

Extended families are the whole network of parents, children, and other relatives who form a family unit. Sometimes extended families, or parts thereof, live together, sharing their labor and economic resources. In some contexts, "kin" may refer to those who are not related by blood or marriage but who are intimately involved in the family support system and are considered part of the family (Stack 1974). *Othermothers*

Diversity can even occur within families, as shown by this photograph of a multiracial, multiethnic family.

may be a grandmother, sister, aunt, cousin, or a member of the local community, but she is someone who provides extensive child care and receives recognition and support from the community around her (Collins 1990: 119). This term emerged from the experience of African American women, whose historically dual responsibilities in the family and work have meant that they have a history of creating alternative means of providing family care for children. Now this is a common practice for many families, including the Obama "first family": Upon moving to the White House, the Obamas brought Michelle Obama's mother, Marian Robinson, with them to assist with raising the Obama daughters, Malia and Sasha.

The system of *compadrazgo* among Chicanos is another example of an extended kinship system. In this system, the family is enlarged by the inclusion of godparents, to whom the family feels a connection that is the equivalent of kinship. The result is an extended system of connections between "fictive kin" (those who are not related by birth but are considered part of the family) and actual kin that deeply affects family relationships among Chicanos (Baca Zinn and Eitzen 2007).

Nuclear Families

In the **nuclear family,** a married couple resides together with their children. Like extended families, nuclear families develop in response to economic and social conditions. The origin of the nuclear family in Western society is tied to industrialization. Before industrialization, families were the basic economic unit of society. Large household units produced and distributed goods, whether in small communities or large plantation or feudal systems where slaves and peasants provided most of the labor. Production took place primarily in households, and all family members were seen as economically vital. Household and production were united, with no sharp distinction between economic and domestic life. Women performed and supervised much of the household work, engaged in agricultural labor, and produced cloth and food. The work of women, men, and children was also highly interdependent. Although the tasks each performed might differ, together they were a unit of economic production.

With industrialization, paid labor was performed mostly away from the home in factories and public marketplaces. The transition to wages for labor created an economy based on cash rather than domestic production. Families became dependent on the wages that workers brought home. The shift to wage labor was accompanied by an assumption that men should earn the "family wage" (that is, be the breadwinner). Thus, men who worked as paid laborers were paid more than women, and women became more economically dependent on men. At the same time, a man's status was enhanced by having a wife who could afford to stay at home—a privilege seldom accorded to working-class or poor families. The *family wage system* has persisted and is reflected still in the unequal wages of men and women.

The unique social conditions that racial–ethnic families have experienced also affect the development of family systems. Disruptions posed by the experiences of slavery, migration, and urban poverty affect how families are formed, their ability to stay together, the resources they have, and the problems they face. For example, historically Chinese American laborers were explicitly forbidden to form families by state laws designed to regulate the flow of labor. Only a small number of merchant families were exempt from the law.

During the westward expansion of the United States, many Mexicans who had settled in the Southwest were displaced. The loss of their land disrupted their families and kinship systems. In the rapidly industrializing United States, many Mexican Americans were able to find work in the mines opening in the new territories or by building the railroads spreading from the east toward the Pacific. Employers apparently thought that they had better control over laborers if their families were not there to distract them, so families were typically prohibited from being with the worker. One result was the development of prostitution camps, which followed workers from place to place (Dill 1988).

Families continue to be influenced by social structural forces. Some families, particularly those with marginal incomes, find it necessary for the entire family to work to meet the economic needs of the household. Migration to a new land and exposure to new customs also disrupts traditional family values. The ability to form and sustain nuclear families is directly linked to the economic, political, and racial organization of society.

Jose Luis Pelaez, Inc./Blend Images/Corbis

Interracial Dating and Marriage

Picture this: A young couple, stars in their eyes, holding hands, intimacy in their demeanor. Newly in love, the couple imagines a long and happy life together. When you visualize this couple, who do you see? If your imagination reflects the sociological facts, odds are that you did not imagine this to be an interracial couple. Although interracial couples are increasingly common (and have long existed), people are more likely to form relationships with those of their same race—as well as social class, for that matter. What do sociologists know about interracial dating and marriage?

First, patterns of interracial dating are influenced by both race and gender. Black men are those most likely to say they would date a person of another race (81 percent say they would), although, along with White women, they are the least likely to say they would marry someone of another race. White women are least likely to say they would date someone of

another race. Both Blacks and Whites in interracial relationships report negative reactions from their families; the majority of these were not extremely hostile, but strong enough to put pressure on the interracial couple (Majete 1999). And although most Blacks and Whites profess to have a color-blind stance toward interracial marriages, when pressed, they raise numerous qualifications and concerns about such pairings (Bonilla-Silva and Hovespan 2000). Among college students, approval of family and friends is the strongest indicator of one's attitude toward interracial dating. Racial minorities are more accepting of interracial dating than are White students; students in the Greek system on college campuses are less accepting than non-Greeks (Khanna et al. 1999).

Regardless of these attitudes, interracial marriage is on the rise, although they are still a small percentage of marriages formed. The most likely interracial marriages are between Black men and White women and between Hispanics and non-Hispanics. National data on Asian American marriage is limited, but studies find that Asian Americans are increasingly likely to marry other Asian Americans, though often someone of a different Asian heritage (Shinagawa and Pang 1996). The data on interracial dating and marriage show how something seemingly "uncontrollable," such as love, is indeed shaped by many sociological factors.

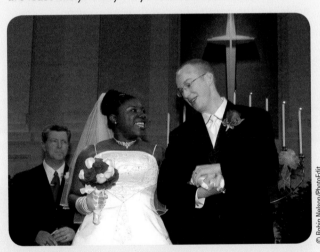

Interracial marriage is increasingly common, although still a small proportion of all marriages.

© Robin Nelson/PhotoEdit

Sources: Khanna, Nikki D., Cherise Harris, and Rana Cullers. 1999. "Attitudes Toward Interracial Dating." Paper presented at the Annual Meetings of the Society for the Study of Social Problems; Bonilla-Silva, Eduardo, and Mary Hovespan. 2000. "If Two People Are in Love: Deconstructing Whites' Views on Interracial Marriage with Blacks." Paper presented at the annual meetings of the Southern Sociological Society; Majete, Clayton A. 1999. "Family Relationships and the Interracial Marriage." Paper presented at the Annual Meetings of the American Sociological Association; Shinagawa, Larry, and Gin Yong Pang. 1996. "Asian American Panethnicity and Intermarriage." *Amerasia Journal* 22 (Spring): 127-152.

thinking sociologically

Identify two popular "family" shows on television. Devise a systematic way to conduct a content analysis of the family ideal portrayed by these shows. What do your observations reveal about how the *family ideal* is communicated through the popular media?

Sociological Theory and Families

Is the family a source of stability or change in society? Are families organized around harmonious interests, or are they sources of conflict and differential power? How do new family forms emerge, and how do people negotiate the changes that affect families? These

table 13.1 — Theoretical Perspectives on Families

	Functionalism	Conflict Theory	Feminist Theory	Symbolic Interactionism
Families	Meet the needs of society to socialize children and reproduce new members	Reinforce and support power relations in society	Are gendered institutions that reflect the gender hierarchies in society	Emerge as people interact to meet basic needs and develop meaningful relationships
	Teach people the norms and values of society	Inculcate values consistent with the needs of dominant institutions	Are a primary agent of gender socialization	Are where people learn social identities through their interactions with others
	Are organized around a harmony of interests	Are sites for conflict and diverse interests of different family members	Involve a power imbalance between men and women	Are places where people negotiate their roles and relationships with each other
	Experience social disorganization ("breakdown") when society undergoes rapid social changes	Change as the economic organization of society changes	Evolve in new forms as the society becomes more or less egalitarian	Change as people develop new understandings of family life

questions and others guide sociological theories of the family (see Table 13.1).

Functionalist Theory and Families

Functionalist theorists interpret the family as filling particular societal needs, including socializing the young, regulating sexual activity and procreation, providing physical care for family members, and giving psychological support and emotional security to individuals. According to functionalism, families exist to meet these needs and to ensure a consensus of values in society. In the functionalist framework, the family is conceptualized as a mutually beneficial exchange, wherein women receive protection, economic support, and status in return for emotional and sexual support, household maintenance, and the production of offspring (Glenn 1986). At the same time, in traditional marriages men get the services that women provide—housework, nurturing, food service, and sexual partnership. Functionalists also see families as providing care for children, who are taught the values that society and the family support. In addition, functionalists see the family as regulating reproductive activity, including cultural sanctions about sexuality.

According to functionalist theory, when societies experience disruption and change, institutions such as the family become disorganized, weakening social cohesion. Currently, some analysts interpret the family as "breaking down" under societal strains. Functionalist theory suggests that this breakdown is the result of the disorganizing forces that rapid social change has fostered.

Functionalists also note that, over time, other institutions have begun to take on some functions originally performed solely by the family. For example, as children now attend school earlier in life and stay in school for longer periods of the day, schools (and other caregivers) have taken on some functions of physical care and socialization originally reserved for the

family. Functionalists would say that the diminishment of the family's functions produces further social disorganization because the family no longer carefully integrates its members into society. To functionalists, the family is shaped by the template of society, and such things as the high rate of divorce and the rising numbers of female-headed and single-parent households are the result of social disorganization.

Conflict Theory and Families

Conflict theory interprets the family as a system of power relations that reinforces and reflects the inequalities in society. Conflict theorists also are interested in how families are affected by class, race, and gender inequality. This perspective sees families as the units through which the advantages, as well as the disadvantages, of race, class, and gender are acquired.

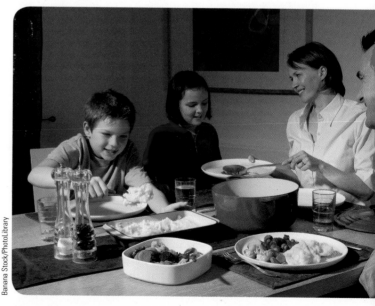

The family is the major institution where socialization of children occurs.

Conflict theorists view families as essential to maintaining inequality in society because they are the vehicles through which property and social status are acquired (Eitzen and Baca Zinn 2007).

The conflict perspective also emphasizes that families in the United States are shaped by capitalism. The family is vital to capitalism because it produces the workers that capitalism requires. Accordingly, within families personalities are shaped to the needs of a capitalist system; thus families socialize children to become obedient, subordinate to authority, and good consumers. Those who learn these traits become the kinds of workers and consumers that capitalism needs. Families also serve capitalism in other ways; giving a child an allowance teaches the child capitalist habits involving money.

Whereas functionalist theory conceptualizes the family as an integrative institution (meaning it has the function of maintaining social stability), conflict theorists depict the family as an institution subject to the same conflicts and tensions that characterize the rest of society. Families are not isolated from the problems facing society as a whole. The struggles brought on by racism, class inequality, sexism, homophobia, and other social conflicts are played out within family life.

Feminist Theory and Families

Feminist theory has contributed new ways of conceptualizing the family by focusing sociological analyses on women's experiences in the family and by making gender a central concept in analyzing the family as a social institution. Feminist theories of the family emerged initially as a criticism of functionalist theory. Feminist scholars argued that functionalist theory assumed that the gender division of labor in the household is functional for society. Feminists have also been critical of functional theory for assuming an inevitable gender division of labor within the family. Feminist critics argue that although functionalists may see the gender division of labor as functional, it is based on stereotypes about men's and women's roles.

Influenced by the assumptions of conflict theory, feminist scholars do not see the family as serving the needs of all members equally. Quite the contrary, feminists have noted that the family is one of the primary institutions producing the gender relations found in society. Feminist theory conceptualizes the family as a system of power relations and social conflict. In this sense, it emerges from conflict theory, but adds that the family is a gendered institution (see Chapter 11).

Symbolic Interaction Theory and Families

Symbolic interaction emphasizes that meanings people give to their behavior and that of others is the basis of social interaction. Those who study families from this perspective tend to take a more microscopic view of families. A symbolic interactionist might ask how different people define and understand their family experience. Symbolic interactionists also study how people negotiate family relationships, such as deciding who does what housework, how they will arrange child care, and how they will balance the demands of work and family life.

To illustrate, when people get married, they form a new relationship and new identities with specific meanings within society. Some changes may seem very abrupt—a change of name certainly requires adjustment, as does being called a husband or wife. Some changes are more subtle—how one is treated by others and the privileges couples enjoy (such as being a recognized legal unit). Symbolic interactionists see the marriage relationship as socially constructed; that is, it evolves through the definitions that others in society give it as well as through the evolving definition of *self* that married partners make for themselves.

The symbolic interactionist perspective understands that roles within families are not fixed, but rather evolve as participants define and redefine their behavior toward each other. Symbolic interaction is especially helpful in understanding changes in the family because it supplies a basis for analyzing new meaning systems and the evolution of new family forms over time. Each theoretical perspective used to analyze families illuminates different features of family experiences.

Diversity Among Contemporary American Families

Today the family is one of the most rapidly changing of all of society's institutions. Families are systems of social relationships that emerge in response to social conditions that, in turn, shape the future direction of society. There is no static or natural form for the family. Change and variation in families are social facts.

Among other changes, families today are smaller than in the past. There are fewer births, and they are more closely spaced, although these characteristics of families vary by social class, region of residence, race, and other factors. Because of longer life expectancy, childbearing and child rearing now occupy a smaller fraction of parents' adult life. During earlier periods, death (often from childbirth) was more likely to claim the mother than the father of small children; thus, men in the past would have been more likely than now to raise children on their own after the death of a spouse. That trend is now reversed; it is now women who are more likely to be widowed with children, and death, once the major cause of early family disruption, has been replaced by divorce (Rossi and Rossi 1990).

Demographic and structural changes have resulted in great diversity in family forms. Compared to thirty years ago, married couples now make up a smaller proportion of households; single-parent households have increased dramatically, and divorced and never-married people make up a larger proportion

a sociological eye on the media

Idealizing Family Life

Cultural norms about motherhood and fatherhood come from many places, but the media is certainly a strong influence on how family ideals—and the ideals for mothers and fathers—are created in society. Media images of the family have certainly changed since the inception of television. In the 1930s, Hollywood Codes—that is, official rules in Hollywood about what could and could not be seen in movies and, later, on television—meant that families were always shown with two parents, marriage intact, father working, and mom staying at home. Parents never talked about sex; indeed, it appeared as though they never had it because the Codes forbade any nudity and required that scenes of passion not excite the audience. As a result, if bedroom scenes between married couples were shown, there were typically only twin beds.

Now, family images on television are more diverse. Shows like *Two and a Half Men, Jon and Kate Plus 8,* and *Desperate Housewives* show very different images of family life. Still, the media continues to construct an ideal for family life—one that continues to stereotype men and women in family

roles. What does research show about these social constructions?

- Most family characters are middle class.
- Men appearing with children are most likely to be shown outside; they are also more likely to be seen with boys, not girls.
- Fathers are infrequently seen with infants.
- Fathers are shown playing with, reading to, talking with, and eating with children, but not preparing meals, cleaning house, changing diapers, and so forth.
- Women are disproportionately shown in family settings in the media.

What gender stereotypes do such images project? How do they influence people's views of ideal family roles? How realistically do they portray the actual gender division of labor in families? What race and class images confound these results?

Sources: Kaufman, Gayle. 1999. "The Portrayal of Men's Family Roles in Television Commercials." *Sex Roles* 41 (September): 439–458. Coltrane, Scott, and Melinda Messineo. 2000. "The Perpetuation of Subtle Prejudice: Race and Gender Imagery in 1990s Television Advertising." *Sex Roles* 42 (March): 363–389.

of the population (see Figure 13.1). Overall, married-couple families make up about half of all households, and single-parent households (typically headed by women), post-childbearing couples, gay and lesbian couples, childless households, and single people are increasingly common (U.S. Census Bureau 2009a). Now people may also spend more years caring for elderly parents than they did raising their children.

Female-Headed Households

One of the greatest changes in family life is the increase in the number of families headed by women. One-quarter of all children live with one parent, the vast majority of whom (87 percent) live with their mother. Although a large number of those living with one parent do so because of divorce, the largest number live with one parent who has never been married. One-quarter of all households are headed by women, although the number of households headed by single fathers has also increased. The odds of living in a single-parent household are even greater for African American and Latino children (U.S. Census Bureau 2008c).

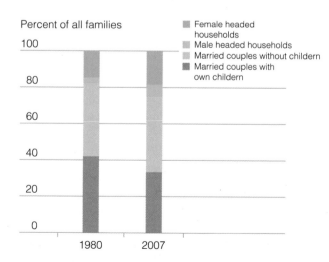

FIGURE 13.1 *The Changing Character of U.S. Families, 1980–2007*

Data: U.S. Bureau of the Census. 2009a. *Statistical Abstract of the United States, 2009.* **www.census.gov**

The two primary causes for the growing number of women heading their own households are the high rate of pregnancy among unmarried teens and the high divorce rate, with death of a spouse also contributing. Although, as we discussed in Chapter 12, the rate of pregnancy among teenagers (of all races), married and unmarried, has declined, the proportion of teen births that occur outside marriage has increased.

Many people see the increase of female-headed households as representing a breakdown of the family and a weakening of social values. An alternate view, however, is that the rise of female-headed households reflects the growing independence of women, some of whom are making decisions to raise children on their own. Not all female-headed households are women who have never married; many are divorced and widowed women whose circumstances may be quite different from those of a younger, never-married woman.

Some claim that female-headed households are linked to problems such as delinquency, the school dropout rate, children's poor self-image, and other social problems. Sometimes the cause of these troubles is attributed explicitly to the absence of men in the family. Sociologists, however, have not found the absence of men as the sole basis for such problems; rather, it is the presence of economic pressure faced by female-headed households, compared with that of male-headed households, that puts female-headed households under great strain—with the threat of poverty being by far the greatest problem they face. Among households headed by women with children, one-third live below the poverty line, with the rates of poverty highest among Black and Hispanic female-headed households (DeNavas-Walt et al. 2008). It is not the makeup of households headed by women that is a problem but the fact that they are most likely to be poor (Brewer 1988). This phenomenon was discussed in Chapter 8 as the *feminization of poverty.*

Debunking Society's Myths

Myth: Absence of fathers is the cause of numerous social problems; if these fathers would just adopt "family values," families would be stronger and children wouldn't get into so much trouble.

Sociological perspective: The mere presence of men in a family does not in itself prevent family problems. Concerns about father-absent families also typically are directed at poor, not middle-class families. Research on never-married, poor African American fathers finds that they typically want to provide for their children, may spend a lot of time with their children, and want to be good fathers, but find that their life opportunities make it difficult to do so (Hamer 2001).

Although the majority of single-parent families are headed by women, families headed by a single father are increasing, but male-headed households are less likely than female-headed households to experience severe economic problems. Unlike female-headed households where a man is not present to help with housework and children, single fathers commonly get domestic help from women—either girlfriends, daughters, or mothers (Popenoe 2001).

Married-Couple Families

Among married-couple families, the increased participation of women in the paid labor force has added new challenges to family life. Families now are able to sustain a median income level only by having both husband and wife in the paid labor force. As a result, families are experiencing substantial speedup (see Figure 13.2).

Women's labor force participation has created other changes in family life. One is the number of married couples who have *commuter marriages,* when work requires one partner in a dual-career couple

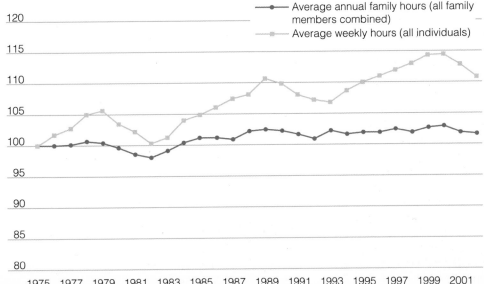

FIGURE 13.2 *Family Work Hours: Social Speedup*

Source: Economic Policy Institute. 2004. "The Rise in Family Work Hours Leads Many Americans to Balance Work and Family." **www.epinet.org.** Reprinted with permission.

to reside in a different city, separated by jobs too distant for a daily commute. The common image of a commuter marriage is one consisting of a prosperous professional couple, each holding important jobs, flashing credit cards, and using airplanes like taxis. However, working-class and poor couples do their share of long-distance commuting: Agricultural workers follow seasonal work; skilled laborers sometimes have to leave their families to find jobs; many families cross national borders in search of work, often separating one or both parents and children. Although their commute may be less glamorous than that of professional spouses, they are commuting nonetheless. When all types of commuter marriages are included, this form of marriage is more prevalent than is typically imagined.

Stepfamilies

Because of the rise in divorce and remarriage, stepfamilies are now fairly common in the United States. They take numerous forms, including married adults with stepchildren, cohabiting stepparents, and stepparents who do not reside together (Stewart 2001). About 40 percent of marriages involve stepchildren. Stepfamilies may face a difficult period in the transition when two families blend, introducing people to a "new" family. Parents and children may have to learn new roles when they become part of a stepfamily. Children accustomed to being the oldest child in the family, or the youngest, may find that their status in the family group is suddenly transformed. New living arrangements may require children to share rooms, toys, and time with people they perceive as strangers.

In stepfamilies, the parenting roles of mothers and fathers suddenly may expand to include more children, each with his or her needs. Jealousy, competition, and demands for time and attention can make the relationships within stepfamilies tense. The problems are compounded by the absence of norms and institutional support systems for stepfamilies. Without norms to follow, people have to adapt by creating new language to refer to family members and new relationships. Many develop strong relationships within this new kinship system; others find the adjustment extremely difficult, resulting in a high probability of divorce among remarried couples with children (Baca Zinn and Eitzen 2007).

Gay and Lesbian Households

The increased visibility of gay and lesbian families has challenged the traditional understanding of families as only heterosexual. The public debate around gay marriage indicates the considerable division around this issue. A small but growing number of states have recognized gay marriages as legal under their state constitutions, though in most states same-sex marriage has neither the official blessings nor supports of social institutions. Nonetheless, many gay and lesbian couples form long-term, primary relationships that they define as marriage. Like other families, gay and lesbian couples share living arrangements and household expenses, make decisions as partners, and in many cases, raise children.

As we saw in Chapter 12, the number of Americans who think same-sex marriage should be legal (40 percent) has increased even since the mid-1990s (Gallup Poll 2009). Gay marriage is also more acceptable in the eyes of younger people than to older groups, raising the question of whether over time social support for gay marriages will increase or whether, as young people age, they will shift their values. In the meantime, in most states and municipalities, gays and lesbians who form strong and lasting relationships do so without formal institutional support and, as a result, have had to be innovative in producing new support systems.

Researchers have found that gay and lesbian couples tend to be more flexible and less gender stereotyped in their household roles than heterosexual couples. Lesbian households, in particular, are more egalitarian than are either heterosexual or gay male couples. Money also has less effect on the balance of power in lesbian relationships than is true for heterosexual couples. However, where one partner is the primary breadwinner and the other the primary caregiver for children, the partner staying at home becomes economically vulnerable and less able to negotiate her needs, just as in heterosexual relationships (Sullivan 1996).

The new family forms that lesbian and gay couples are creating mean that they have to actively construct new meanings of such things as motherhood. Researchers find that they do so in ways that are collaborative, including elaborate networks of family and friends (Dunne 2000; Dalton and Bielby 2000). To date, most gay fathers are those who have children

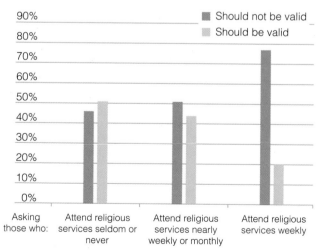

FIGURE 13.3 *Acceptance of Gay Marriage*

The question asked of a national sample was: "Do you think marriages between homosexuals should or should not be recognized by the law as valid, with the same rights as traditional marriages?"

Source: Lyons, Lydia. 2006. "Americans Still Oppose Gay Marriage." Princeton, NJ: The Gallup Organization. **www.gallup.com**

from a previous heterosexual marriage, although the number of gay men adopting children is increasing.

Public debate about gay marriage often centers on the implications for children raised in gay and lesbian families. Research on children in gay and lesbian households finds that for the most part there is little difference in outcomes for children raised in gay and lesbian households compared to those raised in heterosexual households. What differences are found are the result of other factors—not just the sexual orientation of parents. The greatest differences are the result of the homophobia that is directed against children in lesbian and gay families, who are very likely to be stigmatized by others. But such children are also less likely to develop stereotypical gender roles and are more open-minded about sexual matters, although they are no more likely to become gay themselves (Allen 1997; Stacey and Biblarz 2001). If we lived in a society more tolerant of diversity, the differences that emerge might be viewed as strengths, not deficits.

Single People

Single people, including those never married, widowed, or divorced, today constitute 25 percent of the population (over age 15)—an increase from 16 percent in 1970 (U.S. Census Bureau 2009a, 2001). Some of the increase is a result of the rising number of divorced people, but there has also been an increase in the number of those never married. Men and women are also marrying at a later age—25.3 years on average for women, 27.1 for men, compared with 20.8 for women and 23.2 for men in 1970 (Fields 2004).

Among singles, patterns of establishing intimate relationships have changed significantly. "Hooking up"—a phrase referring to a casual, sexual alliance between two people—has supplanted dating as the pattern by which young people get to know each other. Courtship no longer follows preestablished norms. Hooking up is widespread on college campuses and influences the campus culture, although only a minority of students engage in it (about 40 percent of college women). Hooking up carries multiple meanings. For some it means kissing; for others it means sexual-genital play, but not intercourse; for some, it means sexual intercourse. The vagueness of the term contributes to its becoming a shared cultural phenomenon. A majority of college women say that hooking up makes them feel desirable, but also awkward, and they are wary of getting a bad reputation from hooking up too often. The majority of college women still want to meet a spouse while at college (Bogle 2008; Glenn and Marquardt 2001).

The path to a committed relationship, possibly marriage, involves increasing phases of commitment and sexual exclusivity. Many people find the same sexual and emotional gratification in single life as they would in marriage. Being single is also no longer the stigma it once was, especially for women. Nonetheless, studies find that married people are generally happier than singles; this changes in different historical

Popular shows like Friends *gain some of their appeal by the growing proportion of the population who are single.*

periods, however, and depends a lot on other factors, such as financial resources, marital conflict and violence, the extent of other personal ties, and the presence of stressful life problems (Aldous and Ganey 1999; Keith 1997; Lee et al. 1991; Waite 2000).

Cohabitation (living together) has become increasingly common. Some of the increase is the result of better census taking, but the increase is also real. Estimates are that one-quarter of all children will at some time during their childhood live in a family headed by a cohabiting couple (Graefe and Lichtor 1999), indicating how common cohabitation has become. Cohabitation is more common among the young and the less educated and among Black and Native American couples (Cohen 1999; Marquis 2003). Some cohabit because they are critical of the existing norms surrounding marriage (Elizabeth 2000). Cohabitants tend to have more egalitarian attitudes toward gender roles, although they tend to develop a household division of labor similar to that of married heterosexual couples. But they are more likely to remain together when they establish equality within the relationship (Sanchez et al. 1998; Barber and Axinn 1998; Brines and Joyner 1999).

Finally, a growing number of single people are remaining in their parents' homes for longer periods. Known as the "boomerang generation," these young people in their twenties return home when they would normally be expected to live independently. Today, 55 percent of men and 45 percent of women between eighteen and twenty-four years of age live at home with their parents (Fields 2004). The increased cost of living means that many young people find themselves unable to pay their own way, even after marrying or getting an education. They economize by joining the household of their parents.

Marriage and Divorce

Even with the extraordinary diversity of family forms in the United States, the majority of people will still marry at some point in their lives (see Figure 13.4). Indeed, the United States has the highest rate of marriage of any Western industrialized nation, as well as a high divorce rate.

Marriage

The picture of marriage as a consensual unit based on intimacy, economic cooperation, and mutual goals is widely shared, although marital relationships also involve a complex set of social dynamics, including cooperation and conflict, different patterns of resource allocation, and a division of labor. Sociologists must be careful not to romanticize marriage to the point that they miss other significant social patterns within marriage.

Gender roles are a significant reality of family life, shaping power dynamics within marriage, as well as the allocation of work, the degree of marital happiness, the likelihood of marital violence, and even the leisure time that each partner has. Although people do not like to think of marriage as a power relationship, gender shapes the power that men and women have within marriage, as it does in other relationships. For one thing, sociologists have long found that the amount of money a person earns establishes that person's relative power within the marriage, including the ability to influence decisions, the degree of autonomy and independence held by each partner, and the control of expectations about family life. Despite changes in women wanting to work outside the home, in most marriages (64 percent) men are the sole or major earners (Raley et al. 2006). But studies also find that even when wives earn substantially more than their husbands, rare as that is, couples tend to negotiate marital power within the confines of traditional gender expectations (Tichenor 2005).

>doing sociological research

Men's Caregiving

Research Question: Much research has documented the fact that women do the majority of the housework and child care within families. Why? Many have explained it as the result of gender socialization—women learn early on to be nurturing and responsible for others, while men are less likely to do so. Yet, things are changing, and some men are more involved in the "care work" of family life. What explains whether men will be more engaged in family care work?

Research Method: Sociologists Naomi Gerstel and Sally Gallagher studied a sample of 188 married people. They interviewed ninety-four husbands and ninety-four wives, married to each other; the sample was 86 percent White and 14 percent African American but was too small to examine similarities or differences by race.

Research Results: You might expect that men who had attitudes expressing support for men's family responsibilities would be more involved in family care (defined by Gerstel and Gallagher to include elder care, child care, and various household tasks). But this is not what Gerstel and Gallagher found. Gender attitudes did *not* influence men's involvement in caregiving. Rather, the characteristics of the men's families were the most influential determinant of their engagement in housework and child care. Men whose wives spent the most time helping kin and men who had daughters were more likely to help kin. Having sons had no influence. But, in addition, men with more sisters spend less time helping with elder parents than men with fewer sisters. Furthermore, men's employment (measured as hours employed, job flexibility, and job stability) did not affect their involvement in care work.

Conclusions and Implications: It is the social structure of the family, not gender beliefs, that shapes men's involvement in family work. As they put it, "It is primarily the women in men's lives who shape the amount and types of care men provide" (Gerstel and Gallagher 2001: 211). This study shows a most important sociological point: Social structure, not just individual attitudes, is the most significant determinant of social behavior.

Questions to Consider

1. Who does the work in your family? Is it related to the social organization of your family, as Gerstel and Gallagher find in other families?

2. Do you think that men's gender identity changes when they become more involved in care work? What hinders and/or facilitates men's engagement in this kind of work?

Source: Gerstel, Naomi, and Sally Gallagher. 2001. "Men's Caregiving: Gender and the Contingent Character of Care." *Gender & Society* 15 (April): 197–217.

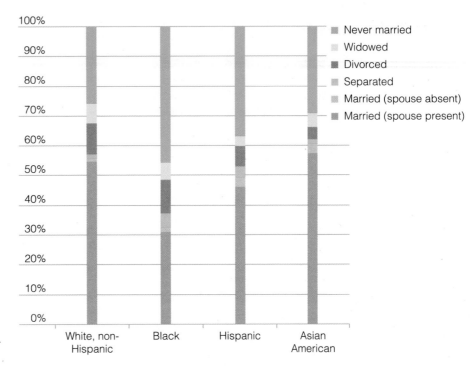

FIGURE 13.4 *Marital Status of the U.S. Population by Race*

Data: U.S. Census Bureau. 2009. *America's Families and Living Arrangements: 2003, Detailed Tables.* Current Population Reports. **www.census.gov**

Within marriage, gender also shapes the division of household labor. Women do far more work in the home and have less leisure time (Baca Zinn and Eitzen 2007). Most employed mothers do two jobs—the so-called *second shift* of housework after working all day in a paid job (Hochschild 1989). With more women in the labor market, wives with children have increased their working hours substantially more than men have (Mishel et al. 2005). Little wonder that people are feeling that they have less and less time.

Are men more involved in housework than in the past? Yes and no. Men report that they do more housework, but they devote only slightly more of their time to housework than in the past. Estimates vary regarding the amount of housework that men do, but studies generally find a large gap between the number of hours women give to housework and child care and the hours men give. Among couples where both partners are employed, only 28 percent equally share the housework. Fathers do more when there is a child in the house under two years of age, but the increase is mostly accounted for by the amount of child care men provide, not the housework they do. The end result is that men have about eleven more hours of leisure per week than women do (Press and Townsley 1998). Interestingly, sociologists have found that the allocation of housework is greatly affected by men's and women's experience in their own families of origin; those from households with a more egalitarian division of labor are likely to carry this into their own relationships (Cunningham 2001).

Despite a widespread belief that young professional couples are the most egalitarian, studies find that there is little difference across social class in the amount of housework that men do (Wright et al. 1992). African American husbands provide a greater

Is It True?*

	True	False
1. Half of all marriages end in divorce.		✓
2. Children who grow up in gay or lesbian families are likely to become gay.		✓
3. Single people are a larger proportion of the population than was true in the past.	✓	
4. Children are better off growing up in a home where mothers are not employed.		✓
5. Women and men find great satisfaction in trying to balance family and work.	✓	
6. Fathers' involvement in child care leads to more stable marriages.	✓	

*The answers can be found on page 326.

Is It True? (Answers)

1. FALSE. The divorce rate is based on the number of divorces in one year per 1000 people in the population; the marriage rate, the number of marriages in one year per 1000 people. This does NOT mean that half of all marriages end in divorce because marriages made in one year can last many years beyond and thus not all marriages are counted.

2. FALSE. There is little difference in outcomes for children growing up in gay or lesbian households relative to those living in heterosexual households, including their later sexual orientation (Stacey and Biblarz 2001).

3. TRUE. The number of never-married people has increased substantially since the 1970s, as have the number of divorced people in the population (U.S. Census Bureau 2009a).

4. FALSE. Researchers find no negative influence of parental employment per se on children; far more significant are other conditions, especially the economic stability of the family and other stresses for children that emerge from conflict and violence (Perry-Jenkins et al. 2000).

5. TRUE. Although balancing the demands of both work and family is stressful for both women and men, both report that being able to do so produces satisfaction. Stress results when women and men feel they have to make trade-offs at work in order to meet family demands (Milkie and Peltola 1999).

6. TRUE. Researchers find that fathers' increased involvement in child care is linked to more stable marriages, in part because mothers are then happier (Kalmijn 1999).

share of housework than do White husbands. Latino households have more diversity in gender roles than stereotypes about machismo would lead us to believe (McLoyd et al. 2000).

Although marriage can be seen as a romantic and intimate relationship between two people, it can also be seen within a sociological context. Marriage relationships are shaped by a vast array of social factors, not just the commitment of two people to each other. You see this especially when examining marital conflicts. Life events, such as the birth of a child, job loss, retirement, and other family commitments, such as elder care or caring for a child with special needs, all influence the degree of marital conflict and stability (Moen et al. 2001; Crowley 1998). As conditions in society change, people make adjustments within their relationships, but how well they can cope within a marriage depends on a large array of sociological not just individual factors.

Divorce

The United States leads the world not only in the number of people who marry, but also in the number of people who divorce. More than sixteen million people have divorced but not remarried in the population today; more women are in this group than men because women are less likely to remarry following a divorce. Since 1960, the rate of divorce has more than doubled, although it has declined recently since its all-time high in 1980.

You will often hear that one in every two marriages ends in divorce, but this is a misleading statistic. The marriage rate is 7.5 marriages per 1000 people and the divorce rate, 3.6 per 1000 people (U.S. Census Bureau 2009a). At first glance, it appears that there are half as many divorces as marriages. But the marriage rate is the number of marriages formed in

a year and does not include the number of continuing marriages; thus, divorce is not as widespread as one in every two of all marriages.

Still, the rate of divorce is high and has risen since 1950, though it has been declining since 1990. The likelihood of divorce is also not equally distributed across all social groups. Divorce is more likely for couples who marry young, while in their teens or early twenties. Second marriages are more likely than first marriages to end in divorce. Divorce is somewhat higher among low-income couples, a fact reflecting the strains that financial problems create. Divorce is also somewhat higher among African Americans than among Whites, partially because African Americans make up a disproportionate part of lower-income groups. Hispanics have a lower rate of divorce than either Whites or Blacks, probably the result of religious influence. Recently, the divorce rate among Asian Americans has also risen, interpreted as the possible shedding of cultural taboos (Armas 2003). This explanation seems supported by the fact that Asian Americans born in the United States are more likely to be divorced than Asians who immigrated (McLoyd et al. 2000).

A number of factors contribute to the current high rate of divorce in the United States. Demographic changes (shifts in the composition of the population) are part of the explanation. The rise in life expectancy, for example, has an effect on the length of marriages. In earlier eras, people died younger, and thus the average length of marriages was shorter. Some marriages that earlier would have ended with the death of a spouse may now be dissolved by divorce. Still, cultural factors also contribute to divorce.

In the United States, individualism is a cultural norm, placing a high value on a person's satisfaction within marriage. The cultural orientation toward individualism may predispose people to terminate a marriage in which they are personally unhappy. In other cultural contexts (including this society years ago), marriage, no matter how difficult, may have been seen as an unbreakable bond, regardless of whether one was unhappy. As we have seen in the previous data, for racial–ethnic groups in the United States, cultural factors can also influence the likelihood of divorce.

Changes in women's roles also are related to the rate of divorce. Women today are now less financially dependent on husbands than in the past, even though they still earn less. As a result, the economic interdependence that bound women and men as a marital unit is no longer as strong. Although most married women would be less well off without access to their husband's income, they could probably still support themselves. This can make it possible for people to end marriages that they find unsatisfactory.

For people in unhappy marriages, divorce, though painful and financially risky, can be a positive option (Kurz 1995). The belief that couples should stay together for the sake of the children is now giving way to a belief, supported by research, that a marriage with protracted conflict is more detrimental to children than divorce. Although there are periodic public outcries about the negative effect of divorce on children, many other factors influence their long-term psychological and social adjustment. Few children feel relieved or pleased by divorce; feelings of sadness, fear, loss, and anger are common, along with desires for reconciliation and feelings of conflicting loyalties. But most children adjust reasonably well after a year or so. Moreover, children's adjustment is influenced most by factors that precede the divorce. The single most important factor influencing children's poor adjustment is marital violence and prolonged discord (Stewart et al. 1997; Arendell 1998; Furstenberg 1998; Cherlin et al. 1998; Amato and Booth 1997). The emotional strain on children is significantly reduced if the couple remains amicable. If both parents remain active in the upbringing of the children, the evidence shows that children do not suffer from divorce; especially important is the ability of the mother to be an effective parent after a divorce. Her ability to be effective can be influenced by the resources she has and her ongoing relationship with the father (Buchanan et al. 1996; Simons 1996; Furstenberg and Nord 1985).

In the aftermath of divorce, many fathers become distant from their children. Sociologists have argued that the tradition of defining men in terms of their role as breadwinners minimizes the attachment they feel for their children. If the family is then disrupted, they may feel that their primary responsibility, as financial provider, is lessened, leaving them with a diminished sense of obligation to their children.

Family Violence

Generally speaking, the family is depicted as a private sphere where members are nurtured and protected, existing away from the influences of the outside world. Although this is the experience of many, families also can be locales for violence, disruption, and conflict. Family violence, hidden for many years, is a phenomenon that has recently been the subject of much sociological research.

Partner Violence. Estimates of the extent of domestic violence are hard to come by and notoriously unreliable because the majority of cases of domestic violence go unreported. The National Violence Against Women Office estimates that 25 percent of women will be raped, physically assaulted, or stalked by an intimate partner in their lifetime. Men also experience partner violence, although far less frequently. Women who experience violence are also twice as likely as men to be injured. Violence also occurs in gay and lesbian relationships, although silence around the issue may be even more pervasive given the marginalized status of gays and lesbians. Men living with male partners are just as likely to be raped, assaulted, or stalked as are women living with men, but the incidence of violence against women by women partners is about half as likely as heterosexual violence. Researchers conclude that this is because most domestic violence is committed by men. Violence is usually accompanied by emotionally abusive and controlling behavior. Jealous and dominating partners are the most likely perpetrators of domestic violence (Tjaden and Thoennes 2000; West 1998; Renzetti 1992).

One of the most common questions asked about domestic violence is why victims stay with their abuser. First, despite the belief that battered women do not leave their abusers, the majority do leave—at least for a period of time—and they seek ways to prevent further victimization (Gelles 1999). But, some

Activists who have spoken out against family violence have raised public awareness about this problem.

Bob Daemmrich/PhotoEdit

do not leave and others leave and then return. Why? The answers are complex and stem from sociological, psychological, and economic problems. Victims tend to believe that the batterer will change, but they also find they have few options; they may perceive that leaving will be more dangerous, because violence can escalate when the abuser thinks he (or she) has lost control. Many women are unable to support their children and meet their living expenses without a husband's income. Mandatory arrest laws in cases of domestic violence can exacerbate this problem because they may, despite their intentions, discourage a woman from reporting violence for fear her batterer will lose his job (Miller 1997). Sociological analyses of violence in the family have led to the conclusion that women's relative powerlessness in the family is at the root of high rates of violence against women. Because most violence in the family is directed against women, the imbalance of power between men and women in the family is the source of most domestic violence. Since women are relatively powerless within the society, they may not have the resources to leave their marriage (Kurz 1989; McCloskey 1996).

Child Abuse. Violence within families also victimizes many children who experience child abuse. Not all forms of child abuse are alike. Some people consider repeated spanking to be abusive; others think of this as legitimate behavior. Child abuse, however, is behavior that puts children at risk and may include physical violence and neglect. As with battering, the exact incidence of child abuse is difficult to know. In 2007, as an example, there were 3.2 million referrals involving 5.8 million children reported to child protective services around the nation. Of these, 25 percent were substantiated, but this is not likely to be an accurate measure of the actual extent of child abuse (U.S. Department of Health and Human Services 2007). Whereas with domestic and sexual violence men are the most likely perpetrators, women are just as likely to be the perpetrators of child abuse as are men.

Research on child abuse finds a number of factors associated with abuse, including chronic alcohol use by a parent, unemployment, and isolation of the family. Sociologists point to the absence of social supports—in the form of social services, community assistance, and cultural norms about the primacy of motherhood—as related to child abuse, because most abusers are those with weak community ties and little contact with friends and relatives (Baca Zinn and Eitzen 2007).

Incest. Incest is a particular form of child abuse involving sexual relations between persons who are closely related. A history of incest has been related to a variety of other problems, such as drug and alcohol abuse, runaways, delinquency, and various psychological problems, including the potential for violent partnerships in adult life. Studies find that fathers and uncles are the most frequent incestuous abusers and that incest is most likely in families where mothers are debilitated (such as by mental illness or alcoholism). In such families, daughters often take on the mothering role, being taught to comply with men's demands to hold the family together. Scholars have linked women's powerlessness within families to the dynamics surrounding incest (Herman 1981).

Elder Abuse. The National Center on Elder Abuse estimates that between one and two million elders are abused in the United States, but it is difficult to gauge the true extent of the problem. Elder abuse tends to be hidden in the privacy of families, and victims are reluctant to talk about their situations, so estimates are only approximations. What is known is that reports of elder abuse have increased. Whether this reflects an actual increase or more reporting is open to speculation (Teaster 2000; National Center on Elder Abuse 2007).

Why are the elderly abused? One explanation is that caring for the elderly is very stressful for the caregiver—usually a daughter who may be employed in addition to caring for the elderly person. Research finds that abusers are most likely to be middle-aged women and (sadly) the daughter of the victim—the person most likely to be caring for the older person. Sons, however, are most likely to be engaged in direct physical abuse, accounting for almost half of the known physical abusers. Sometimes the physical abuser is a husband, where the abuse is a continuation of abusive behavior in the marriage. The same factors that affect family life in any generation contribute to the problem of elder abuse (Teaster 2000).

Changing Families in a Changing Society

Like other social institutions, the family is in a constant state of change, particularly as new social conditions arise and as people in families adapt to the changed conditions of their lives. Some changes affect only a given family—the individual changes that come from the birth of a new child, the loss of a partner, divorce, migration, and other life events. These changes are what C. Wright Mills referred to as "troubles" (see Chapter 1). Some may even be happy events; the point is that they are changes that happen at the individual level, as people adjust to the presence of a new child, adjust to a breakup with a long-term partner, or grieve the loss of a spouse.

As Mills would have pointed out, many microsociological events that people experience in families have their origins in the broader *macrosociological* changes affecting society as a whole.

Global Changes in Family Life

Changes in the institutional structure of families are also being affected by the process of globalization. The increasing global basis of the economy means that people often work long distances from other family members—a phenomenon that occurs at all points on the social class spectrum, although the experience of such global mobility varies significantly by social class. A corporate executive may accumulate thousands—even millions—of first-class flight miles, crossing the globe to conduct business. A regional sales manager may spend most nights away from a family, likely staying in modestly priced motels and eating in fast-food franchises along the way. Truckers may sleep in the cabs of their tractor trailers after logging extraordinary numbers of hours of driving in a given week. Laborers may move from one state to the next, following the pattern of the harvest, living in camps away from families, and being paid by the amount they pick.

These patterns of work and migration have created a new family form, the **transnational family,** defined as families where one parent (or both) lives and works in one country while their children remain in the country of origin. A good example is found in Hong Kong, where most domestic labor is performed by Filipina women who work on multiple-year contracts managed by the government, typically on a live-in basis. They leave their children in the Philippines, usually cared for by a relative, and send money home; the meager wages they earn in Hong Kong far exceed the average income of workers in the Philippines. This pattern is so common that the average Filipino migrant worker supports five people at home; one in five Filipinos directly depends on migrant workers' earnings (Constable 1997; Parreñas 2001).

One need not go to other nations to see such transnational patterns in family life. In the United States, Caribbean women and African American women have had a long history of having to leave their children with others while they sought employment in different regions of the country. Central American and Mexican women may come to work in the United States while their children stay behind. Mothers may return to see their children whenever they can, or alternatively, children may spend part of the year with their mothers, part with other relatives.

Mothers in transnational families have to develop new concepts of their maternal role, because their situation means giving up the idea that biological mothers should raise their own children. Many have expanded their definition of motherhood to include breadwinning, traditionally defined as the role of fathers. Transnational women also create a new sense of home, one not limited to the traditional understanding of "home" as a single place where mothers, fathers, and their children reside (Hondagneu-Sotelo 2001; Hondagneu-Sotelo and Avila 1997; Das Gupta 1997; Alicea 1997).

Families and Social Policy

Family social policies are the subject of intense national debate. Should gay marriages be recognized by the state? What responsibility does society have to help parents balance the demands of work and family? Many issues on the front lines of national social policy engage intense discussions of families. Some claim the family is breaking down. Others celebrate the increased diversity among families. Many blame the family for the social problems our society faces. Drugs, low educational achievement, crime, and violence are often attributed to a crisis in "family values," as if rectifying these attitudes is all it will take to solve our nation's difficulties. The family is the only social institution that typically takes the blame for all of society's problems. Is it reasonable to expect families to solve social problems? Families are afflicted by most of the structural problems that are generated by racism, poverty, gender inequality, and class inequality. Expecting families to solve the problems that are the basis for their own difficulties is like asking a poor person to save us from the national debt.

Balancing Work and Family. Balancing the multiple demands of work and family is one of the biggest challenges for most families. With more parents employed, it is difficult to take time from one's paid job to care for newborn or newly adopted children, tend to sick children, or care for elderly parents or other family members. As more families include two earners, more people feel pulled in multiple directions, always strategizing to find the time to get everything done. Work institutions are structured on a gendered model of the male breadwinner, where family and work are assumed to be separate, nonintersecting spheres. But now there is significant "spillover" between family and work—work seeping into the home and home also affecting people's work (Moen 2003).

The **Family and Medical Leave Act (FMLA),** adopted by Congress in 1993, is meant to provide help for these conflicts. It requires employers to grant employees a total of twelve weeks in unpaid leave to care for newborns, adopted children, or family members with a serious health condition. The FMLA is the first law to recognize the need of families to care for children and other dependents. A number of conditions, however, limit the effectiveness of the FMLA, not the least of which is that the leave is unpaid, making it impossible for many employed parents. Many workers in firms where there are family-friendly policies worry that taking advantage of these policies will harm their prospects for career advancement (Blair-Loy and Wharton 2002). Currently, only 15 percent of workers have child-care benefits available to them from employers (U.S. Bureau of Labor Statistics 2006b). Among industrialized nations, the United States provides the least in support for maternity and child-care policies (see Table 13.2).

Child Care. Family leave policies, much as they are needed, also do not address the ongoing needs for child care. Almost one-half of families with children under age thirteen have child-care expenses, typically taking 9 percent of their earnings. Single-parent families pay an even larger percentage of earnings on child care (Giannarelli and Barsimantov 2000).

table13.2	Maternity Leave Benefits: A Comparative Perspective		
Country	Length of Maternity Leave	Percentage of Wages Paid in Covered Period	Provider of Coverage
Zimbabwe	90 days	60–75%	Employer
Cuba	18 weeks	100%	Social Security
Iran	90 days	66.7% for 16 weeks	Social Security
China	90 days	100%	Employer
Saudi Arabia	10 weeks	50 or 100%	Employer
Canada	17–18 weeks	55% for 15 weeks	Unemployment insurance
Germany	14 weeks	100%	Social Security to a ceiling; employer pays difference
France	16–26 weeks	100%	Social Security
Italy	5 months	80%	Social Security
Japan	14 weeks	60%	Social Security or health insurance
Russian Federation	140 days	100%	Social Security
Sweden	14 weeks	450 days, 100% paid	Social Security
United Kingdom	14–18 weeks	90% for 6 weeks; flat rate thereafter	Social Security
United States	12 weeks	n/a	n/a

Source: United Nations. 2000. *The World's Women 2000: Trends and Statistics.* New York: United Nations, pp. 140–143.

Many parents struggle to find good and affordable child care for their children, some relying on relatives for care; others, on paid providers; and some, a combination of both. In the United States, one-half of three-year-olds and two-thirds of four-year-olds now spend much of their time in child-care centers. But the national approach is one of patching together different programs and primarily relying on private initiatives for care. Compare this with France. Although participation is voluntary, in France almost all parents enroll young children in the *école maternelle* system, where a place is guaranteed to every child aged three to six. These child-care centers are integrated with the school system and are seen as a form of early education. Moreover, in the United States child-care costs match tuition costs at public universities, but in France child care is seen as a social responsibility and is paid by the government. National norms about whether families are a private or public responsibility clearly shape social policy (Clawson and Gerstel 2002; Folbre 2001).

Care work—work that sustains life, including child care, elder care, housework, and other forms of household labor—is increasingly provided to middle- and upper-class families by women of color and immigrant women. Nannies, cleaners, and personal attendants now do much of the domestic work that was once provided by wives and mothers. These trends raise many new questions for sociologists, such as how work is negotiated outside of the public labor market, how "mothering" is defined when it is provided by multiple people, how domestic workers care for their own

families, and what work conditions exist in the lives of domestic laborers (Hondagneu-Sotelo 2001). Clearly, new social policies are needed to address the needs of diverse families (Moen et al. 2004).

Elder Care. The shrinking size of families means that the proportion of elderly people is growing faster than the number of younger potential caretakers. As life expectancy has increased and people live longer, elder care becomes a greater and greater need. Family members provide almost all long-term care for the elderly—work that is often taken for granted (Glazer 1990; Meyer 1994).

Women, who shoulder the work of elder care, can now expect to spend more years as the child of an elderly parent than as the mother of children under eighteen. Indeed, young people now can expect to spend more years caring for an elderly parent than raising their own children. The effects of the burden of care are apparent in the stress that women report from this role. Women also believe they are better at elder care than their husbands and brothers, but with the rapid increase in the older population that lies ahead, these social norms may have to change. As the U.S. population ages, social policies will likely need to respond to this growing need.

Because families are so diverse, different families need different social supports. Family leave policies that give parents time off to care for their children or sick relatives are helpful, but of little use to people who cannot afford to take time off work without pay. Greater employer support for child care can help men and women meet family needs. Some policies will benefit some groups more than others—one reason why policymakers need to be sensitive to the diversity of family experiences. Social policies cannot solve all the problems that families face, but they can go a long way toward creating the conditions under which diverse family units can thrive.

Defining Religion

Sociologists study religion as both a belief system and a social institution. The belief systems of religion have a powerful hold on what people think and how they see the world. The patterns and practices of religious institutions are among the most important influences on people's lives. Sociologists are interested in several questions about religion: How are religious belief and practice related to other social factors, such as social class, race, age, gender, and level of education? How are religious institutions organized? How does religion influence social change? In using sociology to understand religion, what is important is not what one believes about religion, but one's ability to examine religion objectively in its social and cultural context.

thinkingsociologically

For one week keep a daily log, noting every time you see an explicit or implicit reference to religion. At the end of the week, review your notes and ask yourself how religion is connected to other social institutions. Based on your observations, how do you interpret the relationship between the *sacred* and the *secular* in this society.

What is religion? Most people think of it as a category of experience separate from the mundane acts of everyday life, perhaps involving communication with a deity or communion with the supernatural (Johnstone 1992). Sociologists define **religion** as an institutionalized system of symbols, beliefs, values, and practices by which a group of people interprets and responds to what they feel is sacred and that provides answers to questions of ultimate meaning (Glock and Stark 1965: Johnstone 1992). The elements of this definition bear closer examination:

1. **Religion is institutionalized.** Religion is more than just beliefs: It is a pattern of social action organized around the beliefs, practices, and symbols that people develop to answer questions about the meaning of existence. As an institution, religion presents itself as larger than any single individual; it persists over time and has an organizational structure into which members are socialized.

2. **Religion is a feature of groups.** Religion is built around a community of people with similar beliefs. It is a cohesive force among believers because it

is a basis for group identity and gives people a sense of belonging to a community or organization. Religious groups can be formally organized, as in the case of bureaucratic churches, or they may be more informally organized, ranging from prayer groups to cults. Some religious communities are extremely close-knit, as in convents; other communities are more diffuse, such as people who identify themselves as Protestant but attend church only on Easter.

3. **Religions are based on beliefs that are considered sacred.** The **sacred** is that which is set apart from ordinary activity for worship, seen as holy, and protected by special rites and rituals. The sacred is distinguished from the **profane,** which is of the everyday world and specifically not religious (Durkheim 1947/1912; Chalfant et al. 1987). Each religion defines what is to be considered sacred; most religions have sacred objects and sacred symbols. The holy symbols are infused with special religious meaning and inspire awe.

    A **totem** is an object or living thing that a religious group regards with special reverence. A statue of Buddha is a totem and so is a crucifix hanging on a wall. Among the Zuni (a Native American group), fetishes are totems; these are small, intricately carved animal objects representing different dimensions of Zuni spirituality. A totem is important not for what it is, but for what it represents. To a Christian taking communion, a piece of bread is defined as the flesh of Jesus; eating the bread unites the communicant mystically with Christ. To a nonbeliever, the bread is simply that—a piece of bread (McGuire 2001). Likewise, Native Americans hold certain ground to be sacred and are deeply offended when the holy ground is disturbed by industrial or commercial developers who see only potential profit.

4. **Religion establishes values and moral proscriptions for behavior.** A proscription is a constraint imposed by external forces. Religion typically establishes proscriptions for the behavior of believers, some of them quite strict. For example, the Catholic Church defines living together as sexual partners outside marriage as a sin. Often religious believers come to see such moral proscriptions as simply "right" and behave accordingly. At other times, individuals may consciously reject moral proscriptions, although they may still feel guilty when they engage in a forbidden practice. Of course, what people believe and what they do can be very contradictory, perhaps best exemplified by the various scandals recently reported involving sexual abuse of young boys by Catholic priests.

5. **Religion establishes norms for behavior.** Religious belief systems establish social norms about how the faithful should behave in certain situations. Worshippers may be expected to cover

David Silverman/Getty Images
Mario Tama/Getty images

Religious spirituality takes many forms but produces feelings of awe and reverence among believers, as in this Orthodox Christian baptism and this Jain ceremony of soaking in vermillion in recognition of a sacred tradition.

their heads in a temple, mosque, or cathedral, or to wear certain clothes. Such behavioral expectations may be quite strong. The next time you are at a gathering where a prayer is said before a meal, note how many people bow their heads, even though some of those present may not believe in the deity being invoked.

6. **Religion provides answers to questions of ultimate meaning.** The ordinary beliefs of daily life are **secular** beliefs and may be institutionalized, but they are specifically not religious. Science, for example, generates secular beliefs based on particular ways of thinking—logic and empirical observations are at the root of scientific beliefs. Religious beliefs, in contrast, often have a supernatural element. They emerge from spiritual needs and may provide answers to questions that cannot be probed with the profane tools of science and reason. Think of the difference in how religion and science explain the origins of life. Whereas science explains this as the result of biochemical and physical processes, different religions have other accounts of the origin of life.

The Significance of Religion in the United States

The United States is one of the most religious societies in the world. Two-thirds of Americans think religion can solve all or most of society's problems. Forty-two percent of the population describe themselves as "born again" (Winseman 2004). Religion is, for millions of people, the strongest component of their individual and group identity. Much of the world's most celebrated art, architecture, and music has its origins in religion, whether in the classical art of Western Europe, the Buddhist temples of the East, or the gospel rhythms of contemporary rock.

Religion is also strongly related to a number of social and political attitudes. Religious identification is a good predictor of how traditional a person's beliefs will be. People who belong to religious organizations that encourage intolerance of any form are most likely to be racially prejudiced. However, there is not a simple relationship between religious belief and prejudice, because religious principles are also often the basis for lessening racial prejudice. Those with deeper religious involvement tend to have more traditional gender attitudes; *homophobia* has also been linked to religious belief, although some religious congregations have actively worked to encourage the participation of gays and lesbians.

The Dominance of Christianity

Despite the U.S. Constitution's principle of the separation of church and state, Christian religious beliefs and practices dominate U.S. culture. Indeed, Christianity is often treated as if it were the national religion. It is commonly said that the United States is based on a Judeo-Christian heritage, meaning that our basic cultural beliefs stem from the traditions of the pre-Christian Old Testament of the Bible (the Judaic tradition) and the Gospels of the New Testament. The dominance of Christianity is visible everywhere. State-sponsored colleges and universities typically close for Christmas break, not Yom Kippur. Christmas is a national holiday, but not Ramadan, the most sacred holiday among Muslims. Despite the dominance of Christianity, however, the pattern of religion in the United States is a mosaic one.

Measuring Religious Faith

Religiosity is the intensity and consistency of practice of a person's (or group's) faith. Sociologists measure religiosity both by asking people about their religious beliefs and by measuring membership in religious organizations and attendance at religious services. The majority of people in the United States identify themselves as Protestant or Catholic, though there is great religious diversity within the United States (see Figure 13.5).

Forms of Religion

Religions can be categorized in different ways according to the specific characteristics of faiths and how religious groups are organized. In different societies and among different religious groups, the form religion takes reflects differing belief systems and reflects and supports other features of the society. Believing in one god or many, worshiping in small or large groups, and associating religious faith with gender roles all contribute to the social organization of religion and its relationship to the rest of society.

Religiosity varies significantly among different groups in society (Figure 13.6). Church membership and attendance is higher among women than men and more prevalent among older than younger people. African Americans are more likely than Whites to belong to and attend church. On the whole, church membership and attendance fluctuate over time; membership has decreased slightly since 1940, but attendance has remained largely the same since. Large, national religious organizations, such as the mainline Protestant denominations, have lost many members, whereas smaller, local congregations have increased membership (see Figure 13.7).

In recent years, there has been a decrease in the number of people who think that religion can answer

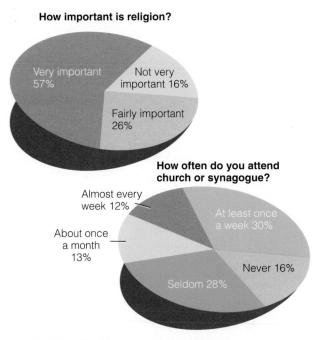

How important is religion?

Very important 57%

Not very important 16%

Fairly important 26%

How often do you attend church or synagogue?

Almost every week 12%

About once a month 13%

At least once a week 30%

Never 16%

Seldom 28%

FIGURE 13.6 *Measuring Religiosity*

Source: Newport, Frank. 2006. "Religion Most Important to Blacks, Women, and Older Americans." Gallup Poll. **www.gallup.org**; Newport, Frank. 2008. "Easter Season Finds a Religious, Largely Christian Nation." Gallup Poll. **www.gallup.com**

all or almost all of today's problems. Changes in immigration patterns have also affected religious patterns in the United States, with Muslims, Buddhists, and Hindus now accounting for several million believers (see Figure 13.8 on page 337; Niebuhr 1998; Haddad et al. 2003). One of the greatest changes has been a tremendous increase in the number identifying as evangelical Protestants, but Islam has also been one of the fastest growing religions in the United States in recent years (Dudley and Roozen 2001; Gallagher 2003).

One basic way to categorize religions is by the number of gods or goddesses adherents worship. **Monotheism** is the worship of a single god. Christianity and Judaism are monotheistic in that both Christians and Jews believe in a single god who created the universe. Monotheistic religions typically define god as omnipotent (all-powerful) and omniscient (all-knowing). **Polytheism** is the worship of more than one deity. Hinduism, for example, is extraordinarily complex, with millions of gods, demons, sages, and heroes—all overlapping and entangled in religious mythology; within Hinduism, the universe is seen as so vast that it is believed to be beyond the grasp of a single individual, even a powerful god (Grimal 1963).

Religions may also be patriarchal or matriarchal. **Patriarchal religions** are those in which the beliefs and practices of the religion are based on male power and authority. Christianity is a patriarchal religion; the ascendancy of men is emphasized by the role of women in the church, the instruction given on relations between the sexes, and even the language of worship itself. **Matriarchal religions** are based on the

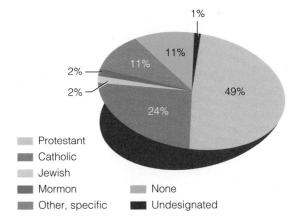

1%

11%

11%

2%

2%

49%

24%

Protestant
Catholic
Jewish
Mormon   None
Other, specific   Undesignated

FIGURE 13.5 *Religious Identification in the United States*

Source: Gallup Poll. 2006. "Religion." Princeton, NJ: Gallup Organization. **www.gallup.com**

Percentage of new congregations started/founded

FIGURE 13.7 *The New Face of Religion*

Source: Dudley, Carl S., and Roozen, David A. 2001. *Faith Communities Today: A Report on Religion in the United States Today.* Hartford, CT: Hartford Institute for Religion Research, Hartford Seminary.

centrality of female goddesses, who may be seen as the source of food, nurturance, and love, or who may serve as emblems of the power of women (McGuire 2001). In societies based on matriarchal religions, women are more likely to share power with men in the society at large. Likewise, in highly sexist, patriarchal societies, religious beliefs are also likely to be patriarchal.

Sociological Theories of Religion

The sociological study of religion probes how religion is related to the structure of society. Recall that one basic question sociologists ask is, "What holds society together?" Coherence in society comes from both the social institutions that characterize society and the beliefs that hold society together. In both instances, religion plays a key role. From the functionalist perspective of sociological theory, religion is an integrative force in society because it has the power to shape collective beliefs. In a somewhat different vein, the sociologist Max Weber saw religion in terms of how it supported other social institutions. Weber thought that religious belief systems provided a cultural framework that supported the development of specific social institutions in other realms, such as the economy. From yet a third point of view, based on the work of Karl Marx and conflict theory, religion is related to social inequality in society (see Table 13.3).

table 13.3	Theoretical Perspectives on Religion		
	Functionalism	**Feminist Theory**	**Symbolic Interaction**
Religion and the social order	Is an integrative force in society	Is the basis for intergroup conflict; inequality in society is reflected in religious organizations, which are stratified by factors such as race, class, or gender	Is socially constructed and emerges with social and historical change
Religious beliefs	Provide cohesion in the social order by promoting a sense of collective consciousness	Can provide legitimation for oppressive social conditions	Are socially constructed and subject to interpretations; they can also be learned through religious conversion
Religious practices and rituals	Reinforce a sense of social belonging	Define in-groups and out-groups, thereby defining group boundaries	Are symbolic activities that provide definitions of group and individual identity

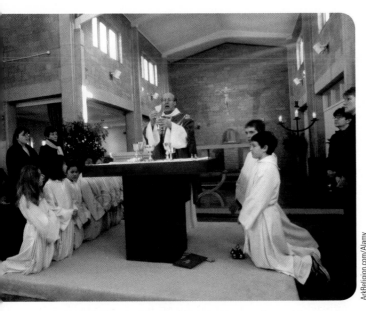

Emile Durkheim theorized that public rituals provide cohesion in society.

Emile Durkheim: The Functions of Religion

Emile Durkheim argued that religion is functional for society because it reaffirms the social bonds that people have with each other, creating social cohesion and integration. Durkheim believed that the cohesiveness of society depends on the organization of its belief system. Societies with a unified belief system are highly cohesive; those with a more diffuse or competing belief system are less cohesive.

Religious **rituals** are symbolic activities that express a group's spiritual convictions. Making a pilgrimage to Mecca, for example, is an expression of religious faith and a reminder of religious belonging. In Durkheim's view, religious rituals are vehicles for the creation, expression, and reinforcement of social cohesion. Groups performing a ritual are expressing their identity as a group. Whether the rituals of a group are highly elaborated or casually informal, they are symbolic behaviors that sustain group awareness of unifying beliefs. Lighting candles, chanting, or receiving a sacrament are behaviors that reunite the faithful and help them identify with the religious group, its goals, and its beliefs (McGuire 2001). Durkheim believed that religion binds individuals to the society in which they live by establishing what he called a **collective consciousness,** the body of beliefs common to a community or society that gives people a sense of belonging. In many societies, religion establishes the collective consciousness and creates in people the feeling that they are part of a common whole.

Durkheim's analysis of religion suggests some of the key ideas in symbolic interaction theory, particularly in the significance he gave to symbols in religious behavior. Symbolic interaction theory sees religion as a socially constructed belief system, one that emerges in different social conditions. From the perspective of symbolic interaction, religion is a meaning system that gives people a sense of identity, defines one's network of social belonging, and confers one's attachment to particular social groups and ways of thinking.

Max Weber: The Protestant Ethic and the Spirit of Capitalism

Theorist Max Weber also saw a fit between the religious principles of society and other institutional needs. In his classic work *The Protestant Ethic and the Spirit of Capitalism,* Weber argued that the Protestant faith supported the development of capitalism in the Western world. He began by noting a seeming contradiction: How could a religion that supposedly condemns extensive material consumption coexist in a society such as the United States with an economic system based on the pursuit of profit and material success?

Weber argued that these ideals were not as contradictory as they seemed. As the Protestant faith developed, it included a belief in predestination—one's salvation is predetermined and a gift from God, not something earned. This state of affairs created doubt and anxiety among believers, who searched for clues in the here and now about whether they were among the chosen—called the "elect." According to Weber, material success was taken to be one clue that a person was among the elect and thus favored by God, which drove early Protestants to relentless work as a means of confirming (and demonstrating) their salvation. As it happens, hard work and self-denial—the key features of the **Protestant ethic**—lead not only to salvation but also to the accumulation of capital. The religious ideas supported by the Protestant ethic therefore fit nicely with the needs of capitalism. According to Weber, these austere religionists stockpiled wealth, had an irresistible motive to earn more (that is, eternal salvation), and were inclined to spend little on themselves, leaving a larger share for investment and driving the growth of capitalism (Weber 1958/1904).

Karl Marx: Religion, Social Conflict, and Oppression

Durkheim and Weber concentrated on how religion contributes to the cohesion of society. Religion can also be the basis for conflict, as we see in the daily headlines of newspapers. In the Middle East, differences between Muslims and Jews have caused decades of political instability. These conflicts are not solely religious, but religion plays an inextricable part. Certainly religious wars, religious terrorism, and religious genocide have contributed to some of the most violent and tragic episodes of world history. The image of religion in history has two incompatible sides: piety and contemplation on the one hand, battle flags on the other. Domestic conflicts over ethical issues such as abortion, assisted suicide, and school prayer evolve from religious values even though they are played out in the secular world of politics and public opinion. Conflict theory illuminates many of the social and political conflicts that engage religious values.

The link between religion and social inequality is also key to the theories of Karl Marx. Marx saw religion as a tool for class oppression. According to Marx, oppressed people develop religion, with the urging of the upper classes, to soothe their distress (Marx, 1843/1972). The promise of a better life hereafter makes the present life more bearable, and the belief that "God's will" steers the present life makes it easier for people to accept their lot. To Marx, religion is a form of *false consciousness* (see Chapter 8) because it prevents people from rising up against oppression. He called religion the "opiate of the people" because it encourages passivity and acceptance.

Marx saw religion as supporting the status quo and being inherently conservative (that is, resisting change and preserving the existing social order). To Marx, religion promotes stratification because it supports a hierarchy of people on Earth and the subordination of humankind to divine authority. Christianity, for example, supported the system of slavery. When European explorers first encountered African people, they regarded them as godless savages, and they justified the slave trade by arguing that slaves were being converted to the Christian way of life. Principles of Christianity thus legitimated the system of slavery in the eyes of the slave owners and allowed them to see themselves as good people, despite their enslavement of other human beings.

At the same time, religion can be the basis for liberating social change. In the civil rights movement in the United States and in Latin American liberation movements, the words and actions of religious organizations have been central in mobilizing people for change. This does not undermine Marx's main point, however, because there remains ample evidence of the role of religion in generating social conflict and resisting social change.

Symbolic Interaction: Becoming Religious

Recall that symbolic interaction theory states that people act toward things on the basis of the meaning things have for them and that those meanings emerge through social interaction. This can explain much about human behavior that is based in religious ideas. Thus, seen from outside the faith, religious practices (kneeling in church, wearing a yamaka, making a pilgrimage to Mecca, or chanting) may seem peculiar or different, but within the faith these and other religious practices carry meaning—meaning that is deeply important to religious believers. But not only does symbolic interaction explain particular religious behaviors, it can also explain how other behaviors may be based in the meanings that religion holds for

people. For example, we have earlier said that even something as reprehensible as suicide bombings can be understood if you understand how religious zealots interpret the meaning of religious texts.

Symbolic interaction theory can also help you understand how people become religious, a process sociologists call *religious socialization*. Religious socialization may be a slow and gradual process, such as in how children learn religious values over time. Religious socialization can also be more dramatic, as when a person joins a cult or some other extreme religious group. And, people may reinterpret religious beliefs when they question their religious faith or even switch to a new religion, such as when a Christian converts to Judaism. The emphasis on meaning that is typical of symbolic interaction helps explain how the same religion can be interpreted differently by different groups or in different times. How, for example, could Christian beliefs be used by some to make slavery seem legitimate, while at the same time for others it provided the belief system that helped them survive slavery as well as to fight against it. And now, religion is very much a factor in shaping people's political behavior—how they think about social issues, as well as how they vote.

Symbolic interaction thus sees religious belief—and its meaning to different people—as essential for understanding many forms of social behavior. Symbolic interaction also helps explain how different religious beliefs and practices emerge in social and historical contexts—contexts that shape what religion means to people.

Diversity and Religious Belief

The world is marked by diverse religious beliefs. Christianity has the largest membership, followed by Islam. But Hindus, Jews, Confucianists, Buddhists, and other folk religions also comprise the world's religions (see Figure 13.8). In the United States, religious identification varies with a number of social factors, including age, income level, education, and political affiliation. Younger people are more likely than older people to express no religious preference. Those in higher income brackets are more likely to identify as Catholic or Jewish than those in lower income brackets, who are more likely to identify as Protestant, although these trends vary among Protestants by denomination. Fundamentalist Protestants, for example, are most likely to come from lower-income groups.

The Influence of Race and Ethnicity

Race is one of the most significant indicators of religious orientation (see Figure 13.9). African Americans are much more likely than Whites, Hispanics, or Asian Americans to say that religion is very important in

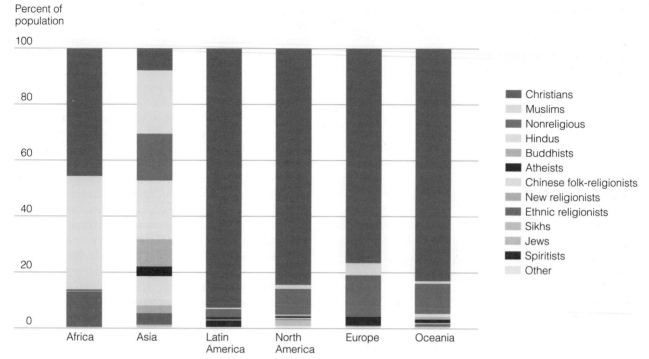

Percent of
population

FIGURE 13.8 *Viewing Society in Global Perspective: The World's Religions*

Data: From the U.S. Census Bureau. 1999. *Statistical Abstract of the United States, 1999.* Washington, DC: U.S. Government Printing Office, p. 831.

their lives. And, though most African American identify as Protestant, a small, but growing number, are Catholic (Newport 2006; Pew Research Center 2003). Many urban African Americans have also become committed Black Muslims, which involves strict regulation of dietary habits and prohibition of many activities, such as alcohol use, drug use, gambling, use of cosmetics, and hair straightening. The emphasis among Black Muslims on self-reliance and traditional African identity has earned it a fervent following, although the actual number of Black Muslims in the United States is relatively small. For many African Americans,

religion has been a defense against the damage caused by racism. Churches have served as communal centers, political units, and sources of social and community support, making churches among the most important institutions within the African American community (Gilkes 2000). Religion also has been a strong force in Latino communities, with the largest number identifying as Catholic. However, there are a growing number of Latino Protestants, both in mainstream Protestant denominations and in fundamentalist groups.

Asian Americans have a great variety of religious orientations, in part because the category "Asian

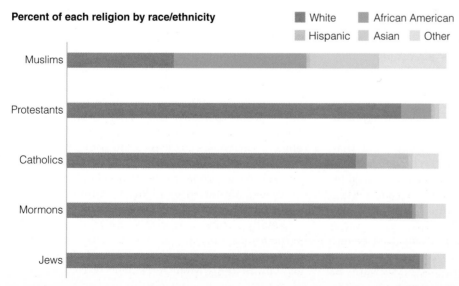

Percent of each religion by race/ethnicity

FIGURE 13.9 *Racial–Ethnic Breakdown of Religious Groups in the United States*

Source: Younis, Mohammed. 2009. "Muslim Americans Exemplify Diversity, Potential." Gallup Poll. **www.gallup.com.** Used with permission.

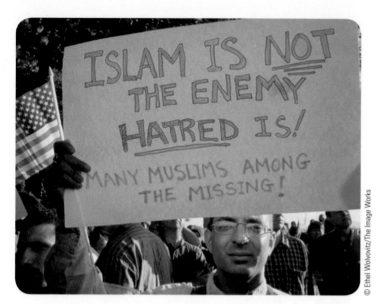

Religion can be interpreted both as producing social conflict and as promoting social justice.

American" is constructed from so many different Asian cultures. Hinduism and Buddhism are common among Asians, but so is Christianity. As with all groups whose family histories include immigration, religious belief and practice among Asian Americans frequently changes between generations. The youngest generation may not worship as their parents and grandparents did, although some aspects of the inherited faith may be retained. Within families, the discontinuity with a religious past brought on by cultural assimilation can be a source of tension between grandparents, parents, and children. Within the United States, Asian Americans often mix Christian and traditionally Buddhist, Confucian, or Hindu beliefs, resulting in new religious practices.

Muslims are a growing segment of U.S. society; two-thirds of Muslims are former immigrants, but a substantial portion (35 percent) are native-born, either African American or others who have converted to this faith or were raised in a Muslim household. Despite stereotypes about Muslim conservatism, on many issues Muslim Americans are actually more liberal than the general public—for example, they are more likely to vote Democratic. At the same time, however, Muslim Americans are generally less tolerant of homosexuality than the public at large (Pew Research Center 2007). And, the majority say that being Muslim in the United States has become more difficult since 9/11. Interestingly, studies find that younger Muslims (those under thirty) tend to be more observant than older Muslims—perhaps explained by the heightened identity that has emerged since 9/11.

Religious Extremism

Religious extremism refers to actions and beliefs that are driven by high levels of religious intolerance. Religious extremists tend to see the world in simplistic either/or terms—dividing people into either good or evil, godly or demonic. Such divisive imagery reduces the complexity of human life into simplistic categories—categories that fuel hate and conflict. Thus, religious extremism usually becomes the basis for extremely violent behavior. And it produces martyrs and enemies, as if the world's people were divided along some "axis of evil," good people on one side and everyone else on the other (Anthony et al. 2002). When such religious fanaticism is intertwined with the power of a state government, religiously inspired leaders can use the power of the military and government propaganda to wield extraordinary power.

Religious extremism is now associated with terrorism in the Middle East, but it has fueled other horrendous acts, including mass executions and genocide, enslavement, and other heinous crimes against humanity. All religions, taken to an extreme, are dangerous social forces because they can drive adherents to think they are doing sacred work even when they are engaging in violent, murderous behavior.

It is easy to see the acts of religious extremists as the work of misled individuals, but those who study religious extremism know that it has social origins. Religious extremism is learned, usually within a narrowly circumscribed social world, such as the *madrassas*—religious camps in Pakistan (and other areas) where young boys are taught a strict interpretation of Islam. For young boys uprooted from families by war, detached from other social contacts, and with no other education, it is easy to be socialized into a narrow worldview that gives them a cause to fight for (Rashid 2000).

Knowing this helps explain how young men in Pakistan and other places could celebrate the death of Americans by cheering and waving guns in the streets—behavior that can easily emerge when all you are taught is to perceive some other group as your enemy.

Religious extremism typically emerges in countries (or subcultures) where people are very poor and without access to education. Young men without an education and with little life opportunity can be attracted to violence as a way of asserting a collective identity (Khashan and Kreidie 2001; Stern 2003). Many of the young suicide bombers who have committed horrendous acts of violence come from needy families who gain status and money from having their children do the footwork of terrorists.

Also, where there is a lack of modernism—and where people perceive that their traditional way of life is being overtaken by Western influences—religious extremism can come from trying to defend a traditional way of life (Pain 2002). Where most people are poor, frustrations can be channeled into extremist movements fueled by highly traditional religious beliefs (Heilman and Kaiser 2002).

Finally, religious extremist movements tend to be highly patriarchal—that is, based on the power of men and the subordination of women. This is true not only in the extremist factions of contemporary Islamic movements, but also in extremist segments

map13.1   *MAPPING AMERICA'S DIVERSITY*

Religiosity Across the Nation

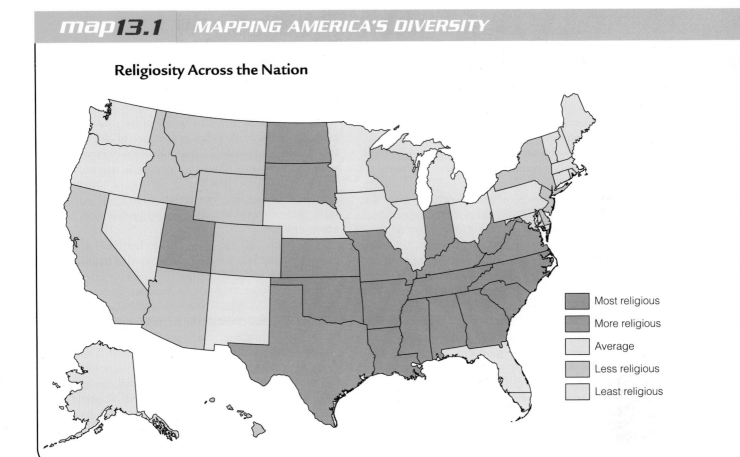

Most religious
More religious
Average
Less religious
Least religious

A survey by the Gallup Organization asked a national sample of people from different states whether religion was an important part of their daily lives. By comparing responses in the 50 different states, the Gallup Organization was able to rank the states according to the most religious (Mississippi, where 85 percent said religion was important in their daily lives) and least religious (Vermont, where 42 percent said religion was important in their daily lives). This map shows the range of states from least to most religious. What social factors do you think might influence the extent of religiosity in these different states and regions of the nation?

Source: Newport, Frank. 2009. "State of the States: Importance of Religion." Michelle Krogmeier/Gallup. Used with permission.

of the Christian right in the United States (Antrobus 2002; Ferber 1998). When religious extremism links with militaristic and patriarchal values, it becomes extremely dangerous.

Religious Organizations

Sociologists have organized their understanding of the various religious organizations into three types: churches, sects, and cults. These are *ideal types* in the sense that Max Weber used the term. That is, the ideal types convey the essential characteristics of some social entity or phenomenon, even though they do not explain every feature of each entity included in the generic category.

Churches are formal organizations that tend to see themselves, and are seen by society, as the primary and legitimate religious institutions. They tend to be integrated into the secular world to a degree that sects and cults are not. They are sometimes closely tied to the state. Many churches are organized as complex bureaucracies with a division of labor and different roles for groups within, including a formally trained clergy and professional staff. However, some churches may be smaller, less formal, with devoted, but less formally trained, clergy. A new phenomenon for churches is the development of *megachurches*—those with memberships numbering into the thousands. These are increasingly common. Not only do megachurches have huge attendance, but they may broadcast on huge screens, possibly even televising church services.

Sects are groups that have broken off from an established church. They emerge when a faction within an established religion questions the legitimacy or purity of the group from which they are separating. Many sects form as offshoots of existing religious organizations. Sects tend to place less emphasis on organization (as in churches) and more emphasis on the purity of members' faith. The Shakers, for example, were formed by departing from the Society of Friends (the Quakers). They retained some Quaker practices, such as simplicity of dress and a belief in pacifism, but departed from Quaker religious philosophy. The Shakers believed that the second coming of Christ was imminent, but that Christ would appear in the form of a woman (Kephart 1993). Sects tend to admit only truly committed members, refusing to compromise their beliefs. Some sects hold emotionally charged worship services, although others, like the Amish, are more stoic. The Shakers, for example, had such emotional services that they shook, shouted, and quivered while "talking with the Lord," earning them their name. The only bodily contact permitted among the Shakers was during the unrestrained religious rituals; they were celibate (did not have sexual relations) and gained new members only through adoption of children or recruitment of newcomers (Kephart 1993).

Cults, which are like sects in their intensity, are religious groups devoted to a specific cause or charismatic leader. Many cults arise within established religions and sometimes continue to peaceably reside within the parent religion simply as a fellowship of people with a particular, often mystical, dogma. As they are developing, it is common for tension to exist between cults and the society around them. Cults tend to exist outside the mainstream of society, arising when believers think that society is not satisfying their spiritual needs and attracting those who feel a longing for meaningful attachments. Internally, cults seldom develop an elaborate organizational structure but are instead close-knit communities held together by personal attachment and loyalty to the cult leader.

Cults form around leaders with great **charisma,** a quality attributed to individuals believed by their followers to have special powers (Johnstone 1992). Typically, followers are convinced that the charismatic leader has received a unique revelation or possesses supernatural gifts. Although there are exceptions, cult leaders are usually men, probably because men are more likely to be seen as having the characteristics associated with charismatic leadership.

Debunking Society's Myths

Myth: People who join extreme religious cults are maladjusted and have typically been quickly brainwashed by cult leaders.

Sociological perspective: Conversion to a religious cult usually involves a gradual process of resocialization wherein the convert voluntarily develops new associations with others and develops a new worldview based on these new relationships.

Religion and Social Change

What is the role of religion in social change? Durkheim saw religion as promoting social cohesion; Weber saw it as culturally linked to other social institutions; Marx assessed religion in terms of its contribution to social oppression. Is religion a source of oppression, or is it a source of personal and collective liberation from worldly problems? There is no simple answer to this question. Religion has had a persistent conservative influence on society, but it has also been an important part of movements for social justice and human emancipation.

The role of religious organizations in social change has of late become a question of public policy. Should faith-based organizations receive government support for work they do in helping people, or does this violate the constitutional separation of church and state? The constitutional issues will ultimately be settled by law, but sociological research sheds light on the implications of such organizations. Though liberals fear that faith-based organizations will infuse religion into government too much and conservatives hope that faith-based initiatives will support their political agenda, research shows that faith-based initiatives enhance the participation of traditionally disadvantaged groups in the democratic process (Wood 2002; Kniss 2003).

The public debate about faith-based initiatives is occurring at a time when evangelical groups have increased membership and influence and have affiliated with conservative political causes, dramatically increasing the influence of religion on politics. At the same time, there has been a decrease in the importance of religion to many people. As a social institution, religion is in transition. Religion, like other aspects of society, is also becoming more commercialized. A large self-help industry has developed in religious publication; religious music is increasingly successful as a form of enterprise. All sorts of religious products are bought and sold in what sociologists now call a "spiritual marketplace" (Roof 1999; Wuthnow 1998). Clearly, religion influences social change, but it is also influenced by the same changes that affect other social institutions.

At the same time, religion continues to have an important role in liberation movements around the world. Throughout the world, liberation theologians have used the prestige and organizational resources of the Catholic Church to develop a consciousness of oppression among poor peasants and working-class people. Likewise, in the United States, churches have had a prominent role in the civil rights movement (Marx 1967/1867; Morris 1984). Churches supplied the infrastructure of the developing Black protest movements of the 1950s and 1960s, and the moral authority of the church was used to reinforce the appeal to Christian values as the basis for racial justice. Now, they continue to be important places for the mobilization of Black politics and provide an important source of community support—often when other institutions have abandoned the Black community (Zuckerman 2002).

The role of women is also changing in most religious organizations. Women have long been denied the right to full participation in many faiths. Some religions still refuse to ordain women as clergy, but the public generally supports the ordination of women. Women now make up a large portion of divinity students. Whereas traditional religious images of women have provided the basis for the subordination of women, those stereotypes are eroding. In sum, religion is a force of both social change and social stability.

Chapter Summary

- ### How are different kinship systems defined?

  All societies are organized around a *kinship system,* varying in how many marriage partners are allowed, who can marry whom, how descent is determined, family residence, and power relations within the family. *Extended family* systems develop when there is a need for extensive economic and social cooperation. The *nuclear family* is the result of the rise of Western industrialization that separated production from the home.

- ### What does sociological theory contribute to our understanding of families?

  Functionalism emphasizes that families have the function of integrating members to support society's needs. *Conflict theorists* see the family as a power relationship, related to other systems of inequality. *Feminist theory* emphasizes the family as a gendered institution and is critical of perspectives that take women's place in the family for granted. *Symbolic interaction* takes a more microscopic look at families, emphasizing how different family members experience and define their family experience.

- ### What changes characterize the diversity in contemporary families?

  One of the greatest changes in families has been the increase in female-headed households, which are most likely to live in poverty. The increase in women's labor force participation has also affected families, resulting in dual roles for women. Stepfamilies face unique problems stemming from

  the blending of two households. Gay and lesbian households are also more common and challenge traditional heterosexual definitions of the family. Single people make up an increasing portion of the population, due in part to the later age when people marry.

- ### Is marriage declining?

  The United States has both the highest marriage rate and the highest divorce rate of any industrialized nation. The high divorce rate is explained as the result of a cultural orientation toward individualism and personal gratification, as well as structural changes that make women less dependent on men within the family.

- ### Why is family violence such a problem?

  Family violence takes several forms, including partner violence, child abuse, incest, and elder abuse. Power relationships within families, as well as gender differences in the division of labor, help explain domestic violence.

- ### What major changes are affecting contemporary families?

  Changes at the global level are producing new forms of families—*transnational families*—where at least one parent lives and works in a different nation from the children. Social policies designed to assist families should recognize the diversity of family forms and the interdependence of the family with other social conditions and social institutions.

- ### *What are the elements of a religion?*
  Sociologists are interested in religion because of the strong influence it has in society. *Religion* is an institutionalized system of symbols, beliefs, values, and practices by which a group of people interprets and responds to what they feel is sacred and that provides answers to questions of ultimate meaning.

- ### *How do sociologists measure the significance of religion for people, and what forms does religion take?*
  The United States is a deeply religious society. Christianity dominates the national culture, even though the U.S. Constitution specifies a separation between church and state. *Religiosity* is the measure of the intensity and practice of religious commitment.

- ### *How do the different sociological theories analyze religion?*
  Durkheim understood religions and religious rituals as creating social cohesion. Weber saw a fit between the ideology of the *Protestant ethic* and the needs of a capitalistic economy. Religion is also related to social conflict. Marx saw religion as supporting societal oppression and encouraging people to accept their lot in life. *Symbolic interaction theory* focuses on the process by which people become religious. Religious conversion involves a dramatic transformation of religious identity and involves several phases through which individuals learn to identify with a new group and lose other existing social ties.

- ### *What diversity exists in religious faith and practice?*
  The United States is a diverse religious society. Protestants, Catholics, Jews, and, increasingly, Muslims, make up the major religious faiths in the United States. *Religious extremism* can emerge in any religion and is generated by certain societal characteristics.

- ### *How is a religion organized?*
  Churches are formal religious organizations. They are distinct from *sects,* which are religious groups that have withdrawn from an established religion. *Cults* are groups that have also rejected a dominant religious faith, but they tend to exist outside the mainstream of society.

- ### *How has religion been affected by social change?*
  In recent years, there has been an enormous growth in conservative religious groups. Religion is a conservative influence in society, but religion also has an important part in movements for human liberation, including the civil rights movement and the move to ordain women in the church.

Key Terms

bilateral kinship   315
charisma   340
church   339
collective consciousness   335
cult   340
extended families   315
family   315
Family and Medical Leave
   Act (FMLA)   329
kinship system   315
matriarchal religion   333

matrilineal kinship   315
matrilocal   315
monogamy   315
monotheism   333
nuclear family   316
patriarchal religion   333
patrilineal kinship   315
patrilocal   315
polygamy   315
polytheism   333
profane   331

Protestant ethic   335
religion   331
religiosity   333
religious extremism   338
ritual   335
sacred   331
sect   339
secular   332
totem   331
transnational family   329

Online Resources

Sociology: The Essentials Companion Website

www.cengage.com/sociology/andersen
Visit your book companion website where you will find more resources to help you study and write your research papers. Resources include web links, learning objectives, Internet exercises, quizzing, and flash cards.

 is an easy-to-use online resource that helps you study in less time to get the grade you want NOW.

www.cengage.com/login
Need help studying? This site is your one-stop study shop. Take a Pre-Test and Cengage NOW will generate a Personalized Study Plan based on your test results. The Study Plan will identify the topics you need to review and direct you to online resources to help you master those topics. You can then take a Post-Test to determine the concepts you have mastered and what you still need to work on.

There are both positive and negative consequences to early tracking of students (a system discussed in detail later in this chapter), but the negatives outweigh the positives. Ollie Taylor's case exemplifies the negative consequences, yet he may also benefit from the positive aspects of getting an education in the United States, even with tracking in place: As an African American, he is six times more likely now than in 1940 to graduate from high school.

His chances of dropping out prior to graduation are, unfortunately, still fairly high. They are considerably higher than for a White youngster of his age, but significantly lower than for a Hispanic young man. If he graduates, he will stand a better chance of avoiding the grim path of many urban Black and Latino male dropouts and many poor White youth—a path leading to unemployment and possibly drug abuse, poor health, perhaps prison, and death at an early age.

Like education, health care in the United States is an institution. And like education, health care as an institution is in crisis. The public wonders: Has the quality of health care in the United States declined over the last two decades? Are doctors and nurses as good as they used to be? Are doctors overpaid? Are hospitals overcharging? Are lawsuits against the medical profession increasing, and if so, why? Why are costs for medical insurance so high, and why do so many people not have health insurance?

Ethnic and cultural diversity in the public high school classroom are increasingly common.

Bob Daemmrich/The Image Works

These public issues may seem to be simple matters of personal responsibility or individual rights, but sociologists show that illness in our society is strongly influenced by social factors. How we treat illnesses is highly influenced by race, social class, gender, and age; the same can be said of how we treat ill people. In different societies, the *same* physical state may be regarded as either sickness or wellness, demonstrating that the definition of illness is at least partly socially constructed. Physical conditions regarded as illness might be treated by prayer or witchcraft, surgery or radiation therapy, depending on the culture or society in question.

Schooling and Society: Theories of Education

Education in a society is concerned with the systematic transmission of the society's knowledge. This includes teaching formal knowledge such as the "three R's," reading, writing, and arithmetic, as well as morals, values, and ethics. Education prepares the young for entry into society and is thus a form of socialization. Sociologists refer to the more formal, institutionalized aspects of education as **schooling.**

The Rise of Education in the United States

Compulsory education is a relatively new idea. During the nineteenth century, many states did not yet have laws requiring education for everyone. Most jobs in the middle of the nineteenth century demanded no education or literacy whatsoever. Education was considered a luxury, available only to children of the upper classes (Cookson and Persell 1985). Education for slaves was prohibited by law until 1900, long after the Emancipation Proclamation passed in 1863.

When compulsory education was established in 1900, the law passed in all states except a few in the South, where Black Americans were still largely denied formal education of any kind (Higginbotham 1978). In the past, state laws in the South and West have also prohibited education for Hispanics, American Indians, and Chinese immigrants. State laws requiring attendance were generally enforced for White Americans at least through eighth grade. Education all the way through high school lagged considerably. In 1910, less than 10 percent of White eighteen-year-olds in the United States had graduated from high school.

Attendance in both high school and college has expanded considerably, such that in the United States today almost 90 percent of those under thirty-five have received at least a high school diploma (see Figure 14.1). High school and college graduation rates, however, are not equal across racial groups. High school completion for Blacks lags behind that of Whites, and that of Hispanics lags significantly behind that of Blacks. In fact the rate of dropout before graduation for Hispanics is

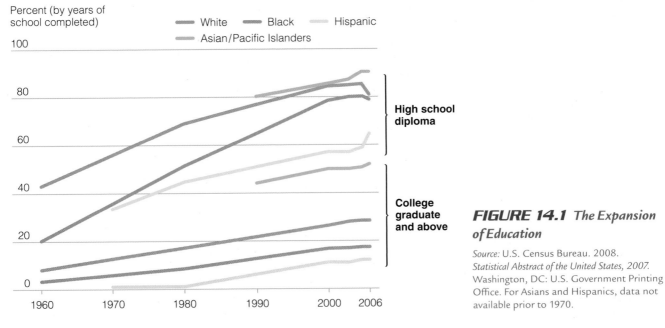

Percent (by years of school completed)

— White   — Black   — Hispanic
— Asian/Pacific Islanders

High school diploma

College graduate and above

FIGURE 14.1 *The Expansion of Education*

Source: U.S. Census Bureau. 2008. *Statistical Abstract of the United States, 2007.* Washington, DC: U.S. Government Printing Office. For Asians and Hispanics, data not available prior to 1970.

twice that of Blacks (about 40 percent versus 20 percent, respectively; Darling-Hammond 2007). As Figure 14.1 shows, similar trends exist for college graduations.

Education in the United States is of four kinds: public, private, parochial, and especially within the last two decades, *home schooling*. Like private schooling, home schooling has been disproportionately for White and wealthier students: As of 2000, over 95 percent of home-schooled students at the elementary school level were White, and almost all have been Protestant. Parents of the home-schooled wish to avoid the negative peer influences and bullying they perceive exist in public education, and the parents seek to teach moral and religious precepts that are generally not taught in public schools. Some parents engage their children in home schooling for a limited number of years, returning them to public or private school afterward. Some studies have shown that home-schooled children score higher than those public-schooled on standardized subject tests, although home schooling is still criticized on the grounds that in general the students learn less, all else constant, than those in public schools. Courts have of late tended to rule that as long as parents can demonstrate to the satisfaction of educators in the community that they are not depriving their children of the right to a quality education, then they can school them at home.

The Functionalist View of Education

All known societies have an education system of some sort. In the United States, as in other industrialized societies, the education system is large and highly formalized. In other societies, such as pastoral societies, it may consist only of parents teaching their children how to till land and gather food. Under these circumstances, the family is both the education system and the kinship institution.

Why does an education institution exist? Functionalist theory in sociology argues that education accomplishes the following consequences, or "functions," for a society.

Socialization is brought about as the cultural heritage is passed on from one generation to the next. This heritage includes a lot more than "book knowledge." It includes moral values, ethics, politics, religious beliefs, habits, and norms—in short, culture.

Debunking Society's Myths

Myth: To get ahead in society, all you need is an education.

Sociological perspective: Education is necessary, but not sufficient, for getting ahead in society. Success depends significantly on one's class origins; the formal education of one's parent or parents; and one's race, ethnicity, and gender.

Occupational training is another function of education, especially in an industrialized society such as the United States. In less complex societies, as was the United States prior to the nineteenth century, jobs and training were passed from father to son, or more rarely, from father or mother to daughter. A significant number of occupations and professions today are still passed on from parent to offspring, particularly among the upper classes (a father passing on a law practice to his son). It also occurs among certain highly skilled occupations (plumbers, ironworkers, and electricians), where both training and union memberships may be passed on. Modern industrialized societies need a system that trains people for jobs. Most jobs today require at least a high school education, and most professions require a college or postgraduate degree.

Social control, or the regulation of deviant behavior, is also a function of education, although a less obvious one. Such indirect, subtle consequences emerging from the activities of institutions are called **latent functions** of the institution. Increased urbanization and immigration beginning in the late nineteenth century were accompanied by rises in crime, overcrowding, homelessness, and other urban ills. One perceived benefit of compulsory education (that is, one latent function) was that it kept young people off the streets and out of trouble (see Table 14.1).

The Conflict View of Education

In contrast to functionalist theory, which emphasizes how education unifies and stabilizes society, conflict theory emphasizes the disintegrative and disruptive aspects of education. Conflict theory focuses on the competition between groups for power, income, and social status, emphasizing the prevailing importance of institutions in the conflict. One intersection of education between group and class competition is embodied in the significant correlations that exist between education and class, race, and gender. The unequal distribution of education allows it to be used to separate groups. The higher the educational attainment of a person, the more likely that person will be middle to upper class, White, and male. Conflict theorists argue that educational level is a mechanism for producing and reproducing inequality in our society.

According to conflict theorists, educational level can be a tool for discrimination by using the mechanism of **credentialism**—the insistence upon educational credentials for their own sake, even if the credentials bear little relationship to the intended job (Collins 1979; Marshall 1997). This device can be used by potential employers to discriminate against minorities, working-class people, or women—that is, those who are often less educated and least likely to be credentialed because discriminatory practices within the education system limited their opportunities for educational achievement.

The Symbolic Interaction View of Education

Symbolic interaction focuses on what arises from the operation of the interaction process during the schooling experience. Through interaction between student and teacher, certain expectations arise on the part of both. As a result, the teacher begins to expect or anticipate certain behaviors, good or bad, from students. Through the operation of the **teacher expectancy effect** (discussed later in this chapter), the expectations a teacher has for a student can actually create the very behavior in question. Thus fulfilled, the behavior is actually caused by the expectation rather than the other way around. For example, if a White teacher expects Latino boys to perform below average on a math test relative to White students, over time the teacher may act in ways that encourage the Latino boys to get below average math test scores. Later we will examine just how the expectancy effect works.

Does Schooling Matter?

How much does schooling really matter? Does more schooling actually lead to a better job, more annual income, and greater happiness? Is the effect of education great or small?

Effects of Education on Occupation and Income

One way that sociologists measure a person's social class or socioeconomic status (SES) is to determine the person's amount of schooling, income, and type of occupation (see Chapter 8 on class stratification). Sociologists call these the *indicators* of SES. In the general population, there is a strong relationship between formal education and occupation, although

	Functionalism	Conflict Theory	Symbolic Interaction
Education in society	Fulfills certain societal needs for socialization and training; "sorts" people in society according to their abilities	Reflects other inequities in society, including race, class, and gender inequality, and perpetuates such inequalities by tracking practices, for example	Emerges depending on the character of social interaction between groups in schools
Schools	Inculcate values needed by the society	Are hierarchical institutions reflecting conflict and power relations in society	Are sites where social interaction between groups (such as teachers and students) influences chances for individual and group success
Social change	Means that schools take on functions that other institutions, such as the family, originally fulfilled	Threatens to put some groups at continuing disadvantage in the quality of education	Can be positive as people develop new perceptions of formerly stereotyped groups

*table*14.1 **Sociological Theories of Education**

the relationship is not perfect. Measuring occupations in terms of social status or prestige, we find that the higher a person's occupational status, the more formal education he or she is likely to have received. Overall, we know that, on average, doctors, lawyers, professors, and nuclear physicists spend many more years in school than unskilled laborers such as garbage collectors and shoe

table 14.2	Median Income, by Education and Gender	
Level of Schooling	**Men**	**Women**
Less than 9th grade	$26,789	$20,499
9th–12th grade (no diploma)	31,434	23,351
High school graduate	42,466	29,410
Some college, no degree	48,431	35,916
Associate degree	51,485	40,463
Bachelor's degree or more	88,843	59,052

Source: U.S. Census Bureau. 2007. Average Earnings Year-Round, Full-time Workers by Educational Attainment and Gender. **www.census.gov**

shiners. This relationship is strong enough that we can often, although not always, guess a person's educational attainment just by knowing his or her occupation. There are indeed instances of semiskilled laborers, such as taxi drivers, who have Ph.D.s, but they are relatively rare. Also exceedingly rare is the reverse: the self-educated, self-made individual who completed only the fourth grade and is now the CEO of a major corporation.

The connection between income (and jobs) and education is not independent of gender. Gender heavily influences the relationship between income and education. Note from Table 14.2 that although the higher one's education, the higher one's (average) income, it is nonetheless true that the average income for women is less than the average income for men at *each* education level. This is because, in general, throughout our society the average woman earns less than a man of the same *or even less* education. Men with professional degrees (law, business, and so forth) earn a median annual income of around $100,000, whereas a woman with that same education earns only about $70,000—about two-thirds of what a man earns. A man with no graduate education but a college-only education earns more than a woman with a master's degree. And men with some college, but no bachelor's degree, earn more than women with a bachelor's degree.

Effects of Social Class Background on Education and Social Mobility

Education has traditionally been viewed in the United States as the way out of poverty and low social standing—the main route to upward social mobility. The assumption has been that a person can overcome modest beginnings, starting by staying in school.

Much sociological research has demonstrated that the effect of education on a person's eventual job and income greatly depends on the *social class* that the person was born into. Among White people of the upper classes, including those who inherited wealth, as well

as professionals and high-level managers, social class origin is *more important than education* in determining occupation and income (Bowles et al. 2005; Jencks 1993; Persell 1990; Cookson and Persell 1985; Bielby 1981; Jencks et al. 1979; Taylor 1973; Jencks et al. 1972).

Class and race work together to "protect" the upper classes from downward social mobility and to block the lower classes from too much upward mobility. Education is used by the upper classes to avoid downward mobility by such means as sending their children to elite private secondary schools. A disproportionate number of upper-class children attend elite boarding schools and day schools compared to working-class children who are considerably underrepresented in such schools (Rendon and Hope 1996; Cookson and Persell 1985).

Among middle-class Whites, education considerably improves the chances of getting middle-class jobs, yet access to upper-class positions is limited. Among those of lower-class origins, such as unskilled laborers or the chronically unemployed, chances of getting a good education as well as a prestigious job are poor. In sum, education is affected by social-class origins, and occupation (and income) is heavily influenced by social class and also by education. Individuals with lower-class origins are less likely to get a college education and thus are less likely to get a prestigious job. These interrelationships are summarized in Figure 14.2, which shows that social class origin affects occupation and income both directly and indirectly by way of education.

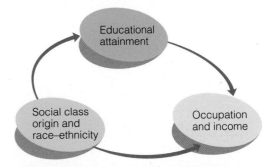

FIGURE 14.2 *Relationship of Social Class, Race–Ethnicity, Education, Occupation, and Income*

Debunking Society's Myths

Myth: Education is more important than social class in determining one's job and income.

Sociological perspective: Although education has an effect on the job one gets and the income one earns, overall, social class origin *is more important* than acquired education in determining one's job and earned income.

Education, Social Class, and Mobility Seen Globally

It is sometimes argued that because of our education system, there is more occupational and income mobility in the United States than in other industrialized nations, particularly England, Germany, and Japan. In general this is true, but not by much. Until a few years ago, students in England were required to take an examination, called the Eleven Plus, at age eleven. A person's score on this examination determined whether he or she was put on a track to prestigious universities such as Oxford or Cambridge or went directly into the labor force from high school. Children of the upper classes stood a *far* better chance of scoring high on this examination than did middle- or working-class children, and the average scores of women and minorities, especially Africans and East Indians, were considerably lower than those of upper-class White males.

A similar situation exists in the United States, as illustrated in this chapter's opening vignette on Ollie Taylor. Students from lower-class families have lower average scores on exams such as the Scholastic

Assessment Test (SAT) and the American College Testing Program (ACT). As shown in Table 14.3, there is a smooth and dramatic increase in average (mean) SAT score as family income increases, for both SAT verbal and math scores. In this sense, one's SAT score is a "proxy" or substitute measure of one's social class: *Within a certain range, you can guess someone's likely SAT score from knowing only the income and social class of his or her parents!* As you can see from Table 14.3, each additional $10,000 in family income is worth about 10 to 15 more points on either the SAT verbal or the SAT math tests (thus 20 or 30 points more for a combined score)! This is truly ironic because the multiple-choice SAT was originally designed in the 1940s as an "objective" test to combat the pattern of children from wealthy families having an advantage for admission to college.

With lower scores comes a diminished chance of getting into the best colleges or universities. African Americans, Latinos, and American Indians score on the average less than Whites, and women tend to score lower than men on the quantitative (mathematical) sections of the SAT (Table 14.4). Asian Americans as a group have scored higher than Whites in recent years on the quantitative sections of the SAT, but lower on the verbal sections. Women of any ethnic group score lower on the quantitative sections than do men of the same ethnic group. These patterns indicate that the SAT has an effect in the United States similar to that of the Eleven Plus test in England, directing the futures of young people according to the results of widely administered exams.

In general, the White-versus-minority test score gap—on "ability" tests as well as achievement (subject) tests—contributes to the low-ranking that

table14.3	Average SAT Scores by Family Income	
Family Income	**SAT Verbal Average Scores[a]**	**SAT Math Average Scores[a]**
<$10,000	429	457
$10,000–$20,000	445	465
$20,000–$30,000	462	474
$30,000–$40,000	478	488
$40,000–$50,000	493	501
$50,000–$60,000	500	509
$60,000–$70,000	505	515
$70,000–$80,000	511	521
$80,000–$90,000	523	534
>$90,000	549	564

[a]SAT scores, on either the verbal or the math sections, range from 200 to 800, with 800 being a perfect score.

Source: College Board. 2008. *College Bound Seniors 2008: A Profile of SAT Program Test Takers.* New York: © 2008 The College Board. Used with permission. **www.collegeboard.com.** Reproduced with permission.

table14.4	Average SAT Scores, by Ethnicity and Gender			
	SAT Verbal Average Scores		SAT Math Average Scores	
SAT test takers who described themselves as:	Male	Female	Male	Female
American Indian or Alaskan Native	489	485	512	478
Asian, Asian American, or Pacific Islander	511	509	594	562
African American or Black	430	437	438	423
Hispanic or Latino:				
• Mexican or Mexican American	459	451	485	450
• Puerto Rican	462	458	472	443
• Latin American, South American, Central American, or other Hispanic or Latino	464	454	484	448
White	529	526	555	520
Other	494	494	534	497

Source: College Board. 2008. *College Bound Seniors 2008: A Profile of SAT Program Test Takers.* New York: © 2008 The College Board. **www.collegeboard.com.** Reproduced with permission.

characterizes the United States globally: In 1960 the United States was number one in both math and reading. By 2007 the United States had fallen to nineteenth out of forty in reading and twenty-eighth out of forty—not far from the bottom—in math (Darling-Hammond 2007).

In Germany, an examination called the *Abitur* is taken during the equivalent of the junior year in high school. A high score on the *Abitur* facilitates admission to a university; a low score inhibits getting into a university. Low-scoring students must take two or three more years of courses and then reapply to a university if they wish to attend.

In Japan, a similar examination given at age 12 determines even more rigidly a child's subsequent educational opportunities. Students who wish to continue their education at a college or university must score high enough to gain admission to prep schools. Especially high scores guarantee admission to prestigious prep schools, which is necessary for later admission to the best universities. Low scorers are virtually shut out from prep school admission, and these students become ineligible for a university education. In recent years, many parents have begun to send their children to weekend "cram" seminars called *jukos* to prepare for this examination, adding a grueling additional regimen to the already stiff requirements of the Japanese school system. Many Japanese, including educators, are becoming concerned that the extreme competitiveness of this system and the great burden of work placed on students are brutalizing the youngsters.

Overall, the educational system in the United States appears to allow for a bit more social mobility than can be achieved in Germany, possibly England, and certainly Japan. The stratification system in those countries and others with similar systems is more rigid, or *castelike,* than that of the United States. However, there is a danger in concluding that the U.S. educational system permits much more social mobility. In general, the similarities tend to be more prominent than the differences.

Nearly ten years ago, the administration of President George W. Bush introduced the No Child Left Behind (NCLB) law and program in an attempt to narrow the achievement and test score gap between Whites and minorities of color in the United States. The program involved restructuring secondary education, replacement of teachers on the basis of their schools' mean test scores, repeated administration of

Japanese students and their parents at this juko, *or "cram" seminar, vow academic success at a meeting in a hotel in Tokyo over the New Year's holiday.*

both verbal and math standardized tests, and other measures. But despite then-President Bush's repeated assertions that this law and program were having a positive and "dramatic" effect, the results have unfortunately shown that not only does a wide gap in both verbal and math test scores continue to exist but that gap has *widened* since the late nineties on both tests and has gotten even wider from 2001 through 2008—precisely the period during which the NCLB law and programs were presumably in effect. The data showing this come from the National Assessment of Educational Progress, often said to be the "nation's report card" (Dillon 2009). Only part of this score gap can be attributed to mismeasurement and cultural bias in the tests (see the following section Cognitive Ability and Its Measurement). Clearly, the failure of programs such as the NCLB law is significantly responsible, and educators continue to search for institutional ways to reduce the score and achievement gap between Whites and minorities of color (Harris and Robinson 2007; Darling-Hammond 2007; Jencks and Phillips 1998).

Education and Inequality

In its original nineteenth-century conception, the education system was to serve as a leveling force in American society—the road to full equality for all citizens regardless of race, social class origin, nationality, religion, or gender. Jew and gentile, Irish and Polish, Black and White, rich and poor, male and female would learn together side by side. Through education, each student would learn the ways of others and thus come to understand and respect them. Full equality for humankind was to follow.

Education has indeed reduced many inequalities in society since compulsory education began at the turn of the twentieth century. The percentage of high school graduates has risen among Whites and minorities, both male and female, as have certain types of social mobility. Despite continuing inequalities in college enrollments comparing African Americans, Hispanics, and Whites, the enrollment of minorities has risen overall—although the high school dropout rate for Hispanics has remained highest among all racial–ethnic groups. Furthermore, as more minorities and women attend and graduate from two- and four-year colleges, the result—at least until the major economic recession of 2008–2009—has been more employment for them in mid-level and high-level jobs. Nonetheless, many inequalities still exist in U.S. education.

Cognitive Ability and Its Measurement

Since as long ago as classical Greece, humans have sought to measure a "mental faculty" or "intelligence." It is now called **cognitive ability,** the capacity for abstract thinking. Since early in the twentieth century, educators in our society, from preschools to universities, have attempted to measure intelligence by means of **standardized ability tests,** such as the SAT or IQ tests, which are intended to measure ability or potential. These are not the same as **achievement tests,** which are intended to measure what has actually been learned, in addition to ability or potential. Advanced Placement (AP) exams are achievement tests taken before entering college. Students who score high demonstrate that they have already mastered certain material and can in many cases skip those courses in college.

The education system in the United States has relied heavily upon the idea that intelligence, or ability, or potential is a single unitary trait. Cognitive ability has been gauged according to the numerical results of the standardized tests. There has been a will to reduce measurements of cognitive ability to a single number, or perhaps two numbers, such as language and math scores of SAT tests and IQ tests in the recent past.

There are three major criticisms made regarding using standardized tests such as the SAT or ACT as measures of cognitive ability. First, the tests tend to measure only limited ranges of abilities (such as quantitative aptitude or verbal aptitude) while ignoring other cognitive endowments such as creativity, musical ability, spatial perception, or even political skill and athletic ability (Zwick 2004; Freedle 2003; Gardner 1999; Lehmann 1999; Sternberg 1988).

Second, the tests possess at least some degree of cultural and gender bias—and also a strong social class bias. As a result, they may perpetuate rather than reduce inequality between different cultural or racial groups and social classes, as well as perpetuate social, economic, and educational inequality between men and women. The tests were designed primarily by middle-class White males, and the "standardization" they strive to achieve mirrors middle-class White male populations. Many studies show that although standardized ability tests are somewhat capable of predicting future school performance for White males, a significant number of studies—in fact *most* studies—show less accurate forecasts for the success of minorities, especially Hispanics, African Americans, and American Indians, and they often predict school performance less accurately for women than for men (Darling-Hammond 2007; Zwick 2004; Epps 2002; Taylor 2009, 2002, 1992a; Jencks and Phillips 1998; Fleming and Garcia 1998; Pennock-Roman 1994; Young 1994; Crouse and Trusheim 1988). In other words, the **predictive validity** of the tests, which is the extent to which the tests accurately predict later college grades, is compromised for minorities, women, and persons of working-class origins.

The third criticism of the SATs is that they do not predict school performance very well, even for Whites. For example, SAT scores are only modestly

Who gets assigned to which tracks? Research shows that track assignment is not solely based on the performance in cognitive ability tests. Social class and race are involved. Students with the *same* test scores often get assigned to different tracks because of differences in their social class and race. Few administrators or teachers consciously and deliberately assign students to tracks based on these criteria, but it occurs nevertheless. Researchers have consistently found that when following two students with identical scores on cognitive ability tests, the student of higher social class is more likely than the student of lower social class status to get assigned to the higher track.

Teacher Expectancy Effect

Similar to the labeling effect of tracking is the teacher expectancy effect, which is the effect of teacher expectations on a student's actual performance—*independent* of the student's actual ability. What the teacher expects students to do affects what they will do. The expectations a teacher has for a student's performance can dramatically influence how much the student learns.

Insights into the teacher expectancy effect come from symbolic interactionist theory. In a classic study, Rosenthal and Jacobson (1968) told teachers of several grades in an elementary school that certain children in their class were academic "spurters," who would increase their performance that year. The rest of the students were called "nonspurters." The researchers selected the "spurters" list completely at random, unbeknownst to the teachers. The distinction had no relation to an ability test the children took early in the school year, although the teachers were told (falsely) that it did. At the end of the school year, it was found that although all students improved somewhat on the achievement test, those labeled spurters made greater gains than those designated nonspurters, especially among first and second graders. This experiment isolates the effect of the label (hence showing a *labeling effect*), because it is the only difference that

distinguishes the two groups. Variations of this clever and revealing study have been conducted many times over, and the results are generally similar.

How are expectations converted into performance? The powerful mechanism of the **self-fulfilling prophecy,** in which merely applying a label has the effect of justifying the label, affects performance (Taylor et al. 2009; Cardenas 1996; Darley and Fazio 1980). Recall the quote from early sociologist W. I. Thomas in Chapter 1: "If men [sic] define situations as real, they are real in their consequences" (Thomas 1928: 572). If a student is defined (labeled) as a certain type, the student becomes that type. The self-fulfilling prophecy is at the core of the practice of tracking.

The process unfolds in stages (see Figure 14.3). First, a teacher is told that a student merits a label such as *spurter*; perhaps the designation originates with administrators or comes from the scoring key of a standardized exam. The teacher's perception of the student is then colored by the label; a student labeled a spurter may be coaxed and praised more often than nonspurters. The student then reacts to the teacher's behavior; students expected to perform well and encouraged to excel actually perform better in class and on exams than other students. Finally, the original prophecy fulfills itself: The teacher observes the behavior of the student, notes the increase in performance, and concludes that the designation spurter is affirmed because the so-called spurters actually perform better by objective measures, such as grades and standardized test scores, than the nonspurters.

Schooling and Gender

Teachers hold different expectations for girls and boys in school, and it comes as no surprise that gender affects teacher expectations and therefore the actual performance of these children. Tracking, particularly in high school, is significantly dependent on gender; for example, a far greater proportion of young men are in the science and math tracks than are young

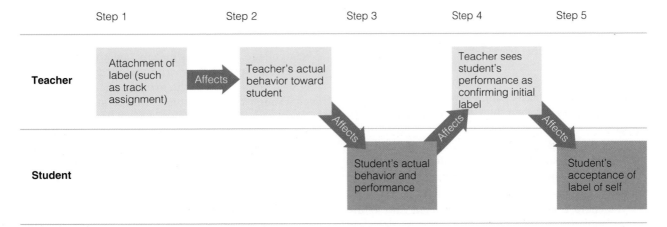

FIGURE 14.3 *The Self-Fulfilling Prophecy*

Source: Adapted from Taylor, Shelley E., Letitia Anne Peplau, and David O. Sears. 2009. *Social Psychology,* 13th ed. Upper Saddle River, NJ: Prentice Hall.

women. Throughout schooling, from preschool through graduate or professional school, women and men are treated differently, with women discriminated against. This consistent and long-term differential treatment has had profound consequences. *Girls and boys start out in school roughly equal in skills and confidence*, a fact that contradicts the often-heard assertion that girls are "inherently" worse than boys in math, spatial, and other kinds of skills.

By the end of high school, however, women trail men, especially in mathematics and the sciences. What happens in between has been documented in a comprehensive report commissioned by the American Association of University Women (1998) as well as by recent research (Zwick 2004).

1. In general, teachers pay less attention to girls and women. In elementary school as well as high school, teachers direct more interaction to boys than to girls. As a result, boys tend to talk more in class, and thus teachers tend to interact with them more.

2. On national tests of reading and writing, girls perform equally to boys. But on advanced mathematics and some science tests, differences emerge. The most dramatic gender differences appear on tests with the highest stakes—the quantitative sections of the Preliminary Scholastic Assessment Tests (PSAT), Scholastic Assessment Tests (SAT), and Advanced Placement (AP) tests. Researchers explain this as the result of several factors, including course-taking patterns, differences in family socialization, and, as we have noted, bias in the standardized tests themselves. With regard to courses, young men and women tend to take the same number of math and science courses, but what they take differs substantially. Girls are more likely to take biology and chemistry, and boys are more likely to take physics, trigonometry, and calculus.

3. As already noted above, some standardized math and science tests still retain gender bias, despite twenty years or more of effort to weed out bias on the part of education specialists and testing organizations such as the Educational Testing Service, manufacturer of the SAT. Bias is especially prevalent in mathematical word problems. Certain mathematics word problems are gender-typed because they employ words and concepts more familiar to men than women (Chipman 1991; American Association of University Women 1998).

4. Standardized tests in math tend to underpredict women's actual grades in mathematics. Women tend to do somewhat better in math courses than their test scores predict (Zwick 2004).

5. Teachers tend to treat Black women and White women differently by rebuffing Black girls and interacting more with White girls. This is particularly true of Black and White girls during the elementary school years. The trend appears to be independent of the teacher's own race; namely, Black teachers tend to interact more with White girls, just as do White teachers.

6. Textbooks still tend to either ignore women or stereotype them. In this respect, textbooks are gender role socializers. In elementary school texts, boys are portrayed as building things, being clever, and leading others. Girls tend to be shown performing dull tasks and following the leadership of boys.

Stereotype Threat

As has already been noted, racial and gender stereotypes can affect actual behavior. To what extent can a negative stereotype about yourself affect your own behavior and academic performance? As with the self-fulfilling prophecy, to what extent do minorities and women internalize negative stereotypes about themselves that show such effects?

An answer has been provided by the groundbreaking research of Claude M. Steele and associates (Steele et al. 2002; Steele 1999, 1997, 1996, 1992; Steele and Aronson 1995; and also Massey and Fisher 2007; Woods et al. 2005; McIntyre et al. 2003; Aronson et al. 2002; Brown and Josephs 1999). They note that two common stereotypes exist in the United States and other countries as well. First, because on average Blacks perform less well than Whites on math and verbal ability tests, Blacks must have, or so it is believed, some "inherent" deficiency in math and verbal abilities relative to Whites. Second, because women perform less well than men on math ability tests, women must therefore have some "inherent" deficiency in math ability.

To the extent that Black students in high school or college may believe (internalize) such stereotypes, they may perform less well on a test if they are told that "this is a genuine test of your true ability," because telling them that would tend to increase their anxiety about the test. The same should apply to women on a math ability test. Either Whites, or males, who are told this should be *less* affected because these societal stereotypes are not *about* Whites or males.

Results in over 100 research studies show that this is just what happens (Kiefer 2007; Steele et al. 2002). Black college students who are told that the test is a "genuine" test of their true verbal ability perform less well than Whites who are told the same thing—even though the Whites and Blacks start out equal in their average test scores. This is the **stereotype threat effect**. If the groups of Blacks and Whites are told nothing, then they perform about the same on the test. Note that nothing is said to the students specifically about Black and White test performance, only that the test was designed to be a "genuine" test of verbal ability. Perhaps Ollie Taylor was told this when he was subjected to a cognitive ability test and thus suffered some stereotype threat effect, which then resulted in his assignment to a low track, even though he was quite bright.

Stereotype threat appears to operate in the same way with regard to the presumed female–male difference in math ability test performance. When both men and women are told that a math test given to them is a "genuine" test of their true math ability, the women do worse than the men. If men and women are told nothing, then women and men perform about the same on the test (Kiefer 2007; Smith 2006; McIntyre et al. 2003; Gonzales et al. 2002; Lau 2002; Brown and Josephs 1999). This suggests that the long-believed female deficit in math ability may stem at least in part simply from what they are told before they take the test and less from so-called inherent differences between men and women in math ability.

Health Care in the United States

Like education, the health care system is one of America's major institutions. Generally speaking, the citizens of the United States are quite healthy in relation to the rest of the world. However, there are very great discrepancies among people within the United States in terms of longevity, general health, and access to health care, with the least advantaged part of the curve consisting primarily of minorities, the lower classes, and for a number of ailments, women. But the fact remains that we are a robust nation.

The Emergence of Modern Medicine

The highly technological, scientific, corporate-based health care that now characterizes modern medicine in the United States is not how health care was delivered in the past. In colonial times, American physicians received their training in Europe. Their competitors in the healing arts included alchemists, herbalists, ministers, faith healers, and even barbers. Treatments were a combination of folk wisdom, superstition, tried-and-true regimens, and often, dangerous quackery. A simple scratch, once infected, could easily cost a limb or a life. Patients were often "bled" as a cure, which involved the removal of "bad" blood from the patient, sometimes by means of drilling holes in the skull. Needless to say, the cure was often worse than the disease.

By the start of the nineteenth century, advances in biology and chemistry ignited a century of explosive growth in medical knowledge. One fruit of the scientific revolution of the mid-1800s was the *germ theory,* the idea that many illnesses were caused by microscopic organisms, or germs. Now considered scientific fact, the claim that something called *germs* caused illness was then hotly debated. Shortly thereafter, germ theory established itself as a foundation of medicine. Doctors were able to show that isolating infected people and sealing infected wounds could stop the spread of illness by stopping the spread of germs. Relentless study and research transformed medicine into a science. Coincidentally, the social prestige of medicine greatly increased, contributing mightily to the status of physicians, who had formerly enjoyed more modest social standing.

The American Medical Association (AMA) was founded in 1847 after half a century of sweeping away rivals in the healing arts, many of whom were women. The AMA successfully campaigned to outlaw or delegitimize alternative therapies and emerged as the most powerful organization in U.S. health care.

In the late 1800s the image of medicine as an upper-class profession took hold. A medical education was expensive, and medical schools drew on White, male, urban populations for their students. Those trained as physicians took their place in the upper

Mystic Seaport Museum, Rosenfeld Collection, Mystic, Connecticut

Rob Crandell, The Image Works

The development of new technologies has transformed the delivery of health care. The "house call" by a physician is now a rarity, although it is being renewed in a few communities. Many new technologies, such as this heart bypass procedure assisted by robots, would have been unimaginable not many years ago.

Some fear that certain viruses will become pandemics, that is infecting people worldwide. Such was the case with the swine flu scare in 2009, shown by this international traveler.

social strata. Herbalists and faith healers came more frequently from the rural lower class and generally remained there. As the ranks of the medical profession swelled with wealthy Whites, African Americans and Hispanics became proportionately more affiliated with older folk practices and midwifery. This overall trend continued through the early part of the twentieth century, although today folk practices continue to have adherents among rural lower-class Whites and rural and urban lower-class Blacks, Hispanics, and Native Americans (Starr 1982).

Health, Diversity, and Social Inequality

Prominent problem areas in the U.S. health care system include the following.

- **Unequal distribution of health care by race–ethnicity, social class, or gender.** Health care is more readily available and more readily delivered to White or middle-class individuals in urban and suburban areas than to minorities. The lack of health care delivery to Native American populations is particularly serious. Likewise, men and women receive unequal treatment for certain types of medical conditions, with women more likely than men to receive truncated treatment.

- **Unequal distribution of health care by region.** Each year, many in the United States die because they live too far away from a doctor, hospital, or emergency room. Doctors and hospitals are concentrated in cities and suburbs; they are much less likely to be situated in isolated rural areas. Rural people in Appalachia and some parts of the South and Midwest may have to travel 100 miles or more to get to the doctor or emergency room.

- **Inadequate health education of inner-city and rural parents.** Many inner-city and rural parents do not understand the importance of immunizing their children against smallpox, tuberculosis, and other illnesses, and they are often suspicious of immunization programs. This hesitancy is reinforced by the depersonalized and inadequate health care ghetto residents often encounter when care is available at all.

The definitions of *sick* and *well* have varied greatly over time in the United States. For example, from the early 1900s until the mid-1940s, thinness was associated with poverty and hunger. If you were skinny, that meant you were in bad health. From the late 1950s through the present, a positive value has been placed on being thin. Female role models in our society, such as movie stars and fashion models, have firmly established that "thin is in." Millions of young women have tried to copy this look, and one result has been an increased incidence of anorexia nervosa (Weitz 2001; Williams and Collins 1995; Taylor et al. 2009).

Anorexia nervosa (*anorexia* for short) is an eating disorder characterized by compulsive dieting. Victims of this illness starve themselves, sometimes to death, even though sufferers do not typically define themselves as ill; they tend to see themselves as overweight, even though they are dangerously thin. A related malady, *bulimia,* is an eating disorder characterized by alternate binge eating and then purging or inducing vomiting to lose weight.

Like many other diseases, anorexia has social as well as biological causes. A recent study of 990,000 individuals indicated that gender, ethnicity, and social class were large influences on the probability of developing anorexia (Lindberg 2003). A majority of people suffering from the disease are young White women, from well-to-do families, most often two-parent families. Many behavioral scientists have noted that generally anorexics have been pressured excessively by their parents to be high achievers. Others have detected a link between anorexia and the socially constructed ideals of beauty in our society, which are fixated on thinness and so-called ideal body types. Images of physical "perfection" are emblazoned across television, magazines, and billboards, with slenderness displayed as the ideal of femininity. Researchers note that these social values, which encourage compulsive dieting, are comparable to the foot-binding once practiced in China and other forms of female mutilation found in some foreign cultures (Wolf 1991; Chernin 1991).

Anorexia is least likely to afflict African American women, Latinos, and lesbians. According to researchers (Logio-Rau 1998; Thompson 1994), many in these groups overeat, rather than self-starve, thus producing a more familiar disorder: *obesity.* Thompson interprets this behavior as a reaction to oppressive life experiences associated with racism, sexism, and homophobia.

Men have not been exempt from the pressure of such values (Logio-Rau 1998). Since the 1940s, and especially from the mid-1970s, one of the most persistent male physical ideals has been the rippling physique of the bodybuilder or weight lifter. Young men have been urged by the media and their peers to "pump iron" for the perfect body. Many athletes, professional and amateur, have been goaded by athletic ambitions to use *anabolic steroids,* powerful

The Americans with Disabilities Act

The Americans with Disabilities Act (ADA), passed by Congress in 1990, prohibits discrimination against disabled persons. In 1999, the U.S. Supreme Court restricted the definition of disability to exclude disabilities that can be corrected with devices such as eyeglasses or with medication. The decision was based on cases in which people had been denied employment because they did not meet a health standard required by the employer, even though, with correction, the standard was met. One case involved two nearsighted

The disability rights movement has opened up new opportunities to those who face the challenge of disability.

© Peter Hvizdak/The Image Works

women denied jobs as airline pilots because, without glasses, their vision did not meet the airline company's standard of 20/40. With glasses, their vision was 20/20. Another case involved a man whose high blood pressure was above the federal standard for driving trucks. He was denied a job as a trucker even though with medication his pressure was regulated to an acceptable level. The Court's decision denied these people the right to claim discrimination.

The decision raises several interesting sociological questions about disabilities and civil rights. Writing for the majority, then-Justice Sandra Day O'Connor argued that the law requires people to be assessed based on each individual's condition, not as members of groups affected in a particular way. The disability rights movement would argue that people with disabilities are a minority group with certain civil rights. The issue being mediated by the courts involves competing definitions of what constitutes a disability. This point intrigues symbolic interactionists who would note the role of socially constructed meanings in legal negotiations. Furthermore, as one lawyer who drafted the original ADA law has pointed out, according to the original decision, someone may be disabled enough to be excluded from a job, but not disabled enough to claim discrimination. This raises interesting questions about the rights of employers to establish physical and medical standards for certain jobs, even if those standards result in the exclusion from employment for some groups of people.

---

hormones that stimulate the growth of muscle. Used widely (despite dire warnings by physicians), steroids not only build muscle as advertised, they can also shrivel the testicles and cause impotence, hair loss, heart arrhythmia, liver damage, strokes, and very possibly some forms of cancer.

Obesity has recently become defined as a public health problem. Obesity has traditionally been considered a matter of individual habit, but in 2004 officials in the Medicare program changed their policy to include obesity as a disease (and thus various weight loss programs are now eligible for Medicare funding). As obesity has come to be defined as a medical problem, analysts tend to emphasize the physical causes of being overweight. But a sociological perspective on obesity would also look to cultural and social structural factors as possible sources of obesity (Czerniawski 2007).

Debunking Society's Myths

Myth: The health care system works with the best interests of clients in mind.

Sociological perspective: The health care system is structured along the same lines as other social institutions, thus reflecting similar patterns of inequality in society.

Race and Health Care

Health and sickness in the United States is influenced by social factors. **Epidemiology** is the study of all the biological, social, economic, and cultural factors associated with disease in society. **Social epidemiology** is the study of the effects of social, cultural, temporal, and regional factors in disease and health. Among the more important social factors that affect disease and health in the United States are race, ethnicity, social class, gender, and age.

Among women, African Americans are more likely than Whites to fall victim to diseases such as cancer, heart disease, stroke, and diabetes. Death of the mother during childbirth is three times higher among African American women than among White women, and African American women are three times more likely to die while pregnant (Williams and Collins 1995; Livingston 1994). In the 45–64 age group, African American women die at twice the rate of White women. The occurrence of breast cancer is lower among African American women than White women, yet the *mortality rate* (death rate) for breast cancer for African American women is considerably higher than for White women (Weitz 2001; National Cancer Institute 2002). This reflects the fact that White women are more likely to get high-quality care, and get it more rapidly, than African American women. African American women also develop cervical cancer at twice the rate of White women. Hypertension (high blood pressure) afflicts 25 percent of all African American women and only 11 percent of White women. Although differences in culture, diet, and lifestyle account for some of these differences, it is nevertheless clear that African American women and men do not receive medical attention as early as Whites, and when they do eventually get treatment, the stage of their illness is more advanced and the treatment they receive is not of the same quality (National Center for Health Statistics 2008; National Cancer Institute 2002).

Hispanics, like African Americans, Native Americans, and other minorities, are significantly less healthy than Whites (Office of Minority Health 2004). Hispanics contract tuberculosis at a rate seven times that of Whites. Other indicators of health, such as infant mortality, reveal a picture for Hispanics similar to that of African Americans and Native Americans. Hispanics are less likely than Whites to have a regular source of medical care, and when they do, it is likely to be a public health facility or an outpatient clinic. Because of language barriers as well as other cultural differences, Hispanics are less likely than other minority groups to use available health services, such as hospitals, doctors' offices, and clinics (National Center for Health Statistics 2008; Weitz 2001).

One of the challenges for the health care system in a society with increasing diversity is responding to the different cultural orientations of various groups in society. Immigrants, for example, who may have limited English language skills and may come from cultures with very different health care practices, may be especially confused by the practices within the U.S. health care system (Suro 2000). Developing the ability for greater cross-cultural administration of health care will likely continue to be a challenge in the future. Also, because the majority of physicians are White, patients of a different racial or ethnic background will likely feel some social distance between themselves and the health care provider, contributing to a reluctance to seek care (Malat 2001). At the same time, researchers find that patients are much more satisfied with their care when their physician is of the same race as they are (Laveist and Nuru-Jeter 2002). Such findings indicate the continuing significance that race and color have in shaping people's health.

Social Class and Health Care

In the United States, social class has a pronounced effect on health and the availability of health services. The lower the social class status of the person or family, the less access they have to adequate health care (Jacobs and Morone 2004). Consequently, the lower one's social class, the less long one will live. People with higher incomes who are asked to rate their own health tend to rate themselves higher than people with lower incomes. Almost 50 percent of those in the highest income bracket rate their health as excellent,

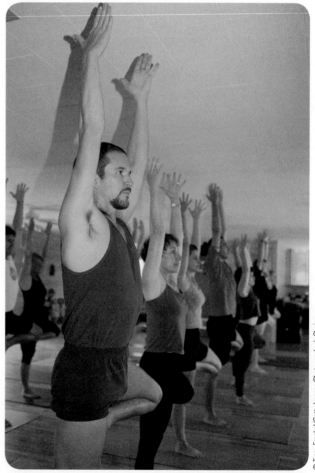

As many people have become aware of the importance of good health, a major industry has developed to provide assorted health services, such as this yoga class. But the higher your social class and income, the more likely you are to use such health services.

whereas only 25 percent in the lowest income bracket do so. In fact, the effects of social class are nowhere more evident than in the distribution of health and disease, showing up dramatically in the rates of infant mortality, stillbirths, tuberculosis, heart disease, cancer, arthritis, diabetes, and a variety of other illnesses. The reasons lie partly in personal habits that are themselves partly dependent on one's social class. For example, those with lower socioeconomic status smoke more often, and smoking is the major cause of lung cancer and a significant contributor to cardiovascular disease.

Social circumstances also have an effect on health. Poor living conditions, elevated levels of pollution in low-income neighborhoods, and lack of access to health care facilities all contribute to the high rate of disease among lower classes. Another contributing factor is the stress caused by financial troubles. Research has consistently shown correlations between psychological stress and physical illness (Taylor et al. 2009; Worchel et al. 2000; Jackson 1992; Thoits 1991; House 1980). The poor are more subject to psychological stress than

the middle and upper classes, and it shows up in their comparatively high level of illness.

With the exception of the elderly, now subsidized by Medicare, low-income people remain largely outside the mainstream of private health care. Nearly 41 million Americans—14.5 percent of the population—have no health insurance at all (see Map 14.1). The main sources for medical care for many of them are hospital emergency rooms in the inner cities, often called the "doctor's office of the poor." Even then, emergency room service is lacking. Treatment is given only for specific critical ailments, and rarely is there any follow-up care or comprehensive treatment.

Sociologists have found that during interactions between health care providers and poor patients, the poor, particularly Black and Hispanic poor, are most likely to be *infantilized* and to receive health counseling that is incorrect, incomplete, or delivered in inappropriate language not easily understood by the patient (Weitz 2001; Gonzalez 1996). The symbolic interaction perspective in sociology has noted the tendency for inadequate care. It has also been attributed

map14.1   *MAPPING AMERICA'S DIVERSITY*

Persons Not Covered by Health Insurance

Persons not covered by health insurance (in percent)

- 7.5–10.7
- 10.8–13.9
- 14.0–17.1
- 17.2–20.3
- 20.4–24.5

This map shows the percentage of the population in each state that was not covered by health insurance in 2007. Texas has the highest percentage of people not covered (24.5%), followed by New Mexico (22.9%), then California (19%). Iowa has the lowest percentage of those not covered by health insurance (2.5%), followed by Rhode Island (8.6%). Were you to design a study to explain state-by-state variation in coverage, what explanatory factors would you want to examine?

Source: From the U.S. Census Bureau. 2008. *Statistical Abstract of the United States, 2007.* Washington, DC: U.S. Department of Commerce.

to an attitude among health counselors that the poor are charity cases who should be satisfied with whatever they get, because they are probably not paying for their own care (Diaz-Duque 1989).

Gender and Health Care

Although women live longer on average than men, older women are more likely than older men to suffer from stress, obesity, hypertension, and chronic illness. National health statistics show that hypertension is more common among men than women until age 55, when the pattern reverses. This may reflect differences in the social environment experienced by men and women, with women finding their situation to be more stressful as they advance toward old age. Under the age of 35, men are more likely to be overweight than women; after that, women are more likely to be overweight. Women have a higher likelihood of contracting chronic disease than men, although men are more likely to be disabled by disease (National Center for Health Statistics 2008; Atchley 2000).

The rate of death caused by infectious diseases is higher for men than women and has declined significantly for both since the early 1900s. Researchers cite differences in male and female roles and cultural practices to explain the differences between the genders. In particular, male occupational roles call for more travel and more exposure to other people, the major sources of infection. In addition, men smoke more, and the overall deleterious effect of smoking on health can account for some of the differences in infectious diseases between men and women (National Cancer Institute 2008; Williams and Collins 1995).

It is typically assumed that the work-oriented, hard-driving lifestyle associated with the traditional role of men tends to produce elevated levels of heart disease and other health problems. In general, this is true. However, the role of women in society is changing, with associated changes in women's health. The health of women and men varies with their social circumstances. People who are "tokens" in the workplace—meaning mainly women and Blacks, especially Black women—suffer more stress in the form of depression and anxiety than do nontokens, or women and Blacks who work in places where there is nothing exceptional about their presence (Jackson et al. 1995). Research also has found that perceiving that there is discrimination against you is significantly related to both physical and mental health (Pavalko et al. 2003). Those who experience discrimination by both gender and race are therefore the most vulnerable.

Housewives have higher rates of illness than women who work outside the home, indicating that employment can have positive effects on women's health. Moreover, housewives who never worked outside the home are healthier than those who once worked outside the home but then returned to homemaking. Employed women are more likely than housewives to get well quickly when sick, and they tend to return to their normal activities sooner after illness than do housewives (Andersen 2009).

AIDS and Sexually Transmitted Diseases (STDs)

Approximately fifty sexually transmitted diseases have been medically diagnosed. The four major STDs are syphilis, gonorrhea, genital herpes, and AIDS (acquired immune deficiency syndrome). Less frequent are the more esoteric diseases such as *lymphogranuloma venereum* (LGV), which untreated devastates the body with open sores. The incidence of all STDs increased during the sexual revolution of the sixties and seventies, when sex was thought of as simply a pleasant way to express affection. Little fear was harbored about contracting a venereal disease because most, such a syphilis or gonorrhea, were known to be medically curable, and the remaining diseases were thought of as too rare to worry about. With the subsequent dramatic rise in STDs, especially AIDS, the late 1980s and early 1990s witnessed a new revolution in contrast to the sexual revolution. The new counterrevolution caused people to reevaluate the nature of sex, the possibly fatal risks involved in unprotected sexual activity, and sexual behavior in general (Laumann et al. 1994; Guttmacher Organization 2008, 1994).

Syphilis and gonorrhea have been around for a long time. Both are caused by microorganisms transmitted through sexual contact involving the mucous membranes of the body. It is virtually impossible to get either syphilis or gonorrhea any other way. If untreated, syphilis causes damage to major body organs, blindness, mental deterioration, and death. Each result was seen all too dramatically in the infamous Tuskegee Syphilis Study, beginning in 1932 and ending in 1972 (see Chapter 3), which was done on a group of Black men who were never told they were infected with syphilis and who were never treated.

Genital herpes (herpes simplex II) is more widespread than either syphilis or gonorrhea and affects roughly 30 million people in the United States alone. That represents one person in seven. Although genital herpes can remain dormant for the life of the infected individual, to date it is nonetheless incurable. Genital herpes began to receive attention in the early 1980s, when concern about the risks of STDs generally was on the rise. A person with genital herpes may have no symptoms or may experience blisters in the genital area and a fever as well. Genital herpes is not fatal to adults, but it can be fatal to infants born through vaginal delivery, but not via cesarean section.

People tend to regard sexually transmitted diseases as not merely diseases but as punishment for being immoral. As a result, people who contract an STD become negatively stigmatized, sometimes resulting in their not seeking treatment. This shows the power that social influence can have on the treatment of disease.

A **stigma** occurs when an individual is socially devalued because of some malady, illness, misfortune, or similar attribute. A stigma is viewed as a relatively permanent characteristic of the stigmatized individual; the negative attribute of the stigmatized person is expected to persist, with no cure in sight (Goffman 1963; Jones et al. 1986; Epstein 1996). When the disease AIDS first appeared in the early 1980s, it was mostly associated with gay men and was heavily stigmatized. The result was that the federal government (during the Reagan administration) largely ignored this spreading problem and devoted little in research funds to identify its causes. The stigma associated AIDS with gay men, and the resulting delay by the medical and scientific community in researching treatment likely cost many lives (Shilts 1988).

AIDS is the term for a category of disorders that result from a breakdown of the body's immune system. HIV (human immuno deficiency virus), the virus that causes AIDS, was first identified in 1981. The incubation period between infection with HIV and the development of AIDS can stretch longer than ten years. Thus, one can be infected with HIV and yet not have full-blown AIDS for up to ten years. Consequently, one can have it, not know it, and thus spread it to others over a long period of time. It is not the actual HIV infection that causes death; rather, death is caused by a complex of severe illnesses that thrive in the absence of a working immune system, such as pneumonia, certain cancers, and a number of other illnesses rare enough that their presence is judged to be diagnostic of AIDS.

Over 40 million adults and children worldwide are infected with HIV, 17 million of whom are women. The global AIDS epidemic among women is overwhelmingly the result of heterosexual contact, almost entirely so in Africa and South and Southeast Asia. Analysts

have argued that the high rate of transmission to women worldwide results from women's financial dependence on men. Because of such dependence, women may have little control over when and with whom they have sex. Many women have to exchange sex for financial support, and in highly patriarchal cultures, women are not expected, nor allowed, to make decisions about sex. If they refuse sex or request condom use, they risk abuse and violence or may be suspected of infidelity, which can also put them at great risk, sometimes the risk of death. These facts mean that treating the worldwide AIDS epidemic requires that an analysis of gender relations in different cultures might be a part of the solution to this health epidemic.

The AIDS disease is transmitted through the exchange of bodily fluids, particularly blood and semen. A little more than one-third of all new cases of AIDS are the result of male-to-male sexual contact. Although AIDS initially affected primarily White, gay men, now one-third of new AIDS cases are women, the largest share of whom acquire AIDS through heterosexual contact. The second most frequent cause of transmission for women and men is intravenous drug use (U.S. Census Bureau 2008b). In the United States, AIDS has hit inner-city minority communities disproportionately hard. The contextual problems of poverty, poor health, inadequate health care, drug and alcohol abuse, and violence must be taken into account in explaining the high rate of AIDS in these communities (see Figure 14.4).

AIDS can be understood in medical terms, and new treatment protocols promise to bring more effective treatment for those who contract the disease. But to reduce the incidence of AIDS requires a sociological perspective that takes into account the *social networks* and *social norms* that contribute to the transmission of this disease. Social norms affecting the age of first intercourse, number of sexual partners, drug use, and homosexual and bisexual sexual practices influence the spread of AIDS.

In recent years, the public has mobilized in the fight against HIV/AIDS.

Theoretical Perspectives on Health Care

A deeper understanding of the nation's health care system and its problems can be achieved by applying the three major theoretical paradigms of sociology: functionalism, conflict theory, and symbolic interaction theory (see Table 14.5).

The Functionalist View of Health Care

Functionalism argues that any institution, group, or organization can be interpreted by looking at its positive and negative functions in society. Positive

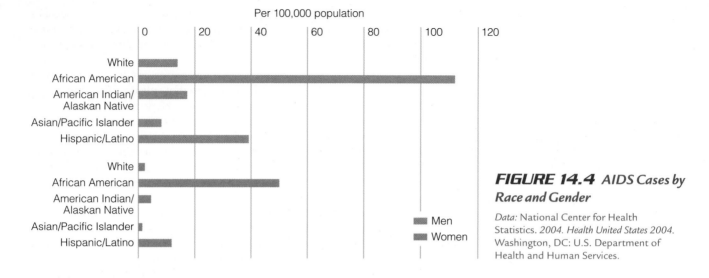

FIGURE 14.4 *AIDS Cases by Race and Gender*

Data: National Center for Health Statistics. *2004. Health United States 2004.* Washington, DC: U.S. Department of Health and Human Services.

functions contribute to the harmony and stability of society. The positive functions of the health care system are the prevention and treatment of disease. Ideally, this would mean the delivery of health care to the entire population without regard to race, ethnicity, social class, gender, age, or any other characteristic. At the same time, the health care system is notable for a number of negative functions, those that contribute to disharmony and instability of society (some will be reviewed in the section titled The Health Care Crisis in the United States).

Functionalism also emphasizes the systematic way that various social institutions are related to each other, together forming the relatively stable character of society. You can see this with regard to how the health care system is entangled with government through such things as federal regulation of new drugs and procedures. The government is also deeply involved in health care through scientific institutions such as the National Institutes of Health, a huge government agency that funds new research on various

matters of health and health care policy. As a social institution, health care is also one of the nation's largest employers and, thus, is integrally tied to systems of work and the economy.

The Conflict Theory View of Health Care

Conflict theory stresses the importance of social structural inequality in society. From the conflict perspective, the inequality inherent in our society is responsible for the unequal access to medical care. Minorities, the lower classes, and the elderly, particularly elderly women, have less access to the health care system in the United States than Whites, the middle and upper classes, and the middle-aged. (To the contrary, functionalists argue that relatively greater access of the middle and upper classes to medical care is good for society because the upper classes are more beneficial—"functional"—to society.) Restricted access is further exacerbated by the high

table14.5	**Theoretical Perspectives on the Sociology of Health**		
	Functionalism	**Conflict Theory**	**Symbolic Interaction**
Central point	The health care system has certain functions, both positive and negative.	Health care reflects the inequalities in society.	Illness is partly socially constructed.
Fundamental problem uncovered	The health care system produces some negative functions.	Excessive bureaucratization of the health care system and privatization lead to excess cost.	Patients are patronized and infantilized.
Policy implications	Policy should decrease negative functions of health care system for minority groups, the poor, and women.	Policy should improve access to health care for minority racial-ethnic groups, the poor, and women.	Doctors, nurses, and other medical personnel should periodically take the sick role of the patient, as an instructional device.

costs of medical care, stemming from high fees and the abuses of the fee-for-service and third-party payment systems (explained later in the chapter). The exceptionally high incomes of some medical professionals amplify the social chasm between medical practitioners and an increasingly resentful public.

Excessive bureaucratization is another affliction of the health care system that adds to the alienation of patients. The U.S. health care system is burdened by endless forms for both physician and patient, including paperwork to enter individuals into the system, authorize procedures, dispense medicines, monitor progress, and process payments. Long waits for medical attention are normal, even in the emergency room. Prolonged waits have reached alarming proportions in the emergency rooms of many urban hospitals in the United States and can only deepen the alienation of patients. One study showed that up to 15 percent of emergency room patients give up and leave before receiving care because of long waits—some stretching to an abysmal fifteen hours.

Symbolic Interactionism and Health Care

Symbolic interactionists hold that illness is partly (although obviously not totally) socially constructed (Armstrong 2003). The definitions of illness and wellness are culturally relative—sickness in one culture may be wellness in another. It is time dependent as well. A condition considered optimal in one era (such

as being thin) may be defined as sickness at another time in the same culture (at the turn of the twentieth century in this country, a healthy woman was supposed to be plump). Similarly, the health care system itself has a socially constructed aspect. The ways we behave toward the ill, toward doctors, and toward innovative ventures such as HMOs are all social creations.

The symbolic interaction perspective highlights a number of socially constructed problems in the health care system. Medical practitioners frequently subject patients to *infantilization* (treat them like children, even if adult). The patient is assigned a role that depends heavily on the physician and the health care system, much as an infant is dependent on its parents. Doctors and nurses may begin patronizing the patient from the initial greeting, a condescending "How are *we* today?" Physicians commonly address patients by their first names, yet patients virtually always address physicians as "Doctor." Women patients are far more likely than men to be addressed by their first names. Such patronizing and infantilizing of patients is common in emergency rooms, where minority patients are infantilized most often (Armstrong 2003; Weitz 2001; Gonzalez 1996).

The symbolic interactionist analysis of the health care system allows us to see these problems more clearly. One solution is to give health care professionals training about such matters in medical school. This is just starting to happen in a few U.S. medical schools. For example, a social issue addressed in some medical school courses on gynecology is how to manage

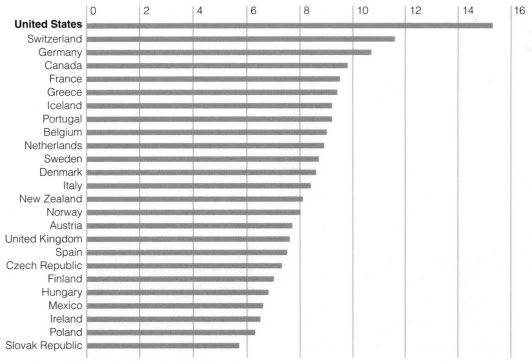

FIGURE 14.5 _Health Care Expenditures as a Percentage of Gross National Product (GNP)_

Source: U.S. Census Bureau. 2008. *Statistical Abstract of the United States, 2007.* Washington, DC: U.S. Printing Office.

the interaction when a male gynecologist treats his woman patient. Women patients feel uncomfortable and vulnerable when they lie partially naked on an examination table with their heels in elevated stirrups and their legs apart. Furthermore, strongly fixed social norms say that when a man touches a woman's genitals, it is an act of intimacy. Yet the gynecological examination is supposed to be completely impersonal. Male gynecologists are notorious for their failure to appreciate the discomfort of their female patients (Scully 1994; Emerson 1970).

The Health Care Crisis in the United States

Currently, the cost of medical care in the United States is approximately 14 percent of our gross domestic product, making health care the nation's third leading industry. The cost of health care has at times risen at twice the annual rate of inflation. The United States tops the list of all countries in per person expenditures for health care (see Figure 14.5). Despite all that spending, the U. S. health care system is not the best in the world. Other countries spend considerably less money and deliver a level of health care at least as good. For example, Sweden and France spend roughly one-half as much per capita as the United States, and Great Britain spends a bit more than one-third as much, yet these countries have national health insurance programs that cover virtually their entire populations and deliver a level of care not much different overall from that achieved in the United States.

The battlefronts of the health care crisis extend to the cost of health care, the epidemic of malpractice suits, the development of HMOs, and the debate

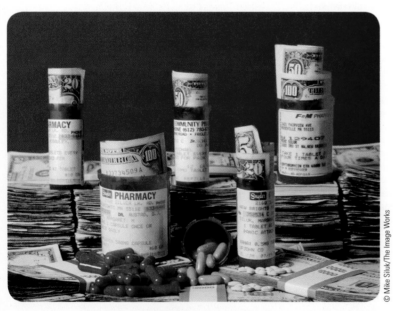

The high cost of prescription drugs is indicative of the problems generated in a profit-based health care system.

over universal health care (Starr 1995; Scarpitti et al. 1997). Here we examine each.

The Cost of Health Care

Health care must be paid for, and in the United States the structures in place for paying the doctor's bill are in a state of chaos. The central element in the system is the fee-for-service principle. Under this arrangement, the patient is responsible for paying the fees charged by the physician or hospital. Fortunate patients will be adequately insured and able to pass on their health expenses to the insurance company. Many people are not insured, however—especially people with lower incomes. This group must reckon with large bills by drawing on their own limited resources.

Overall, 31 million people in the United States are without health insurance. Most severely hit are the southern and southwestern states, where people without health insurance reach more than 15 percent of the population (see Map 14.1 on page 361). A third source of payment is the government, which is increasingly the ultimate payer of health care expenses.

The greatest contributors to skyrocketing health care costs are the soaring costs of hospital care and the rise in fees for the services of physicians. The services of specialists in particular are hugely expensive. Another culprit is the third-party payment system. With hospitals interested in making a profit and patients simply passing bills on to insurance providers, there is little pressure to keep down the cost of care. Patients may be treated on an in-patient basis when a hospital stay is not actually necessary, unnecessary treatments may be given, and some diagnostic procedures may be ordered simply because expensive equipment is available that needs to be used to pay for itself.

The U.S. government has for some time sought to have some form of widespread guaranteed health service, at least for certain categories of people, such as veterans, the poor, and the elderly. The **Medicare** program, begun in 1965 under the administration of President Lyndon Johnson, provides medical insurance covering hospital costs for all individuals age 65 or older. The Medicare program does not cover physician costs incurred outside the hospital, but other programs do, although the patient must pay a portion of the cost.

Medicaid is a governmental program that provides medical care in the form of health insurance for the poor, welfare recipients, and the disabled. The program is funded through tax revenues. The costs covered per individual vary from state to state because the state must provide funds to the individual in addition to the funds that are provided by the federal government. The Medicare and Medicaid programs together are as close as the United States has come to the ideal of universal health insurance. More recent attempts to establish universal coverage have not been successful (Weitz 2001), although the administration of President

Obama has promised to reintroduce a system of universal health care.

Medical Malpractice

A rising problem is malpractice insurance, a problem forcing many physicians to abandon some services. Annual insurance premiums (costs) for physicians have risen tenfold in the past two decades and range now from about $5000 per year for physicians in family practice to well over $170,000 per year for physicians in specialties such as radiology, anesthesiology, and surgery. The cost of these insurance premiums is simply passed along to consumers (patients) and has contributed to the rise in the overall cost of health care.

The U.S. public has traditionally accorded physicians a high social status and high incomes. The recent increase in malpractice suits suggests that the public is beginning to question the privileged status of doctors (Pescosolido et al. 2001). Several specific reasons have been put forward for client revolts against the medical profession. Attorneys claim that the primary cause of the malpractice dilemma is a declining standard of medical care and a rising incidence of medical negligence. Another argument is that, as a result of specialization, doctors now fail to establish old-fashioned rapport with patients, and thus patients are less attached to their physicians and more likely to turn hostile. A third notion is that the high cost of medical care and resentment about the outsized incomes of physicians has generated animosity in patients that is quick to show itself.

Adding to the crisis is the eagerness with which lawyers take on malpractice suits against doctors. To protect themselves against potential malpractice suits, doctors increasingly practice **defensive medicine,** which entails ordering excessively thorough tests, X-rays, and so on at the least indication that something might be amiss. This is done partly to ensure that nothing is missed and partly to document

that the highest possible level of care was given. Although this extra attentiveness may contribute marginally to favorable medical outcomes, it contributes mightily to the overall cost of health care.

A Response to the Problem: HMOs

Health maintenance organizations (HMOs) are an innovation in health care first begun shortly after the Second World War but becoming widespread only within the last twenty years. The staffs in HMOs can include anywhere from several doctors to several hundred doctors. People who join HMOs are assigned to a physician who administers care and when necessary gives referrals to specialists affiliated with the HMO. The doctors in an HMO earn salaries rather than fees. All services are ultimately paid for by the membership fee paid by subscribers to the HMO. Eliminating both the fee-for-service system and the third-party insurer drives health care costs down.

The Universal Health Care Debate

Periodically, in the United States plans for the reorganization of the health care system are suggested in order to solve problems of high costs and other problems. Universal health care for all Americans is suggested as part of such intended programs. As we noted above, the administration of President Barack Obama has promised to introduce such a program.

The core of such programs is called **managed care.** Under this plan, all individuals in the United States

Elder care for Wonder Woman, Superman, Captain America—can you name the rest?

would belong to a complex of managed care organizations, rather like HMOs, that would use their collective bargaining force to drive down the cost of health insurance, while accepting the responsibility for operating their own facilities in an economical manner and continuing to provide high-quality care. Everyone would join the managed care complex. However, individuals would still be free to retain their personal physicians if they so choose, as long as their physicians meet government-stipulated criteria. The plan was intended to achieve the advantages of socialized medicine systems as administered in Great Britain, Canada, and other places where everyone is entitled to see a doctor when they need one, while retaining elements of the profit motive in the system.

Chapter Summary

- ## What is the importance of the education institution?

  Education is the social institution that is concerned with the formal transmission of society's knowledge. It is therefore part of the socialization process. More money is spent in the United States on education than on any other activity except health care.

- ## How does education affect, or not affect, the likelihood of social mobility?

  The number of years of formal education for individuals has important, but in many ways modest, effects on their ultimate occupation and income. Social class origin affects the extent of educational attainment (the higher the social class origins, the more education is ultimately attained) as well as occupation and income (higher social class origin likely means a more prestigious occupation and more income).

- ## How does the educational system perpetuate inequality, instead of reduce it?

  Although the education system in the United States has traditionally been a major means for reducing racial, gender, and class inequalities among people, it is also true that the educational institution has perpetuated these inequalities. Test biases based on culture, language, race, gender, and class have been only slightly reduced in standardized tests, if at all. *Tracking* continues to disproportionately and negatively affect minorities, women, and the working classes. Students are subject to *teacher expectancy effect,* which affects both teacher expectations and student perfor-

mance. The *self-fulfilling prophecy* and *stereotype threat effect* work to the detriment of minorities, working-class people, and women.

- ## What are the social factors in illness?

  Race–ethnicity, social class, and gender are major factors in the state of health care in the United States: Whites live longer than Blacks and Hispanics, due partly to worse health care and less access to health care for minorities. Black women are more likely than White women to fall victim to cancer, heart disease, strokes, and diabetes. The mortality rate among Native Americans is one and one-half times that of the general population.

- ## How serious are the effects of these social factors?

  The lower one's social class status, the greater are one's chances of tuberculosis, heart disease, cancer, and arthritis. The poor are more likely to be depersonalized and treated inadequately by the health care system than those of higher social class status. Men and women tend to be treated differently in the health care system because there is still a tendency for the male-dominated profession to regard the problems of women as non-mainstream and "special."

- ## How have AIDS and other sexually transmitted diseases developed?

  A person who is known to be HIV positive, or who has full-blown AIDS, is partly blamed, negatively *stigmatized,* and treated as a social outcast. Although once thought to be primarily a disease among gay men, AIDS has increased among intravenous drug users and among women who are infected from contact with heterosexual men. Among new cases of AIDS worldwide, half are women.

- ***What is the health care crisis in the United States?***

High costs and malpractice suits have resulted in a policy crisis today in the U.S. health care system. The growth of *health maintenance organizations (HMOs)* shows promise for easing the crisis, although programs for large HMOs and national universal health insurance (including programs for managed care) have had only limited success to date.

Key Terms

achievement test 352
anorexia nervosa 358
cognitive ability 352
cognitive elite 354
credentialism 348
defensive medicine 367
epidemiology 359
health maintenance
   organization (HMO) 367

latent functions 348
managed care 367
Medicaid 366
Medicare 366
predictive validity 352
schooling 346
self-fulfilling prophecy 355
social epidemiology 359
standardized ability test 352

stereotype threat effect 356
stigma 363
teacher expectancy effect 348
tracking 354

Online Resources

Sociology: The Essentials Companion Website

www.cengage.com/sociology/andersen
Visit your book companion website where you will find more resources to help you study and write your research papers. Resources include web links, learning objectives, Internet exercises, quizzing, and flash cards.

 is an easy-to-use online resource that helps you study in less time to get the grade you want NOW.

www.cengage.com/login
Need help studying? This site is your one-stop study shop. Take a Pre-Test and Cengage NOW will generate a Personalized Study Plan based on your test results. The Study Plan will identify the topics you need to review and direct you to online resources to help you master those topics. You can then take a Post-Test to determine the concepts you have mastered and what you still need to work on.

15

Chapter fifteen
CHAPTER FIFTEEN
Chapter

Politics and the Economy

Power, Politics, and the State

Power and Authority

Theories of Power

Government: Power and Politics in a Diverse Society

Economy and Society

The Changing Global Economy

Sociological Perspectives on Work

Characteristics of the Labor Force

Power in the Workplace

Chapter Summary

[*Picture a couple* lounging together on a public beach on a national holiday. This couple is probably giving no thought to how much their life is influenced by the government at that moment. Permission to walk on the public beach comes from the government; the day off to celebrate a holiday is sanctioned by the government; the drive to the beach took place in a car registered with the state and inspected by the state, driven by a motorist licensed by the state.

Perhaps the couple imagine themselves married someday, maybe having children they will send to public school. Does the government have anything to say about the couple's children? They may decide to use birth control until they are ready for children; if so, their choice of birth control methods will be limited to those the government allows. What if they cannot have children? They might consider adopting if state agencies judge them acceptable as parents.

continued

Kevin Dietsch/Corbis

As you imagined this couple, whom did you picture? Were they the same race? Not so long ago, the law would have forbidden the marriage if it were interracial. Is this a man and a woman, or is this a same-sex couple? If they are lesbian or gay, despite being in love and wanting to form a lifelong relationship, in most states they are prevented by law from marrying. The range of things regulated by the government is simply enormous, and yet many are never noticed.

The reach of government—and the politics associated with it—is extensive, indicating a sociological lesson you have learned throughout this book: Every day the things we do and how we behave are influenced by social structures that extend beyond our immediate day-to-day life. Sometimes this becomes apparent to people; other times, it is less tangible, but the influence of social structure and social forces is everywhere. You see this, too, in the work that people do—or the work they do not do. As the economic recession of 2009 evolved, people became increasingly aware of how social and economic forces far beyond their immediate control shaped people's lives. Thousands of people lost their jobs, many lost their homes, and those who had retirement funds found themselves losing financial resources they worked a lifetime to save. Young people entering the labor market for the first time, even as college graduates, found themselves facing possible joblessness—through no fault of their own—merely because they exited school at a time when broader social and economic forces were shaping their life opportunities.

All of this can be understood in a sociological framework. Sociologists study the social forces that shape politics and government, as well the social changes that are affecting work and the economy.

Power, Politics, and the State

Sociologists use the term **state** to refer to the organized system of power and authority in society. The state is an abstract concept that includes the institutions that represent official power in society, including the government, the legal system (law, courts, and the prison system), the police, and the military. Theoretically, the state exists to regulate social order, ranging from individual behavior and interpersonal conflicts to international affairs. Less powerful groups in the society may see the state more as an oppressive force than as a protector of individual rights. However, they may still turn to the state to rectify injustice, for example, by advocating for civil rights or seeking state-based rights and protections for people with disabilities.

The state has a central role in determining the rights and privileges of various groups. The state determines who is a citizen and who is not; the state may be called upon to resolve conflicts between management and labor (such as in airline strikes); and the state may pass legislation determining the benefits of different groups or make decisions that extend rights to various groups, such as the right for same-sex marriage.

Numerous institutions make up the state, including the government, the legal system, the police, and the military. The *government* creates laws and procedures that regulate and guide a society. The *military* is the branch of government responsible for defending the nation against domestic and foreign conflicts. The *court system* is designed to punish wrongdoers and adjudicate disputes. Court decisions also determine the guiding principles or laws of human interaction. *Law* is a fundamental type of formal social control that outlines what is permissible and what is forbidden. The *police* are responsible for enforcing law in the community and for maintaining public order. The *prison system* is the institution responsible for punishing those who have broken the law. Under the U.S. Constitution, these state institutions treat people equally, although sociologists have documented how often this is not the case (see also Chapter 7).

Sociological analyses of the state focus on several different questions. One important issue is the relationship between the state and inequality in society. State policies can have very different impacts on different groups, as we will see later in this chapter. Another issue explored by sociological theory is the connection between the state and other social institutions—the state and religion or the state and the family.

The State and Social Order

Throughout this book we have seen that a variety of social processes contribute to order in society, but none so explicitly and unambiguously as the official system of power and authority in society. In making laws, the state decrees which actions are legitimate and which are not. Punishments for illegitimate actions are enforced, and systems for administering punishment are maintained. The state also influences public opinion through its power to regulate the media. In some cases the state circulates **propaganda,** which is disseminated with the intention to justify the state's power. Censorship is another means by which the state can direct public opinion and uphold dominant group values, such as the Federal Communications Commission requirement for all TVs with a 13-inch screen or greater to have a V-chip, thus permitting parents to control what their children can view.

The state's role in maintaining public order is also apparent in how it manages dissent. Protest movements perceived by those in power to challenge state authority or disrupt society may be repressed through state action, such as by surveillance, imprisonment, or police or military force. For example, throughout the 1960s the FBI conducted illegal surveillance operations targeting Martin Luther King Jr. and other civil rights leaders (Garrow 1981). Now, the increased security screenings at airports, the power to intercept email and voice mail via the Patriot Act of 2001, and the presence of cameras at traffic intersections are evidence of increased surveillance by the state in U.S. society.

Global Interdependence and the State

The global character of modern society means that political systems, along with economic systems, are now elaborately entangled—a phenomenon that can be observed daily in the newspaper. The global economy profoundly affects the character of different nation-states and their relationships. One example is the World Trade Organization (WTO), created in 1994 to monitor and resolve trade disputes. Member nations may challenge the decisions made by local and national governments, with the WTO Council in Geneva resolving disagreements. Another is the European Union (EU), an alliance of separate nations established to promote a common economic market and develop political unions with Western Europe. The creation of these organizations has produced a system that transcends individual nation-states by formalizing and strengthening the rules that govern many aspects of world trade.

At the same time, this world interdependence is also occurring at a time when in some nations there is increasing nationalism. **Nationalism** is the strong identity associated with an extreme sense of allegiance to one's culture or nation, often to the exclusion of interdependent relations with others. Nationalism can become a political movement, such as when groups subordinated by external nations use their original national culture as the basis for resisting oppression. It can also be a political movement when a group identifying itself as a nation (regardless of its official status as such) tries to become a dominant force in the world, such as is the case with the Taliban in Afghanistan. Nationalism is often associated with ethnocentrism because, taken to the extreme, nationalists define others in hostile terms; it is thus often a dangerous force in world affairs.

Power and Authority

The concepts of power and authority are central to sociological analyses of the state. **Power** is the ability of one person or group to exercise influence and control over others. The exercise of power can be seen in relationships ranging from the interaction of two people (husband and wife, police officer and suspect) to a nation threatening or dominating other nations. Sociologists are most interested in how power is structured in society: who has it, how it is used, and how it is built into institutions such as the state. In the United States, a society that is heavily stratified by race, class, and gender, power is structured into basic social institutions in ways that reflect these inequalities. Moreover, institutionalized power in society influences the social dynamics within individual and group relationships.

The exercise of power may be persuasive or coercive. For example, a strong political leader may persuade the nation to support a military invasion or a social policy through popular appeal. Or, power may be exerted by sheer force. Generally speaking, groups with the greatest material resources have the greatest power, but not always. A group may by sheer size be able to exercise power, through, for example, organized social protests. But smaller groups may also be able to exercise power, such as in armed uprisings.

Power can be legitimate—accepted by the members of society as right and just—or it can be illegitimate. **Authority** is power perceived by others as legitimate. Authority emerges not from the exercise of power, but from the belief of constituents that the power is legitimate. In the United States, the source of the president's domestic power is not just his status as commander of the armed forces but also the belief by most people that his power is legitimate. The law is also perceived by most as a legitimate system of authority. In contrast, *coercive power* is achieved through force, often against the will of the people being forced. A dictatorship often relies on its ability to exercise coercive power through its control of the military or state police.

Types of Authority

Max Weber (1864–1920), the German classical sociologist, postulated that three types of authority exist in society: traditional, charismatic, and rational–legal (Weber 1978/1921). **Traditional authority** stems from long-established patterns that give certain people or groups legitimate power in society. A monarchy is an example of a traditional system of authority. Within a monarchy, kings and queens rule, not necessarily because of their appeal, or because they have won elections, but because of long-standing traditions within the society.

Charismatic authority is derived from the personal appeal of a leader. Charismatic leaders are often believed to have special gifts, even magical powers, and their presumed personal attributes inspire devotion and obedience. Charismatic leaders often emerge from religious movements, but they come from other realms also. President Obama is for many a charismatic leader, admired not only for being the first African American president of the United States, but also for his ability to inspire so many people, especially young people.

Rational–legal authority stems from rules and regulations, typically written down as laws, procedures, or codes of conduct. This is the most common form of authority in the contemporary United States. People obey not because national leaders are charismatic or because of social traditions, but because there is a legal system of authority established by formalized rules and regulations. Rulers gain legitimate authority by election or appointment in accordance with society's rules. In systems based on rational–legal authority, the rules are upheld by state agents such as the police, judges, social workers, and other state functionaries to whom power is delegated.

thinkingsociologically

Observe the national evening news for one week, noting the people featured who have some kind of *authority*. List each of them and note their area of influence. What form of authority would you say each represents: *traditional, charismatic,* or *rational–legal*? How is the kind of authority that a person has reflected in his or her position in society (that is, race, class, gender, occupation, education, and so on)?

The Growth of Bureaucratic Government

According to Weber, rational–legal authority leads inevitably to the formation of bureaucracies. As we noted in Chapter 6, a **bureaucracy** is a formal organization characterized by an authority hierarchy, a clear division of labor, explicit rules, and impersonality. Bureaucratic power comes from the accepted legitimacy of the rules, not personal ties to individuals. The rules may change, but they do so through formal, bureaucratic procedures. People who work within bureaucracies are selected, trained, and promoted based on how well they apply the rules. Those who establish the rules are unlikely to be the same people who administer them. Bureaucracies are hierarchical, and the bureaucratic leadership may be quite remote. Power in bureaucracies is dispersed downward through the system to those who actually carry out the bureaucratic functions. It is an odd feature of bureaucracy that those with the least power to influence how the rules are formulated—those at the bottom of the hierarchy—are often the most adamant about strict adherence to the rules; their job evaluation may rest on their enforcement.

Is the bureaucratic growth of government in the United States becoming unwieldy? Many would say yes, judging from the sheer size of bureaucratic governmental institutions. Bureaucracy has become the modern system of administration. In principle,

Bureaucracies are organized according to hierarchical and rule-driven forms of social organization.

bureaucracies are highly efficient modes of organization, but the reality is often very different. Bureaucracies tend to proliferate rules, ensnaring the organization in its own bureaucratic requirements. For example, government procurement of computers is so overburdened by bureaucratic rules for review, comparison, bidding, and approval that by the time a purchase is approved, the approved models may be obsolete.

Within bureaucracies, personal temperament and individual discretion are not supposed to influence the application of rules. But, bureaucracy has another face. Bureaucratic workers frequently exercise discretion in applying rules and procedures, "working the system," perhaps by personalizing the interaction or dodging bureaucratic stipulations. But, most of the time, dealing with an elaborate bureaucracy—even an electronic one like voice mail—can be very frustrating.

Theories of Power

Does the state act in the interests of its different constituencies or does it merely reflect the needs of a the most powerful? In other words, how is power exercised in society? This question has spawned much sociological study and debate and has resulted in several theoretical models of state power. Four theoretical models have been developed by sociologists to answer this question: the *pluralist model,* the *power-elite model,* the *autonomous state model,* and *feminist theories of the state.* Each begins with a different set of assumptions and arrives at different conclusions (see Table 15.1).

table 15.1	**Theories of Power in Society**			
	Pluralism	**Power Elite**	**Autonomous State**	**Feminist Theory**
Interprets the state	As representing diverse and multiple groups in society	As representing the interests of a small, but economically dominant, class	As taking on a life of its own, perpetuating its own form and interests	As masculine in its organization and values (that is, based on rational principles and a patriarchal structure)
Interprets political power	As derived from the activities of interest groups and as broadly diffused throughout the public	As held by the ruling class	As residing in the organizational structure of state institutions	As emerging from the dominance of men over women
Interprets social conflict	As the competition between diverse groups that mobilize to promote their interests	As stemming from the domination of elites over less powerful groups	As developing between states, as each vies to uphold its own interests	As resulting from the power men have over women
Interprets social order	As the result of the equilibrium created by multiple groups balancing their interests	As coming from the interlocking directorates created by the linkages among those few people who control institutions	As the result of administrative systems that work to maintain the status quo	As resulting from the patriarchal control that men have over social institutions

The Pluralist Model

The **pluralist model** interprets power in society as derived from the representation of diverse interests of different groups in society. This model assumes that in democratic societies, the system of government works to balance the different interests of groups in society. An **interest group** can be any constituency in society organized to promote its own agenda, including large, nationally based groups such as the American Association of Retired Persons (AARP) and the National Rifle Association (NRA). Some interest groups are organized around professional and business interests, such as the American Medical Association (AMA) and the Tobacco Institute. Others concentrate on one political or social goal, such as NORML, working to reform marijuana laws. According to the pluralist model, interest groups achieve power and influence through their organized mobilization of concerned people and groups.

The pluralist model has its origins in functionalist theory. This model sees the state as benign and representative of the whole society. The pluralist model sees power as broadly diffused across the public with people who want to effect a change or express their points of view needing only to mobilize to do so. The pluralist model suggests that members of diverse groups can participate equally in a representative and democratic government. As seen from this model, various special interest groups compete for government attention and action. The pluralist model sees special interest groups as an integral part of the political system, even though they are not an official part of government. In the pluralist view, special interest groups make government more responsive to the needs and interests of different people, an especially important function in a highly diverse society.

The pluralist model helps explain the importance of **political action committees (PACs),** groups of people who organize to support candidates they feel will represent their views. In 1974, Congress passed legislation enabling employees of companies, members of unions, professional groups, and trade associations to support political candidates with money they raise collectively. The number of political action committees has now grown to almost 4000. PACs have enormous influence on the political process. They are now so powerful that many are critical of the influence they have on elections. In the 2008 presidential election, counting all PAC contributions, PACs spent over 400 million dollars to influence voters (Federal Election Commission 2008).

The Power Elite Model

The **power elite model** originated in the work of **Karl Marx** (1818–1883) and developed from the framework of conflict theory. According to Marx, the dominant or ruling class controls all the major institutions in society; the state itself is simply an instrument by which the ruling class exercises its power. The Marxist view of the state emphasizes the power of the upper class over the lower classes, the small group of elites over the rest of the population. The state, according to Marx, is not a representative, rational institution, but an expression of the will of the ruling class (Marx 1972/1843).

Marx's theory was elaborated much later by **C. Wright Mills** (1956), who popularized the term *power elite*. Mills attacked the pluralist model, arguing

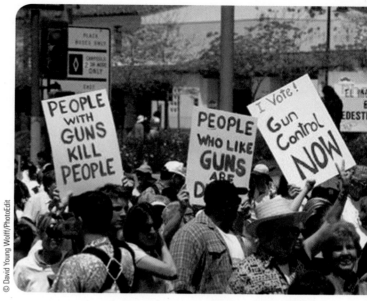

The pluralist model of the state sees diverse interest groups as mobilizing to influence policy and gain political power.

that the true power structure consists of people well positioned in three areas: the economy, the government, and the military. These three institutions are considered the bastions of the power elite, although some have argued that Mills overemphasized the role of the military (Domhoff 2002). While sharing common beliefs and goals, the power elite shape political agendas and outcomes in the society along the narrow lines of their particular collective interests. This small and influential group at the top of the pyramid of power holds the key positions in the major state institutions, giving them extensive power and control over the rest of society.

The power elite model posits a strong link between government and business, a view supported by the strong hand government takes in directing the economy and by the role of military spending as a principal component of U.S. economic affairs. The power elite model also emphasizes how power overlaps between influential groups.

Interlocking directorates are organizational linkages created when the same people sit on the board of directors for numerous corporations. People in elite circles may serve on the boards of several major companies, universities, and foundations at the same time. People drawn from the same elite group receive most of the major government appointments; thus, the same relatively small group of people tends to the interests of all these organizations and the interests of the government. These interests naturally overlap and reinforce each other.

The power elite model sees the state as part of the structure of domination in society, one in which the state is simply a piece of the whole. Members of the upper class do not need to occupy high office themselves to exert their will, as long as they are in a position to influence people who are in power (Domhoff 2002). The majority of the power elite are White men,

which means that the interests and outlooks of White men dominate the national agenda.

The Autonomous State Model

A third view of power developed by sociologists, the **autonomous state model,** interprets the state as its own major constituent. From this perspective, the state develops interests of its own, which it seeks to promote independently of other interests and the public that it allegedly serves. The state does not reflect the needs of the dominant groups, as Marx and power elite theorists would contend. It is an administrative organization with its own needs, such as maintenance of its complex bureaucracies and protection of its special privileges (Evans et al. 1985; Skocpol 1992; Rueschmeyer and Skocpol 1996).

Autonomous state theorists note that states tend to grow over time, including the possible expansion beyond their original boundaries by military incursions. Another example of expansion is the North American Free Trade Agreement (NAFTA), in which the United States is expanding its state interests by regulating business not only in the United States but also in Mexico and Canada.

The huge government apparatus now in place in the United States is a good illustration of autonomous state theory. The government provides a huge array of social support programs, including Social Security, unemployment benefits, agricultural subsidies, public assistance, and other economic interventions intended to protect citizens from the vagaries of a capitalist market system (Collins 1988). The purpose of these programs is to serve people in need. Autonomous state theory argues that the government has grown into a massive, elaborate bureaucracy, run by bureaucrats more absorbed in their own interests than in meeting the needs of the people. As a consequence, government can become paralyzed in conflicts between revenue-seeking state bureaucrats and those who must fund them. This can lead to revolt against the state, as in the tax revolts appearing sporadically throughout the country (Lo 1990; Collins 1988).

Feminist Theories of the State

Feminist theorists diverge from the preceding theoretical models by seeing men as having the most important power in society. The pluralists see power as widely dispersed through the class system, power elite theorists see political power directly linked to upper-class interests, and autonomous state theorists see the state as relatively independent of class interests.

Some feminist theorists argue that all state institutions reflect men's interests; they see the state as fundamentally patriarchal, its organization embodying the fixed principle that men are more powerful than women. Feminist theories of the state conclude that despite the presence of a few powerful women, the state is devoted primarily to men's interests, and moreover, the actions of the state will tend to support gender inequality (Haney 1996; Blankenship 1993). One historical example would be laws denying women the right to own property once they married. Such laws protected men's interests at the expense of women.

Evidence that "the state is male" (MacKinnon 2006, 1983) is easy to observe by looking at powerful political circles. Despite the recent inclusion of more women in powerful circles and the presence of some notable women as major national figures, most of the powerful are men. The U.S. Senate is 86 percent men; groups that exercise state power, such as the police and military, are predominantly men. Moreover, these institutions are structured by values and systems that can be described as culturally masculine—that is,

based on hierarchical relationships, aggression, and force. Feminist theory begins with the premise that an understanding of power cannot be sound without a strong analysis of gender (Haney 1996).

Government: Power and Politics in a Diverse Society

The terms *government* and *state* are often used interchangeably. More precisely, the government is one of several institutions that make up the state. The **government** includes those institutions that represent the population, making rules that govern the society. The government of the United States is a **democracy**; therefore, it is based on the principle of representing all people through the right to vote.

The actual makeup of the government, however, is far from representative of society. Not all people participate equally in the workings of government, either as elected officials or as voters, nor do their interests receive equal attention. Women, the poor and working class, and racial–ethnic minorities are less likely to be represented by government than are White middle- and upper-class men. Sociological research on political power has concentrated on inequality in government affairs and demonstrated large, persistent differences in the political participation and representation of various groups in society.

Diverse Patterns of Political Participation

One would hope that in a democratic society all people would be equally eager to exercise their right to vote and be heard. That is far from the case. Among democratic nations, the United States has one of the *lowest* voter turnouts (see Figure 15.1). In the 2008 presidential election, the percentage of eligible voters who went to the polls was only 64 percent of the population, one of the highest voter turnouts ever for a presidential election. A turnout of 50 percent or less is typical of national elections, although elections in 1960, 1964, and 1968 had turnouts just over 60 percent of the population. Voter turnout in congressional and local elections is even lower.

Generally, older, better educated, and financially better off people are the most likely to vote. Although young people are generally not as politically engaged as other groups, the presidential election of 2008 saw the highest turnout ever of youth voters (those aged 19–29); estimates are the 52 percent of young people voted in the 2008 election, with a two to one margin for now President Obama. Voter turnout is not, however, the only measure of youth participation. In the 2008 elections, young people were very active in the campaign, attending campaign rallies, following

THE LONDON SUMMIT 2009
STABILITY | GROWTH | JOBS

Stefan Rousseau/PA Photos/Landov

The leadership of the world's most powerful nations, represented by the G20 and gathered here at a summit meeting in 2009. Where are the women?

understanding diversity

Diversity in the Power Elite

As society has become more diverse, has it made a difference in the makeup of the power elite? Various groups—women, racial–ethnic groups, lesbians, and gays—have vied for more representation in the halls of power, but have their efforts succeeded? If they make it to power, does this change the corporations, military, or government—the major institutions composing the power elite?

Sociologists Richard L. Zweigenhaft and G. William Domhoff examined these questions by analyzing the composition of boards of directors and chief executive officers (CEOs) of the largest banks and corporations in the United States, as well as analyzing Congress, presidential cabinets, and the generals and admirals who form the military elite. In addition, they examined the political party preferences and the political positions of people found among the power elite. Do women and minorities bring new values into power, thereby changing society as they move into powerful positions, or do their values match those of the traditional power elite or become absorbed by a system more powerful than they are? Zweigenhaft and Domhoff's study looks as well at whether those who do make it into the power elite are within the innermost circles or whether they are marginalized.

They find that women, Jews, gays, lesbians, Black Americans, and Hispanics have become more numerous within the power elite, but only to a small degree. The power elite is still overwhelmingly White, wealthy, Christian, and male. Women and other minorities who make it into the power elite also tend to come from already privileged backgrounds, as measured by their social class and education. Among African Americans and Latinos, skin color continues to make a difference, with darker-skinned Blacks and Hispanics less likely to achieve prominence compared with lighter-skinned people. Furthermore, Zweigenhaft and Domhoff find that the perspectives and values of women and minorities who rise to the top do not differ substantially from their White male counterparts. Some of this is explained by the common class origins of those in the power elite. The researchers also attribute the managing of one's identity to avoid challenging the system as a sorting factor that perpetuates the dominant worldview and practices of the most powerful.

The authors of this study conclude that "the irony of diversity" is that greater diversity may have strengthened the position of the power elite, because its members appear to be more legitimate through their inclusion of those previously left out. But, by including only those who share the perspectives and values of those already in power, little is actually changed. Condoleeza Rice and Clarence Thomas may make it seem that women and African Americans have it made, but as long as they support the positions of White male elites, the power elite goes undisturbed.

Sources: Zweigenhaft, Richard L, and G. William Domhoff. 2006. *Diversity in the Power Elite: Have Women and Minorities Reached the Top?* New Haven: Yale University Press; G. William Domhoff. 2002. *Who Rules America?* New York: McGraw-Hill.

---

the election via the Internet, and contacting other voters. The most striking thing about young voters in the 2008 election was that they are more racially diverse than older groups of voters; 32 percent of young voters in 2008 were African American or Hispanic; among older voters, there was far less racial and ethnic diversity (Keeter et al. 2008).

It is too soon to tell whether the diversity within the youth vote will affect elections in the future. The 2008 election saw record turnouts of African Americans, Latinos, and Asian Americans—all of whom had been less likely to vote than Whites prior to the 2008 election. In fact, the 2008 election was the first time that turnout for African Americans nearly matched that for White Americans (65.3 percent for African Americans, 66.1 percent for Whites; see also Figure 15.2 p. 381). And, for the first time in history, Black women had a higher turnout rate than any other group—68.8 percent (Pew Research Center 2009).

Not only do social factors influence the likelihood of voting, they also influence how people vote. African Americans and Latinos, with the exception of Cuban Americans, tend to be markedly Democratic. Ninety-five percent of Black voters voted for Obama in 2008, as did 67 percent of Latinos and 62 percent of Asian Americans. Traditionally, Latinos and African Americans have been more attracted to the Democratic Party, in part because of party policies, but also because of socioeconomic factors such as age, gender, income, education, and religion (see Figure 15.3 p. 382).

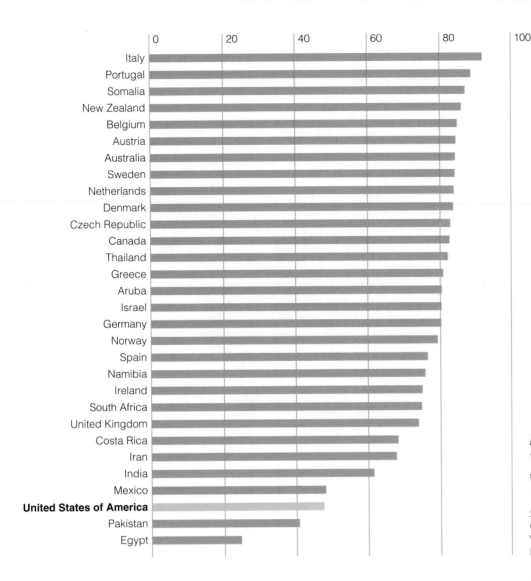

FIGURE 15.1 *International Voter Turnout (as percent of voting age population, 1945–2001).*

Source: Pintor, Rafael López, Maria Gratschew, and Kate Sullivan. 2007. Voter Turnout Rates from a Comparative Perspective. **www.iknowpolitics.gov**

Gender is also a major social factor influencing political attitudes and behavior. The **gender gap** refers to the differences between men and women in political attitudes and behavior. Women tend to have more liberal views than men on a variety of social and political issues and are more likely to vote Democratic. And, women are now more likely to vote than men; in the 2008 elections, this was even more true as women's turnout was higher in every racial–ethnic group and exceeded men overall by 10 million voters; in fact, in the 2008 election, White men dropped in their voter turnout. This election was particularly interesting for women because of the candidacies of both Hillary Clinton and Sarah Palin; in the end, 58 percent of women voted for Obama and 43 percent for McCain (Pew Research Center 2009; www.nytimes.com).

Political Power: Who's in Charge?

A democratic government is supposed to be representative of the people in the nation. Is this the case in the United States? Hardly. As Figure 15.4 p. 383 shows, women and racial–ethnic minorities are vastly underrepresented in our government. Moreover, most members are from upper middle-class or upper-class backgrounds; the vast majority have law, politics, and business as their prior occupation; very few were blue-collar workers before coming to Congress. The majority identify themselves as Protestants, although Roman Catholic is the largest single religious denomination (Congressional Research Service 2008). Many in Congress have large financial interests in the industries that they regulate.

Simply getting into politics requires a substantial investment of money. The average cost of a Senate campaign in 2008 was $2 million; in the House of Representatives, $710,000. Candidates depend on contributions from individuals and groups to finance their election campaigns, with wealthy individuals among the largest campaign contributors, especially to presidential elections (Allen and Broyles 1991). The 2008 presidential election was the most costly in U.S. history. Candidates spent a total of $5.3 billion dollars! President Barack Obama spent a record $750 million and John McCain, the Republican candidate, spent $333 million. For both candidates, media expenditures were the largest piece of campaign expenses,

map15.1 MAPPING AMERICA'S DIVERSITY

Youth and the Obama Election

2008 general election results, all voters

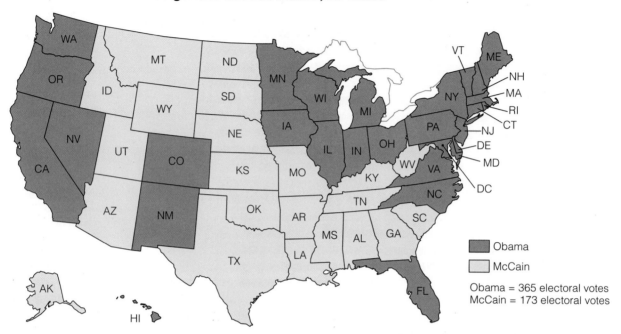

Obama
McCain

Obama = 365 electoral votes
McCain = 173 electoral votes

**What the electoral map would look like if the election
were decided by 18-29 years olds**

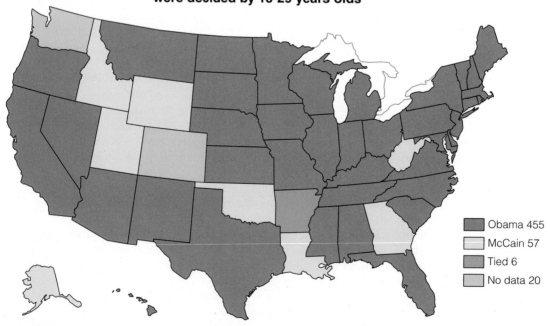

Obama 455
McCain 57
Tied 6
No data 20

Source: The New York Times; The Center for Information and Research on Civic Learning and Engagement. **www.civicyouth.org**

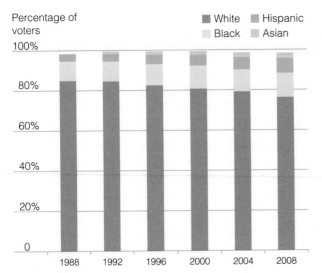

FIGURE 15.2 *The Changing Electorate? Demographic Composition of Voters by Race and Ethnicity, 1988 to 2008*

Source: Lopez, Mark. 2009. "Dissecting the 2008 Electorate: Most Diverse in History." Pew Research Center. **www.pewresearch.com**

Note: Whites include only non-Hispanic Whites. Blacks include only non-Hispanic Blacks. Asian include only non-Hispanic Asians. Native Americans and mixed-race groups are not shown. Asian share not available prior to 1990.

$298 million for Obama (38 percent of expenses) and $119 million for McCain (31 percent of expenses; Center for Responsive Politics 2009).

The largest contributors to political campaigns are typically PACs. Much of the money given by individuals and PACs goes to incumbents, who have an overwhelming edge in elections and already sit on the committees where public policy is hammered out. This picture of elites and business interests funneling money to candidates, who return to the same donors for more money when the next campaign rolls around, has shaken the faith of many Americans in their political system. There is little belief that the political process is a democratic and populist mechanism by which the "little people" can select political leaders to represent them. In fact, national surveys show that only 6 percent of U.S. citizens have a great deal of confidence in Congress, and only 13 percent have some confidence in the Supreme Court (Jones 2008).

Women and Minorities in Government

Although there have been some gains in the number of women and minorities in government, they are still underrepresented—both at the federal and state level. There are now 17 women in the Senate (out of 100) and 75 in the House of Representatives (out of 435), an all-time high in both houses of government. Among those in the House of Representatives, 40 are African Americans, 23 are Latino/a, only 3 are Asian American, and only 1 is Native American.

Although 13 percent of the U.S. population is African American, there is only one African American senator. Hispanics are 13 percent of the U.S. population, but there are only two in the Senate. There are only two Asian Americans in the Senate, and no Native Americans (Ben Nighthorse Campbell, elected in 1992, was for a while the only Native American in the Senate). The House of Representatives is more diverse, but still woefully lacking in reflecting the actual diversity of the nation (see Figure 15.4). The 111th Congress, convened in 2009, has more religious diversity than any previous Congress, with thirty-one Jewish representatives, two Muslims, two Buddhists, one Quaker, and one self-identified atheist. There are only three openly gay members of the House. Clearly, there is a long way to go before we have a representative government in this country.

Researchers offer several explanations for why women and racial–ethnic minorities continue to be underrepresented in government. Certainly, prejudice plays a role. It was not long ago, in the 1960 Kennedy–Nixon election, that Kennedy became the first Catholic president elected. In the 2000 presidential election, Joseph Lieberman was the first Jewish candidate to appear on a major national ticket. Gender and racial prejudice run just as deep in the public mind. Although the percentage of Americans who say they would vote for a woman for president has climbed to 92 percent (53 percent in 1969), a substantial number (42 percent) also say they think a man would make a better president than a woman (Simmons 2001).

Individual prejudice alone, however, cannot account for the lack of representation. Societal causes are a major factor in the successful elections of women and people who are better represented in local political office. Women and minority candidates receive a great deal of political support from local groups, but at the national level, they do not fare as well. The power of incumbents, most of whom are White men, is a disadvantage to any new office seeker.

The Military

The military arm of the state is among the most powerful and influential social institutions in almost all societies. In the United States, the military is the largest single employer. Approximately 3 million men and women serve in the U.S. military, 1.4 million on active duty, including those in Iraq and Afghanistan, and the rest are in the reserves. This does not include the many hundreds of thousands who are employed in industries that support the military, nor does it include civilians who work for the Department of Defense and other military-affiliated agencies (U.S. Census Bureau 2009a).

The Military as a Social Institution. Institutions are stable systems of norms and values that fulfill certain functions in society. The military is a social institution whose function is to defend

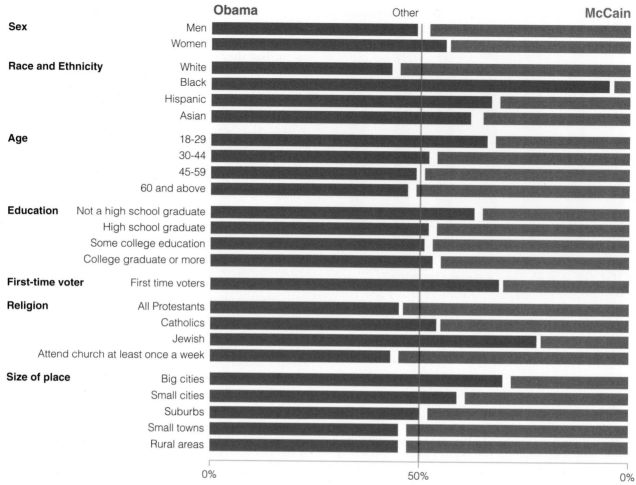

FIGURE 15.3 *Who Voted How? The 2008 Presidential Election*

Source: www.nytimes.com

the nation against external (and sometimes internal) threats. A strong military is often considered an essential tool for maintaining peace, although the values that promote preparedness in the armed forces are perilously close to the warlike values that lead to military aggression against others.

The military is one of the most hierarchical social institutions, and its hierarchy is extremely formalized. People who join the military are explicitly labeled with rank, and if promoted, they pass through a series of well-defined levels, each with clearly demarcated sets of rights and responsibilities. An explicit line exists between officers and enlisted personnel, and officers have many privileges that others do not. Higher ranks are also entitled to absolute obedience from the ranks below them, with elaborate rituals created to remind both dominants and subordinates of their status. As in other social institutions, military enlistees are carefully socialized to learn the norms of the culture they have joined. Military socialization places a high premium on conformity and eliminates individuality. All new recruits have their heads shaved, are issued identical uniforms,

and are allowed to retain very few of their personal possessions. They must quickly learn new, strictly enforced codes of behavior.

Most of the military is a part of the institution of government, but there has also been *privatization of the military*—meaning that an increasing number of military functions have been paid on a contract basis to private, for-profit employers. Under this development, the military becomes like a business, with people and corporations reaping profits on activities that once would have been not for profit. An example of this is the work of Halliburton, a company that amassed millions of dollars in contracts during the war in Iraq—even as then Vice President Dick Cheney was its CEO. The privatization of the military can include companies that provide specific services (such as security), as well as engineering and building contracts. Critics of this trend warn that it will sacrifice safety and national security for the sake of corporate and individual profit and that could lure the brightest people away from traditional military service if they see economic gains from private military service (Singer 2007).

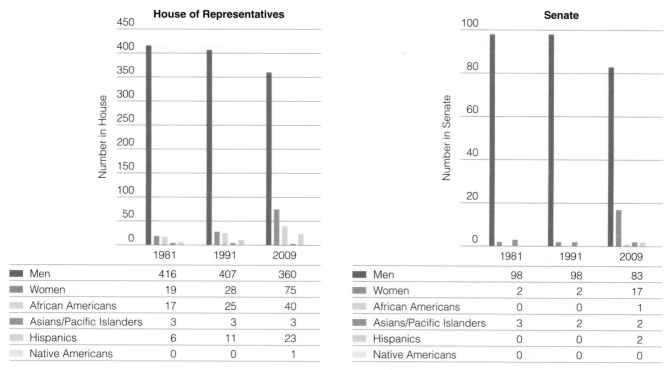

House of Representatives	1981	1991	2009
■ Men	416	407	360
■ Women	19	28	75
▨ African Americans	17	25	40
■ Asians/Pacific Islanders	3	3	3
▒ Hispanics	6	11	23
░ Native Americans	0	0	1

Senate	1981	1991	2009
■ Men	98	98	83
■ Women	2	2	17
▨ African Americans	0	0	1
■ Asians/Pacific Islanders	3	2	2
▒ Hispanics	0	0	2
░ Native Americans	0	0	0

FIGURE 15.4 *A Representative Government?*

Source: U.S. Census Bureau, 2007. *Statistical Abstract of the United States, 2006.* Washington, DC: U.S. Government Printing Office. http://www.census.gov/compendia/statab/

Race and the Military. The greatest change in the military as a social institution is the representation of racial minority groups and women within the armed forces. Picture a U.S. soldier. Whom do you see? At one time you would have almost certainly pictured a young, White male, possibly wearing army green camouflage and carrying a weapon. Today the image of the military is much more diverse. Drawing on the cultural images you have stored in your mind, you are just as likely to picture a young African American man in military blues with a stiffly starched shirt and a neat and trim appearance, or perhaps a woman wearing a flight helmet in the cockpit of a fighter plane.

African Americans have served in the military for almost as long as the U.S. armed forces have been in existence. Except for the Marines, which desegregated in 1942, the armed forces were officially segregated until 1948, when President Harry Truman signed an executive order banning discrimination in the armed services. Although much segregation continued after this order, the desegregation of the armed forces is often credited with promoting more positive interracial relationships and increased awareness among Black Americans of their right to equal opportunities. Until that time the widespread opinion among Whites was that to allow Black and White soldiers to serve side by side would destroy soldiers' morale.

Currently 22 percent of active military personnel are African American (20 percent of men and 34 percent of women); 10 percent are Hispanic. Data on Asian Americans and Native Americans are not available

(Segal and Segal 2004). For groups with limited opportunities in civilian society, joining the military seems to promise an educational and economic boost. Is this realized? Within the military today, there is a policy of equal pay for equal rank. African Americans and Latinos, however, are overrepresented in lower-ranking support positions. Often they are excluded from the higher-status, technologically based positions—those most likely to bring advancement and higher earnings both in the military and beyond. Most minorities remain in positions with little supervisory responsibility, such as service and supply jobs. Although the number of racial minorities in officers' positions has been increasing, they are still underrepresented and are less likely to get there via the route of military academies, as is the case for White officers (Segal and Segal 2004). Still, for both Whites and racial minorities, serving in the military leads to higher earnings relative to one's nonmilitary peers.

Women in the Military. The military academies did not open their enrollment to women until 1976. Since then, the armed services have profoundly changed their admission policies, and in 1996 the Supreme Court ruled (in *United States* v. *Virginia*) that women cannot be excluded from state-supported military academies such as the Citadel and the Virginia Military Institute (VMI). This was a landmark decision that opened new opportunities for women who want the rigorous physical and academic training that military academies provide (Kimmel 2000).

With the introduction of the all-volunteer armed services, more military personnel are from working class and/or racial-ethnic backgrounds. Should there be a military draft so that all have an equal chance of being called for military service? Is your answer shaped by your class and/or race background?

The involvement of women in the military has reached an all-time high in recent years. Women are 14 percent of enlisted military personnel. Now, almost 200,000 women are on active duty, with an additional 151,000 in the reserves, not including the Coast Guard and its reserves. The Air Force has the highest proportion of women (20 percent), followed by the Navy (14 percent), the Army (13 percent), and the Marines (6 percent). Women are 12 percent of the U.S. Coast Guard (Woman's Research and Education Institute 2009).

The former exclusion of women from military service was rationalized by the popular conviction that women should not serve in combat. Despite this attitude, still prevailing and technically the policy of the military, women have been fighting in active combat to defend the nation.

The presence of women in the military has transformed the armed forces, but it also has raised new issues for military personnel. Half of military personnel are married, and there has been an increase in the number of dual-military couples (12 percent of all military marriages). Family separations, frequent moves (on average every three years), risk of injury or death, living in a foreign country—these are only some of the challenges that military personnel face in trying to manage their lives (Segal and Segal 2004).

For women in the military, the highly gendered organization of which they are a part is also a challenge. Indeed, recent reports have documented an alarmingly high rate of sexual assault and sexual harassment against women in the military—by other military personnel. A recent Pentagon report found that one-third of women in the military (and 6 percent of men) experienced sexual harassment, including unwanted crude and offensive behavior, unwanted sexual attention, and sexual coercion; the same report found that 5 percent of women in the military experienced some form of unwanted sexual contact, such as rape, unwanted sodomy, or indecent assault (Defense Manpower Data Center 2008). Periodic scandals involving rape, sexual harassment, and other forms of intimidation against women in the military (including the military academies) reveal that, although certainly not all military men engage in these behaviors, institutions organized around such masculine characteristics as aggression, domination, and hierarchy put women at risk.

Gays and Lesbians in the Military. Gays and lesbians have long served in military duty, despite the policies that have attempted to exclude them. The military has admitted that there always have been gays and lesbians in all branches of the U.S. armed forces, but homophobia is a pervasive part of military culture (Myers 2000; Becker 2000).

President Clinton introduced a "don't ask, don't tell" policy in the military by which recruiting officers

The men and women who serve in the Armed Forces, such as this young woman returning from Iraq, are often separated from families and loved ones for long periods of time.

could not ask about sexual preference, and individuals in the military who keep their sexual preference a private matter could not be discriminated against. While this policy did not explicitly permit the military to seek out and discharge service people on the grounds of homosexuality, critics have argued that the policy does not clearly assert the civil rights of gays and lesbians, and gay soldiers and potential recruits have filed several lawsuits against the military. President Obama promised to end this policy, but then delayed it. It remains unclear whether gays and lesbians will ultimately be permitted to live openly as homosexuals while also pursuing careers in the armed services.

Supporters of the ban on gays in the military often use arguments similar to the arguments used before 1948 to defend the racial segregation of fighting units. They claim that the morale of soldiers will drop if forced to serve alongside gay men and women, national security will be threatened, and known homosexuals serving in the military will upset the status quo and destroy the fighting spirit of military units.

Economy and Society

We move next to an examination of work and the economy in society. To understand *work,* you first must see it in the context of the broad social institution known as the economy. All societies are organized around an economic base. The **economy** of a society is the system by which goods and services are produced, distributed, and consumed. We first look at the economic significance of the historic transformation from agriculturally based societies to industrial and now postindustrial societies.

The Industrial Revolution

In Chapter 5, we discussed the evolution of different types of societies. One of the most significant of these changes was, first, the development of agricultural societies and, later, the far-ranging impact of the Industrial Revolution. Now, the Industrial Revolution is giving way to the growth of postindustrial societies—a development in the economic system with far-reaching consequences for how society is organized.

The Industrial Revolution is usually pinpointed as beginning in mid-eighteenth–century Europe, soon thereafter spreading throughout other parts of the world. The Industrial Revolution led to numerous social changes because Western economies were organized around the mass production of goods. The Industrial Revolution led to the creation of factories, which separated work and family by relocating the place where most people were employed.

We still live in a society that is largely industrial, but that is quickly giving way to a new kind of social organization: postindustrial society. Whereas industrial societies are primarily organized around the

The Industrial Revolution transformed labor, moving it from the household to the factory or other sites where work was mechanized and oriented to mass production.

production of goods, **postindustrial societies** are organized around the provision of services. Thus, in the United States, we have moved from being a manufacturing-based economy to an economy centered on the provision of services. *Service industry* is a broad term meant to encompass a wide range of economic activities now common in the labor market. It includes banking and finance, retail sales, hotel and restaurant work, and health care; it also includes parts of the information technology industry—not electronics assembly, but areas such as software design and the exchange of information (through publishing, video production, and the like).

Comparing Economic Systems

The three major economic systems found in the world today are *capitalism, socialism,* and *communism.* These are not totally distinct, that is, many societies have a mix of these economic systems. **Capitalism** is an economic system based on the principles of market competition, private property, and the pursuit of profit. Within capitalist societies, stockholders own corporations—or a share of the corporation's wealth. Under capitalism, owners keep a surplus of what is generated by the economy; this is their *profit,* which may be in the form of money, financial assets, and other commodities.

Socialism is an economic institution characterized by state ownership and management of the basic industries; that is, the means of production are the property of the state not of individuals. Modern socialism emerged from the writings of Karl Marx, who predicted that capitalism would give way to egalitarian, state-dominated socialism, followed by a transition to stateless, classless communism.

Many European nations, for example, have strong elements of socialism that mix with the global forces of capitalism. Sweden supports an extensive array of state-run social services, such as health care, education, and social welfare programs, but Swedish industry is capitalist. Other world nations are more strongly socialist, although they are not immune from the penetrating influence of capitalism. The People's Republic of China was formerly a strongly socialist society that is currently undergoing transformation to a mix of socialist and capitalist principles, including state encouragement of a market-based economy, the introduction of privately owned industries, and increased engagement in the international capitalist economy.

Communism is sometimes described as socialism in its purest form. In pure communism, industry is not the private property of owners. Instead, the state is the sole owner of the systems of production. Communist philosophy argues that capitalism is fundamentally unjust because powerful owners take more from laborers (and society) than they give and use their power to maintain the inequalities between the worker and owner classes. Communist theorists in the nineteenth century declared that capitalism would inevitably be overthrown as workers worldwide united against owners and the system that exploited them. Class divisions were supposed to be erased at that time, along with private property and all forms of inequality. History has not borne out these predictions.

The Changing Global Economy

One of the most significant developments of modern times is the creation of a global economy, affecting work in the United States and worldwide. The concept of the **global economy** acknowledges that all dimensions of the economy now cross national borders, including investment, production, management, markets, labor, information, and technology (Altman 2001; Carnoy et al. 1993). Economic events in one nation now can have major reverberations throughout the world. When the economies of any major nation are unstable, the effects are felt worldwide.

Multinational corporations—those that draw a large share of their revenues from foreign investments

and conduct business across national borders—have become increasingly powerful, spreading their influence around the globe. The global economy links the lives of millions of Americans to the experiences of other people throughout the world. You can see the internationalization of the economy in everyday life: Status symbols such as high-priced sneakers are manufactured for just a few cents in China. The Barbie dolls that young girls accumulate are inexpensive by U.S. standards, yet it would require one month's wages for the Indonesian worker who makes the doll to buy it for her child.

In the global economy, the most developed countries control research and management, and assembly-line work is performed in nations with less privileged positions in the global economy. A single product, such as an automobile, may be assembled from parts made all over the world—the engine assembled in Mexico, tires manufactured in Malaysia, electronic parts constructed in China. The relocation of manufacturing to wherever labor is cheap has led to the emergence of the *global assembly line,* a new international division of labor in which research and development is conducted in the United States, Japan, Germany, and other major world powers, and the assembly of goods is done primarily in underdeveloped and poor nations—mostly by women and children.

Within the United States, the development of a global economy has also created anxieties about foreign workers, particularly among the working class. Because it is easier to blame foreign workers for unemployment in the United States than it is to understand the complex processes that have produced this phenomenon, U.S. workers have been prone to **xenophobia,** the fear and hatred of foreigners. Campaigns to "buy American" reflect this trend, although the concept of buying American is increasingly antiquated in a global economy.

When buying a product from a U.S. company, it is likely that the parts, if not the product itself, were built overseas. In a global economy, distinctions between U.S. and foreign businesses blur. Moreover, the label "Made in U.S.A." does not necessarily mean that the product was made by well-paid workers in the United States. In the garment industry, sweatshop workers, many of whom are recent immigrants and primarily women, are likely to have stitched the clothing that bears such a label. Moreover, these workers are likely to be working under exploitative conditions. For example, recent surveys of the garment industry in Los Angeles, where 25 percent of all women's outerwear is made, find that 96 percent of the garment firms are in violation of health and safety regulations for workers, and 61 percent are violating wage regulations (Bonacich and Applebaum 2000; Louie 2001).

The development of a global economy is part of the broad process of **economic restructuring,** which refers to the contemporary transformations in the basic structure of work that are permanently altering the workplace. This process includes the changing composition of the workplace, deindustrialization, and use of enhanced technology. Some changes are *demographic*—that is, resulting from changes in the population. The labor force is becoming more diverse, with women and people of color becoming the majority of those employed. Other changes are driven by *technological developments*. For example, the economy is based less on its earlier manufacturing base and more on service industries—those in which the primary business is not the production of goods, but the delivery of services (banking, health care, provision of food, or the like). All these developments are happening within a global context.

thinkingsociologically

Identify a job you once held (or currently hold) and make a list of all the ways that workers in this segment of the labor market are being affected by the various dimensions of *economic restructuring*: demographic changes, globalization of the economy, and technological change. What does your list tell you about how people's individual work experiences are shaped by social structure?

A More Diverse Workplace

A more diverse workplace is becoming a common result of economic restructuring. In a few years, women and racial minorities are expected to increase to more than one-half of the labor force. Another upcoming change in the labor market will be an increase in the number of older people (those over age fifty-five) as the population bulge of the Baby Boomer generation passes through late middle age.

These changes in the social organization of work and the economy are creating a more diverse labor force, but much of the growth in the economy is projected to be in service industries, where, for the better jobs, education and training are required. People without these skills will not be well positioned for success. Manufacturing industries, where racial minorities have in the past maintained a foothold on employment, are now in decline. New technologies and corporate layoffs have reduced the number of entry-level corporate jobs that recent college graduates have always used as a starting point for career mobility. Many college graduates are employed in jobs that do not require a college degree. College graduates, however, do still have higher earnings than those with less education.

Deindustrialization

Deindustrialization refers to the transition from a predominantly goods-producing economy to one based on the provision of services (Harrison and Bluestone 1982). This does not mean that goods are no longer produced, but that fewer workers in the United States are required to produce goods because machines can do the work people once did and many goods-producing jobs have moved overseas.

Deindustrialization is most easily observed by looking at the decline in the number of jobs in the manufacturing sector of the U.S. economy since the Second World War. At the end of the war in 1945, the majority of workers (51 percent) in the United States were employed in manufacturing-based jobs. Now the majority (at least 70 percent) are employed in the *service sector* (U.S. Department of Labor 2009). The service sector includes two segments: service delivery (such as food preparation, cleaning, or child care) and information processing (such as banking and finance, computer operation, or clerical work). Service delivery consists of many low-wage, semi-skilled, and unskilled jobs and employs many women and people of color.

The human cost of deindustrialization can be severe. Deindustrialization has led to **job displacement,** the permanent loss of certain job types that occurs when employment patterns shift. When a manufacturing plant shuts down, many people may lose their jobs at the same time, and whole communities can be affected. Among the areas hardest hit by deindustrialization are communities that were heavily dependent on a single industry, such as steel towns or automobile-manufacturing cities such as Detroit and Cleveland. But this also hits rural communities hard because those communities often have only one major employer, such as a textile plant. Job displacement hits people in both rural and inner cities hard because emerging new industries tend to be located in suburban, not urban or rural areas—a phenomenon called **spatial mismatch.** It is then no surprise that poverty is highest in central cities and rural America. This process has also been especially hard on young people, especially African American and Latino youth. Unless young people have the educational and technical skills for employment in a new economy, or if they live in areas hard hit by job displacement, they have little opportunity for getting a good start in the now global economy. You see this in the extremely high unemployment rates for Black and Hispanic teens (Wilson 1996, 2009).

Technological Change

Coupled with deindustrialization, rapidly changing and developing technologies are bringing major changes in work, including how it is organized, who does it, and how much it pays. One of the most influential technological developments of the twentieth century has been the invention of the semiconductor. Computer technology has made possible workplace transactions that would have seemed like science fiction just a few years ago. Electronic information can be transferred around the world in less than a second. Employees can provide work for corporations located on another continent; thus, a woman in Southeast Asia or the Caribbean can type a book manuscript for a publishing house in New York. Some argue that the computer chip has as much significance for social change as the earlier inventions of the wheel and the steam engine.

Increasing reliance on the rapid transmission of electronic data has produced *electronic sweatshops,* a term referring to the back offices found in many industries, such as airlines, insurance firms, and mail-order houses where workers at computer terminals process thousands of transactions in a day. Workers' performance is likely monitored by a computer, conjuring up images of "Big Brother" invisibly watching. Computers can measure how fast cashiers ring up groceries and how fast ticket agents book reservations. Records derived from computer monitoring then become the basis for job performance evaluation.

Technological innovation in the workplace is a mixed blessing. **Automation**—the process by which human labor is replaced by machines—eliminates many repetitive and tiresome tasks, and it makes rapid communication and access to information possible. But our increasing dependence on technology may make workers subservient to machines. Robots can do the spot-welding on automobiles; some are even used for human surgery. Sophisticated robots are capable of highly complex tasks, enabling them to assemble finished products or flip burgers in fast-food restaurants. Will robots replace human workers? Robots are expensive to buy, but employers can see other savings because "the robot hamburger-flipper would need no lunch or bathroom breaks, would not take sick days, and most certainly would neither strike nor quit" (Rosengarten 2000: 4).

Deskilling is a process whereby the level of skill required to perform certain jobs declines over time.

Automation means that machines can now supply the labor originally provided by human workers, such as these soccer-playing robots.

Deskilling may result when a job is automated or when a more complex job is divided into a sequence of easily performed units. With deskilling, workers are paid less and have less control over their tasks. Jobs may become routine and boring. Deskilling contributes to polarization of the labor force. The best jobs require increasing levels of skill and technological knowledge, whereas at the bottom of the occupational hierarchy, people stuck in dead-end positions become alienated from their work (Apple 1991).

Along with deskilling has come an increasing reliance on temporary or **contingent workers** who do not hold regular jobs, but whose employment is dependent on demand. This includes contract workers, temporary workers, on-call workers (those called only when needed), the self-employed, and day laborers. Women are more likely than men to be employed in these jobs, and women are concentrated in the least desirable jobs—those with the lowest pay and least likelihood of providing benefits. Considering race, Whites are more likely to be independent contractors or self-employed, whereas Blacks and Hispanics are more likely to be found in temporary and part-time work (Kalleberg et al. 2000; Cook 2000; Hudson 1999).

Immigration

One of the most significant changes in the U.S. labor force has been the increased presence of immigrant labor. There are approximately 37 million immigrants in the United States—one-eighth of all U.S. residents. Included in this number are about 11 million undocumented immigrants, so-called illegal immigrants. Immigrants constitute more than one-third of the labor force in fields such as building cleaning and maintenance, agriculture, meat/poultry and fish production, and construction. Eighty-two percent of immigrant households have at least one worker present, more than is true for native-born households (where the figure is 73 percent). Immigrants have far lower wages than native-born citizens, and they also experience higher rates of poverty. Although immigrants have been concentrated in particular states (California, Florida, and Texas, among others), the recent trend has been that immigrants have moved into geographic regions and smaller towns, thus transforming the character of hundreds of American communities (Camaroto 2007).

Not only has immigration changed the composition of the work force, but it has also stimulated intense political debate. Should the nation restrict immigration? Should a guest worker program be created? What rights do immigrants have? These issues engage different interest groups—including immigrants themselves and the organizations that support them, business leaders, nonimmigrant workers, and local communities where immigrants are employed. These interests often diverge, often creating conflict and certainly creating questions for policymakers about the best way to handle immigration.

Popular wisdom holds that the bulk of new immigrants are illegal, poor, and desperate, but the data show otherwise (Rumbaut 1996a, 1996b; Pedraza and Rumbaut 1996; Lamphere et al. 1994). In fact, the proportion of professionals and technicians among legal immigrants exceeds the proportion of professionals in the labor force as a whole. This conclusion, though, is based on formal immigration data that exclude undocumented immigrants, most of whom are working class. Still, those who migrate are usually not the most downtrodden in their home country; seldom are the poorest able to migrate. Even undocumented immigrants tend to have higher levels of education and occupational skills than the typical workers in their homeland. Immigrants include both the most educated and the least educated segments of the population.

Some European nations have created *guest worker programs* to allow the labor that immigrants provide. Guest worker programs allow immigrants to work in a nation for a limited period of time without fear of deportation, but they must return to their nation of origin at the end of their time limit. Critics argue that such programs threaten jobs for U.S. workers and allow employers to exploit this unequal class of workers. Supporters say guest worker programs allow for better regulation of immigration and avoid some of the problems of illegal immigration. It is unclear what direction the United States will take in immigration policy, but what is clear is that the economy is highly dependent on immigrant labor.

Sociological Perspectives on Work

Most people think of work as an activity for which a person gets paid, but work includes the labor that people do without pay. Unpaid jobs such as housework, child care, and volunteer activities make up much of the work done in the world. Sociologists define **work** as productive human activity that creates something of value, either goods or services. Given this definition of work, housework, though unpaid, is defined as work, even though it is not included in the official measures of productivity used to indicate national work output.

Arlie Hochschild (1983) has introduced the concept of emotional labor to address some forms of work that are common in a service-based economy. **Emotional labor** is work specifically intended to produce a desired state of mind in a client and often involves putting on a false front before clients. Many jobs require some handling of other people's feelings, and emotional labor is performed where inducing or suppressing a feeling in the client is one of the primary work tasks. Airline flight attendants perform emotional labor—their job is

to please the passenger and, as Hochschild suggests, to make passengers feel as though they are guests in someone's living room.

The Division of Labor

The **division of labor** is the systematic *interrelatedness* of different tasks that develop in complex societies. When different groups engage in different economic activities, a division of labor is said to exist. In a relatively simple division of labor, one group may be responsible for planting and harvesting crops, whereas another group is responsible for hunting game. As the economic system becomes more complex, the division of labor becomes more elaborate.

In the United States, the division of labor is affected by gender, race, class, and age—the major axes of stratification. The *class division of labor* can be observed by looking at the work done by people with different educational backgrounds, because education is a fairly reliable indicator of class. People with more education tend to work in higher-paid, higher-prestige occupations. Class also leads to perceived distinctions in the value of manual labor versus mental labor. Those presumed to be doing mental labor (management and professional positions) tend to be paid more and have more job prestige than those presumed to be doing manual labor. Class thus produces stereotypes about the working class; manual labor is presumed to be the inverse of mental labor, meaning it is presumed to require no thinking. By extension, workers who do manual labor may be assumed not to be very smart, regardless of their intelligence.

Race and gender segregation in the labor market mean that women of color are concentrated in occupations where most other workers are also women of color.

thinkingsociologically

Think about the labor market in the region where you live. What racial and ethnic groups have historically worked in various segments of this labor market and how would you now describe the *racial and ethnic division of labor*?

SEE *FOR YOURSELF*

ANATOMY OF WORKING-CLASS JOBS

Identify an occupation that you think of as a working-class job. Then, find someone who does this kind of work and who would be willing to talk with you. Ask this person to tell you exactly what he or she does and what is most difficult and most rewarding about the work. Where does this job fit in the *division of labor*? Who does it and how are they rewarded (or not)? How would each of the three major theoretical perspectives in sociology (functionalism, conflict theory, and symbolic interaction) interpret this work in the context of the broader society?

The *gender division of labor* refers to the different work that women and men do in society. In societies with a strong gender division of labor, the belief that some activities are women's work (for example, secretarial work) and other activities men's work (for example, construction) contributes greatly to the propagation of inequality between women and men, especially because cultural expectations usually place more value (both social and economic) on men's work.

This helps explain why librarians and social workers are typically paid less than electricians despite the likelihood the women have higher education.

Debunking Society's Myths

Myth: Because of programs such as affirmative action, women and minorities are taking jobs away from White men.

Sociological perspective: Although women and minorities have made many gains as a result of programs such as affirmative action, women and minority workers still are clustered in occupations that are segregated by race and gender. There is little actual evidence that the progress of these groups has been at the expense of White men.

Similarly, the *racial division of labor* is seen in the pattern of workers from different racial groups working, in general, in different jobs. The labor performed by racial minority groups has often been the lowest paid, least prestigious, and most arduous work. The racial division of labor found today is rooted in the past.

Functionalism, Conflict Theory, and Symbolic Interaction

The major theoretical perspectives identified in this book also provide the frameworks for the social structure of work. Each viewpoint—conflict theory, functionalist theory, and symbolic interaction—offers a unique analysis of work and the economic institution of which it is a part (see Table 15.2).

table 15.2 *Theoretical Perspectives on Work*

	Functionalism	Conflict Theory	Symbolic Interaction
Defines work	As functional for society because work teaches people the values of society and integrates people within the social order; more "talented" people rank higher	As generating class conflict because of the unequal rewards associated with different jobs	As organizing social bonds between people who interact within work settings
Views work organizations	As functionally integrated with other social institutions	As producing alienation, especially among those who perform repetitive tasks	As interactive systems within which people form relationships and create beliefs that define their relationships to others
Interprets changing work systems	As an adaptation to social change	As based in tensions arising from power differences between different class, race, and gender groups	As the result of the changing meanings of work resulting from changed social conditions
Explains wage inequality	As motivating people to work harder	As reflecting the devaluation of different classes of workers	As producing different perceptions of the value of different occupations

Conflict theorists view the transformations taking place in the workplace as the result of inherent tensions in the social systems, tensions that arise from the power differences between groups vying for social and economic resources. Class conflict is then a major element of the social structure of work. Conflict theorists see the class division of labor as the source of unequal rewards for workers. Conflict theorists analyze the fact that some forms of work are more highly valued than others, both in how the work is perceived by society and how it is rewarded. As noted long ago by social–economic theorist **Thorstein Veblen** (1931), mental labor has always been more highly valued than manual labor.

Functionalist theorists, in contrast, interpret the work and the economy as a functional necessity for society. Certain tasks must be done to sustain society, and how work is organized reflects the values and other characteristics of a given social order. Functionalists argue that society "sorts" people into occupations, with the more able sorted into prestigious occupations that pay more because they are more valuable—more "functional"—for society. Functionalist theory also explains that when society changes too rapidly, work institutions generate social disorganization—perhaps creating **alienation,** a feeling of powerlessness and separation from society.

Symbolic interaction brings a different perspective to the sociology of work. *Symbolic interaction theorists* would be interested in what work means to people and how social interactions in the workplace form social bonds. Some symbolic interaction studies examine how new workers learn their roles and how workers' identities are shaped by the social interactions in the workplace. Symbolic interaction also notes the creative ways that people deal with routinized jobs, perhaps performing exaggerated displays of routine tasks to humanize otherwise boring work (Leidner 1993).

Characteristics of the Labor Force

Data on characteristics of the U.S. labor force typically are drawn from official statistics reported by the U.S. Department of Labor. The labor force now includes approximately 150 million people (U.S. Department of Labor 2009). However, who works, where, and how varies considerably for different groups in the population.

One of the most dramatic changes in the labor force since the Second World War has been the increase in the number of women employed. Since 1948, the employment of women has increased from 35 to 60 percent of all women; women now constitute almost half (47 percent) of all workers. Other changes in the labor force include that racial minorities are the fastest growing segment of the labor force. Although White, non-Hispanic workers are still 70 percent of the workforce, Hispanics, Asians, and African Americans are the

fastest growing segments of the labor force. Hispanics and Asians have the fastest growth in the labor force, largely because of immigration.

These trends, however, do not mean—as popularly believed—that minorities and women are routinely taking jobs from White men. Jobs where White men have predominated are precisely those that have been declining because of economic restructuring, such as in the manufacturing sector. It is jobs in areas of the labor market that are race- and gender-segregated that have been increasing, such as fast-food work and other low-wage service jobs. The conflicts that exist about work—such as that immigrant and foreign workers are taking U.S. jobs or that women and minorities are taking away White men's jobs—stem from social structural transformations in the economy, not the individual behaviors of people affected by these changes.

Unemployment and Joblessness

The U.S. Department of Labor regularly reports the **unemployment rate,** defined as the percentage of those not working but officially defined as looking for work. At the end of 2008, about nine million people were officially unemployed. The unemployment rate was 5.8—that is, 5.8 percent of the labor force (Bureau of Labor Statistics 2009).

But, the official unemployment rate does not include all people who are jobless. It includes only those who meet the official definition of unemployment—those who do not have a job and who have looked for work in the previous four weeks. Thus, the official measure excludes people who earned money at any job during the week prior to the data being collected, people who have given up looking for full-time work (so-called *discouraged workers*), those who have settled for part-time work, people who are ill or disabled, and those who cannot afford the child care, transportation, or other necessities for getting to work. Migrant workers and other transient populations are also undercounted in the official statistics, as are people who work only a few hours a week are excluded even though their economic position may differ little from that of someone with no work at all. Workers on strike are also counted as employed, even if they receive no income while they are on strike. Because the number of people excluded from the official definition of unemployment, the official unemployment rate seriously underestimates actual joblessness.

Moreover, the people most likely to be left out of the unemployment rate are those for whom unemployment runs the highest—the youngest and oldest workers, women, and racial minority groups. These groups are also those most likely to have left jobs that do not qualify them for unemployment insurance, because to be eligible for unemployment you have to have worked a certain period of time and earned enough wages to qualify for a claim. Current estimates are that only about one-third of all workers qualify for unemployment insurance (Wenger 2001)—a percentage that has been falling.

> doing sociological research

All in the Family: Children of Immigrants and Family Businesses

A relatively common pattern among Asian American immigrants is to establish a small family business, using the labor of family members to establish and run it. Owning a family business is part of the American dream, thus idealized as promoting entrepreneurial and family values, including the value of hard work, devotion to family, and self-sacrifice.

Research Question: What kind of life does this produce for the children of immigrant families? This is what sociologist Lisa Park wanted to know in her research on the children of Asian American entrepreneurs.

Research Methods: Park, a child of such entrepreneurs herself, conducted her study by interviewing Korean American and Chinese American high school and college students who grew up in small family businesses. She utilized several qualitative research methods, including (1) in-depth individual interviews, (2) focus group interviews, and (3) participant observation in family stores.

Research Results: Although many teens hold jobs often to help support their families, sometimes for their own spending money, Park found that the life of Asian American teenagers who are sons and daughters of entrepreneurs was difficult. Her research subjects reported wanting to have the freedom to be bored that they associated with White, middle-class teens. They were critical of the popular images of teens in families as only worrying about their hair, their clothes, and their friends.

In contrast to the stereotype of Asian American families as close-knit and conflict-free, Park found

that teens in these families report feeling overworked and with little family "quality time." Some of these young people have worked since they were six years old. Often parents set up a small "home away from home" in the family restaurant or business—a small table in the corner where children can read, color, or watch television. But her subjects also reported being burdened and not able to have a normal childhood. They often had to do all the household chores while parents worked, as well as serve customers or help in other ways.

Conclusions and Implications: Park's research shows immigration shapes family structures and affects children's lives. She challenges simplistic views of the family business as an ideal way of life and helps us understand some of the social forces that are part of the immigrant experience and that shape the life of some Asian American youth.

Questions to Consider

1. Are there businesses in your community run by immigrant families? If so, what evidence of children's involvement in the business do you see and how do you think this influences the children?

2. In addition to immigrant businesses, how are children involved in the work that their parents do? In what ways does this influence the children's identities, values about work, future opportunities?

Source: Park, Lisa Sun-Hee. 2002. "A Life of One's Own." *Contexts* 1 (Summer): 56–57.

Official unemployment rates also ignore **underemployment**—the condition of being employed at a skill level below what would be expected given a person's training, experience, or education. This condition can also include working fewer hours than desired. A laid-off autoworker flipping hamburgers at a fast-food restaurant is underemployed and so is a person with a law degree who drives a taxi.

Even given the problems in measuring unemployment, the highest rates of unemployment are among Black men, followed by Hispanic men. When the government reports the national unemployment rate, it is a safe bet that unemployment among African

Americans will be at least twice the national rate—a pattern that has persisted over time. Although not regularly reported by the Department of Labor, unemployment among Native Americans is also staggeringly high—the greatest barrier being the simple absence of jobs. For Native Americans living on reservations (about one-quarter of the population), there are few jobs available; those that do exist are often held by Whites (Snipp 1996). In urban areas, Native Americans face many of the same problems as urban African Americans and Hispanics. Were the unemployment rate that is found among Native Americans, African Americans, and Hispanics (with the exception of Cuban Americans) the national unemployment rate, it would be called a major economic depression (see Figure 15.5).

People often attribute unemployment to the individual failings of workers, claiming that unemployed people do not try hard enough to find jobs or prefer a welfare check to hard work and a paycheck. This viewpoint reflects the common myth that anyone who works hard enough can succeed. But, as the economic recession the nation entered in 2009 clearly showed, it is usually conditions beyond an individual's control that result in unemployment and joblessness. During the recession, hundreds of thousands of workers either lost jobs or found themselves in highly precarious working situations. The Department of Labor itself estimated that the ranks of discouraged workers (those not counted in the unemployment rate) rose by three-quarters of a million in just one quarter of the year (U.S. Department of Labor 2009).

In times like a major recession, it is easier to see how structural changes in the economy affect people at the individual level. (Recall the distinction between *troubles and issues* made by C. Wright Mills and examined in Chapter 1 of this book.) But at all times, sociologists examine how changes in the social organization of the economy trickle into the day to day reality of people's lives. Although people do not ordinarily interpret their situations as shaped by processes such as deindustrialization, corporate downsizing, or spatial mismatch, the fact is that these structural conditions have enormous influence on how we live.

Diversity in the U.S. Occupational System

Jobs are organized into an *occupational system*—the array of jobs that together constitute the labor market. Within the occupational system, people are distributed in patterns that reflect the race, class, and gender organization of society. Jobs vary in their economic rewards, their perceived value and prestige, and the opportunities they hold for advancement.

The Dual Labor Market. The labor market can be seen as comprising two major segments, the *primary labor market* and the *secondary labor market*, a phenomenon known as the **dual labor market** (see also Chapter 11). The primary labor market offers jobs with relatively high wages, benefits, stability, good working conditions, opportunity for promotion, job protection, and due process for workers (meaning workers are treated according to established rules and procedures that are allegedly fairly administered). Blue-collar and service workers in the primary labor market are often unionized, which leads to better wages and job benefits. High-level corporate jobs and unionized occupations fall into this segment of the labor market.

The secondary labor market is characterized by low wages, few benefits, high turnover, poor working conditions, little opportunity for advancement, no job protection, and arbitrary treatment of workers. Many service jobs, such as waiting tables, nonunionized assembly work, and domestic work, are in the secondary labor market. Women and minority workers are the most likely groups to be employed in the secondary labor market. This particular structure in the labor market can explain much about race, gender, and class inequalities in work.

Occupational Distribution. Occupational distribution describes the pattern by which workers are located in the labor force. Workers are dispersed throughout the occupational system in patterns that vary greatly by race, class, and gender, revealing a certain **occupational segregation** on the basis of such characteristics. Women are most likely to work in technical, sales, and administrative support, primarily because of their heavy concentration in clerical work. This is now true for both White women and women of color. White men are most likely found in managerial and professional jobs, whereas African American and Hispanic men are most likely employed as operators and laborers—among the least well paid and least prestigious in the occupational system (Bureau of Labor Statistics 2009).

Changes in occupational segregation are noticeable over time. For example, in 1960, 38 percent of all Black women were employed as private domestic workers; by the 1990s, this had declined to less than 2 percent. Over time, there has also been some increase in the number of women employed in working-class jobs traditionally held only by men. Today, women are 12 percent of precision production, craft, and repair workers, compared with 1 percent in 1970 (U.S. Department of Labor 2009). This is a significant increase in the number of women in these jobs, although women are still a small proportion of all such workers.

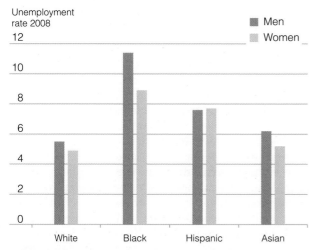

FIGURE 15.5 *Unemployment by Race and Gender*

Data: Bureau of Labor Statistics. 2009. **www.bls.gov**

Occupational Prestige and Earnings.

Occupational prestige is the perceived social value of an occupation in the eyes of the general public (see Chapter 8). Sociologists have found a strong correlation between occupational prestige and the race and gender of people employed in given jobs. African American and Latino men are disproportionately found in jobs that have the lowest occupational prestige scores; White and Asian American men hold the jobs with the highest occupational prestige, followed by White women, Asian American women, African American women, and Latinas (Xu and Leffler 1992; Stearns and Coleman 1990). The gender composition of jobs and their occupational prestige is also linked: Jobs that employ mostly women are lower in prestige than those that employ more men. Indeed, jobs often lose their prestige as many women enter a given profession. Likewise, the prestige of jobs increases as more men enter the field.

Beyond what people think about jobs, what really matters is what they pay. Sociologists have extensively documented that earnings from work are highly dependent on race, gender, and class, as shown in Figure 15.6. White men earn the most, with a gap between men's and women's earnings among all groups. African American women and Hispanic men and women earn the least. Occupations in which White men are the numeric majority tend to pay more than occupations in which women and minorities are a majority of the workers.

Why are there such disparities? Again, we have to turn to theory to put the facts into perspective. According to functionalist theories, workers are paid according to their value, which is derived from the characteristics they bring to the job: education, experience, training, and motivation to work. As we saw in Chapter 8, functionalist theorists see inequality

as what motivates people to work. From this point of view, the high wages and other rewards associated with some jobs are the incentive for people to spend long years in training and garnering experience; otherwise, the jobs would go unfilled. To functionalists, then, differential wages are a source of motivation and a means to ensure that the most talented workers fill jobs essential to society and that different wages reflect the differently valued characteristics (education, years experience, training, and so forth) that workers bring to a job.

Conflict theorists strongly disagree with the functionalist point of view, arguing that many talented people are thwarted by the systems of inequality they encounter in society. Thus, far from ensuring that the most talented will fill the most important jobs, conflict theorists see that some of the most essential jobs are, in fact, the most devalued and underrewarded. From a conflict perspective, wage inequality is one way that systems of race, class, and gender inequality are maintained. Factors such as the dual labor market, overt race and gender discrimination, and persistent and unequal judgments about what work is appropriate for what groups shape the inequalities that are found in work.

Power in the Workplace

Pay, benefits, and prestige are important dimensions of people's work experience. But also important are opportunities for advancement and the culture one experiences at work. The **glass ceiling** is the term used to describe the limits to advancement that women, as well as racial–ethnic people and minorities, experience at work. Many barriers to the advancement of women and minorities have been removed, yet invisible barriers still persist. The existence of the glass ceiling is well documented. Although there has been an increase in the number of managers who are women and minorities, most top management jobs are still held by White men, who are far more likely than other groups to control a budget, participate in hiring and promotion, and have subordinates who report to them. Women and racial minorities remain clustered at the bottom of managerial hierarchies.

Sexual Harassment

Women workers also are more likely to experience **sexual harassment** in the workplace. Sexual harassment is legally defined as unwanted physical or verbal sexual behavior that occurs in the context of a relationship of unequal power and that is experienced as a threat to the victim's job or educational activities (Saguy 2003). Sexual harassment is of two forms, according to the law. *Quid pro quo sexual harassment* forces sexual compliance in exchange for an

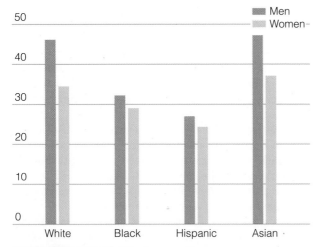

FIGURE 15.6 *The Income Gap*

Data: U.S. Census Bureau. 2009. *Work Experience in 2007—People 15 Years Old and Over by Total Money Earnings in 2007, Age, Race, Hispanic Origin, and Sex.* Table PINC-05. **www.census.gov**

employment or educational benefit. A professor who suggests to a student that going out on a date or having sex would improve the student's grade is engaging in quid pro quo sexual harassment. The other form of sexual harassment recognized by law is the creation of a *hostile working environment* in which unwanted sexual behaviors are a continuing condition of work. This kind of sexual harassment may not involve outright sexual demands, but includes unwanted behaviors such as touching, teasing, sexual joking, and other kinds of sexual comments.

Sexual harassment was first made illegal by Title VII of the 1964 Civil Rights Act, which identifies sexual harassment as a form of sex discrimination. In 1986, the U.S. Supreme Court upheld the principle that sexual harassment violates federal laws against sex discrimination (*Meritor Savings Bank* v. *Vinson*). The law defines sexual harassment as discriminatory because it places workers at a disadvantage on the basis of their sex. Same-sex harassment also falls under the law, as does harassment directed by women against men. The law also makes employers liable for financial damages if they do not have policies appropriate for handling complaints or have not educated employees about their paths of redress.

Fundamentally, sexual harassment is an abuse of power when perpetrators use their position to exploit subordinates. The true extent of sexual harassment is difficult to estimate because it tends to be underreported. Surveys indicate though that as many as one-half of all employed women experience some form of sexual harassment at some time in their working lives. Men are sometimes the victims of sexual harassment, although far less frequently than women—about three percent of all cases. There is some evidence that women of color are more likely to be harassed than White women (Gruber 1982). Same-gender harassment also occurs; when men are harassed by other men, they react with more severe consequences than do men who have been harassed by women (DuBois et al. 1998). Most studies find that typically neither women nor men are aware of the proper channels for reporting sexual harassment. Women are less likely than men to report sexual harassment primarily because they believe that nothing will be done to stop the behavior (VanRoosmalen and McDaniel 1998; Cortina et al. 1998).

Gays and Lesbians in the Workplace

The increased willingness of lesbians and gay men to be open about their sexual identity has resulted in more attention paid to their experience in the workplace.

The Glass Ceiling
Whimsically depicted as an architectural drawing by Norman Andersen, the glass ceiling refers to the structural obstacles still inhibiting upward mobility for women workers. Most employed women remain clustered in low status, low-wage jobs that hold little chance for mobility. Those who make it to the top often report being blocked and frustrated by patterns of exclusion and gender stereotyping.

Surveys find that a large majority of the U.S. population (89 percent) endorse the general concept that lesbians and gay men should have equal rights in job opportunities, but when asked about specific occupations, a significant proportion hold prejudices that gays should not be elementary school teachers (43 percent), high school teachers (36 percent), or clergy (47 percent; Gallup Poll 2008).

Thus, although there is increasing acceptance of lesbians and gays generally, persistent homophobia can affect the self-esteem, productivity, and general well-being of gay and lesbian workers. Gays and lesbians often fear they will suffer adverse career consequences if heterosexual coworkers know they are gay. Shielding themselves from antagonism or rejection may make them appear distant and isolate them from social networks. These behaviors can actually have a negative effect on their performance reviews because fellow workers may find them unfriendly and withdrawn. Research finds, however, that the relationships of gay employees with their coworkers are less stressful when the employees are open about their identity (Schneider 1984). The work organization is also improved for lesbian, transgender, and gay employees when employee assistance programs encourage open communication, policies are sensitive to lesbian and gay needs, and discriminatory practices can be identified and stopped (Sussal 1994).

Disability and Work

Not too many years ago, people with disabilities were not thought of as a social group; rather, disability was thought of as an individual frailty or perhaps a stigma. Sociologist **Irving Zola** (1935–1994) was one of the first to suggest that people with disabilities face issues similar to those of minority groups. Instead of using a medical model that treats disability like a disease and sees individuals as impaired, conceptualizing people with disabilities as a minority group enabled people to think about the social, economic, and political environment that this population faces. Instead of seeing people with disabilities as pitiful victims, this approach emphasizes the group rights of the disabled, illuminating things such as access to employment and education (Zola 1989, 1993).

Now those with disabilities have the same legal protections afforded to other minority groups. Key to these rights is the Americans with Disabilities Act (ADA), adopted by Congress in 1990. Building on the Civil Rights Act of 1964 and earlier rehabilitation law, in particular the Rehabilitation Act of 1973, the Americans with Disabilities Act protects people with disabilities from discrimination in employment and stipulates that employers and other providers (such as schools and public transportation systems) must

The Americans with Disabilities Act provides legal protection for workers with disabilities, giving them rights to reasonable accommodations by employers and access to education and jobs.

Owen Franken/Corbis

provide "reasonable accommodation." People with disabilities must be qualified for the jobs or activities for which they seek access, meaning that they must be able to perform the essential requirements of the job or program without accommodation to the disability. For students, reasonable accommodation includes provision of adaptive technology, exam assistants, and accessible buildings.

The law prohibits employers with fifteen or more employees from discriminating against job applicants who are disabled or current employees who become disabled, including job application, wages and benefits, advancement, and employer-sponsored social activities. The law applies to state and local governments, as well as employers. ADA also legislates that public buses, trains, and light rail systems must be accessible to riders with disabilities; airlines are excluded from this requirement. The law also requires businesses and public accommodations to be accessible and requires telephone companies to provide services that allow people with speech or hearing impairments to communicate by telephone.

Not every person with a disability is covered by ADA. To be considered disabled, a person must have a condition or the history of a condition that impairs a major life activity. The law specifically excludes the use of illegal drugs as a cause of disability and excludes pregnancy as a disability—which is ironic because pregnancy is treated as a disability by other federal laws for purposes of maternity leave.

Chapter Summary

• **What is the state?**

The *state* is the organized system of power and authority in society. It comprises different institutions, including the government, the military, the police, the law and the courts, and the prison system. The state is supposed to protect its citizens and preserve society, but it often protects the status quo, sometimes to the disadvantage of less powerful groups in the society.

• **How do sociologists define power and authority?**

Power is the ability of a person or group to influence another. *Authority* is power perceived to be legitimate. There are three kinds of authority: *traditional authority,* based on long-established patterns; *charismatic authority,* based on an individual's personal appeal or charm; and *rational–legal authority,* based on the authority of rules and regulations (such as law).

• **What theories explain how power operates in the state?**

Sociologists have developed four theories of power. The *pluralist model* sees power as operating through the influence of diverse interest groups in society. The *power elite model* sees power as based on the interconnections between the state, industry, and the military. *Autonomous state theory* sees the state as an entity in itself that operates to protect its own interests. *Feminist theorists* argue that the state is patriarchal, representing primarily men's interests.

• **How well does the government represent the diversity of the U.S. population?**

An ideal democratic government would reflect and equally represent all members of society. The makeup of the U.S. government does not reflect the diversity of the general population. African Americans, Latinos, Native Americans, Asians, and women are underrepresented within the government. Political participation also varies by a number of social factors, including income, education, race, gender, and age. African Americans and Latinos, however, are overrepresented in the military, in part because of the opportunity the military purports to offer groups otherwise disadvantaged in education and the labor market; however, both are underrepresented at the levels of high-level commissioned officers. There is an increased presence of women in the military; however, prejudice and discrimination continue against lesbians and gays in the military.

• **How are societies economically organized?**

Societies are organized around an economic base. The *economy* is the system on which the production, distribution, and consumption of goods and services are based. *Capitalism* is an economic system based on the pursuit of profit, market competition, and private property. *Socialism* is characterized by state ownership of industry; *communism* is the purest form of socialism.

• **How do sociologists define work?**

Sociologists define *work* as human activity that produces something of value. Some work is judged to be more valuable than other work. *Emotional labor* is work that is intended to produce a desired state of mind in a client. The *division of labor* is the differentiation of work roles in a social system. In the United States there is a class, gender, and racial division of labor.

• **How has the global economy changed?**

As capitalism has spread throughout the world, *multinational corporations* conduct business across national borders. A number of countries have undergone *deindustrialization,* or changeover from a goods-producing economy to a services-producing one. This has caused many heavy-industry jobs in U.S. cities to vanish, thus increasing the unemployment rate in those cities. Changes in information technology, plus increased *automation,* have resulted in the further elimination of jobs in both the United States and abroad.

• **What are some of the characteristics of the contemporary labor force?**

In recent years, the employment of women has been increasing whereas that of men has been decreasing. Official *unemployment rates* underestimate the actual extent of joblessness. Women and racial minorities are most likely to be unemployed.

Recent immigrants also have a unique status in the labor market, often in the lowest paying and least prestigious jobs. The U.S. occupational system is characterized by a *dual labor market*. Jobs in the primary sector of the labor market carry better wages and working conditions, whereas those in the secondary labor market pay less and have fewer job benefits. Women and minorities are disproportionately employed in the secondary labor market. Patterns of occupational distribution also show tremendous segregation by race and gender in the labor market. Race and gender also affect the occupational prestige, as well as the earnings, of given jobs.

• *How does power in the workplace affect women, racial–ethnic groups, and gays and lesbians?*

Women and minorities often encounter the *glass ceiling*—a term used to describe the limited mobility of women and minority workers in male-dominated organizations. In addition, women more often than men face *sexual harassment* at work—defined as the unequal imposition of sexual requirements in the context of a power relationship. Homophobia in the workplace also negatively affects the working experience of gays and lesbians.

Key Terms

alienation   391
authority   373
automation   388
autonomous state model   376
bureaucracy   374
capitalism   385
charismatic authority   373
communism   386
contingent worker   388
deindustrialization   387
democracy   377
division of labor   389
dual labor market   393
economic restructuring   386
economy   385

emotional labor   389
gender gap   379
glass ceiling   394
global economy   386
government   377
interest group   375
interlocking directorate   376
job displacement   387
multinational corporations   386
nationalism   373
occupational segregation   393
pluralist model   375
political action committee
   (PAC)   375
postindustrial society   385

power   373
power elite model   375
propaganda   372
rational–legal authority   374
sexual harassment   394
socialism   386
spatial mismatch   387
state   372
traditional authority   373
underemployment   392
unemployment rate   391
work   389
xenophobia   386

Online Resources

Sociology: The Essentials Companion Website

www.cengage.com/sociology/andersen
Visit your book companion website where you will find more resources to help you study and write your research papers. Resources include web links, learning objectives, Internet exercises, quizzing, and flash cards.

 is an easy-to-use online resource that helps you study in less time to get the grade you want NOW.

www.cengage.com/login
Need help studying? This site is your one-stop study shop. Take a Pre-Test and Cengage NOW will generate a Personalized Study Plan based on your test results. The Study Plan will identify the topics you need to review and direct you to online resources to help you master those topics. You can then take a Post-Test to determine the concepts you have mastered and what you still need to work on.

16

Chapter sixteen
CHAPTER SIXTEEN
Chapte

Population, the Environment, and Social Change

Demography and the U.S. Census

Diversity and the Three Basic Demographic Processes

Population Characteristics

Theories of Population Growth Locally and Globally

Checking Population Growth

Urbanization

Ecology and the Environment

What Is Social Change?

Theories of Social Change

Global Theories of Social Change

Modernization

The Causes of Social Change

Collective Behavior and Social Movements

Diversity, Globalization, and Social Change

Chapter Summary

[*The study of* population and environment is a most important topic as we begin the new millennium. A great many issues of central concern to the entire country are driven by population. Population growth and density are responsible for the major policy issues of the day, such as overcrowding in cities, traffic jams, long lines at markets and other stores,

continued

environmental pollution, family planning, diminishing resources, and food shortages. The federal as well as the state governments endlessly debate the population issue of who should or should not be permitted to immigrate into the United States.

The estimated population of the United States is presently more than 280 million. At the current rate of growth, the country will reach almost 300 million by the year 2025. The population has more than doubled since 1946, when it stood at approximately 132 million. In that year, a spike in the number of births began that lasted until 1964—from the end of the Second World War to the beginning of the Vietnam War. The *Baby Boom* is the name given to that crop of 75 million babies, who are now middle-aged and older and consist of nearly one-third of all the people in the United States. The generation that in the tumultuous 1960s warned, "Don't trust anyone over thirty," is now hovering in its mid-fifties and early sixties and starting to have grandchildren.

The education and occupation structure of the country has been greatly affected by the decisions by so many people to have children during the previous years. Today, as globalization, economic restructuring, terrorism, and war cause young people to feel insecure about their future, many twenty-something couples are deciding, at least for the present, not to have children. Their decisions will ripple forward in the years to come, as fewer youngsters are enrolled in grade school and then in college two decades down the road. Inevitably, there will follow a decline in the number of middle-aged people to take care of others in their old age.

Technological innovation can transform the population of an entire society—in other words, result in social change. Consider the following: A gigabyte of information can travel from China to the United States in less time than it takes you to read this paragraph. At the same time, someone can send a fax in the opposite direction while also sipping coffee grown in Colombia, roasted in Seattle, distributed across the nation, and delivered via express mail. Was this imaginable fifty years ago? Even thirty years ago? Not really. What will the next thirty to fifty years bring?

We are quick to think of change as progress, but sometimes people in a changing society wonder if the things gained from progress are not overbalanced by the things lost. The Alaskan pipeline and related innovations have brought considerable wealth to the communities of northern Alaska, and many Eskimo entrepreneurs have prospered from the changes. Others have fared less well. Some researchers have attributed increases in suicide and alcohol-related deaths to changes that simply came too rapidly in the social structure and culture of the Eskimos.

Social movements often produce major social change in a society. Fads and fashions often serve as a basis for relatively permanent social change. For example, blue jeans have been around since the late 1800s and show no signs of becoming less popular. Skateboarding, begun in the early 1960s, has endured for nearly fifty years. Other social movements have resulted in deeper and more far-reaching cultural and structural changes in society, such as the civil rights movement, the social movement to protect the environment, or the animal rights movement—examples of true social movements.

Demography and the U.S. Census

The scientific study of population is called **demography,** which is the study of the current state and changes over time in the size, distribution, and composition of human populations. This field of sociology draws on huge bodies of data generated by a variety of sources. One source is the U.S. Census Bureau.

A **census** is a head count of the entire population of a country, usually done at regular intervals. The census conducted every ten years by the U.S. Census Bureau, as required by the U.S. Constitution, attempts not only to enumerate every individual, but also to obtain information such as gender, race, ethnicity, age, education, occupation, and other social factors.

The 2000 U.S. Census is estimated to have undercounted a small percentage of the country's population. Among those most likely to have been undercounted by the census are the homeless, immigrants, minorities living in ghetto neighborhoods, and other residents of low social status. The constitutional requirement for a census was included to ensure fair apportionment of representatives in the federal government. Undercounting minorities and the underclass tends to leave these groups underrepresented in government. The estimated undercount for the entire U.S. population overall is only about 2 percent, yet the undercount for African Americans nationally has been estimated to be as high as 20 percent, and for Hispanics, as high as 25 percent (NAACP Legal Defense and Education Fund 2006; Harrison 2000). Things may improve with the 2010 census.

The U.S. Congress and the U.S. Census Bureau have fiercely debated the following issue: Should people be allowed to select multiracial (or "mixed race") responses on the census questionnaire, which it now does, by allowing someone to check more than one racial–ethnic category from among several—such as African American, Hispanic, non-Hispanic White, Native American, Asian, Eskimo, and Aleut—or, should individuals be limited to checking one category only? The use of the multiracial response option gives individuals an opportunity to define themselves as mixed race. One argument against this option is that it subtracts from the number of people who would have otherwise indicated only one category, especially a racial–ethnic minority category, thus further undercounting African Americans, Hispanics, and Native Americans. One study estimates, however, that only about 2 percent of the persons responding actually indicate a multiracial response (Harrison 2000; Morning 2008, 2007).

Diversity and the Three Basic Demographic Processes

The total number of people in a society at any given moment is determined by only three variables: births, deaths, and migrations. These three variables show different patterns for different racial and ethnic groups, different social strata, and both genders. Births add to the total population, and deaths subtract from it. Migration into a society from outside, called **immigration,** adds to the population, whereas **emigration,** the departure of people from a society (also called *out-migration*), subtracts from the population.

The total population of the world is increasing at a rate of approximately 270,000 people per day, or just less than 200 people per minute. The world's population does not increase in a linear fashion—the line on a graph of population does not rise in a straight line, with the same number of people added each year. Instead, the population grows exponentially, with an upward accelerating curve, as shown in Figure 16.1. An ever increasing number of people are added each year. At the present rate of growth, the world's population will double in forty years, barring some major catastrophe such as world war, an international epidemic of an untreatable disease, or some such global calamity.

Birthrate

The **crude birthrate** of a population is the number of babies born each year for every 1000 members of the population or, alternatively, the number of births divided by the total population, multiplied by 1000:

$$\text{Crude birth (CBR)} = \frac{\text{number of births}}{\text{total population}} \times 1000$$

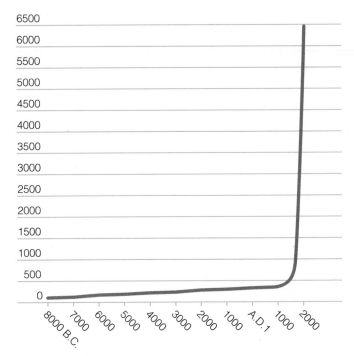

FIGURE 16.1 *World Population Growth (in millions)*

Data: From the Population Reference Bureau. 2008. Washington, DC. **www.prb.org.** Reprinted with permission.

The crude birthrate for the entire world population is approximately 28 births per 1000 people. Various countries and different subgroups within a country can have dramatically different birthrates. For example, the country with the highest birthrate in the world is Niger, with 49.6 births per 1000 people. The lowest birthrate is found in Japan, with only 7.9 births per 1000 people (U.S. Census Bureau 2008).

The overall birthrate for the United States is approximately 14.2 births per 1000 people, compared with the all-time high rate of 27 births per 1000 people in 1946, the start of the so-called Baby Boom that followed the Second World War. The rate varies according to racial–ethnic group, region, socioeconomic status, religion, and other factors. Overall, we find that for different racial–ethnic groups, the birthrates in the United States vary—Latinos (all subgroups) with the highest (23.0 per thousand) down to Whites with the lowest (13.4 per thousand).

The effects of birthrates are somewhat cumulative. For example, minorities tend to be overrepresented at the lower end of the socioeconomic scale, compounding the likelihood of a high birthrate. Similarly, religious and cultural differences can make themselves felt. Catholics, for example, have a higher birthrate than non-Catholics of the same socioeconomic status. Hispanic Americans have a high likelihood of being Catholic, another factor that contributes to the higher birthrate among Hispanic Americans. Because minorities tend to have higher birthrates, assuming that present migration rates continue and assuming that death rates do not outstrip the birth-rates, in the coming years the United States will have a significantly greater proportion of minorities, thus a relatively lower proportion of Whites.

Death Rate

The **crude death rate** of a population is the number of deaths each year per 1000 people, or the number of deaths divided by the total population, times 1000:

$$\text{Crude death rate (CDR)} = \frac{\text{number of deaths}}{\text{total population}} \times 1000$$

The crude death rate can be an important measure of the overall standard of living for a population. In general, the higher the standard of living enjoyed by a country, or a group within the country, the lower the death rate. The death rate of a population also reflects the quality of medicine and health care. Poor medical care, which goes along with a low standard of living, will correlate with a high death rate.

Another measure that can reflect the standard of living in a population is the **infant mortality rate,** which is the number of deaths per year of infants less than one year old for every 1000 live births. In the United States, the overall infant mortality rate is on average 7.0 infant deaths for every 1000 live births—but much higher than that for Blacks (13.7) but lower for Whites (5.7), a big difference. The highest infant mortality rates among seventy-seven countries throughout the world are Angola, with a rate of 200, and Afghanistan, with a rate of 150 (U.S. Census Bureau 2008).

Infant mortality rates, a measure of the chances of the very survival of members of the population, are important to compare across racial–ethnic groups and across social class strata. They are a good indicator of the overall quality of life, as well as the survival chances for members of that racial or class group. There are also many other causes of higher infant mortality, such as presence of toxic wastes, malnutrition of the mother, inadequate food, and outright starvation.

The **life expectancy** of a population or group is defined as the average number of years a member of the group can expect to live. In the United States, life expectancy has gone from forty years of age in 1900 to over seventy-seven years of age for people born now. Although a life expectancy of seventy-seven years might seem high, the truth is that the United States does not compare very well to other developed nations in either life expectancy or infant mortality. Table 16.1 shows that the United States ranks near the *bottom* among *industrialized nations* in life expectancy, behind Japan, the Netherlands, Canada, and several other countries. The picture is similar regarding infant mortality. Interestingly, Russia, with whom the United States was engaged in a long-running, expensive Cold War, also has a low life expectancy and an exceptionally high infant mortality rate.

As shown in this photo, in the 1920s from virtually every southern town, African Americans left by the thousands to go north to enter northern industry. This has come to be known as the Great Black Migration in America.

Migration

Joining the birthrate and death rate as factors in determining the size of a population is the migration of people into and out of the country. Migration affects society in many ways. Israel, since its establishment in 1948, has experienced considerable growth in its population, primarily due to a tremendous migration of Jews from Europe and the United States. These migrants tend to be younger on average than the rest of the Israeli population, therefore their arrival has certain direct consequences, such as increasing the birthrate, which is higher among the young than the older Israelis.

Migration can also occur within the boundaries of a country. In the 1980s, internal migration by African Americans, Hispanics, Asians, and Pacific Islanders within the borders of the United States has occurred at a rate unmatched since the First World War and the Great Black Migration from the South to the North in the United States early in the twentieth century. In that era, Blacks migrated from the South to major industrial urban areas in the North, such as Chicago,

Debunking Society's Myths

Myth: High infant mortality and a high death rate for young adults are problems in underdeveloped countries, but not in the United States.

Sociological perspective: Both infant mortality and adult death rates are significantly greater among lower-class than among middle-class persons in the United States, greater among people of color (Latinos, Native Americans, and African Americans) than among Whites of the same social class, and greater among men than women.

table16.1	Life Expectancy and Infant Mortality (by country, industrialized countries only)	
Country	**Life Expectancy**	**Infant Mortality Rate**[a]
Japan	80.8	3.9
Australia	79.8	5.0
Canada	79.4	5.0
Italy	79.1	5.8
France	78.9	4.5
Spain	78.9	4.9
Netherlands	78.4	4.4
United Kingdom	77.8	5.5
Germany	77.6	4.7
United States	**76.6**	**6.8**
China	71.6	28.1
Russia	67.3	20.1

[a]Per 1000 live births

Source: U.S. Census Bureau. 2009a. *Statistical Abstract of the United States, 2008.* Washington, DC: U.S. Census Bureau.

New York, Detroit, and Cleveland, where jobs were available. The recent pattern of migration for African Americans has been not only from the South, but also from the major northern urban centers to the West, the Southwest, and New England. This migration has carried many African Americans into areas that were previously all White and has frequently been associated with the presence of the migrants at institutions such as military bases or universities.

Among Hispanics, migration patterns have traditionally been loosely linked to the agriculture industry. More recently, these patterns are linked to other industries such as meatpacking and textiles and to industries centered in urban areas (Portes and Rumbaut 1996; Edmondson 2000). Although Mexican Americans have traditionally settled in the West and Southwest, recent movement has taken them to Michigan, Washington State, and New England. Farm workers from Mexico have settled near the beet fields of northern Minnesota, tripling the Hispanic population there since 1980. Puerto Rican Americans have migrated northward in increasing numbers to rural, urban, and suburban areas. In New England towns that were once predominantly White, such as Lynn and Lowell, Massachusetts, the Hispanic population has increased 180 percent since 1980 (Rumbaut and Portes 2001; Portes and Rumbaut 2001).

As noted in Chapter 15, the structure of the U.S. labor force is being shaped by the employment of recent Hispanic immigrants. Contrary to present stereotypes, these new immigrants are not poor and unmotivated, but in contrast show a proportion of professionals and technicians higher than that in the labor force as a whole (Portes et al. 2008; Portes and Rumbaut 2001).

Population Characteristics

The composition of a society's population can reveal a tremendous amount about the society's past, present, and future. The populations of many nations show a striking imbalance in the number of men and women of certain ages, with many fewer men than would be expected. Quite often, the explanation can be found by looking back in time to when men in this category were the right age for military service—the demographic vacancy usually coincides with a major war from which many young men failed to return. The demographic data are thus a record of national history. The Second World War was responsible for killing many men, and some women, in the early 1940s—people in their twenties. As a result, this age category in the United States revealed a shortage of people, and considerably fewer men than women, as the entire population of the country aged.

Sociologists put together data about population characteristics to develop pictures of the population in slices or as a whole. Important characteristics of the population are sex ratio, age composition, the age–sex population pyramid, and age cohorts. We examine these main approaches to describing the population in the next section.

Sex Ratio and the Population Pyramid

Two factors that affect the composition of a population are its sex ratio and its age–sex pyramid. The **sex ratio** (also called *gender ratio*) is the number of males per 100 females, or the number of males divided by the number of females, times 100.

$$\text{Sex ratio} = \frac{\text{number of males}}{\text{number of females}} \times 100$$

A sex ratio above 100 indicates there are more males than females in the population; below 100 indicates there are more females than males. The sex ratio could just as easily have been defined as the number of females per 100 males. A ratio of exactly 100 indicates the number of males equals the number of females. In almost all societies, there are more boys born than girls, but because males have a higher infant mortality rate and a higher death rate after infancy, there are usually more females in the overall population. In the United States, approximately 105 males are born for every 100 females, thus giving a sex ratio for live births of 105. After factoring in male mortality, the sex ratio for all ages for the entire country ends up being 94; there are 94 males for every 100 females.

The *age composition* of the U.S. population is presently undergoing major changes. More and more people are entering the sixty-five and older age bracket. This trend is known as the *graying of America*. The elderly will soon become the largest population category in our society. As our society gets grayer, its older members will have more influence on national policy and a greater say in matters such as health care, housing, and other areas where the elderly traditionally experience age discrimination.

Gender and age data are often combined into a graphic format called an **age–sex pyramid** (or age–gender pyramid) that represents the age and gender structure of a society (see Figure 16.2). The left side of the figure shows the age–sex pyramid for the United States. Note that there are slightly more males than females in the younger age ranges (due to the higher birthrate for males), a trend that reverses in the middle-age brackets. At the upper range of the age scale, women outnumber men. The pronounced bulge around the middle of the pyramid represents the Baby Boom generation, people now in their late fifties and sixties. As these Baby Boomers age, the bulge will rise toward the top of the pyramid, to be replaced underneath by whatever birth trends occur in the coming years. This restructuring of the population pyramid will necessitate restructuring of society's institutions. Marketing will have to be directed more often toward those over age 65, considerably more funds will be needed for their health care, and recreational facilities for the elderly will have to be greatly expanded.

The right side of Figure 16.2 shows the age–sex pyramid for Mexico. It is quite wide at the bottom and very narrow at the top—exactly pyramidal in shape. This suggests a high birthrate and a death rate that increases rapidly with age.

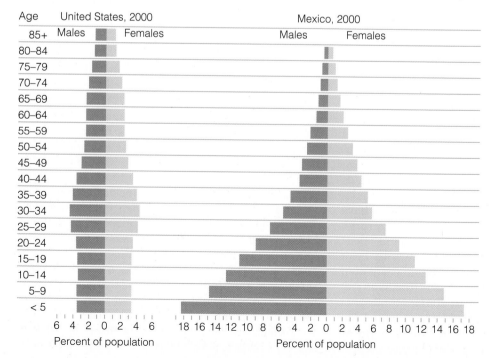

FIGURE 16.2 *A Comparison of Two Age–Sex Pyramids*

Source: John R. Weeks. 2008. *Population: An Introduction to Concepts and Issues,* 10th ed. Belmont, CA: Wadsworth Publishing Co.

Cohorts

A birth cohort, or more simply, a **cohort,** consists of all the persons born within a given period. A cohort can include all persons born within the same year, decade, or other time period. Over time, cohorts either stay the same in size or get smaller owing to deaths, but can never grow larger. If we have knowledge of the death rates for this population, we can predict quite accurately the size of the cohort as it passes through the stages of life from infancy to old age. This enables us to predict things such as how many people will enter the first grade at age 6 between the years 2015 and 2020, how many are likely to enroll in college, and how many will arrive at retirement decades down the road. Administrators of social entities such as schools and pension funds can make preparations on the basis of cohort predictions.

To see how dramatically a single cohort can affect a society over time, consider the effect of the cohort born in the United States between 1946 and 1964—the Baby Boom cohort. Many parents of students reading this book are Baby Boomers. The Baby Boom has been one of the most significant demographic events in all of U.S. history, along with the Great Black Migration from the South to the North in the late 1880s through the 1920s, the simultaneous massive immigration of Europeans and other groups, and current migration of Hispanic groups (Portes and Rumbaut 2001; Weeks 2004; Kennedy 1989).

The Baby Boom cohort, now comprising nearly one-third of the entire U.S. population, has had a major impact on the practices, politics, habits, preferences, and culture of our society. Raised in the relatively permissive late 1950s and 1960s, watching the likes of *Howdy Doody* and the *Mickey Mouse Club* on television, they became a large part of the "greed generation" of the 1990s. Some argue this greed generation cohort was at least in part responsible for the financial excesses of recent years and the accompanying failures of major banks, brokerage houses, and the major automakers.

Theories of Population Growth Locally and Globally

Among the major problems facing modern-day civilization is the specter of uncontrolled population growth. As noted earlier, the world population increases by approximately 270,000 people every day. Some view overpopulation as an epochal catastrophe about to roll over us like a tidal wave. Others dispute whether the problem exists at all, explaining that there is no scientific consensus on the *carrying capacity* of the planet, the number of people the planet can support on a sustained basis, and technological advances that have dependably met our needs in the past, can be counted on to do so in the future as the number of mouths to feed continues to grow.

Is the world overpopulated? Is the United States? What can we expect from the future? If the less-optimistic scenarios turn out to be accurate, these could be the most important questions facing humankind.

Malthusian Theory

Humans, like animals, can survive and reproduce only when they have access to the means of *subsistence,* meaning the necessities of life, such as food and shelter. Human populations and animal populations are both prone to decline and die off when encountering checks on population growth such as famine, disease, and war. In the face of a daunting environment, humans have managed to thrive, and their population has doubled many times over. The period of doubling gets shorter and shorter.

Thomas R. Malthus, a clergyman born in 1766 in Scotland and educated in England, pondered the realities of life on earth and assembled his observations into a chilling depiction of disastrous population growth called **Malthusian theory,** the idea that a population tends to grow faster than the subsistence needed to sustain it. Malthus propounded his gloomy views in his *First Essay on Population,* published in 1798 (Malthus 1798), and predicted wide spread catastrophe and famine.

Malthus noted that populations tend to grow not by *arithmetic increase,* adding the same number of new individuals each year, but by *exponential increase,* in which the number of individuals added each year grows, with the larger population generating an even larger number of births with each passing year. Arithmetic increase would cause a population to double in size at a decelerating rate. Exponential increase, in contrast, causes a population to double ever faster.

Unchecked doubling of any population must swiftly spawn enough people to carpet the earth many times over. Clearly, something prevents populations, human and others, from doubling every generation. Malthus reasoned that two forces were at work to keep population growth in check: (1) the growth in the amount of food produced tends to be only arithmetical and not exponential; and, (2) his surmise that there were three major *"positive"* checks on population growth—famine, disease, and war. In Malthus's time, disease could reach apocalyptic scales. The outbreak of bubonic plague in Europe from 1334 to 1354 eliminated one-third of the population; a smallpox epidemic in 1707 wiped out *three-fourths* of the populations of Mexico and the West Indies. Wars took a toll on European men, with deaths in battle causing semipermanent gaps in the population pyramids of European populations. Along with "positive" checks on population growth, Malthus acknowledged what he called *preventive checks,* such as sexual abstinence, but he knew that sexual abstinence was unlikely to be the behavior change that halted uncontrolled population growth.

Malthusian theory actually predicted rather well the population of many agrarian societies such as Egypt from A.D. 500 through Malthus's own lifetime. However, Malthus failed to foresee three revolutionary developments that derailed his predictions of growth and catastrophe. In agriculture, technological advances have permitted farmers to work larger plots of land and grow more food per acre, resulting in subsistence levels higher than Malthus would have predicted. In medicine, science has fought off diseases such as the bubonic plague that Malthus expected to periodically wipe out entire nations. Finally, the development and widespread use of contraceptives in many countries have kept the birthrate at a level lower than Malthus would have thought possible.

The technological victories of this century have not completely erased the specter of Malthus. The worldwide AIDS epidemic warns us that nature can still hurl catastrophes our way. Heartrending pictures of swollen, starving babies remind us that famine continues to destroy human populations in some parts of the world just as it has for thousands of years. Overall, Malthus's theory has served as a warning that subsistence and natural resources are limited. The Malthusian doomsday has not yet occurred, but some believe that Malthus's warning was not in error, just premature.

Demographic Transition Theory

An alternative to Malthusian theory is demographic transition theory, developed initially in the early 1940s by Kingsley Davis (Davis 1945) and extended by Ansley Coale (1974, 1986) and others (Weeks 2004). *Demographic transition theory* proposes that countries pass through a consistent sequence of population patterns linked to the degree of development in the society and end with a situation in which the birthrates and death rates are both relatively low. Overall, the population level is predicted to eventually stabilize, with little subsequent increase or decrease over the long term.

There are three main stages to population change, according to demographic transition theory (see Figure 16.3). Stage 1 is characterized by a high birthrate and high death rate. The United States during its colonial period was in this stage. Women began bearing children at younger ages, and it was not uncommon for a woman to have twelve or thirteen children—a very high birthrate. However, infant mortality was also high, as was the overall death rate owing to primitive medical techniques and unhealthy sanitary conditions.

Stage 2 in the demographic transition is characterized by a high birthrate but a declining death rate. Hence, the overall level of the population increases. The United States entered stage 2 in the second half of the nineteenth century as industrialization took

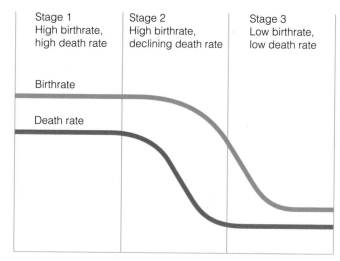

FIGURE 16.3 *Demographic Transition Theory*

hold in earnest. The norms of the day continued to encourage large families, thereby causing high birthrates, while advances in medicine and public sanitation whittled away at the infant mortality rate and the overall death rate. Life expectancy increased, and the population grew in size.

The characteristics of stage 2 did not apply across all social groups or social classes, however. Minorities at the time (Blacks, and especially Native Americans and Chinese in the Midwest and West) were less likely to benefit from medical advances. The infant mortality and overall mortality rates for Blacks, Native Americans, and Chinese remained high, while their life expectancy remained considerably shorter than life expectancy for the White population. Lack of access to quality medical care was particularly devastating to Native Americans, who had very high death rates for all ages during this period. Demographic transition theory is not completely accurate for different racial–ethnic groups within the same society.

Stage 3 of the demographic transition is characterized by a low birthrate and low death rate. The overall level of the population tends to stabilize in stage 3. Medical advances continue, the general prosperity of the society is reflected in lowered death rates, and cultural changes take place, such as a reduction in family size. The United States entered this stage prior to the Second World War, and with the notable exception of the Baby Boom, has exhibited stage 3 demographics since then.

Demographic transition theory has received some criticism primarily because it tends to be based on heavily industrialized countries with a White majority (Coale 1986). In this respect the theory is *ethnocentric* (Weeks 2004). It argues that industrialized countries in the past have gone from high birthrates and death rates to low birthrates and death rates, with the demographic "transition" phase in between. The theory carried with it the presupposition that nations and ethnic groups of color were considerably less industrialized and were thus at stage 1, with high birthrates and death rates. Yet, it is not true that such groups are necessarily nonindustrialized. Thus, such nations and groups can be industrialized yet still be at stage 1.

The "Population Bomb" and Zero Population Growth

In 1968, a modern Malthus appeared in the form of Paul Ehrlich, a biology professor at Stanford University. His book, *The Population Bomb* (Ehrlich 1968), was the first in a series of writings in which Ehrlich argued that many of the dire earlier predictions of Malthus were not far from wrong. The growth in world population, according to Ehrlich, was a time bomb ready to go off in the near future, with dismal consequences. Ehrlich supported his argument with scientific research. This commodity was used rather loosely in the work of Malthus, whose reasoning depended heavily on speculative guesswork about how many people the planet could and should bear. Ehrlich stated that the sheer mathematics of population growth worldwide were sufficient to demonstrate that world population could not possibly continue to expand at its present rates.

Ehrlich pointed out that worldwide population growth has outgrown food production and that massive starvation must inevitably follow. He went beyond Malthusian theory, however, to state that the problem transcended food production and was in fact a problem of the environment. Ehrlich was among the earliest modern thinkers to argue that the quality of the environment, especially the availability of clean air and water, was a critical factor in the growth and health of populations.

Many of the dire predictions Ehrlich made back in 1968 have come true. In a later book Ehrlich wrote with his wife, Anne, *The Population Explosion* (1991), they pointed to a variety of disasters that confirmed the predictions of the 1968 work and even that of Malthus: mass starvation in parts of Africa, starvation in the United States among Black and Hispanic populations, increased homelessness in cities among minorities in particular, acid rain, rampant extinctions of plant and animal species, the irrecoverable destruction of environments such as the rain forests, and, particularly, **global warming,** the systematic increase in worldwide surface temperatures. The effect of global warming was underscored in former Vice President Al Gore's book and movie, *An Inconvenient Truth* (2006) (see Table 16.2).

Checking Population Growth

As early as the 1950s, most countries had accepted population growth as a problem that must be addressed. By the 1980s, countries representing 95 percent of the world's population had formulated some policies aimed at stemming population growth; however, no consensus exists on how to control population. For example, many religious and political authorities have argued against the use of birth control. Efforts to encourage the use of contraceptives among rapidly growing populations have therefore had mixed support, whereas other attempts to curb population growth, such as encouraging changes in social habits, generally meet with little success.

Family Planning and Diversity

Many governments, including that of the United States, make contraceptives available to individuals

table 16.2 A Comparison of Demographic Theories

	Malthusian Theory	Demographic Transition Theory	Zero Population Growth (ZPG) Theory
Main point	A population grows faster than the subsistence (food supply) needed to sustain it.	Populations go through predictable stages (transitions) from high birth and death rates to a stable population with low birth and death rates.	Achievement of zero population growth solves the Malthusian problem of unchecked population growth.
Positive checks on population growth	Famine, disease, and war are likely.	Famine, disease, and war are moderately likely.	Famine, disease, and war are unlikely.
Preventive checks on population growth	Sexual abstinence.	Sexual abstinence, birth control, and contraceptive methods.	Sexual abstinence, birth control, and contraceptive methods.
Predictions for the future	Pessimistic; despite positive and preventive checks, a population will ultimately outstrip its food supply.	Optimistic; given technology and medical advances in a population.	Very optimistic; zero population growth has already been achieved in the United States and other countries.

A field worker in Bangladesh explains birth control pills and their use to village women.

and families. Doing so, however, is not always consistent with the beliefs and cultural practices of all groups in the society. Catholics are taught that it is acceptable to use natural means of birth control (such as the "rhythm method") but are forbidden to use contraceptive devices such as the "pill." Many Catholics choose to use contraceptives anyway; some studies have shown that the majority of Catholics in the United States practice forms of birth control forbidden by their church.

Governmental programs that advocate contraception can be successful only if couples themselves choose smaller families over larger ones. This is most likely to occur if the wider culture supports that decision. Creating a new image of the ideal family size is a central concern of those involved in the family planning movement.

Before contraceptive devices such as condoms and the pill are widely adopted, cultural values in favor of undisturbed fertility must be countered (Vanlandingham 1993; Westoff et al. 1990), although some studies have shown that certain countries, such as Bangladesh, in stage 2 of demographic transition, have become very receptive to birth control programs and contraceptive devices (Stevens 1994).

Urbanization

The growth and development of cities—centers of human activity with high degrees of population density—is a relatively recent occurrence in the course of human history. Scholars locate the development of the first city around 3500 B.C. (Flanagan 1995). The study of the urban, the rural, and the suburban is the task of *urban sociology,* a subfield of sociology that examines the social structure and cultural aspects of the city compared with rural and suburban centers. These comparisons involve what urban sociologist Gideon Sjøberg (1965) calls the *rural–urban continuum,* those structural and cultural differences that exist as a consequence of differing degrees of urbanization. **Urbanization** is the process by which a community acquires the characteristics of city life and the "urban" end of the rural–urban continuum.

Urbanization as a Lifestyle

Early German sociological theorist **Georg Simmel** (1905/1950) argued that urban living had profound social psychological effects on the individual. Thus he was among the early theorists who argued that social structure could affect the individual. He argued that urban life has a quick pace and is stimulating, but as a consequence of this intense style of life, individuals become insensitive to people and events that surround them. The urban dweller tends to avoid the emotional involvement that, according to Simmel, was more likely found in rural communities. Interaction tends to be characterized as economic rather than social, and close, personal interaction is frowned upon and discouraged. Yet, urban dwelling can increase the likelihood of other ills: Early theorist **Emile Durkheim** noted that the suicide rate per 10,000 people was greater in more urbanized areas than in rural areas (Durkheim 1897/1951).

The sociologist **Louis Wirth** (1938), focusing on Chicago in the 1930s, also argued that the city was a center of distant, cold interpersonal interaction, and as a result the urban dweller experienced alienation, loneliness, and powerlessness. One positive consequence of all this, according to Wirth and Simmel, was the liberating effect that arises from the relative absence of close, restrictive ties and interactions. Thus, city life offered individuals a certain feeling of freedom.

A contrasting view of urban life is offered by Herbert Gans (1962/1982), who studied people in Boston in the late 1950s and concluded that many city residents develop strong loyalties to others and are characterized by a sense of community. Such subgroupings he referred to as the *urban village,* which is characterized by several "modes of adaptation," among them *cosmopolites*—typically students, artists, writers, and musicians, who together form a tightly knit community and choose urban living to be near the city's cultural facilities. A second category are the *ethnic villagers,* people who live in ethnically and racially segregated neighborhoods. Such "urban enclaves," as they are called by researcher Mark Abrahamson (2006), tend to develop their own unique identities, such as San Francisco's Chinatown or Miami's Little Havana.

The New Suburbanites

The United States ended the twentieth century as it began it—in a great wave of immigration. The 1924 National Origins Quota Law encouraged immigration from northern and western Europe (England, France, Germany, Switzerland, and the Scandinavian countries), but discouraged immigration from eastern and southern Europe (Greece, Italy, Poland, Turkey, and eastern European Jews generally, among others) (see Chapter 10). Despite this openly discriminatory law, millions of eastern Europeans successfully made the journey to Ellis Island and thence the U.S. mainland, only to face prejudice, discrimination, and the accusation that they were taking jobs that would have otherwise gone to the already-present White majority.

At the end of the twentieth century, once again American shores received millions from abroad, and once again prejudice and discrimination were part of the picture, but often in more subtle forms and at times in clearly and fortunately decreased forms. Neighborhoods are now invigorated and culturally enriched by mosques or Buddhist temples; by whole neighborhoods of Vietnamese Catholics, Koreans, or Asian Indians; or by war refugees from Somalia and Bosnia. The most prominent immigrants in suburban neighborhoods are Hispanic Americans and Asian Americans, the groups that presently make up most of the country's foreign-born population (Edmondson 2000).

One long-term consequence of the current immigration settlement is what at least one demographer (Frey 2009) calls the new *demographic divide*. Because today's immigrants settle in a relatively small number of big cities and their suburbs, mostly located on either the East or West Coast, many parts of the United States are still not feeling the effects of these new immigrants. This distinction has a number of consequences. For example, for immigrant groups who are youthful and have a relatively high birthrate, schooling has become a major political issue. But in other areas with a higher proportion of older immigrants, tax cuts for the elderly and Medicare are the major issues.

Ecology and the Environment

It should be apparent by now that population size has an important social dimension. Social forces can cause changes in the size of the population, and population changes can transform society. Values are acquired from one's culture, including values about what family size is considered ideal or what degree of crowding in a city is considered tolerable. The values people hold will affect the number of children they want, the place they want to live, and the degree to which they consider population control an urgent problem.

Population density is the total number of people per unit of area, usually per square mile. As population density rises to high levels, as it has in today's cities, the familiar problems of urban living appear, including high rates of crime and homelessness. Interacting with these problems are crises of the physical environment, such as air and water pollution, acid rain, and the growing output of hazardous wastes, and the resultant global warming. Humans sometimes forget that, like all other creatures, we are intimately dependent on the physical environment.

Human ecology is the scientific study of the interdependencies that exist between humans and our physical environment. A *human ecosystem* is any system of interdependent parts that involves human beings in interaction with one another and the physical environment. A city is a human ecosystem; so is a rural farmland community. In fact, the entire world is a human ecosystem. Two fundamental and closely related problems confront our present ecosystems: overpopulation and the destruction or exhaustion of natural resources (Weeks 2004; Hawley 1986).

Vanishing Resources

In all ecosystems, whether human, animal, or plant, organisms depend for their survival on one another as well as on the physical environment. Plants use carbon dioxide and give off oxygen, which all animals need to survive. Terrestrial creatures metabolize oxygen and produce carbon dioxide, which is then used by plants, completing a cycle. Humans and other animals require nutrients they can get only by eating plants. When they die, they decompose and provide nutrients to the soil that are taken up by plants, completing another cycle.

The examination of ecosystems has demonstrated two things: The supply of many natural resources is finite, and if one element of an ecosystem is disturbed, the entire system is affected. For much of the history of humankind, the natural resources of the earth were so abundant compared with the amounts used by humans that they may as well have been infinite. No more. Some resources, such as certain fossil fuels, are simply nonrenewable and will be gone soon. Other resources, such as timber or seafood, are renewable as long as we do not plunder the sources of supply so recklessly that they disappear. This is an ecological blunder we have made many times before. Finally, some natural resources are so abundant that they still seem infinite, such as the planet's stock of air and water; but at this stage of our technological development, we are learning that our powers extend to such heights and depths that we can even destroy the near-infinite resources (Gore 2006).

On the planetary scale, our gaseous wastes are gnawing away at the ozone layer, and our buried chemical wastes are trickling into the water table creating underground pools of poison. One study (Barlow and Clarke 2002) actually notes that pollution has

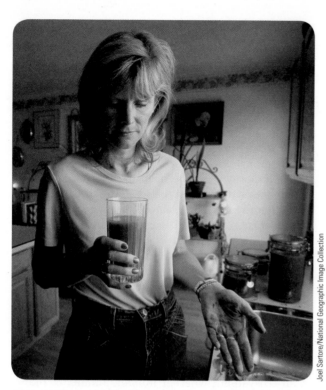

This woman was told by a gas company in Powder River Basin, Wyoming, that the drilling for gas near her house "would never cause you to lose your water." Shortly after drilling began near her, her well water turned into a muddy methane slurry, which she unhappily holds in a glass in this photo.

SEE *FOR YOURSELF*

THE WASTEFUL SOCIETY

For just one day (a full 24-hour period), make a list of everything that you use up or discard. Include everything that you throw away, including garbage, waste from cooking and eating, gasoline in your car, and so forth. At the end of the day, list the things you discarded. Indicate whether there were any alternatives to discarding these things. How might one reduce the amount of waste produced in society generally?

Environmental Pollution

It would be a bitter irony if we managed to avoid exhausting the resources of the planet only to find that what we had conserved was too degraded by pollution to be useful. The most threatening forms of pollution are the poisoning of the planet's air and water. Air pollution is not only ugly and uncomfortable, it is deadly. The skies of all major cities around the world are stained with pollution hazes, and in cities that rest within geological basins, such as Mexico City and Los Angeles, the concentrations of pollutants can rise so high that pollution-sensitive individuals cannot leave their homes. The occurrence of respiratory cases in the hospitals rises and falls with the passing of weather systems that cause the pollutants to concentrate or disperse.

The leading air and water polluters are the United States, Japan, Russia, China, and Poland. When the reign of secrecy ended in the former Soviet Union, bloodcurdling stories of environmental vandalism emerged. Nuclear-powered ships at the ends of their useful lives were blithely sunk—contaminated reactors, spent fuel, and all—in Arctic seas. In the farther reaches of the giant Russian territory (which crosses eleven time zones; see Map 16.1), areas were designated as open dumping sites for toxic wastes and then sealed off. Similarly, a toxic quarantine area exists downwind of the Chernobyl nuclear power plant that blew up in 1986, spewing radioactive poisons over Europe in amounts never accurately determined, but now appearing to be *far in excess* of what was once thought to be the worst case imaginable. Incredibly, the Chernobyl nuclear complex was not shut down until November 2000.

A huge portion of the pollutants released into the air come from the exhaust pipes of motor vehicles. The major component of this exhaust is carbon monoxide, a highly toxic substance. Also found in exhaust fumes are nitrogen oxides, the substances that give smog its brownish-yellow tinge. The action of sunlight causes these oxides to combine with hydrocarbons also emitted from exhausts, forming a host of health-threatening substances (see Figure 16.4).

A daunting international issue has grown around a group of chemicals called chlorofluorocarbons

damaged the earth's surface water so badly that worldwide underground water reserves are being mined faster than nature can replenish them.

If growing population is a problem of the developing world, shrinking resources are a problem of the industrialized world. The United States consumes more than 40 percent of the world's aluminum and coal as well as about 30 percent of its platinum and copper (Ehrlich et al. 1977; Ehrlich and Ehrlich 1991). According to some observers, within forty years we will reach the end of the world's supply of lead, silver, tungsten, and mercury—mainstays of heavy and high-tech manufacturing, including the computer hardware industry.

In the United States alone, real estate development takes over millions of acres of farmland each year. In the western and southwestern United States, the groundwater supply is being depleted at a rapid pace. We are racing through our nonrenewable natural resources and destroying much that should be renewable; however, some activists have begun to claim that the environmental situation, though perilous, is improving (Simon 2007). There is evidence to support this view, but even optimists who wish to show that things are getting better are quick to point out that "better" is not sufficient if the situation is bad enough to begin with (for example, Montagne 1995).

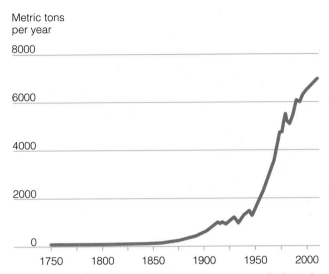

Metric tons
per year

FIGURE 16.4 *Worldwide Carbon Dioxide (CO₂)
Emissions from Burning Fossil Fuels, 1751–2000*

Source: U.S. Census Bureau. 2008b. *Statistical Abstract of the United
States, 2007.* Washington, D.C.: U.S. Department of Commerce.

(CFCs), used as a coolant in refrigerators, in the manufacture of plastics, and as an aerosol can propellant. CFCs released into the air find their way to the ozone layer in the upper atmosphere, where they eliminate the highly reactive ozone. The ozone layer is a shield that blocks dangerous ultraviolet light, and as this shield is destroyed, more ultraviolet light gets through, causing an increase in sunburn, skin cancer, and other illnesses.

Related to the problem of ozone depletion is the **greenhouse effect.** As the sun's energy pours onto the earth, some is reflected from the earth's surface to the earth's atmosphere. Of the reflected energy, a portion is captured by carbon dioxide in the earth's atmosphere, while the rest radiates into space. If the amount of solar energy trapped by carbon dioxide rises, the temperature increases, producing, as we have already noted, *global warming.*

Rather small changes in the average temperature of the earth can have dramatic consequences. A few degrees of difference can cause greater melting in the arctic regions, which raises the level of the sea, affecting water, land, and weather systems worldwide. Today, we see on television images of polar bears drowning because of the breakup of ice floes, which is their habitat, caused by global warming. Some researchers have noted that at least until recently, public opinion in countries outside the United States, including less-developed nations, reveals more concern about global warming than that in the United States (Brechen 2003). It is a sad commentary that the administration of President George W. Bush and Vice President Richard Cheney (2000–2008) consistently denied these realities of global warming, despite overwhelming scientific evidence of it.

The nation's rivers and lakes have long been dumping grounds for heavy industry. Yet these same industries—paper, steel, automobile, and chemical—depend on clean water for their production processes, during which they take water from the rivers and lakes and return it heated and polluted. The difference in temperature can alter aquatic habitats and kill aquatic life, earning this change the name of *thermal pollution.* The chemical pollutants that industries discharge into rivers, lakes, and the oceans include solid wastes, sewage, nondegradable by-products, synthetic materials, toxic chemicals, and radioactive substances. Add the polluting effects of sewage systems of towns and large cities, detergents, oil spills, pesticide runoff, and runoff from mines, and the enormity of the problem is clear.

The Environmental Protection Agency (EPA) estimates that 63 percent of rural Americans may be drinking water contaminated by agricultural runoff and the improper disposal of toxic substances in landfills. Thousands of rural water wells have been abandoned due to contamination. Households served by municipal water systems are also endangered; fully 20 percent of the country's public water systems do not meet the minimum toxicity standards set by the government.

Federal and state statutes now prohibit industry from polluting the nation's water, but the pollution continues. Why? The answer is economic, political, and sociological. Industries that contribute to a vigorous economy have traditionally met little interference from the government. Public awareness and outrage can force the government to crack down on major polluters. Nevertheless, the federal government has often chosen a look-the-other-way attitude.

Many argue that of all environmental problems facing the United States today, the most urgent is the dumping of hazardous wastes, if only for the sheer noxiousness of the materials being dumped (see Map 16.1). It is estimated that since 1970, the production of toxic wastes increased ninefold (Weeks 2008). This dramatic increase in the amount and variety of hazardous waste production is traceable to new and profitable industrial technologies.

Environmental Inequality

As noted earlier in this book, the brutally devastating hurricanes Katrina and Rita hit New Orleans, Louisiana, as well as Beaumont, Texas, and other locations along the country's southern Gulf Coast early in the fall of 2005. Because of slowness of response by the federal government, by FEMA (the Federal Emergency Management Administration), and by the president (President Bush) himself, over 1000 people died from drowning, from direct hits from flying debris, and from lack of medical attention. The most negatively affected neighborhoods were the lowest-lying ones, which were primarily poor and

map16.1　*MAPPING AMERICA'S DIVERSITY*

Total Number of Waste Sites by State

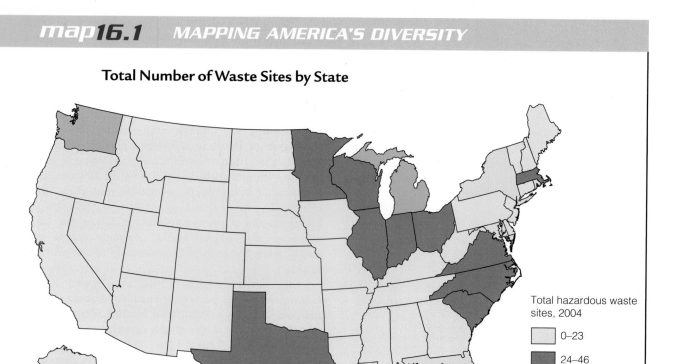

Total hazardous waste sites, 2004

- 0–23
- 24–46
- 47–69
- 70–92
- 93–115

The total number of waste sites in a state varies greatly from state to state. Pick out your home state on this map. What is the approximate number of hazardous waste sites in it?

Data: From the U.S. Census Bureau. 2008b. *Statistical Abstract of the United States, 2007.* Washington, D.C.: U.S. Census Bureau. **www.census.gov**

Black or Hispanic, thus *hypersegregated*. News commentators and some social scientists argued that if the White and wealthy had been so concentrated in such neighborhoods, the response of the federal government and the president would probably have been much more rapid and perhaps more effective (Dyson 2006).

Adding to the problem is the fact that toxic wastes are dumped with disproportionate frequency in areas that have high concentrations of minorities, particularly American Indians, Hispanics, and African Americans, as well as people of lower socioeconomic status (Mohai and Saha 2007; Downy 2005; Pellow 2003; Holmes 2000; Boer et al. 1997; Bullard 1994a, 1994b). Studies determined that it was "virtually impossible" that dumps were being placed so often in minority and lower socioeconomic status communities by chance alone (Mohai and Saha 2007; Bullard 1994a, 1994b). Such communities also tended to fall below the national average economically and educationally. **Environmental racism** consists of the dumping of toxic wastes with disproportionate frequency at or very near areas with high concentrations of minorities.

Debunking Society's Myths

Myth: Environmental pollution is in fact more common in or near economically poor areas, but race has nothing to do with it.

Sociological perspective: Even when comparing areas of the same low economic status but different racial compositions, the areas with a higher percentage of minorities are on average closer to polluted areas than those with a lower percentage of minorities.

One study of households throughout Florida found that Native American, Hispanic, and particularly African American populations reside disproportionately closer to toxic sources than do Whites. Such patterns are *not* explainable by social class differences alone. That is, when communities of the same socioeconomic characteristics but different racial–ethnic compositions are compared, Native Americans, Hispanics, and African Americans of a given socioeconomic level live closer to toxic dumps than do

Whites of the *same* socioeconomic level (Mohai and Saha 2007; Holmes 2000).

One of the largest commercial hazardous waste landfills in the nation is located in Emelle, Alabama, where Blacks make up nearly 80 percent of the population. In Scotlandville, Louisiana, the site of the fourth-largest toxic landfill in the country, Blacks comprise 93 percent of the population. In Kettleman City, California, the site of the fifth-largest toxic landfill, more than 78 percent of the residents are Hispanic. In many cases, the siting of the landfills was linked to the economic interests of both residents and corporations. Companies with profit in mind negotiated favorable deals with the residents of these communities to permit the dumping of wastes, *often misrepresented as nontoxic*, in exchange for jobs and other economic incentives, which were often slow in coming (Bullard 1994b).

From the 1950s through the 1970s, the Navajo population of Shiprock, New Mexico, was exposed to waste from uranium mining and dumping that was 90 to 100 times more radioactive than the level permissible by law. Kerr-McGee, the corporation responsible, was forced out of the area in the early 1970s. When it left, it simply abandoned the site, leaving 70 acres of radioactive mine tailings (the residue from the separation of ores). An even worse situation developed during the same time in Laguna, New Mexico, involving the Pueblo Indians. Anaconda Copper, a subsidiary of the Atlantic-Richfield Corporation, virtually wrecked the traditional Pueblo economy by recruiting the community's youth for hazardous jobs even as it contaminated their environment with the wastes from uranium mining. A high rate of cancer deaths in these communities serves as testimony to the horrible consequences of carelessly discarding toxic wastes (Churchill 1992).

Feminism and the Environment

Women and men do not regard environmental issues equally. In general, women tend to be more concerned with issues of environmental risk, and this has important policy implications. Lack of attention on the part of local and federal governments to environmental issues can be interpreted as lack of attention to policy that differentially affects women. In this respect, it is a feminist issue.

Women consistently show more concern than men for environmental issues and perceived themselves to be at considerably more risk from environmental hazards than did men. For example, women are more likely than men to believe that abandoned waste sites cause cancer and produce miscarriages and other health problems. They are also more likely than men to feel that waste sites posed dangers to trees, fish, and other wildlife. Women are also more likely than men to perceive dangers in global warming; they were more likely to predict, as a result, coastal flooding from polar ice meltdown; loss of forests; increased killing off of certain animal species (such as polar bears); increases in hurricanes and tornadoes; and increased air and water pollution.

Environmental Policy

Environmental policy of the U.S. government over the last thirty years or so has been affected, to greater or lesser degrees, by what has come to be called the environmental movement, a social movement (see following)) consisting of various loosely organized groups such as the Earth Day groups, NIMBY ("Not in my backyard"), the Sierra Club, the Berkeley, California, tree-sitters, and other similar organizations. Biologist Rachel Carson's seminal book, *The Silent Spring* (1962), served as a significant impetus to the environmental movement. The book was a scathing indictment concerning the pesticide chemical DDT and its far-reaching polluting effects from toxic runoff, including the severe reduction of many bird populations and the contamination of human breast milk. The Earth Day mobilization in 1970, now noted every year, received tremendous support from the U.S. public (nearly 70 percent according to one survey; Dunlop and Mertig 1992).

The Love Canal debacle of the 1980s is often cited as one of the historical spurs to the environmental movement. During this period, in an area of upstate New York near Niagara Falls called Love Canal, new homes were discovered to have been built on land previously used as large toxic waste dumps and landfills. The pollution was so severe that many

Environmental racism refers to the pattern whereby people living in predominantly minority communities are more likely exposed to toxic dumping and other forms of pollution. Nuclear waste and testing in the American Southwest, for example, have been located in areas predominantly inhabited by Native Americans. In other areas, African Americans and Latinos are exposed to the effects of industrial waste.

Visual concept by Norman Andersen

[seeing society in *everyday life*

The Hurting Environment

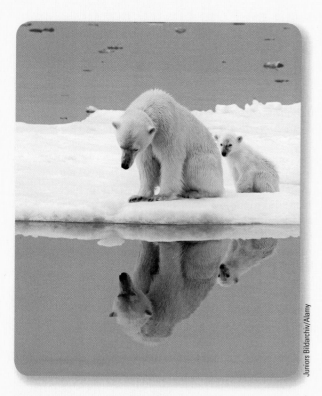

Juniors Bildarchiv/Alamy

As a direct result of global warming, the ice masses at both the North and South Poles are melting, and thus forcing polar bears and their offspring from the floes that are their natural habitat. This has also increased the distance between such ice floes, hence making it more difficult for the bears to swim from one to the other. Polar bears have drowned trying to do so.

Michael Penn/The New York Times/Redux Pictures

A peculiarity of global warming is as glaciers melt, the land rises, thus causing the sea to retreat, not advance. The person in this photo holds a piece of such a glacier. Glaciers are exceptionally heavy (billions upon billions of tons), and their melting relieves the land underneath of their immense weight, thus causing the land to rise—so geologists tell us—much as a cushion regains its shape after someone gets up from a couch! As shown in this photograph, this has been happening in and near Juneau, Alaska for thirty years or more, causing ponds and once-navigable streams and rivers to become shallow or completely dry.

Roland Emmerlich/Picture Desk

Global warming, a direct result of the greenhouse effect (see text), is upon us. The scientific evidence is indisputable. This causes warming and melting of the glaciers at the poles. Melting of the glaciers causes a rise in the level of the globe's oceans. An ocean rise of only a few feet would wipe out large parts of low-lying coastal cities across the entire world, as suggested in this rendering.

Major environmental mishaps, such as the near meltdown at the nuclear plant at Three Mile Island, Pennsylvania, in 1979; the explosion of a nuclear reactor at Chernobyl, Ukraine, in 1986; and the running aground of the oil tanker *Exxon Valdez* off the coast of Alaska in 1989, have resulted in organized attempts to fight such massively polluting mishaps. In 2005, an agreement among more than 160 countries, known as the Kyoto Protocol, went into effect. This agreement focused on the fight against macrolevel problems such as pollution of the oceans, global warming, ozone depletion, and rain forest loss. Countries ratifying this protocol all agreed to reduce emissions of carbon dioxide and other "greenhouse" gasses. The United States, a major producer of greenhouse gasses, has to date not ratified this protocol.

Only in the last thirty years or so has the tremendous energy potential of the wind has been realized. Wind technology is the driving force for many wind farms that have sprung up across the country. Privately owned wind farms have enjoyed some success, but state or federal governmental plans to erect such windmills in large fields or waterways have met with public opposition on the grounds that they represent eyesores.

This person is a "tree sitter" in Berkeley, California, so named because the woman in this picture occupied the tree platform for several days in order to protest the severe logging (cutting down) of trees in parts of the state. Loggers have stripped certain lands of large trees — even some of the state's prized very large and ancient Redwood trees — and this has caused erosion of the land which had been formerly held in place by the trees' roots. Thus the need for lumber has come into conflict with the environmentalists' campaign to save the trees and inhibit erosion of soil.

of the homeowners discovered a dark, gray chemical ooze issuing from their backyard lawns (Levine 1982). The homeowners abandoned their properties, which remain unoccupied and unsellable to this day.

Antipollution laws have been resisted by industry because the laws require expensive adaptations of manufacturing processes. They have also been resisted by unions that fear the added expense to industry would cost jobs. Despite these points of opposition, many antipollution laws have been passed since the late 1960s and have won great public support. However, in the early 1980s, President Ronald Reagan's administration relaxed antipollution standards based on the claim that they were too costly for industry, and these relaxed policies were continued under the administration of President George H.W. Bush.

The development of new technologies in the last twenty years has played a major role in the reduction of certain kinds of pollution. Since the early 1970s, emissions controls for automobile exhaust have been widely installed. Electric cars are once again in vogue, as is the search for alternative fuels to replace gasoline, such as methanol or methane from the fermentation of human and animal waste products.

What Is Social Change?

Social change is the alteration of social interactions, institutions, stratification systems, and elements of culture over time. Societies are in a constant state of flux. Some changes are rapid, such as those brought about by desktop computers in little more than ten years. Other changes are more gradual, such as the increasing urbanization that characterizes the contemporary world. Sometimes people adapt quickly to change, such as in response to the development of electronic communication. Other times people resist change or are slow to adapt to new possibilities. Despite decades of effort to promote contraceptive methods, in some developing nations there are only the most sluggish gains (Menken and Rahman 2001; Goldman et al. 1989). The speed of social change varies from society to society and from time to time within the same society.

Microchanges are subtle alterations in the day-to-day interactions between people. A fad "catching on" is an example of a microchange. Take the popularity of bungee jumping. Although not as widespread as some previous fads, this highly dangerous recreation is one of a group of "extreme sports" that have become popular across the country. Bungee jumping has caused quite a few serious injuries and deaths, but has also provided thrilling footage for soft-drink commercials. This may account for why a large number of youths have suddenly developed a taste for putting themselves in bone-smashing danger.

Macrochanges are gradual transformations that occur on a broad scale and affect many aspects of society. In the process of *modernization,* societies absorb the changes that come with new times and shed old ways. One frequently noted trend accompanying modernization is that societies develop greater differentiation in social rank, divisions of labor, and so on. The effects of the fast-food industry and its impact on social structure exemplify a macrochange (see the discussion of "McDonaldization" in Chapter 6). In the United States, the rise of the computer through all its generations from vacuum tube to microchip has dramatically changed society, another example of a macrochange. Not many years ago, who would have imagined that you could "surf the Web" to the extent possible today? As recently as 1990, few people had heard of the Internet, much less used it (Friedman 1999).

Social norms about dress and human activity sometimes make social change more evident when there is historical contrast.

Large or small, fast or slow, social change generally has in common the following characteristics (Lenksi and Nolan 2005):

1. **Social change is uneven.** The various parts of a society do not all change at the same rate; some parts lag behind others. This is the principle of *culture lag,* a term coined by sociological theorist **William F. Ogburn** (1922) and first described in Chapter 2. Recall that culture lag refers to the delay between when social conditions change and when cultural adjustments are made. Often the first change is a development in material culture (such as a technological change in computer hardware), which is followed by a change in nonmaterial culture (meaning the habits and mores of the culture).

2. **The onset and consequences of social change are often unforeseen.** The inventors of the atomic bomb in the early 1940s could not predict the vast changes in the character of international relations, including a Cold War that lasted until the demise of the Soviet Union in the early 1990s. Television pioneers, who envisioned a mode of mass communication more compelling than radio, could not predict television would become such a dominant force in determining the interests and habits of youth and the activities and structure of the family. Culture lag is present in both examples: A change in material culture (invention of the atomic bomb, invention of television) precedes later changes in nonmaterial culture (international relations, youth culture, and family structure).

3. **Social change often creates conflict.** Change often triggers conflicts along racial–ethnic lines, social class lines, and gender lines. Terrorism—both in the United States and abroad—focuses attention on the deep conflicts that exist worldwide in political, ethnic, and religious division. These conflicts not only produce international tension, they often drive the world events that generate social change.

4. **The direction of social change is not random.** Change has "direction" relative to a society's history. A populace may want to make a good society better, or it may rebel against a status quo regarded as unendurable. Whether change is wanted or resisted, when it occurs it takes place within a specific social and cultural context.

Social change cannot erase the past. As a society moves toward the future, it carries along its past, its traditions, and its institutions (Lenski and Nolan 2005; Lenski 2005). A generally satisfied populace that strives to make a good society better obviously wishes to preserve its past, but even when a society is in revolt against a status quo that is intolerable, the social change that occurs must be understood in the context of the past as much as the future.

Theories of Social Change

As we have seen, social change may occur for different reasons. It may occur quickly or slowly, may be planned or unplanned, and may represent microchange or macrochange. Different theories of social change emphasize different aspects of the change process. The three main lines of contention in social change theory are functionalist theories, conflict theories, and cyclical theories (see Table 16.3). Later in this chapter we consider three additional global theories: modernization theory, world systems theory, and dependency theory. We will consider each individually.

Functionalist and Evolutionary Theories

Recall from previous chapters that functionalist theory builds on the postulate that all societies, past and present, possess basic elements and institutions that perform certain functions permitting a society to survive and persist. A *function* is a consequence of a social element that contributes to the continuance of a society. For example, the function of an institution such as the family is to provide the society with sufficient population to assure its continuance.

The early theorists **Herbert Spencer** (1882) and **Emile Durkheim** (1964/1895) both argued that as societies move through history, they become more complex. Spencer argued that societies moved from "homogeneity to heterogeneity." Durkheim similarly argued that societies moved from a state of *mechanical solidarity,* a cohesiveness based on the similarity among its members, to *organic solidarity* (also called *contractual solidarity*), a cohesiveness based on difference; a division of labor that exists among its members joins them together, because each depends on the others to perform specialized tasks (see Chapter 5). Through the creation of specialized roles, structures, and institutions, societies thus move from a condition of relative undifferentiation to higher social differentiation.

According to functional theorists, societies that are structurally simple and homogeneous, such as foraging or pastoral societies, where all members engage in similar tasks, move to societies more structurally complex and heterogeneous, such as agricultural, industrial, and postindustrial societies, where great social differentiation exists in the division of labor among people who perform many specialized tasks. The consequence (or function) of increased differentiation and division of labor is a higher degree of stability and cohesiveness in the society, brought about by mutual dependence (Parsons 1951a, 1966).

Evolutionary theories of social change are a branch of functionalist theory. One variety, called *unidimensional evolutionary theory,* argued that societies follow a single evolutionary path from simple, undifferentiated societies to more complex and highly differentiated societies. The more differentiated societies are then perceived as more "civilized." Early theorists such as Lewis Morgan (1877) labeled the distinctions between societies as "primitive" and "civilized," an antiquated notion that has been severely criticized; there is no reason to suppose that an undifferentiated society is necessarily more primitive than a more differentiated one. Furthermore, these earlier theories offered no firm definitions for the terms *primitive* or *civilized*. Nevertheless, the notion that some societies are primitive continues to persist today.

Unidimensional theories of social change fell out of favor because social change occurs in several dimensions and affects a variety of institutions and cultural elements. Meeting the need for a theory that better matches what is actually observed, *multidimensional evolutionary theory* (also called *neoevolutionary* theory) argues that the structural, institutional, and cultural development of a society can simultaneously follow many evolutionary paths, with the different paths all emerging from the circumstances of the society in question.

One formulation of multidimensional evolutionary theory is that of Lenski (Lenski 2005; Lenski et al. 1998). Lenski gives a central role to technology, arguing that technological advances are significantly (though not wholly) responsible for other changes, such as alterations in religious preference, the nature of law, the form of government, and relations between races and genders. Although the role of technology is presented as central, other relationships among institutions continue to be important. For example, advances in computer hardware and software can produce changes in the legal system by creating a need for new laws to deal with computer crimes, such as identity theft.

In support of the overall argument that social change is in fact evolutionary—cumulative and not easily reversible—Lenski and his associates point out that many agricultural societies have transformed into industrial societies throughout history, but few have made the reverse trip from industrial to agricultural, although certain countercultural groups have tried, such as hippie communes of the 1960s and 1970s.

Debunking Society's Myths

Myth: Societies change in linear, directed fashion from primitive to civilized.

Sociological perspective: Social change can occur in several directions at roughly the same time. Furthermore, the terms *primitive* and *civilized* are out of favor as concepts, in that they imply an ethnocentric value judgment about the relative sophistication of diverse cultures.

Conflict Theories

Karl Marx, the founder of conflict theory (Marx 1967/1867), was influenced by the early functionalist and evolutionary theories of Herbert Spencer. Marx agreed that societies change and social change has

direction, the central principle in Spencer's social evolutionary theory, but Marx placed greater emphasis on the role of economics. He argued that societies could indeed "advance" and that advancement was to be measured by the movement from a class society to a society with no class structure. Marx believed that, along the way, class conflict was inevitable.

As noted earlier in this book, the central notion of conflict theory is that conflict is inherently built into social relations (Dahrendorf 1959). For Marx, social conflict, particularly between the two major social classes—working class versus upper class, proletariat versus bourgeoisie—was not only inherent in social relations but was indeed the driving force behind all social change. Marx believed that the most important causes of social change were the tensions between social groups, especially those defined along social class lines. Different classes have different access to power, with the relatively lower class carrying less power. Although the groups Marx originally referred to were indeed social classes, subsequent interpretations include conflict between any socially distinct groups that receive unequal privileges and opportunities (G. Marx 1967; Rodney 1974). However, be aware that the distinction between class and other social variables is necessarily murky. For example, conflict between Whites and minorities is at least partly (but not wholly) class conflict, because minorities are disproportionately represented among the less well-off classes.

Racial and ethnic conflict in the United States involves far more than class differences alone: Many *cultural* differences exist between Whites and Native Americans, Latinos, Blacks, and Asians. Furthermore, cultural differences exist *within* broadly defined ethnic groups as well. We have pointed out earlier in this book that there are broad differences in norms and heritage among Chinese Americans, Japanese Americans, Vietnamese Americans, and so on—all often grouped rather coarsely as Asian Americans. The central idea of conflict theory is the notion that social groups will have competing interests regardless of how they are defined. Conflict is an inherent part of the social scene in any society.

Cyclical Theories

Cyclical theories of social change invoke patterns of social structure and culture that are believed to recur at regular intervals. Cyclical theories build on the idea that societies have a life cycle, like seasonal plants, or at least a life span, like humans. **Arnold J. Toynbee,** a social historian and a principal theorist of cyclical social change, argues that societies are born, mature, decay, and sometimes die (Toynbee 2006; Toynbee and Caplan 1972). For at least part of his life, Toynbee believed that Western society was fated to self-destruct as energetic social builders were replaced by entrenched elite minorities who ruled by force, and society would wither under the sterile regimes. Some believe that societies become more decrepit, only to be replaced by more youthful societies. This belief is typified in Oswald Spengler's famous work, *The Decline* *of the West* (1932), which held that Western European culture was already deeply in decline, following a path Spengler believed was observable in all cultures.

Sociological theorists **Pitrim Sorokin** (1941) and, more recently, Theodore Caplow (1991), have argued that societies proceed through three phases or cycles. In the first phase, the *idealistic culture,* the society wrestles with the tension between the ideal and the practical. An example would be the situation captured in **Gunnar Myrdal's** classic work, *An American Dilemma* (1944), in which our nation declared a belief in equality for all, despite intractable racial, class, and gender stratification.

The second phase, *ideational culture,* emphasizes faith and new forms of spirituality as a phase in social change. The current New Age spirituality movement stresses nontraditional techniques of meditation and the use of such things as crystals, yoga, and chanting in a journey toward self-fulfillment and spiritual peace (Wuthnow 2007: 1994).

The third phase is *sensate culture,* which stresses practical approaches to reality, and involves the hedonistic and the sensual ("sex, drugs, and rock and roll"). Sorokin may have foreseen the hedonistic elements of popular culture in the 1960s and 1970s as indicative of sensate culture. According to the theory, when a society tires of the sensate, the cyclical process begins again with the society seeking refuge in idealistic culture. The emphasis beginning in the late 1980s and continuing now for a return to "family values," meaning older and more traditional values, is an example of a return to idealistic culture, presumably as a response to a perceived sensate culture.

Global Theories of Social Change

Globalization is the increased interconnectedness and interdependence of numerous societies around the world. No longer can the nations of the world be viewed as separate and independent societies. The irresistible current trend has been for societies to develop deep dependencies on each other, with interlocking economies and social customs. In Europe, this trend proceeded as far as developing a common currency, the *euro,* for all nations participating in the newly constructed common economy.

If the world is becoming increasingly interconnected, does this mean we are moving toward a single, homogeneous culture? In such a culture, electronic communications, computers, and other developments would erase the geographic distances between cultures and, eventually, the cultural differences. However, greater interconnectedness among societies may *magnify* the cultural differences between interacting groups by emphasizing their incompatibilities.

In fact, both processes take place. As societies become more interconnected, cultural diffusion between them creates common ground, while cultural differences may become more important as the relationships among nations becomes more intimate. The different perspectives on globalization are represented by three main theories that we will review: modernization theory, world systems theory, and dependency theory, which are included in Table 16.3.

Modernization Theory

Modernization theory states that global development is a worldwide process including nearly all societies affected by technological change. As a result, societies are more homogeneous in terms of differentiation and complexity. Modernization theory traces the beginnings of globalization to technological advances in Western Europe and the United States that propelled them ahead of the less-developed nations of the world, which were left to adopt the new technologies years later. Homogenization resulted, with developing nations being shaped in the mold of the Western nations that had modernized first.

Proponents of modernization theory, such as William McCord and Arline McCord, reject the assumption that

table16.3	Theories of Social Change				
	Modernization Theory	**Dependency Theory**	**Functionalist/ Evolutionary Theory**	**Conflict Theory**	**Cyclical Theory**
How do societies change?	Societies become more homogenous as a result of technological change.	The most successful nations control the development of less powerful nations, which become dependent on them.	Societies change from simple to complex and from an undifferentiated to a highly differentiated division of labor.	Conflict is inherent in social relations, and society changes from a class-based to a classless society.	Societies develop in cycles from idealistic to sensate culture.
What is the primary cause of social change?	Technology and global development	Economic inequality in the global economy	Technology	Economic conflict between social classes	Necessary for growth

only Western European countries and the United States have led the technological globalization (McCord 1991; McCord and McCord 1986). The McCords argue that non-Western societies, most notably Japan, have also been leaders in modernization. As a result, Japanese culture has profoundly influenced other countries and cultures with its emphasis on the importance of small friendship groups in the workplace and a traditional work ethic. According to the McCords, Japan and other technological leaders, such as Taiwan and South Korea, have added to the impetus of global economic growth.

Dependency Theory

Dependency theory maintains that highly industrialized nations tend to imprison developing nations in dependent relationships rather than spurring the upward mobility of developing nations with transfers of technology and business acumen (Rodney 1981, 1974; Reich 1991). Dependency theory sees the highly industrialized *core nations* as transferring only those narrow capabilities that it serves them to deliver. Once these unequal relationships are forged, core nations seek to preserve the status quo because they benefit from the cheap raw materials and cheap labor from the noncore nations. In this sense the core nations actively *prevent* upward mobility within the developing noncore nations. In the meantime, the developing nations remain dependent on the core nations for markets and support in maintaining what industry they have acquired, while they experience minimal social development, limited economic growth, and increased income stratification among their own population. Rodney (1981) has argued that this pattern of dependency is to blame for the exceptional underdevelopment of a number of African countries.

Borrowing dependency is a form of a dependent relationship. Former Secretary of Labor Robert Reich (1991) has noted that core nations have been willing to lend money to noncore nations, but often at high interest rates that put severe economic strain on these nations, sometimes requiring interventions such as wage and price freezes in the developing societies to maintain solvency. The hardship produced falls disproportionately on the lower social classes; the upper classes are less affected, and occasionally they benefit extravagantly.

Modernization

As societies grow and change, in a general sense they become more modern. As already noted, sociologists use the term *modernization* in a specific sense: **Modernization** is a process of social and cultural change initiated by industrialization and followed by increased social differentiation and division of labor. Societies can, of course, experience social change without industrialization. Modernization is a specific type of social change that industrialization tends to bring

about. The change toward an industrialized society can have positive consequences, such as improved transportation and a higher gross national product, or negative consequences, such as pollution, elevated stress, and increases in certain job discrimination.

Modernization has three general characteristics (Berger et al. 1974). First, *modernization is typified by the decline of small, traditional communities.* The individuals in foraging or agrarian societies live in small-scale settlements with their extended families and neighbors. The primary group is prominent in social interaction. Industrialization causes an overall decline in the importance of primary group interactions and an increase in the importance of secondary groups, such as colleagues at work. Second, *with increasing modernization, a society becomes more bureaucratized:* Interactions come to be shaped by formal organizations. Traditional ties of kinship and neighborhood feeling decrease, and members of the society tend to experience feelings of uncertainty and powerlessness. Third, *there is a decline in the importance of religious institutions,* and with the mechanization of daily life, people begin to feel that they have lost control of their own lives: People may respond by building new religious groups and communities (Wuthnow 1994).

From Community to Society

The German sociologist **Ferdinand Tönnies,** who died in 1936, formulated a theory of modernization that still applies to today's societies (Tönnies 1963/1887). Tönnies viewed the process of modernization as a progressive loss of *gemeinschaft* (German for "community"), a state characterized by a sense of common feeling, strong personal ties, and sturdy primary group memberships, along with a sense of personal loyalty to one another. Tönnies argued the Industrial Revolution, which emphasized efficiency and task-oriented behavior, destroyed the sense of community and personal ties associated with an earlier rural life. At the crux of this was a society organized on the basis of self-interest, which caused the condition of *gesellschaft* (German for "society"), a kind of social organization characterized by a high division of labor, less prominence of personal ties, the lack of a sense of community among the members of society, and the absence of a feeling of belonging—maladies often associated with modern urban life.

The United States since the 1940s has become a gesellschaft, with social interaction less intimate and less emotional, although certain primary groups such as the family and the friendship group still permit strong emotional ties. However, Tönnies noted that the role of the family is considerably less prominent in a gesellschaft than in a gemeinschaft. Patriarchy is less prominent, yet more public, and more women are employed outside the home. In the large cities that characterize the gesellschaft, people live among strangers and pass people on the street who are unfamiliar.

In a gemeinschaft, most of the people one encounters one has seen before. The level of interpersonal trust is considerably less in a gesellschaft. Social interaction tends to be even more confined within ethnic, racial, and social class groups. To find personal contact and to satisfy the need for intimate interaction, individuals often join small church groups, training groups, or personal awareness groups (Wuthnow 1994).

Social Inequality, Powerlessness, and the Individual

Another product of modernization, along with mass society, is pronounced social stratification, according to theorists such as Karl Marx (1967/1867) and Jurgen Habermas (1970). In their view, the personal feelings of powerlessness that accompany modernization are due to social inequalities related to race, ethnicity, class, and gender stratification. Marx argued that inequalities are the inevitable product of the capitalist system. Habermas argued that inequalities are the cause of social conflict.

The social structural conditions that arise from modernization, such as increased social stratification, are influenced by the individuals. Building a stable personal identity is difficult in a highly modernized society that presents the individual with complex and conflicting choices about how to live. Many individuals flounder among lifestyles while searching for personal stability and a sense of self. According to Habermas, individuals in highly modernized environments are more likely than their less modernized peers to experiment with new religions, social movements, and lifestyles in search of a fit with their conception of their own "true self." These individual responses to social structural conditions reveal how the social structure can affect personality.

The influential social theorist Herbert Marcuse (1964) has argued that modernized society fails to meet the basic needs of people, among them the need for a fulfilling identity. In this respect, modern society and its attendant technological advances are not stable and rational, as is often argued, but unstable and irrational. The technological advances of modern society do not increase the feeling of having control over one's life, but instead reduce that control and lead to feelings of powerlessness.

This powerlessness leads to the *alienation* of the individual from society—the individual experiences feelings of separation from the group or society. This alienation is most likely to affect people traditionally denied access to power, such as racial minorities, women, and the working class. This alienation from the highly modernized, technological society is, in Marcuse's view, one of the most pressing problems of civilization today. Marcuse argues that, despite the popular view that technology is supposed to yield efficient solutions to the world's problems, it may be more accurate to say that technology is a primary cause of many problems in modern society.

The Causes of Social Change

The causes of social change are many and varied but fall into several broad areas, including cultural diffusion, inequality, changes in population, war, technological innovation, and the mobilization of people through collective behavior and social movements. We examine each of these topics in the following sections.

Revolution

A **revolution** is the overthrow of state or the total transformation of central state institutions. A revolution thus results in far-reaching social change. Numerous sociologists have studied revolutions and identified the conditions under which revolutions are likely to occur. Revolutions can sometimes break down a state and various disenfranchised groups. An array of groups in a society may be dissatisfied with the status quo and organize to replace established institutions. Dissatisfaction alone is not enough to produce a revolution, however. The opportunity must exist for the group to mobilize en masse. Thus, revolutions can result when structured opportunities are created, such as through war or an economic crisis or mobilization through a *social movement,* as we will see later in this chapter.

Social structural conditions that often lead to revolution can include a highly repressive state—so repressed that a strong political culture develops out of resistance to state oppression. A major economic crisis can also produce revolution—as can the development of a new economic system, such as capitalism—that transforms the world economy.

Cultural Diffusion

Cultural diffusion (as noted in Chapter 2) is the transmission of cultural elements from one society or cultural group to another. Cultural diffusion can occur by trade, migration, mass communications media, and social interaction. Anthropologist **Ralph Linton** (1937) long ago alerted us to the fact that much of what many people regard as "American" originally came from other lands—cloth (developed in Asia), clocks (invented in Europe), coins (developed in Turkey), and much more.

Expressions and cultural elements found in the English-speaking United States have been harvested from all over the world. Barbecued ribs, originally eaten by Black slaves in the South after the ribs were discarded by White slave owners who preferred meatier parts of the pig, are now a delicacy enjoyed throughout the United States by virtually all ethnic and racial groups. One theorist, Robert Ferris Thompson (1993), points out that an exceptionally large range of elements in material and nonmaterial culture that originated in Africa have diffused throughout virtually all groups and subcultures in the United States, including aspects of language, dance, art, dress, decorative styles, and even forms of greeting. These examples illustrate cultural diffusion not only from one place to another (such as from West Africa to the United States), but also diffusion across time from a community in the past to many diverse ethnic groups in the present.

Similarly, the immigration of Latino groups into the United States over time has dramatically altered U.S. culture by introducing new food, music, language, slang, and many other cultural elements (Muller and Espenshade 1985). By a similar token, popular culture in the United States has diffused into many other countries and cultures: Witness the adoption of American clothing styles, rock, rap, and Big Macs in countries such as Japan, Germany, Russia, and China. In grocery shops worldwide, from the rain forests of Brazil to the ice floes of Norway, can be found the Coca-Cola logo.

At one time, it was thought that slavery killed off most of the institutions and cultural elements that the slaves brought to the Americas (Herskovits 1941). Extensive research over the past three decades now demonstrates that elements of culture carried from Africa by Black slaves continue to survive among African Americans and, thanks to cultural diffusion,

among many other groups as well. A *step show* is an energetic, highly rhythmic, group choreography performed as a special event by predominantly Black fraternities and sororities. Researchers have traced these displays to traditional West African and central African group dances (Thompson 1993; Gates 1988, 1992). The step show has recently been noticed by non-Black students at universities and colleges all over the country, with a few White groups, fraternal, sororal, and otherwise, taking up "steppin'."

Inequality and Change

Inequalities between people on the basis of class, ethnicity, gender, or other social structural characteristics can be a powerful spur toward social change. Social movements may blossom into full-blown revolution if the underlying tension is great enough. An example of the mechanism of change can be seen when inequalities between the middle class and the urban underclass produce governmental initiatives, such as increased education for the poor, which are designed to reduce this inequality.

Culture itself can sometimes contribute to the persistence of social inequality and thus becomes a source of discontent among the individuals in the society. Inequalities within the education system often have a cultural basis. For example, a poor child in the United States, who adopts a language useful in the ghetto, is at a disadvantage in the classroom where standard English is used. Culturally specific linguistic systems such as urban Black English, or "Ebonics," which is not merely a different dialect (as many people assume) but a distinct linguistic system with its own grammar and syntax (Dillard 1972; Harrison and Trabasso 1976; DiAcosta 1998), are generally not adopted by schools. As a result, this may serve to strengthen the inequalities between the poor and the privileged—unless the child is bicultural and can speak both standard English and Ebonics, which many African Americans can do.

Compounding the problem, a female student may shrink from studying mathematics because she has received the cultural message that to be adept at mathematics is not feminine (Kiefer 2007; Lau 2002; Shih et al. 1999; Sadker and Sadker 1994). The perpetuation of the inequalities of class, race, and gender can stoke the desire for social change on the part of the disadvantaged groups.

War and Social Change

War and severe political conflict result in large and far-reaching changes for both the conquering society, or a region within a society (as in civil war), and for the conquered. The conquerors can impose their will on the conquered and restructure many of their institutions, or the conquerors can exercise only minimal changes.

The U.S. victory over Japan and Germany in the Second World War resulted in societal changes in each country. The war transformed the United States into a mass-production economy and affected family structure

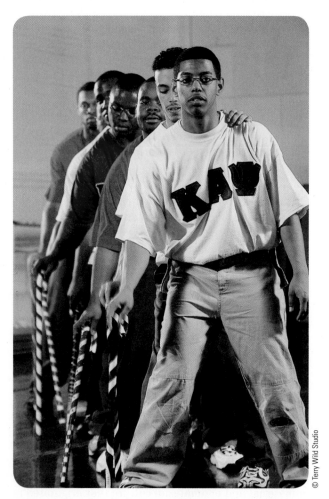

Here members of a predominantly African American fraternity, Kappa Alpha Psi (ΚΑΨ) put on a step show, a highly rhythmic and energetic dance form with roots in slave society as well as in West Africa.

© Terry Wild Studio

(father's absence increasing and women not previously employed joining the labor force) and education (men of college age went off to war in large numbers). Many in the armed forces who returned from the war were educated under a scholarship plan called the GI Bill.

The war also transformed Germany in countless ways, given the vast physical destruction brought on by U.S. bombs and the worldwide attention brought to anti-Semitism and the Nazi Holocaust. The cultural and structural changes in Japan were extensive, as well. The decimation of the Jewish population in Germany, and other nations throughout Europe, resulted in the massive migration of Jews to the United States. The Vietnam War also resulted in many social changes, including the migration of Vietnamese to the United States. If this history of war is any indication, we might shortly expect a wave of migration of Iraqis to the United States.

Technological Innovation and the Cyberspace Revolution

Technological innovations can be strong catalysts of social change. The historical movement from agrarian societies to industrialized societies has been tightly linked to the emergence of technological innovations and inventions (see Chapter 5). Inventions often come about because they answer a need in the society that promises great rewards. The waterwheel promised agrarian societies greater power to raise crops despite dry weather, while also saving large amounts of time and labor. It is possible to trace a timeline from the use of the waterwheel to the use of the large hydroelectric dams that power industrialized societies and along the way find evidence of how each major advance changed society.

In today's world, the most obvious technological change transforming society is the rise of the computer and the subsequent development of desktop computing since the 1980s. The invention and development of the Internet and the resulting communication is now called *cyberspace,* which includes the use of computers for communication between persons and communication between persons and computers. Unique in its vastness and lack of a required central location, the Internet has very rapidly become so much a part of human communication and social reality that it pervades and has transformed literally every social institution—educational, economic, political, familial, and religious.

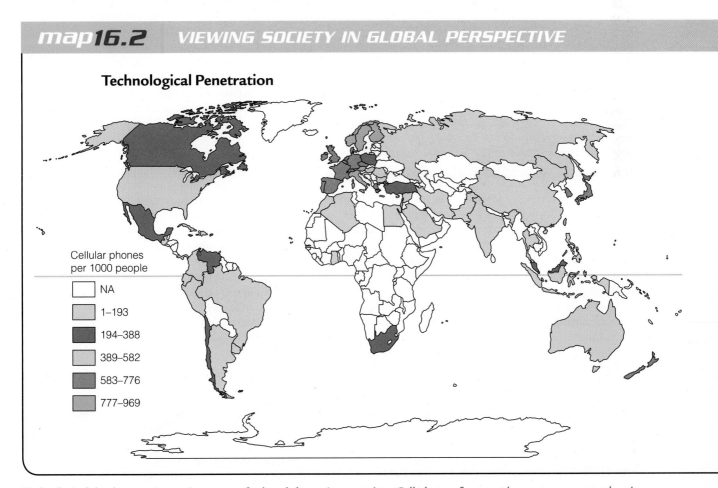

map 16.2 — VIEWING SOCIETY IN GLOBAL PERSPECTIVE

Technological Penetration

Cellular phones per 1000 people

- NA
- 1–193
- 194–388
- 389–582
- 583–776
- 777–969

Technological development is a major source of cultural change in any society. Cell phones, for example, are now commonplace in the United States and other nations. What cultural changes inspire the use of cell phones? And what cultural changes does the increased use of cell phones then create?

Data: U.S. Census Bureau. 2008b. *Statistical Abstract of the United States, 2007.* Washington, DC: U.S. Government Printing Office. **www.census.gov**

© "Disappearing World," Granada TV/the Hutchison Library

The Kaipo people of Brazil wear colorful formal dress. Technology from outside this society (TV, guns) presently threatens the persistence of such cultural practices. Recently, the Kaipo have mobilized to oppose outside intervention such as oil drilling.

The cyberspace revolution began with vacuum tube mainframe computers in the 1950s and early 1960s, followed by the transistorized computer of the mid-1960s and the integrated circuit computers of the late 1960s and 1970s, and was accelerated by the advent of the PC (personal computer) and the invention of the microchip. What can now be stored in a microchip memory the size of a wristwatch, in the late 1960s would have required a transistorized computer the size of a small auditorium.

The path by which technology is introduced into society often reflects the predominant cultural values in that society. Some cultural values may prevent a technological innovation from changing a society. For example, anthropologists have noted that new technologies introduced into an agrarian society very often meet with resistance even though the new technology might greatly benefit the society. The *Yanomami*, an agrarian society existing deep in the rain forests of South America, live without electricity, automobiles, guns, and other items of material culture associated with industrialized societies. The Yanomami place great positive cultural value on their way of hunting and engaging in war. The recent introduction of steel into Yanomami culture, however, may have introduced major social changes and changed them into a more warlike society (Tierney 2000).

Mobilizing People for Change

Social change does not develop in the abstract. Change comes from the actions of human beings. Collective behavior and social movements are ways that people organize to promote, or in some cases, to resist change. *Collective behavior* occurs when normal conventions cease to guide people's behavior, and people establish new patterns of interaction and social structure. *Social movements* are organized and persistent forms of collective behavior. The purpose of a social movement is often to initiate or vigorously resist social change. Examples abound: the civil rights movement, the women's movement, the environmental movement, and the militia movement, just to name a few. So significant are the changes that result from collective behavior and social movements that we examine them in detail now.

Collective Behavior and Social Movements

Social change is propelled by the actions of people who engage in collective behavior and social movements. As we have seen, **collective behavior** occurs when the usual conventions that guide behavior are suspended and people establish new norms of behavior in response to an emerging situation (Turner and Killian 1993). Although collective behavior may emerge spontaneously, it can be predicted. Some of the phenomena defined as collective behavior are whimsical and fun, such as fads and fashions and certain crowds. Other collective behaviors can be terrifying, as in panics or riots. Whether whimsical or awesome, collective behavior is innovative, sometimes revolutionary, and it is this feature that links collective behavior to social change.

Social movements are led by groups that act with some continuity and organization to promote or resist change in society (Turner and Killian 1988). Social movements tend to persist over time more than other forms of collective behavior. Evidence of social movements can be seen throughout society, in movements to protect the environment, promote racial justice, defend the rights of diverse groups (including animal rights), attack the government (such as militia groups), or advocate particular beliefs (such as the Christian Coalition).

Characteristics of Collective Behavior

Collective behavior exhibits certain common characteristics.

1. **Collective behavior always represents the actions of groups of people, not individuals.** The action of a lone gunman who opens fire in a high school cafeteria is not collective behavior because it is one person acting alone. However, groups that gather at the scene of the incident to observe the emergency response are engaged in collective behavior.

2. **Collective behavior involves new or emergent relationships in groups that arise in unusual or unexpected circumstances.** The behavior of people who commute to work together is not considered collective behavior because commuting is an ordinary part of their everyday life. But when a community is struck by a natural disaster, such as a flood, earthquake, or hurricane such as Hurricane Katrina, suddenly nothing can be taken for granted—not food, water, transportation, electricity, nor shelter. Collective behavior emerges to meet the new needs that people in the community face.

3. **Collective behavior captures the more novel, dynamic, and changing elements of society.** An example is a fad that introduces something new into everyday social life. Collective behavior may mark the beginnings of more organized social behavior and often precedes the establishment of formal social movements. People who spontaneously organize to protest something may develop structured ways of sustaining their protest. The environmental justice movement includes a wide array of Native American, African American, Latino, and other communities that have organized to protest the dumping and pollution that imperil their neighborhoods.

4. **Collective behavior is patterned and not the irrational, overly emotional behavior of crazed individuals.** People may follow new guidelines of social behavior, but they do follow guidelines. Even episodes of panic, which appear to be asocial and disorganized, follow a relatively orderly pattern. Episodes of collective behavior often exhibit the more emotional side of life.

5. **During collective behavior, people may communicate extensively through rumors.** Lacking communication channels or distrusting the ones available, people use *rumors* to help define an otherwise ambiguous situation. Rumors are transmitted by people who are piecing together information about a story whose facts are partly obscured. Rumoring is most common when there is inadequate information to interpret a problematic situation or event (Dahlhamer and Nigg 1993; Shibutani 1966).

The Organization of Social Movements

Social movements are both spontaneous and structured, although within movements tension typically exists between spontaneity and structure. Unlike everyday organizations, movements thrive on spontaneity and often must swiftly develop new strategies and tactics in the quest for change. During the civil rights movement, students improvised the technique of sit-ins, which quickly spread because they succeeded in gaining attention for the activists' concerns. Social movements differ from other forms of collective behavior, however, by containing routine elements of organization and lasting for longer periods.

There are three broad types of social movements: personal transformation movements, political or social change movements, and reactionary movements. **Personal transformation movements** aim to change the individual. Rather than pursuing social change, participants adopt a new identity, one they use to redefine their life, both in its current and former state. They focus on the development of new meaning within individual lives (Klapp 1972). The contemporary New Age movement defines mainstream life as stressful and overly rational and promotes relaxation and spiritualism as an emotional release and route to expanded perceptions. New Age music, crystals, massage therapy, and meditation are intended to restore the New Age person to a state of unstressed wholeness. Religious and cult movements are also personal transformation movements. The recent rise in evangelical religious movements can be explained in part by the need people have to give clear meaning to their lives in a complex and sometimes perplexing society.

Social/political change movements aim to change some aspect of society, such as the environmental movement, the gay and lesbian movement, the civil rights movement, the animal rights movement, and the religious right movement. All seek social change, although in distinct and sometimes oppositional ways. Some movements want radical change in existing social institutions, and others want a retreat to a former way of life or even a move to an imagined past (or future) that does not exist. Social movements use a variety of tactics, strategies, and organizational forms to achieve their goals. The civil rights movement, starting in the mid 1950s, used collective action during sit-ins and organizational activity to overturn statutes that supported the "separate but equal" principle of segregation. These efforts culminated in the U.S. Supreme Court decision *Brown v. Board of Education of Topeka, Kansas,* which declared the "separate but equal" doctrine unconstitutional (Morris 1999).

Social change movements may be either reformist or radical (Turner and Killian 1988). **Reform movements** seek change through legal or other mainstream political means, typically working *within* existing institutions. *Radical movements* seek broader-based fundamental change in the basic institutions of society. Instead of working within society's institutions, radical social movements tend to work outside of society's institutions. Although most movements are primarily reformist or radical, within a given movement both factions may exist. In the environmental movement, for example, the Sierra Club is a classic reform movement that lobbies within the existing political system to promote legislation protecting the environment. Greenpeace, in contrast, is a more radical group that sometimes uses dramatic tactics to disrupt activities the group finds objectionable, such as the killing of whales.

Reactionary movements are organized to resist change or to reinstate an earlier social order that participants perceive to be better, and they are reacting against contemporary changes in society. The Aryan Nation, a White supremacist coalition, is a reactionary movement that wants to suppress Jews and minorities and institute a "racially pure" decentralized state (Ferber 1998). The militia movement is also a reactionary movement, seeking to resist government authority and reinstate the perceived lost power of White people (Esterberg 2003). Reactionary movements represent an organized backlash against social values and societal changes that participants find deeply objectionable. Rather than adjust to the future, they may instead hark back to some mythical past that their movement ideology defines as ideal.

thinkingsociologically

Identify a *social movement* in your community or school that you find interesting. By talking to the movement leaders and participants, and if possible, by examining any written material from the movement, how would you describe this movement in sociological terms? Is it *reform, radical,* or *reactionary*? How have the available resources helped or hindered the movement's development? What tactics does the movement use to achieve its goals? How is it connected to other social movements?

Origins of Social Movements

Social movements do not typically develop out of thin air. For a movement to begin, there must be a preexisting communication network (Freeman 1983). The importance of a preexisting communication network is well illustrated by the beginning of the civil rights movement, which most people date to December 1, 1955, the day Rosa Parks was arrested in Montgomery, Alabama, for refusing to give up her seat on a municipal bus to a White man. Although she is typically understood as simply too tired to give up her seat that day, Rosa Parks had been an active member of the movement against segregation in Montgomery. Parks was secretary of the local NAACP chapter that had already been organizing to boycott the Montgomery buses. She was the one chosen to be the plaintiff in a test case against the bus company because she was soft-spoken, middle-aged, and a model citizen. NAACP leaders believed her to make a credible and sympathetic plaintiff in what they had long hoped would be a significant legal challenge to segregation in public transportation.

When, according to plan, Rosa Parks refused to give up her seat, the movement stood ready to mobilize. News of her arrest spread quickly via networks of friends, kin, church, and school organizations. A small group, the Women's Political Council, had previously discussed plans to announce a boycott of the bus system. The arrest of Rosa Parks presented an opportunity for implementation. One member of the Women's Political Council, Jo Ann Gibson Robinson, called a friend who had access to the mimeograph machine at Alabama State College. In the middle of the night, Robinson and two of her students duplicated 52,500 leaflets calling for the boycott to be distributed among Black neighborhoods (Morris 1999; Robinson 1987).

Celebrities can advance the cause of social movements by bringing visibility to the movement, although sociologists have found that celebrities typically involve themselves in less controversial movements that already have widespread popular support. The entry of celebrities into social movements can unintentionally hurt a movement's development, because celebrities tend to depoliticize it (Meyer and Gamson 1995). Their endorsement and presence, nonetheless, brings visibility to a movement.

Another cause for a movement to begin must be a perceived sense of injustice among the potential participants. The environmental justice movement has emerged largely from grassroots organizations formed in communities where residents have organized to clean up a toxic waste site, close a polluting industry, or protect children from a perceived threat to their health and safety.

As movements develop, they quickly establish an organizational structure. The shape of the movement's organization may range from formal bureaucratic structures to decentralized, interpersonal, and egalitarian arrangements. Many movements combine both. Examples of large bureaucracies are the National Organization for Women, the National Association for the Advancement of Colored People (NAACP), Amnesty International, Greenpeace, the Jewish Defense League, and the National Rifle Association. As social movements become institutionalized, they are most likely to take on bureaucratic form.

© Thomas Frey/EPA/Landov

Social movements such as the disability rights movement can raise public awareness and result in new forms of social behavior.

Susan Welch/AP Photos

Social movements often use highly visible tactics to promote their causes, as in the March for Women's Lives in Washington, DC.

Theories of Social Movements

Sociologists have developed several theories to explain the development of social movements, including resource mobilization theory, political process theory, and new social movements theory.

Resource mobilization theory is an explanation of the development of social movements that focuses on how movements gain momentum by successfully garnering resources, competing with other movements, and mobilizing the resources available to them (Marx and McAdam 1994; McCarthy and Zald 1973). Money, communication technology, special technical or legal knowledge, and people with organizational and leadership skills are all examples of resources that can be put to use in organizing a social movement (Zald and McCarthy 1975). Interpersonal contacts are one of the most important resources a group can mobilize because the contacts provide a continuous supply of new recruits, as well as money, knowledge, skills, and other assistance. Sometimes social movement organizations acquire the resources of other organizations.

Aldon Morris, an African American sociologist, has used resource mobilization theory to explain the development of the civil rights movement. That nationwide movement relied heavily on Black churches and colleges for resources such as money, leadership, meeting

Debunking Society's Myths

Myth: Social movements develop usually as a result of extremists who are single-minded in their interests.

Sociological perspective: Social movements often develop from the everyday concerns of ordinary people who mobilize to address conditions in their lives that they find unacceptable.

space, and administrative support (Morris 1984). In other words, it mobilized existing resources on behalf of its own cause.

Resource mobilization theory, among other things, notes how movements are connected to each other. The gay and lesbian movement has borrowed many strategies used by the anti–violence against women movement in developing its own campaign to halt hate crimes against gay men and lesbian women (Jenness and Broad 1997; Jenness 1995).

Political process theory posits that movements achieve success by exploiting a combination of internal factors, such as the ability of organizations to mobilize resources, and external factors, such as changes occurring in the society (McAdam 1982). Some structural conditions provide opportunities for collective action. A war, pressure from international parties, demographic shifts, or an economic crisis may create the possibility for those who challenge the social order to mobilize a movement (McAdam 1999). Political process theory explicitly acknowledges the effects that larger social and political processes have on the mobilization of social movements. Large-scale changes such as industrialization, urbanization, or ending repression may provide opportunities for the mobilization of social movements not present previously. Charles Tilly argued that as capitalism developed after popular uprisings in seventeenth-century France, markets emerged in cities for grain from the countryside. To meet the needs of the new markets, grain had to be stored and transported. Rural peasants, angry that their food supply was being transported away, saw the opportunity for collective action and attacked and sabotaged storage and transportation facilities (Tilly 1986).

Political process theory also stresses the vulnerability of the political system to social protest. For example, several groups in Watsonville, California, used the Loma Prieta earthquake in 1989 as an opportunity to protest. Organized Latino groups had demanded better low-income housing in that community for approximately a decade (Simile 1995). Among Mexican farm laborers in the valley, there was already a long tradition of labor organizing, including a cannery workers' strike in 1986 (Bardacke 1988; Amott and Matthaei 1996). After the 1989 earthquake, these groups mobilized around the issues of temporary and permanent housing and the way that aid had been distributed to Latinos. In other words, groups already organized for protest on other issues were ready to squeeze an opportunity out of disaster.

Resource mobilization theory and political process theory are social structural as opposed to individual-level explanations. Other explanations of social movements are more cultural in focus (see Table 16.4). People will not organize within social movements unless they develop a shared definition of the situation that gives meaning to their action, in other words, a culture.

Sociologists use the concept of *framing* to explain this process. **Frames** are specific schemes of interpretation that allow people to perceive, identify, and label events within their lives that can become the basis for collective

Chapter Summary

- ### What is demography?

  Demography is the study of population, a field that focuses on three fundamental processes, all of which determine the level of population at a given moment: births, deaths, and migrations. We noted that the United States ranks very low in *life expectancy* and high in *infant mortality,* relative to other Western countries.

- ### How is diversity relevant?

  Diversity is of great significance since both *infant mortality* and *life expectancy* are not equal across all races, social classes, or for men and women. Women have a greater life expectancy than men in virtually all countries and at all social class levels. However, the lower one's social class, the lower one's life expectancy, regardless of gender, and the greater the infant mortality. Minority group individuals, especially African Americans, Hispanics, and American Indians, all have lower life expectancies and higher infant mortality, than Whites.

- ### What about current migration patterns?

  Current migration patterns show that in addition to movement from city to suburb, not only by Whites but by people of color as well, including the "new suburbanite" populations from other countries and cultures, large portions of populations, such as the underclass, remain stuck in central cities. Yet, recent populations of Hispanics have become upwardly occupationally mobile and have become new suburbanites. The study of a *cohort,* the Baby Boomers, showed how a demographic cohort can both use and produce cultural change.

- ### What is the Malthusian problem and why is it important?

  Malthusian theory, still relevant today, warns us about the dangers of exponential population growth along with only arithmetic increases in food and natural resources. To avoid the so-called Malthusian positive checks of famine and war, population control can be, and is being, instituted by programs for family planning and birth control.

- ### What are the current problems pertaining to urbanization, human ecology, and the environment?

  Any society is a *human ecosystem* with interacting and interdependent forces, consisting of human populations, natural resources, and the state of the environment. *Urbanization* has caused such forces to become more prominent. Depletion of one natural resource affects many other things in the ecosystem. The dumping of toxic wastes is a very major problem in our society, especially when toxic dumps are found more frequently, as they are, in or very near African American, Hispanic, and Native American communities. Such practices constitute what some researchers call *environmental racism.* Surveys have shown that environmental risks are of more concern to women than to men. As a consequence, environmental policy has more impact on women than on men.

- ### What is social change and what are its types?

  Social change is that process by which social interaction, the social stratification system, and entire institutions in a society change over time. Social change can take place quickly or it can take longer, sometimes involving *microchanges* (such as a fad that "catches on"), and at other times, *macrochanges* (such as a technological innovation).

- ### How do sociologists explain social change?

  Functionalist theories and *evolutionary theories* predict that societies move or evolve from the structurally simple to the structurally complex. *Unidimensional evolutionary theory* predicts that societies evolve along a path from simpler socially undifferentiated societies to more complex highly differentiated ones. *Multidimensional evolutionary theory* predicts that societies follow not one but several different paths in the process of social change. *Conflict theories* predict that social conflict is an inherent part of any social structure and that conflict between social class strata or racial–ethnic groups can bring about social change. *Cyclical theories,* such as those of Arnold J. Toynbee and Pitirim Sorokin, predict that certain patterns of social structure and culture recur in a society at different times. *Modernization theory* states that global development among societies is a worldwide process not confined to any one society. *Dependency theory* notes that industrialized core nations tend to keep noncore nations economically dependent, and this retards the economic development and upward mobility of the noncore nations.

- ***What are the causes of social change?***

  Sources of social change include revolution, cultural diffusion, social inequalities, war, technological innovation, collective behavior, and social movements.

- ***What are collective behavior and social movements?***

  Collective behavior occurs when the usual conventions to guide behavior are suspended and people establish new norms of behavior. *Social movements* are organized social groups that have some continuity and promote or resist social change.

Social movements start when there is a preexisting communication network, a collective sense of grievance, a precipitating factor initiating the movement, and mobilization of a group of people. *Resource mobilization theory* suggests that social movements develop when people can compete for and gain resources needed for mobilization. *Political process theory* and *framing* suggest that large-scale social changes, such as industrialization or urbanization, provide the conditions that spawn social movements. *New social movement theory* adds that social movements are places where people construct their identities.

Key Terms

age–sex pyramid   406
census   402
cohort   407
collective behavior   426
crude birthrate   403
crude death rate   404
cultural diffusion   423
cyclical theory   420
demography   402
dependency theory   422
emigration   403
environmental racism   414
frames   429
global warming   409

globalization   421
greenhouse effect   413
immigration   403
infant mortality rate   404
life expectancy   404
macrochange   418
Malthusian theory   407
microchange   418
modernization   422
modernization theory   421
new social movement theory   430
personal transformation
   movements   427
political process theory   429

population density   411
reactionary movements   428
reform movements   427
resource mobilization
   theory   429
revolution   423
sex ratio   406
social change   418
social movements   426
social/political change
   movements   427
urbanization   410

Online Resources

Sociology: The Essentials Companion Website

www.cengage.com/sociology/andersen
Visit your book companion website where you will find more resources to help you study and write your research papers. Resources include web links, learning objectives, Internet exercises, quizzing, and flash cards.

 is an easy-to-use online resource that helps you study in less time to get the grade you want NOW.

www.cengage.com/login
Need help studying? This site is your one-stop study shop. Take a Pre-Test and Cengage NOW will generate a Personalized Study Plan based on your test results. The Study Plan will identify the topics you need to review and direct you to online resources to help you master those topics. You can then take a Post-Test to determine the concepts you have mastered and what you still need to work on.

Glossary

absolute poverty: the situation in which individuals live on less than $365 a year, or $1.00 a day

achieved status: a status attained by effort

achievement test: test intended to measure what is actually learned rather than potential

adult socialization: the process of learning new roles and expectations in adult life

affirmative action: a method for opening opportunities to women and minorities that specifically redresses past discrimination by taking positive measures to recruit and hire previously disadvantaged groups

age cohort: an aggregate group of people born during the same time period

age discrimination: different and unequal treatment of people based solely on their age

age prejudice: a negative attitude about an age group that is generalized to all people in that group

age stereotype: preconceived judgments about what different age groups are like

age stratification: the hierarchical ranking of age groups in society

ageism: the institutionalized practice of age prejudice and discrimination

age–sex (or age–gender) pyramid; also population pyramid: a graphic representation of the age and gender structure of a society

alienation: the feeling of powerlessness and separation from one's group or society

altruistic suicide: the type of suicide that can occur when there is excessive regulation of individuals by social forces

anomic suicide: the type of suicide occurring when there are disintegrating forces in the society that make individuals feel lost or alone

anomie: the condition existing when social regulations (norms) in a society break down

anorexia nervosa: a condition characterized by compulsive dieting resulting in self-starvation

anticipatory socialization: the process of learning the expectations associated with a role one expects to enter in the future

anti-Semitism: the belief or behavior that defines Jewish people as inferior and that targets them for stereotyping, mistreatment, and acts of hatred

ascribed status: a status determined at birth

assimilation: process by which a minority becomes socially, economically, and culturally absorbed within the dominant society

attribution error: error made in attributing the causes for someone's behavior to their membership in a particular group, such as a racial group

attribution theory: the principle that dispositional attributions are made about others (what the other is "really like") under certain conditions, such as out-group membership

authoritarian personality: a personality characterized by a tendency to rigidly categorize people and to submit to authority, rigidly conform, and be intolerant of ambiguity

authority: power that is perceived by others as legitimate

automation: the process by which human labor is replaced by machines

autonomous state model: a theoretical model of the state that interprets the state as developing interests of its own, independent of other interests

aversive racism: subtle, nonovert, and nonobvious racism

beliefs: shared ideas held collectively by people within a given culture

bilateral kinship: a kinship system where descent is traced through the father and the mother

biological determinism: explanations that attribute complex social phenomena to physical characteristics

bioterrorism: a form of terrorism involving the dispersion of chemical or biological substances intended to cause widespread disease and death

bureaucracy: a type of formal organization characterized by an authority hierarchy, a clear division of labor, explicit rules, and impersonality

capitalism: an economic system based on the principles of market competition, private property, and the pursuit of profit

caste system: a system of stratification (characterized by low social mobility) in which one's place in the stratification system is determined by birth

census: a count of the entire population of a country

charisma: a quality attributed to individuals believed by their followers to have special powers

charismatic authority: authority derived from the personal appeal of a leader

church: a formal organization that sees itself and is seen by society as a primary and legitimate religious institution

class: see social class

class consciousness: the awareness that a class structure exists and the feeling of shared identification with others in one's class with whom one perceives common life chances

coercive organization: organizations for which membership is involuntary; examples are prisons, mental hospitals

cognitive ability: the capacity for abstract thinking

cognitive elite: the idea that there is a presumably genetically based elite class in the United States containing those with high IQs, high incomes, and prestigious jobs

cohort (birth cohort): see age cohort

collective behavior (action): behavior that occurs when the usual conventions are suspended and people collectively establish new norms of behavior in response to an emerging situation

collective consciousness: the body of beliefs that are common to a community or society and that give people a sense of belonging

colonialism: system by which Western nations became wealthy by taking raw materials from other societies (the colonized) and reaping profits from products finished in the homeland

color-blind racism: ignoring legitimate racial, ethnic, and cultural differences between groups, thus denying the reality of such differences

coming out: the process of defining oneself as gay or lesbian

commodity chain: the network of production and labor processes by which a product becomes a finished commodity. By following the commodity chain, it is evident which countries gain profits and which ones are being exploited

communism: an economic system where the state is the sole owner of the systems of production

concept: any abstract characteristic or attribute that has the potential to be measured

conflict theory: a theoretical perspective that emphasizes the role of power and coercion in producing social order

contact theory: the theory that prejudice will be reduced through social interaction with those of different race or ethnicity but of equal status

content analysis: the analysis of meanings in cultural artifacts like books, songs, and other forms of cultural communication

contingent worker: those who do not hold regular jobs, but whose employment is dependent upon demand

controlled experiment: a method of collecting data that can determine whether something actually causes something else

core countries (core nations): within world systems theory, those nations that are more technologically advanced

correlation: a statistical technique that analyzes patterns of association between pairs of sociological variables

counterculture: subculture created as a reaction against the values of the dominant culture

creationism: the belief that humans have evolved not via a Darwinian natural selection process, but via the action of a god

credentialism: the insistence upon educational credentials only for their own sake

crime: one form of deviance; specifically, behavior that violates criminal laws

criminology: the study of crime from a scientific perspective

cross-tabulation: a table that shows how the categories of two variables are related

crude birthrate: the number of babies born each year for every 1000 members of the population

crude death rate: the number of deaths each year per 1000 members of the population

cult: a religious group devoted to a specific cause or charismatic leader

cultural capital: (also known as *social capital*) cultural resources that are socially designated as being worthy (such as knowledge of elite culture) and that give advantages to groups possessing such capital

cultural diffusion: the transmission of cultural elements from one society or cultural group to another

cultural hegemony: the pervasive and excessive influence of one culture throughout society

cultural relativism: the idea that something can be understood and judged only in relationship to the cultural context in which it appears

culture: the complex system of meaning and behavior that defines the way of life for a given group or society

culture lag: the delay in cultural adjustments to changing social conditions

culture of poverty: the argument that poverty is a way of life and, like other cultures, is passed on from generation to generation

culture shock: the feeling of disorientation that can come when one encounters a new or rapidly changed cultural situation

cyberspace interaction: interaction occurring when two or more persons share a virtual reality experience via electronic communication and interaction with each other

cyberterrorism: the use of the computer to commit one or more terrorist acts

cyclical theory: the idea that societies go through a "life cycle," for example, from idealistic through hedonistic culture and back to idealistic

data: the systematic information that sociologists use to investigate research questions

data analysis: the process by which sociologists organize collected data to discover what patterns and uniformities are revealed

debriefing: a process whereby the researcher explains the true purpose of a research study to the subject (respondent); usually done after completion of the study

debunking: looking behind the facades of everyday life

deductive reasoning: the process of creating a specific research question about a focused point, based on a more general or universal principle

defensive medicine: practiced by physicians who order extra tests on a patient in an effort to fend off a lawsuit by the patient

deindividuation: the feeling that one's self has merged with a group

deindustrialization: the transition from a predominantly goods-producing economy to one based on the provision of services

democracy: system of government based on the principle of representing all people through the right to vote

demography: the scientific study of population

dependency theory: the global theory maintaining that industrialized nations hold less-industrialized nations in a dependent, thus exploitative, relationship that benefits the industrialized nations at the expense of the less-industrialized ones

dependent variable: the variable that is a presumed effect (*see* independent variable)

deviance: behavior that is recognized as violating expected rules and norms

deviant career: continuing to be labeled as deviant even after the initial (primary) deviance may have ceased

deviant community: groups that are organized around particular forms of social deviance

deviant identity: the definition a person has of himself or herself as a deviant

differential association theory: theory that interprets deviance as behavior one learns through interaction with others

discrimination: overt negative and unequal treatment of the members of some social group or stratum solely because of their membership in that group or stratum

disengagement theory: theory predicting that as people age, they gradually withdraw from participation in society and are simultaneously relieved of responsibilities

diversity: the variety of group experiences that result from the social structure of society

division of labor: the systematic interrelation of different tasks that develops in complex societies

doing gender: a theoretical perspective that interprets gender as something accomplished through the ongoing social interactions people have with one another

dominant culture: the culture of the most powerful group in society

dominant group: the group that assigns a racial or ethnic group to subordinate status in society

dual labor market: the division of the labor market into two segments: the primary and secondary labor markets

dyad: a group consisting of two people

economic restructuring: contemporary transformations in the basic structure of work that are permanently altering the workplace, including the changing composition of the workplace, deindustrialization, the use of enhanced technology, and the development of a global economy

economy: the system on which the production, distribution, and consumption of goods and services is based

educational attainment: the total years of formal education

ego: the part of the self representing reason and common sense

egoistic suicide: the type of suicide that occurs when people feel totally detached from society

elite deviance: the wrongdoing of powerful individuals and organizations

emigration (vs. immigration): migration of people from one society to another (also called out-migration)

emotional labor: work that is explicitly intended to produce a desired state of mind in a client

empirical: refers to something that is based on careful and systematic observation

Enlightenment: the period in eighteenth- and nineteenth-century Europe characterized by faith in the ability of human reason to solve society's problems

environmental racism: the dumping of toxic wastes with disproportionate frequency at or very near areas with high concentrations of minorities

epidemiology: the study of all factors—biological, social, economic, and cultural—that are associated with disease and health

Equal Rights Amendment: a constitutional principle, never passed, guaranteeing that equality of rights under the law shall not be denied or abridged on the basis of sex

estate system: a system of stratification in which the ownership of property and the exercise of power is monopolized by an elite or noble class, that has total control over societal resources

ethnic group: a social category of people who share a common culture, such as a common language or dialect, a common religion, or common norms, practices, and customs

ethnocentrism: the belief that one's in-group is superior to all out-groups

ethnomethodology: a technique for studying human interaction by deliberately disrupting social norms and observing how individuals attempt to restore normalcy

eugenics: a social movement in the early twentieth century that sought to apply scientific principles of genetic selection to "improve" the offspring of the human race

evaluation research: research assessing the effect of policies and programs

expressive needs: needs for intimacy, companionship, and emotional support

extended families: the whole network of parents, children, and other relatives who form a family unit and often reside together

extreme poverty: the situation in which people live on less than $275 a year, or 75 cents a day

false consciousness: the thought resulting from subordinate classes internalizing the view of the dominant class

family: a primary group of people—usually related by ancestry, marriage, or adoption—who form a cooperative economic unit to care for any offspring (and each other) and who are committed to maintaining the group over time

Family and Medical Leave Act (FMLA): federal law requiring employers of a certain size to grant leave to employees for purposes of family care

feminism: a way of thinking and acting that advocates a more just society for women

feminist theory: analyses of women and men in society intended to improve women's lives

feminization of poverty: the process whereby a growing proportion of the poor are women and children

feral children: children who come of age with very little, if any, interaction with humans; "wild children"

first-world countries: industrialized nations based on a market economy and with democratically elected governments

folkways: the general standards of behavior adhered to by a group

formal organization: a large secondary group organized to accomplish a complex task or set of tasks

forms of racism: types or versions of racism, such as traditional racism, aversive racism, institutional racism, and others

frames: specific schemes of interpretation that allow people to perceive, identify, and label events within their lives that can become the basis for collective action

functionalism: a theoretical perspective that interprets each part of society in terms of how it contributes to the stability of the whole society

game stage: the stage in childhood when children become capable of taking a multitude of roles at the same time

game theory: a mathematical theory that regards human interaction as a game, thus characterized by strategies, rewards and punishments, and winners and losers

gemeinschaft: German for *community,* a state characterized by a sense of common feeling among the members of a society, including strong personal ties, sturdy primary group memberships, and a sense of personal loyalty to one another; associated with rural life

gender: socially learned expectations and behaviors associated with members of each sex

gender apartheid: the extreme segregation and exclusion of women from public life

gender gap: gender differences in behavior, such as voting behavior

gender identity: one's definition of self as a woman or man

gender segregation: the distribution of men and women in different jobs in the labor force

gender socialization: the process by which men and women learn the expectations associated with their sex

gender stratification: the hierarchical distribution of social and economic resources according to gender

gendered institutions: the total pattern of gender relations that structure social institutions, including the stereotypical

expectations, interpersonal relationships, and the different placement of men and women that are found in institutions

generalization: the ability to make claims that a finding represents something greater than the specific observations on which the finding is based

generalized other: an abstract composite of social roles and social expectations

gesellschaft: a type of society in which increasing importance is placed on the secondary relationships people have—that is, less intimate and more instrumental relationships

glass ceiling: popular concept referring to the limits that women and minorities experience in job mobility

global culture: the diffusion of a single culture throughout the world

global economy: term used to refer to the fact that all dimensions of the economy now cross national borders

global stratification: the systematic inequalities between and among different groups within nations that result from the differences in wealth, power, and prestige of different societies relative to their position in the international economy

global warming: the systematic increase in worldwide surface temperatures

globalization: increased economic, political, and social interconnectedness and interdependence among societies in the world

government: those state institutions that represent the population and make rules that govern the society

greenhouse effect: a rise in the earth's surface temperature caused by heat trapped by excess carbon dioxide in the atmosphere; global warming

gross national income (GNI): the total output of goods and services produced by residents of a country each year plus the income from nonresident sources, divided by the size of the population

group: a collection of individuals who interact and communicate, share goals and norms, and who have a subjective awareness as "we"

group size effect: the effect upon the person of groups of varying sizes

groupthink: the tendency for group members to reach a consensus at all costs

hate crime: assaults and other malicious acts (including crimes against property) motivated by various forms of bias, including that based on race, religion, sexual orientation, ethnic and national origin, or disability

health maintenance organization (HMO): a cooperative of doctors and other medical personnel who provide medical services in exchange for a set membership fee

hermaphroditism: a condition produced when irregularities in chromosome formation or fetal differentiation produces persons with biologically mixed sex characteristics

heterosexism: the institutionalization of heterosexuality as the only socially legitimate sexual orientation

HMO: see Health Maintenance Organization

homophobia: the fear and hatred of homosexuality

human capital theory: a theory that explains differences in wages as the result of differences in the individual characteristics of the workers

human poverty index: a multidimensional measure of poverty meant to indicate the degree of deprivation in four dimensions: life expectancy, education, economic well-being, and social inclusion

hypersegregation: a pattern of extreme racial, ethnic, and/or social class residential segregation, such that nearly all individuals in an area are of one such group

hypothesis: a statement about what one expects to find in research

id: the part of the personality that includes various impulses and drives, including sexual passions and desires, biological urges, and human instincts

identity: how one defines oneself

ideology: a belief system that tries to explain and justify the status quo

imitation stage: the stage in childhood when children copy the behavior of those around them

immigration (vs. emigration): the migration of people into a society from outside it (also called in-migration)

impression management: a process by which people attempt to control how others perceive them

imprinting: a process whereby a newly hatched or newborn member of a species attaches itself to the first object "seen" by it, whether it is the mother or not, and whether it is an animal, human, or a physical object

income: the amount of money brought into a household from various sources during a given year (wages, investment income, dividends, etc.)

independent variable: a variable that is the presumed cause of a particular result (*see* dependent variable)

index crimes: the FBI's tallying of violent crimes of murder, manslaughter, rape, robbery, and aggravated assault, plus property crimes

indicator: something that points to or reflects an abstract concept

inductive reasoning: the process of arriving at general conclusions from specific observations

infant mortality rate: the number of deaths per year of infants under age 1 for every 1000 live births

informed consent: a formal acknowledgment by the research subject (respondent) that she/he understands the purpose of the research and agrees to be studied

institutional racism: racism involving notions of racial or ethnic inferiority that have become ingrained into society's institutions

instrumental needs: emotionally neutral, task-oriented (goal-oriented) needs

interest group: a constituency in society organized to promote its own agenda

interlocking directorate: organizational linkages created when the same people sit on the boards of directors of a number of different corporations

internalization: a process by which a part of culture becomes incorporated into the personality

international division of labor: system of labor whereby products are produced globally, while profits accrue only to a few

intersexed person: a person born with the physical characteristics of both sexes

issues: problems that affect large numbers of people and have their origins in the institutional arrangements and history of a society

job displacement: the permanent loss of certain job types when employment patterns shift, as when a manufacturing plant shuts down

kinship system: the pattern of relationships that define people's family relationships to one another

labeling theory: a theory that interprets the responses of others as most significant in understanding deviant behavior

labor force participation rate: the percentage of those in a given category who are employed

laissez-faire racism: maintaining the status quo of racial groups by persistent stereotyping and blaming of minorities themselves for achievement and socioeconomic gaps between groups

language: a set of symbols and rules that, when put together in a meaningful way, provides a complex communication system

latent functions: subtle, unintended consequences of an institutional element of which the participants (the people) are usually unaware

law: the written set of guidelines that define what is right and wrong in society

liberal feminism: a feminist theoretical perspective asserting that the origin of women's inequality is in traditions of the past that pose barriers to women's advancement

life chances: the opportunities that people have in common by virtue of belonging to a particular class

life course: the connection between people's personal attributes, the roles they occupy, the life events they experience, and the social and historical context of these events

life expectancy: the average number of years individuals and particular groups can expect to live

looking glass self: the idea that people's conception of self arises through reflection about their relationship to others

macroanalysis: analysis of the whole of society, how it is organized and how it changes

macrochange: gradual transformations in a society that occur on a broad scale and affect many aspects of a society

Malthusian theory: after T. R. Malthus, the principle that a population tends to grow faster than the subsistence (food) level needed to sustain it

managed care: use of collective bargaining as part of a large collection of HMOs

mass media: channels of communication that are available to very wide segments of the population

master status: some characteristic of a person that overrides all other features of the person's identity

material culture: the objects created in a given society

matriarchal religion: a religion based on the centrality of a female goddess or goddesses

matriarchy: a society or group in which women have power over men

matrilineal kinship: kinship systems in which family lineage (or ancestry) is traced through the mother

matrilocal: a pattern of family residence in which married couples reside with the family of the wife

McDonaldization: the increasing and ubiquitous presence of the fast-food model in vast numbers of organizations

mean: the sum of a set of values divided by the number of cases from which the values are obtained; an average

mechanical solidarity: unity based on similarity, not difference, of roles

median: the midpoint in a series of values that are arranged in numerical order

median income: the midpoint of all household incomes

Medicaid: a governmental assistance program that provides health care assistance for the poor, including the elderly

medicalization of deviance: explanations of deviant behavior that interpret deviance as the result of individual pathology or sickness

Medicare: a governmental assistance program established in the 1960s to provide health services for older Americans

meritocracy: a system in which one's status is based on merit or accomplishments

microanalysis: analysis of the smallest, most immediately visible parts of social life, such as people interacting

microchange: subtle alterations in the day-to-day interaction between people, such as a fad "catching on"

minority group: any distinct group in society that shares common group characteristics and is forced to occupy low status in society because of prejudice and discrimination

mode: the value (or score) that appears most frequently in a set of data

modernization: a process of social and cultural change that is initiated by industrialization and followed by increased social differentiation and division of labor

modernization theory: a view of globalization in which global development is a worldwide process affecting nearly all societies that have been touched by technological change

monogamy: the marriage practice of a sexually exclusive marriage with one spouse at a time

monotheism: the worship of a single god

mores: strict norms that control moral and ethical behavior

multinational corporation: corporations that conduct business across national borders

multiracial feminism: form of feminist theory noting the exclusion of women of color from other forms of theory and centering its analysis in the experiences of all women

nationalism: the strong identity associated with an extreme sense of allegiance to one's culture or nation

neocolonialism: a form of control of poor countries by rich countries, but without direct political or military involvement

net worth: the value of one's financial assets minus debt

new social movement theory: a theory about social movements linking culture, ideology, and identity conceptually to explain how new identities are forged within social movements

newly industrializing countries (NICs): countries that have shown rapid growth and have emerged as developed countries

nonmaterial culture: the norms, laws, customs, ideas, and beliefs of a group of people

nonverbal communication: communication by means other than speech, as by touch, gestures, use of distance, eye movements, and so on

normative organization: an organization having a voluntary membership and that pursues goals; examples are the PTA or a political party

norms: the specific cultural expectations for how to act in a given situation

nuclear family: family in which a married couple resides together with their children

occupational prestige: the subjective evaluation people give to jobs as better or worse than others

occupational segregation: a pattern in which different groups of workers are separated into different occupations

old-fashioned racism: overt and obvious expressions of racism, such as physical assaults, lynchings, and other such acts against a minority

organic metaphor: refers to the similarity early sociologists saw between society and other organic systems

organic (contractual) solidarity: unity based on role differentiation, not similarity

organized crime: crime committed by organized groups, typically involving the illegal provision of goods and services to others

out-group homogeneity effect: the tendency for an in-group member to perceive members of any out-group as similar or identical to each other

participant observation: a method whereby the sociologist becomes both a participant in the group being studied and a scientific observer of the group

patriarchal religion: religion in which the beliefs and practices of the religion are based on male power and authority

patriarchy: a society or group where men have power over women

patrilineal kinship: a kinship system that traces descent through the father

patrilocal: a pattern of family residence in which married couples reside with the family of the husband

peers: those of similar status

percentage: the number of parts per hundred

peripheral countries (nations): poor countries, largely agricultural, having little power or influence in the world system

personal crimes: violent or nonviolent crimes directed against people

personal transformation movements: social movements that aim to change the individual

personality: the cluster of needs, drives, attitudes, predispositions, feelings, and beliefs that characterize a given person

play stage: the stage in childhood when children begin to take on the roles of significant people in their environment

pluralism: pattern whereby groups maintain their distinctive culture and history

pluralist model: a theoretical model of power in society as coming from the representation of diverse interests of different groups in society

political action committees (PACs): groups of people who organize to support candidates they feel will represent their views

political process theory: explanation of social movements positing that movements achieve success by exploiting a combination of internal organizational factors as well as external changes in society

polygamy: a marriage practice in which either men or women can have multiple marriage partners

polytheism: the worship of more than one deity

popular culture: the beliefs, practices, and objects that are part of everyday traditions

population: a relatively large collection of people (or other unit) that a researcher studies and about which generalizations are made

population density: the number of people per square mile

positivism: a system of thought that regards scientific observation to be the highest form of knowledge

postindustrial society: a society economically dependent upon the production and distribution of services, information, and knowledge

postmodernism: a theoretical perspective based in the idea that communication in all its forms is reality, such that understanding society requires studying all forms of communication, including cultural ideas, language, texts, and self-conceptions

poverty line: the figure established by the government to indicate the amount of money needed to support the basic needs of a household

power: a person or group's ability to exercise influence and control over others

power elite model: a theoretical model of power positing a strong link between government and business

predictive validity: the extent to which a test accurately predicts later college grades, or some other criterion such as likelihood of graduating

preindustrial society: one that directly uses, modifies, and/or tills the land as a major means of survival

prejudice: the negative evaluation of a social group, and individuals within that group, based upon conceptions about that social group that are held despite facts that contradict it

prestige: the value with which different groups or people are judged

primary group: a group characterized by intimate, face-to-face interaction and relatively long-lasting relationships

profane: that which is of the everyday, secular world and is specifically not religious

propaganda: information disseminated by a group or organization (such as the state) intended to justify its own power

property crimes: crimes involving theft of or harm to property without bodily harm to the victim(s)

Protestant ethic: belief that hard work and self-denial lead to salvation

proxemic communication: meaning conveyed by the amount of space between interacting individuals

psychoanalytic theory: a theory of socialization positing that the unconscious mind shapes human behavior

qualitative research: research that is somewhat less structured than quantitative research but that allows more depth of interpretation and nuance in what people say and do

quantitative research: research that uses numerical analysis

queer theory: a theoretical perspective that recognizes the socially constructed nature of sexual identity

race: a social category, or social construction, that we treat as distinct on the basis of certain characteristics, some biological, that have been assigned social importance in the society

racial formation: process by which groups come to be defined as a "race" through social institutions such as the law and the schools

racial profiling: the use of race alone as the criterion for deciding whether to stop and detain someone on suspicion of their having committed a crime

racialization: a process whereby some social category, such as a social class or nationality, is assigned what are perceived to be race characteristics

racism: the perception and treatment of a racial or ethnic group, or member of that group, as intellectually, socially, and culturally inferior to one's own group

radical feminism: feminist theoretical perspective that interprets patriarchy as the primary cause of women's oppression

random sample: a sample that gives everyone in the population an equal chance of being selected

rate: parts per some number (e.g., per 10,000; per 100,000)

rational–legal authority: authority stemming from rules and regulations, typically written down as laws, procedures, or codes of conduct

reactionary movements: social movements organized to resist change or to reinstate an earlier social order that participants perceive to be better

reference group: any group (to which one may or may not belong) used by the individual as a standard for evaluating her or his attitudes, values, and behaviors

reflection hypothesis: the idea that the mass media reflect the values of the general population

reform movements: social movements that seek change through legal or other mainstream political means, by working within existing institutions

relative poverty: a definition of poverty that is set in comparison to a set standard

reliability: the likelihood that a particular measure would produce the same results if the measure were repeated

religion: an institutionalized system of symbols, beliefs, values, and practices by which a group of people interprets and responds to what they feel is sacred and that provides answers to questions of ultimate meaning

religiosity: the intensity and consistency of practice of a person's (or group's) faith

religious extremism: actions and beliefs that are driven by high levels of religious intolerance

replication study: research that is repeated exactly, but on a different group of people at a different point in time

research design: the overall logic and strategy underlying a research project

residential segregation: the spatial separation of racial and ethnic groups in different residential areas

resocialization: the process by which existing social roles are radically altered or replaced

resource mobilization theory: theory of how social movements develop that focuses on how movements gain momentum by successfully garnering organizational resources

revolution: the overthrow of a state or the total transformation of central state institutions

risky shift (also polarization shift): the tendency for group members, after discussion and interaction, to engage in riskier behavior than they would while alone

rite of passage: ceremony or ritual that symbolizes the passage of an individual from one role to another

ritual: symbolic activities that express a group's spiritual convictions

role: behavior others expect from a person associated with a particular status

role conflict: two or more roles associated with contradictory expectations

role modeling: imitation of the behavior of an admired other

role set: all roles occupied by a person at a given time

role strain: conflicting expectations within the same role

sacred: that which is set apart from ordinary activity, seen as holy, and protected by special rites and rituals

salience principle: categorizing people on the basis of what initially appears prominent about them

sample: any subset of units from a population that a researcher studies

Sapir–Whorf hypothesis: a theory that language determines other aspects of culture because language provides the categories through which social reality is defined and perceived

scapegoat theory: argument that dominant group aggression is directed toward a minority as a substitute for frustration with some other problem

schooling: socialization that involves formal and institutionalized aspects of education

scientific method: the steps in a research process, including observation, hypothesis testing, analysis of data, and generalization

secondary group: a group that is relatively large in number and not as intimate or long in duration as a primary group

second-world countries: socialist countries with state-managed economies and typically without a democratically elected government

sect: groups that have broken off from an established church

secular: the ordinary beliefs of daily life that are specifically not religious

segregation: the spatial and social separation of racial and ethnic groups

self: our concept of who we are, as formed in relationship to others

self-concept: a person's image and evaluation of important aspects of oneself

self-fulfilling prophecy: the process by which merely applying a label changes behavior and thus tends to justify the label

semiperipheral countries: semi-industrialized countries that represent a kind of middle class within the world system

serendipity: unanticipated, yet informative, results of a research study

sex: used to refer to biological identity as male or female

sex ratio (gender ratio): the number of males per 100 females

sex trafficking: refers to the practice whereby women, usually very young women, are forced by fraud or coercion into commercial sex acts

sexual harassment: unwanted physical or verbal sexual behavior that occurs in the context of a relationship of unequal power and that is experienced as a threat to the victim's job or educational activities

sexual identity: the definition of oneself that is formed around one's sexual relationships

sexual orientation: the attraction that people feel for people of the same or different sex

sexual politics: the link feminists argue exists between sexuality and power and between sexuality and race, class, and gender oppression

sexual revolution: the widespread changes in men's and women's roles and a greater public acceptance of sexuality as a normal part of social development

sexual scripts: the ideas taught to us about what is appropriate sexual behavior for a person of our gender

sexual tourism: practice whereby people travel to particular parts of the world specifically to engage in commercial sexual activities

significant others: those with whom we have a close affiliation

social capital: see cultural capital

social change: the alteration of social interaction, social institutions, stratification systems, and elements of culture over time

social change movements: movements that aim to change some aspect of society

social class: the social structural hierarchical position groups hold relative to the economic, social, political, and cultural resources of society

social construction perspective: a theoretical perspective that explains sexual identity (or any other identity) as created and learned within a cultural, social, and historical context

social control: the process by which groups and individuals within those groups are brought into conformity with dominant social expectations

social control agents: those who regulate and administer the response to deviance, such as the police or mental health workers

social control theory: theory that explains deviance as the result of the weakening of social bonds

social Darwinism: the idea that society evolves to allow the survival of the fittest

social differentiation: the process by which different statuses in any group, organization, or society develop

social epidemiology: the study of the effects of social and cultural factors upon disease and health

social facts: social patterns that are external to individuals

social institution: an established and organized system of social behavior with a recognized purpose

social interaction: behavior between two or more people that is given meaning

social learning theory: a theory of socialization positing that the formation of identity is a learned response to social stimuli

social mobility: a person's movement over time from one class to another

social movement: a group that acts with some continuity and organization to promote or resist social change in society

social network: a set of links between individuals or other social units such as groups or organizations

social organization: the order established in social groups

social/political change movement: a type of social movement that intends to change some status quo aspect of society, such as the civil rights movement or the environmental movement

social sanctions: mechanisms of social control that enforce norms

social stratification: a relatively fixed hierarchical arrangement in society by which groups have different access to resources, power, and perceived social worth; a system of structured social inequality

social structure: the patterns of social relationships and social institutions that make up society

socialism: an economic institution characterized by state ownership and management of the basic industries

socialist feminism: a feminist theoretical perspective that interprets the origins of women's oppression as lying in the system of capitalism

socialization: the process through which people learn the expectations of society

socialization agents: those who pass on social expectations

society: a system of social interaction that includes both culture and social organization

socioeconomic status (SES): a measure of class standing, typically indicated by income, occupational prestige, and educational attainment

sociological imagination: the ability to see the societal patterns that influence individual and group life

sociology: the study of human behavior in society

spatial mismatch: the pattern whereby jobs are located in geographic area located away from groups who need work

spurious correlation: a false correlation between X and Y, produced by their relationship to some third variable (Z) rather than by a true causal relationship to each other

standardized ability test: tests given to large populations and scored with respect to population averages

state: the organized system of power and authority in society

status: an established position in a social structure that carries with it a degree of prestige

status attainment: the process by which people end up in a given position in the stratification system

status inconsistency: exists when the different statuses occupied by the individual bring with them significantly different amounts of prestige

status set: the complete set of statuses occupied by a person at a given time

stereotype: an oversimplified set of beliefs about the members of a social group or social stratum that is used to categorize individuals of that group

stereotype interchangeability: the principle that negative stereotypes are often interchangeable from one racial group (or gender or social class) to another

stereotype threat effect: the effect of a negative stereotype about one's self upon one's own test performance

stigma: an attribute that is socially devalued and discredited

Stockholm Syndrome: a process whereby a captured person identifies with the captor as a result of becoming inadvertently dependent upon the captor

structural strain theory: a theory that interprets deviance as originating in the tensions that exist in society between cultural goals and the means people have to achieve those goals

subculture: the culture of groups whose values and norms of behavior are somewhat different from those of the dominant culture

superego: the dimension of the self representing the cultural standards of society

symbolic interaction theory: a theoretical perspective claiming that people act toward things because of the meaning things have for them

symbols: things or behavior to which people give meaning

taboos: those behaviors that bring the most serious sanctions

tactile communication: patterns of touch, influenced by gender, that express emotional support, assert power, or express sexual interest

taking the role of the other: the process of imagining oneself from the point of view of another

teacher expectancy effect: the effect of the teacher's expectations on the student's actual performance, independent of the student's ability

Temporary Assistance to Needy Families: federal program by which grants are given to states to fund welfare

tenure: a guarantee of continuing employment in an organization

terrorism: the unlawful use of force or violence against persons or property to intimidate or coerce a government or population in furtherance of political or social objectives

third-world countries: countries that are poor, underdeveloped, largely rural, and with high levels of poverty; typically governments in such countries are autocratic dictatorships and wealth is concentrated in the hands of a small elite

Title IX: legislation that prohibits schools that receive federal funds from discriminating based on gender

total institution: an organization cut off from the rest of society in which individuals are subject to strict social control

totem: an object or living thing that a religious group regards with special awe and reverence

tracking: grouping, or stratifying, students in school on the basis of ability test scores

traditional authority: authority stemming from long-established patterns that give certain people or groups legitimate power in society

transgendered: those who deviate from the binary (that is, male or female) system of gender

transnational family: families where one parent (or both) lives and works in one country while the children remain in their country of origin

triad: a group consisting of three people

triadic segregation: the tendency for a triad to separate into a dyad and an isolate

troubles: privately felt problems that come from events or feelings in one individual's life

underemployment: the condition of being employed at a skill level below what would be expected given a person's training, experience, or education

unemployment rate: the percentage of those not working, but officially defined as looking for work

urban underclass: a grouping of people, largely minority and poor, who live at the absolute bottom of the socioeconomic ladder in urban areas

urbanization: the process by which a community acquires the characteristics of city life

utilitarian organization: a profit or nonprofit organization that pays its employees salaries or wages; examples are Ford Motor Co. (for profit); Microsoft Corp. (for profit); a college or university (nonprofit)

validity: the degree to which an indicator accurately measures or reflects a concept

values: the abstract standards in a society or group that define ideal principles

variable: something that can have more than one value or score

verstehen: the process of understanding social behavior from the point of view of those engaged in it

victimless crimes: violations of law not listed in the FBI's serious crime index, such as gambling or prostitution

wealth: the monetary value of everything one actually owns

work: productive human activity that produces something of value, either goods or services

world cities: cities that are closely linked through the system of international commerce

world systems theory: theory that capitalism is a single world economy and that there is a worldwide system of unequal political and economic relationships that benefit the technologically advanced countries at the expense of the less technologically advanced

xenophobia: the fear and hatred of foreigners

References

Aberle, David F., Albert K. Cohen, A. Kingsley Davis, Marion J. Levy Jr., and Francis X. Sutton. 1950. "The Functional Prerequisites of a Society." *Ethics* 60 (January): 100–111.

Abrahamson, Mark. 2006. *Urban Enclaves: Identity and Place in the World,* 2nd ed. New York: Worth.

Acker, Joan. 1992. "Gendered Institutions: From Sex Roles to Gendered Institutions." *Contemporary Sociology* 21 (September): 565–569.

Acker, Joan, Sandra Morgen, and Lisa Gonzales. 2002. "Welfare Restructuring, Work & Poverty: Policy from Oregon." Working Paper, Eugene, OR: Center for the Study of Women in Society.

ACORN Fair Housing. 2007. "Foreclosure Exposure: A Study of Racial and Income Disparities in Home Mortgage Lending in 172 American Cities." New Orleans, LA: ACORN Fair Housing. **www.acorn.org**

Acs, Gregory, and Seth Zimmerman. 2008. *U.S. Intergenerational Mobility from 1984 to 2004.* Pew Charitable Trust Economic Mobility Project. **www.pewtrusts.org**

Adorno, T. W., Else Frenkel-Brunswik, D. J. Levinson, and R. N. Sanford. 1950. *The Authoritarian Personality.* New York: Harper and Row.

Alba, Richard. 1990. *Ethnic Identity: The Transformation of Ethnicity in the Lives of Americans of European Ancestry.* New Haven, CT: Yale University Press.

Alba, Richard, and Gwen Moore. 1982. "Ethnicity in the American Elite." *American Sociological Review* 47 (June): 373–383.

Alba, Richard D., and Victor Nee. 2003. *Remaking the American Mainstream: Assimilation and Contemporary Immigration.* Cambridge, MA: Harvard University Press.

Albas, Daniel, and Cheryl Albas. 1988. "Aces and Bombers: The Post-Exam Impression Management Strategies of Students." *Symbolic Interaction* 11: 289–302.

Albelda, Randy, and Chris Tilly. 1996. "It's a Family Affair: Women, Poverty, and Welfare." Pp. 79–86 in *For Crying Out Loud: Women's Poverty in the United States,* edited by Diane Dujon and Ann Withorn. Boston, MA: South End Press.

Albelda, Randy, and Ann Withorn (eds.). 2002. *Lost Ground: Welfare Reform, Poverty, and Beyond.* Boston, MA: South End Press.

Aldag, Ramon J,. and Sally R. Fuller. 1993. "Beyond Fiasco: A Reappraisal of the Groupthink Phenomenon and a New Model of Group Decision Processes." *Psychological Bulletin* 113: 533–552.

Alden, Helena L. 2001. "Gender Role Ideology and Homophobia." Paper presented at the Annual Meeting of the Southern Sociological Society.

Aldous, Joan, and Rodney F. Ganey. 1999. "Family Life and the Pursuit of Happiness: The Influence of Gender and Race." *Journal of Family Issues* 20(2): 155–180.

Aldrich, Howard, and Martin Ruef. 2006. *Organizations Evolving.* Thousand Oaks, CA: Sage.

Alicea, Marixsa. 1997. "'What Is Indian About You?' A Gendered, Transnational Approach to Ethnicity." *Gender & Society* 11 (October): 597–626.

Allen, Katherine R. 1997. "Lesbian and Gay Families." Pp. 196–218 in *Contemporary Parenting: Challenges and Issues,* edited by Terry Arendell. Thousand Oaks, CA: Sage.

Allen, Michael Patrick, and Philip Broyles. 1991. "Campaign Finance Reforms and the Presidential Campaign Contribu-

tions of Wealthy Capitalist Families." *Social Science Quarterly* 72 (December): 738–750.

Allison, Anne. 1994. *Nightwork: Sexuality, Pleasure, and Corporate Masculinity in a Tokyo Hostess Club.* Chicago, IL: University of Chicago Press.

Allport, Gordon W. 1954. *The Nature of Prejudice.* Reading, MA: Addison-Wesley.

Alon, Sigal, and Marta Tienda. 2007. "Diversity, Opportunity, and the Shifting Meritocracy in Higher Education." *American Sociological Review* 72 (August): 487–511.

Altemeyer, Bob. 1988. *Enemies of Freedom: Understanding Right-Wing Authoritarianism.* San Francisco, CA: Jossey-Bass.

Altman, Dennis. 2001. *Global Sex.* Chicago, IL: University of Chicago Press.

Amato, Paul R., and Alan Booth. 1996. "A Prospective Study of Divorce and Parent-Child Relationships." *Journal of Marriage and the Family* 58(2): 356–365.

American Association of University Women. 1992. *How Schools Shortchange Girls.* Washington, DC: American Association of University Women.

American Association of University Women. 1998. *Gender Gaps: Where Schools Still Fail Our Children.* Washington, DC: American Association of University Women.

American Psychological Association. 2007. *Report of the APA Task Force on the Sexualization of Girls.* Washington, DC: American Psychological Association.

American Sociological Association. 2008. "Communal Grief and Solidarity After 4/16/07 at Virginia Tech." *Footnotes* (April): 3–4.

Amott, Teresa L., and Julie A. Matthaei. 1996. *Race, Gender, and Work: A Multicultural History of Women in the United*

States, 2nd ed. Boston, MA: South End Press.

Andersen, Margaret L. 2003. *Thinking About Women: Sociological Perspectives on Sex and Gender,* 6th ed. Boston, MA: Allyn and Bacon.

Andersen, Margaret L. 2009. *Thinking About Women: Sociological Perspectives on Sex and Gender*, 8th ed. Boston, MA: Allyn and Bacon.

Andersen, Margaret L., and Patricia Hill Collins. 2010. *Race, Class and Gender: An Anthology,* 7th ed. Belmont, CA: Wadsworth/Cengage.

Anderson, David A., and Mykol Hamilton. 2005. "Gender Role Stereotyping of Parents in Children's Picture Books: The Invisible Father." *Sex Roles: A Journal of Research* 52(3–4): 145–151.

Anderson, Elijah. 1976. *A Place on the Corner.* Chicago, IL: University of Chicago Press.

Anderson, Elijah. 1990. *Streetwise: Race, Class, and Change in an Urban Community.* Chicago, IL: University of Chicago Press.

Anderson, Elijah. 1999. *Code of the Street: Decency, Violence, and the Moral Life of the Inner City.* New York: W. W. Norton.

Anderson, Kristin J., and Campbell Leaper. 1998. "Meta-Analyses of Gender Effects of Conversational Interruption: Who, What, When, Where, and How." *Sex Roles* 39 (August): 225–252.

Andreasse, Arthur. 1997. "Evaluating the 1995 Industry Employment Projections." *Monthly Labor Review* 120 (September): 9–31.

Angotti, Joseph. 1997. "Content Analysis of Local News Programs in Eight U.S. Television Markets." Miami, FL: University of Miami Center for the Advancement of Modern Media.

Anthony, Dick, Thomas Robbins, and Steven Barrie-Anthony. 2002. "Cult and Anticult Totalism: Reciprocal Escalation and Violence." *Terrorism and Political Violence* 14 (Spring): 211–239.

Antrobus, Peggy. 2002. "Feminism as Transformational Politics: Towards Possibilities for Another World." *Development* 45 (June): 46–52.

Apple, Michael W. 1991. "The New Technology: Is It Part of the Solution or Part of the Problem in Education?" *Computers in the Schools* 8 (April-October): 59–81.

Arendell, Terry. 1998. "Divorce American Style." *Contemporary Sociology* 27 (May): 226–228.

Arendt, Hannah. 1963. *Eichmann in Jerusalem: A Report on the Banality of Evil.* New York: Viking Press.

Argyris, Chris. 1990. *Overcoming Organizational Defenses: Facilitating Organizational Learning.* Boston, MA: Allyn and Bacon.

Armas, Genaro C. 2003. "Asian American Divorce Rate Up." *San Francisco Chronicle:* A4.

Armstrong, E. A., L. Hamilton, and B. Sweeney. 2006. "Sexual Assault on Campus: A Multilevel, Integrative Approach to Party Rape." *Social Problems* 53: 483–499.

Armstrong, Elizabeth M. 2003. *Conceiving Risk, Bearing Responsibility: Fetal Alcohol Syndrome and the Diagnosis of Moral Disorder.* Baltimore, MD: Johns Hopkins University Press.

Aronson, Joshua, Carrie B. Fried, and Catherine Good. 2002. "Reducing the Effects of Stereotype Threat on African American College Students by Shaping Theories of Intelligence." *Journal of Experimental Social Psychology* 38: 113–125.

Asch, Solomon. 1951. "Effects of Group Pressure upon the Modification and Distortion of Judgments." In *Groups, Leadership, and Men,* edited by H. Guetzkow. Pittsburgh, PA: Carnegie Press.

Asch, Solomon. 1955. "Opinions and Social Pressure." *Scientific American* 19 (July): 31–35.

Asch, Solomon. 1956. "Opinion and Social Pressure." *Scientific American* 19 (July): 31–36.

Atchley, Robert C. 2000. *Social Forces and Aging,* 9th ed. Belmont, CA: Wadsworth.

Avakame, Edem F., James J. Fyfe, and Candace McCoy. 1999. "'Did You Call the Police? What Did They Do?' An Empirical Assessment of Black's Theory of Mobilization of Law." *Justice Quarterly* 16 (December): 765–792.

Babbie, Earl. 2007. *The Practice of Social Research.* Belmont, CA: Thompson/Wadsworth.

Baca Zinn, Maxine. 1995. "Chicano Men and Masculinity." Pp. 33–41 in *Men's Lives,* 3rd ed., edited by Michael S. Kimmel and Michael A. Messner. Boston, MA: Allyn and Bacon.

Baca Zinn, Maxine, and Bonnie Thornton Dill. 1996. "Theorizing Difference from Multiracial Feminism." *Feminist Studies* 22 (Summer): 321–331.

Baca Zinn, Maxine, and D. Stanley Eitzen. 2008. *Diversity in Families,* 8th ed. Boston, MA: Allyn and Bacon.

Baca Zinn, Maxine, Pierrette Hondagneu-Sotelo, and Michael Messner, eds. 2005. *Through the Prism of Difference.* New York: Oxford University Press.

Bachman, Daniel, and Ronald K. Esplin. 1992. "Plural Marriage." *Encyclopedia of Mormonism,* Vol. 3. New York: Macmillan.

Bailey, Garrick. 2003. *Humanity: An Introduction to Cultural Anthropology,* 6th ed. Belmont, CA: Wadsworth.

Bandura, Albert, and R. H. Walters. 1963. *Social Learning and Personality Development.* New York: Holt, Reinhart, and Winston.

Banks, W. Curtis. 1976. "White Preference in Blacks: A Paradigm in Search of a Phenomenon." *Psychological Bulletin* 83: 1179–1186.

Barber, Benjamin R. 1995. *Jihad vs. McWorld: How Globalism and Tribalism Are Reshaping the World.* New York: Random House.

Barber, Jennifer S., and William G. Axinn. 1998. "Gender Role Attitudes and Marriage among Young Women." *Sociological Quarterly* 39 (Winter): 11–31.

Bardacke, Frank. 1988. "Watsonville: A Mexican Community on Strike." Pp. 149–182 in *Reshaping the US Left: Popular Struggles in the 1980s,* edited by Mike Davis and Michael Sprinker. London: Verso.

Barlow, Maude and Tony Clarke. 2002. "Who Owns Water?" *The Nation* (Sept. 2/9): 11–14.

Barton, Bernadette. 2006. *Stripped: Inside the Lives of Exotic Dancers.* New York: New York University Press.

Basow, Susan A. 1992. *Gender: Stereotypes and Roles,* 3rd ed. Pacific Grove, CA: Brooks/Cole.

Basow, Susan A., and Kelly Johnson. 2000. "Predictors of Homophobia in Female College Students." *Sex Roles* 42 (March): 391–404.

Baumeister, Roy F. 1998. "The Self." Pp. 680-740 in *The Handbook of*

Social Psychology, vol. 1, edited by D. T. Gilbert, S. F. Fiske, and G. Lindzey. New York: McGraw Hill.

Baumeister, Roy F., and Brad J. Bushman. 2008. *Social Psychology and Human Nature.* Belmont, CA: Thompson/Wadsworth.

Bean, Frank D., and Marta Tienda. 1987. *The Hispanic Population of the United States.* New York: Russell Sage Foundation.

Becker, Elizabeth. 2000. "Harassment in the Military Is Said to Rise." *The New York Times* (March 10): 14.

Becker, Howard S. 1963. *Outsiders: Studies in the Sociology of Deviance.* New York: Free Press.

Becker, Penny Edgell, and Heather Hofmeister. 2001. "Work, Family, and Religious Involvement for Men and Women." *Journal for the Scientific Study of Religion* 40 (4): 707–722.

Belknap, Joanne. 2001. *The Invisible Woman.* Belmont, CA: Wadsworth.

Belknap, Joanne, Bonnie S. Fisher, and Francis T. Cullen. 1999. "The Development of a Comprehensive Measure of the Sexual Victimization of College Women." *Violence Against Women* 5 (February): 185–214.

Bellah, Robert (ed.). 1973. *Emile Durkheim on Morality and Society: Selected Writings.* Chicago, IL: University of Chicago Press.

Beller, Emily, and Michael Hout. 2006. "Intergenerational Social Mobility: The United States in Comparative Perspective." *The Future of Children* 16 (Fall): 19–36.

Benedict, Ruth. 1934. *Patterns of Culture.* Boston, MA: Houghton Mifflin.

Ben-Yehuda, Nachman. 1986. "The European Witchcraze of the Fourteenth-Seventeenth Centuries: A Sociologist's Perspective." *American Journal of Sociology* 86: 1–31.

Berger, Joseph, and Morris Zelditch, Jr. (eds.). 1985. *Status, Rewards, and Influence.* San Francisco, CA: Jossey-Bass.

Berger, Peter L. 1963. *Invitation to Sociology: A Humanistic Perspective.* Garden City, NY: Doubleday Anchor.

Berger, Peter L., and Thomas Luckmann. 1967. *The Social Construction of Reality: A Treatise in the Sociology of Knowledge.* Garden City, NY: Anchor Books.

Berger, Peter L., Brigitte Berger, and Hansfried Kellner. 1974. *The Homeless Mind: Modernization and Consciousness.* New York: Vintage Books.

Berkman, Lisa F., Thomas Glass, Ian Brissette, and Teresa E. Seeman. 2000. "From Social Integration to Health: Durkheim in the New Millennium." *Social Science and Medicine* 51 (September): 843–857.

Bernstein, Nina. 2002. "Side Effect of Welfare Law: The No-Parent Family." *The New York Times,* July 29: A1.

Berscheid, Ellen, and Harry R. Reis. 1998. "Attraction and Class Relationships." Pp. 193–281 in *The Handbook of Social Psychology,* 4th ed., edited by Daniel T. Gilbert, Susan T. Fiske, and Gardner Lindzey. New York: Oxford University Press/McGraw-Hill.

Best, Joel. 1999. *Random Violence: How We Talk About New Crimes and New Victims.* Berkeley, CA: University of California Press.

Best, Joel. 2001. *Damned Lies and Statistics: Untangling Numbers from the Media, Politicians, and Activists.* Berkeley, CA: University of California Press.

Best, Joel. 2007. *Social Problems.* New York: W. W. Norton.

Bielby, William T. 1981." Models of Status Attainment." Pp. 3–26 in *Research in Social Stratification and Mobility,* edited by Donald Treiman and Robert Robinson. Greenwich, CT: JAI Press.

Binns, Allison. 2003. *White Gold, Weed, and Blow: The Drug Trades of Afghanistan, Columbia, and Mexico in Comparative Historical Perspective.* Senior Thesis, Princeton University, Princeton, NJ.

Blair-Loy, Mary, and Amy S. Wharton. 2002. "Employees' Use of Work-Family Policies and the Workplace Social Context." *Social Forces* 80 (3): 813–845.

Blake, C. Fred. 1994. "Footbinding in Neo-Confucian China and the Appropriation of Female Labor." *Signs* 19 (Spring): 676–712.

Blalock, Hubert M., Jr. 1991. *Understanding Social Inequality.* Newbury Park, CA: Sage.

Blankenship, Kim. 1993. "Bringing Gender and Race In: U.S. Employment Discrimination Policy." *Gender & Society* 7 (June): 204–226.

Blaskovitch, J. 1973. "Blackjack and the Risky Shift." *Sociometry* 36 (March): 42ff.

Blassingame, John. 1973. *The Slave Community: Plantation Life in the Antebellum South.* New York: Oxford University Press.

Blau, Francine, and Lawrence Kahn. 2004. "The U.S. Gender Pay Gap in the 1990s: Slowing Convergence." *National Bureau of Economics Research Working Papers,* #10853. **www.nber.org**

Blau, Peter M. 1986. *Exchange and Power in Social Life,* revised ed. New Brunswick, NJ: Transaction.

Blau, Peter M., and Otis Dudley Duncan. 1967. *The American Occupational Structure.* New York: Wiley.

Blau, Peter M., and W. Richard Scott. 1974. *On the Nature of Organizations.* New York: Wiley.

Blee, Kathleen M., and Ann R. Tickamyer. 1995. "Racial Differences in Men's Attitudes About Women's Gender Roles." *Journal of Marriage and the Family* 57 (February): 21–30.

Blinde, Elaine M., and Diane E. Taub. 1992a. "Homophobia and Women's Sport: The Disempowerment of Athletes." *Sociological Focus* 25 (May): 151–166.

Blinde, Elaine M., and Diane E. Taub. 1992b. "Women Athletes as Falsely Accused Deviants: Managing the Lesbian Stigma." *Sociological Quarterly* 33 (Winter): 521–533.

Blinde, Elaine M., Diane E. Taub, and Lingling Han. 1993. "Sport Participation and Women's Personal Empowerment: Experiences of the College Athlete." *Journal of Sport and Social Issues* 17 (April): 47–60.

Blinde, Elaine M., Diane E. Taub, and Lingling Han. 1994. "Sport as a Site for Women's Group and Social Empowerment: Perspectives from the College Athlete." *Sociology of Sport Journal* 11 (March): 51–59.

Blumer, Herbert, 1969. *Studies in Symbolic Interaction.* Englewood Cliffs, NJ: Prentice Hall.

Bobo, Lawrence D. 1999. "Prejudice as Group Position: Microfoundations of a Sociological Approach to

Racism and Race Relations." *Journal of Social Issues* 55 (3): 445–492.

Bobo, Lawrence D. 2006. "The Color Line, the Dilemma, and the Dream: Race Relations in America at the Close of the Twentieth Century." Pp. 87–95 in *Race and Ethnicity in Society: The Changing Landscape*, edited by Elizabeth Higginbotham and Margaret L. Andersen. Belmont, CA: Thompson/Wadsworth.

Bobo, Lawrence D., and Michael Dawson. 2006." Katrina: Unmasking Race, Poverty, and Politics in the 21st Century." *DuBois Review* 3 (Spring): 1–6.

Bobo, Lawrence, and James R. Kluegel. 1991. "Modern American Prejudice: Stereotypes, Social Distance, and Perceptions of Discrimination Toward Blacks, Hispanics, and Asians." Paper presented before the American Sociological Association, Cincinnati, OH.

Bobo, L., and R. A. Smith. 1998. "From Jim Crow Racism to Laissez-Faire Racism: An Essay on the Transformation or Racial Attitudes in America." Pp. 182–220 in *Beyond Pluralism: Essays on the Conception of Groups and Group Identities in America*, edited by W. Katkin, N. Landsmand, and A. Tyree. Urbana, IL: University of Illinois Press.

Bobo, Lawrence D., and Camille Zubrinsky Charles. 2009. "Race in the American Mind: From the Moynihan Report to the Obama Candidacy." *Annals of the American Academy of Political and Social Science* 621 (1): 243–259.

Boer, J. Tom, Manuel Pastor Jr., James L. Sadd, and Lori D. Snyder. 1997. "Is There Environmental Racism? The Demographics of Hazardous Waste in Los Angeles County." *Social Science Quarterly* 78 (December): 793–810.

Boeringer, Scott B. 1999. "Associations of Rape Supportive Attitudes with Fraternal and Athletic Participation." *Violence against Women* 5 (January): 81–90.

Boggs, Vernon W. 1992. *Salsiology: Afro-Cuban Music and the Evolution of Salsa in New York City*. Westport, CT: Greenwood Press.

Bogle, Kathleen. 2008. *Hooking Up: Sex, Dating, and Relationships on Campus*. New York: New York University Press.

Bonacich, Edna, and Richard P. Applebaum. 2000. *Behind the Label: Inequality in the Los Angeles Apparel Industry*. Berkeley, CA: University of California Press.

Bonilla-Silva, Eduardo. 1997. "Rethinking Racism: Toward a Structural Interpretation." *American Sociological Review* 62 (3): 465–480.

Bonilla-Silva, Eduardo. 2001. *White Supremacy and Racism in the Post-Civil Rights Era*. Boulder, CO: Lynne Rienner.

Bonilla-Silva, Eduardo, and Gianpaolo Baiocchi. 2001. "Anything But Racism: How Sociologists Limit the Significance of Racism." *Race and Society* 4 (No. 2): 117–131.

Bonilla-Silva, Eduardo, and Mary Hovespan. 2000. "If Two People Are in Love: Deconstructing Whites' Views on Interracial Marriage with Blacks." Paper presented at the annual meeting of the Southern Sociological Society.

Bontemps, Arna (ed.). 1972. *The Harlem Renaissance Remembered*. New York: Dodd, Mead.

Boonstra, Heather, Rachel Benson Gold, Cory L. Richards, and Lawrence B. Finer. 2006. *Abortion in Women's Lives*. New York: Guttmacher Institute.

Bostic, Raphael W., and Kwan Ok Lee. 2008. "Mortgages, Risk, and Home Ownership Among Low-Income and Moderate Income Americans." *American Economic Review* 98 (May): 310–314.

Bouchard, Thomas J., Jr., and Matt McGue. 2003. "Genetic and Environmental Influences on Human Psychological Differences." *Journal of Neurobiology* 54(1): 4–45.

Bouchard, Thomas J., Jr., David T. Lykken, Matthew McGue, Nancy Segal, and Auke Tellegen. 1990. "Sources of Human Psychological Differences: The Minnesota Study of Twins Reared Apart." *Science* 250 (October 12): 223–228.

Bourdieu, Pierre. 1984. *Distinction: A Social Critique of the Judgement of Taste,* translated by Richard Nice. Cambridge, MA: Harvard University Press.

Bowditch, Christine. 1993. "Getting Rid of Troublemakers: High School Disciplinary Procedures and the Production of Dropouts." *Social Problems* 40 (November): 493–510.

Bowen, William G., and Derek Bok. 1998. *The Shape of the River: Long-Term Consequences of Considering Race in College and University Admissions*. Princeton, NJ: Princeton University Press.

Bowles, Samuel, and Herbert Gintis. 1976. *Schooling in Capitalist America*. New York: Basic Books.

Bowles, Samuel, Herbert Gintis, and Melissa Osborn Groves (eds.) 2005. "Introduction." Pp. 1–22 in *Unequal Chances: Family Background and Economic Success*. Princeton, NJ: Princeton University Press.

Branch, Taylor. 1988. *Parting the Waters: America in the King Years, 1954–1963*. New York: Simon & Schuster.

Branch, Taylor. 1998. *Pillar of Fire: America in the King Years, 1963–1965*. New York: Simon and Shuster.

Branch, Taylor. 2006. *At Canaan's Edge: America in the King Years, 1965–1968*. New York: Simon and Shuster.

Brechin, Steven R. 2003. "Comparative Public Opinion and Global Climatic Change and the Kyoto Protocol: The U.S. Versus the World?" *International Journal of Sociology and Social Policy* 23: 106–134.

Brehm, S. S., R. S. Miller, D. Perlman, and S. M. Campbell. 2002. *Intimate Relationships,* 3rd ed. Boston, MA: McGraw-Hill.

Brewer, Rose. 1988. "Black Women in Poverty: Some Comments on Female-Headed Families." *Signs* 13 (Winter): 331–333.

Bridges, George, and Robert Crutchfield. 1988. "Law, Social Standing and Racial Disparities in Imprisonment." *Social Forces* 66 (June): 699–724.

Brignall, Thomas Wells III, and Thomas Van Valey. 2005. "The Impact of Internet Communications on Social Interaction." *Sociological Spectrum* 25: 335–348.

Brines, Julie, and Kara Joyner. 1999. "The Ties That Bind: Principles of Cohesion in Cohabitation and Marriage." *American Sociological Review* 64 (June): 333–355.

Brislen, William, and Clayton D. Peoples. 2005. "Using a Hypothetical Distribution of Grades to

Introduce Social Stratification." *Teaching Sociology* 33 (January): 74–80.

Britt, Chester L. 1994. "Crime and Unemployment among Youths in the United States, 1958–1990: A Time Series Analysis." *American Journal of Economics and Sociology* 53 (January): 99–109.

Brodkin, Karen. 2006. "How Did Jews Become White Folks?" Pp. 59–66 in Elizabeth Higginbotham and Margaret L. Andersen, eds., *Race and Ethnicity in Society: The Changing Landscape.* Belmont, CA: Wadsworth.

Brooke, James. 1998. "Utah Struggles with a Revival of Polygamy." *The New York Times* (August 23): A12.

Brookey, Robert Alan. 2001. "Bio-Rhetoric, Background Beliefs, and the Biology of Homosexuality." *Argumentation and Advocacy* 37: 171–183.

Brooks-Gunn, Jeanne, Wen-Jui Han, and Jane Waldfogel. 2002. "Maternal Employment and Child Cognitive Outcomes in the First Three Years of Life: The NICHD Study of Early Child Care." *Child Development* 73 (July-August): 1052–1072.

Brown, Cynthia. 1993. "The Vanished Native Americans." *The Nation* 257 (October 11): 384–389.

Brown, Elaine. 1992. *A Taste of Power: A Black Woman's Story.* Pantheon: New York.

Brown, Michael K., Martin Carnoy, Troy Duster, Elliott Currie, David B. Oppenheimer, Marjorie Schulz, and David Wellman. 2005. *White-washing Race: The Myth of a Color-Blind Society.* Berkeley, CA: University of California Press.

Brown, Roger. 1986. *Social Psychology,* 2nd ed. New York: Free Press.

Brown, Ryan P., and Robert A. Josephs. 1999. "A Burden of Proof: Stereotype Relevance and Gender Differences in Math Performance." *Journal of Personality and Social Psychology* 76 (2): 246–257.

Browne, Irene (ed.). 1999. *Latinas and African American Women at Work: Race, Gender, and Economic Inequality.* New York: Russell Sage Foundation.

Broyard, Bliss. 2007. *One Drop: My Father's Hidden Life.* New York: Little, Brown.

Bryant, Alyssa N. 2003. "Changes in Attitudes toward Women's Roles: Predicting Gender-Role Traditionalism among College Students." *Sex Roles* 48 (February): 131–142.

Bryant, Susan L., and Lillian M. Range. 1997. "Type and Severity of Child Abuse and College Students' Lifetime Suicidality." *Child Abuse and Neglect* 21 (December): 1169–1176.

Buchanan, Christy M., Eleanor Maccoby, and Sanford M. Dornsbusch. 1996. *Adolescents After Divorce.* Cambridge, MA: Harvard University Press.

Budig, Michelle J. 2002. "Male Advantage and the Gender Composition of Jobs: Who Rides the Glass Escalator?" *Social Problems* 49 (May): 257–277.

Bullard, Robert D. 1994a. *Dumping in Dixie: Race, Class, and Environmental Quality.* Boulder, CO: Westview.

Bullard, Robert D. 1994b. *Unequal Protection: Environmental Justice and Communities of Color.* San Francisco, CA: Sierra Club Books.

Bullock, Heather E., Karen Fraser, and Wendy R. Williams. 2001. "Media Images of the Poor." *The Journal of Social Issues* 57 (Summer): 229–246.

Bureau of Labor Statistics. 2009. **www.bls.gov**

Burgess, Samuel H. 2001. "Gender Role Ideology and Anti-Gay Opinion in the United States." Paper presented at the annual meeting of the Southern Sociological Association, Atlanta, GA.

Burkholder, Richard. 2003. "Iraq and the West: How Wide Is the Morality Gap?" The Gallup Poll, Princeton, NJ. **www.gallup.com**

Burris, Val. 2000. "The Myth of Old Money Liberalism: The Politics of the Forbes 400 Richest Americans." *Social Problems* 47 (Summer): 360–378.

Burt, Cyril. 1966. "The Genetic Determination of Differences in Intelligence: A Study of Monozygotic Twins Reared Together and Apart." *British Journal of Psychology* 57: 137–153.

Bushman, Brad P. 1998. "Primary Effects of Media Violence on the Accessibility of Aggressive Constructs in Memory." *Personality and Social Psychology Bulletin* 24: 537–545.

Butsch, Richard. 1992. "Class and Gender in Four Decades of Television Situation Comedy: Plus ça Change . . ." *Critical Studies in Mass Communication* 9: 387–399.

Buunk, Bram and Ralph B. Hupka. 1987. "Cross-Cultural Differences in the Elicitation of Sexual Jealousy." *Journal of Sex Research* 23 (February): 12–22.

Calhoun, Craig. 1994. *Social Theory and the Politics of Identity.* Cambridge, MA: Blackwell.

Camaroto, Steven. 2007. "Immigrants in the United States, 2007: A Profile of America's Foreign-Born Population." Washington, DC: Center for Immigration Studies. **www.cis.org**

Cameron, Deborah. 1998. "Gender, Language, and Discourse: A Review Essay." *Signs* 23 (Summer): 945–974.

Campbell, Anne. 1987. "Self-Definition by Rejection: The Case of Gang Girls." *Social Problems* 34 (December): 451–466.

Campenni, C. Estelle. 1999. "Gender Stereotyping of Children's Toys: A Comparison of Parents and Nonparents." *Sex Roles* 40 (January): 121–138.

Cantor, Joanne. 2000. "Media Violence." *Journal of Adolescent Health* 27 (August): 30–34.

Caplow, Theodore. 1968. *Two Against One: Coalitions in Triads.* Englewood Cliffs, NJ: Prentice-Hall.

Cardenas, Jose A. 1996. "Ending the Crisis in the K-12 System." Pp. 51–70 in *Educating a New Majority: Transforming America's Educational System for Diversity,* edited by Laura I. Rendon and Richard O. Hope. San Francisco, CA: Jossey-Bass.

Carmichael, Stokely, and Charles V. Hamilton. 1967. *Black Power: The Politics of Liberation in America.* New York: Vintage.

Carnoy, Martin, Manuel Castells, Stephen S. Cohen, and Fernando Henrique Cardoso. 1993. *The New Global Economy in the Information Age.* University Park, PA: The Pennsylvania State University Press.

Carroll, J. B., 1956. *Language, Thought, and Reality: Selected Writings of Benjamin Lee Whorf.* Cambridge, MA: MIT Press.

Carson, Rachel. 1962. *The Silent Spring.* New York: Knopf.

Carter, Timothy S. 1999. "Ascent of the Corporate Model in Environmental-Organized Crime." *Crime, Law and Social Change* 21: 1–30.

Cashmore, Ellis. 1991. "Black Cops Inc." Pp. 87–108 in *Out of Order: Policing Black People,* edited by Ellis Cashmore and Eugene McLaughlin. New York: Routledge.

Cassen, Robert. 1994. "Population and Development: Old Debates, New Conclusions." *U.S.-Third World Policy Perspectives* 19: 282.

Cassidy, J., and P. R. Shaver (eds.). 1999. *Handbook of Attachment: Theory, Research, and Critical Applications.* New York: Guilford.

Catanzarite, Lisa, and Vilma Ortiz. 1996. "Family Matters, Work Matters? Poverty among Women of Color and White Women." Pp. 121–140 in *For Crying Out Loud: Women's Poverty in the United States,* edited by Diane Dujon and Ann Withorn. Boston, MA: South End Press.

Cavalier, Elizabeth. 2003. "'I Wear Dresses, I Wear Muscles': Media Images of Women Soccer Players." Paper presented at the annual meeting of the Southern Sociological Society, New Orleans, Louisiana.

Centeno, Miguel A. and Eszter Hargittai. 2003. "Defining a Global Geography." *The American Behavioral Scientist* 44 (10).

Center for Responsive Politics. 2009. **www.opensecrets.org**

Centers, Richard. 1949. *The Psychology of Social Classes.* Princeton, NJ: Princeton University Press.

Chafetz, Janet. 1984. *Sex and Advantage.* Totowa, NJ: Rowman and Allanheld.

Chalfant, H. Paul, Robert E. Beckley, and C. Eddie Palmer. 1987. *Religion in Contemporary Society,* 2nd ed. Palo Alto, CA: Mayfield.

Chambliss, William J., and Howard F. Taylor. 1989. *Bias in the New Jersey Courts.* Trenton, NJ: Administrative Office of the Courts.

Chang, Jung. 1991. *Wild Swans: Three Daughters of China.* New York: Simon & Schuster.

Chen, Elsa, Y. F. 1991. "Conflict Between Korean Greengrocers and Black Americans." Unpublished senior thesis, Princeton University.

Cherlin, Andrew J., P. Lindsay Chase-Lansdale, and Christine McRae. 1998. "Effects of Parental Divorce on Mental Health Throughout the Life Course." *American Sociological Review* 63 (April): 219–249.

Chernin, Kim. 1991. *The Obsession.* New York: Harper Colophon.

Chesney-Lind, Meda. 1992. "Women's Prisons: Putting the Brakes on the Building Binge." *Corrections Today* 54 (August): 30–34.

Chia, Rosina C., Jamie L. Moore, Ka Nei Lam, C. J. Chuang, and B. S. Cheng. 1994. "Cultural Differences in Gender Role Attitudes between Chinese and American Students." *Sex Roles* 31 (July): 23–30.

Childs, Erica Chito. 2005. *Navigating Interracial Borders: Black-White Couples and Their Social Worlds.* New Brunswick, NJ: Rutgers University Press.

Chipman, Susan. 1991. "Word Problems: Where Bias Creeps In." Unpublished Research Report, Office of Naval Research, Arlington, VA.

Chipuer, H. M., M. J. Rovine, and R. Plomin. 1990. "LISREL Modeling: Genetic and Environmental Influences on IQ Revisited." *Intelligence* 14: 11–29.

Chiricos, Ted, Sarah Eschholz, and Marc Gertz. 1997. "Crime, News and Fear of Crime: Toward an Identification of Audience Effects." *Social Problems* 44 (August): 342–357.

Chodorow, Nancy. 1978. *The Reproduction of Mothering: Psychoanalysis and the Study of Gender.* Berkeley, CA: University of California Press.

Churchill, Ward. 1992. *Struggle for the Land.* Toronto: Between the Lines.

Cialdini, R. B. 1993. *Influence: Science and Practice,* 3rd ed. New York: Harper Collins.

Cicourel, Aaron V. 1968. *The Social Organization of Juvenile Justice.* New York: Wiley.

Clark, Kenneth B., and Mamie P. Clark. 1947. "Racial Identification and Preference in Negro Children." Pp. 602–611 in *Readings in Social Psychology,* edited by T. M. Newcomb and E. L. Hartley. New York: Holt.

Clawson, Dan, and Naomi Gerstl. 2002. "Caring for Our Young: Child Care in Europe and the United States." *Contexts* 1 (Fall-Winter): 28–35.

Clawson, Rosalee A., and Rakuya Trice. 2000. "Poverty As We Know It: Media Portrayals of the Poor." *The Public Opinion Quarterly* 64 (Spring): 53–64.

Clifford, N. M., and Elaine Walster. 1973. "Research Note: The Effects of Physical Attractiveness on Teacher Expectations." *Sociology of Education* 46: 248–258.

Coale, Ansley. 1974. "The History of the Human Population." *Scientific American* 231 (3): 40–51.

Coale, Ansley. 1986. "Population Trends and Economic Development." Pp. 96–104 in *World Population and the U.S. Population Policy: The Choice Ahead,* edited by J. Menken. New York: W. W. Norton.

Cohen, Philip N. 1999. "Racial-Ethnic and Gender Differences in Returns to Cohabitation and Marriage: Evidence from the Current Population Survey." Paper presented at Annual Meeting of the Population Association of America, **www.census.gov**

Cole, David. 1999. "The Color of Justice." *The Nation* (October 11): 12–15.

College Board. 2008. *College Bound Seniors 2008: A Profile of SAT Program Test Takers.* New York: The College Board. **www.collegeboard.com**

Colley, Ann, and Zazie Todd. 2002. "Gender-Linked Differences in the Style and Content of E-mails to Friends." *Journal of Language and Social Psychology* 21 (December): 380–393.

Collins, Patricia Hill. 1990. *Black Feminist Theory: Knowledge, Consciousness and the Politics of Empowerment.* Cambridge, MA: Unwin Hyman.

Collins, Patricia Hill. 1998. *Fighting Words: Black Women and the Search for Justice.* Minneapolis, MN: University of Minnesota Press.

Collins, Patricia Hill. 2004. *Black Sexual Politics: African Americans, Gender, and the New Racism.* New York: Routledge.

Collins, Patricia Hill, Lionel A. Maldonado, Dana Y. Takagi, Barrie Thorne, Lynn Weber, and

Howard Winant. 1995. "On West and Fenstermaker's 'Doing Difference.'" *Gender & Society* 9 (August): 491–505.

Collins, Randall. 1979. *The Credential Society.* New York: Academic Press.

Collins, Randall. 1988. *Theoretical Sociology.* San Diego, CA: Harcourt Brace Jovanovich.

Collins, Randall. 1994. *Four Sociological Traditions.* New York: Oxford.

Collins, Randall, and Michael Makowsky. 1972. *The Discovery of Society.* New York: Random House.

Collins, Sharon M. 1989. "The Marginalization of Black Executives." *Social Problems* 36: 317–331.

Collins-Lowry, Sharon M. 1997. *Black Corporate Executives: The Making and Breaking of a Black Middle Class.* Philadelphia, PA: Temple University Press.

Congressional Research Service. 2008. "Membership of the 110th Congress: A Profile." **www. senate.gov**

Connell, R. W. 1992. "A Very Straight Gay: Masculinity, Homosexual Experience, and the Dynamics of Gender." *American Sociological Review* 57 (December): 735–751.

Conrad, Peter, and Joseph W. Schneider. 1992. *Deviance and Medicalization: From Badness to Sickness,* expanded edition. Philadelphia, PA: Temple University Press.

Constable, Nicole. 1997. *Maid to Order in Hong Kong: Stories of Filipina Workers.* Ithaca, NY: Cornell University Press.

Cook, Christopher D. 2000. "Temps Demand a New Deal." *The Nation* (March 27): 13–19.

Cook, Karen S., and Alexandra Gervasi. 2006. "Homans and Emerson on Power: Out of the Skinner Box." In *George C. Homans: History, Theory, and Method,* edited by Javier Trevino. New York: Paradigm Publishers.

Cook, S. W. 1988. "The 1954 Social Science Statement and School Segregation: A Reply to Gerard." Pp. 237–256 in *Eliminating Racism: Profiles in Controversy,* edited by D. A. Taylor. New York: Plenum.

Cooksey, Elizabeth C., and Ronald R. Rindfuss. 2001. "Patterns of Work and Schooling in Young Adulthood." *Sociological Forum* 16 (December): 731–755.

Cookson, Peter W., Jr., and Caroline Hodges Persell. 1985. *Preparing for Power: America's Elite Boarding Schools.* New York: Basic Books.

Cooley, Charles Horton. 1902. *Human Nature and Social Order.* New York: Scribner's.

Cooley, Charles Horton. 1967 [1909]. *Social Organization.* New York: Schocken Books.

Coontz, Stephanie. 1992. *The Way We Never Were.* New York: Basic Books.

Cooper, Alvin, and Coralie R. Scherer. 1999. "Sexuality on the Internet: From Sexual Exploration to Pathological Expression." *Professional Psychology Research and Practice* 30. **www.sex-centre.com/ Internetsex_Folder/MSNBC_ Study_pp.htm**

Cortina, Lilia M., Suzanna Swan, Louise F. Fitzgerald, and Craig Walo. 1998. "Sexual Harassment and Assault: Chilling the Climate for Women in Academia." *Psychology of Women Quarterly* 22 (September): 419–441.

Coser, Lewis. 1977. *Masters of Sociological Thought.* New York: Harcourt Brace Jovanovich.

Costello, Mark. 2004. "Throwing Away The Key." *The New York Times Magazine* (June 6): 41–43.

Craig, Maxine Leeds. 2002. *Ain't I a Beauty Queen? Black Women, Beauty and the Politics of Race.* New York: Oxford University Press.

Crane, Diana (ed.). 1994. *The Sociology of Culture: Emerging Theoretical Perspectives.* Cambridge, England: Blackwell.

Crawford, Mary. 1995. *Talking Difference: On Gender and Language.* Thousand Oaks, CA: Sage.

Crittenden, Ann. 2002. *The Price of Motherhood: Why the Most Important Job in the World Is the Least Valued.* New York: Holt.

Croll, Elisabeth. 1995. *Changing Identities of Chinese Women.* London: Zed Books.

Crouse, James, and Dale Trusheim. 1988. *The Case Against the SAT.* Chicago, IL: University of Chicago Press.

Crowley, Sue M. 1998. "Men's Self-Perceived Adequacy as the Family Breadwinner: Implications for Their Psychological, Marital, and Family Well-Being." *Journal of Family and Economic Issues* 19 (Spring): 7–23.

Csikszentmihalyi, Mihaly, and Barbara Schneider. 2000. *Becoming Adult: How Teenagers Prepare for the World of Work.* New York: Basic Books.

Cunningham, Mick. 2001. "The Influence of Parental Attitudes and Behaviors on Children's Attitudes toward Gender and Household Labor in Early Adulthood." *Journal of Marriage and Family* 63 (February): 111–122.

Currie, Dawn. 1997. "Decoding Femininity: Advertisements and Their Teenage Readers." *Gender & Society* 11 (August): 453–477.

Curtiss, Susan. 1977. *Genie: A Psycholinguistic Study of a Modern-Day "Wild Child."* New York: Academic Press.

Czerniawski, Amanda. 2007. "From Average to Ideal: The Evolution of the Height and Weight Table in the United States." *Social Science History* 31 (2): 273–296.

Dahlhamer, James, and Joanne M. Nigg. 1993. "An Empirical Investigation of Rumoring: Anticipating Disaster under Conditions of Uncertainty." Paper presented at the Annual Meetings of the Southern Sociological Society, Chattanooga, TN.

Dahrendorf, Rolf. 1959. *Class and Class Conflict in Industrial Society.* Stanford, CA: Stanford University Press.

D'Alessio, Stewart J., and Lisa Stolzenberg. 2003. "Race and the Probability of Arrest." *Social Forces* 81 (June): 1381–1397.

Dalton, Susan E., and Denise D. Bielby. 2000. "'That's Our Kind of Constellation': Lesbian Mothers Negotiate Institutionalized Understandings of Gender Within the Family." *Gender & Society* 14 (February): 36–61.

Darley, John M., and R. H. Fazio. 1980. "Expectancy Confirmation Processes Arising in the Social Interaction Sequence." *American Psychologist* 35: 867–881.

Darling-Hammond, Linda. 2007. "The Flat Earth and Education: How America's Commitment to Equality Will Determine Our Future." Paper presented before the conference Embracing

Diversity, Center for Comparative Studies in Race and Ethnicity, Stanford University, Palo Alto, California.

Das Gupta, Monisha. 1997. "'A Chambered Nautilus': The Contradictory Nature of Puerto Rican Women's Role in the Construction of a Transnational Community." *Gender & Society* 11 (October): 627–655.

Davis, Angela. 1981. *Women, Race, and Class.* New York: Random House.

Davis, James A., and Tom Smith. 1984. *General Social Survey Cumulative File, 1972–1982.* Ann Arbor, MI: Inter-University Consortium for Political and Social Research.

Davis, Kingsley. 1945. "The World Demographic Transition." *Annals of the American Academy of Political and Social Sciences* 237: 1–11.

Davis, Kingsley, and Wilbert E. Moore. 1945. "Some Principles of Stratification." *American Sociological Review* 10 (April): 242–247.

DeCarlo, Scott. 2006. "CEO Compensation." *Forbes,* April 20. **www. forbes.com**

Deegan, Mary Jo. 1988. "W.E.B. Du Bois and the Women of Hull-House, 1895–1899." *The American Sociologist* 19 (Winter): 301–311.

Defense Manpower Data Center. 2006. *2006 Gender Relations Survey of Active Duty Members.* Ft. Belvoir, VA: Defense Manpower Data Center.

DeFleur, Lois. 1975. "Biasing Influences on Drug Arrest Records: Implications for Deviance Research." *American Sociological Review* 40 (March): 88–103.

Dellinger, Kristen, and Christine L. Williams. 1997. "Makeup at Work: Negotiating Appearance Rules in the Workplace." *Gender & Society* 11 (April): 151–177.

D'Emilio, John. 1998. *Sexual Politics, Sexual Communities: The Making of a Homosexual Minority in the United States, 1940–1970,* rev. edition. Chicago, IL: University of Chicago Press.

Demeny, Paul. 1991. "Tradeoffs between Human Numbers and Material Standards of Living." Pp. 408–421 in *Resources, Environment, and Population:*

Present Knowledge, Future Options, edited by Kingsley Davis and Michail S. Bernstam. New York: Population Council.

DeNavas-Walt, Carmen, and Robert W. Cleveland. 2002. *Money Income in the United States 2002.* Washington, DC: U.S. Census Bureau.

DeNavas-Walt, Carmen, Bernadette D. Proctor, and Jessica C. Smith. 2008. *Income, Poverty, and Health Insurance Coverage in the United States: 2007,* P60-235. Washington, DC: U.S. Census Bureau.

DeSteno, D., R. E. Petty, D. D. Rucker, D. T. Wegener, and J. Braverman. 2004. "Discrete Emotions and Persuasion: The Role of Emotion-Induced Expectancies." *Journal of Personality and Social Psychology* 78: 397–416.

DeVault, Marjorie. 1991. Feeding the Family: *The Social Organization of Caring as Gendered Work.* Chicago, IL: University of Chicago Press.

Devlin, Bernie, Michael Daniels, and Kathryn Roeder. 1997. "The Heritability of IQ." *Nature* 388 (July 31): 468–471.

DeWitt, Karen. 1995. "Blacks Prone to Job Dismissal in Organizations." *The New York Times* (April 20): A19.

DiAcosta, Diego. 1998. "A Sociolinguistic Study of Two Baptist Churches." Senior thesis, Princeton University, Department of Sociology, Princeton, NJ.

Diaz-Duque, Ozzie F. 1989. "Communication Barriers in Medical Settings: Hispanics in the United States." *International Journal of the Sociology of Language* 79: 93–102.

Dill, Bonnie Thornton. 1988. "Our Mothers' Grief: Racial Ethnic Women and the Maintenance of Families." *Journal of Family History* 13 (October): 415–431.

Dillard, J.L. 1972. *Black English: Its Historical Usage in the United States.* New York: Random House.

Dillon, Sam. 2009. "No Child Law Is Not Closing a Racial Gap." *The New York Times* (April 26): A15–16.

DiMaggio, Paul J., and Walter W. Powell. 1991. "Introduction." Pp. 1–38 in *The New Institutionalism in Organizational Analysis,* edited by W. W. Powell and P. J.

DiMaggio. Chicago, IL: University of Chicago Press.

Dines, Gail, and Jean M. Humez. 2002. *Gender, Race, and Class in Media,* 2nd ed. Thousand Oaks, CA: Sage.

Dion, K. K. 1972. "Physical Attractiveness and Evaluating Children's Transgressions." *Journal of Personality and Social Psychology* 24: 285–290.

Dixit, Avinash K., and Susan Sneath. 1997. *Games of Strategy.* New York: W. W. Norton.

Dolan, Jill. 2006. "Blogging on Queer Connections in the Arts and the Five Lesbian Brothers." *GLQ* 12: 491–506.

Dollard, John, Neal E. Miller, Leonard W. Doob, O. H. Mowrer, and Robert R. Sears. 1939. *Frustration and Aggression.* New Haven, CT: Yale University Press.

Domhoff, G. William. 2002. *Who Rules America?* New York: McGraw-Hill.

Dominguez, Silvia, and Celeste Watkins. 2003. "Creating Networks for Survival and Mobility: Social Capital Among African-American and Latin-American Low-Income Mothers." *Social Problems* 50 (1): 111–135.

Dovidio, J. F., and S. L. Gaertner (eds.). 1986. *Prejudice, Discrimination, and Racism.* New York: Academic Press.

Dowd, James J., and Laura A. Dowd. 2003. "The Center Holds: From Subcultures to Social Worlds." *Teaching Sociology* 31 (January): 20–37.

Downey, Liam. 2005. "The Unintended Significance of Race: Environmental Racial Inequality in Detroit." *Social Forces* 83 (March): 971–1008.

Doyal, I. 1990. "Hazards of Hearth and Home." *Women's Studies International Forum* 13: 501–517.

Driggs, Ken. 1990. "After The Manifesto: Modern Polygamy and Fundamentalist Mormons." *Journal of Church and State* 32 (Spring): 367–389.

Du Bois, W.E.B. 1901. "The Freedmen's Bureau." *Atlantic Monthly* 86: 354–365.

Du Bois, W.E.B. 1903. *The Souls of Black Folk: Essays and Sketches.* Chicago: A. C. McClurg and Co.

DuBois, Cathy L., Deborah E. Knapp, Robert H. Faley, and Gary A. Kustis. 1998. "An

Empirical Examination of Same and Other Gender Sexual Harassment in the Workplace." *Sex Roles* 9–10 (November): 731–749.

Dubrofsky, Rachel. 2006. "The Bachelor: Whiteness in the Harem." *Critical Studies in Media Communication* 23: 39–56.

Dudley, Carl S., and David A. Roozen. 2001. *Faith Communities Today: A Report on Religion in the United States Today.* Hartford, CT: Hartford Institute for Religion Research, Hartford Seminary.

Due, Linnea. 1995. *Joining the Tribe: Growing Up Gay & Lesbian in the '90s.* New York: Doubleday.

Duneier, Mitchell. 1999. *Sidewalk.* New York: Farrar, Strauss and Giroux.

Dunlop, Riley E., and Angela G. Mertig. 1992. *American Environmentalism: The U.S. Environmental Movement, 1970–1990.* Philadelphia, PA: Taylor and Francis.

Dunne, Gillian A. 2000. "Opting into Motherhood: Lesbians Blurring the Boundaries and Transforming the Meaning of Parenthood and Kinship." *Gender & Society* 14 (February): 11–35.

Durkheim, Emile. 1897 [1951]. *Suicide.* Glencoe, IL: Free Press.

Durkheim, Emile. 1912 [1947]. *Elementary Forms of Religious Life.* Glencoe, IL: Free Press.

Durkheim, Emile. 1938 [1950]. *The Rules of Sociological Method.* Glencoe, IL: Free Press.

Durkheim, Emile. 1964 [1895]. *The Division of Labor in Society.* New York: Free Press.

Dworkin, Shari L., and Michael A. Messner. 1999. "Just Do . . . What?" Pp. 341–361 in *Revisioning Gender,* edited by Myra Marx Ferree, Judith Lorber, and Beth B. Hess. Thousand Oaks, CA: Sage.

Dyson, Michael Eric. 2006. *Come Hell or High Water: Hurricane Katrina and the Color of Disaster.* New York: Basic Books.

Eckberg, Douglass Lee. 1992. "Social Influences on Belief in Creationism." *Sociological Spectrum* 12 (April-June): 145–165.

Edin, Kathryn. 2000. "What Do Low-Income Single Mothers Say About Marriage?" *Social Problems* 47 (February): 112–133.

Edin, Kathryn, and Laura Lein. 1997. *Making Ends Meet: How Single Mothers Survive Welfare and Low-Wage Work.* New York: Russell Sage Foundation.

Edin, Kathryn, and Maria Kefalas. 2005. *Promises I Can Keep: Why Poor Women Put Motherhood before Marriage.* Berkeley, CA: University of California Press.

Edmondson, Brad. 2000. "Immigration Nation." *Preservation* (January-February): 31–38.

Ehrlich, Paul. 1968. *The Population Bomb.* New York: Ballantine Books.

Ehrlich, P. 1990. *The Population Explosion.* New York: Ballantine Books.

Ehrlich, Paul R. Anne H. Ehrlich, and John P. Holdren. 1977. *Ecoscience: Population, Resources, Environment.* San Franciso, CA: Freeman.

Ehrlich, Paul R., and Jianguo Liu. 2002. "Some Roots of Terrorism." *Population and Environment,* vol. 24, no. 2, Human Sciences Press.

Eichenwald, Kurt. 1996. "The Two Faces of Texaco." *The New York Times* (November 10): Section 3: 1ff.

Eichenwald, Kurt. 2002. "Two Ex-Officials at WorldCom Are Charged in Huge Fraud." *The New York Times* (August 2): A1 and A11.

Eitzen, D. Stanley, and Maxine Baca Zinn. 2006. *Globalization: The Transformation of Social Worlds,* 2nd ed. Belmont, CA: Wadsworth.

Eitzen, D. Stanley, and Maxine Baca Zinn. 2007. *In Conflict and Order,* 11th ed., Boston: Allyn and Bacon.

Elizabeth, Vivienne. 2000. "Cohabitation, Marriage, and the Unruly Consequences of Difference." *Gender & Society* 14 (February): 87–110.

Ellison, Christopher G., and John P. Bartkowski. 2002. "Conservative Protestantism and the Division of Household Labor among Married Couples." *Journal of Family Issues* 23 (November): 950–985.

Emerson, Joan P. 1970. "Behaviors in Private Places: Sustaining Definitions of Reality in Gynecological Examinations." Pp. 74–97 in *Recent Sociology,* vol. 2, edited by H. P. Dreitzel. New York: Collier.

Emerson, Rana A. 2002. "Where My Girls At?: Negotiating Black Womanhood in Music Videos." *Gender & Society* 16 (February): 115–135.

Enloe, Cynthia. 2001. *Bananas, Beaches, and Bases: Making Feminist Sense of International Politics,* updated edition. Berkeley, CA: University of California Press.

Epps, Edgar T. 2002. "Race, Class and Educational Opportunity:" In *2001 Race Odyssey: African Americans and Sociology,* edited by Bruce R. Hare. Syracuse, NY: Syracuse University Press. 164–177.

Epstein, Steven. 1996. *Impure Science: AIDS, Activism, and the Politics of Knowledge.* Berkeley, CA: University of California Press.

Erikson, Eric. 1980. *Identity and the Life Cycle.* New York: W.W. Norton.

Erikson, Kai. 1966. *Wayward Puritans: A Study in the Sociology of Deviance.* New York: Wiley.

Esbensen-Finn, Aage, Elizabeth Piper Deschenes, and Thomas L. Winfree Jr. 1999. "Differences between Gang Girls and Gang Boys: Results from a Multisite Survey." *Youth and Society* 31 (September): 27–53.

Eschholz, Sarah, Jana Bufkin, and Jenny Long. 2002. "Symbolic Reality Bites: Women and Racial/Ethnic Minorities in Modern Film." *Sociological Spectrum* 22 (July-August): 299–334.

Espenshade, Thomas J., Lauren E. Hale, and Chang Y. Chung. 2005. "The Frog Pond Revisited: High School Academic Context, Class Rank, and Elite College Admission." *Sociology of Education* 78 (October): 269–293.

Essed, Philomena. 1991. *Understanding Everyday Racism.* Newbury Park, CA: Sage.

Esterberg, Kristin G. 2003. "New Right." *Encyclopedia of Lesbian, Gay, Bisexual and Transgendered History in America.* New York: Charles Scribner's Sons.

Etzioni, Amatai. 1975. *A Comparative Analysis of Complex Organization: On Power, Involvement, and Their Correlates,* rev. ed. New York: Free Press.

Etzioni, Amatai, John Wilson, Bob Edwards, and Michael W. Foley. 2001. "A Symposium on Robert D. Putnam's *Bowling Alone: The Collapse and Revival of American Community.*" *Contemporary Sociology* 30 (May): 223–230.

Evans, Peter B., Dietrich Ruesschemeyer, and Theda Skocpol. 1985. *Bringing the State Back In.* Cambridge, MA: Cambridge University Press.

Farber, Naomi. 1990. "The Significance of Race and Class in Marital Decisions Among Unmarried Adolescent Mothers." *Social Problems* 37 (February): 51–63.

Fattah, Ezzat A., 1994. "The Interchangeable Roles of Victim and Victimizer." *HEUNI* Papers 3: 1–26.

Fausto-Sterling, Anne. 1992. *Myths of Gender: Biological Theories About Women and Men.* New York: Basic Books.

Fausto-Sterling, Anne. 2000. *Sexing the Body: Gender Politics and the Construction of Sexuality.* New York: Basic Books.

Feagin, Joe R. 2000. *Racist America: Roots, Future Realities, and Racial Reparations.* New York: Routledge.

Feagin, Joe R., and Clairece B. Feagin. 1993. *Racial and Ethnic Relations,* 4th ed. Englewood Cliffs, NJ: Prentice Hall.

Feagin, Joe R., and Hernan Vera 1995. *White Racism.* New York: Routledge.

Federal Bureau of Investigation. 2006. "Uniform Crime Reports." Washington, DC: U.S. Department of Justice.

Federal Bureau of Investigation. 2008. "Uniform Crime Reports." Washington, DC: U.S. Department of Justice.

Federal Bureau of Investigation. 2009. "Uniform Crime Reports." Washington DC: U.S. Department of Justice. **www.fbi.gov**

Federal Election Commission. 2008. "Growth in PAC Financial Activity Slows." **www.fec.gov**

Fein, Melvyn L. 1988. "Resocialization: A Neglected Paradigm." *Clinical Sociology.* 6: 88–100.

Ferber, Abby. 1998. *White Man Falling: Race, Gender, and White Supremacy.* Lanham, MD: Rowman and Littlefield.

Ferber, Abby. 1999. "What White Supremacists Taught a Jewish Scholar about Identity." *The Chronicle of Higher Education* (May 7): 86–87.

Fernald, Anne, and Hiromi Morikawa. 1993. "Common Themes and Cultural Variations in Japanese and American Mothers' Speech to Infants." *Child Development* 64 (June): 637–656.

Festinger, Leon, Stanley Schachter, and Kurt Back. 1950. *Social Pressures in Informal Groups: A Study of Human Factors in Housing.* Stanford, CA: Stanford University Press.

Fields, Jason. 2004. *America's Families and Living Arrangements: 2003.* Current Population Reports, P20–553. Washington, DC: U.S. Bureau of the Census.

Fischer, Claude S., Michael Hout, Mártin Sánchez Jankowski, Samuel R. Lucas, Ann Swidler, and Kim Voss. 1996. *Inequality by Design: Cracking the Bell Curve Myth.* Princeton, NJ: Princeton University Press.

Fisher, Bonnie S., Francis T. Cullen, and Michael G. Turner. 2000. *Sexual Victimization of College Women.* Washington, DC: Bureau of Justice Statistics.

Fitzgibbon, Marian, and Melinda Stolley. 2002. "Minority Women: The Untold Story," NOVA Online. **http://www.pbs.org/wgbh/nova/thin/minorities.html**

Flanagan, William G. 1995. *Urban Sociology: Images and Structure.* Boston, MA: Allyn and Bacon.

Fleisher, Mark. 2000. *Dead End Kids: Gang Girls and the Boys They Know.* Madison, WI: University of Wisconsin Press.

Fleming, Jacqueline, and Nancy Garcia. 1998. "Are Standardized Tests Fair to African Americans? Predictive Validity of the SAT in Black and White Institutions." *Journal of Higher Education* 69 (September-October): 471–495.

Flowers, M.L. 1977. "A Laboratory Test of Some Implications of Janis' Groupthink Hypothesis." *Journal of Personality and Social Psychology* 35 (December): 888–896.

Folbre, Nancy. 2001. *The Invisible Heart: Economics and Family Values.* New York: New Press.

Frankenberg, Erica, and Chungmei Lee. 2002. *Race in American Public Schools: Rapidly Resegregating School Districts.* Cambridge, MA: The Civil Rights Project, Harvard University. Website: **www.civilrightsproject.harvard.edu**

Frankenberg, Erica, and Chungmei Lee. 2006. "Rapidly Resegregating School Districts." Pp. 247–353 in Elizabeth Higginbotham and Margaret L. Andersen, eds., *Race and Ethnicity in Society: The Changing Landscape.* Belmont, CA: Wadsworth.

Frazier, E. Franklin. 1957. *The Black Bourgeoisie.* New York: Collier Books.

Fredrickson, George M. 2003. *Racism: A Short History.* Princeton, NJ: Princeton University Press.

Freedle, Roy O. 2003. "Correcting the SATs Ethnic and Social Class Bias: A Method of Re-Estimating SAT Scores." *Harvard Educational Review* 73 (Spring): 1–43.

Freedman, Estelle B., and John D'Emilio. 1988. *Intimate Matters: A History of Sexuality in America.* New York: Harper & Row.

Freeman, Jo. 1983. "A Model for Analyzing the Strategic Options of Social Movement Organizations." Pp. 193–210 in *Social Movements of the Sixties and Seventies,* edited by Jo Freeman. New York: Longman.

Freese, Jeremy. 2008. "Genetics and the Social Science Explanation of Individual Outcomes." *American Journal of Sociology* 114 (Suppl.): S1–S35.

Freese, Jeremy, Jui-Chung Allen Li, and Lisa D. Wade. 2003. "The Potential Relevances of Biology to Social Inquiry." *Annual Review of Sociology* 29: 233–256.

Freud, Sigmund. 1901 [1965]. *The Psychopathology of Everyday Life,* translated by Alan Tyson and edited by James Strachey. New York: W.W. Norton.

Freud, Sigmund. 1923 [1960]. *The Ego and the Id,* translated by Joan Riviere. New York: W.W. Norton.

Freud, Sigmund. 1930 [1961]. *Civilization and Its Discontents,* translated by James Strachey. New York: W.W. Norton.

Frey, William. 2009. *Getting Current: Recent Demographic Trends in Metropolitan America.* Washington, DC: The Brookings Institution.

Fried, Amy. 1994. "'It's Hard to Change What We Want to Change': Rape Crisis Centers as Organizations." *Gender & Society* 4 (December): 562–583.

Friedman, Thomas L. 1999. *The Lexus and the Olive Tree.* New York: Farrar, Strauss, and Giraux.

Fryberg, Stephanie. 2003. "Really: You Don't Look Like an American Indian: Social Representations and Social Group Identities." Ph.D. Dissertation, Department of Psychology, Stanford University.

Frye, Marilyn. 1983. *The Politics of Reality*. Trumansburg, NY: The Crossing Press.

Fukuda, Mari. 1994. "Nonverbal Communication within Japanese and American Corporations." Unpublished manuscript, Princeton University.

Fullerton, Howard F. 1999. "Labor Force Projections to 2008: Steady Growth and Changing." *Monthly Labor Review* 122: 19–32.

Furstenberg, Frank. 1998. "Relative Risk: What Is the Family Doing to Our Children?" *Contemporary Sociology* 27 (May): 223–225.

Furstenberg, Frank F., Jr., and Christine Winquist Nord. 1985. "Parenting Apart: Patterns of Childrearing After Marital Disruption." *Journal of Marriage and the Family* 47 (November): 898–904.

Gagné, Patricia, and Richard Tewksbury. 1998. "Conformity Pressures and Gender Resistance among Transgendered Individuals." *Social Problems* 45 (February): 81–101.

Gallagher, Charles A. 2006. "Color-Blind Privilege: The Social and Political Functions of Erasing the Color Line in Post Race America." Pp. 95–99 in *Race, Class and Gender: An Anthology*, 7th ed, edited by Margaret Andersen and Patricia Hill Collins. Belmont, CA: Wadsworth/Cengage.

Gallagher, Sally K. 2003. *Evangelical Identity and Gendered Family Life*. New Brunswick, NJ: Rutgers University Press.

Gallup Organization. 2006. "Tobacco and Smoking." The Gallup Poll. Princeton, NJ: The Gallup Organization. **www.gallup.com**

Gallup Poll. 2008. "Homosexual Relations." Princeton, NJ: Gallup Poll. **www.gallup.com**

Gallup Poll. 2009. "Homosexual Relations." Princeton, NJ: Gallup Poll. **www.gallup.com**

Gamson, Joshua. 1995. "Featured Essay." *Contemporary Sociology* 24 (May): 294–298.

Gamson, Joshua. 1998a. *Freaks Talk Back: Tabloid Talk Shows and Sexual Nonconformity*. Chicago, IL: University of Chicago Press.

Gamson, Joshua. 1998b. "Publicity Traps: Television Talk Shows and Lesbian, Gay, Bisexual, and Transgender Visibility." *Sexualities* 1: 11–41.

Gamson, William A. 1992. "The Social Psychology of Collective Action." Pp. 53–76 in *Frontiers in Social Movement Theory*, edited by Aldon D. Morris and Carol McClurg Mueller. New Haven, CT: Yale University Press.

Gamson, William, and Andre Modigliani. 1974. *Conceptions of Social Life: A Text-Reader for Social Psychology*. Boston, MA: Little, Brown.

Gardner, Howard. 1999 [1993]. *Frames of Mind: The Theory of Multiple Intelligences*. New York: Basic Books.

Garfinkel, Harold. 1967. *Studies in Ethnomethodology*. Englewood Cliffs, NJ: Prentice Hall.

Garrow, David J. 1981. *The FBI and Martin Luther King*. New York: W. W. Norton.

Gastil, John. 1990. "Generic Pronouns and Sexist Language: The Oxymoronic Character of Masculine Generics." *Sex Roles* 23 (December): 629–643.

Gates, Henry Louis, Jr. 1988. *The Signifying Monkey: A Theory of African-American Literary Criticism*. New York: Oxford University Press.

Gates, Henry Louis, Jr. 1992. "Integrating the American Mind." Pp. 105–120 in *Loose Canons: Notes on the Culture Wars*, edited by Henry Louis Gates Jr. New York: Oxford University Press.

Gelles, Richard J. 1999. "Family Violence." Pp. 1–24 in *Family Violence: Prevention and Treatment*, edited by Robert J. Hampton, Thomas P. Gallota, Gerald R. Adams, Earl H. Potter III, and Roger P. Weissberg. Newbury Park, CA: Sage.

Genovese, Eugene. 1972. *Roll, Jordan, Roll: The World the Slaves Made*. New York: Pantheon.

Gerami, Shahin, and Melodye Lehnerer. 2001. "Women's Agency and Houshold Diplomacy: Negotiating Fundamentalism." *Gender & Society* 15 (August): 556–573.

Gersch, Beate. 1999. "Class in Daytime Talk Television." *Peace Review* 11 (June): 275–281.

Gerson, Kathleen. 1993. *No Man's Land: Men's Changing Commitments to Family and Work*. New York: Basic Books.

Gerstel, Naomi, and Sally Gallagher. 2001. "Men's Caregiving: Gender and the Contingent Character of Care." *Gender & Society* 15 (April): 197–217.

Gerth, Hans, and C. Wright Mills (eds.). 1946. *From Max Weber: Essays in Sociology*. New York: Oxford University Press.

Giannarelli, Linda, and James Barsimantov. 2000. *Child Care Expenses of America's Families*. Washington. DC: Urban Institute.

Giddings, Paula. 1994. *In Search of Sisterhood: Delta Sigma Theta and the Challenge of the Black Sorority Movement*. New York: William Morrow.

Giddings, Paula. 2008. *A Sword among Lions: Ida B. Wells and the Campaign against Lynching*. New York: Amistad.

Gilbert, Daniel R., Susan T. Fiske, and Gardner Lindzey (eds.). 1998. *The Handbook of Social Psychology*, 4th ed. New York: Oxford University and McGraw-Hill.

Gilbert, D. T., and P. S. Malone. 1995. "The Correspondence Bias." *Psychological Bulletin* 117: 21–38.

Gilens, Martin. 1996. "Race and Poverty in America: Public Misperceptions and the American News Media." *The Public Opinion Quarterly* 60 (Winter): 515–541.

Gilkes, Cheryl Townsend. 2000. *"If It Wasn't For the Women . . ." Black Women's Experience and Womanist Culture in Church and Community*. Maryknoll, NY: Orbis Books.

Gilligan, Carol. 1982. *In a Different Voice: Psychological Theory and Women's Development*. Cambridge, MA: Harvard University Press.

Gimlin, Debra. 1996. "Pamela's Place: Power and Negotiation in the Hair Salon." *Gender & Society* 10 (October): 505–526.

Gimlin, Debra. 2002. *Body Work: Beauty and Self-Image in American Culture*. Berkeley, CA: University of California Press.

Gitlin, Todd. 2002. *Media Unlimited: How the Torrent of Images and Sounds Overwhelms Our Lives*. New York: Metropolitan Books.

Glassner, Barry. 1999. *Culture of Fear: Why American Are Afraid of the Wrong Things*. New York: Basic Books.

Glaze, Lauren E., and Laura M. Maruschak. 2008. "Parents in Prison and Their Minor Children." Washington, DC: U.S. Bureau of Justice Statistics. **www.ojp.usdoj.gov**

Glazer, Nathan. 1970. *Beyond the Melting Pot: The Negroes, Puerto Ricans, Jews, Italians, and Irish of New York City.* Cambridge, MA: MIT Press.

Glazer, Nona. 1990. "The Home as Workshop: Women as Amateur Nurses and Medical Care Providers." *Gender & Society* 4: 479–499.

Glenn, Evelyn Nakano. 1986. *Issei, Nisei, War Bride: Three Generations of Japanese American Women in Domestic Service.* Philadelphia, PA: Temple University Press.

Glenn, Evelyn Nakano. 2002. *Unequal Freedom: How Race and Gender Shaped American Citizenship and Labor.* Cambridge, MA: Harvard University Press.

Glenn, Norval, and Elizabeth Marquardt. 2001. "Hooking Up, Hanging Out, and Hoping for Mr. Right." New York: Institute for American Values.

Glock, Charles, and Rodney Stark. 1965. *Religion and Society in Tension.* Chicago, IL: Rand McNally.

Gluckman, Amy, and Betsy Reed (eds.). 1997. *Homo Economics: Capitalism, Community, and Lesbian and Gay Life.* New York: Routledge.

Goffman, Erving. 1961. *Asylums: Essays on the Social Situation of Mental Patients and Other Inmates.* Garden City, NY: Anchor.

Goffman, Erving. 1963. *Stigma: Notes on the Management of Spoiled Identity.* Englewood Cliffs, NJ: Prentice Hall.

Goffman, Irving. 1959. *The Presentation of Self in Everyday Life.* Garden City, NY: Doubleday.

Goldberger, Arthur S. 1979. "Heritability." *Econometrica* 46: 327–347.

Goldman, Noreen, Yorenzo Moreno, and Charles F. Westoff. 1989. *Peru Experimental Study: An Evaluation of Fertility and Child Health Information.* Princeton, NJ: Office of Population Research.

Gonzales, P., H. Blanton, and K. Williams. 2002. "The Effect of Stereotype Threat and Double-Minority Status on Test Performance in Latino Women." *Personality and Social Psychology Bulletin* 28: 659–670.

González, Tina Esther. 1996. "Social Control of Medical Professionals, Cultural Notions of Pain, and Doctors as Social Scientists." Unpublished junior thesis, Princeton University.

Gordon, Linda. 1977. *Woman's Body/Woman's Right.* New York: Penguin.

Gore, Albert. 2006. *An Inconvenient Truth: The Planetary Emergency of Global Warming and What We Can Do About It.* Emmaus, PA: Rodale.

Gottfredson, Michael R., and Travis Hirschi. 1990. *A General Theory of Crime.* Stanford, CA: Stanford University Press.

Gottfredson, Michael R., and Travis Hirschi. 1995. "National Crime Control Policies." *Society* 32 (January-February): 30–36.

Gould, Stephen Jay. 1981. *The Mismeasure of Man.* New York: W.W. Norton.

Gould, Stephen J. 1994. "Curveball." *The New Yorker* 70 (November 28): 139–149.

Gould, Stephen Jay. 1999. "The Human Difference." *The New York Times* (July 2).

Graefe, Deborah Roempke, and Daniel T. Lichtor. 1999. "Life Course Transitions of American Children: Parental Cohabitation, Marriage, and Single Motherhood." *Demography* 36 (May): 205–217.

Graham, Lawrence Otis. 1999. *Our Kind of People: Inside America's Black Upper Class.* New York: Harper Collins.

Gramsci, Antonio. 1971. *Selections from the Prison Notebooks of Antonio Gramsci,* edited by Quintin Hoare and Geoffrey Nowell. London: Lawrence and Wishart.

Granovetter, Mark. 1973. "The Strength of Weak Ties." *American Journal of Sociology* 78 (May): 1360–1380.

Granovetter, Mark. 1974. *Getting a Job: A Study of Contacts and Careers.* Cambridge, MA: Harvard University Press.

Granovetter, Mark S. 1995. "Afterward 1994: Reconsiderations and a New Agenda." Pp. 139–182 in *Getting A Job,* 2nd ed. by Mark S. Granovetter. Chicago, IL: University of Chicago Press.

Grant, Don Sherman, II, and Ramiro Martinez Jr. 1997. "Crime and the Restructuring of the U.S. Economy: A Reconsideration of the Class Linkages." *Social Forces* 75 (March): 769–799.

Green, Gary Paul, Leam M. Tigges, and Daniel Diaz. 1999. "Racial and Ethnic Differences in Job Search Strategies In Atlanta, Boston, and Los Angeles." *Social Science Quarterly* 80 (June): 263–290.

Greenhouse, Linda. 1996. "Gay Rights Laws Can't Be Banned, High Court Rules." *The New York Times* (May 21): A1ff.

Greenhouse, Steven. 2000. "Poll of Working Women Finds Them Stressed." *The New York Times,* April 3.

Grimal, Pierre (ed.). 1963. *Larousse World Mythology.* New York: Putnam.

Grindstaff, Laura. 2002. *The Money Shot: Trash Class, and the Making of TV Talk Shows.* Chicago, IL: University of Chicago Press.

Gruber, James E. 1982. "Blue-Collar Blues: The Sexual Harassment of Women Autoworkers." *Work and Occupations* 3 (August): 271–298.

Guinther, J. 1988. *The Jury in America.* New York: Facts on File Publications.

Guo, Guang, and Elizabeth Stearns. 2002. "The Social Influences on the Realization of Genetic Potential for Intellectual Development." *Social Forces* 80: 881–910.

Guttmacher Institute. 2006a. "Contraceptive Use." New York: Guttmacher Institute. **www.guttmacher.org**

Guttmacher Institute. 2006b. "Facts on American Teens' Sexual and Reproductive Health." New York: Guttmacher Institute. **www.guttmacher.org**

Guttmacher Organization. 2008. "Facts on Young Men's Sexual and Reproductive Health." **www.guttmacher.org**

Habermas, Jürgen. 1970. *Toward a Rational Society: Student Protest, Science, and Politics.* Boston, MA: Beacon Press.

Hadaway, C. Kirk, and Penny Long Marler. 1993. "All in the Family: Religious Mobility in America." *Review of Religious Research* 35 (December): 97–116.

Haddad, Yvonne Yazbeck, Jane I. Smith, and John L. Esposito

(eds.). 2003. *Religion and Immigration: Christian, Jewish, and Muslim Experiences in the United States.* Walnut Creek, CA: AltaMira Press.

Hagan, John. 1993. "The Social Embeddedness of Crime and Unemployment." *Criminology* 31 (November): 465–491.

Hall, Edward T. 1966. *The Hidden Dimension.* New York: Doubleday.

Hall, Edward T., and Mildred Hall. 1987. *Hidden Differences: Doing Business with the Japanese.* New York: Anchor Press/Doubleday.

Hall, Elaine J. 1993. "Waitering/Waitressing: Engendering in the Work of Table Servers." *Gender & Society* (September): 329–346.

Hallinan, Maureen T. 2003. *Ability Grouping and Student Learning.* Washington, DC: Brookings Papers on Educational Policy.

Hamer, Dean H., Stella Hu, Victoria L. Magnuson, Nan Hu, and Angela M. L. Pattatucci. 1993. "A Linkage Between DNA Markers on the X Chromosome and Male Sexual Identification." *Science* 261 (July): 321–327.

Hamer, Jennifer. 2001. *What It Means to Be Daddy.* New York: Columbia University Press.

Hamilton, Mykol C. 1988. "Using Masculine Generics: Does Generic He Increase Male Bias in the User's Imagery?" *Sex Roles* 19 (December): 785–799.

Handlin, Oscar. 1951. *The Uprooted.* Boston, MA: Little, Brown.

Haney, C., C. Banks, and P. G. Zimbardo. 1973. "Interpersonal Dynamics in a Simulated Prison." *International Journal of Criminology and Penology* 1: 69–97.

Haney, Lynne. 1996. "Homeboys, Babies, Men in Suits: The State and the Reproduction of Male Dominance." *American Sociological Review* 61 (October): 759–778.

Hans, Valerie (ed.). 2006. *The Jury System: Contemporary Scholarship.* Burlington, VT: Ashgate Publishing Co.

Hans, Valerie, and Ramiro Martinez. 1994. "Intersections of Race, Ethnicity, and the Law." *Law and Human Behavior* 18 (June): 211–221.

Hargittai, Eszther. 2008. "The Digital Reproduction of Inequality." Pp. 936–944 in *Social Stratification*, edited by David Grusky. Boulder, CO: Westview Press.

Harp, Dustin, and Mark Tremayne. 2006. "The Gendered Blogosphere: Examining Inequality Using Network and Feminist Theory." *Journalism & Mass Communication Quarterly* 83: 247–264.

Harris, Allen C. 1994. "Ethnicity as a Determinant of Sex Role Identity: A Replication Study of Item Selection for the Bern Sex Role Inventory." *Sex Roles* 31 (August) 241–273.

Harris, Angel L. 2006. "I Don't Hate School: Revisiting 'Oppositional Culture' Theory of Blacks' Resistance to Schooling." *Social Forces* 85: 797–834.

Harris, Angel L., and Keith Robinson. 2007. "Schooling Behavior or Prior Skills? A Cautionary Tale of Omitted Variable Bias within Oppositional Culture Theory." *Sociology of Education* 80: 139–157.

Harris, Marvin. 1974. *Cows, Pigs, Wars, and Witches: The Riddles of Culture.* New York: Vintage.

Harrison, Bennett, and Harry Bluestone. 1982. *The Deindustrialization of America: Plant Closings, Community Abandonment, and the Dismantling of Basic Industry.* New York: Basic Books.

Harrison, Roderick. 2000. "Inadequacies of Multiple Response Race Data in the Federal Statistical System." Manuscript. Joint Center for Political and Economic Studies and Howard University, Department of Sociology, Washington, DC.

Harry, Joseph. 1979. "The Marital 'Liaisons' of Gay Men." *The Family Coordinator* 28 (October): 622–629.

Hartman, Chester, and Gregory D. Squires (eds.). 2006. *There Is No Such Thing as a Natural Disaster: Race, Class, and Hurricane Katrina.* New York: Routledge.

Hastorf, Albert, and Hadley Cantril. 1954. "They Saw a Game: A Case Study." *Journal of Abnormal and Social Psychology* 40 (2): 129–134.

Hauan, Susan M., Nancy S. Landale, and Kevin T. Leicht. 2000. "Poverty and Work Effort among Urban Latino Men." *Work and Occupations* 27 (May): 188–222.

Hauser, Robert M., Howard F. Taylor, and Troy Duster. 1995. "The Bell Curve." *Contemporary Sociology* 24 (March): 149–161.

Hawley, Amos H. 1986. *Human Ecology: A Theoretical Essay.* Chicago, IL: University of Chicago Press.

Hays, Sharon. 2003. *Flat Broke with Children: Women in the Age of Welfare Reform.* New York: Oxford University Press.

Hearnshaw, Leslie. 1979. *Cyril Burt: Psychologist.* Ithaca, NY: Cornell University Press.

Heider, Fritz. 1958. *The Psychology of Interpersonal Relations.* New York: John Wiley and Sons, Inc.

Heilman, Brice E., and Paul J. Kaiser. 2002. "Religion, Identity, and Politics in Tanzania." *Third World Quarterly* 23 (August): 691–709.

Heimer, Karen. 1997. "Socioeconomic Status, Subcultural Definitions and Violent Delinquency." *Social Forces* 75 (March): 799–833.

Hendy, Helen M., Cheryl Gustitus, and Jamie Leitzel-Schwalm. 2001. "Social Cognitive Predictors of Body Images in Preschool Children." *Sex Roles* 44 (May): 557–569.

Henry, Kathy. 2008. "Warrior Woman—Ida B. Wells Barnett." *Ezinearticles.* **ezinearticles.com**

Henslin, James M. 1993. "Doing the Unthinkable." Pp. 253–262 in *Down to Earth Sociology,* 7th ed., edited by James M. Henslin. New York: Free Press.

Herman, Judith. 1981. *Father-Daughter Incest.* Cambridge, MA: Harvard University Press.

Herrnstein, Richard J., and Charles Murray. 1994. *The Bell Curve: Intelligence and Class Structure in American Life.* New York: Free Press.

Herskovits, Melville J. 1941. *The Myth of the Negro Past.* New York: Harper and Brothers.

Hesse-Biber, Sharlene Hagy. 2007. *The Cult of Thinness,* 2nd ed. New York: Oxford University Press.

Higginbotham, A. Leon. 1978. *In the Matter of Color: Race and the American Legal Process.* New York: Oxford University Press.

Higginbotham, Elizabeth. 2001. *Too Much to Ask: Black Women in the Era of Integration.* Chapel Hill, NC: University of North Carolina Press.

Hill, C.T., Z. Rubin, and L.A. Peplau. 1976. "Breakups Before Marriage: The End of 103 Affairs."

Journal of Social Issues 32: 147–168.

Hill, Lori Diane. 2001. *Conceptualizing Educational Attainment Opportunities of Urban Youth: The Effects of School Capacity, Community Context and Social Capital.* Ph.D. dissertation, University of Chicago.

Hirschi, Travis. 1969. *Causes of Delinquency.* Berkeley: University of California Press.

Hirschman, Charles. 1994. "Why Fertility Changes." *Annual Review of Sociology* 20: 203–223.

Hochschild, Arlie. 1983. *The Managed Heart: Commercialization of Human Feelings.* Berkeley, CA: University of California Press.

Hochschild, Arlie Russell. 1997. *The Time Bind: When Work Becomes Home and Home Becomes Work.* New York: Metropolitan Books.

Hochschild, Arlie Russell, with Anne Machung. 1989. *The Second Shift: Working Parents and the Revolution at Home.* New York: Viking.

Hoffman, Saul D., and Greg J. Duncan. 1988. "What *Are* the Economic Consequences of Divorce?" *Demography* 25 (November): 641–645.

Hoffnung, Michele. 2004. "Wanting It All: Career, Marriage, and Motherhood During College-Educated Women's 20s." *Sex Roles* 50(May): 711–723.

Hofstadter, Richard. 1944. *Social Darwinism in American Thought.* Philadelphia, PA: University of Pennsylvania Press.

Hollander, Jocelyn A. 2002. "Resisting Vulnerability: The Social Reconstruction of Gender in Interaction." *Social Problems* 49 (November): 474–496.

Hollingshead, August B., and Frederick C. Redlich. 1958. *Social Class and Mental Illness: A Community Study.* New York: Wiley.

Holmes, Schuyler. 2000. "Environmental Racism and Classism in Toxic Waste Dumping in Cleveland, Ohio." Junior thesis. Princeton University.

Homans, George. 1974. *Social Behavior: Its Elementary Forms,* revised ed. New York: Harcourt Brace Jovanovich.

Hondagneu-Sotelo, Pierrette, and Ernestine Avila. 1997. "'I'm Here, but I'm There': The Meanings of Latina Transnational Motherhood." *Gender & Society* 11(5): 548–571.

Hondagneu-Sotelo, Pierrette. 2001. *Doméstica: Immigrant Workers Cleaning and Caring in the Shadows of Affluence.* Berkeley, CA: University of California Press.

Hong, Laurence K. 1978. "Risky Shift and Cautious Shift: Some Direct Evidence on the Culture-Value Theory." *Social Psychology* 41 (December): 342–346.

Hornung, Carlton A. 1977. "Social Status, Status Inconsistency and Psychological Stress." *American Sociological Review* 42 (August): 623–638.

Horowitz, Ruth. 1995. *Teen Mothers: Citizens or Dependents?* Chicago, IL: University of Chicago Press.

Horton, Hayward Derrick, Beverlyn Lundy Allen, Cedric Herring, and Melvin E. Thomas. 2000. "Lost in the Storm: The Sociology of the Black Working Class, 1850 to 1990." *American Sociological Review* 65 (February): 128–137.

House, James S. 1980. *Occupational Stress and the Mental and Physical Health of Factory Workers.* Ann Arbor, MI: Survey Research Center.

Hoyt, Wendy, and Lori R. Kogan. 2001. "Satisfaction with Body Image and Peer Relationships for Males and Females in a College Environment." *Sex Roles* 45 (August): 199–215.

Hudson, Ken. 1999. "No Shortage of 'Nonstandard' Jobs." *Briefing Paper.* Washington, DC: Economic Policy Institute, December.

Huesmann, L. R., J. Moise, C. D. Podolski, and J. D. Eron. 2003. "Longitudinal Relations Between Childhood Exposure to Media Violence and Adult Aggression and Violence, 1977–1992." *Developmental Psychology* 39: 201–221.

Hughes, Langston. 1967. *The Big Sea.* New York: Knopf.

Hunt, Ruth. 2005. "Subjective Rating by Skin Color Gradation, Gender, and Other Characteristics." Junior Project, Princeton University, manuscript.

Hunter, Andrea, and Sherrill L. Sellers. 1998. "Feminist Attitudes Among African American Women and Men." *Gender & Society* 12 (February): 81–99.

Hunter, Margaret. 2002. "Rethinking Epistemology, Methodology, and Racism: or, Is White Sociology Really Dead?" *Race and Society* 5 (No. 2): 119–138.

Ignatiev, Noel. 1995. *How the Irish Became White.* New York: Routledge.

Inglehart, Ronald, and Wayne E. Baker. 2000. "Modernization, Cultural Change, and the Persistence of Traditional Values." *American Sociological Review* 65 (February): 19–51.

International Labour Organization. 2002. *Every Child Counts: Estimates on Child Labour.* **www.ilo.org**

Irwin, Katherine. 2001. "Legitimating the First Tattoo: Moral Passage through Informal Interaction." *Symbolic Interaction* 24 (March): 49–73.

Jackson, Pamela Brayboy. 1992. "Specifying the Buffering Hypothesis: Support, Strain, and Depression." *Social Psychology Quarterly* 55: 363–378.

Jackson, Pamela Braboy. 2000. Unpublished. "Stress and Coping among Black Elites in Organizational Settings." manuscript.

Jackson, Pamela B., Peggy A. Thoits, and Howard F. Taylor. 1994. "The Effects of Tokenism on America's Black Elite." Paper read before the American Sociological Association. August, Los Angeles, CA.

Jackson, Pamela B., Peggy A. Thoits, and Howard F. Taylor. 1995. "Composition of the Workplace and Psychological Well-Being: The Effects of Tokenism on America's Black Elite." *Social Forces* 74 (December): 543–557.

Jacobs, David, Zhenchao Qian, Jason T. Charmichael, and Stephanie L. Kent. 2007. "Who Survives Death Row: An Individual and Contextual Analysis." *American Sociological Review* 72 (August): 610–632.

Jacobs, Jerry A., and Kathleen Gerson. 2004. *The Time Divide: Work, Family, and Gender Inequality.* New York: Oxford University Press.

Jacobs, Lawrence R., and James A. Morone. 2004. "Health and Wealth." *The American Prospect* (June): A20–21.

Jang, Kerry L., W. J. Lively, and Philip A. Vernon. 1996. "Heritability of the Big Five Personality

Dimensions and Their Facets: A Twin Study." *Journal of Personality* 64: 577–589.

Janis, Irving L. 1982. *Groupthink: Psychological Studies of Policy Decisions and Fiascos,* 2nd ed. Boston, MA: Houghton Mifflin.

Janus, Samuel S., and Cynthia L. Janus. 1993. *The Janus Report on Sexual Behavior.* New York: Wiley.

Jencks, Christopher. 1993. *Rethinking Social Policy: Race, Poverty, and the Underclass.* New York: Harper Collins.

Jencks, Christopher, and Meredith Phillips (eds.). 1998. *The Black-White Test Score Gap.* Washington, DC: Brookings Institution Press.

Jencks, Christopher, Susan Bartlett, Mary Corcoran, James Crouse, David Eaglesfield, Gregory Jackson, Kent McClelland, Peter Mueser, Michael Olneck, Joseph Schwartz, Sherry Ward, and Jill Williams. 1979. *Who Gets Ahead? The Determinants of Economic Success in America.* New York: Basic Books.

Jencks, Christopher, Marshall Smith, Henry Ackland, Mary Jo Bane, David Cohen, Herbert Gintis, Barbara Heyns, and Stephan Michelson. 1972. *Inequality: A Reassessment of the Effect of Family and Schooling in America.* New York: Basic Books.

Jenness, Valerie. 1995. "Social Movement Growth, Domain Expansion, and Framing Processes: The Gay/Lesbian Movement and Violence Against Gays and Lesbians as a Social Problem." *Social Problems* 42 (February): 145–170.

Jenness, Valerie, and Kendal Broad. 1997. *Hate Crimes: New Social Movements and the Politics of Violence.* New York: Aldine de Gruyter.

Jenness, Valerie, and Kendal Broad. 2002. *Hate Crimes: New Social Movements and the Politics of Violence.* New York: Aldine de Gruyter.

Jennings, M. K., and R. G. Niemi. 1974. *The Political Character of Adolescence.* Princeton, NJ: Princeton University Press.

Johnson, Norris R., James G. Stember, and Deborah Hunter. 1977. "Crowd Behavior as Risky Shift: A Laboratory Experiment." *Sociometry* 40 (2): 183–187.

Johnston, David Cay. 2000. "Corporations' Taxes are Falling Even as Individual Burden Rises." *The New York Times* (February 20): A1ff.

Johnstone, Ronald I. 1992. *Religion in Society: A Sociology of Religion,* 4th ed. Englewood Cliffs, NJ: Prentice Hall.

Jones, Diane Carlson. 2001. "Social Comparison and Body Image: Attractiveness Comparison to Models and Peers among Adolescent Girls and Boys." *Sex Roles* 45 (November): 645–664.

Jones, E. E., and R. E. Nisbett. 1972. "The Actor and the Observer: Divergent Perceptions of Causes of Behavior." Pp. 79–94 in *Attribution: Perceiving the Causes of Behavior,* edited by E. E. Jones, D. E. Kanouse, H. H. Kelley, R. E. Nisbett, S. Valins, and B. Weiner. Morristown, NJ: General Learning Press.

Jones, Edward E., Amerigo Farina, Albert H. Hastorf, Hazel Markus, Dale T. Miller, and Robert A. Scott. 1986. *Social Stigma: The Psychology of Marked Relationships.* New York: W. H. Freeman.

Jones, James M. 1997. *Prejudice and Racism,* 2nd ed. New York: McGraw-Hill.

Jones, Jeffrey M. 2008. "Confidence in Congress: Lowest Ever for Any U.S. Institution." **www.gallup.com**

Jordon, Winthrop D. 1968. *White Over Black: American Attitudes Toward the Negro 1550–1812.* Chapel Hill, NC: University of North Carolina Press.

Jordan, Winthrop D. 1969. *The White Man's Burden: Historical Origins of Racism in the United States.* New York: Oxford University Press.

Joseph, Jay. 2001. "Separated Twins and the Genetics of Personality Differences: A Critique." *American Journal of Psychology* 114 (1): 1–30.

Joseph, Jay. 2004. *The Gene Illusion: Genetic Research in Psychiatry and Psychology Under the Microscope.* New York: Algora.

Jucha, Robert. 2002. *Terrorism.* Belmont, CA: Wadsworth.

Kadushin, Charles. 1974. *The American Intellectual Elite.* Boston, MA: Little, Brown.

Kalleberg, Arne L., Barbara F. Reskin, and Ken Hudson. 2000. "Bad Jobs in America: Standard and Nonstandard Employment Relations and Job Quality in the United States." *American Sociological Review* 65 (April): 256–278.

Kalmijn, Matthijs. 1991. "Status Homogamy in the United States." *American Journal of Sociology* 97 (September): 496–523.

Kalmijn, Matthijs. 1999. "Father Involvement in Childrearing and the Perceived Stability of Marriage." *Journal of Marriage and the Family* 61 (May): 409–421.

Kalof, Linda. 1999. "The Effects of Gender and Music Video Imagery on Sexual Attitudes." *Journal of Social Psychology* 139 (June): 378–385.

Kamin, Leon J. 1974. *The Science and Politics of IQ.* Potomac, MD: Lawrence Erlbaum.

Kamin, Leon J. 1995. "Behind the Curve." *Scientific American* 272 (February): 99–103.

Kamin, Leon J., and Arthur S. Goldberger. 2002. "Twin Studies in Behavioral Research: A Skeptical View." *Theoretical Population Biology* 61: 83–95.

Kane, Emily W. 2006. "'No Way My Boys Are Going to Be Like That!'": Parents' Responses to Children's Gender Nonconformity." *Gender & Society* 20(2):149–176.

Kanter, Rosabeth Moss. 1977. *Men and Women of the Corporation.* New York: Basic Books.

Kaplan, Elaine Bell. 1996. *Not Our Kind of Girl: Unraveling the Myths of Black Teenage Motherhood.* Berkeley, CA: University of California Press.

Kasarda, Jack. 1999. "Industrial Restructuring and the Changing Location of Jobs." In *State of the Union: America in the 1990s: Economic Trends,* edited by Reynolds Farley. New York: Russell Sage Foundation.

Katz, I., J. Wackenhut, and R. G. Hass. 1986. "Racial Ambivalence, Value Duality, and Behavior." Pp. 35–60 in *Prejudice, Discrimination, and Racism,* edited by J. F. Dovidio and S. L. Gaertner. New York: Academic Press.

Kaufman, Gayle. 1999. "The Portrayal of Men's Family Roles in Television Commercials." *Sex Roles* 41 (September): 439–458.

Keeter, Scott, Juliana Horowitz, and Alec Tyson. 2008. *Young Voters in the 2008 Election.* Washington, DC: Pew Research Center.

Keith, Verna M. 1997. "Life Stress and Psychological Well-Being among Married and Unmarried Blacks." Pp. 95–116 in *Family Life in Black America.* Thousand Oaks, CA: Sage.

Kelley, J. R., J. W. Jackson, and S. L. Huston-Comeaux. 1999. "The Effects of Time Pressure and Task Differences on Influence Models and Accuracy in Decision-Making Groups." *Personality and Social Psychology Bulletin* 23: 10–22.

Kendall, Lori. 2002. "'Oh No! I'm a Nerd!': Hegemonic Masculinity on an Online Forum." *Gender & Society* 14 (April): 256–274.

Kennedy, Randall. 2003. *Interracial Intimacies: Sex, Marriage, Identity and Adoption.* New York: Pantheon.

Kennedy, Robert E. 1989. *Life Choices: Applying Sociology,* 2nd ed. New York: Holt, Rinehart and Winston.

Kephart, W. H. 1993. *Extraordinary Groups: An Examination of Unconventional Life,* rev. ed. New York: St. Martin's Press.

Kerr, N. L. 1992. "Issue Importance and Group Decision Making." Pp. 68–88 in *Group Process and Productivity,* edited by S. Worchel, W. Wood, and J. A. Simpson. Newbury Park, CA: Sage.

Kershaw, Sarah. 2009. "Talk About Race?" *The New York Times* (January 15): E1, E5.

Kessler, Suzanne J. 1990. "The Medical Construction of Gender: Case Management of Intersexed Infants." *Signs* 16 (Autumn): 3–26.

Khanna, Nikki D., Cherise Harris, and Rana Cullers. 1999. "Attitudes toward Interracial Dating." Paper presented at the annual meeting of the Society for the Study of Social Problems, Chicago, IL.

Khashan, Hilal, and Lina Kreidie. 2001. "The Social And Economic Correlates of Islamic Religiosity." *World Affairs* 1654 (Fall): 83–96.

Kiefer, Amy K. 2007. "Implicit Stereotypes and Women's Math Performance: How Implicit Gender-Math Stereotypes Influence Women's Susceptibility to Stereotype Threat." *Journal of Experimental Social Psychology* 43 (5): 825–832.

Kim, Elaine H. 1993. "Home Is Where the *Han* Is: A Korean American Perspective on the Los Angeles Upheavals." Pp. 215–235 in *Reading Rodney King / Reading Urban Uprising,* edited by Robert Gooding-Williams. New York: Routledge.

Kimmel, Michael. 2000. "Saving the Males: The Sociological Implications of Virginia Military Institute and the Citadel." *Gender & Society* 14 (August): 494–516.

Kimmel, Michael. 2008. *Guyland: The Perilous World Where Boys Become Men.* New York: Harper.

Kimmel, Michael S., and Michael A. Messner. 2004. *Men's Lives,* 6th ed. Boston, MA: Allyn and Bacon.

Kitano, Harry. 1976. *Japanese Americans: The Evolution of a Subculture,* 2nd ed. New York: Prentice Hall.

Kitsuse, John I., and Aaron V. Cicourel. 1963. "A Note on the Uses of Official Statistics." *Social Problems* 11 (Fall): 131–139.

Klapp, Orin. 1972. *Currents of Unrest: Introduction to Collective Behavior.* New York: Holt, Rinehart, and Winston.

Kleinfeld, Judith S. 2002. "The Small World Problem." *Society* 39: 61–66.

Klinger, Lori J., James A. Hamilton, and Peggy J. Cantrell. 2001. "Children's Perceptions of Aggression and Gender-Specific Content in Toy Commercials." *Social Behavior and Personality* 29: 11–20.

Kluegel, J. R., and Lawrence Bobo. 1993. "Dimensions of Whites' Beliefs About the Black-White Socioeconomic Gap." Pp. 127–147 in *Race and Politics in American Society,* edited by P. Sniderman, P. Tetlock, and E. Carmines. Stanford, CA: Stanford University Press.

Kniss, Fred. 2003. "Church and State." *Contexts* 2 (Spring) 62–63.

Knoke, David. 1992. *Political Networks: The Structural Perspective.* New York: Cambridge University Press.

Kochen, M. (ed.). 1989. *The Small World.* Norwood, NJ: Ablex Press.

Kocieniewski, David, and Robert Hanley. 2000. "Racial Profiling Was Routine, New Jersey Says." *The New York Times* (November 28): 1.

Kohlberg, Lawrence. 1969. "'Stage and Sequence': The Cognitive-Developmental Approach to Socialization." Pp. 347–480 in *Handbook of Socialization and Research,* edited by D.A. Goslin. Chicago, IL: Rand McNally.

Kohut, Andrew. 2007. "Muslim Americans: Middle Class and Mostly Mainstream." New Pew Research Center.

Konishi, Hideo, and Debraj Ray. 2003. "Coalition Formation as a Dynamic Process." *Journal of Economic Theory* 110 (May): 1–41.

Kovel, Johnathan. 1970. *White Racism: A Psychohistory.* New York: Pantheon.

Krasnodemski, Memory. 1996. "Justified Suffering: Attribution Theory Applied to Perceptions of the Poor in America." Unpublished senior thesis, Princeton University.

Kuhn, Harold W., and Sylvia Nasar. 2002. *The Essential John Nash.* Princeton, NJ: Princeton University Press.

Kung, H. C., D. L. Hoyert, J. Q. Xu, and S. L. Murphy. 2008. *Deaths: Final Data for 2005. National Vital Statistics Reports* 56, No 10. Hyattsville, MD: National Center for Health Statistics.

Kurz, Demi. 1989. "Social Science Perspectives on Wife Abuse: Current Debates and Future Directions." *Gender & Society* 3 (December): 489–505.

Kurz, Demie. 1995. *For Richer for Poorer: Mothers Confront Divorce.* New York: Routledge.

Lacy, Karen R. 2007. *Blue-Chip Black: Race, Class and Status in the New Black Middle Class.* Berkeley, CA: University of California Press.

Ladner, Joyce A. 1986. "Teenage Pregnancy: Implications for Black Americans." Pp. 65–84 in *The State of Black America 1986,* edited by James D. Williams. New York: National Urban League.

LaFlamme, Darquise, Andree Pomerrleau, and Gerard Malcuit. 2002. "A Comparison of Fathers' and Mothers' Involvement in Childcare and Stimulation Behaviors during Free-Play with Their Infants at 9 and 15 Months." *Sex Roles* 11–12 (December): 507–518.

LaFrance, Marianne, 2002. "Smile Boycotts and Other Body Politics." *Feminism & Psychology* 12 (August): 319–323.

Lamanna, Mary Ann, and Agnes Riedman. 2008. *Marriage and*

Families: Making Choices in a Diverse Society, 10th ed. Belmont, CA: Wadsworth.

Lamb, Michael. 1998. "Cybersex: Research Notes on the Characteristics of the Visitors to Online Chat Rooms." *Deviant Behavior* 19 (April–June): 121–135.

Lamont, Michèle. 1992. *Money, Morals, and Manners: The Culture of the French and the American Upper-Middle Class.* Chicago, IL: University of Chicago Press.

Lamphere, Louise, Alex Stepick, and Guillermo Grenier (eds.). 1994. *Newcomers in the Workplace: Immigrants and the Restructuring of the U.S. Economy.* Philadelphia, PA: Temple University Press.

Langhinrichsen-Rohling, Jennifer. Peter Lewinsohn, Paul Rohde, John Seeley, Candice M. Monson, Kathryn A. Meyer, and Richard Langford. 1998. "Gender Differences in the Suicide-Related Behaviors of Adolescents and Young Adults." *Sex Roles* 39 (December): 839–854.

Langman, Lauren, and Douglas Morris. 2002. "Internetworked Social Movements: The Promises and Prospects for Global Justice." Paper presented at the International Sociological Association, Brisbane, Australia.

Lareau, Annette. 2003. *Unequal Childhoods: Class, Race, and Family Life.* Berkeley, CA: University of California Press.

Larana, Enrique, Hank Johnston, and Joseph R. Gusfield (eds.). 1994. *New Social Movements: From Ideology to Identity.* Philadelphia, PA: Temple University Press.

Lau, Bonnie. 2002. "Stereotype Threat: Minority Education, Intelligence Theory, Affirmative Action, and an Empirical Study of Women in the Quantitative Domain." Senior thesis, Princeton University.

Laumann, Edward O., John H. Gagnon, Robert T. Michael, and Stuart Michaels. 1994. *The Social Organization of Sexuality: Sexual Practices in the United States.* Chicago, IL: University of Chicago Press.

Laveist, Thomas A., and Amani Nuru-Jeter. 2002. "Is Doctor-Patient Race Concordance Associated with Greater Satisfaction with Care?" *Journal of*

Health and Social Behavior 43 (September): 296–306.

Lavelle, Louis. 2001. "Executive Pay." *Business Week,* April 16, **www.businessweek.com**

Lawson, Helene M., and Kira Leck. 2006. "Dynamics of Internet Dating." *Social Science Computer Review* 24 (Summer): 189–208.

Lawson-Borders, Gracie, and Rita Kirk. 2005. "Blogs in Campaign Communication." *American Behavioral Scientist* 49: 548–559.

Lederman, Douglas. 1992. "Men Outnumber Women and Get Most of the Money in Big-Time Sports Programs." *The Chronicle of Higher Education* 38 (April): A1ff.

Lee, Gary R., Karen Seccombe, and Constance L. Shehan. 1991. "Marital Status and Personal Happiness: An Analysis of Trend Data." *Journal of Marriage and the Family* 53(4): 839–844.

Lee, Matthew T., and M. David Ermann. 1999. "Pinto 'Madness' as a Flawed Landmark Narrative: An Organizational and Network Analysis." *Social Problems* 46 (February): 30–47.

Lee, Sharon M. 1993. "Racial Classification in the U.S. Census: 1890–1990." *Ethnic and Racial Studies* 16 (1): 75–94.

Lee, Sharon M. 1994. "Poverty and the U.S. Asian Population." *Social Science Quarterly* 75 (September): 541–559.

Lee, Stacey J. 1996. *Unraveling the "Model Minority" Stereotype: Listening to Asian American Youth.* New York: Teacher's College Press.

Lehmann, Nicholas. 1999. *The Big Test: The Secret History of the American Meritocracy.* New York: Farrar, Straus, and Giroux.

Leidner, Robin. 1993. *Fast Food, Fast Talk: Service Work and the Routinization of Everyday Life.* Berkeley, CA: University of California Press.

Lengermann, P. M., and Niebrugge-Brantley, J. 1998. *The Woman Founders: Sociology and Social Theory, 1830–1930.* Boston: McGraw-Hill.

Lenski, Gerhard. 2005. *Ecological-Evolutionary Theory: Principles and Applications.* Boulder, CO: Paradigm Publishers.

Lenski, Gerhard, Jean Lenski, and Patrick Nolan. 1998. *Human Societies: An Introduction to*

Macro-Sociology, 8th ed. New York: McGraw-Hill.

Leonhardt, David. 2006. "Gender Pay Gap, Once Narrowing, Is Stuck in Place." *The New York Times* (December 24): A1.

Lever, Janet. 1978. "Sex Differences in the Complexity of Children's Play and Games." *American Sociological Review* 43 (August): 471–483.

Levine, Adeline. 1982. *Love Canal: Science, Politics, and People.* Lexington, MA: Lexington Books.

Levine, Felice J., and Katherine J. Rosich. 1996. *Social Causes of Violence: Crafting a Science Agenda.* Washington, DC: American Sociological Association.

Levine, Lawrence. 1984. "William Shakespeare and the American People: A Study in Cultural Transformation." *American Historical Review* 89 (1): 34–66.

Levy, Ariel. 2007. *Female Chauvinist Pigs: Women and the Rise of Raunch Culture.* New York: Free Press.

Levy, Marion J. 1949. *The Structure of Society.* Princeton, NJ: Princeton University Press.

Lewin, Tamar. 2002. "Study Links Working Mothers to Slower Learning." *The New York Times* (July 17): A14.

Lewis, Amanda. 2003. *Race in the Schoolyard: Negotiating the Color Line in Classrooms and Communities.* New Brunswick, NJ: Rutgers University Press.

Lewis, Dan A., Amy Bush Stevens, and Kristen Shook Slack. 2002. *Illinois Families Study, Welfare Reform in Illinois: Is the Moderate Approach Working?* Evanston, IL: Institute for Poverty Research, Northwestern University.

Lewis, Oscar. 1960. *Five Families: Mexican Case Studies in the Culture of Poverty.* New York: Basic Books.

Lewis, Oscar. 1966. "The Culture of Poverty." *Scientific American* 215 (October): 19–25.

Lewontin, Richard C. 1970. "Race and Intelligence." *Bulletin of the Atomic Scientists* 26: 2–8.

Lewontin, Richard. 1996. *Human Diversity.* New York: W. H. Freeman.

Lieberson, Stanley. 1980. *A Piece of the Pie: Black and White Immigrants Since 1880.* Berkeley, CA: University of California Press.

Light, Ivan, Rebecca Kim, and Connie Hunt. 1998. "Globalization, Vacancy Chains, or Migration Networks?: Immigrant Employment and Income in Greater Los Angeles, 1970–1990." Paper presented at the International Sociological Association, Montreal.

Lin, Nan. 1989. "The Small World Technique as a Theory Construction Tool." Pp. 231–238 in *The Small World,* edited by M. Kochen. Norwood, NJ: Ablex Press.

Lindberg, L. 2003. "Risk Factors for Anorexia Nervosa: A National Cohort Study." *International Journal of Eating Disorders* 334 (4): 397–408.

Livingston, I. L. 1994. *Handbook of Black American Health: The Mosaic of Conditions, Issues and Prospects.* Westport, CT: Greenwood.

Lo, Clarence. 1990. *Small Property Versus Big Government: Social Origins of the Property Tax Revolt.* Berkeley, CA: University of California Press.

Locklear, Erin M. 1999. "Where Race and Politics Collide: The Federal Acknowledgement Process and Its Effects on Lumsee and Pequot Indians." Unpublished senior thesis, Princeton University.

Logio-Rau, Kim. 1998. *Here's Looking at You, Kid: Race, Gender and Health Behaviors among Adolescents.* Unpublished doctoral dissertation, University of Delaware.

Lombardo, William K., Gary A. Cretser, and Scott C. Roesch. 2001. "For Crying Out Loud— The Differences Persist into the '90s." *Sex Roles* 45 (December): 529–547.

Lorber, Judith. 1994. *Paradoxes of Gender.* New Haven, CT: Yale University Press.

Lorenz, Conrad. 1966. *On Aggression.* New York: Harcourt Brace Jovanovich.

Louie, Miriam Ching Yoon. 2001. *Sweatshop Warriors: Immigrant Women Workers Take on the Global Factory.* Cambridge, MA: South End Press.

Lovejoy, Meg. 2001. "Disturbances in the Social Body: Differences in Body Image and Eating Problems Among African American and White Women." *Gender & Society* 15 (April): 239–261.

Lucal, Betsy. 1994. "Class Stratification in Introductory Textbooks:

Relational or Distributional Models?" *Teaching Sociology* 22 (April): 139–150.

Lucas, Samuel R. 1999. *Tracking Inequality: Stratification and Mobility in American High Schools.* New York: Teachers College Press.

Luckenbill, David F. 1986. "Deviant Career Mobility: The Case of Male Prostitutes." *Social Problems* 33 (April): 283–296.

Luker, Kristin. 1975. *Taking Chances: Abortion and the Decision Not to Contracept.* Berkeley, CA: University of California Press.

Luker, Kristin. 1984. *Abortion and the Politics of Motherhood.* Berkeley, CA: University of California Press.

Luker, Kristin. 1996. *Dubious Conceptions: The Politics of Teenage Pregnancy.* Cambridge, MA: Harvard University Press.

Lyons, Linda. 2002. "Teen Attitudes Contradict Sex-Crazed Stereotype." *The Gallup Poll.* January 29, **www.gallup.com**

Lyons, Linda. 2004. "Teens: Sex Can Wait." Princeton, NJ: The Gallup Poll. **www.gallup.com**

Macdougall, Robert. 2005. "Identity, Electronic Ethos, and Blogs: A Technologic Analysis of Symbolic Exchange on the New News Medium." *American Behavioral Scientist* 49: 575–599.

Machel, Graca. 1996. *Impact of Armed Conflict on Children.* New York: UNICEF/United Nations.

MacKinnon, Catherine. 1983. "Feminism, Marxism, Method, and the State: An Agenda for Theory." *Signs* 7 (Spring): 635–658.

MacKinnon, Catherine A. 2006. "Feminism, Marxism, Method, and the State: An Agenda for Theory." Pp. 829–868 in D. Kennedy and W.F. Fisher, *The Canon of American Legal Thought.* Princeton: Princeton University Press.

MacKinnon, Neil J., and Tom Langford. 1994. "The Meaning of Occupational Prestige Scores: A Social Psychological Analysis and Interpretation." *Sociological Quarterly* 35 (May): 215–145.

Mackintosh, N. J. 1995. *Cyril Burt: Fraud or Framed?* Oxford, England: Oxford University Press.

Madriz, Esther. 1997. *Nothing Bad Happens to Good Girls: Fear of*

Crime in Women's Lives. Berkeley, CA: University of California Press.

Maimon, David, and Danielle C. Kuhl. 2008. "Social Control and Youth Suicidality: Situating Durkheim's Ideas in a Multilevel Framework." *American Sociological Review* 73 (December): 921–943.

Majete, Clayton A. 1999. "Family Relationships and the Interracial Marriage." Paper presented at the annual meeting of the American Sociological Association, Chicago, IL.

Malat, Jennifer. 2001. "Social Distance and Patients' Rating of Healthcare Providers." *Journal of Health and Social Behavior* 42 (December): 36–372.

Malcomson, Scott L. 2000. *The American Misadventure of Race.* New York: Farrar, Strauss, and Giroux.

Maldonado, Lionel, A. 1997. "Mexicans in the American System: A Common Destiny." In *Ethnicity in the United States: An Institutional Approach,* edited by William Velez. Bayside, NY: General Hall.

Maldonado, Lionel A., and Charles V. Willie. 1996. "Developing a 'Pipeline' Recruitment Program for Minority Faculty." Pp. 330–371 in *Educating a New Majority: Transforming America's Educational System for Diversity,* edited by Laura I. Rendon and Richard O. Hope. San Francisco, CA: Jossey-Bass.

Malthus, Thomas Robert. 1798 [1926]. *First Essay on Population 1798.* London: Macmillan.

Mandel, Daniel. 2001. "Muslims on the Silver Screen." *Middle East Quarterly* 8 (Spring): 19–30.

Manning, Winton H., and Rex Jackson. 1984. "College Entrance Examinations: Objective Selection or Gatekeeping for the Economically Privileged," In *Perspectives on Bias in Mental Testing,* ed. by Cecil R. Reynolds and Robert T. Brown. New York: Plenum.

Marciniak, Liz-Marie. 1998. "Adolescent Attitudes Toward Victim Precipitation of Rape." *Violence and Victims* 12 (Fall): 287–300.

Marcuse, Herbert. 1964. *One-Dimensional Man.* Boston, MA: Beacon Press.

Margolin, Leslie. 1992. "Deviance on Record: Techniques for Labeling

Child Abusers in Official Documents." *Social Problems* 39 (February): 58–70.

Marks, Carole. 1989. *Farewell, We're Good and Gone: The Great Black Migration.* Bloomington, IN: Indiana University Press.

Marks, Carole, and Deana Edkins. 1999. *The Power of Pride: Stylemakers and Rulebreakers of the Harlem Renaissance.* New York: Crown.

Marquis, Christopher. 2003. "Total of Unmarried Couples Surged in 2000 Census." *The New York Times* (March 13): A32.

Marshall, Susan. 1997. *Splintered Sisterhood: Gender & Class in the Campaign against Woman Suffrage.* Madison, WI: University of Wisconsin Press.

Martin, Patricia Yancey, and Robert Hummer. 1989. "Fraternities and Rape on Campus." *Gender & Society* 3 (December): 457–473.

Martineau, Harriet. 1837. *Society in America.* London: Saunders and Otley.

Martineau, Harriet. 1838. *How to Observe Morals and Manners.* London: Charles Knight and Co.

Martinez, Ramiro. 2002. *Latino Homicide: Immigration, Violence and Community.* New York: Routledge.

Marx, Anthony. 1997. *Making Race and Nation: A Comparison of the United States, South Africa, and Brazil.* New York: Cambridge University Press.

Marx, Gary T. 1967. "Religion: Opiate or Inspiration of Civil Rights Militancy among Negroes." *American Sociological Review* 32 (February): 64–72.

Marx, Gary T., and Douglas McAdam. 1994. *Collective Behavior and Social Movements: Process and Structure.* Englewood Cliffs, NJ: Prentice Hall.

Marx, Karl. 1967 [1867]. *Capital.* F. Engels (ed.). New York: International Publishers.

Marx, Karl. 1972 [1843]. "Contribution to the Critique of Hegel's *Philosophy of Right.*" Pp. 11–23 in *The Marx-Engels Reader,* edited by Robert C. Tucker. New York: W. W. Norton.

Massey, D., and M. Fischer. 2007. "Stereotype Threat and Academic Performance: New Findings from a Racially Diverse Sample of College Freshmen." *Du Bois Review* 2 (1): 45–67.

Massey, Douglas S. 1993. "Latino Poverty Research: An Agenda for the 1990s." *Social Science Research Council Newsletter* 47 (March): 7–11.

Massey, Douglas S. 2005. *Strangers in a Strange Land: Humans in an Urbanizing World.* New York: Norton.

Massey, Douglas S., and Nancy A. Denton. 1993. *American Apartheid: Segregation and the Making of the Underclass.* Cambridge, MA: Harvard University Press.

Mathews, Linda. 1996. "More Than Identity Rests on a New Racial Category." *The New York Times* (July 6): 1–7.

Mauer, Marc. 1999. *Race to Incarcerate.* New York: The New Press.

Mazunder, Bhashkar. 2008. "Upward Intergenerational Mobility in the United States." Pew Charitable Trust Economic Mobility Project. **www. pewtrusts.org**

McAdam, Doug. 1982. *Political Process and the Development of Black Insurgency 1930–1970.* Chicago, IL: University of Chicago Press.

McAdam, Doug. 1999. *Political Process and the Development of Black Insurgency. 1930–1970,* 2nd ed. Chicago, IL: University of Chicago Press.

McCabe, Janice. 2005. "What's in a Label? The Relationship between Feminist Self-Identification and 'Feminist' Attitudes among U.S. Women and Men." *Gender & Society* 19(4): 480–505.

McCall, Leslie. 2001. *Complex Inequality: Gender, Class, and Race in the New Economy.* New York: Routledge.

McCarthy, John, and Mayer Zald. 1973. *The Trend of Social Movements in America: Professionalism and Resource Mobilization.* Morristown, NJ: General Learning Press.

McCauley, C. 1989. "The Nature of Social Influence in Groupthink: Compliance and Internalization." *Journal of Personality and Social Psychology* 57 (August): 250–260.

McClelland, Susan. 2003. "A Grim Toll on the Innocent." *Maclean's* (May 12): 20.

McCloskey, Laura Ann. 1996. "Socioeconomic and Coercive Power Within the Family." *Gender & Society* 10 (August): 449–463.

McCollum, Chris. 2002. "Relatedness and Self-Definition: Two Dominant Themes in Middle-Class Americans' Life Stories." *Ethos* 30: 113–139.

McComb, Chris. 2001 (May 4). "Few Say It's Ideal for Both Parents to Work Full Time Outside of Home." Princeton, NJ: The Gallup Poll.

McCord, William, and Arline McCord. 1986. *Paths to Progress: Bread and Freedom in Developing Societies.* New York: W.W. Norton.

McGuire, Gail M., and Barbara F. Reskin. 1993. "Authority Hierarchies at Work: The Impacts of Race and Sex." *Gender & Society* 7 (December): 487–506.

McGuire, Meredith. 2001. *Religion: The Social Context,* 5th ed. Belmont, CA: Wadsworth.

McIntyre, Rusty B., Rene M. Paulson, and Charles G. Lord. 2003. "Alleviating Women's Mathematics Stereotype Threat Through Salience of Group Achievements." *Journal of Experimental Social Psychology* 39: 83–90.

McLoyd, Vonnie C., Ana Mari Cauce, David Takeuchi, and Leon Wilson. 2000. "Marital Processes and Parental Socialization in Families of Color: A Decade Review of Research." *Journal of Marriage and the Family* 62 (November): 1070–1093.

Mead, George Herbert. 1934. *Mind, Self, and Society.* Chicago. IL: University of Chicago Press.

Mediamark Research and Intelligence. 2007. New York. **www. census.gov**

Meng, Susan. 2001. "Pet Sanctuary." *Forbes* (October 8): 236.

Menken, Jane, and M. Omar Rahman. 2001. "Reproductive Health." In *International Public Health,* edited by Michael K. Merson, Robert E. Black, and Anne J. Mills. Gaithersberg, MD: Aspen Publishers.

Meredith, Martin. 2003. *Elephant Destiny: Biography of an Endangered Species in Africa.* New York: HarperCollins.

Mernissi, Fataima. 1987. *Beyond the Veil: Male-Female Dynamics in Modern Muslim Society.* Bloomington, IN: Indiana University Press.

Merton, Robert K. 1957. *Social Theory and Social Structure.* New York: Free Press.

Merton, Robert K. 1968. "Social Structure and Anomie." *American Sociological Review* 3: 672–682.

Merton, Robert, and Alice K. Rossi. 1950. "Contributions to the Theory of Reference Group Behavior." Pp. 279–334 in *Continuities in Social Research Studies, Scope and Method of "The American Soldier,"* edited by Robert K. Merton and Paul F. Lazarsfeld. New York: Free Press.

Messerschmidt, James W. 1997. *Crime as Structured Action: Gender, Race, Class and Crime in the Making.* Thousand Oaks, CA: Sage.

Messner, Michael A. 1992. *Power at Play: Sports and the Problem of Masculinity.* Boston, MA: Beacon Press.

Messner, Michael A. 2002. *Taking the Field: Women, Men, and Sports.* Minneapolis, MN: University of Minnesota Press.

Metz, Michael A., B. R. Rosser-Simon, and Nancy Strapko. 1994. "Differences in Conflict-Resolution Styles Among Heterosexual, Gay, and Lesbian Couples." *Journal of Sex Research* 31: 293–308.

Meyer, David S., and Joshua Gamson. 1995. "The Challenge of Cultural Elites: Celebrities and Social Movements." *Sociological Inquiry* 65 (Spring): 181–206.

Meyer, Madonna Harrington. 1994. "Gender, Race, and the Distribution of Social Assistance: Medicaid Use Among the Elderly." *Gender & Society* 8 (March): 8–28.

Mickelson, Roslyn Arlin (ed.). 2000. *Children on the Streets of the Americas: Globalization, Homelessness, and Education in the United States, Brazil, and Cuba.* New York: Routledge.

Milgram, Stanley. 1974. *Obedience to Authority: An Experimental View.* New York: Harper & Row.

Milkie, Melissa A. 1999. "Social Comparisons, Reflected Appraisals, and Mass Media: The Impact of Pervasive Beauty Images on Black and White Girls' Self-Concepts." *Social Psychology Quarterly* 62 (June): 190–210.

Milkie, Melissa A. 2002. "Contested Images of Femininity: An Analysis of Cultural Gatekeepers' Struggles with the 'Real Girl' Critique." *Gender & Society* 16 (December): 839–859.

Milkie, Melissa A., Suzanne M. Bianchi, Marybeth J. Mattingly, and John P. Robinson. 2002. "Gendered Division of Childrearing: Ideals, Realities, and the Relationship to Parental Well-being." *Sex Roles* 47 (July): 21–38.

Milkie, Melissa, and Pia Peltola. 1999. "Playing All the Roles: Gender and the Work–Family Balancing Act." *Journal of Marriage and the Family* 61 (May): 476–490.

Miller, Eleanor. 1991. "Jeffrey Dahmer, Racism, Homophobia, and Feminism." *SWS Network News* 8 (December): 2.

Miller, Matthew, and Duncan Greenberg. 2008. "The Forbes 400." **www.forbes.com**

Miller, Susan L. 1997. "The Unintended Consequences of Current Criminal Justice Policy." Talk presented at Research on Women Series, University of Delaware, Newark, DE.

Mills, C. Wright. 1956. *The Power Elite.* New York: Oxford University Press.

Mills, C. Wright. 1959. *The Sociological Imagination.* New York: Oxford University Press.

Milner, Murray, Jr. 2004. *Freaks, Geeks, and Cool Kids: American Teenagers, Schools, and the Culture of Consumption.* New York: Routledge.

Miner, Horace. 1956. "Body Ritual Among the Nacirema." *American Anthropologist* 58: 503–507.

Mintz, Beth, and Michael Schwartz. 1985. *The Power Structure of American Business.* Chicago, IL: University of Chicago Press.

Mirandé, Alfredo. 1979. "Machismo: A Reinterpretation of Male Dominance in the Chicano Family." *The Family Coordinator* 28: 447–449.

Mirandé, Alfredo. 1985. *The Chicano Experience.* Notre Dame, IN: Notre Dame University Press.

Mishel, Lawrence, Jared Bernstein, and Sylvia Allegretto. 2005. *The State of Working America 2004/2005.* Washington, DC: Economic Policy Institute.

Mishel, Lawrence, Jared Bernstein, and Sylvia Allegretto. 2007. *The State of Working America 2006/2007.* Ithaca: NY: ILR Press.

Mishel, Lawrence, Jared Bernstein, and Heidi Shierholz. 2008. *State of Working America 2008/2009.* Washington, DC: Economic Policy Institute.

Misra, Joy, Stephanie Moller, and Maria Karides. 2003. "Envisioning Dependency: Changing Media Depictions of Welfare in the 20th Century." *Social Problems* 50 (November): 482–504.

Mitchell, G., Stephanie Obradovich, Fred Herring, Chris Tromberg, and Alysson L. Burns. 1992. "Reproducing Gender in Public Places: Adults' Attention to Toddlers in Three Public Locales." *Sex Roles* 26 (September): 323–330.

Mizruchi, Ephraim H. 1983. *Regulating Society: Marginality and Social Control in Historical Perspective.* New York: Free Press.

Mizruchi, Mark S. 1992. *The Structure of Corporate Political Action: Interfirm Relations and Their Consequences.* Cambridge, MA: Harvard University Press.

Moen, Phyllis, Jungmeen E. Kim, and Heather Hofmeister. 2001. "Couples' Work/Retirement Transitions, Gender, and Marital Quality." *Social Psychology Quarterly* 64 (March): 55–71.

Moen, Phyllis. 2003. *It's About Time: Couples and Careers.* Ithaca, NY: Cornell University Press.

Moen, Phyllis, with Donna Dempster-McClain, Joyce Altobelli, Wipas Wimonsate, Lisa Dahl, Patricia Roehling, and Stephen Sweet. 2004. *The New "Middle" Workforce.* The Bronfenbrenner Life Course Center and Cornell Careers Institute. **www.lifecourse.cornell**

Mohai, Paul, and Robin Saha. 2007. "Racial Inequality in the Distribution of Hazardous Waste: A National-Level Reassessment." *Social Problems* 54 (3): 343–370.

Moller, Laura C., Shelley Hymel, and Kenneth H. Rubin. 1992. "Sex Typing in Play and Popularity in Middle Childhood." *Sex Roles* 26 (April): 331–353.

Montada, L., and M. Lerner Jr. (eds.). 1998. *Responses to Victimization and Beliefs in a Just World.* New York: Plenum.

Montagne, Peter. 1995. "Review of Julian L. Simon, ed., *The State of Humanity.*" Cambridge, MA: Blackwell.

Montgomery, James D. 1992. "Job Search and Network Composition: Implications of the Strength of Work Ties Hypothesis." *American Sociological Review* 57 (October): 586–596.

Moore, David W. 2005. (August 15). "Gender Stereotypes Prevail on Working outside the Home." Princeton, NJ: The Gallup Poll.

Moore, Gwen. 1979. "The Structure of a National Elite Network." *American Sociological Review* 44 (October): 673–692.

Moore, Joan. 1976. *Hispanics in the United States.* Englewood Cliffs, NJ: Prentice Hall.

Moore, Joan W., and John M. Hagedorn. 1996. "What Happens to Girls in Gangs?" Pp. 205–218 in *Gangs in America,* edited by C. Ronald Huff. Thousand Oaks, CA: Sage.

Moore, Robert B. 1992. "Racist Stereotyping in the English Language." Pp. 317–328 in *Race, Class, and Gender: An Anthology,* 2nd ed., edited by Margaret L. Andersen and Patricia Hill Collins. Belmont, CA: Wadsworth.

Moore, Valerie A. 2001. "'Doing' Racialized and Gendered Age to Organize Peer Relations: Observing Kids in Summer Camp." *Gender & Society* 15 (December): 835–858.

Moreland, Richard L., and Scott R. Beach. 1992. "Exposure Effects in the Classroom: The Development of Affinity among Students." *Journal of Experimental Social Psychology* 28: 255–276.

Morgan, Lewis H. 1877. *Ancient Society, or Researches in the Lines of Human Progress, from Savagery Through Barbarism to Civilization.* Cambridge, MA: Harvard University Press.

Morlan, Patricia A. 2005. "Are We Taught to Blame the Poor? Attributions for Poverty: Sociodemographic Predictors and the Effects of Higher Education." Senior thesis, Princeton University.

Morning, Ann. 2005. "Race." *Contexts* 4 (Fall): 44–46.

Morning, Ann. 2007. "Everyone Knows It's A Social Construct: Contemporary Science and the Nature of Race." *Sociological Focus* 40 (4): 436–454.

Morning, Ann. 2008. "Reconstructing Race in Science and Society: Biology Textbooks, 1952–2002." *American Journal of Sociology* 114 (Suppl.): S106–S137.

Morris, Aldon. 1984. *The Origins of the Civil Rights Movement: Black Communities Organizing for Change.* New York: Free Press.

Morris, Aldon D. 1999. "A Retrospective on the Civil Rights Movement: Political and Intellectual Landmarks." *Annual Review of Sociology* 25: 517–539.

Morris, Aldon, and Carol McClurg Mueller (eds.). 1992. *Frontiers in Social Movement Theory.* New Haven, CT: Yale University Press.

Moskos, Peter. 2008. *Cop in the Hood: My Year in Policing Baltimore's Eastern District.* Princeton, NJ: Princeton University Press.

Moynihan, Daniel P. 1965. *The Negro Family: The Case for National Action.* Washington, DC: Office of Policy Planning and Research, U.S. Department of Labor.

Muller, Thomas, and Thomas J. Espenshade. 1985. *The Fourth Wave: California's Newest Immigrants.* Washington, DC: The Urban Institute Press.

Myers, Steven Lee. 2000. "Survey of Troops Finds Antigay Bias Common in Service." The *New York Times* (March 24): 1.

Myers, Walter D. 1998. *Amistad Affair.* New York: NAL/Dutton.

Myerson, Allen R. 1998. "Rating the Bigshots: Gates vs. Rockefeller." *The New York Times* (May 24): 4.

Myrdal, Gunnar. 1944. *An American Dilemma: The Negro Problem and Modern Democracy.* 2 Vols. New York: Harper and Row.

NAACP Legal Defense and Education Fund. 2006. *Annual Report.* Washington, DC: NAACP Legal Defense Fund.

Nagel, Joane. 1996. *American Indian Ethnic Renewal: Red Power and the Resurgence of Identity and Culture.* New York: Oxford University Press.

Nagel, Joane. 2003. *Race, Ethnicity, and Sexuality: Intimate Intersections, Forbidden Frontiers.* New York: Oxford University Press.

Nanda, Serena. 1998. *Neither Man Nor Woman: The Hijras of India.* Belmont, CA: Wadsworth.

Nardi, Peter M., and Beth Schneider. 1997. *Social Perspectives on Lesbian and Gay Studies.* New York: Routledge.

Nash, John F. 1951. "Non-Cooperative Games." *Annals of Mathematics* 54: 286–295.

National Cancer Institute. 2002. *Cancer among Blacks and Other Minorities: Statistical Profiles.* Washington, DC: U. S. Department of Health and Human Services, National Institutes of Health.

National Center for Health Statistics. 2002. *National Vital Statistics Report* 50 (No. 6, March 21). Rockville, MD: U.S. Department of Health and Human Services.

National Center for Health Statistics. 2004. *Health United States.* Rockville, MD: U.S. Department of Health and Human Services.

National Center for Health Statistics 2007. *Health United States.* Hyattsville, MD: U.S. Department of Health and Human Services.

National Center for Health Statistics 2008. "Health United States 2008." Hyattsville, MD: U.S. Department of Health and Human Services. **www.cdc.gov/ nchs**

National Center on Elder Abuse. 2007. NCEA Fact Sheet. Elder Abuse Prevention and Incidence. Washington, DC: National Center on Elder Abuse. **www.elderabusecenter.org**

National Coalition against Domestic Violence. 2001. **www.ncadv.org**

National Coalition for the Homeless. 2008. "Fact Sheets." Washington, DC. **www.nationalhomeless.org**

National Law Center on Homelessness and Poverty. 2008. *Homelessness in the United States and the Human Right to Housing.* **www.nlchp.org**

National Opinion Research Center. 2004. *General Social Survey.* **www.norc.uchicago.edu**

National Priorities Project. 2004. "Half of Low-Income People Not Receiving Food Stamps." **www. nationalpriorities.org**

National Television Violence Study. 1997. *National Television Violence Study.* Thousand Oaks, CA: Sage.

Nee, Victor. 1973. *Longtime Californ': A Documentary Study of an American Chinatown.* New York: Pantheon Books.

Newman, Katherine. 1999. *No Shame in My Game: The Working Poor in the Inner City.* New York: Russell Sage Foundation/Vintage Books.

Newman, Katherine S. 2006. *Rampage: The Social Roots of School Shootings.* New York: Basic Books.

Newport, Frank. 2000. "Women's Most Pressing Concerns Today are Money, Family, Health, and Stress." *The Gallup Poll Monthly* (March): 40–41.

Newport, Frank. 2006. "Religion Most Important to Blacks, Women, and Older Americans." Gallup Organization. **www.gallup.com/poll/25585/Religion-Most-Important-Blacks-Women-Older-Americans.aspx**

Niebuhr, Gustav. 1998. "Makeup of American Religion Is Looking More Like Mosaic, Data Say." *The New York Times* (April 12): 14.

Nolan, Patrick, and Gerhard Lenski. 2005. *Human Societies.* New York: Paradigm Publishers.

Nolan, Patrick, and Gerhard Lenski. 2008. *Human Societies.* New York: Paradigm Publishers.

Norris, Pippa, and Ronald Inglehart. 2002. "Islamic Culture and Democracy: Testing the 'Clash of Civilizations' Thesis." *Comparative Sociology* 1: 235–263.

Oakes, Jeannie. 1985. *Keeping Track: How Schools Structure Inequality.* New Haven, CT: Yale University Press.

Oakes, Jeannie. 1992. *Educational Matchmaking. Academic and Vocational Tracking in Comprehensive High Schools.* Santa Monica: Rand.

Oakes, Jeannie, and Martin Lipton. 1996. "Developing Alternatives to Tracking and Grading." Pp. 168–200 in *Educating A New Majority: Transforming America's Educational System for Diversity,* edited by Laura I. Rendon and Richard O. Hope. San Francisco, CA: Jossey-Bass.

Oakes, Jeannie, Karen Hunter Quartz, Steve Ryan, and Martin Lipton. 2000. *Becoming Good American Schools. The Struggle for Civic Virtue in Education Reform.* San Francisco: Jossey-Bass.

Ogburn, William F. 1922. *Social Change with Respect to Cultural and Original Nature.* New York: B. W. Huebsch.

O'Leary, Carol. 2002. "The Kurds of Iraq: Recent History, Future Prospects." *Middle East Review of International Affairs* 6 (December 2002).

Oliver, Melvin, and Thomas M. Shapiro. 1995. *Black Wealth/White Wealth: A New Perspective on Racial Inequality.* New York: Routledge.

Oliver, Melvin L., and Thomas M. Shapiro. 2001. "Wealth and Racial Stratification." Pp. 222–240 in *America Becoming: Racial Trends and Their Consequences,* edited by Neil Smelser, William Julius Wilson, and Faith Mitchell. Washington, DC: National Academies Press.

Oliver, Melvin L., and Thomas M. Shapiro. 2006. *Black Wealth/White Wealth: A New Perspective on Racial Inequality,* 10th ed. New York: Routledge.

Oliver, Melvin L., and Thomas M. Shapiro. 2008. "Subprime as a Black Catastrophe." *The American Prospect,* September 22. **www.prospect.org**

Ollivier, Michele. 2000. "'Too Much Money off Other People's Backs': Status in Late Modern Societies." *Canadian Journal of Sociology* 25 (Fall): 441–470.

Omi, Michael, and Howard Winant. 1994. *Racial Formation in the United States,* 2nd ed. New York: Routledge.

Orzechowski, Shawna, and Peter Sepielli. 2003. *Net Worth and Asset Ownership of Households: 1998 and 2000.* Washington, DC: U.S. Census Bureau, P70–88.

Ouchi, William. 1981. *Theory Z: How American Business Can Meet the Japanese Challenge.* Reading, MA: Addison-Wesley.

Owens, Sarah E. 1998. "The Effects of Race, Gender, Their Interactions, and Selected School Variables upon Educational Aspirations and Achievements." Senior thesis, Department of Sociology, Princeton University.

Padavic, Irene, and Barbara Reskin. 2002. *Women and Men at Work,* 2nd ed. Thousand Oaks, CA: Sage.

Page, Charles H. 1946. "Bureaucracy's Other Face." *Social Forces* 25 (October): 89–94.

Pager, Devah. 2007. *Marked: Race, Crime, and Finding Work.* Chicago: University of Chicago Press.

Pain, Emil. 2002. "The Social Nature of Extremism and Terrorism." *Social Sciences* 33: 55–68.

Park, Lisa Sun-Hee. 2005. *Consuming Citizenship: Children of Asian Immigrant Entrepreneurs.* Palo Alto, CA: Stanford University Press.

Park, Robert E., and Ernest W. Burgess. 1921. *Introduction to the Science of Society.* Chicago, IL: University of Chicago Press.

Parmelee, L. F. 2001. "A Roper Center Data Review: Mending the Fabric: Race Relations in Black and White." *Public Perspective* 12: 22–31.

Parreñas, Rhacel Salazar. 2001. *Servants of Globalization: Women, Migration, and Domestic Work.* Stanford, CA: Stanford University Press.

Parsons, Talcott (ed.). 1947. *Max Weber: The Theory of Social and Economic Organization.* New York: Free Press.

Parsons, Talcott. 1951a. *The Social System.* Glencoe, IL: Free Press.

Parsons, Talcott. 1951b. *Toward a General Theory of Action.* Cambridge, MA: Harvard University Press.

Parsons, Talcott. 1966. *Societies: Evolutionary and Comparative Perspectives.* Englewood Cliffs, NJ: Prentice Hall.

Pascoe, C. J. 2007. *Dude, You're a Fag: Masculinity and Sexuality in High School.* Berkeley, CA: University of California Press.

Paternoster, Raymond, and Robert Brame. 2003. "An Empirical Analysis of Maryland's Death Sentencing System with Respect to the Influence of Race and Legal Jurisdiction." **www.urhome.umd.edu/newsdesk**

Pathways to Peace. 2009. "Impact on War on Poverty." **www.icrc.org**

Pattillo-McCoy, Mary. 1999. *Black Picket Fences: Privilege and Peril among the Black Middle Class.* Chicago, IL: University of Chicago Press.

Paulus, P. B., T. S. Larey, and M. T. Dzindolet. 2001. "Creativity in Groups and Teams." Pp. 319–338 in M. E. Turner, ed., *Groups At Work: Theory and Research.* Mahwah, NJ: Earlbaum.

Pavalko, Eliza K., Krysia N. Mossakowski, and Vanessa J. Hamilton. 2003. "Does Perceived Discrimination Affect Health? Longitudinal Relationships between Work Discrimination and Women's Physical and Emotional Health." *Journal of Health and Social Behavior* 43 (March): 18–33.

Pedraza, Silvia. 1996a. "Cuba's Refugees: Manifold Migrations." Pp. 263–279 in *Origins and Destinies: Immigration, Race, and Ethnicity in America,* edited by Silvia Pedraza and Rubén

Rumbaut. Belmont, CA: Wadsworth.

Pedraza, Silvia, and Rubén G. Rumbaut. (eds.). 1996. *Origins and Destinies: Immigration, Race, and Ethnicity in America.* Belmont, CA: Wadsworth.

Pelham, Brett, and Steve Crabtree. 2009. "Religiosity and Perceived Intolerance of Gays and Lesbians." Princeton, NJ: Gallup Poll, March 10. **www.gallup.com**

Pellow, David N. 2003. *Garbage Wars: The Struggle for Environmental Justice in Chicago.* Cambridge, MA: MIT Press.

Pennock-Roman, Maria. 1994. "College Major and Gender Differences in the Prediction of College Grades." College Board Research Report No. 94-2, ETS Research Reports No. 94-24.

Peoples, James, and Garrick Bailey. 2003. *Humanity: An Introduction to Cultural Anthropology,* 6th ed. Belmont, CA: Wadsworth.

Peralta, Robert L. 2002. *Getting Trashed in College: Doing Alcohol, Doing Gender, Doing Violence.* Ph.D. dissertation, University of Delaware.

Perez, Denise N. 1996. "A Case for Derailment: Tracking in the American Public School System as an Obstacle to the Improvement of Minority Education." Senior thesis, Department of Sociology, Princeton University.

Perlmutter, David. 2008. *Blogwars: The New Political Battleground.* New York: Oxford University Press.

Perrow, Charles. 1986. *Complex Organization: A Critical Essay,* 3rd ed. New York: Random House.

Perrow, Charles. 1994. "The Limit of Safety: The Enhancement of a Theory of Accidents." *Journal of Contingencies and Crisis Management* 22: 212–220.

Perrow, Charles. 2007. *Organizing America: Wealth, Power, and Origins of American Capitalism.* Princeton: Princeton University Press.

Perry-Jenkins, Maureen, Rena L. Repetti, and Anne C. Crouter. 2000. "Work and Family in the 1990s." *Journal of Marriage and the Family* 62 (November): 981–998.

Persell, Caroline Hodges. 1990. *Understanding Society: An Introduction to Sociology.* New York: Harper & Row.

Pescosolido, Bernice A., Elizabeth Grauerholz, and Melissa A. Milkie. 1997. "Culture and Conflict: The Portrayal of Blacks in U.S. Children's Picture Books Through the Mid- and Late-Twentieth Century." *American Sociological Review* 62 (June): 443–464.

Pescosolido, Bernice A., Steven A. Tuch, and Jack K. Martin. 2001. "The Profession of Medicine and the Public: Examining Americans' Changing Confidence in Physician Authority from the Beginning of the 'Health Care Crisis' to the Era of Health Care Reform." *Journal of Heath and Social Behavior* 42 (March): 1–16.

Petersen, Trond, Ishak Saporta, and Mark-David L. Seidel. 2000. "Offering A Job: Meritocracy and Social Networks." *American Journal of Sociology* 106 (November): 763–816.

Pettigrew, Thomas F. 1992. "The Ultimate Attribution Error: Extending Allport's Cognitive Analysis of Prejudice." Pp. 401–419 in *Readings About the Social Animal,* edited by Elliott Aronson. New York: Freeman.

Pew Internet and American Life Project. 2009. "Adults and Social Network Websites." **www. pewinternet.org**.

Pew Research Center. 2003. "The 2004 Political Landscape." Pew Research Center for People and the Press. **www.people-press.org**

Pew Research Center. 2007. "Muslim Americans: Middle Class and Mostly Mainstream." **www. pewresearch.org**

Pew Research Center. 2009. *Dissecting the 2008 Election: Most Diverse in U.S. History.* Washington, DC: Pew Research Center.

Piaget, Jean. 1926. *The Language and Thought of the Child.* New York: Harcourt.

Polce-Lynch, Mary, Barbara J. Myers, Wendy Kliewer, and Christopher Kilmartin. 2001. "Adolescent Self-Esteem and Gender: Exploring Relations to Sexual Harassment, Body Image, Media Influence, and Emotional Expression." *Journal of Youth and Adolescence* 30 (April): 225–244.

Popenoe, David. 2001. "Today's Dads: A New Breed?" *The New York Times* (June 19): A22.

Population Reference Bureau. 2008. Washington, DC. **www.prb.org**

Portes, Alejandro. 2002. "English-Only Triumphs, But the Costs Are High." *Contexts* 1 (February): 10–15.

Portes, Alejandro, Christina Escobar, and Renelinda Arana. 2008. "Bridging the Gap: Transnational and Ethnic Organizations in the Political Incorporation of Immigrants in the United States." *Ethnic and Racial Studies* 31 (September): 1056–1090.

Portes, Alejandro, and Rubén G. Rumbaut. 1996. *Immigrant America: A Portrait,* 2nd ed. Berkeley, CA: University of California Press.

Portes, Alejandro, and Rubén G. Rumbaut. 2001. *Legacies: The Story of the Immigrant Second Generation.* Berkeley, CA: University of California Press.

Press, Andrea. 2002. "The Paradox of Talk." *Contexts* 1 (Fall–Winter): 69–70.

Press, Eyal. 1996. "Barbie's Betrayal." *The Nation* (December 30): 11–16.

Press, Julie E., and Eleanor Townsley. 1998. "Wives and Husbands' Reporting: Gender, Class, and Social Desirability." *Gender & Society* 12 (April): 188–218.

Preves, Sharon E. 2003. *Intersex and Identity: The Contested Self.* New Brunswick, NJ: Rutgers University Press.

Prostitutes Education Network. 2009. **www.bayswan.org**

Pruitt, D. G. 1971. "Choice Shifts in Group Discussion." *Journal of Personality and Social Psychology* 20 (December): 339–360.

Prus, Robert, and C. R. D. Sharper. 1991. *Road Hustler.* New York: Kauffman and Greenbery.

Punch, Maurice. 1996. *Dirty Business: Exploring Corporate Misconduct; Analysis and Cases.* Thousand Oaks, CA: Sage.

Putnam, Robert D. 2000. *Bowling Alone: The Collapse and Revival of American Community.* New York: Simon and Schuster.

Raboteau, Albert J. 1978. *Slave Religion: The "Invisible Institution" in the Antebellum South.* New York: Oxford University Press.

Raley, Sara B., Marybeth J. Mattingly, and Suzanne M. Bianchi. 2006. "How Dual Are

Dual-Income Couples? Documenting Change from 1970 to 2001." *Journal of Marriage and Family* 68(1): 11–28.

Rampersad, Arnold. 1986. *The Life of Langston Hughes: Vol. I: 1902–1941. I, Too, Sing America.* New York: Oxford University Press.

Rampersad, Arnold. 1988. *The Life of Langston Hughes: Vol. II: 1941–1967. I Dream A World.* New York: Oxford University Press.

Rashid, Ahmed. 2000. *Taliban: Militant Islam, Oil, and Fundamentalism in Central Asia.* New Haven, CT: Yale University Press.

Read, Jen'nan Ghazal. 2003. "The Sources of Gender Role Attitudes among Christian and Muslim Arab-American Women." *Sociology of Religion* 64 (Summer): 207–222.

Read, Piers Paul. 1974. *Alive: The Story of the Andes Survivors.* Philadelphia, PA: Lippincott.

Reich, Robert. 1991. *The Work of Nations: Preparing Ourselves for 21st Century Capitalism.* New York: Knopf.

Reiman, Jeffrey. 2004. "Executive Pay." *BusinessWeek* (April 17). **www.businessweek.com**

Reiman, Jeffrey H. 2007. *The Rich Get Richer and the Poor Get Prison,* 8th ed. Boston, MA: Allyn and Bacon.

Rendon, Laura I., and Richard O. Hope (eds.). 1996. *Educating a New Majority: Transforming America's Educational System for Diversity.* San Francisco, CA: Jossey-Bass.

Renzetti, Claire. 1992. *Violent Betrayal: Partner Abuse in Lesbian Relationships.* Newbury Park, CA: Sage.

Reskin, Barbara. 1988. "Bringing the Men Back In: Sex Differentiation and the Devaluation of Women's Work." *Gender & Society* 2 (March): 58–81.

Reynolds, Ariel Catherine. 2003. "Religiosity as a Predictor of Attitudes toward Homosexuality." Paper presented at the Annual Meetings of the Southern Sociological Society, New Orleans, Louisiana.

Rheault, Magail, and Dalia Mogahed. 2008. "Moral Issues Divide Westerners from Muslims in the West." Princeton, NJ: Gallup Poll. **www.gallup.com**

Rich, Adrienne. 1980. "Compulsory Heterosexuality and Lesbian Existence." *Signs* 5 (Summer): 631–660.

Ridley, Matt. 2003. *Nature via Nurture: Genes, Experience and What Makes Us Human.* New York: Harper Collins.

Rieff, David. 1993. "A Global Culture?" *World Policy Journal* 10 (Winter): 73–81.

Rifkin, Jeremy. 1998. "The Biotech Century: Human Life as Intellectual Property." *The Nation* 266 (April): 11–89.

Rindfuss, Ronald R., Elizabeth C. Cooksey, and Rebecca L. Sutterlin. 1999. "Young Adult Occupational Achievement: Early Expectations versus Behavioral Reality." *Work and Occupations* 26 (May): 220–263.

Risman, Barbara, and Pepper Schwartz. 2002. "After the Sexual Revolution: Gender Politics in Teen Dating." *Contexts* 1 (Spring): 16–24.

Ritzer, George. 2007. *The McDonaldization of Society,* 5th ed. Thousand Oaks, CA: Sage Publications.

Rivera, Ammad, et al. 2008. "Foreclosed: State of the Dream." Boston: United for a Fair Economy. **www.faireconomy. org**

Robbins, Thomas. 1988. *Cults, Convents, and Charisma: The Sociology of New Religious Movements.* Beverly Hills, CA: Sage.

Roberts, Dorothy. 1997. *Killing the Black Body: Race, Reproduction and the Meaning of Liberty.* New York: Vintage Books.

Robertson, Tatsha, and Garrance Burke. 2001. "Fighting Terror: Concerned Family;" "Shock, Worry for Family of U.S. Man Captured with Taliban." *The Boston Globe* (December 4): A1.

Robinson, Dawn T., and Lynn Smith-Lovin. 2001. "Getting a Laugh: Gender, Status, and Humor in Task Discussions." *Social Forces* 80 (September): 123–158.

Robinson, Jo Ann Gibson. 1987. *The Montgomery Bus Boycott and the Women Who Started It.* Knoxville, TN: The University of Tennessee Press.

Rodney, Walter. 1974. *How Europe Underdeveloped Africa.* Washington, DC: Howard University Press.

Rodney, Walter. 1981. *How Europe Underdeveloped Africa,* rev. ed. Washington, DC: Howard University Press.

Rodriguez, Clara E. 1989. *Puerto Ricans: Born in the U.S.A.* Boston, MA: Unwin Hyman.

Rodriguez, Clara E. 2006. "Changing Race." Pp. 22–25 in *Race and Ethnicity in Society: The Changing Landscape.* Belmont, CA: Wadsworth.

Rodriquez, Nestor P. 1999. "Globalization, Autonomy, and Transnational Migration: Impacts on U.S. Intergroup Relations." *Research in Politics and Society* 6: 65–84.

Roethlisberger, Fritz J., and William J. Dickson. 1939. *Management and the Worker.* Cambridge, MA: Harvard Univesity Press.

Romain, Suzanne. 1999. *Communicating Gender.* Mahwah, NJ: Erlbaum.

Roof, Wade Clark. 1999. *Spiritual Marketplace.* Princeton, NJ: Princeton University Press.

Rose, Tricia. 1994. *Black Noise: Rap Music and Black Culture in Contemporary America.* Middletown, CT: Wesleyan University Press.

Rosenau, Pauline Marie. 1992. *Post-Modernism and the Social Sciences: Insights, Inroads, and Intrusions.* Princeton, NJ: Princeton University Press.

Rosenbaum Janet Elise. 2009. "Patient Teenagers? A Comparison of the Sexual Behavior of Virginity Pledgers and Matched Nonpledgers." *Pediatrics* 123 (January): 110–120.

Rosenfeld, Michael J., and Byung-Soo Kim. 2005. "The Independence of Young Adults and the Rise of Interracial and Same-Sex Unions." *American Sociological Review* 70 (August): 541–562.

Rosengarten, Danielle. 2000. "Modern Times." *Dollars & Sense* (September): 4.

Rosenhan, David L. 1973. "On Being Sane in Insane Places." *Science* 179 (January 19): 250–258.

Rosenthal, Robert, and Lenore Jacobson. 1968. *Pygmalian in the Classroom: Teacher Expectations and Pupils' Intellectual Development.* New York: Holt, Rinehart and Winston.

Ross, L. 1977. "The Intuitive Psychologist and His Shortcoming: Distortions in the Attribution Process." Pp. 174–221 in

Advances in Experimental Social Psychology, vol 10, edited by L. Berkowitz. New York: Academic Press.

Rossi, Alice S., and Peter H. Rossi. 1990. *Of Human Bonding: Parent-Child Relations Across the Life Course.* New York: Aldine de Gruyter.

Rostow, W.W. 1978. *The World Economy: History and Prospect.* Austin, TX: University of Texas Press.

Rubin, Gayle. 1975. "The Traffic in Women." Pp. 157–211 in *Toward an Anthropology of Women,* edited by Rayna Reiter. New York: Monthly Review Press.

Ruef, Martin, Howard Aldrich, and N. Carter. 2003. "The Structure of Foundation Teams: Homophily, Strong Ties, and Isolation Among U.S. Entrepreneurs." *American Sociological Review* 68: 195–222.

Rueschmeyer, Dietrich, and Theda Skocpol. 1996. *States, Social Knowledge, and the Origins of Modern Social Policies.* Princeton, NJ: Princeton University Press.

Rumbaut, Rubén G. 1996a. "Origins and Destinies: Immigration, Race, and Ethnicity in Contemporary America." Pp. 21–42 in *Origins and Destinies: Immigration, Race, and Ethnicity in America,* edited by Sylvia Pedraza and Rubén G. Rumbaut. Belmont, CA: Wadsworth.

Rumbault, Rubén. 1996b. "Prologue." Pp. xvi–xix in *Origins and Destinies: Immigration, Race, and Ethnicity in America,* edited by Silvia Pedraza and Rubén Rumbaut. Belmont, CA: Wadsworth.

Rupp, Leila J., and Verta Taylor. 2003. *Drag Queens at the 801 Cabaret.* Chicago, IL: University of Chicago Press.

Rust, Paula C. 1993. "'Coming Out' in the Age of Social Constructionism: Sexual Identity Formation among Lesbian and Bisexual Women." *Gender & Society* 7 (March): 50–77.

Rust, Paula. 1995. *Bisexuality and the Challenge to Lesbian Politics.* New York: New York University Press.

Ryan, Charlotte. 1996. "Battered in the Media: Mainstream News Coverage of Welfare Reform." *Radical America* 26 (August): 29–41.

Ryan, Suzanne, Jennifer Manlove, and Sandra L. Hofferth. 2006. "State-Level Welfare Policies and Nonmarital Subsequent Childbearing." *Population Research and Policy Review* 25 (1): 103–126.

Ryan, William. 1971. *Blaming the Victim.* New York: Pantheon.

Rymer, Russ. 1993. *Genie: A Scientific Study.* New York: HarperCollins.

Saad, Lydia. 2002. "There's No Place Like Home to Spend an Evening, Say Most Americans." *The Gallup Poll.* Princeton, NJ: Gallup Organization. **www.gallup.com**

Saad, Lydia. 2003. "Pondering 'Women's Issues,' Part II." The Gallup Poll. Princeton, NJ: The Gallup Organization. **www.gallup.com**

Saad, Lydia. 2007. "Women Slightly More Likely to Prefer Working to Homemaking." Princeton, NJ: Gallup News Service. **www.gallup.com**

Sabol, William J., and Heather Couture. 2008. "Prison Inmates at Mid Year 2007." U.S. Bureau of Justice Statistics. **www.ojp.usdoj.gov**

Sadker, Myra, and David Sadker. 1994. *Failing at Fairness: How America's Schools Cheat Girls.* New York: Scribner's.

Saguy, Abigail. 2003. *What Is Sexual Harassment? From Capitol Hill to the Sorbonne.* Berkeley, CA: University of California Press.

Saks, M. J., and M. W. Marti. 1997. "A Meta-Analysis of the Effects of Jury Size." *Law and Human Behavior* 21: 451–467.

Sampson, Robert J. 1987. "Urban Black Violence: The Effect of Male Joblessness and Family Disruption." *American Journal of Sociology* 93: 348–382.

Sanchez, Laura, Wendy D. Manning, and Pamela J. Smock. 1998. "Sex-Specialized or Collaborative Mate Selection? Union Transitions among Cohabitors." *Social Science Research* 27 (September): 280–304.

Sanchez, Lisa Gonzalez. 1999. "Reclaiming Salsa." *Cultural Studies* 13 (April): 237–250.

Sanchez-Jankowski, Martin. 1991. *Islands in the Street: Gangs and American Urban Society.* Berkeley: University of California Press.

Sanday, Peggy. 2002. *Women at the Center: Life in a Modern Matriarchy.* Ithaca, NY: Cornell University Press.

Sandnabba, N. Kenneth, and Christian Ahlberg. 1999.

"Parents' Attitudes and Expectations About Children's Cross-Gender Behavior." *Sex Roles* 40 (February): 249–263.

Santelli, John, et al. 2007. "Explaining Recent Declines in Adolescent Pregnancy in the United States: The Contribution of Abstinence and Increased Contraceptive Use." *American Journal of Public Health American Journal of Public Health* 97(1): 150–156.

Sapir, Edward. 1921. *Language: An Introduction to the Study of Speech.* New York: Harcourt Brace.

Sawhill, Isabel, and Sara McClanahan (eds.). 2006. "Introducing the Issue." *The Future of Children* 16 (Fall): 3–17.

Sax, L. J., et al. 2003. *The American Freshman: National Norms for Fall 2003.* Higher Education Research Institute. Los Angeles, CA: University of California, Los Angeles.

Scarpitti, Frank R., Margaret L. Andersen, and Laura O'Toole. 1997. *Social Problems,* 3rd ed. New York: Addison Wesley Longman.

Schacht, Steven P. 1996. "Misogyny on and off the 'Pitch': The Gendered World of Male Rugby Players." *Gender & Society* 10 (October): 550–565.

Schaffer, Kay, and Song Xianlin. 2007. "Unruly Spaces: Gender, Women's Writing and Indigenous Feminism in China." *Journal of Gender Studies* 16 (1): 17–30.

Scheper-Hughes, Nancy. 1992. *Death without Weeping: The Violence of Everyday Life in Brazil.* Berkeley, CA: University of California Press.

Schlosser, Eric. 2001. *Fast Food Nation: The Dark Side of the All-American Meal.* New York: Houghton Mifflin.

Schmitt, Eric. 2001. "Segregation Growing among U.S. Children." *The New York Times* (May 6): 28.

Schmitt, Frederika E., and Patricia Yancey Martin. 1999. "Unobtrusive Mobilization by an Institutionalized Rape Crisis Center: 'It Comes From the Victims.'" *Gender & Society* 13: 364–384.

Schneider, Barbara, and David Stevenson. 1999. *The Ambitious Generation: Motivated but Directionless.* New Haven, CT: Yale University Press.

Schneider, Beth. 1984. "Peril and Promise: Lesbians' Workplace

Participation." Pp. 211–230 in *Women Identified Women,* edited by Trudy Darty and Sandee Potter. Mountain View, CA: Mayfield.

Schumann, Howard, Charlotte Steeh, Lawrence Bobo, and Maria Krysan. 1997. *Racial Attitudes in America: Trends and Interpretations.* Cambridge, MA: Harvard University Press.

Schur, Edwin M. 1984. *Labeling Women Deviant.* New York: Random House.

Schwartz, John, and Matthew L. Wald. 2003. "NASA's Failings Go Far Beyond Foam Hitting Shuttle, Panel Says." *The New York Times* (June 7): 1 and 12.

Schwartz, Pepper, and Virginia Rutter. 1998. *The Gender of Sexuality.* Thousand Oaks, CA: Sage.

Scott, Ellen K., Andrew S. London, and Nancy A. Myers. 2002. "Dangerous Dependencies: The Intersection of Welfare Reform and Domestic Violence." *Gender & Society* 16 (December): 878–897.

Scott, Tracy Lee. 2000. "What's God Got to Do with It? Protestantism, Gender, and the Meaning of Work in the U.S." *Dissertation Abstracts International, A: The Humanities and Social Sciences* (April): 3813A–3814A.

Scully, Diana. 1994. *Men Who Control Women's Health: The Miseducation of Obstetrician-Gynecologists.* New York: Teachers College Press.

Segal, David R., and Mady Wechsler Segal. 2004. "America's Military Population." *Population Bulletin* 59 (December).

Seidman, Steven. 2003. *The Social Construction of Sexuality.* New York: Oxford University Press.

Sen, Amartya. 2000. "Population and Gender Equity." *The Nation* (July 24–31): 16–18.

Sen, Amartya. 2002. "How to Judge Globalism." *The American Prospect,* Special Supplement (Winter): A2–A6.

Settersten, Richard A., Jr., and Loren D. Lovegreen. 1998. "Educational Experiences throughout Adult Life: New Hopes or No Hope for Life-Course Flexibility?" *Research on Aging* 20 (July): 506–538.

Sherkat, Darren E. 2002. "Sexuality and Religious Commitment in the United States: An Empirical Examination." *Journal for the Scientific Study of Religion* 41 (June): 313–323.

Shibutani, Tomatsu. 1961. *Society and Personality: An Interactionist Approach to Social Psychology.* Englewood Cliffs, NJ: Prentice Hall.

Shibutani, Tomatsu. 1966. *Improvised News.* Indianapolis, IN: Bobbs-Merrill.

Shih, Margaret, Todd L. Pittinsky, and Nalini Ambady. 1999. "Stereotype Susceptibility: Identiy Salience and Shifts in Quantitative Performance." *Psychological Science* 10 (January): 80–83.

Shilts, Randy. 1988. *And the Band Played On.* New York: Penguin.

Shinagawa, Larry, and Gin Yong Pang. 1996. "Asian American Panethnicity and Intermarriage." *American Journal* 22 (Spring): 127–152.

Sigall, H., and N. Ostrove. 1975. "Beautiful but Dangerous: Effects of Offender Attractiveness and Nature of Crime on Judicial Judgment." *Journal of Personality and Social Psychology* 31: 410–414.

Sigelman, Lee, and Paul J. Wahlbeck. 1999. "Gender Proportionality in Intercollegiate Athletics: The Mathematics of Title IX Compliance." *Social Science Quarterly* 80 (September): 518–538.

Signorielli, Nancy. 1991. *A Sourcebook on Children and Television.* New York: Greenwood Press.

Signorielli, Nancy, and Aaron Bacue. 1999. "Recognition and Respect: A Content Analysis of Prime-Time Television Characters across Three Decades." *Sex Roles* 40 (April): 527–544.

Signorielli, Nancy, Douglas McLeod, and Elaine Healy. 1994. "Gender Stereotypes in MTV Commercials: The Beat Goes On." *Journal of Broadcasting and Electronic Media* 38 (Winter): 91–101.

Silverthorne, Zebulon A., and Vernon L Quinsey. 2000. "Sexual Partner Age Preferences of Homosexual and Heterosexual Men and Women." *Archives of Sexual Behavior* 29 (No. 1): 67–76.

Simile, Catherine. 1995. *Disaster Settings and Mobilization for Contentious Collective Action: Case Studies of Hurricane Hugo and the Loma Prieta Earthquake.* Unpublished dissertation, University of Delaware.

Simmel, Georg. 1902. "The Number of Members as Determining the Sociological Form of the Group." *The American Journal of Sociology* 8 (July): 1–46.

Simmons, Wendy W. 2001. "Majority of Americans Say More Women in Political Office Would Be Positive for the Country." The Gallup Poll. Princeton, NJ.: Gallup Organization. **www.gallup.com**

Simon, David R. 2007. *Elite Deviance,* 8th ed. Boston, MA: Allyn and Bacon.

Simons, Ronald L. 1996. "The Effect of Divorce on Adult and Child Adjustment." Pp. 3–20 in *Understanding Differences between Divorced and Intact Families.* Thousand Oaks, CA: Sage.

Simpson, George Eaton, and J. Milton Yinger. 1985. *Racial and Cultural Minorities: An Analysis of Prejudice and Discrimination,* 5th ed. New York: Plenum.

Singer, Peter W. 2007. *Corporate Warriors: The Rise of the Privatized Military Industry.* Ithaca, NY: Cornell University Press.

Sirvananadan, A. 1995. "La trahision des clercs. (Racism)." *New Statesman and Society* 8: 20–22.

Skocpol, Theda. 1992. *Protecting Soldiers and Mothers: The Origins of Social Policy in the United States.* Cambridge, MA: Belknap Press.

Smelser, Neil J. 1992a. "Culture: Coherent or Incoherent." Pp. 3–28 in *Theory of Culture,* edited by R. Münch and N. J. Smelser. Berkeley, CA: University of California Press.

Smith, Barbara. 1983. "Homophobia: Why Bring It Up?" *Interracial Books for Children Bulletin* 14: 112–113.

Smith, Jesse L. 2006. "The Interplay Among Stereotypes, Performance-Avoidance Goals, and Women's Math Performance Expectations." *Sex Roles* 54 (3/4): 287–295.

Smith, M. Dwayne, Joel A. Devine, and Joseph F. Sheley. 1992. "Crime and Unemployment: Effects Across Age and Race Categories." *Sociological Perspectives* 35 (Winter): 551–572.

Smith, Richard B., and Robert A. Brown. 1997. "The Impact of

Social Support on Gay Male Couples." *Journal of Homosexuality* 33: 39–61.

Smith, Sandra. 2007. *Lone Pursuit: Distrust and Defensive Individualism Among the Black Poor.* New York: Russell Sage Foundation.

Smith, Tom W. 1999. "The Religious Right and Anti-Semitism." *Review of Religious Research* 40 (March): 244–258.

Snipp, C. Matthew. 1989. *American Indians: The First of This Land.* New York: Russell Sage Foundation.

Snipp, C. Matthew. 1996. "The First Americans: American Indians." Pp. 390–404 in *Origins and Destinies: Immigration, Race, and Ethnicity in America,* edited by Sylvia Pedraza and Rubén G. Rumbaut. Belmont, CA: Wadsworth.

Snow, David A., and Robert D. Benford. 1988. "Ideology, Frame Resonance, and Participant Mobilization." *International Social Movements Research* 1: 197–217.

Sorokin, Pitrim. 1941. *The Crisis of Our Age.* New York: Dutton.

Sotirovic, Mira. 2000. "Effects of Media Use on Audience Framing and Support for Welfare." *Mass Communication & Society* 2–3 (Spring–Summer): 269–296.

Sotirovic, Mira. 2001. "Media Use and Perceptions of Welfare." *Journal of Communication* 51 (December): 750–774.

Spangler, Eve, Marsha A. Gordon, and Ronald Pipkin. 1978. "Token Women: An Empirical Test of Kanter's Hypothesis." *American Journal of Sociology* 84: 160–170.

Speed, A., and S.W. Gangestad. 1997. "Romantic Popularity and Mate Preferences: A Peer-Nomination Study." *Personality and Social Psychology Bulletin* 23: 928–936.

Spitzer, Steven. 1975. "Toward a Marxian Theory of Deviance." *Social Problems* 22: 638–651.

Squires, Gregory. 2007. "Demobilization of the Individualistic Bias: Housing Market Discrimination as a Contributor to Labor Market and Economic Inequality." *The Annals of the American Academy of Political and Social Science* 609: 200–214.

Stacey, Judith, and Timothy J. Biblarz. 2001. "(How) Does the Sexual Orientation of Parents Matter?" *American Sociological Review* 66 (April): 159–183.

Stack, Carol. 1974. *All Our Kin: Strategies for Survival in a Black Community.* New York: Harper Colophon Books.

Starr, Paul. 1982. *The Social Transformation of American Medicine.* New York: Basic Books.

Starr, Paul. 1982. What Happened to Health Care Reform?" *The American Prospect* (20/Winter): 20–31.

Stearns, Linda Brewster, and Charlotte Wilkinson Coleman. 1990. "Industrial and Labor Market Structures and Black Male Employment in the Manufacturing Sector." *Social Science Quarterly* 71 (June): 285–298.

Steele, Claude M. 1992. "Race and the Schooling of Black Americans." *The Atlantic Monthly* 269 (April): 68–78.

Steele, Claude M. 1996. "A Burden of Suspicion: How Stereotypes Shape the Intellectual Identities and Performance of Women and African-Americans." Paper read before the Princeton Conference on Higher Education, Princeton, NJ, February.

Steele, Claude. 1997. A Threat in the Air: How Stereotypes Shape Intellectual Identity and Performance." *American Psychologist* 52 (June): 613–629.

Steele, Claude M. 1999. "Thin Ice: Stereotype Threat and Black College Students." *The Atlantic Monthly* (August): 44–54.

Steele, Claude M., and Joshua Aronson. 1995. "Stereotype Vulnerability and African American Intellectual Performance." Pp. 409–421 in *Readings About the Social Animal,* 7th ed., edited by Elliot Aronson. New York: W. H. Freeman.

Steele, Claude M., S. Spencer, and Joshua Aronson. 2002. "Contending With Group Image: The Psychology of Stereotype and Social Identity Threat." In *Advances in Experimental Social Psychology,* vol 37, edited by Mark Zanna. New York: Academic Press.

Steffensmeier, Darrell, and Stephen Demuth. 2000. "Ethnicity and Sentencing Outcomes in U.S. Federal Courts: Who Is Punished More Harshly?" *American Sociological Review* 65 (October): 705–729.

Steffensmeier, Darrell, Jeffery Ulmer, and John Kramer. 1998. "The Interaction of Race, Gender, and Age in Criminal Sentencing: The Punishment Cost of Being Young, Black and Male." *Criminology* 36 (November): 763–797.

Stein, Arlene, and Ken Plummer. 1994. "'I Can't Even Think Straight': Queer Theory and the Missing Sexual Revolution in Sociology." *Sociological Theory* 12 (July): 178–187.

Steinheimer, Karen. 2007. "Do Video Games Kill?" *Contexts* 6 (Winter): 13–17.

Stern, Jessica. 2003. *Terror in the Name of God: Why Religious Militants Kill.* New York: Ecco.

Stern, Kenneth S. 1996. *A Force upon the Plain.* New York: Simon and Schuster.

Sternberg, Robert J. 1988. *The Triarchic Mind: A New Theory of Human Intelligence.* New York: Penguin.

Stevens, William K. 1994. "Poor Lands Success in Cutting Birthrate Upsets Old Theories." *The New York Times* (January 2): 1, 8.

Stewart, Abigail J., Anne P. Copeland, Nia Lane Chester, Janet E. Malley, and Nicole B. Barenbaum. 1997. *Separating Together: How Divorce Transforms Families.* New York: Guilford Press.

Stewart, Susan D. 2001. "Contemporary American Stepparenthood: Integrating Cohabiting and Nonresident Stepparents." *Population Research and Policy* 20 (August): 345–364.

Stjerna, Marie-Louise, Sonja Olin Lauritzen, and Per Tillgren. 2004. "Social Thinking and Cultural Images: Teenagers' Notions of Tobacco Use." *Social Science & Medicine* 59: 573–83.

Stoller, Eleanor Palo, and Rose Campbell Gibson. 2000. *Worlds of Difference: Inequality in the Aging Experience.* Thousand Oaks, CA: Pine Forge Press.

Stombler, Mindy, and Irene Padavic. 1997. "Sister Acts: Resisting Men's Domination in Black and White Fraternity Little Sister Programs." *Social Problems* 44 (May): 257–275.

Stone, Pamela. 2007. *Opting Out: Why Women Really Quit Careers and Head Home.* Berkeley, CA: University of California Press.

Stoner, J. A. F. 1961. "A Comparison of Individual and Group Decisions Involving Risk." Unpublished M. A. thesis, MIT.

Stotik, Jeffrey, Thomas E. Shriver, and Sherry Cable. 1994. "Social Control and Movement Outcome: The Case of AIM." *Sociological Focus* 27 (February): 53–66.

Stover, R. G., and C. A. Hope. 1993. *Marriage, Family and Intimate Relationships*. New York: Harcourt Brace Jovanovich.

Stryker, Susan, and Stephen Whittle (eds.). 2006. *The Transgender Studies Reader*. New York: Routledge.

Sudnow, David N. 1967. *Passing On: The Social Organization of Dying*. Englewood Cliffs, NJ: Prentice Hall.

Sullivan, Maureen. 1996. "Rozzie and Harriet? Gender and Family Patterns of Lesbian Coparents." *Gender & Society* 12 (December): 747–767.

Sullivan, Teresa A., Elizabeth Warren, and Jay Lawrence Westbrook. 2000. *The Fragile Middle Class: Americans in Debt*. New Haven, CT: Yale University Press.

Sumner, William Graham. 1906. *Folkways: A Study of the Sociological Importance of Usages, Manners, Customs, Mores, and Morals*. Boston: Ginn.

Suro, Roberto. 2000. "Beyond Economics." *American Demographics* (February): 48–55.

Surratt, Hilary L., and James A. Inciardi. 1998. "Unraveling the Concept of Race in Brazil." *Journal of Psychoactive Drugs* 30 (3): 255–260.

Sussal, Carol M. 1994. "Empowering Gays and Lesbians in the Workplace." *Journal of Gay and Lesbian Social Services* 1: 89–103.

Sussman, N. M., and D. H. Tyson. 2000. "Sex and Power: Gender Differences in Computer-Mediated Interactions." *Computers in Human Behavior* 16 (July): 381–394.

Sutherland, Edwin H. 1940. "White Collar Criminality." *American Sociological Review* 5 (February): 1–12.

Sutherland, Edwin H., and Donald R. Cressey. 1978. *Criminology*, 10th ed. New York: Lippincott.

Swidler, Ann. 1986. "Culture in Action: Symbols and Strategies." *American Sociological Review* 51: 273–286.

Switzer, J.Y. 1990. "The Impact of Generic Word Choices: An Empirical Investigation of Age- and Sex-Related Differences." *Sex Roles* 22: 69–82.

Szasz, Thomas S. 1974. *The Myth of Mental Illness*. New York: Harper & Row.

Takaki, Ronald. 1989. *Strangers from a Different Shore: A History of Asian Americans*. New York: Penguin.

Tannen, Deborah. 1990. *You Just Don't Understand: Women and Men in Conversation*. New York: William Morrow.

Tatum, Beverly.1997. *Why Are All the Black Kids Sitting Together at the Cafeteria?* New York: Basic Books.

Taylor, Howard F. 1973. "Linear Models of Consistency: Some Extensions of Blalock's Strategy." *American Journal of Sociology* 78 (March): 1192–1215.

Taylor, Howard F. 1980. *The IQ Game: A Methodological Inquiry into the Heredity–Environment Controversy*. New Brunswick, NJ: Rutgers University Press.

Taylor, Howard F. 1992a. "Intelligence." Pp. 941–949 in *Encyclopedia of Sociology*, edited by E. F. Borgatta and M. L. Borgatta. New York: Macmillan.

Taylor, Howard F. 1992b. "The Structure of a National Black Leadership Network: Preliminary Findings." Unpublished manuscript, Princeton University.

Taylor, Howard F. 1995. "Symposium on 'The Bell Curve'." *Contemporary Sociology* 24 (March): 153–158.

Taylor, Howard F. 2002. "Deconstructing the Bell Curve: Racism, Classism, and Intelligence Measurement in America. "In *2001: Race Odyssey: African Americans and Sociology,* edited by Bruce R. Hare. Syracuse, NY: Syracuse University Press. 60–76.

Taylor, Howard F. 2006. "Defining Race." Pp. 47–54 in *Race and Ethnicity in U.S. Society: The Changing Landscape,* edited by Elizabeth Higginbotham and Margaret L. Andersen. Belmont, CA: Wadsworth.

Taylor, Howard F. 2008. "The Nature-Nurture Debate." Pp. 612–616 in *Encyclopedia of Social Problems*, vol. 2, edited by Vincent N. Parrillo. Los Angeles: Sage Reference Publications.

Taylor, Howard F. 2009. *Race and Class and the Bell Curve in America*. Unpublished manuscript, Princeton University, Princeton, NJ.

Taylor, Howard F., and Carlton A. Hornung. 1979. "On A General Model for Social and Cognitive Consistency." *Sociological Methods and Research* 7 (February): 259–287.

Taylor, Shelley E., Letitia Anne Peplau, and David O. Sears. 2006. *Social Psychology,* 12th Ed. Upper Saddle River, NJ: Prentice Hall.

Taylor, Shelley E., Letitia Ann Peplau, and David O. Sears. 2009. *Social Psychology*, 13th edition. Upper Saddle River, NJ: Pearson/Prentice Hall.

Teaster, Pamela B. 2000. "A Response to the Abuse of Vulnerable Adults: The 200 Survey of State Adult Protective Services." Washington, DC: National Center on Elder Abuse. **www.elderabusecenter.org**

Telles, Edward E. 1992. "Residential Segregation and Skin Color in Brazil." *American Sociological Review* 57 (April): 186–197.

Telles, Edward E. 2004. *Race in Another America: The Significance of Skin Color in Brazil*. Princeton, NJ: Princeton University Press.

Telles, Edward E., and Vilma Ortiz. 2008. *Generations of Exclusion: Mexican Americans, Assimilation, and Race*. New York: Russell Sage Foundation Press.

Thakkar, Reena R., Peter M. Gutierrez, Carly L. Kuczen, and Thomas R. McCanne. 2000. "History of Physical and/or Sexual Abuse and Current Suicidality in College Women." *Child Abuse and Neglect* 24 (October): 1345–1354.

The Sentencing Project. 2009. Washington, DC. **www.sentencingproject.org**

Thoits, Peggy A. 1991. "On Merging Identity Theory and Stress Research." *Social Psychology Quarterly* 54: 101–112.

Thoits, Peggy A. 2009. "Sociological Approaches to Mental Illness." In *A Handbook for the Study of Mental Health*, edited by Teresa L. Scheid and Tony L. Brown.

Cambridge: Cambridge University Press.

Thomas, William I. 1931. *The Unadjusted Girl.* Boston, MA: Little, Brown.

Thomas, William I. 1966 [1931]. "The Relation of Research to the Social Process." Pp. 289–305 in *W. I. Thomas on Social Organization and Social Personality,* edited by Morris Janowitz. Chicago, IL: University of Chicago Press.

Thomas, William I., with Dorothy Swaine Thomas. 1928. *The Child in America.* New York: Knopf.

Thompson, Becky. 1994. *A Hunger So Wide and So Deep: American Women Speak Out on Eating Problems.* Minneapolis, MN: University of Minnesota Press.

Thorne, Barrie. 1993. *Gender Play: Girls and Boys in School.* New Brunswick, NJ: Rutgers University Press.

Thornton, Russell. 1987. *American Indian Holocaust and Survival: A Population History.* Norman, OK: University of Oklahoma Press.

Thornton, Russell. 2001. "Trends among American Indians in the United States." Pp. 135–169 in *America Becoming: Racial Trends and Their Consequences,* vol. I, edited by Neil J. Smelser, William Julius Wilson, and Faith Mitchell. Washington, DC: National Academy Press.

Tichenor, Veronica. 2005. "Maintaining Men's Dominance: Negotiating Identity and Power When She Earns More." *Sex Roles* 53(3–4): 191–205.

Tienda, Marta, and Haya Stier. 1996. "Generating Labor Market Inequality: Employment Opportunities and the Accumulation of Disadvantage." *Social Problems* 43 (May): 147–165.

Tierney, Patrick. 2000. *Darkness in El Dorado: How Scientists and Journalists Devastated the Amazon.* New York: W.W. Norton.

Tilly, Charles. 1986. *The Contentious French.* Cambridge, MA: Harvard University Press.

Tilly, Louise, and Joan Scott. 1978. *Women, Work, and Family.* New York: Holt, Rinehart, and Winston.

Tjaden, Patricia, and Nancy Thoennes. 2000. *Extent, Nature, and Consequences of Intimate Partner Violence.* Washington, DC: National Institute of Justice and the Centers for Disease Control and Prevention.

Tönnies, Ferdinand. 1963 [1887]. *Community and Society (Gemeinschaft and Gesellschaft).* New York: Harper & Row.

Toubia, Nahid, and Susan Izett. 1998. "Female Genital Mutilation: An Overview." Geneva: World Health Organization.

Toynbee, Arnold J., and Jane Caplan. 1972. *A Study of History.* New York: Oxford University Press.

Travers, Jeffrey, and Stanley Milgram. 1969. "An Experimental Study of the Small World Problem." *Sociometry* 32: 425–443.

Treiman, Donald J. 2001. "Occupations, Stratification, and Mobility." Pp. 297–313 in *The Blackwell Companion to Sociology,* edited by Judith R. Blau. Malden, MA: Blackwell.

Trent, Katherine, and Sharon L. Harlan. 1990. "Household Structure Among Teenage Mothers in the United States." *Social Science Quarterly* 71 (September): 439–457.

Tuan, Yi-Fu. 1984. *Dominance and Affection: The Making of Pets.* New Haven, CT: Yale University Press.

Tuchman, Gaye. 1979. "Women's Depiction by the Mass Media." *Signs* 4 (Spring): 528–542.

Tumin, Melvin M. 1953. "Some Principles of Stratification." *American Sociological Review* 18 (August): 387–393.

Turner, Jonathan. 1974. *The Structure of Sociological Theory.* Homewood, IL: Dorsey Press.

Turner, Ralph, and Lewis Killian. 1988. *Collective Behavior,* 3rd ed. Englewood Cliffs, NJ: Prentice Hall.

Turner, Ralph, and Lewis Killian. 1993. *Collective Behavior,* 4th ed. Englewood Cliffs, NJ: Prentice Hall.

Turner, Ralph, Joanne M. Nigg, and Denise Paz. 1986. *Waiting for Disaster: Earthquake Watch in California.* Berkeley, CA: University of California Press.

Turner, Terence. 1969. "Tchikrin: A Central Brazilian Tribe and Its Symbolic Language of Body Adornment." *Natural History* 78 (October): 50–59.

U.S. Bureau of Justice Statistics. 2005. *Criminal Victimization in the United States 2005, Statistical Tables.* Washington, DC: U.S. Bureau of Justice Statistics.

U.S. Bureau of Justice Statistics. 2007. *Criminal Victimization in the United States 2007, Statistical Tables.* Washington, DC: U.S. Bureau of Justice Statistics.

U.S. Bureau of Justice Statistics, 2008. "Prison and Jail Inmates." **www.ojp.usdoj.gov/bjs/ prisons.html**

U.S. Bureau of Labor Statistics 2006. *Percent of All Workers with Access to Employer Asisstance for Child Care.* **www.bls.gov**

U.S. Bureau of Labor Statistics. 2006a. *Employment Characteristics of Families in 2005,* USDL 06-731. Washington, DC: U.S. Department of Labor.

U.S. Bureau of Labor Statistics. 2008. "A Profile of the Working Poor, 2006, August." Washington, DC: U.S. Department of Labor. **www.bls.gov**

U.S. Census Bureau. 2001. *Statistical Abstract of the United States.* **www.census.gov**

U.S. Census Bureau. 2002. *Statistical Abstract of the United States.* **www.census.gov**

U.S. Census Bureau. 2003. "Racial and Ethnic Classification Used in Census 2000 and Beyond." **www. census.gov**

U.S. Census Bureau. 2004. "The American Community—Asians 2004." Washington, DC: U.S. Department of Commerce.

U.S. Census Bureau. 2004. *Detailed Income Tables, Table P2.* Washington, DC: U.S. Department of Commerce, **www.census.gov**

U.S. Census Bureau. 2006. Community Survey. **hfactfinder. census.gov**

U.S. Census Bureau. 2006. *Statistical Abstract of the United States 2005.* **www.census.gov**

U.S. Census Bureau. 2007. *Statistical Abstract of the United States: 2006.* **www.census.gov**

U.S. Census Bureau. 2007a. "Current Population Survey, March 2007." **www.census.gov**

U.S. Census Bureau. 2007b. "Current Population Reports," Figure 10.1. **www.census.gov**

U.S. Census Bureau. 2008a. "Current Population Survey, Table FINC-02. Age of Reference Person, by Total Money Income in 2007, Type of Family, Race and Hispanic Origin of Reference Person." **www.census.gov**

U.S. Census Bureau. 2008b. *Statistical Abstract of the United States.* **www.census.gov**

U.S. Census Bureau. 2008c. "America's Families and Living Arrangements." **www.census.gov**

U.S. Census Bureau. 2009a. *Statistical Abstract of the United States.* **www.census.gov**

U.S. Census Bureau. 2009b. "American FactFinder." **www.census.gov**

U.S. Census Bureau. 2009c. "Detailed Income Tabulations from the Current Population Survey: Educational Attainment People 25 Years Old and Over, by Total Money Earnings in 2005, Work Experience in 2005, Age, Race, Hispanic Origin, and Sex," Table PINC-03.

U.S. Department of Health and Human Services. 2007. "Child Maltreatment 2007." **www.acf.hhs.gov**

U.S. Department of Health and Human Services. 2009. "Sex Trafficking Fact Sheet." **www.acf.hhs.gov**

U.S. Department of Justice. 2009. "Child Sex Tourism." **www.usdoj.gov**

U.S. Department of Labor. 2009. "Ranks of Discouraged Workers and Others Marginally Attached to the Labor Force Rise During Recession." *Issues in Labor Statistics* (April), Summary 90–04.

Uleman, J. S., L. S. Newman, and G. B. Moskowitz. 1996. "People as Flexible Interpreters: Evidence and Issues From Spontaneous Trait Inference." Pp. 211–279 in *Advances in Experimental Social Psychology,* vol. 28, edited by Mark Zanna. Boston, MA: Academic Press.

Ullman, Sarah E., George Karabatsos, and Mary P. Koss. 1999. "Alcohol and Sexual Assault in a National Sample of College Women." *Journal of Interpersonal Violence* 14 (June): 603–625.

UNICEF. 2000. *Domestic Violence against Women and Girls.* New York: United Nations.

United Nations. 2000. *Human Development Report.* New York: United Nations.

United Nations. 2003. "Human Annual Development Report 2003." **www.undp.org**

United Nations. 2005a. *The World's Women: Progress in Statistics.* New York: United Nations.

United Nations. 2005b. *Human Development Report.* New York: United Nations.

United Nations. 2006. "Human Development Report 2006." **www.undp.org**

United Nations. 2006a. *In Depth Study on All Forms of Violence Against Women.* New York: United Nations.

United Nations. 2006b (August). United Nations Statistics Division: *Statistics and Indicators on Women and Men.* **www.unstats.un.org/demographic/products/default.htm**

United Nations News Center. 2007. "One Third of All Iraqis Live in Poverty, Study Says." February 13. New York: United Nations.

Uvin, Peter. 1998. *Aiding Violence.* West Hartford, CT: Kumarian Press.

Van Ausdale, Deborah, and Joe R. Feagin. 2000. *The First R: How Children Learn Race and Racism.* Lanham, MD: Rowan and Littlefield.

Van Ausdale, Debra, and Joe R. Feagin. 1996. "The Use of Racial and Ethnic Concepts by Very Young Children." *American Sociological Review* 61 (October): 779–793.

Vanlandingham, Mark. 1993. *Two Perspectives on Risky Sexual Practices among Northern Thai Males: The Health Belief Model and the Theory of Reasoned Action.* Unpublished doctoral dissertation, Princeton University.

Vanneman, Reeve, and Lynn Weber Cannon. 1987. *The American Perception of Class.* Philadelphia: Temple University Press.

VanRoosmalen, Erica, and Susan A. McDaniel. 1998. "Sexual Harassment in Academia: A Hazard to Women's Health." *Women and Health* 28: 33–54.

Varelas, Nicole, and Linda A. Foley. 1998. "Blacks' and Whites' Perceptions of Interracial and Intraracial Date Rape." *Journal of Social Psychology* 138 (June): 392–400.

Vaughan, Diane. 1996. *The Challenger Launch Decision: Risky Technology, Culture, and Deviance at NASA.* Chicago, IL: The University of Chicago Press.

Vazquez, Carmen. 1992. "Appearances." Pp. 157–166 in *Homophobia: How We All Pay the Price,* edited by Warren J. Blumenfeld. Boston, MA: Beacon Press.

Veblen, Thorstein. 1899 [1953]. *The Theory of the Leisure Class: An Economic Study of Institutions.* New York: The New American Library.

Velminckx, Koen, and Timothy Smeeding (eds.). 2001. *Child Well-Being, Child Poverty, and Child Policy in Modern Nations.* Tonawanda, NY: University of Toronto Press.

Waite, Linda J. 2000. "Trends in Men's and Women's Well-Being in Marriage." Pp. 368–392 in *The Ties that Bind: Trends in Men's and Women's Well-Being in Marriage.* New York: Aldine de Gruyter.

Waldfogel, Jane, Wen-Jui Han, and Jeanne Brooks-Gunn. 2002. "The Effects of Early Maternal Employment in Child Cognitive Development." *Demography* 39: 369–392.

Walker-Barnes, Chanequa J., and Craig A. Mason. 2001. "Perceptions of Risk Factors for Female Gang Involvement among African American and Hispanic Women." *Youth and Society* 32 (March): 303–336.

Wallerstein, Immanuel M. 1974. *The Modern World System: Capitalist Agriculture and the Origins of the European World Economy in the Sixteenth Century.* New York: Academic Press.

Wallerstein, Immanuel M. 1980. *The Modern World-System II.* New York: Academic Press.

Walters, Suzanna Danuta. 1999. "Sex, Text, and Context: (In) Between Feminism and Cultural Studies." Pp. 222–260 in *Revisioning Gender,* edited by Myra Marx Ferree, Judith Lorber, and Beth B. Hess. Thousand Oaks, CA: Sage.

Warner, Rebecca L., and Brent S. Steel. 1999. "Child Rearing as a Mechanism for Social Change: The Relationship of Child Gender to Parents' Commitment to Gender Equity." *Gender & Society* 13 (August): 503–517.

Washington, Scott. 2006. "Racial Taxonomy." Unpublished manuscript, Princeton University, Princeton, NJ.

Wasserman, Stanley, and Katherine Faust (eds.). 1994. *Social Network Analysis: Methods and Applications.* Cambridge, MA: Cambridge University Press.

Waters, Mary C., and Peggy Levitt (eds.). 2002. *The Changing Face of Home: The Transnational Lives of the Second Generation.* New York: Russell Sage Foundation Press.

Waters, Mary C. 1990. *Ethnic Options: Choosing Identities in America.* Berkeley, CA: University of California Press.

Watkins, Susan. 1987. "The Fertility Transition: Europe and the Third World Compared." *Sociological Forum* 2 (Fall): 645–673.

Watt, Toni Terling, and Susan F. Sharp. 2001. "Gender Differences in Strains Associated with Suicidal Behavior among Adolescents." *Journal of Youth and Adolescence* 30 (June): 333–348.

Watts, Duncan J. 1999. "Networks, Dynamics, and the Small-World Phenomenon." *American Journal of Sociology* 105 (September): 493–527.

Watts, Duncan J., and Stephen H. Strogatz. 1998. "Collective Dynamics of 'Small World' Networks." *Nature* 393: 440–442, **www.sentencing project.org**

Weber, Max. 1925 [1947]. *The Theory of Social and Economic Organization.* New York: Free Press.

Weber, Max. 1958 [1904]. *The Protestant Ethic and the Spirit of Capitalism.* New York: Scribner's.

Weber, Max. 1962 [1913]. *Basic Concepts in Sociology.* New York: Greenwood.

Weber, Max. 1978 [1921]. *Economy and Society: An Outline of Interpretive Sociology,* Guenther Roth and Claus Wittich (eds.). Berkeley, CA: University of California Press.

Webster, Murray, Jr., and Stuart J. Hysom. 1998. "Creating Status Characteristics." *American Sociological Review* 63 (June): 351–378.

Weeks, John R. 2004. *Population: An Introduction to Concepts and Issues,* 9th ed. Belmont, CA: Wadsworth.

Weeks, John R. 2008. *Population: An Introduction to Concepts and Issues,* 10th ed. Belmont, CA: Wadsworth.

Weisburd, David, Cynthia M. Lum, and Anthony Petrosino. 2001. "Does Research Design Affect Study Outcomes in Criminal Justice?" *The Annals of the American Academy of Political and Social Science* 578 (November): 50–70.

Weisburd, David, Stanton Wheeler, Elin Waring, and Nancy Bode. 1991. *Crimes of the Middle Class: White Collar Defenders in the Courts.* New Haven, CT: Yale University Press.

Weitz, Rose. 2001. *The Sociology of Health, Illness, and Health Care: A Critical Approach.* Belmont, CA: Wadsworth.

Wenger, Jeffrey. 2001. *Divided We Fall: Deserving Workers Slip Through America's Patchwork Unemployment Insurance System.* Washington, DC: Economic Policy Institute. **www.epi.org**

West, Candace, and Don Zimmerman. 1987. "Doing Gender." *Gender & Society* 1 (June): 125–151.

West, Candace, and Sarah Fenstermaker. 1995. "Doing Difference." *Gender & Society* 9 (February): 8–37.

West, Carolyn M. 1998. "Leaving a Second Closet: Outing Partner Violence in Same-Sex Couples." Pp. 163–183 in *Partner Violence: A Comprehensive Review of 20 Years of Research,* edited by Jana L. Jasinksi and Linda M. Williams. Thousand Oaks, CA: Sage.

Western, Bruce. 2006. *Punishment and Inequality in America.* New York: Russell Sage Foundation.

White, Deborah Gray. 1985. *Ar'n't I a Woman?: Female Slaves in the Plantation South.* New York: W. W. Norton.

White, Jonathan R. 2002. *Terrorism: An Introduction.* Belmont, CA: Wadsworth.

White, Paul. 1998. "The Settlement Patterns of Developed World Migrants in London." *Urban Studies* 35 (October): 1725–1744.

Whiten, A., J. Goodall, W. C. McGrew, T. Nishida, V. Reynolds, Y. Sugiyama, C. E. G. Tutin, R. W. Wrangham, and C. Boesch. 1999. "Cultures in Chimpanzees." *Nature* 399 (June 17): 682–684.

Whorf, Benjamin. 1956. *Language, Thought, and Reality: Selected Writings.* Cambridge, MA: Technology Press, MIT.

Whyte, William F. 1943. *Street Corner Society.* Chicago, IL: University of Chicago Press.

Wickelgren, Ingrid. 1999. "Discovery of 'Gay Gene' Questioned." *Science* 284 (April 23): 571.

Wilder, Esther I., and Toni Terling Watt. 2002. "Risky Parental Behavior and Adolescent Sexual Activity at First Coitus." *The Milbank Quarterly* 80 (September): 481–524.

Wilkerson, Isabel. 2005. "In a Strange Land: From New Orleans to Sallisaw; Scattered in a Storm's Wake and Caught in a Clash of Cultures." *The New York Times,* October 9.

Wilkie, Jane. 1993. "Changes in U.S. Men's Attitudes Toward the Family Provider Role, 1972 to 1989." *Gender & Society* 7 (June): 261–279.

Williams, Christine L. 1992. "The Glass Escalator: Hidden Advantages for Men in the 'Female' Professions." *Social Problems* 39 (August): 253–267.

Williams, Christine L. 1995. *Still a Man's World: Men Who Do Women's Work.* Berkeley, CA: University of California Press.

Williams, David R., and Chiquita Collins. 1995. "U.S. Socioeconomic and Racial Differences in Health: Patterns and Explanations." *Annual Review of Sociology* 21: 349–386.

Willie, Charles Vert. 1979. *The Caste and Class Controversy.* Bayside, NY: General Hall.

Wilson, Barbara J., Stacy L. Smith, W. James Potter, Dale Kunkel, Daniel Linz, Carolyn M. Colvin, and Edward Donnerstein. 2002. "Violence in Children's Television Programming: Assessing the Risks." *Journal of Communication* 52 (March): 5–35.

Wilson, James Q., and Richard J. Herrnstein. 1985. *Crime and Human Nature.* New York: Simon and Schuster.

Wilson, John F. 1978. *Religion in American Society: The Effective Presence.* Englewood Cliffs, NJ: Prentice Hall.

Wilson, John. 1994. "Returning to the Fold." *Journal for the Scientific Study of Religion* 33 (June): 148–161.

Wilson, William Julius. 1978. *The Declining Significance of Race: Blacks and Changing American Institutions.* Chicago, IL: University of Chicago Press.

Wilson, William Julius. 1987. *The Truly Disadvantaged: The Inner City, the Underclass and Public Policy.* Chicago, IL: University of Chicago Press.

Wilson, William Julius. 1996. *When Work Disappears: The World of the New Urban Poor.* New York: Knopf.

Wilson, William Julius. 2009. *More Than Just Race: Being Black and*

Poor in the Inner City. New York: W. W. Norton.

Winant, Howard. 2000. "Race and Race Theory." *Annual Review of Sociology* 26: 169–185.

Winnick, Louis. 1990. "America's 'Model Minority.'" *Commentary* 90 (August): 222–229.

Winseman, Albert L. 2004. "'Born Agains' Wield Political, Economic Influence." The Gallup Poll. Princeton, NJ: Gallup Organization. **www.gallup.com**

Wirth, Louis. 1938. "Urbanism as a Way of Life." *American Journal of Sociology* 40: 1–24.

Wolcott, Harry F. 1996. "Peripheral Participation and the Kwakiutl Potlatch." *Anthropology and Education Quarterly* 27 (December): 467–492.

Wolf, Naomi. 1991. *The Beauty Myth: How Images of Beauty Are Used Against Women*. New York: William Morrow.

Woman's Research and Education Institute. 2009. *Women in the Military and Women Veterans*. **www.weri.org**

Woo, Deborah. 1998. "The Gap Between Striving and Achieving." Pp. 247–256 in *Race, Class, and Gender: An Anthology*, 3rd ed., edited by Margaret L. Andersen and Patricia Hill Collins. Belmont, CA: Wadsworth.

Wood, Julia T. 1994. *Gendered Lives: Communication, Gender and Culture*. Belmont, CA: Wadsworth.

Wood, Richard L. 2002. *Faith in Action: Religion, Race, and Democratic Organizing in America*. Chicago, IL: University of Chicago Press.

Woods, T., B. Kurts-Costes, and S. Rowley. 2005. "The Development of Stereotypes About the Rich and Poor: Age, Race, and Family Income Differences in Beliefs." *Journal of Youth and Adolescence* 34 (5): 437–445.

Worchel, Stephen, Joel Cooper, George R. Goethals, and James L. Olsen. 2000. *Social Psychology*. Belmont, CA: Wadsworth.

World Bank. 2004. *Water Resources Sector Strategy: Strategic Directions for World Bank Engagement*. **www.worldbank. org**

World Bank. 2007. "GNI per Capita 2005." *Quick References Tables*. **www.worldbank.org**

World Bank. 2009. "GNI per Capita 2005." Quick References Tables. **www.who.int**

World Health Organization. 2000. *The State of Food Insecurity in the World: 2000*. Geneva: World Health Organization. **www.who.int**

World Health Organization. 2002. **www.worldbank.org**

Wright, Erik Olin. 1979. *Class Structure and Income Determination*. New York: Academic Press.

Wright, Erik Olin. 1985. *Classes*. London: Verso.

Wright, Erik Olin, Karen Shire, Shu-Ling Hwang, Maureen Dolan, and Janeen Baxter. 1992. "The Non-Effects of Class on the Gender Division of Labor in the Home: A Comparative Study of Sweden and the United States." *Gender & Society* 6 (June): 252–282.

Wright, Laurence. 1994. "One Drop of Blood." *The New Yorker* 70 (July 25): 46–55.

Wuthnow, Robert (ed.). 1994. *I Come Away Stronger: How Small Groups Are Shaping American Religion*. Grand Rapids, MI: Eerdmans.

Wuthnow, Robert. 1998. *After Heaven: Spirituality in America Since the 1950s*. Berkeley, CA: University of California Press.

Wuthnow, Robert. 2007. *After the Baby Boomers: How Twenty-Somethings and Thirty-Somethings Are Shaping the Future of American Religion*. Princeton, NJ: Princeton University Press.

Wysocki, Diane Kholos. 1998. "Let Your Fingers Do the Talking: Sex on an Adult Chat-Line." *Sexualities* 1 (November): 425–452.

Xu, Wu, and Ann Leffler. 1992. "Gender and Race Effects on Occupational Prestige, Segregation, and Earnings." *Gender & Society* 6 (September): 376–392.

Xuewen, Sheng, Norman Stockman, and Norman Bonney. 1992. "The Dual Burden: East and West (Women's Working Lives in China, Japan, and Great Britain)." *International Sociology* 7 (June): 209–223.

Yoder, Janice. 1991. "Rethinking Tokenism: Looking Beyond Numbers." *Gender & Society* 5 (June): 178–192.

Young, J. W. 1994. "Differential Prediction of College Grades by Gender and by Ethnicity: A Replication Study." *Educational and Psychological Measurement* 54: 1022–1029.

Zajonc, Robert B. 1968. "Attitudinal Effects of Mere Exposure." *Journal of Personality and Social Psychology*. Monograph Supplement, Part 2: 1–29.

Zald, Mayer, and John McCarthy. 1975. "Organizational Intellectuals and the Criticism of Society." *Social Service Research* 49: 344–362.

Zhou, Min, and Carl L. Bankston. 2000. "Immigrant and Native Minority Groups, School Performance, and the Problem of Self-Esteem." Paper presented at the Southern Sociological Society, April, New Orleans, LA.

Zimbardo, Phillip G., Ebbe B. Ebbesen, and Christina Maslach. 1977. *Influencing Attitudes and Changing Behavior*. Reading, MA: Addison-Wesley.

Zola, Irving Kenneth. 1989. "Toward the Necessary Universalizing of a Disability Policy." *Milbank Quarterly* 67 (Suppl. 2): 401–428.

Zola, Irving Kenneth. 1993. "Self, Identity and the Naming Question: Reflections on the Language of Disability." *Social Science and Medicine* 36 (January): 167–173.

Zorza, J. 1991. "Woman Battering: A New Cause of Homelessness." *Clearinghouse Review* 25 (4).

Zuberi, Tukufu. 2001. *Thicker Than Blood: How Racial Statistics Lie*. Minneapolis, MN: University of Minneapolis Press.

Zuckerman, Phil. 2002. "The Sociology of Religion of W.E.B. Du Bois." *Sociology of Religion* 63: 239–253.

Zweigenhaft, Richard L., and G. William Domhoff. 2006. *Diversity in the Power Elite: How It Happened, Why It Matters*. Lanham, MD: Rowman and Littlefield.

Zwerling, Craig, and Hilary Silver. 1992. "Race and Job Dismissals in a Federal Bureaucracy." *American Sociological Review* 57 (October): 651–660.

Zwick, Rebecca, (ed.). 2004. *Rethinking the SAT: The Future of Standardized Testing in University Admissions*. New York: Routledge Falmer.

Photo Credits

Chapter 1

p.1 Richard G. Bingham II/Alamy; p.2 Jeff Greenberg/The Image Works; p.4 Lara Jo Regan/Liaison/Getty Images; p.6 Spencer Platt/Getty Images; p.7BL William Thomas Cain/Getty Images; p.7 Lindsay Hebberd/Corbis; p.10 courtesy of Maggie Andersen; p.11 courtesy of Howard Taylor; p.12T Robert Brenner/PhotoEdit; p.12B Warren Miller/The New Yorker Collection/cartoonbank.com; p.13 Eugene Hoshiko/AP Photo; p.14BR Bettmann/Corbis; p.15TL Frontline Photography/Alamy; p.15CL P-59 Photos/Alamy; p.15BL Alexis C. Glenn/UPI/Landov; p.15CR Owen Franken/Corbis; p.15TR David Grossman/Alamy; p.15BR Karel Prinsloo/AP Photo; p.16TR Bettmann/Corbis; p.16BR AKG London; p.17 Bettmann/Corbis; p.18BL Special Collections and Archives W. E.B. Du Bois Library University of Massachusetts at Amherst; p.18TL Schomburg Center for Research in Black Culture/New York Public Library; p.24 David Young-Wolff/Photographer's Choice/Getty Images

Chapter 2

p.27 Zave Smith/PhotoLibrary; p.28TL David Reed/Corbis; p.28TR Paul Chesley/Stone/Getty Images; p.28BL Lindsay Hebberd/Corbis; p.28BR Jack Kurtz/The Image Works; p.30 NBAE/Getty Images; p.31 Mark Duncan/AP Photo; p.32 Paul Sakuma/AP Photo; p.37TL Charles Bennett/AP Photo; p.37TR Don Ryan/AP Photo; p.39 Amy Sancetta/AP Photo; p.40 Konstantynov/Shutterstock.com; p.43 M Becker/American Idol 2009/Getty Images

Chapter 3

p.57 Sonda Dawes/The Image Works; p.58T Bob Daemmrich/PhotoEdit; p.58B Emmanuel Dunand/AFP/Getty Images; p.62 Rhoda Sidney/PhotoEdit; p.66 Mark Richards/ PhotoEdit; p.67 Dwayne Newton/PhotoEdit; p.69 Laura Noel/

Sygma/Corbis; p.72 Jim Erickson/Corbis

Chapter 4

p.74 Keith Meyers/The New York Times/Redux; p.78 Thinkstock/Jupiter Images; p.79 Dima Gavrysh/AP Photo; p.80L Amy Etra/PhotoEdit; p.80R AP Photo; p.84 Paul Chesley/Stone/Getty Images; p.87 Kate Powers/Taxi/Getty Images; p.89 Caroline Penn/Corbis; p.92L Bob E. Daemmrich/Sygma/Corbis; p.92R Brooks Kraft/Corbis; p.95 piluhin/Alamy; p.96TL Juliet Highet/The Hutchison Library; p.96TR Judy Griesedieck/Corbis; p.96BL E.A. Heiniger/Photo Researchers Inc.; p.96BR Boris Grdanoski/AP Photo; p.97 Angela Hampton Picture Library/Alamy; p.100 Darrin Klimek/Digital Vision/Getty Images

Chapter 5

p.102 RayArt Graphics/Alamy; p.103L Shout/Alamy; p.103R Mark Edwards/Peter Arnold Inc.; p.106L BlueMoon Stock/Alamy; p.106R Sergei Grits/AP Photo; p.110 Simon Jarratt/Corbis; p.111 Lon C. Diehl/PhotoEdit; p.112 David Woolfall/Getty Images; p.113 Michael Newman/PhotoEdit; p.115L Nina Leen//Time Life Pictures/Getty Images; p.115R Alex Mares-Manton/Asia Images/Jupiter Images; p.119 Rich Abrahamson/The Fort Collins Coloradoan/AP Photo; p.122CL Matt Sayles/AP Photo; p.122TR Bob Daemmrich/The Image Works; p.122BR Joanne Carole/AP Photo; p.123TL Kevork Djansezian/AP Photo; p.123CR Jeff Greenberg/PhotoEdit; p.123BL Michelle Gabel/Syracuse Newspapers/The Image Works; p.126 Scott Olson/Reportage/Getty Images

Chapter 6

p.128 Michael Kelley/Getty Images; p.129 Momatiuk Eastcott/ The Image Works; p.131 DoD/ Staff Sgt. Dawn Finniss; p.132 tomazl/istockphoto; p.133 Bebeto Matthews/AP Photo; p.136 From the film

OBEDIENCE © 1965 by Stanley Milgram and distributed by Penn State Media Sales; p.139 Sutcliffe/SIPA Press; p.143L Michael Soforonski/ SIPA Press; p.143TR NASA/AP Photo; p.143BR Dr. Scott Lieberman/AP Photo; p.144 Tony Dejak/AP Photo; p.145 Bananastock/Jupiter Images; p.150 Nic Cleave Photography/Alamy

Chapter 7

p.153 RAWA/WorldPicture News; p.154 Jeff Greenberg/PhotoEdit; p.156 Paul Chesley/Stone/Getty Images; p.160 David Young-Wolff/PhotoEdit; p.163 Jim West/Alamy; p.164 Keith Erskine/Alamy; p.168 Brendan McDermid/Reuters/Landov; p.169 Abbot Genser/HBO/courtesy Everett Collection; p.175 Moshe Bursuker/AP Photo; p.178 Beth A. Keiser/AP Photo

Chapter 8

p.181L Tony Freeman/PhotoEdit; p.181R SUNNYphotography.com/Alamy; p.182L Carolyn Kaster/AP Photo; p.182R Eric Gay/AP Photo; p.183 Jeff Greenberg/Alamy; p.184TL Thomas Barwick/Getty Images; p.184TR Don Ryan/AP Photo; p.184BL Rob Kim/Landov; p.184BR Richard Vogel/AP Photo; p.185TL Michael Newman/PhotoEdit; p.185TL Satchan/zefa/Corbis; p.185BL Andrew Holbrooke/Corbis; p.185BR Kareem Black/Getty Images; p.187TL Radius Images/PhotoLibrary; p.187TR Ric Feld/AP Photo; p.188BL Jesse Grant/Getty Images; p.188BR Sean Garnsworthy/Getty Images; p.190 Jim West/ The Image Works; p.192 Justin Sullivan/Getty Images; p.201TL Tom Hauck/Getty Images; p.201TR Visions LLC/PhotoLibrary; p.201BL Jed Jacobsohn/Getty Images; p.201BR Natalie Fobes/Corbis; p.210 Paula Bronstein/Reportage/Getty Images

Chapter 9

p.213TL EPA/ Wilson Wen/ Landov; p.213TR WoodyStock/Alamy; p.214TL Ryan Pyle/Corbis; p.215

Name Index

A

Aberle, David F., 103
Abma, J. C., 306
Abrahamson, Mark, 410
Acker, Joan, 207, 274
Acs, Gregory, 196
Addams, Jane, 15, 18
Adorno, T. W., 241, 245
Aguilar-San Juan, Karin, 75
Alba, Richard, 134, 236, 254
Albas, Daniel, 119
Albelda, Randy, 205, 207
Aldag, Ramon J., 138
Alden, Helena L., 274
Aldous, Joan, 323
Aldrich, Howard, 133
Alicea, Marixsa, 329
Allen, Katherine R., 323
Allen, Michael Patrick, 379
Allison, Anne, 275, 300
Allport, Gordon W., 241, 242, 245
Alon, Sigal, 353
Altemeyer, Bob, 245
Altman, Dennis, 295, 300, 386
Amato, Paul R., 327
Amott, Teresa L., 249, 250, 429
Andersen, Margaret L., 10–1, 110,
    111, 140, 194, 246, 252, 256, 265,
    267, 276, 283, 362
Andersen, Norman, 85, 415
Anderson, David A., 270
Anderson, Elijah, 66, 114
Anderson, Kristin J., 263
Andreasse, Arthur, 205
Angotti, Joseph, 42
Anthony, Dick, 338
Antrobus, Peggy, 339
Apple, Michael W., 388
Arendell, Terry, 327
Arendt, Hannah, 137
Argyris, Chris, 147
Armas, Genaro C., 326
Armstrong, E.A., 308
Armstrong, Elizabeth M., 365
Aronson, Joshua, 356

Asch, Solomon, 135
Atchley, Robert C., 362

B

Babbie, Earl, 163
Baca Zinn, Maxine, 205, 265, 274,
    283, 316, 322, 325, 328
Bachman, Daniel, 315
Bacon, Sir Francis, 56
Bailey, Garrick, 224
Bandura, A., 83
Banks, W. Curtis, 132
Barber, Benjamin R., 42
Barber, Jennifer S., 323
Bardacke, Frank, 429
Barlow, Maude, 411
Barton, Bernadette, 7, 292
Basow, Susan A., 263, 274
Baumeister, Roy F., 67, 79, 132,
    133
Bean, Frank D., 249
Becker, Elizabeth, 384
Becker, Howard S., 152, 161,
    162, 163
Becker, Penny Edgell, 80
Belknap, Joanne, 171, 172
Bellah, Robert, 16
Beller, Emily, 196, 207
Ben-Yehuda, Nachman, 160
Benedict, Ruth, 37
Berger, Joseph, 128
Berger, Peter L., 7–8, 20, 76, 85,
    117, 422
Berkman, Lisa F., 156
Bernstein, Nina, 207
Berscheid, Ellen, 115, 116
Best, Joel, 9, 163, 167, 169, 170
Bielby, William T., 349
Binns, Allison, 175
Blair-Loy, Mary, 329
Blake, C. Fred, 8
Blalock, Hubert M., Jr., 110
Blankenship, Kim, 377
Blaskovitch, J., 139
Blassingame, John, 248

Blau, Francine, 3
Blau, Peter M., 119, 140, 187
Blinde, Elaine M., 81
Blumenbach, Alfred, 34
Blumer, Herbert, 20, 86, 117, 163
Bobo, Lawrence D., 133, 202, 241,
    243, 244, 245
Boer, J. Tom, 414
Boeringer, Scott B., 172
Boggs, Vernon W., 39
Bogle, Kathleen, 323
Bonacich, Edna, 386
Bonilla-Silva, Eduardo, 7, 243, 244,
    246, 317
Bontemps, Arna, 249
Boonstra, Heather, 304, 305, 307
Bostic, Raphael W., 193
Bouchard, Thomas J., Jr., 353
Bourdieu, Peirre, 48
Bowditch, Christine, 82
Bowles, Samuel, 82, 349
Branch, Taylor, 258
Brechin, Steven R., 413
Brehm, S. S., 116
Brewer, Rose, 321
Bridges, George, 173
Brignall, Thomas Wells, III, 124
Brines, Julie, 323
Brislen, William, 186
Britt, Chester L., 170
Brodkin, Karen, 238
Brooke, James, 315
Brookey, Robert Alan, 291
Brooks-Gunn, Jeanne, 60
Brown, Cynthia, 238
Brown, Elaine, 256, 258
Brown, Roger, 114
Brown, Ryan P., 356, 357
Browne, Irene, 205, 275, 278, 286
Broyard, Bliss, 237
Bryant, Alyssa N., 284
Bryant, Susan Leslie, 156
Buchanan, Christy M., 327
Budig, Michelle J., 280
Bullard, Robert D., 414, 415

Bullock, Heather E., 197
Burgess, Samuel H., 274
Burkholder, Richard, 41
Burris, Val, 191
Burt, Cyril, 65
Bushman, Brad P., 67
Butsch, Richard, 46
Buunk, Bram, 299

C

Calhoun, Craig, 430
Camaroto, Steven, 388
Cameron, Deborah, 263
Campbell, Anne, 154
Campenni, C. Estelle, 88
Cantor, Joanne, 67, 79
Caplow, Theodore, 129, 421
Cardenas, Jose A., 355
Carmichael, Stokely, 258
Carnoy, Martin, 386
Carroll, J. B., 33
Carson, Rachel, 415
Carter, Timothy S., 169
Cashmore, Ellis, 173
Cassen, Robert, 222
Cassidy, J., 114
Catanzarite, Lisa, 205, 206
Cavalier, Elizabeth, 81
Centeno, Miguel A., 133
Centers, Richard, 196
Chafetz, Janet, 275
Chalfant, H. Paul, 331
Chambliss, William J., 173
Chang, Jung, 8
Chen, Elsa Y. F., 251
Cherlin, Andrew J., 327
Chernin, Kim, 358
Chesney-Lind, Meda, 140
Chia, Rosina C., 273
Chipman, Susan, 356
Chipuer, H. M., 353, 354
Chiricos, Ted, 42
Chodorow, Nancy, 83
Churchill, Ward, 415
Cialdini, R. B., 135
Cicourel, Aaron V., 162
Clark, Kenneth B., 132
Clawson, Dan, 330
Clawson, Rosalee A., 197
Clifford, N. M., 116
Coale, Ansley, 222, 408
Cohen, Philip N., 323
Cole, David, 173
Colley, Ann, 263
Collins, Patricia Hill, 45, 246, 256, 258, 274, 282, 283, 316

Collins, Randall, 14, 17, 19, 348, 376, 377
Collins, Sharon M., 145
Collins-Lowry, Sharon M., 134
Coltrane, Scott, 320
Comte, August, 13–4
Connell, R. W., 291, 296
Conrad, Peter, 154
Constable, Nicole, 329
Cook, Christopher D., 388
Cook, Karen S., 119
Cook, S. W., 245
Cooksey, Elizabeth C., 91
Cookson, Peter W., Jr., 346, 349
Cooley, Charles Horton, 17, 85–6, 129
Coontz, Stephanie, 314
Cooper, Alvin, 308
Cortina, Lilia M., 395
Coser, Lewis, 16, 17
Costello, Mark, 170
Craig, Maxine Leeds, 44
Crane, Diana, 49
Crawford, Mary, 263
Crittenden, Ann, 276
Croll, Elisabeth, 315
Crouse, James, 352
Crowley, Sue M., 326
Csikzentmihalyi, Mihaly, 90
Cunningham, Mick, 325
Currie, Dawn, 46
Curtiss, Susan, 74
Czerniawski, Amanda, 359

D

Dahlhamer, James, 427
Dahrendorf, Rolf, 20, 420
D'Alessio, Stewart J., 170
Dalton, Susan E., 322
Darley, John M., 355
Darling-Hammond, Linda, 347, 351, 352, 354
Darwin, Charles, 17, 27
Das Gupta, Monisha, 329
Davis, Angela, 248
Davis, James A., 187
Davis, Kingsley, 199, 408
Deegan, Mary Jo, 18
DeFleur, Lois, 163
Dellinger, Kristen, 280
Demeny, Paul, 222
D'Emilio, John, 297
DeNavas-Walt, Carmen, 93, 181, 186, 187, 195, 202, 203, 207, 225, 264, 277, 321
Denton, Nancy A., 255, 256
DeVault, Marjorie, 267

Devlin, Bernie, 353
DeWitt, Karen, 145, 146
DiAcosta, Diego, 424
Diaz-Duque, Ozzie F., 362
Dill, Bonnie, 316
Dillard, J. L., 424
Dillon, Sam, 352
DiMaggio, Paul J., 126, 139
Dines, Gail, 46
Dion, K. K., 116
Dixit, Avinash K., 120
Dolan, Jill, 51
Dollard, John, 245
Domhoff, G. William, 134, 376, 378
Dominguez, Silvia, 134
Dovidio, J. F., 243
Dowd, James J., 39
Downey, Liam, 414
Dreier, Peter, 5
Driggs, Ken, 315
Du Bois, W. E. B., 15, 18, 49, 158, 201, 298
DuBois, Cathy L., 395
Dubrofsky, Rachel, 46
Dudley, Carl S., 333, 334
Due, Linnea, 33, 91
Duneier, Mitchell, 55–6, 66
Dunlop, Riley E., 415
Dunne, Gillian A., 322
Durkheim, Emile, 14, 15, 16, 18–9, 49, 102, 104–5, 153, 155–6, 158, 201, 244, 298, 331, 335, 410, 419
Duster, Troy, 18
Dworkin, Shari L., 81
Dyson, Michael, 414

E

Eckberg, Douglass Lee, 79
Edin, Kathryn, 203, 206, 207, 208, 306
Edmondson, Brad, 405, 411
Ehrlich, Anne, 409, 412
Ehrlich, Paul R., 222, 224, 409, 412
Eichenwald, Kurt, 145, 170
Eitzen, D. Stanley, 19, 199, 230, 319
Elizabeth, Vivienne, 323
Ellison, Christopher G., 80
Emerson, Juan P., 366
Emerson, Rana A., 45
Engels, Friedrich, 16
Enloe, Cynthia, 295, 300
Epps, Edgar T., 352
Epstein, Steven, 363
Erikson, Erik, 88–9
Erikson, Kai, 152, 153, 155, 160, 163
Esbensen-Finn, Aage, 154
Eschholz, Sarah, 271

Espenshade, Thomas J., 260
Essed, Philomena, 253
Esterberg, Kristin G., 428
Etzioni, Amatai, 47, 140
Evans, Peter B., 376

F

Farber, Naomi, 306
Fattah, Ezzat A., 88
Fausto-Sterling, Anne, 266
Feagin, Joe R., 89, 140, 242, 245, 246
Fein, Melvyn L., 96
Ferber, Abby, 40, 253, 339, 428
Fernald, Anne, 78
Festinger, Leon, 115
Fields, Jason, 323
Fischer, Claude S., 354
Fisher, Bonnie S., 172, 307
Fitzgibbon, Marian, 116
Flanagan, William G., 410
Fleisher, Mark, 154
Fleming, Jacqueline, 352, 353
Flowers, M. L., 138
Folbre, Nancy, 330
Frankenberg, Erica, 243, 256
Frazier, E. Franklin, 132, 194
Fredrickson, George M., 217
Freedle, Roy O., 352
Freedman, Estelle B., 291, 299, 308
Freeman, Jo, 428
Freese, Jeremy, 76
Freud, Sigmund, 82–3, 86
Frey, William, 411
Fried, Amy, 140
Friedman, Thomas L., 187, 418
Fryberg, Stephanie, 28
Frye, Marilyn, 104
Fukuda, Mari, 113
Furstenberg, Frank F., Jr., 327

G

Gagné, Patricia, 267
Gallagher, Charles, 44–5
Gallagher, Charles A., 243, 244
Gallagher, Sally, 324, 333
Gamson, Joshua, 46, 293, 430
Gamson, William, 118
Gans, Herbert, 410
Gardner, Howard, 352
Garfinkel, Harold, 36, 117–8
Garrow, David J., 373
Gastil, John, 35
Gates, Henry Louis, Jr., 424
Gelles, Richard J., 327
Genovese, Eugene, 248, 299
Gerami, Shahin, 271
Gersch, Beate, 197

Gerson, Kathleen, 134
Gerstel, Naomi, 324
Gerth, Hans, 17
Giannarelli, Linda, 329
Giddings, Paula, 18, 140
Gilbert, D. T., 132–3
Gilbert, Daniel R., 116
Gilens, Martin, 197
Gilkes, Cheryl Townsend, 337
Gilligan, Carol, 84
Gimlin, Debra, 44, 66, 120–1, 268–9
Gitlin, Todd, 26, 42
Glassner, Barry, 167
Glaze, Lauren E., 7
Glazer, Nathan, 254, 330
Glenn, Evelyn Nakano, 249, 251, 318
Glenn, Norval, 323
Glock, Charles, 331
Gluckman, Amy, 301
Goffman, Erving, 118–9, 124, 140, 165, 363, 430
Goldberger, Arthur S., 353
Goldman, Noreen, 418
Gonzales, P., 357
González, Tina Esther, 361, 365
Gordon, Linda, 302
Gore, Albert, 411
Gottfredson, Michael R., 159
Gould, Stephen J., 27, 254, 353, 354
Graefe, Deborah Roempke, 323
Graham, Lawrence Otis, 132
Gramsci, Antonio, 43
Granovetter, Mark, 133
Grant, Don Sherman, III, 159
Green, Gary Paul, 134
Greenhouse, Linda, 302
Greenhouse, Steven, 284
Grimal, Pierre, 333
Grindstaff, Laura, 46
Gruber, James E., 395
Guinther, J., 128
Guo, Guang, 76

H

Habermas, Jürgen, 423
Hadaway, C. Kirk, 80
Haddad, Yvonne Yazbeck, 333
Hagan, John, 170
Hall, Edward T., 113
Hall, Elaine J., 279
Hallinan, Maureen T., 354
Hamer, Dean H., 291, 321
Hamilton, Mykol C., 35
Handlin, Oscar, 252
Haney, C., 137
Haney, Lynne, 377

Hans, Valerie, 127, 128
Hargittai, Eszther, 121
Harp, Dustin, 51
Harris, Allen C., 272
Harris, Angel L., 132, 352
Harris, Marvin, 37
Harrison, Bennett, 387, 402, 403
Harrison, Deborah, 424
Harrison, Roderick, 237
Harry, Joseph, 302
Hartman, Chester, 202
Hastorf, Albert, 117
Hauan, Susan M., 202
Hauser, Robert M., 354
Hawley, Amos H., 411
Hays, Sharon, 206, 207, 208
Hearnshaw, Leslie, 65
Heider, Fritz, 117
Heilman, Brice E., 338
Heimer, Karen, 159
Hendy, Helen M., 267
Henry, Kathy, 18
Henslin, James. M., 152
Herman, Judith, 328
Herrnstein, Richard J., 353–4
Herskovitz, Melville J., 424
Hesse-Biber, Sharlene Hagy, 271, 272
Higginbotham, A. Leon, 194, 238, 346
Higginbotham, Elizabeth, 237
Hill, C. T., 116
Hill, Lori Diane, 48
Hirschi, Travis, 159
Hirschman, Charles, 222
Hochschild, Arlie R., 112, 280, 281, 325, 389
Hoffnung, Michele, 285
Hofstadter, Richard, 17
Hollander, Jocelyn A., 46
Hollingshead, August B., 164
Holmes, Schuyler, 414, 415
Homans, George, 119
Hondagneu-Sotelo, Pierrette, 329, 330
Hong, Laurence, 139
Hornung, Carlton A., 110
Horowitz, Ruth, 307
Horton, Hayward Derrick, 195
House, James S., 361
Hoyt, Wendy, 270
Hudson, Ken, 388
Huesmann, L. R., 67, 79
Hughes, Langston, 249
Hunt, Ruth, 240
Hunter, Andrea, 273, 284
Hunter, Margaret, 243

I

Ignatiev, Noel, 238
Inglehart, Ronald, 41
Irwin, Katherine, 7, 30, 162

J

Jackson, Pamela Braboy, 134, 146, 361, 362
Jacobs, David, 173
Jacobs, Jerry A., 280
Jacobs, Lawrence R., 360
Jang, Kerry L., 353
Janis, Irving L., 137–8
Janus, Samuel S., 294
Jencks, Christopher, 349, 352, 353, 354
Jenness, Valerie, 168, 429
Jennings, M. K., 135
Johnson, Norris R., 139
Johnston, David Cay, 194
Johnstone, Ronald L., 331, 340
Jones, Diane Carlson, 270
Jones, E. E., 117
Jones, Edward E., 117, 363
Jones, James M., 241, 243
Jones, Jeffrey M., 381
Jordan, Winthrop D., 248, 299
Joseph, Jay, 354
Jucha, Robert, 175

K

Kadushin, Charles, 134
Kahn, Lawrence, 3
Kalleberg, Arne L., 388
Kalmijn, Matthijs, 315, 326
Kalof, Linda, 46
Kamin, Leon J., 65, 253, 254, 353, 354
Kane, Emily W., 270
Kanter, Rosabeth Moss, 146
Kaplan, Elaine Bell, 8–9
Katz, I., 243
Kaufman, Gayle, 320
Keeter, Scott, 378
Keith, Verna M., 323
Kelley, J. R., 138
Kendall, Lori, 7
Kennedy, Robert E., 315, 407
Kephart, William H., 340
Kerr, N. L., 139
Kershaw, Sarah, 244
Kessler, Suzanne J., 266
Khanna, Nikki D., 317
Khashan, Hilal, 338
Kiefer, Amy K., 356, 357, 424
Kim, Elaine H., 252
Kimmel, Michael, 7, 383

Kimmel, Michael S., 77
Kirby, Douglas, 306
Kitano, Harry, 251
Kitsuse, John I., 162, 163
Klapp, Orin, 427
Kleinfeld, Judith S., 134
Klinger, Lori J., 272
Kluegel, J. R., 133, 241
Kniss, Fred, 340
Knoke, David, 133
Kochen, M., 134
Kocieniewski, David, 173
Kohlberg, Lawrence, 84
Kohut, Andrew, 252
Konishi, Hideo, 129
Kovel, Jonathan, 243
Krasnodemski, Memory, 241
Kuhn, Harold W., 120
Kung, H.C., 264
Kurz, Demie, 327, 328

L

Lacy, Karen R., 195
Ladner, Joyce A., 307
LaFlamme, Darquise, 270
LaFrance, Marriane, 263
Lamanna, Mary Ann, 315
Lamb, Michael, 308
Lamont, Michèle, 67
Lamphere, 389
Langhinrichsen-Rohling, Jennifer, 156
Langman, Lauren, 223
Larana, Enrique, 430
Lareau, Annette, 87–8
Lau, Bonnie, 357, 424
Laumann, Edward O., 291, 294, 362
Lavelle, Louis, 181
Laviest, Thomas A., 360
Lawson, Helene M., 115
Lawson-Borders, Gracie, 51
Lederman, Douglas, 285
Lee, Gary R., 323
Lee, Matthew T., 170
Lee, Sharon M., 237, 239, 254
Lee, Stacey J., 195, 202, 254
Lehmann, Nicholas, 352
Leidner, Robin, 391
Lemert, Edwin M., 162
Lengermann, P. M., 18
Lenski, Gerhard, 420
Leonhardt, David, 3
Lever, Janet, 82
Levine, Adeline, 418
Levine, Felice J., 9
Levy, Ariel, 308
Levy, Marion J., 103

Lewin, Tamar, 60
Lewis, Amanda, 8
Lewis, Dan A., 207
Lewis, Oscar, 256
Lewontin, Richard, 238, 354
Lieberson, Stanley, 279
Light, Ivan, 221
Lin, Nan, 134
Lindberg, L., 358
Linton, Ralph, 423
Livingston, I. L., 360
Lo, Clarence, 377
Locklear, Erin M., 238
Logio-Rau, Kim, 358
Lombardo, William K., 263
Lopez, Mark, 381
Lorber, Judith, 265, 291, 296
Lorenz, Conrad, 114
Louie, Miriam Ching Yoon, 386
Lovejoy, Meg, 116, 273
Lucal, Betsy, 183
Luker, Kristin, 304, 306
Lyons, Linda, 291, 322

M

Macdougall, Robert, 51
Machel, Graca, 228
MacKinnon, Catherine, 377
MacKinnon, Neil J., 187
Mackintosh, N. J., 65
Maimon, David, 156
Majete, Clayton A., 317
Malat, Jennifer, 360
Malcomson, Scott L., 237, 238
Maldonado, Lionel A., 249, 354
Malthus, Thomas R., 407–9
Mandel, Daniel, 45
Manning, Winton H., 353
Mantsios, Gregory, 197
Marciniak, Liz-Marie, 308
Marcuse, Herbert, 423
Margolin, Leslie, 162
Marks, Carole, 248, 249
Marquis, Christopher, 323
Marshall, Susan, 348
Martin, Patricia Yancey, 308
Martineau, Harriet, 14
Martinez, Ramiro, Jr., 170
Marx, Anthony, 217
Marx, Gary T., 420, 429
Marx, Karl, 14, 15, 16, 18, 20, 49, 158, 159–60, 189, 196, 197–8, 199, 201, 218, 298, 334, 335–6, 340, 375, 386, 420, 423
Massey, D., 356
Massey, Douglas S., 194, 242, 243, 255, 256

Mathews, Linda, 237
Mauer, Marc, 173, 174
Mazunder, Bhashkar, 196
McAdam, Doug, 429
McCabe, Janice, 282
McCall, Leslie, 205
McCarthy, John, 429
McCauley, C., 138
McClelland, 228
McCloskey, Laura Ann, 328
McCollum, Chris, 115
McComb, Chris, 284
McCord, Arline, 421–2
McCord, William, 218, 421–2
McGuire, Gail M., 145
McGuire, Meredith, 331, 334, 335
McIntyre, Rusty B., 356, 357
McLoyd, Vonnie C., 326
Mead, George Herbert, 17, 85, 86
Menken, Jane, 418
Meredith, Martin, 27
Mernissi, Fatima, 299
Merton, Robert, 19, 110, 132, 156–7, 158, 199
Messerschmidt, James W., 154
Messner, Michael A., 58, 81
Metz, Michael A., 302
Meyer, David S., 428
Meyer, Madonna Herrington, 330
Mickelson, Roslyn Arlin, 227
Milgram, Stanley, 135–7
Milkie, Melissa, 326
Milkie, Melissa A., 46, 266, 270
Miller, Eleanor, 152, 161, 171
Miller, Matthew, 187
Miller, Susan L., 328
Mills, C. Wright, 5, 6, 12, 20, 88, 93, 134, 211, 302, 328, 375–6, 393
Milner, Murray, Jr., 40
Miner, Horace, 26
Mintz, Beth, 133
Mirandé, Alfredo, 249, 274
Mishel, Lawrence, 181, 189, 190, 192, 282, 325
Misra, Joy, 197
Mitchell, G., 88
Mizruchi, Mark S., 133, 163
Moen, Phyllis, 326, 329, 330
Mohai, Paul, 414, 415
Moller, Lora C., 88
Montada, L., 162
Montagne, Peter, 412
Montgomery, James D., 133
Moore, David W., 284
Moore, Gwen, 154
Moore, Joan, 249
Moore, Robert B., 33, 35

Moore, Valerie A., 270
Moreland, Richard L., 115
Morgan, Lewis H., 420
Morlan, Patricia A., 241
Morning, Ann, 237, 403
Morris, Aldon D., 258, 340, 427, 428, 429, 430
Moskos, Peter, 56, 70
Moynihan, Daniel P., 256
Muller, Thomas, 424
Myers, Steven Lee, 384
Myers, Walter D., 248
Myerson, Allen R., 191
Myrdal, Gunnar, 421

N
Nagel, Joane, 247, 299
Nakao and Treas, 187
Nanda, Serena, 265
Nardi, Peter M., 296
Nash, John F., 120
Nee, Victor, 251
Newman, Katherine, 256
Newman, Katherine S., 156, 159
Newport, Frank, 281, 333, 337
Niebuhr, Gustav, 333
Nolan, Patrick, 106
Norris, Pippa, 224

O
Oakes, Jeannie, 354
Ogburn, William F., 50, 419
O'Leary, Carol, 48
Oliver, Melvin L., 192, 193, 194, 207, 242, 244, 256
Ollivier, Michele, 187
Omi, Michael, 237, 238, 249
O'Neil, John, 60
Orzechowski, Shawna, 191
Ouchi, William, 147
Owens, Sarah E., 354

P
Padavic, Irene, 134, 275, 286
Page, Charles, 142
Pager, Devah, 162
Pain, Emil, 338
Park, Lisa Sun-Hee, 7, 392
Park, Robert, 17
Parmelee, L. F., 243
Parreñas, Rhacel Salazar, 216, 329
Parsons, Talcott, 17, 19, 103, 199, 419
Pascoe, C.J., 270, 274
Paternoster, Raymond, 4
Pattillo-McCoy, Mary, 194, 195, 256
Paulus, P. B., 138

Pavalko, Eliza K., 362
Pedraza, Sylvia, 250, 389
Pelham, Brett, 294
Pellow, David N., 414
Pennock-Roman, Marcia, 352
Peoples, James, 93
Peralta, Robert L., 282
Perez, Denise N., 354
Perlmutter, David, 51
Perrow, Charles, 141, 142, 145
Perry-Jenkins, Maureen, 326
Persell, Caroline Hodges, 349
Pescosolido, Bernice A., 67, 366
Petersen, Trond, 133
Pettigrew, Thomas F., 132
Piaget, Jean, 84
Polce-Lynch, Mary, 270
Popenoe, David, 321
Portes, Alejandro, 33, 250, 254, 405, 407
Press, Andrea, 46
Press, Eyal, 212
Press, Julie E., 325
Preves, Sharon E., 266
Pruitt, D. G., 139
Prus, Robert, 161
Punch, Maurice, 170
Putnam, Robert, 47

R
Raboteau, Albert J., 248
Raley, Sara B., 324
Rampersad, Arnold, 249
Rashid, Ahmed, 338
Read, Jen'nan Ghazal, 45
Read, Piers Paul, 152
Reich, Robert, 422
Reiman, Jeffrey, 162, 169, 170, 171
Rendon, Laura I., 349, 354
Renzetti, Claire, 327
Reskin, Barbara F., 279
Reynolds, Ariel Catherine, 81
Rheault, Magail, 293
Rich, Adrienne, 297
Ridley, Matt, 76
Rieff, David, 41
Rifkin, Jeremy, 303
Rindfuss, Ronald R., 91
Risman, Barbara, 295, 306
Ritzer, George, 142, 144
Rivera, Ammad, 194
Robbins, Thomas, 98
Roberts, Dorothy, 250, 303
Robertson, Tatsha, 97
Robinson, Dawn T., 263
Robinson, Jo Ann Gibson, 428
Rodney, Walter, 420, 422

Rodriquez, Clara, 237, 238
Rodriquez, Nestor P., 221, 250
Roethlisberger, Fritz J., 142
Romain, Suzanne, 114
Roof, Wade Clark, 340
Rose, Tricia, 39
Rosenau, Pauline Marie, 21
Rosenbaum Janet Elise, 305
Rosenfeld, Michael J., 7
Rosengarten, Danielle, 388
Rosenhan, David L., 140
Rosenthal, Robert, 355
Ross, L., 117
Rossi, Alice S., 319
Rostow, W. W., 217
Rubin, Gayle, 299
Ruef, 134
Ruef, Martin, 133
Rueschmeyer, Dietrich, 376
Rumbaut, Rubén G., 254, 389, 405
Rupp, Leila J., 48, 297
Rust, Paula, 296
Ryan, Charlotte, 197
Ryan, William, 205
Rymer, Russ, 74

S
Saad, Lydia, 42, 264, 284
Sabol, William J., 7
Sadker, Myra, 8, 82, 271, 424
Saguy, Abigail, 394
Saks, M. J., 128
Sampson, Robert J., 256
Sanchez, Laura, 323
Sanchez, Lisa Gonzalez, 39
Sanchez-Jankowski, Martin, 66
Sanday, Peggy Reeves, 276
Sandnabba, N. Kenneth, 270
Santelli, John, 305
Sapir, Edward, 32–3
Sawhill, Isabel, 196
Sax, L. J., 77
Scarpitti, Frank R., 170, 366
Schacht, Steven P., 58
Schaffer, Kay, 51
Scheff, Thomas J., 162
Scheper-Hughes, Nancy, 93
Schlosser, Eric, 50
Schmitt, Eric, 256
Schmitt, Frederica E., 140
Schneider, Barbara, 90
Schneider, Beth, 280, 395
Schuman, Howard, 243
Schur, Edwin M., 152
Schwartz, John, 143
Schwartz, Pepper, 291, 292, 295, 297
Scott, Ellen K., 203

Scott, Tracy Lee, 80
Scully, Diana, 366
Segal, David R., 383, 384
Seidman, Steven, 296
Sen, Amartya, 227, 231
Settersten, Richard A., Jr., 91
Sharp, Susan F., 158
Sherkat, Darren E., 81
Shibutani, Tomatsu, 20, 427
Shih, Margaret, 424
Shilts, Randy, 363
Shinagawa, Larry, 317
Sigall, H., 116
Sigelman, Lee, 285
Signorielli, Nancy, 45, 47, 271
Silverthorne, Zebulon A., 116
Simile, Catherine, 429
Simmel, Georg, 9, 129, 410
Simmons, Wendy W., 381
Simon, David R., 159, 162, 164, 169, 412
Simons, Ronald, 327
Simpson, George Eaton, 239, 253
Singer, Peter W., 382
Sirvanandan, A., 217
Sjøberg, Gideon, 410
Skocpol, Theda, 376
Smelser, Neil J., 47
Smith, Barbara, 301
Smith, Bonnie, 357
Smith, Heather, 145
Smith, M. Dwayne, 170
Smith, Richard B., 302
Smith, Sandra, 134, 146
Smith, Tom W., 35
Snipp, C. Matthew, 52, 140, 238, 247, 248, 392
Snow, David A., 430
Sorokin, Pitrim, 421
Sotirovic, Mira, 197
Spangler, Eve, 146
Speed, A., 116
Spencer, Herbert, 419, 420
Spengler, Oswald, 420–1
Spitzer, Steven, 160
Squires, Gregory, 194
Stacey, Judith, 301, 323, 326
Stack, Carol, 315
Starr, Paul, 358, 366
Stearns, Linda Brewster, 394
Steele, Claude M., 132, 356
Steffensmeier, Darrell, 173
Stein, Arlene, 296
Steinhamer, Karen, 67, 79
Stern, Jessica, 225, 338
Stern, Kenneth S., 40
Sternberg, Robert J., 352

Stevens, William K., 410
Stewart, Abigail J., 327
Stewart, Susan D., 322
Stjerna, Marie Louise, 20
Stoller, Eleanor Palo, 88, 92, 93
Stombler, Mindy, 308
Stone, Pamela, 7
Stoner, James, 138
Stotik, Jeffrey, 431
Stover, R. G., 116
Stryker, Susan, 267
Sudnow, David N., 117
Sullivan, Maureen, 322
Sullivan, Teresa A., 193
Sumner, William Graham, 34–5
Suro, Roberto, 360
Surratt, Hilary L., 237
Sussal, Carol M., 395
Sussman, N. M., 263
Sutherland, Edwin H., 160, 161, 168
Swidler, Ann, 48
Switzer, J. Y., 35
Szasz, Thomas S., 164

T
Takaki, Ronald, 251, 254, 315
Tannen, Deborah, 32, 113, 114
Tatum, Beverly, 80
Taylor, Howard F., 10, 11, 65, 110, 134, 237, 249, 254, 349, 352, 353, 354
Taylor, Shelley E., 67, 110, 114, 115, 116, 117, 133, 135, 139, 172, 237, 241, 242, 243, 355, 358, 361
Teaster, Pamela B., 328
Telles, Edward E., 217, 237, 249, 254
Thakkar, Reena R., 156
Thoits, Peggy A., 80, 110, 361
Thomas, 20
Thomas, William I., 117, 132, 161, 355
Thompson, Becky, 358
Thompson, Robert Ferris, 424
Thorne, Barrie, 82
Thornton, Russell, 48, 247
Tichenor, Veronica Jaris, 324
Tienda, Marta, 202
Tierney, Patrick, 426
Tilly, Charles, 429
Tilly, Louise, 108
Tjaden, Patricia, 308, 327
Tocqueville, Alexis de, 14
Tönnies, Ferdinand, 105, 422–3
Toubia, Nahid, 41
Touré, Kwame, 258
Toynbee, Arnold, 420
Travers, Jeffrey, 134

Treiman, Donald J., 48
Trent, Katherine, 306
Tuan, Yi-Fu, 299
Tuchman, Gaye, 46
Tumin, Melvin M., 200
Turner, Jonathan, 19
Turner, Ralph, 426, 427
Turner, Terence, 26

U
Uleman, J. S., 133
Ullman, Sarah E., 172
Uvin, Peter, 216

V
Vail, D. Angus, 162
Van Ausdale, Debra, 89, 242
Vanlandingham, Mark, 410
Vanneman, Reeve, 189
VanRoosmalen, Erica, 395
Varelas, Nicole, 308
Vaughan, Diane, 142–3
Vázquez, Carmen, 301
Veblen, Thorstein, 37, 391

W
Waite, Linda J., 323
Waldfogel, Jane, 60
Walker-Barnes, Chanequa J., 154
Wallerstein, Immanuel, 219–20
Walters, Suzanna, 49
Warner, Rebecca L., 79
Warren, Elizabeth, 193
Washington, Scott, 237
Wasserman, Stanley, 133, 134
Waters, Mary C., 236, 237

Watkins, Susan, 222
Watt, Toni Terling, 156
Watts, Duncan J., 134
Weber, Max, 14, 15, 16–7, 18, 47, 49, 141, 147, 158, 182, 198–9, 201, 218, 298, 334, 335, 373–4
Webster, Murray, Jr., 111
Weeks, John R., 406, 407, 408, 411, 413
Weisburd, David, 160
Weitz, Rose, 358, 360, 361, 365, 366
Wells-Barnett, Ida B., 15, 18
Wenger, Jeffrey, 391
West, Candace, 110, 282
West, Carolyn M., 327
Westbrook, Jay Lawrence, 193
Western, Bruce, 162, 171, 173
Westoff, Charles F., 410
White, Deborah Gray, 248
White, Jonathan R., 175, 224
White, Paul, 221
Whiten, A., 27
Whorf, Benjamin, 32–3
Whyte, William Foote, 66
Wickelgren, Ingrid, 291
Wilder, Esther, 60
Wilkerson, Isabel, 50
Williams, Christine L., 184, 280
Williams, David R., 358, 360, 362
Willie, Charles Vert, 246
Wilson, Barbara J., 79
Wilson, John, 80
Wilson, William Julius, 189, 205, 206, 246, 256, 387
Winnick, Louis, 251, 252
Winseman, Albert L., 332

Wirth, Louis, 410
Wolcott, Harry F., 37
Wolf, Naomi, 358
Woo, Deborah, 254
Wood, Julia T., 113, 114, 133
Wood, Richard L., 340
Woods, T., 356
Worchel, Stephen, 67, 113, 116, 135, 139, 241, 361
Wright, Erik Olin, 183, 189, 325
Wright, Laurence, 237
Wuthnow, Robert, 80, 130–1, 340, 421, 423
Wysocki, Doana Kholos, 308

X
Xu, Wu, 394
Xuewen, Sheng, 284

Y
Yoder, Janice, 146
Young, J. W., 352
Younis, Mohammed, 337

Z
Zajonc, Robert, 115
Zald, Mayer, 429
Zhou, Min, 132
Zimbardo, Phillip G., 135
Zola, Irving, 395
Zorza, J., 204
Zuberi, Tukufu, 7
Zuckerman, Phil, 341
Zweigenhaft, Richard L., 191, 378
Zwerling, Craig, 145, 146
Zwick, Rebecca, 352, 353, 356

Subject Index

Page numbers in **bold** refer to definitions/explanations of key terms.
Page numbers in *italic* refer to boxed material or illustrations.

Abortion, *303*, 303–4, *304*
Abu Ghraib prison, 137, 138
Achievement tests, **352**. *See also* Standardized ability tests
Acquaintance rape, 152, 172, 307–8
Addams, Jane, *15*, *17*, **18**
Adolescence/adolescents
  alcohol, drugs, smoking, 20, 165, *165*
  characterized, 88–90, *91*
  contraceptive use, 305, 306, *306*
  media images of, *266*
  online profiles, *124*
  pregnancy/motherhood, *8–9*, 303, 305–7, *307*, 321
  rites of passage, 96
  sex during, *60*, 291, *291*, 294, *295*, 305–7
  socialization during, 88–90, 159
  stereotypes, 90, 92
  violence and, 156, 159
Adulthood, 90–1, *91*, 94. *See also* Aging/elderly
Affiliation, 114
Affirmative action, **258–9**, 260, **285–6**, *286*, *390*
Afghanistan, 97, *153*, 214, *215*, 275
Africa, *96*, *226*, 226–7, 299, 423–4
African Americans. *See also* Discrimination; Race and ethnicity; Racism; Segregation
  body image/weight, 116
  body image/weight/eating disorders, *273*
  class, 194–5, 256
  as crime victims, 170, *172*
  culture of, 29, 37, 39, *49*, 50–1, 249, 424, *424*
  death penalty, 4, 173
  defining, 237, 238
  "double consciousness," *49*
  families, *8–9*, 316, *321*
  health, 69, *273*, 358, 360, 362
  history, 248–9, 254–5
  home mortgages, 192, 193–4
  Hurricane Katrina and, *257*
  leadership networks, 134
  in media, 44, 45, *197*, 242
  migration to North, 248, *404*, 404–5
  in military, *15*, 383
  murder of, 258, 299
  population statistics, *12*, *13*
  religion, 258, 336–8, 340–1
  rights, 256–8, *284*, 340–1, 427, 429
  self-esteem, 132
  slavery, 238, 248, 299, 336, 424
  social movements, 52, 256–8
  in sociology, 18, *18*
  sororities/fraternities, 424, *424*
  stereotypes, 45, 240, 241, *298*, 299, 356
  stratification, 248–9
  terminology, 33, *35*, 235
  wealth levels, 191–2
  work, 146, 248, *286*, 392, *393*, 394
Age. *See also* Adolescence/adolescents; Aging/elderly; Children; Life course
  as ascribed status, 93
  division of labor, 105
  gender and health, 362
  mass media and, 92
  sexual attitudes and, 294
  as source of stratification, 195
  voting and, 377–8, *380*, *382*
Age cohort, **93**, *95*
Age discrimination, **92**
Ageism, **93**
Age prejudice, **92**
Age-sex pyramid, *94*, **406**, *406*
Age stereotypes, 90, **92**
Age stratification, **93–4**, *95*
Aggression, 79, 245, *265*, 265–6, 272, *281*
Aging/elderly
  death and dying, 117, 156
  elder abuse, 328
  elder care, 330–1
  population statistics, *94*, 402, 403, *406*, 406–8
  poverty, *93*, 94, 195, *195*
  social dimensions of, 91–5
  theoretical perspectives, 94–5, *95*
Agricultural societies, 107, *107*, 181
AIDS, 303, 362–3, *364*
Airplane crashes, 130, 151–2
Alcoholism, 154, 165
Alcohol use, *165*
Alienation, *130*, 143–4, *158*, **391**, 410, 423
Al Qaeda, 40, 175, 224–5, 431
Altruistic suicide, **156**
American Dream, 191, 192–4, *392*
American Indian Movement (AIM), 258, 430–1
American Medical Association (AMA), 357
American Sociological Association (ASA), *22*, 69
Americans with Disabilities Act (ADA), *359*, 396
Amish, *39*, 255
Anabolic steroids, 358–9
Andersen, Margaret L., *10–1*
Animals, 27, 107, 114, *115*
Anomic suicide, **156**
Anomie, 16, **155–6**
Anorexia nervosa, 77, 116, *268*, *273*, **358**
Anti-Semitism, 14, 136–7, **253**
Apache people, *28*
Apartheid, 182, 217, 239, *255*
Apartheid, gender, **275**
Appearance. *See* Beauty/physical ideals
Arabs/Arab Americans, 45, 239, 252
Asch Conformity Experiment, 135, *135*, *137*
Asian Americans. *See also* Race and ethnicity; *specific cultural groups*
  diversity, 250, 338
  family businesses, *392*
  history, 250–2

as "model minority," 195, 254
population statistics, *12, 13*
religion, 337–8
socialization, *75*
stereotypes, 45, 195, 240, 242
terminology, 33, 235
wealth/net worth, 192
Assimilation, **245**
Assimilation perspective, 254–5
Attachment, 83, 93, 114
Attraction, interpersonal, 114–6
Attribution error, 132–3
Attribution theory, **132**
Audiences, 109–10, 128
Authoritarian personality, **245**
Authority, **373**. *See also* Power
   in bureaucracies, 141
   charismatic, **373–4**
   obedience to, 135–7
   rational-legal, **374**
   of the state, 372–3, 377–85
   traditional, **373**
Automation, **388**
Autonomous state model, *375,*
   376–7

Baby Boomers, *94,* 402, 403, 406,
   406, 407
The Bachelor, 46
Bacon, Sir Francis, **56**
Bangladesh, 410
Bankruptcy, *193*
Barbie dolls, 212, *213*
Barr, Doug, *301*
Bay of Pigs invasion, 138
Beauty/physical ideals
   aging and, 92
   body image, 269, 270, 272, *273,*
      358–9
   culture and, 7, 44, *120–1, 268–9*
   eating disorders, 116, *268,* 272,
      273, 358
   gender identity and, 267, *268–9,*
      270, 272
   health and, 77, 358–9
   ideals of, 5–6, 44, 77, *120–1,*
      268–9, 358–9
   mass media and, 44, 45
   self-esteem and, 270
   social interaction and, 115–6
"Beauty work," *120–1*
Beliefs, *31, 36, 334. See also* Ideology; Myths; Religion; Social
   construction of reality; Stereotypes; Values
Bell Curve, The, 353–4
Berdaches, 265
Bilateral kinship, **315**
bin Laden, Osama, 175
Biological determinism, **265–6**
Biology and behavior, 27–8, 74, 76,
   265–7, 290–1, 353–4

Bioterrorism, **175**
Birth control, 250, 292, 302–3,
   305, 306, *306,* 308, 409–10
Birth, culture and, 103, *103*
Birthday celebrations, *27*
Birthrate, crude, **403–4**
Birthrates, 221–2, *408,* 408–9
Bisexuality, 296
Black feminism, 258
Black Muslims, 337
Black Panther Party, 258
Black Power movement, 258
Blogging, *51*
Body image, 116, *268–9,* 269, 270,
   272, *273,* 358–9
"Boomerang generation," 323
Borrowing dependency, 422
Bourgeoisie, 16
Brainwashing, 97–8
Braxton, Ashleigh, *286*
Brazil, 26, 93, 217, 237, *426*
Brown v. Board of Education, 257,
   427
Buffet, Warren, 187
Bureaucracy, **141–4, 374,** 376–7,
   422
Bureaucracy's other face, 141–2,
   374
Bush, George H. W., 418
Bush, George W., *92,* 137, 413, *416*
Business. *See* Corporations

Calculability, 144
Campbell, Ben Nighthorse, 381
Cancer, 360, 415
Cannibalism, 151–2
Capitalism, **385–6**
   classical theorists on, *158*
   class/power, 16, 82, 189–90
   in conflict theory, 189–90, 318–9
   crime and deviance, 159
   culture and, 41–2, 47, *49*
   exploitation and, 16, 198, 216–7
   inequality and, 16, 181, 197–8,
      212, 216–7, 230
   Marx on, *15,* 16, 189–90
   music culture and, *49*
   religion and, 47, 335
   sex and, 292
   socialization/families and,
      82, 319
   working women and, 282–3
Capitalism, Spirit of, 335
Capitalist class, 16, 197, 223
Care work, *324,* 330
Carmichael, Stokely, 258
Carrying capacity, 407
Caste system, 107, **181,** 248
Catholic Church, 403, 410
Caucasian (origin of term), 34. *See*
   also Whites
Causation, *60, 64, 65,* 66, 67

Cell phones, 32, *32, 122–3*
Censorship, 372
Census, 237, *239,* **402–3**
CEOs, income of, *180, 181*
Chaney, James, 258
Change. *See* Social change
Charisma, **340**
Charismatic authority, **373–4**
Cheney, Richard (Dick), 382, 413,
   416
Cherokee, 248
Chicago School, 17, 18, 20, 161
Child abuse, 74, 88, 156, 328
Child care, 206, 270, 280, 281, 316,
   326, 329–30
Child labor, 211, 212, 227, *392*
Children. *See also* Adolescence;
   Birthrates; Family
   effects of divorce on, 327
   in gay/lesbian families, 323, *326*
   gender/race relations, 48, *89*
   global perspective, *90, 221,* 222
   of immigrants, 31, *33, 75, 392*
   learning prejudice, 242
   parental employment and, *326*
   population statistics, *90*
   poverty and, 93, *93, 180,* 195,
      195, 202–4, 227, *229*
   socialization of, *75,* 76, 78–88, *89,*
      242, 318
   socialization of, gender, 47, 264,
      270–2
   in stepfamilies, 322
   violence in media, *67,* 79
   war, effects of, *228*
   of working mothers, *60*
China, 8, 265, 283–4, 303, 386
Chinese Americans, 250–1, 316.
   See also Asian Americans; Race
   and ethnicity
Cho, Seung-Hui, 156
Christianity, 331, 332–3, 335, 336
Chromosomes, 266, 291
Churches, **339**
Cities, 17, 189, 221, 255–6, *257,*
   410–1
Civic engagement, 47
Civil Rights Act (1964), 285, 395
Civil rights movement, 256–8,
   340–1, 373, 427, 429
Class, **182**. *See also* Status;
   Stratification, social; *specific*
   classes
   as achieved *vs.* ascribed status,
      111, 182
   birth control and, 302–3
   classical theorists on, *158, 201*
   crime and deviance, 159–60, 161,
      167, 169–70, *171*
   cultural capital and, 48
   defining, 182–3, 197, 198–9
   division of labor, 105, 389

Class (*continued*)
  education and, 48, 82, 187,
    348–51, *349*, *350*, 352, 355
  employment and, 146
  in everyday life, *184–5*
  gender and, 195
  gender, race interplay, 146, 194,
    240–1, 246, 297–9
  hair stylists and, *120–1*
  health care and, 358, 360–2,
    364–5
  Hurricane Katrina and, *182*, *257*
  income and, 146, 348–50, *349*,
    350, 394
  income *vs.* wealth, 190–4
  indicators of, 183, 186–7, 190
  inequality, growing, 190–4, 256
  inequality, reproducing, 48, *197*,
    200
  intelligence tests and, 353–4
  Internet and, *45*
  intersection perspective, 246, 256
  laddered model of, 186
  layers (categories) of, 187–9
  legal system and, 160
  life expectancy and, *404*
  mass media and, *45*, 46
  mental illness and, 164
  race and, 194–5, 240–1, 246, 256,
    297–9
  race, education, income, 348–50,
    349
  religion and, 336
  socialization and, 87–8
  social mobility, 186, 191–2,
    196–8, *200*, 349–51
  as social structure, 104
  standardized test scores and, 350
  status symbols and, 183
  stereotypes of, 46, *197*, 241
  taxation and, 190, 194
  theoretical perspectives, 159–60,
    189–90, 197–202, *200*
  toxic waste sites and, 413–4, *415*
  in U.S., 183–95, *188*, *191*, *193*,
    196–7
  values and, *30*
  voting, 379
  work *vs.* play, 90
Class conflict, 105, 189–90, 198,
  246, 420
Class consciousness, **196–7**, *200*
Clinton, Hillary, 379
Cliques, 80
Coalition formation, 129
Coercive organizations, **140–1**
Cognitive ability, **352–3**, 354, *354*
Cognitive development, 84
Cognitive elite, **354**
Cohabitation, 323
Cohorts, **407**
Collective behavior, **426–31**

Collective consciousness, **104**, 335
Colonialism, **218–9**
Coming out, 91, **296**
Commodity chain, **220**
Communication. *See also* Inter-
    net; Language; Mass media;
    Television
  across cultures, *34–5*
  by animals, 27
  gender styles, 263
  nonverbal, **113–4**, 263
  proxemic, **113–4**
  rumors, 427
  tactile (touch), **113**
  technology and, 101–2, *122–3*, *437*
Communism, **386**
Community
  sense of, *130–1*, 410, 422–3 (*See
    also* Gemeinschaft)
  transnational, 221
  virtual, *51*, 109–10, 120
Compadrazgo, 316
Computers, 51, 175, 387–8, 425–6.
    See also Internet
Comte, Auguste, **13–4**
Concept, defined, **59**
Confederate flag, 29
Conflict theory, **20**, *21*
  aging, 94, *95*
  class/inequality, 189–90,
    199–202, *200*
  crime, 166, *166*
  culture, *47*, 47–8
  deviance, *155*, 159–61
  education, 82, 348, *348*
  family, *318*, 318–9
  gender inequality, 281–2
  health care, *364*, 364–5
  mental illness, 164
  organizations, 147, *147*
  racism/prejudice, 246, *246*
  religion, 335–6
  sexuality, 295–6, *296*
  social change, 420, *421*
  social institutions, 104
  socialization, *83*, 84–5
  work, *390*, 391, 395
Conformity
  in bureaucracies, 137–8, 142
  cultural hegemony and, 42–3
  deviance and, 157–8, 159
  to gender roles, 272, 274
  groupthink, **137–8**, *138*
  individuality and, 77
  "nonconformists," 40, 134, *293*
  prison guards, 137
  ritualism, 142–3
  schools and, 82
  studies on, *135*, 135–7
Con games, 168–9
Conspicuous consumption, 37
Consumerism, 42, 161

Contact theory, **245–6**
Content analysis, 44–5, **67**, *68*,
    197
Contingent workers, **388**
Controlled experiments, **66–7**, *68*
Cooley, Charles Horton, **85–6**, **129**
Core countries, **215**, 216–7, 220,
    221, 422
Corporate crime, 160, 169–70
Corporate culture, 139
Corporations
  child labor, 211
  control of culture, *43*, 48
  environmental degradation, *412*,
    413, 415, 418
  family policies, 112, *330*
  as formal organizations, 139,
    144–5
  in global economy, *184*, 211,
    216–7, 219, 220, 386
  government and, 194, 376, 381,
    382, 418
  income of CEOs, *180*, *181*
  in Japan, 113
  mass media and, 42–3, *43*
  military and, 382
  multinational, 217, **219**, **386**
  race and, 216–7, *378*
  sports and, *201*
  taxation of, 169, 190, 194
Correlation, *60*, **64**
Correlation, spurious, **64**
Cosmopolites, 410
Countercultures, **40**
Courts. *See* Criminal justice
    system
Creationism, **80**
Credentialism, **348**
Credit cards, *193*
Crime, **165–6**. *See also* Deviance;
    Murder; Rape; Terrorism
  class and, 161, 167
  class, race, gender, 170–2, *171*,
    172
  corporate, 160, 169–70
  costs of, 168, 169, 173
  drugs and, 173, 174–5
  elite, 167, 168–9
  by gangs/Mafia, 154, *169*
  hate, **168**, 235, *301*
  index, **166–7**
  mass media and, 42, 166, *167*
  organized, **169**
  personal, **168**
  poverty and, 159
  property, **168**
  statistics on/reporting of, 163,
    166–9, *171*, 172, *172*, 284,
    307
  theoretical perspectives, *155*,
    155–64, 166, *166*
  victimless, **168**

victims of, 166, 168–9, 170, *171*, 172, *172*
violent, *167*, *171*, *172*
white-collar, 160, 168–9, 170
by women, 170–1
women as victims, 172, *172*
Criminal justice system
class and, 159–60
death penalty and, 4, *153*, 173
jury system, 127–8
minorities and, 4, 160, 170, *171*, *172*, 172–5
police, *78*, 163, 172–3, 244
poverty and, 159
prisons, 7, 137, *137*, 140–1, *171*, 173–5, *174*
racism and, 163, 244
sentencing, 170, *172*, 173
state power and, 372
Criminology, **166**
Cross-tabulation, *64–5*
Cubans, 250. *See also* Latino Americans
Cults, 97, 98, **340**, *340*
Cultural bias, 352
Cultural capital, **48**
Cultural diffusion, **50–1**, 421, **423–4**
Cultural diversity, 37–42, *38*. *See also* Diversity
Cultural gatekeepers, *266*
Cultural hegemony, **42**
Cultural pluralism, **245**, 254–5, 258
Cultural relativism, **41**
Cultural stereotypes, 240–1
Cultural studies, 48–9
Culture, **27–8**. *See also* Mass media
American, 25–6, 37–9, *49*, 49–50, 423–4
animals and, 27
beauty and, *43*, 116, *120–1*, *268–9*, *273*
biology and, 27–8
birth control/family planning, 409–10
change in, 48, 49–52, *51*, *52*, 423–4 (*See also* Social change)
characteristics of, 28–31
class and, *46*, 183
computers and, 425–6
content analysis, **67**
corporate control of, *43*, 48
in corporations, 139
countercultures, **40**
dominant, **38–9**, 50
education and, 350–2
education, multicultural, 132
elements of, *31*, 31–7
elite, 43–4

gender/sex and, 8, 33, *34*, *35*, 43, 265, *266*, 289–92
global, **41–2**
globalization of, 41–2, 223–4, 421
group solidarity, 47
health/health care, 360, 365
idealistic, 421
ideational, 421
kinship/marriage, 315–6
material, **27**, 47, 49, 419
McDonaldization of society, 144–5, 212
media's power and, 42–6, *43*, 48
multidimensional view of, *49*
nonmaterial, **27**, 47, 48–9, 419
of organizations, 139
personal space, 113–4
popular, *43*, **43–4**, 48, 271–2, *298*
power and conflict, 47–8
rites of passage, *28*, **95–6**, *96*
sensate, 421
sex trade, international, 299–300
sexualization/pornification of, 304–5
socialization and, 29, 75, *75*, 78
subcultures, **39**, 96
symbols and, 29–30
technology and, 25–6, *122–3*
theoretical perspectives, 46–9, *47*
Western, 223–4
in workplace/bureaucracy, 142
Culture lag, **50**, 419
Culture of poverty, **204–5**, 218, *255*, 256
Culture shock, **50**
Cyberspace interaction, *51*, **120–1**, *122–3*, 124, 425–6. *See also* Internet
Cyberterrorism, **175**
Cyclical theories, **420–1**, *421*

Dahmer, Jeffrey, 152
Darfur, 299
Darwin, Charles, 17, 27
Data, **59**, 61–3
Data analysis, *60–1*, **63**, *64–5*
Date rape. *See* Acquaintance rape
Dating, 115, *317*, 323
DDT, 415
Death and dying, 36, *37*, 93, 117, *257*. *See also* Murder; Suicide
Death penalty, 4, *153*, 173
Death rates, 222, 360, 362, *404*, *408*, 408–9. *See also* Infant mortality
Death rates, crude, **404**
Debriefing, **69**
Debt, 189, 190, 191, *193*, 219, 229, 422. *See also* Mortgages
Debunking, **7–9**, *8–9*, 9–10. *See also* Myths

Decision making, 127–8, 134–9, *138*, 142–3
Deductive reasoning, **56–7**
Defensive medicine, **367**
Definition of the situation, 20, 117, 355
Dehumanization, 145, *158*, 305
Deindividuation, **139**, *139*
Deindustrialization, **387**
Democracy, 14, **377**, 379
Demographic divide, 411
Demographic transition theory, *408*, 408–9, *409*
Demography, **402**
Dependency theory, *217*, **218–9**, **421**, *421*
Desegregation, 132, 257, 258, 427
Deskilling, 388
Deviance, **152**. *See also* Crime
attribution error and, 132–3
cannibalism, 151–2
context of, 152–3
Durkheim on, 153, *298*
elite, **160**
formal *vs.* informal, 152
forms of, 157, 158, *158*, 164–5
medicalization of, **154–5**
normalization of, 143
norms of, 159
punishment and, 153
social construction of, 143, 153–4
society creating, *153*
sociological perspectives on, 152–4
solidarity and, 153, 155
theoretical perspectives, *155*, 155–64, 166, *166*
Deviant careers, **162–3**
Deviant communities, **163**
Deviant identity, **162**
Differential association theory, **161**
"Digital divide," *122*
Disabilities, *34*, 164–5, *359*, 396
Disasters, *182*, 235, *257*, 427
Discrimination, **242**, **279**. *See also* Income gap; Prejudice; Racism; Segregation; Sexism
aging and, 92–3
class, gender, race, 145–6
disabilities and, *359*
education and, 348
gays and lesbians, 302, 384–5
health and, 362
housing, 192, 194, 243
immigration quotas, 251, 253, 410–1
legislative change, 285–6
in the military, 383, 384–5
myths about, *146*
in organizations, 145–6
police and, 163

Discrimination (*continued*)
in school sports, 81
sexual harassment, **394–5**
wealth accumulation and, 192
in workplace, 145–6, 275–6,
285–6, 348
Disengagement theory, **94**
Dissent, discouraging, 138
Diversity, **10–1.** *See also* Immi-
grants/immigration; Minority
groups; Race and ethnicity
cultural, 37–42, *38*
global, 12–3, 41–2
in government, *377, 378*
language/labeling, *34–5*
nonverbal interactions, 114
in organizations, 145–6
in power elite, *377, 378*
in religion, *333, 334,* 336–9, *337*
significance of, 10–3
social change and, 431
socialization and, 87–8, *89*
sources of, 11
Diversity training, 146
Division of labor. *See* Labor,
division of
Divorce, *325, 326,* 326–7
Domestic work, *216*
Dominant culture, **38–9,** 50
Dominant groups, **239**
Double deprivation, 227
Downsizing, 140
Dramaturgy, *118,* 118–9
Drugs/drug abuse, 165, *165,* 173,
174–5, 358–9, 363
Drug trade, *78,* 175
Dual labor market theory, **278, 393**
Du Bois, W.E.B., *15,* **18,** *18, 49,*
140, *158, 201, 298*
Durkheim, Emile, **14,** *14, 15,* 18,
49, 102, 104, 105, 153, 155–6,
158, 201, 298, 335, *410, 419*
Dyads, **129**
Dysfunctions, 19, 147

Eating disorders, 116, 158, *268,*
272, *273,* 358
Economic development, *217,*
217–21, 222, 230–1
Economic problems
home foreclosures, 192–4
sociological dimensions of, 192–4,
372, 393
unemployment and, 109, 352, 393
youth generation and, 90, 93, 372
Economic restructuring, **192,** 205,
386–7
Economic systems, *107,* 108, 215,
385–6. *See also* Capitalism;
Socialism
Economy, **385.** *See also* Global
economy

Education. *See also* Schools
bilingual, *33*
class and, 48, 82, 187, 348–51,
349, 350, 355
gender and, 8, 82, 223, *277,* 349,
349, 352, 355–6
global perspective on, 223, 350–2
home schooling, 347
income and, 3, *277,* 348–9, *349,*
350
inequality and, *221,* 223, 227,
352–7, *355,* 424
occupation and, 348–9, *349, 350*
in postindustrial societies, 109
poverty and, 8, 227
race/ethnicity, 244, 251, 260,
345–6, 346–7, *347, 349,* 355,
356, 424
social control and, 52, 348
standardized testing, *350,* 350–2,
351, 352–3, 356
teacher expectancy effect, 82,
348, 355, *355*
theoretical perspectives, 346–8,
348
tracking, **354–5**
U.S. history of, 346–7
voting, *382*
Educational attainment, **187**
Efficiency, **141,** 144
Ego, **83**
Egoistic suicide, **156**
Eichmann, Adolf, 136–7
Eisenstadt v. Baird, 302
Elections, 375, 377–9, *379,*
379–81. *See also* Voting
Elite crime, 167, 168–9
Elite deviance, **160**
Elites. *See* Power elites; Upper class
Embarrassment, 119
Emigration, **403**
Emotional labor, **389**
Empirical approach, **7,** 56
Energy use, *222, 417*
England, 350
Enlightenment, **13–4**
Environmental issues, 222–3, 409,
411–8, *412, 413, 414, 416–7*
Environmental justice movement,
428
Environmental racism, **414–5,** *415*
Epidemiology, **359**
Equal Pay Act (1963), 277
Equal Rights Amendment, **285**
Estate system, **181**
Ethics, 68–9, 73
Ethnic cleansing, 48
Ethnic conflict, 105
Ethnic enclaves, 252
Ethnic groups, **236.** *See also*
Immigrants/immigration;
Race and ethnicity

Ethnic identity, 236
Ethnic villagers, 410
Ethnocentrism, **40–1, 241–2,** 373,
408
Ethnomethodology, **36, 117–8,** *118,*
282
Eufunctions, 147
Eugenics, **303**
European Union (EU), 373
Everyday life, society in, *122–3,*
184–5, 268–9, 416–7
Evolutionary theories, 17, 80,
419–20, *421*
Exchange theory, *118,* 119–20
Experimental groups, 66–7
Experiments. *See* Research
Eye contact, 114

False consciousness, **196,** 336
Family, **314–5.** *See also* Children;
Marriage
balancing work and, *61,* 112,
280–1, 321, 325, 329–30
child care, 280, 281, 316, 329–30
as economic unit, 108, 316
elder care, 330–1
extended, **315–6**
functions of, 19, 78–9, 88, *318*
"ideal," 314, *320*
men's care work, *324*
nuclear, **316**
poverty and, 203–4, 206
religion and, 314
size of, 319
social policy and, 329–31
stereotypes, *320*
theoretical perspectives, 317–9,
318
transnational, *216,* **329**
violence in, 74, 88, 156, *224,*
327–8
women's unpaid labor, 283
Family and Medical Leave Act
(FMLA), **329**
Family diversity
changing society and, 314,
319–20, 328–31
gay and lesbian households, *322,*
322–3
interracial, 7, 315, *317*
kinship systems, **315**
married-couple families, *320,*
321, 321–2
race, 315, 316, 321, *321*
single parents, 8–9, 202–4, *203,*
305–6, 320–1
singles/cohabiters, 323
stepfamilies, 322
Family planning, 409–10. *See also*
Birth control
Family–wage economy, 108, 316
Fashion, *30, 268–9*

Fast food, 49–50, 144–5
Fathers, 270, 321, *321*, *324*, *326*
FEMA, *257*, 413
Female-headed households, *186*, 305–6, 320–1. *See also* Single-parent households
Feminism, **282–3**
 Black, 258
 liberal, **282**, *283*
 multiracial, **283**, *283*
 radical, **282–3**, *283*
 socialist, **282**, *283*
Feminist theory, 21, **282–3**
 environmental issues, 415
 family, *318*, 319
 gender roles, 281, *281*
 inequality and gender identity, *283*
 religion, *334*
 sexual violence, 308
 social change, *283*, 284
 state/power, *375*, 377
 women's status, 8, 172, 282–3, *283*
Feminization of poverty, **202–3**, 321
Feral children, **74**
Fertility rates, *91*, 221–2, 227
Fictive kin, 316
Field research. *See* Participant observation
Filipinos, *216*, 251, 329
First-world countries, **215**
Folkways, *31*, **34–5**, 154
Foot binding, 8
Foraging societies, 106–7, *107*
Frames, **429–30**
Fraternities, 38, 97, 308
Freud, Sigmund, 82–3, 86
Friendships, 110, 115, *122–3*, 272. *See also* Peers
Functionalism, **18–20**, 21
 aging, 94, *95*
 class, 199, *200*, 200–1
 crime, 166, *166*
 culture, 47, *47*
 deviance, *155*, 155–9
 education, 347–8, *348*
 family, 19, 318, *318*, 319
 health care, 363–4, *364*
 inequality, 20, 281
 mental illness, 164
 organizations, 147, *147*
 racism/prejudice, 245–6, *246*
 religion, *334*, 335
 sexuality, 294–6, *296*
 social change, 419–20, *421*
 social institutions, 103–4
 socialization, *83*, 84
 work, *390*, 391, 395
Functions
 defined, 419
 latent, **348**
 manifest, 19
Fundamentalism, 271

Games, *67*, 88
Game stage, **86**
Game theory, *118*, **120**
Gandhi, Mohandas, 256
Gangs, 66, 129, 154, *169*
Garvey, Marcus, 258
Gates, Bill, 187, 191
Gays and lesbians
 activism, 48, 297, 429
 attitudes toward, 270, 274, *293*, 293–4, 323
 biology/genetics, 290–1
 coming out, 91, **296**
 crimes against, *301*
 domestic violence, 327
 families/households, 301, 322–3
 HIV/AIDS, 363
 homophobia, 81, **274**, **300**, *301*, 332, 384–5
 interracial couples, 7
 labeling, 33, 155
 marriage, 292, *302*, 322, *322*
 media portrayals of, 46, *60*, 293
 in the military, 384–5
 queer theory, **296–7**
 rights of, 46, 293, 293–4, 297, 299, 302, 396
 sexual identity, 290–1
 stereotypes, 46, 301, 302
 support systems, 302
 terminology, 33
 understanding, 300–2
 in workplace, 280, 395–6
Gemeinschaft, **105**, 106, 422–3
Gender, **265**. *See also* Men; Women
 AIDS and, 363, *364*
 attitude gap, *264*
 attribution error and, 133
 biological determinism, **265–6**
 class and, 195, 241
 class, race interplay, 194, 195, 240–1, 298–9
 communication styles, 263
 crime and, 170–2, *172*
 "doing gender," 110–1, 282
 education and, 8, *277*, 348, 349, *349*, 355–6
 eye contact, 114
 health care, 358, 362
 income and (*See* Income gap)
 inequality (*See* Gender stratification)
 Internet interactions, 121
 intersection perspective, 246, 256
 language and assumptions, *34*, *35*
 in mass media, 43, 45–6, 270, 271–2, *293*, *320*
 mental illness and, 164
 morality and, 84

political power and, *375*, 377, *377*, *383*
 race and, 194, 195, 241, 246, 272–4
 in research, *65*
 self-concepts and, 77
 sexual/sexual politics, 297–9
 social change and, 284–6
 social construction of, 110–1, *111*, 117, 264–7, *266*, *273*, 296–7
 socialization and (*See* Gender socialization)
 standardized tests, 350, *351*, 352, 356
 stereotypes, 46, 92, 240–1, 264, *266*, 270, 271–2, 298–9, *320*, 357
 theoretical perspectives, 281–3, *283*
 touching and, 113
 voting and, 380, *382*
Gender apartheid, **275**
Gender betrayal, *301*
Gender bias, 352–3, 355–6
Gender division of labor, 105, 107, 280–1, 283–4, 316, 319, 325–6, 389
Gendered institutions, **274**, 319
Gender gap (in income). *See* Income gap
Gender gap (in politics), **379**
Gender identity, 265, **267**, 274, 280, *283*
Gender roles, 77, 112, 272–4, 284–5, 316, *324*, 324–6. *See also* Gender socialization
Gender segregation, **278–9**
Gender socialization, **267**
 conformity and, 77, 272
 in families, 83, 88, 264, 270
 homophobia and, **274**, 300–1
 race, gender, and identity, 272–4, *273*
 schools and, 82, 356
 sexual scripts and, 291
 sources of, *266*, 270–2, 270–3, *281*
 sports and, 81, *81*
Gender stratification, **275**
 global, 223, *224*, 227, 283–4
 in income (*See* Income gap)
 in industrialized societies, 108
 in labor force (*See* Women, labor and)
 legislative change, 285–6
 as multidimensional, 275
 in organizations, 145–6
 power and, 276, 286, *375*, 377, *377*, *378*
 sexism and patriarchy, 227, 275–6, 282–3, 338–9

Gender stratification (*continued*)
slavery and, 248
theoretical perspectives, 281–3, *283*
Generalization, *60–1*, **63**, *65*, 66
Generalized other, **86**
Genetic engineering, 31
Genetics
alcoholism and, 154
behavior and, 27–8, 76, 265–7, 290–1
human genome project, 73
intelligence and, 353–4, *354*
sexual identity and, 266–7, 290–1
Genie (feral child), 74
Genital herpes, 362
Genocide, 47–8, 248. *See also* Holocaust
Germany, 351, 424–5
Gerrymandering, 242
Gesellschaft, **105**, 106, 422–3
Gestures, 114
Glass ceiling, 145–6, **394**, *395*
Global assembly line, *184*, 386
Global culture, **41–2**
Global economy, **385–9**. *See also* Global stratification
deindustrialization, **387**
inequality and, *184*, 212
manufacturing, *184*, 192, 211, 212, 220, 386–7
postindustrialism and, 109
poverty and, 215
power and, 215
sex trade, 299–300
the state and, 373
technology, 387–8
Globalization, **421**
of culture, 41–2, 421
family life and, 328–9
social change and, 230–1, 421–2, 431
theoretical perspectives, *421*, 421–2
Global perspectives
overview, 12–3
education, class, mobility, 350–2
gender, 283–4
health/health care, *365*
human development index, *62*
incarceration rates, *174*
interdependence of nations, 373
maternity leave, *330*
population, *90*, 221–2, *403*, 406, 407–10, *408*
religion, *337*
sexuality, 299–300
societies, types of, 106–9
technology, *437*
terrorism, 175
voter participation, *379*

work and family, 328–9
Global stratification, **212**
consequences of, 221–5
education and, 227
health and, *221*, 222–3, 227–8, *228*
hunger and, 227–8
life expectancy, *405*
quality of life and, *221*, 222–3, 225–30
race and, 216–7
rich and poor countries, 212–7, *213*, *214*
social change and, 230–1
theoretical perspectives, *217*, 217–21
war and terrorism, 223–5
women in government, 284
world poverty, *225*, 225–30, *226*, *227*, *228*
Global warming, 413, 415, *416–7*
Goldman, Emma, 302
Goodman, Andrew, 258
Gore, Albert (Al), 409
Government, **377**. *See also* Military; Politics; Power; Social policy; Taxation
bureaucracy, **374**, 376–7
contraception and, 409–10
corporations and, 194, 376, 381, 382, 418
decision making, 137–8
democracy, 14, **377**, 379
dissent, managing, 372–3
in first-, second-, third-world, 215
global economy and, 373
health care, 364, 366–7
minorities in, *378*, 381, *383*
poverty and, 229
power/authority, 373–4
power/politics, 372–3, 377–85
power, theories of, 374–7, *375*
rights and security, 372
state, defined, **372–8**
women in, 284, 381
Gratz v. Bollinger, 260
Greenhouse effect, 222, **413**, *413*, *416–7*
Green Party, 430
Grizwold v. Connecticut, 302
Gross national income (GNI), per capita, **212–5**, *213*, *214*
Group dynamics, 129, 138–9
Groups, **109–10**, **128**. *See also* Collective behavior; Organizations
class, gender, race interplay, 145–6
conformity/obedience, 134–9, *136*, *138*
derogatory terms for, 33–4, 235
dominant, 33, **239**
hierarchies, *66*
in- *vs.* out-, 132–3, 239, 241

jury decision making, 127–8
labeling of, 33–4, *34–5*
minority (*See* Minority groups)
primary, **129–31**, 422
reference, **131–2**
secondary, **129–31**
small group movement, *130–1*
theoretical perspectives, 146–7, *147*
types of, 128–34
Group size effects, 128, **129**, 135, 139
Groupthink, **137–8**, *138*
Grutter v. Bollinger, 260
Gynecology, 365–6

Habermas, Jurgen, 423
Hamilton, Charles V., 258
Harlem Renaissance, 249
Harris, Eric, 156
Hate crimes, **168**, 235, *301*
Hate groups, 40, *163*, 428
Hawthorne effect, 60–1
Hawthorne Studies, 142
Hazing, *97*
Health. *See also* Life expectancy
beauty/physical ideals, 77, 272, *273*, 358–9
conformity and, 77
eating disorders, 116, 158, *268*, 272, *273*, 358
global stratification and, 216, *221*, 222–3, 227–8, *228*, 365
infant mortality rate, *221*, 227, *228*, **404**, *404*, *405*, 408
life expectancy, **404**, *405*
mental, 162, 164, *164*, 204
pollution and, 412, 415
"sick" role, 164
social construction of illness, 346, 358, 365
waterborne illnesses, 222–3
Health care system
access to, 360, 366–7, 404, 408
class and, 360–2
costs of, *365*, 366–8
culture and, 360
gender and, 164, 358, 362
inequality and, 358, *359*
infantilization, 361, 365
insurance, *361*, 362, 366–7
modern medicine, emergence of, 357–8
race and, 358, 359–60
theoretical perspectives, 363–6, *364*
universal care debate, 367–8
Helmsey, Leona, *180*
Heredity. *See* Genetics; Nature *vs.* nurture
Hermaphroditism, **266**
Heterosexism, **301**

Hidden curriculum, 82, *82*
Hip-hop, 39, 40, *49*
Hitler, Adolf, 136, 237
HMOs (health maintenance organizations), **367**, 368
Holocaust, 136–7, 237
Homelessness, 55–6, 203–4, 227
Home schooling, 347
Homophobia, 81, 270, **274**, **300**, *301*, 323, 332, 384–5
Homosexuals. *See* Gays and lesbians
"Hooking up," 298, 323
Hopi Indians, 32
Horticultural societies, 107, *107*
Housework, 112, 280–1, *324*, 325–6. *See also* Domestic work
Housing. *See* Mortgages; Segregation
Hughes, Langston, 249
Human capital theory, **277–8**
Human development index, *62*
Human ecology, 411
Human ecosystem, 411
Human genome project, 73
Human poverty index, **226**, *226*
Hunger, 216, *227*, 227–8
Hunting and gathering, 106, *107*
Hurricane Katrina, *35*, 50, *182*, 202, 235, *257*, 413–4
Hussein, Saddam, 48
Hypersegregation, **255**, *255*, *257*, 414
Hypothesis, **59**

Id, 82–3
Identity, **75**. *See also* Self; Socialization
development of, 77–8, 83
deviant, 155, **162**, 163
ethnic, 236
gender, 266–7, **267**, 272–4, 280, *283*
impression management, 118–9, **118–9**, *120–1*, 121, 124
Internet and, 121, 124
labeling and, 155, 162
language and, *34–5*
master status and, 111, 165
in modernized society, 423
music culture and, *49*
play and games, 88
primary groups and, 129
religion and, 331, 332, 335
rites of passage and, 95–6
sexual, 110–1, 266, **290**, *296*, 296–7
social learning theory on, 84, 111
social movements and, 427, 430
"spoiled," 165
sports and, 81, *201*
stigmas and, 164–5

Ideology, **198**, 275–6
Illiteracy, *221*, 223, 275
Imitation stage, **86**
Immigrants/immigration, **403**. *See also* Migration; *specific groups*
assimilation, *33*, 253, 254–5
cities and, 410–1
culture, 31, 37, 50
family and, 31, *75*, *392*
globalization and, *216*, 220–1
group identity and, 236
guest worker programs, 389
harassment of, 252
in labor force, 220–1, 249, 250, 255, 388–9, *392*, 405
prejudice against, 250–1, 252–3, 303
quotas, 251, 253, 410–1
socialization and, 31, *75*
status inconsistency, 111
stereotypes of, 249, 250–1, 252–3, 405
undocumented, 388, 389
U.S. diversity, 37, *38*, 50–1, *247*, 248–54, 410–1
Imperialism, 218–9, 231
Impression management, **118–9**, *120–1*, 121, 124
Imprinting, **114**, *115*
Incest, 328
Income, **186**, **190**. *See also* Poverty; Wealth
of CEOs, *180*, *181*
class and, 146, 186, 348–50, *349*, *350*, 394
of crime victims, *171*
distribution of, *180*, *186*, *187*, *188*, 190–4, *191*, *207*
education and, *277*, 348–9, *349*, *350*
in factories, 212, 386
global poverty and, 225, 226
living wage campaign, 205
median, *64*, **186**, *186*, *188*
military service and, 383
minimum wage, *180*, 205
occupation and, 348–9, *349*, *350*, 394
per capita GNI, **212–5**, *213*, *214*
SAT scores and, *350*
vs. wealth, and class, 190–2
wages declining, 206, *281*
workfare and, 207
Income gap
education, 3, *277*, 349, *349*
growth of, 108, 190–4, *191*
industrial societies and, 108, 316
between nations, 212–5, *213*, *214*
occupation and, 195, 278, 349, *349*, 394
race/ethnicity, *186*, 191–2, 242, *242*, *243*, *277*, *281*, 349, 394, *394*

theoretical perspectives, 277–9, *390*
women, 3, 108, 264, *277*, 278, 349, *349*, *394*
women's median income, *186*, 195, 206, 276–7, *277*, *281*, 349
women's wages, global, 276
Index crimes, **166–7**
India, 28, 218, *265*
Indicators, **59**
Individualism, 14, 212, 326–7
Individuality, 77. *See also* Conformity; Identity; Self
Individuals, *21*, *137*, 423
Inductive reasoning, **57**
Industrialization, 218, 316, 422
Industrial Revolution, 107–8, 385
Industrial societies, *107*, 107–8, 217–22, 316
Inequality. *See* Gender stratification; Global stratification; Income gap; Stratification, social
Infant mortality rate, *221*, 227, *228*, **404**, *404*, *405*, 408
Informed consent, **69**
Inside Toyland, 184–5
Institutional racism, **244**, *244*
Institutions. *See* Social institutions
Intelligence, 353–4, *354*
Intelligence testing, 352–4
Interest groups, **375**
Interlocking directorates, **376**
Internalization, **76**
Internet
blogging culture, *51*
class and, *45*
cybersex, 308
cyberterrorism, **175**
dating and, 115
identity/impression management, 121, 124
messaging language, *32*
online profiles, *124*
pornography on, 308
social interaction on, 109–10, 120–1, *122–3*, 124, 425–6
transforming society, *51*, *122–3*, 425–6
Interpersonal attraction, 114–6
Intersection perspective, 246, 256
Intersexed people, **266**
Interviews, 63–4
IQ (intelligence quotient), 352–4. *See also* Achievement tests; Standardized ability tests
Iraq, *15*, 137, 138, 225, *228*, 252, 382
Irish Americans, 238, 240, 252, 254
Islam, 40, *41*, 299, 337, 338
Isolate, defined, 129

Isolation, *45*, 74, 114, 156
Israel, 404
Issues (*vs.* troubles), **6**, 13, 393

Jacobs, Carla, *286*
Japan
  communication in, 113
  culture of, *96*, 113, 421
  economic development, 218
  education in, 351
  infant socialization in, 78–9
  organization model, 147
  women in, 275, 283
Japanese Americans, 251. *See also*
  Asian Americans; Race and
  ethnicity
Jazz, 37
Jews. *See also* Race and ethnicity
  anti-Semitism, 14, 136–7, **253**
  Bar Mitzvah, *28*
  as ethnic group *vs.* race, 237
  Holocaust, 136–7, 237
  as immigrants, 252, 253
  in Israel, 404
  Orthodox, 271
  stereotypes, 45, 240
Jihad, 40
Job displacement, **387**
Johnson, Lyndon B., 138
Juries, 127–8

Kennedy, John F., 138, 381
Kim, Bong Hwan, *75*
King, Dr. Martin Luther, Jr., 256,
  257, 373
Kinship systems, **315**, 316, 322
Klebold, Dylan, 156
Koreans, *75*, 251. *See also* Asian
  Americans
Koufax, Sandy, 169
Kurdish people, 48
Kwakiutl society, 37
Kyoto Protocol, *417*

Labeling. *See also* Tracking
  of deviance, 152, 160, 161–4, *163*
  of gays and lesbians, 155
  mental illness, 164, *164*
  racial-ethnic terms, *34–5*, 235–6
Labeling theory, **161–2**
Labor. *See also* Occupation
  definition of work, **389**
  emotional, **389**
  exploitation of, 16, 198, 217, 282
  job training, 347
  poverty and, *180*
  in preindustrial through postin-
    dustrial societies, 106–9
  sexism and patriarchy, 275–6
  technology and, 106–9, 387–8
  theoretical perspectives, 277–9,
    389–91, *390*, 395

Labor, division of, **105**, **389**
  age and, 105
  in bureaucracies, 141
  class, gender, race interplay, 105,
    389–90
  gender, 107, 224, 280–1, 283–4,
    316, 319, *324*, 324–6
  industrialization, 108, 316
  international, **217**, 220, 386
  in preindustrial societies, 107,
    316
Labor force
  age discrimination, 92, 94
  alienation, 143–4, 388, **391**
  bureaucracies, 142–4
  characteristics of, 391–4
  child labor, 211, 227, *392*
  commuter marriages, 321–2
  contingent workers, **388**
  dehumanization, 145
  deindustrialization, **387**
  deskilling, 388
  disabled workers, 396
  discrimination, 162, 275–6, 279,
    285–6, 348
  diversity, 145–6, 387
  domestic workers, *216*, 329, 330
  economy and, 93
  family duties, balancing, 280–1,
    316, 325–6, *326*, 328–31
  family, transnational, 329
  gays and lesbians, 280, 395–6
  in global economy, 211, 212, *216*,
    328–9, 386–7
  immigrants, 220–1, 249, 255,
    388–9, *392*, 405
  in industrial societies, 105
  job displacement, **387**
  job loss, 192, 205, 212, 387
  jobs, finding, 133–4
  at McDonald's, 144–5
  migrant workers, 249
  in military service, 381
  promotion/firing, 145–6
  race, class, gender networks,
    133–4
  race/ethnicity, 258, 260, 387, 390,
    392, 392–3, *393*, 394
  race/ethnicity and gender, 273–4,
    275, *286*, 362, *390*, 391, 393,
    394
  ritualism, 142–3
  sexual harassment, 279, **394–5**
  social speedup and, *321*
  unemployment, 6, 13, 189, 205,
    206, 250, 391–3, *393*
  women (*See* Women, labor and)
  working poor, 189, 205
Labor force participation rate,
  277
Labor market, dual, **278**, **393**
Labor unions, *190*, 279

Language
  bilingual education, *33*
  defined, *31*
  as element of culture, **31**–4
  messaging, *32*
  race/ethnicity and, *29*, *34*, 235–6,
    424
  social meaning of, *34–5*
  U.S. diversity, *38*
La Raza Unida, 52, 140, 258
Latent functions, **348**
Latino Americans. *See also* Race
  and ethnicity; *specific cultural
  groups*
  class and, 194–5
  cultures, *28*, 28–9, 39, 52, 424
  diversity among, 194, 249–50
  families, 316
  gender roles, 272, 326
  gestures, 114
  health, 360
  history, 249–50
  in media, 45, 242
  migration of, 405
  military service, *15*, 383
  population statistics, *12*, *13*
  religion, 337
  rites of passage, *28*, *96*
  social movements, 429
  stereotypes, 45, 240, 242, 272
  terminology, *34*, 235
  wealth/net worth, 192
Lautenberg, Frank, 169
Lawrence v. Texas, 293
Laws, **35**
  anti-pollution, 418
  Civil Rights Act (1964), 285–6, 395
  class and, 160
  defining "Black," 237
  disabilities and, *359*, 396
  discrimination, 277, *281*, 285–6,
    395
  education, 346
  equal pay, 277
  Family and Medical Leave Act
    (FMLA), **329**
  functions of, 372
  gay marriage/partnerships, 299,
    322
  immigration, 251, 411–2
  race and marriage, 315, 316
  sexual harassment, **394–5**
  sexual relations, 293
  social control and, 372
Legal Defense Fund (LDF), 260
"Legitimation techniques," *30*
Lieberman, Joseph, 381
Life chances, 146, *182*, 192–3, 196,
  200
Life course, **88–96**
Life expectancy, *221*, 222, 227,
  404, *404*, *405*, 408

Lindh, John Walker, 97
Linton, Ralph, **423**
Literacy, 275
Literature reviews, 57
Liuzzo, Viola, 258
Looking glass self, *85*, **85–6**
Lorenz, Konrad, *115*
Love, 114–6, *116*
Lower class, 189, 196, 241. *See also* Class; Underclass
Lower middle class, 189
Lumpenproletariat, 16, 198

Machismo, 274
Macroanalysis, **102**, 139, 159
Macrochanges, **418**
Macrosociology, 18, 20, 328
Madoff, Bernard, *168*, 168–9
Madrassas, 338
Mafia, *169*
Malcolm X, 258
Malkovitch, John, 169
Malnutrition, 216, *221*, *227*, 227–8
Malpractice, 367
Malthusian theory, **407–8**, *409*
Managed care, **367–8**
Manifest functions, 19
Marijuana, *165*
Marriage
   age at first, *91*, 323
   battered women, 98, *203*, 205
   commuter marriages, 321–2
   culture and, *96*, 315–6, 327
   divorce, *325*, 326–7
   gender roles, *324*, 324–6
   housework, 112, 280–1, *324*, 325–6
   intermarriage, 315, *317*
   power and, 324–6, 328
   same-sex, 292, 293, *302*, 322, *322*
   statistics on, 320, *320*, *325*
   theoretical perspectives, 318, *319*
   welfare and, *203*, 207, *207*
Marshall, Thurgood, 260
Martineau, Harriet, **14**, *14*
Marx, Karl, 15, **16**, *16*, 18, 20, *49*, *158*, 159, 189, 196, **197–8**, *201*, 218, *298*, 334, 335–6, 375, 386, **420**, 423
Mass media, **42**. *See also* Internet; Television
   accuracy of reporting, 4, 48, *60–1*
   as big business, 42–3, *43*
   blackout experiment, *44–5*, *45*
   class, gender, race, 44–6, *49*
   content analysis, 44–5, **67**, *68*
   control of, 16, *43*, 48, *49*, *266*
   control of ideas, 16, 42–6, *43*, 48, *197*, 198, 372
   crime/criminals in, 42, 166, *167*
   culture and, 42–6, *43*, 48
   families in, *320*

gender in, 45–6, 240–1, *266*, 270, 271–2, 308–9
   pornography, 304–5
   postmodernist theory and, 21
   public opinion, 42–3, *197*
   race in, 45, *197*, 242, 251
   sex in, 45, 46, *293*, 308–9
   social commentary in, 4
   socialization and, 79
   stereotypes, 45–6, 67, 92, *197*, 240–1, 242, *266*, 270
   theoretical perspectives, 46–9
   values, 46–7, 79, 358
   violence in, *67*, 79, 167
Material culture, **27**, 47, 419
Maternity leave, 283, *330*
Matriarchal religions, **333–4**
Matriarchy, **276**
Matrilineal kinship, **315**
Matrilocal, defined, **315**
Matrix of domination, 256
McCain, John, 379, 381
McDonaldization of society, *144*, 144–5, 212
McDonald's, 50, *144*, 144–5, 224
Mead, George Herbert, **85**, 86
Mean (statistical), *64*
Means of production, 197
Mechanical solidarity, **104**
Media. *See* Mass media
Median (statistical), *64*
Medicaid, 203, **366–7**
Medicalization of deviance, **154–5**
Medicare, **366**
Men. *See also* Gender
   aggression and, 77, 265–6, 272
   conformity/gender roles, 77, 272, 300–1
   fathers, 270, *321*, *324*, *326*
   in female-dominated jobs, 280
   housework, 281, *324*, 325–6
   male-headed households, 321
   physical ideals, *268*, 358–9
   race and gender identity, 272–4
   socialization of, 84, 87, 270, *298*, 300–1
   in sports, 81, 285, 358–9
   stereotypes, 240–1, 270, 272, 274
   support for working women, *284*, 284–5
   violence and, 272
Mental illness, 162, 164, *164*, 204
Mere exposure effect, 115
Meritocracy, **196**
Merton, Robert, **19**, **156–7**, *158*
Mexican Americans, 249. *See also* Latino Americans
Mexican-American War, 249
Mexico, 406, *406*
Microanalysis, **102**, 161
Microchanges, **418**
Microsociology, 18, 20, 109

Middle class. *See also* Gays and lesbians
   Blacks in, 194–5, 256
   characterized, 189
   childhood socialization, 87–8
   conflict theory, 189–90
   deviance and, 161
   financial fragility of, 192–4, *193*
   income, 186, 189
   Latinos in, 194–5
   Marx on, 198
   statistics on, 196
   stereotypes, 241
   tattoos and, *30*
   values, *30*
Middle East. *See specific countries*
Middle Easterners, 46, 114, 223–4, 235, 239, 251
Middle East oil-rich countries, 215, 223–4
Migration, *216*, 248, 404–5
Milgram Obedience Studies, 135–7, *136*, *137*
Military, 96–7, 129, 137, 372, 381–5. *See also* War
Military academies, 383
Militia movement, 428
Mills, C. Wright, 6–7, 12, 88, 93, 328, 375–6
Minangkabau (Indonesia), 276
Minority groups, **239**. *See also* Immigrants/immigration; Race and ethnicity; Women
   affirmative action, *258*, **258–9**, 260, **285–6**, *286*, *390*
   characteristics of, 239, *240*
   cultural awareness programs, 132
   disabled people, *34*, 164–5, *359*, 396
   in dominant group context, 80
   in government, *377*, *378*, 381
   income of (*See* Income gap)
   in media, 45–6, 242
   in power elite, *378*
   scapegoat theory, **245**
   "tax" on, in U.S., *259*
Mobility. *See* Social mobility
Mobilization, 426, 429
Mode (statistical), *64*
Modernization, 418, **422–3**
Modernization theory, *217*, *217*, *421*, **421–2**
Monogamy, **315**
Monotheism, **333**
Moral development, 84
Morality, 135–7, *136*, 264, 297–8, 331. *See also* Ethics; Values
Mores, *31*, **35**
Mormons, 315
Mortgages, *192*, 192–4, 243
Motherhood, *8–9*, 307, 316, 329

Mothering, 330
Movies, *44*
Muhammed, Elijah, 258
Multidimensional evolutionary theory, 420
Murder
  cannibalism and, 152
  of civil rights workers, 258
  classical theorists on school shootings, *158*
  of minorities, 235, 248, 258, 299, *301*
  in schools, 156, *158*
  statistics, *9*, 163, *167*
Music, 37, 39, *49*, 50–1, 249, 424
Muslims, 252, 258, 299, 337
Myrdal, Gunnar, **421**
Myths. *See also* Stereotypes
  about police work, *78*
  affirmative action, *286*, *390*
  bilingual education, *33*
  Black teenage motherhood, *8–9*
  Black women workers, 275, *286*
  class, *404*
  crime, *172*
  death rate, *404*
  decisions by experts, *138*
  deviance, *153*
  discrimination, *146*
  economic development, *421*
  education, *347*, *350*
  fathers, effect of absence of, *321*
  gays/lesbians, 301
  gender as ascribed status, *111*
  group labels, *29*
  group organization, *66*
  health care and equality, *359*
  hunger, *227*
  individuals and group pressure, *137*
  intelligence, *354*
  language, *29*
  love, *116*
  mental illness, *164*
  men who support equality of women, *284*
  minority groups, *240*, *378*, *415*
  model minority, 195
  occupation/income, *350*
  pollution, *415*
  poverty, *8–9*, 207, *207*
  prejudice, *244*
  racial differences, *237*
  rags-to-riches stories, 191
  rape, *172*
  religious conversion, *340*
  schools, *82*
  sex education, *306*
  sexuality, *291*
  social change, *408*
  social mobility, 195–7
  social movements, *429*

  sports, *29*, *81*
  success, *347*, 393
  teen pregnancy, *8–9*, *307*
  unemployment, *390*
  violence, *9*
  welfare, *205*, *207*, 207–8

NAACP, 18, 140, 257, 260, 428
Nader, Ralph, 430
NAFTA, 376
NASA, 142–3, *143*
Nash, John, 120
Nationalism, **373**
National Urban League, 140
Native Americans. *See also* Race and ethnicity
  AIM, 258, 430–1
  culture, *28*, *29*, *32*, 37, *48*, 104, 248
  defining, 238
  government policy, 238, 248
  history, 247–8
  in labor force, 392
  in media, 242
  National Museum, *245*
  population statistics, 247, 248
  poverty, 248
  religion, 331
  repatriation movement, 48
  sexual abuse of, 299
  sports mascots, *29*, *29*
  stereotypes, 242
  suicide among, *156*
  terminology, 33, *34–5*
  toxic waste and, 414–5
Nature *vs.* nurture, 27–8, 74, 76, 265–7, 353–4
Navajo people, *156*, 265, 415
Needs, expressive, **131**, 141
Needs, instrumental, **131**, 141
Neocolonialism, **219**
Neoevolutionary theory, 420
Networks, social, *124*, **133–4**, 142. *See also* Online profiles
Net worth, **190**, 191, *193*
New Ageism, 427
New cultural studies, *47*
Newly industrializing countries (NICs), **230**
New Orleans. *See* Hurricane Katrina
News, 48, *60–1*. *See also* Mass media
New social movement theory, **430**, *430*
Nike, 211, 220
Nixon, Richard M., 138
Nonmaterial culture, **27**, 47, 419
Norms, *31*, **34–6**
  in bureaucracies, 142
  culture and, 35–7, 39, 47, *47*
  deviance and, *30*, 152, *153*, 154, 155, 159

  ethnomethodology and, 117–8
  gender roles, 112, 270
  military, 96–7
  of nonconformists, 134
  religion and, 331–2
  sexual, 289–92, *295*, *296*, 299, 363
  values and, 37
Not-me syndrome, 135, 136
Nouveau riche, 187–8

Obama, Barack
  blogging culture and campaign, *51*
  as charismatic leader, 373–4
  election cost, 379, 381
  election, meaning of, 244
  family of, 316
  photos of, 58, 236, *238*
  policies, 243, 256, 367, 373–4, 385, 431
  voting for, 377, 378, 379, *380*
Obama, Malia, 316
Obama, Michelle, *5*, 316
Obama, Sasha, 316
Obedience and conformity, *135*, 135–7, *136*
Obesity, 358, 359
Objectivity, 61, *61*, 66, 69
Occupational distribution, 393
Occupational prestige, **187**, 195, 279, 349, 394
Occupational segregation, **278–9**, **393**
Occupational training, 347
Occupation, education and income, 348–9, *349*, *350*
O'Connell, John, *301*
O'Connor, Sandra Day, *359*
Ogburn, William F., **419**
Old age. *See* Aging/elderly
Oppression, 85, 104, 246, 336, 340–1
Organic metaphor, **17**
Organic solidarity, **104–5**
Organizations. *See also* Bureaucracy
  class, gender, race interplay, 145–6
  coercive, **140–1**
  crime within, 160, 169–70
  diversity training, 146
  formal, **139**
  in industrial societies, 108
  McDonaldization of society, 144–5
  normative, **140**
  ritualism in, 142–3
  social, **103**, *107*, 109–12
  theoretical perspectives, 146–7, *147*
  types of, 110, 139–45

utilitarian, **141**
voluntary, 140
Organized crime, **169**, *169*
Othermothers, 316
Out-group homogeneity effect, **239**
Outsourcing, 140

Palin, Sarah, 379
Park, Robert, **17**
Parks, Rosa, 257, 428
Parsons, Talcott, **19**
Participant observation, **56**, 65–6, *68*, *78*
Pastoral societies, 106–7, *107*
Patriarchal religions, **333–4**, 338–9
Patriarchy, 227, 248, **276**, 282–3
Patrilineal kinship, **315**
Patrilocal, defined, **315**
Patriot Act, 373
Peers, **79–80**, 129, 161, 270
Percentage, *64*
Peripheral countries, **215**, 220, 221
Personality, **75**
Personality, authoritarian, **245**
Personal space, 113–4
Personal transformation movements, **427**
Petty bourgeoisie, 16, 197–8
Piaget, Jean, **84**
Play, 88, 90, 270
Play stage, **86**
Pluralism, **245**, 254–5, 258
Pluralist model, **375**, *375*
Polarization shift, 138–9
Police, *78*, 163, 172–3, 244
Policy research, 68. *See also* Social policy
Political action committees (PACs), **375**, 380, 381
Political power, 191, *375*, 375–84
Political process theory, **429**, *430*
Political protest, 48, 52, 212
Politics. *See also* Government; Social policy
    abortion, *303*, 303–4
    campaign costs, 375, 379, 381
    culture and, *43*, 48
    elections, 377–9, *379*, 379–81, *383*
    family and, 134–5
    leadership networks, 134
    political participation patterns, 377–9, *379*
    power elite and, 191, 375–6, *378*
    racial segregation and, 243
    religion and, 340–1
    sexual, *297*, 298–9, 304
    sports and, *201*
Polls, 63–4
Pollution, 222, 411–8, *412*, *413*, *414*, 415, *415*, 418

Polygamy, **315**
Polytheism, **333**
Ponzi, Carlo, 168
Ponzi scheme, **168–9**
Popular culture, *43*, **43–4**, 48, 271–2, *298*
Population (in research), **62–3**
Population (of societies)
    age–sex ratios, 406, *406*
    characteristics of, 405–7
    children, *90*, *94*
    cohorts, **407**
    demography, **402**
    elderly, *94*, 402, 406, *406*
    environment and, 409, 411
    growth of, 221–2, *403*, 403–5, 407–10, *409*
    of U.S., 10, *12*, *13*, 37, *38*, *90*, *94*, 108, 222, 402–5, *409*
Population density, **411**
Population pyramids, *94*, 406, *406*
Pornography, 304–5, 308
Positivism, **14**
Postindustrial societies, *107*, **109**, **385**
Postmodernism, **21**, 48
Potlatch ceremony, 37
Poverty. *See also* Income gap; Stratification, social
    absolute, **225**, *226*
    blaming the poor for, 204–5, 218
    causes of, 204–6, 215, 229–30, 256
    characteristics of, 202–8, 225–8
    children in, 93, *93*, *180*, 195, *195*, 202–4, *229*, 321
    class, race, gender interplay, 202
    crime and, 159, 162, 170, *171*
    culture of, **204–5**, 218, *255*, 256
    divorce and, 203, 206, 321
    duration of, 205
    education and, 8, 227
    elderly people, *93*, 94, 195, *195*
    exploitation, 215, 218–9
    extreme, **225**
    extremism and violence, 338
    family and, 203–4, 205–6, 306, 321
    feminization of, **202–3**, 321
    gender and, *180*, 189, 202–4, *203*, 206, 223, 227, 306, 321
    global (*See* Global stratification)
    government definition of, 202
    health care and, 361, 366
    homelessness, 55–6, 203–4, 227
    indices of, 226, *226*
    manufacturing job loss, 192, 205, 212, 387
    media portrayals of, *197*
    mental illness and, 164
    myths about, *8–9*, 207, *207*
    per capita GNI, **212–5**, *213*, *214*

population growth, 221–2
    race/ethnicity, *93*, *180*, 189, 202, *203*, 242, *242*, *243*, 248, 256, 321
    relative, **225**
    segregation and, *255*
    stereotypes of, *197*, 205
    teen pregnancy and, *8–9*, 306
    theoretical perspectives, *200*, 217, 217–21
    in U.S., *195*, 200–8, *203*, *204*, 225, *226*
    war and, 228, 229–30
    welfare system, *8–9*, *197*, *203*, 206–8, *207*
    working poor, 189, 205
Poverty line, **202**, 207, 225
Power, **215**, **373**. *See also* Conflict theory; Social control; State
    authority, 372–3, **373–4**
    in bureaucracies, 141, 374
    in capitalism, 16
    coercive, 373
    culture and, 47–8
    family violence and, 328
    gender and, 272, 275–6, 279, 286, *375*, 377, *378*
    global conflict, 223–5
    inequality and, 189–90, 215, 218–9
    institutional racism and, 244
    of labeling, 33–4, 160
    legitimate, 373
    in marriage/relationships, 322, 324–6, 328
    of mass media, 42–6, *43*, 48, *197*
    of nations, economic, *213*, 215
    in organizations, 145
    political, 191, *375*, 377–85
    sexual harassment, **394–5**
    sexuality and, 295, 297
    sexual violence and, 307
    social institutions and, 104
    social order and, 20
    stratification and, 199
    theoretical perspectives, 20, 199–202, 374–7, *375*
    in workplace, 394–6, *395*
Power elite model, *375*, **375–6**
Power elites
    control, 181, 187, 196, 198, 200
    gender in, *377*, *378*
    legal system and, 160
    minorities in, *378*
    political power of, 191, 375–6, 381
    stratification and, *200*
    wealth of, 187–8, 190–1, *191*
Powerlessness, 423
Predictability, 144
Predictive validity, **352**
Pregnancy, *8–9*, 303, 305–7, *307*, 321, *330*, 360, 396

Preindustrial society, **106–7**, *107*, 316

Prejudice, **241**. *See also* Discrimination; Hate crimes; Hate groups; Racism; Segregation; Sexism; Stereotypes
 aging and, 92–3
 courts, racial bias in, 173
 homophobia, 81, **274**, **300**, *301*, 323, 332, 384–5
 immigrants and, 411
 media and minorities, 242
 minorities in government, 381
 myths about, *244*
 reducing, 245–6
 religion and, 332
 socialization and, 242
 theoretical perspectives, 244–6, *246*

Prestige, **187**
 occupational, **187**, 195, 279, 349, 394
 "status" (Weber), 198, 199

Prisoners of war, 98, 137, 138

Prison records and job market, 162

Prisons
 class/poverty, *171*
 functions of, 162, 173–5
 incarceration rates, 173, *174*
 "new color caste" and, 173
 prison guard study, 137, *137*
 as total institutions, 140–1
 women in, 7, 173–4

Private property, 16

Profane, defined, **331**

Proletariat, 16, 189

Propaganda, **372–3**

Prostitution, 157, 159, 227, 292, *298*, 299–300

Protestant ethic, 47, 218, **335**

Proxemic communication, **113–4**

Psychoanalytic theory, **82–3**, *83*

Psychology, 164, 245

Public opinion
 gay marriage, *322*
 gender and morality, *264*
 global warming, 413, 415
 government and, 372
 mass media and, 42–6, *197*, 372
 on sexuality, 291, *291*, 292–4, *293*

Public policy. *See* Social policy

Pueblo Indians, 415

Puerto Ricans, 39, 202, 249–50, *250*. *See also* Latino Americans

Punishment, 119–20, 153, 163

Pyramid schemes, 168–9

Quakers, 340

Qualitative research, **58–9**

Quality of life indicators, *221*, 222–3

Quantitative research, **58**

Queer theory, **296–7**

Questionnaires, 63–4

Quinceañera, *28*, 96

Race, **236–7**
 as ascribed status, 111
 biology and, 237–8
 categories of, 217, 237–9, *239*
 children's understanding of, *89*
 feminism and, 283
 mixed-race people, 110, 217, 237, *239*, 403
 social construction of, 111, 117, 217, 236–9, *237*, 245–6
 vs. class, 194–5, 246

Race and ethnicity. *See also* Diversity; Immigrants/immigration; Minority groups; *specific groups*
 affirmative action, **258–9**, 260, **285–6**, *286*, *390*
 AIDS, *364*
 arrest rates, 163, 170, *172*, 244, 299
 assimilation *vs.* pluralism, 254–5
 attribution error and, 132–3
 birth control, 302–3, *306*
 birthrates, 403
 census and, 237, *239*, 402–3
 class and, 194–5, 237, 241, 246, 256, 297–9
 class, education, income, 349
 class/gender interplay, 194, 195, 240–1
 classical theorists on school shootings, *158*
 in conflict theory, 420
 crime victims, 172, *172*, 299, 308
 criminal justice system and, 4, 160, 163, 170, *172*, 172–5, 244, *244*, 299
 cultural awareness programs, 132
 death penalty, 4, 173
 death rates, 360, 404, *404*
 defining groups by, 217, 236–9, *237*, *239*
 diversity training, 146
 division of labor, 105, 390
 eating disorders, *268*, *273*
 education, *89*, 132, 251, 258, 260, 346–7, *347*, *351*, 355, 356, 424
 equality, attaining, 254–60, *255*, *259*
 families and, 315, 316, 321, *321*
 gender and, 194–5, 240–1, 270, 272–4
 global inequality and, 216–7, 350–1
 in government, 381, *383*
 health/health care, 69, 358, 359–60, 362, 404, 409, 415
 Hurricane Katrina and, *182*, 257

income gap, *186*, 191–2, 242, *242*, *243*, 277, *349*, *394*
 infant mortality, 404, *404*
 intelligence/ability, 352, 353–4
 intermarriage, 315, *317*
 labeling, 33–4, *34–5*, *75*, 235–6
 labor, 145–6, 189, 387, *390*, 391, *392*, 392–3, *393*, 394
 life expectancy, *404*, 409
 in mass media, 44–6, 243, 251
 "matrix of domination," 256
 in the military, 383
 mortgages, *192*, 192–4
 in organizations, 145–6
 personal space, 114
 population statistics, 10, *12*, *13*
 poverty, *93*, *180*, 189, 202, *203*, 242, *242*, *243*, 248, 256, 321
 religion and, 333, 336–8, *337*
 self-esteem, 132
 sexuality and class, 297–9
 sexual violence, 308
 socialization, *75*, *89*
 social mobility, 196
 sports, *201*
 standardized tests, 350–1, *351*, 352–3
 stereotypes, *8–9*, 45, 240–1, 242, 249, 250, 272, 274, 298–9, 356
 "tax" on, in U.S., *259*
 theoretical perspectives, 244–6, *246*
 U.S. history of, 246–54, *247*
 voluntary organizations and, 140
 voting, 378, 379, *381*, *382*
 wealth distribution, 191–2

Racial formation, **238–9**, 249

Racialization, **237**

Racialized mass incarceration, 172–3

Racial profiling, **172–3**, *244*, 252

Racism, **243–4**. *See also* Anti-Semitism; Discrimination; Hate crimes; Hate groups; Prejudice; Segregation
 aversive, **243**
 color-blind, **243**
 environmental, **414–5**, *415*
 forms of, **243**
 institutional, **244**, *244*
 intersection perspective, 246
 in labor market, 275–6
 laissez-faire (symbolic), **243**
 in language, 33–4, *34–5*
 old-fashioned (traditional or Jim Crow), **243**
 theoretical perspectives, 244–6, *246*
 in U.S., 235, 243–4, 251, *255*

Radical movements, 427

Random samples, *60*, **63**

Rape

date/acquaintance, 152–3, 172, 307–8
  power and, 172, 307–8
  race and, 172, *172*, 299, 308
  reporting/statistics on, 163, 166, *167*, *171*, 307
Rape crisis centers, 140
Rates, defined, *64*
Reactionary movements, **428**
Reagan, Ronald, 418
Reality. *See* Definition of the situation; Social construction of reality
Reason, 13, 56–7, 84
Rebellion, 158, *158*
Recidivism, 162
Reference groups, **131–2**
Reflection hypothesis, **46–7**
Reform movements, **427**. *See also* Social movements
Reliability, **59–60**
Religion, **331**. *See also specific religions*
  conversion, 97, *340*
  cults, 97, **340**, *340*
  culture and, *31*, *41*
  diversity, *333*, *334*, 336–9
  elements of, 331–2
  family and, 314
  forms of, 333–4
  fundamentalism, 271
  gender and, 271, 341
  matriarchal, **333–4**
  opinions and, 294
  patriarchal, **333–4**, 338–9
  politics and, 340–1
  race/ethnicity and, 333, 336–8, *337*
  significance in U.S., 332–4, *333*
  social change and, 336, 340–1, 427
  socialization and, 80–1, 97, 98, 336, 338
  statistics on, *333*, *334*, *337*
  theoretical perspectives, *334*, 334–6
  types of organizations, 339–40
  violence and, 338
  voting, *382*
Religiosity, **333**, *333*, *339*
Religious extremism, **338–9**
Religious socialization, 336
Replication study, **57**, *61*
Reproductive technologies, 303
Research
  bias in, *60–1*, 63, *65*
  debunking, **7–9**, *8–9*, 9–10
  ethics/values, 68–9
  media reports, assessing, *60–1*
  reliability, **59**
  variables, **59**, *59*
Research designs, **58**

Research process
  overview, *56*
  data analysis, **63**, *64–5*
  data gathering, 59, 60, 61–3
  developing questions, 57, 58
  empirical approach, **7**
  reporting results, 63
  scientific method, **56**
  statistics, *64–5*, *68*
  tools, 63–8, *68*
Research, types of
  content analysis, 44–5, **67**, *68*
  controlled experiments, **66–7**, *68*
  ethnomethodology, 36, **117–8**
  evaluation (policy), **68**, *68*
  historical, 67–8, *68*
  participant observation, **56**, 65–6, *68*, *78*
  qualitative, **58–9**, *68*
  quantitative, **58**, *68*
  replication study, **57**
  surveys, 63–4, *68*
Resocialization, **96–8**
Resource mobilization theory, **429**, *430*
Revolution, *122–3*, **423**
Rewards, 119–20, 154, 199–200, *200*
Rights
  abortion and, *303*, 303–4
  birth control and, 302–3
  Civil Rights Act (1964), 285, 395
  civil rights movement, 256–8, 340–1, 373, 427, 429
  of disabled persons, *359*, 396
  Equal Rights Amendment, **285**
  of gays/lesbians, 46, *293*, 293–4, 299, 302, 396
  government and, 372
  legislative change, 285–6
  of women, 223, 275, 285 (*See also* Feminism)
Risky shift, **138–9**, 142
Rites of passage, *28*, **95–6**, *96*
Ritualism, 142–3
Rituals, 14, 27, *334*, **335**. *See also* Rites of passage
Robinson, Marian, 316
Rockefeller, John D., 191
Roe v. Wade, 303–4
Role change, 95–6
Role conflict, 87, 91, **111–2**
Role differentiation, 105, 107
Role modeling, **111**, 132
Roles, **74–5**, 86, **111–2**. *See also* Gender roles
Role set, **111**, *112*
Role strain, **112**
Role-taking, 77–8, 86
Romer v. Evans, 302
Roosevelt, Franklin D., 251
Ruling class. *See* Power elites

Rumors, 427
Rumsfeld, Donald H., 137
Russia, 412

Sacred, *131*, **331**
Salience principle, **240**
Samples, *60*, **62–3**, *65*
Sanctions, social, **35–6**
Sanger, Margaret, 302
Sapir-Whorf hypothesis, **32**
SATs (Scholastic Aptitude Tests), 350, *350*, *351*, 352–3, 356
Scapegoat theory, **245**
Schema, 84
Schiavo, Michael, 36
Schiavo, Terry, 36, *37*
Schooling, **346**
Schools. *See also* Education
  culture and, 132
  dropouts, 346–7, *349*
  gender and, 8, 271, 356
  race/ethnicity, *89*, 347
  resegregation, 256
  segregation/desegregation, 132, 257, 258, 427
  socialization and, 81–2, *82*, 318, 346, 347, 356
  sports in, *29*, 81, 285, *286*
  violence in, 67, 156, *158*
Schwerner, Michael, 258
Scientific method, 14, **56**
"Screenagers," 26
"Second shift," 112, 280–1, 325–6
Second-world countries, **215**
Sects, **339–40**
Secular, defined, **332**
Sedimentation of racial inequality, 192
Segregation, **255–6**
  apartheid, 182, 217, 239, *255*
  de jure and de facto, 256
  gender, **278–9**
  hypersegregation, *255*, **256**, *257*, 414
  in the labor force, 275, 278–9
  occupational, **278–9**, **393**
  residential, *192*, 192–4, 195, 243, **244**, *255*
  schools, 132, 256, 257, 258, 427
  triadic, **129**
  urban underclass and, 255–6
Self, **85**. *See also* Identity
  deindividuation and, 139
  looking glass self, *85*, **85–6**
  in psychoanalytic theory, 83
  socialization theories and, *83*
  as socially constructed, 85–7
  in symbolic interaction theory, 85–7
Self-concept/self-esteem, **77–8**
  beauty/appearance and, 77, 270
  child abuse and, 88

Self-concept/self-esteem (*continued*)
gender and, 77, *77*, *266*, 270
media's effect on, 79, *266*
race and, *29*, 85, 132
sexualization and, 309
socialization and, 77–8, *80*, 267
Self-fulfilling prophecies, 82, **355**, *355*. *See also* Teacher expectancy effect
Semiperipheral countries, **215**, 220
Sept. 11, 2001 terrorist attacks
as altruistic suicide, 156
crime statistics and, 163
formation of groups, 130
as global crime, 175
hate crimes and, 235
racial profiling and, 252
sociological perspective and, 6–7
Serial monogamy, 315
Sex, **265**. *See also* Gender; Men; Women
adolescence and, *60*, *291*, 294, *295*, 305–7
attitudes toward, 291, *291*, 292–4, *295*, 304–5
biology and, 265–7
commercialization of, 308–9
culture and, 289–92, 299–300
cybersex, 308
"double standard," 297–8
exploitation and, *298*
"hooking up," 323
popular culture and classical theory, *298*
pornography, 304–5
prostitution, 157, 159, 299–300
reproductive technologies, 303
social change and, 308–9, 323
social issues and, 302–8
on talk shows, *293*
Sex education, *306*
Sexism, *34*, 275–6. *See also* Gender stratification
Sex ratio, **406**
Sexual abuse, 137, 272, 299
Sexual assault. *See* Rape
Sexual harassment, 140, 279, **394–5**
Sexual identity, 110–1, 266, **290**, 291, *296*, 296–7
Sexuality
class, gender, race, 297–9
cultural and social basis of, 289–92
diversity and inequality, 297–302
global perspective on, 299–300
heterosexism, **301**
norms, *295*, *296*, 299
sociological theory, 294–7, *296*, *298*
stereotypes, 298–9
violence and, 295

Sexually transmitted diseases, 69, 362–3, *364*
Sexual orientation, **290**. *See also* Gays and lesbians; Transgendered people
Sexual politics, **297**, 298–9, 304
Sexual practices, 294, 298
Sexual revolution, *295*, **308**
Sexual scripts, **291**
Shakers, 340
Shopping, 179–80, *184–5*
Sick role, 164
Significant others, **86**
Simmel, Georg, 9, **129**, **410**
Single-parent households, 8–9, 202–4, *203*, 305–6, 320–1
Single people, 323, *325*, *326*
Sioux, 248
Six degrees of separation, 134
Slavery, 238, 248, 299, 336, 424
Small group movement, *130–1*
Small world problem, 134
Smoking, 20, *60*, 153, *165*, 272, 361, 362
Social action, 17–8. *See also* Social movements
Social categories, 109, 128, 237–9
Social change, *19*, **418–9**. *See also* Social movements
capitalism and, 16
causes of, 50–2, *421*, 423–6, *429*
characteristics of, 419
culture and, 48, 49–52
education and, *348*
families and, 314, 318, 319–20, 329–31
fast food, 49–50
gender and, *283*, 284–6
global, 230–1, 386–9, 431
language and, 32, *122*
micro- *vs.* macrochanges, **418**
modernization, 418
myths, *408*
race relations, 256–8
religion and, 336, 340–1
sex and, 291, *291*, *296*, 308–9
social evolution and, 17
technology and, 51, 106–9, *122–3*, 218, 425–6, *437*
theoretical perspectives, 19, *21*, *217*, *246*, 419–22, *421*
war/terrorism, 424–5
Social class. *See* Class
Social construction of reality
overview, 117
beauty/physical ideals, 44, *268–9*, 358
culture, *34–5*, 48
"definition of the situation," 20, 117, 355
deviance, 143, 153–4
family/marriage, 319

gender/sexuality, 110–1, 117, 264–7, *266*, *273*, 290–2, 296–7
health/illness, 346, 365
identity, 85–7
language and, 32–3, *34–5*
nature-nurture debate and, 76
race/ethnicity, 110, 117, 217, 236–9, *237*, 245–6
symbolic interaction and, 21
as theory of social interaction, *118*
Social construction perspective, **296**
Social control, **76–7**, **160**. *See also* Conformity
in conflict theory, 20
deviance and, 159, 160
education and, 52, 348
in gemeinschafts, 105
norms and, 34–6
socialization as, *76*, 76–7
the state and, 372–3, *375*, 375–7
in total institutions, 140–1
Social control agents, **160**
Social control theory, **159**
Social Darwinism, **17**
Social differentiation, **180**
Social epidemiology, **359–60**
Social exchange model, *118*, 119–20
Social facts, **14**, *15*, 170, 244
Social institutions, *19*, **103**. *See also specific institutions*
as agents of socialization, 78–82, 103
culture and, 47
gender embedded in, 274–5, 319
the military as, 381–3
race and, 244
sexuality and, 292, 294, 295, 297
theories of, 103–4
total institutions, **140–1**
Social interaction, *19*
in bureaucracies, 141
in cyberspace, 109–10, 120–1, *122–3*, 124, *124*
everyday, 112–4
gemeinschaft *vs.* gesellschaft, 105
interpersonal attraction, 114–6
social institutions and, 103–4
society and, 109–16
solidarity, 104–5
theories of, 116–20, *118*
Socialism, **386**
Socialization, **74–9**
adult, **90–1**
agents of, **78–82**, 270–2, 270–3
anticipatory, **91**
of children, *75*, 76, 78–82, 83–8, *89*, 242, 318
class and, 87–8

class, gender, racial inequalities, 81–2, 275
conformity/individuality, 77
consequences of, 77–8
deviance and, 161, 162
diversity and, *75*, 87–8, *89*
education as, 346, 347
family and, 78–9, 318
gender and (*See* Gender socialization)
immigrants and, *75*
of inmates, 162, *171*
institutions and, 78–82, 103
across life course, 88–96
military norms, 96–7, 382
peers and, 79–80
prejudice and, 242
religion and, 80–1, 97, 336, 338, *340*
resocialization, **96–8**, *340*
rites of passage, *28*, **95–6**, *96*
schools and, 81–2, *82*, *89*, 318
self-concept and, 77, *77*
as social control, *76*, 76–7
sports and, 81
theoretical perspectives, 82–7, *83*
Social learning theory, *83*, **83–4**, 111
Social location, 3, 4, 256
Social mobility, 186, 191–2, **196–8**, *200*, *207*, 349–51
Social movements, **426**
  "Black is beautiful," 52
  Black power, 258
  civil rights, 256–8, 340–1, 427, 429
  environmental issues, 415, 428
  gays and lesbians, 48, 429
  La Raza Unida, 52
  myths about, *429*
  organization of, 427–8
  origins of, 428
  perception of deviance and, 153
  religion and, 340–1
  sexual violence and, 307–8
  social/political change, **427**
  state power and, 372–3
  theoretical perspectives, 429–31, *430*
  trade policies and, 212
  transnational, 431
  women's movement, 284
Social networks, **133–4**, 142
Social organization, **103**, *107*, 109–12
Social policy
  affirmative action, **258–9**, 260, **285–6**, *286*, *390*
  birth control, 292, 302–3, 409–10
  care work, 329–31
  child care, 329–30
  environment, 415, 418

families, 329–31, *330*
gender equality, 285–6
health care, 364, *364*, 366–7, 367–8
mass media and, *61*, *197*
poverty/welfare, *197*, 203, 205, 206–8
prisons, 173–5
race and, 194, 243, 258, 260
religion and, 340
research on, 68
sex and reproduction, 292, 302–8
sexual violence, 307–8
stratification and, *200*, 201
Social/political change movements, **427**
Social sanctions, **35–6**, 119–20
Social sciences, 5
Social solidarity. *See* Solidarity
Social speedup, *321*
Social stratification. *See* Stratification, social
Social structure, **6**, *19*, **104**, 105–9. *See also* Class; Social institutions
Society, **102–4**. *See also* Groups; Social interaction
  in everyday life, *122–3*, *184–5*, *268–9*, *416–7*
  functions of, 14, 18–9
  gesellschaft and gemeinschaft, **105**, 422–3
  institutions/functions of, 103–4
  McDonaldization of, *144*, 144–5
  micro- and macroanalysis, **102–3**, 159, 161
  theories of, *118*
  types of, 105–9, *107*
  Weber's basic dimensions of, 16–7
Socioeconomic status (SES), **186**, 348. *See also* Class; Income; Poverty
Socioemotional needs, *130–1*, 131
Sociological imagination, **5**, 9, 87, 93
Sociological theory. *See also* specific theories
  classical theorists on contemporary topics, *15*, *49*, *158*, *201*, *298*
  history of, 13–8
  major theories, 18–21, *21*
  testing of (*See* Research process)
Sociology, **4**
  careers in, 9–10, *10–1*, 22
  debunking in, **7–9**, *8–9*, *9–10*
  famous sociology majors, 5
  key concepts, 19
  purpose of, 6–7, 16, 17, 87, 93, 211
  urban, 410
  value-free, 17, 68–9

Solidarity
  contractual, 419
  culture and, 47
  deviance and, 153, 155
  Durkheim on, 14, 104–5, 153, 335
  gemeinschafts and gesellschafts, **105**, 422–3
  mechanical, **104**, 419
  organic, **104–5**, 419
  religion and, *334*, 335
  resocialization and, 97
Sorokin, Pitrim, **421**
Sororities, 38, 97
South African apartheid, 182, 217, 239
Space shuttles, 142–3, *143*
Spacial mismatch, **387**
Special-interest groups, **375**
Spencer, Herbert, **419**
Spengler, Oswald, 420–1
Sports
  class and, *201*
  color line in, *201*
  myths, *81*
  Native American names and mascots, 29
  perception of facts study, 117
  socialization and, 81
  steroid use, 358–9
  women in, 267, 285, *286*
Standardized ability tests, 260, *350*, 350–2, *351*, **352**, 356
Starbuckization, 144
State, **372–8**, *375*, *377*. *See also* Government; Power
Statistics, *64–5*
Status, **110–1**, **180**. *See also* Class; Prestige; Stratification, social
  achieved, **110–1**, *111*, 182
  age and, 92–3
  ascribed, 93, **110–1**, *111*, 181–2
  culture and, 37, 48
  language/labels and, 33
  master, **111**, **164**
  tattoos and, *30*
  touching behavior and, 113
  Weber on, 182, 198, 199
  of women, 8, 33, 172, 223, 267, 282–3
Status attainment, **183**
Status inconsistency, **110**, *216*, 251
Status set, **110**
Status symbols, *30*, 183, *184*
Stepfamilies, 322
Step shows, 424, *424*
Stereotype interchangeability, **241**
Stereotypes, **240**. *See also* Labeling; Myths
  adolescents, 90, 92
  age, **92**
  class and, 46, *197*

Stereotypes (*continued*)
  class, gender, race interplay,
    240–1
  criminals, 167
  deviant labels, 160
  families, *320*
  gays and lesbians, 46, 301, 302
  gender, 45–6, 81, 92, 206, 240–1,
    264, *266*, 270, 271, 272, 298–9,
    320, 357
  immigrants, 252–3, 405
  internalizing, 356
  language and, *34–5*
  in the media, 45–6, 67, 92, *197*,
    240–1, 242, *266*, 270
  occupational segregation and, 280
  poverty, *197*, 205, 206
  prejudice and, 241
  racial/ethnic, *8–9*, 45–6, 167, 195,
    240–1, 242, 249, 250, 272,
    274, 298–9, 356
  salience principle, **240**
  teacher expectations, 82, *355*
Stereotype threat effect, **356–7**
Sterilization, 250, 303
Stigmas, **164–5**, *171*, **362–3**
Stockholm Syndrome, **98**
Stock ownership, *207*
Strangers, 9
Stratification, social, **180–1**. *See
   also* Class; Income gap; Social
   mobility; Status
  age and, 93–4, *94*, 195
  in capitalism, 197–8
  caste system, 107, **181**, 248
  class, race, gender interplay,
    194–5
  components/dimensions of, 198–9
  diverse sources of, 194–5
  education and, 348, *350*, 350–8,
    351, *355*, 424
  gender (*See* Gender stratifica-
    tion)
  global (*See* Global stratification)
  in health care, 358–63, *359*,
    364–5
  ideas supporting, 195–7
  language and, 33–4, *34–5*, 424
  powerlessness, 423
  racial, 194–5, 248–9, 258 (*See
    also* Race and ethnicity)
  religion and, 336
  sexuality and, 297–9
  social change and, 108, 424
  systems of, 180–3
  theoretical perspectives, 19–20,
    21, 197–202, *200*
Street Corner Society, 66
"Strength of weak ties," 133
Stress, 111, 280–1, 361, 362
Structural strain theory, **156–8**,
  158

Subcultures, **39**, 96, 154
Suburbs, 410–1
Success, 8, 157–8, 254–5, *347*,
  352–3, 393
Suicide
  altruistic, **156**
  anomic, **156**
  Durkheim on, 16, 155–6
  egoistic, **156**
  rates of, *157*
Suicide bombings, 156, 225, 338
Sui generis, 14, 102
Suleman, Nadya, 299
Sullenberger, Chesley B. "Scully,"
  130
Sumner, William Graham, **34–5**
Superego, **82–3**
Support groups, *130–1*
Surveillance, 373
Surveys, 63–4, *68*
Sweden, 275, 386
Symbolic interaction theory, **20–1**,
  21
  aging, 94–5, *95*
  crime, 166, *166*
  culture, 47, 48–9
  deviance, *155*, 161–4
  education, 348, *348*, 355
  family, *318*, 319
  health care, 361–2, *364*, 365–6
  mental illness, 164
  organizations, 147, *147*
  racism and prejudice, 245–6,
    246
  religion, *334*, *335*, 336
  sexuality, *296*, 296–7
  social construction of reality, 21,
    117, 124
  social interaction and, 117, 124
  socialization, *83*, *85*, 85–7
  work, *390*, 391
Symbols, **29**, *201*
Syphilis, 69, 362

Taboos, **35**
Tabula rasa (blank slate), 66
Tactile communication, **113**
Taken-for-granted assumptions,
  29, 40
Taking the role of the other, **86**
Taliban, 97, *153*, 275
Task-oriented needs, 131
Tattoos, 7, *30*, 163
Taxation, 169, 190, 194
Taylor, Howard F., *11*
Taylor, Murtis, *11*
Tchikrin people, 26, *26*
Teacher expectancy effect, 82, 271,
  348, 355, *355*
Technology. *See also* Computers;
   Industrialization; Internet
  bureaucracy and, 142–3

  change and, *51*, 51–2, 106–9,
    122–3, 218, 425–6, *437*
  culture and, 25–6, 50, 51, *51*,
    122–3
  economy and, 217–8, 387
  "green" energy production, *417*
  industrial societies, 107–9
  labor, impact on, 109, 387–8
  modernization and, 421–2
  pollution and, 418
  population growth and, 408
  postindustrial societies, 109
  preindustrial societies, 106–7
  reproduction and, 303
  sex and cybersex, 308
Teenagers. *See* Adolescence
Television. *See also* Mass media
  American culture and, 26, 42
  gender and, 45–6, 271–2, *320*
  race/ethnicity, 45–6
  sex on talk shows, *293*
  values and, 46–7
  violence and, 79
Temporary Assistance for Needy
  Families (TANF), **206**
Tenure, **141**
Terrorism, **224**. *See also* Sept. 11,
  2001
  Al Qaeda, 40, 224–5, 431
  bin Laden, Osama, 175
  ethnocentrism and, 40–1
  global perspective on, 175, 223–5
  inequality and, 224–5
  religion and, 40, 97, 224
  social change and, 431
  suicide bombers, 156, 225, 338
Tertius gaudens, 129
Testing. *See* Standardized ability
  tests
Testosterone, 265
Text messaging, 32, *32*
Thailand, sex trade in, 300
Theoretical perspectives. *See*
  Sociological theory
Thermal pollution, *413*
Third-world countries, **215**
Thomas, W. I., 20, 117, **132**, 161,
  355
Tibet, 41
Till, Emmett, 258
Time, social meaning of, 32–3
Title IX, **271**
Tocqueville, Alexis de, **14**
Tokenism, 80, 146, 362
Tönnies, Ferdinand, **422–3**
Torture, 98, 137, 138
Total institutions, **140–1**
Totem, **331**
Touch, 113
Touré, Kwame, 258
Toynbee, Arnold J., **420**
Tracking, 244, 345–6, **354–5**, *355*

Transgendered people, **266–7**
Transnational capitalist class, 223
Transnational communities, 221
Transnational families, *216*, **329**
Transnational social movements, 431
Triadic segregation, **129**
Triads, **129**
Troubles (*vs.* issues), **6**, 13, 328, 393
Truman, Harry S., 138
Tuskegee Syphilis Study, 69, 362
Tyranny of the majority, 14

Underclass, global, 226
Underclass, urban, **189**, **255**
Underemployment, **392**
Unemployment, 6, 13, 109, 205, 250, *393*
Unemployment rate, **391–3**
Unidimensional evolutionary theory, 419–20
United States
    age composition of, *90*, *94*, 405–7, *406*
    class in, 183–95
    cultural diffusion, 423–4
    cultural diversity, *38*, 52
    culture and technology, 25–6
    elections, 377–9, *379*
    Hurricane response, *257*
    immigration (*See* Immigrants/immigration)
    language diversity, *38*
    life expectancy, 404, *404*, *405*
    per capita GNI, 214, *214*
    population statistics, *12*, 41, *90*, *94*, *108*, 222, 402–5, 408, 409
    religion in, 332–4, *333*
    sociology in, 17–8
University of California Regents v. Bakke, 260
Upper class (elites), 160, 187–8, 191, 196, 241. *See also* Power elites
Upper middle class, 188–9
Urbanization, **410–1**. *See also* Cities
Urban underclass, **189**, **255**
Urban village, 410

Validity, **59**, *61*, 65
Values, *31*, **36–7**
    class and, *30*
    culture and, *30*, 36–7, 39–40, *41*, 41–2, 46–7, *47*
    deviance and, *30*, 159
    economic achievement and, 218
    "family values," 105, *321*, 329
    globalization and, 41–2, 223–4
    mass media and, 46–7, 79
    militarism, 382

physical perfection, 358–9 (*See also* Beauty/physical ideals)
    of power elite, *378*
    religion and, 331
    sexuality and, 291, 292–4, *295*, *296*, 302–8
    in sociological research, 17, *61*, 68–9
    sports and, 81
    support groups and, *130–1*
Variables, **59**, *59*
Variables, dependent, **59**, *59*
Variables, independent, **59**, *59*, *64*
Variables, intervening, *59*, *59*
Verstehen, **17**
Vietnamese, 251–2. *See also* Asian Americans
Violence. *See also* Aggression; Murder; Suicide; Terrorism; War
    child abuse, 74, 88, 156, 328
    classical theorists on school shootings, *158*
    elder abuse, 328
    gender socialization and, 272
    homophobic, *301*
    incest, 328
    media and, *67*, 79, *167*
    partner violence, 98, *203*, 204, 327–8
    police brutality, 173
    race and, 235, 258, 299
    rates of, *9*
    religion and, 338
    in schools, *67*, 156, *158*
    socialization, 79, 159
    violent crime statistics, *167*, *171*, *172*
    against women, 98, 272, 283, 327–8 (*See also* Rape)
    against women: global analysis, *153*, 223, *224*, 284
Virginia Tech University, 156
Voting, 377–9, *379*, *380*, *381*, *382*

War, *15*, 223–5, *228*, 229–30, 251, 299, 424–5. *See also* Military
War, prisoners of, 98, 137, 138
Watergate affair, 138
Water shortages/pollution, *221*, 222–3, 411–2, *412*, 413
Wealth, **186**, **190**. *See also* Income; Upper class
    of countries, 212–5, *213*, *214*
    distribution of, *180*, 187–9, 190–4, *191*, *193*
    intergenerational effects, 190, 191–2
Weber, Max, *15*, *16*, **16–7**, 18, 47, *49*, **141**, 144, *158*, 198–9, *201*, 218, *298*, 334, 335, 373–4
Weight, *268–9*

Welfare system, *8–9*, *197*, *203*, *205*, 206–8, *207*
Wells-Barnett, Ida B., *15*, **18**, *18*
White-collar crime, 160, 168–9, 170
Whites. *See also* Race and ethnicity
    as crime victims, *172*
    eating disorders, 116, *268*, *273*, 358
    ethnic groups, 236, 238, 240, 252–4
    men: power in workplace, 279
    in organizations, 145
    population statistics, *12*, *13*
    privilege, 243, 259
    stereotypes of, 240
    unemployment, *390*, *391*, *393*
White supremacists, 40, 235, 428
White supremacy, 248
Wilson, William Julius, 246, 256
Winfrey, Oprah, 79
Wirth, Louis, **410**
"Witch hunts," 160
Women. *See also* Gender; Pregnancy
    AIDS, 363, *364*
    beauty, 44, 45, 116, *120–1*, *268–9*, 358
    conformity, 77, 272, *273*
    crime by, 170–1
    as crime victims, 172, *172*
    elder care, 330–1
    empowering, 227
    equality, societies with, 275
    gender apartheid, **275**
    in government, 284, 377, *377*, *378*, 381, *383*
    labels/terminology, 33
    media images of, 44, 45–6, *266*, 270, 308–9
    mental illness, 164
    military service, 383–4
    motherhood, *8–9*, 307, 315–6, 329
    poverty, *180*, 189, 202–4, *203*, 206, 227, 306, 321
    in prisons, 7, 173–4
    religion and, 341
    as sex objects/sexualization, *269*, 271, 272, 308–9
    sexual exploitation of (*See* Prostitution)
    single parents, *8–9*, 202–4, *203*, 305–6, 320–1
    socialization of, 84 (*See also* Gender socialization)
    in sociology, *17*, 17–8
    sports and, 81, *81*, 267, 285, *286*
    status of, 33, 172, 223, 267, 282–3 (*See also* Gender stratification)

Women (*continued*)
  stereotypes, 81, 92, 206, 241, *266*, 357
  Taliban and, *153*, 275
  violence against, 98, 152–3, *153*, 223, *224*, 272, 283, 284, 327–8 (*See also* Rape)
Women, labor and
  commuter marriages, 321–2
  domestic workers, *216*, 329, 330
  gender division of labor, 105, 107, 280–1, 283–4, 316, 319, 325–6, 390
  housework/"second shift" for, 112, 280–1, *324*, 325–6
  immigrants, *216*
  income of (*See* Income gap)
  job segregation, *34*, 275, 393–5
  in male-dominated jobs, 274, 279–80
  maternity leave, *330*
  men's support for, 284–5
  "mommy tax," 276
  promotion/glass ceiling, 145–6, 394, *395*

  race, 273–4, 275, *286*, 362, *390*, 391, 393, *393*, *394*
  role conflict/strain, 112
  sexual harassment, **394–5**
  statistics on, 277, 279
  stress and health, 362
  "taking jobs from men," 275, *390*, 391
  theoretical perspectives, 277–9
  unpaid labor, 283
  "women's work" devalued, 108, 276, 279–80
  working mothers, *60*, *61*, 112, 207, 276, 280–1, *326*, 329, *330*
Women's movement. *See* Feminism
Work, **389**. *See also* Corporations; Labor; Labor force; Occupation; Women, labor and
Workers. *See* Labor force
Work ethic, 33, 47
Workfare, 206, 207
Working class
  characterized, 189, 190, 197
  childhood socialization, 82, 87–8

  crime and, 159–60
  declining fortunes of, 192, 205
  exploitation of, 16, 198
  in the media, 46, *197*
  schools and, 82, 349
  women in, 195
World cities, **221**
WorldCom Corporation, 170
World systems theory, *217*, **219–21**, *421*
World Trade Center. *See* Sept. 11, 2001 terrorist attacks
World Trade Organization (WTO), 373

Xenophobia, **386**

Yanomami people, 426

Zero population growth (ZPG) theory, 409, *409*
Zola, Irving, **396**

TO THE OWNER OF THIS BOOK:

I hope that you have found *Sociology: The Essentials,* Sixth Edition, useful. So that this book can be improved in a future edition, would you take the time to complete this sheet and return it? Thank you.

School and address: _____

_____

Department: _____

Instructor's name: _____

1. What I like most about this book is: _____

_____

_____

2. What I like least about this book is: _____

_____

_____

3. My general reaction to this book is: _____

_____

_____

4. The name of the course in which I used this book is:

_____

5. Were all of the chapters of the book assigned for you to read? _____

   If not, which ones weren't? _____

6. In the space below, or on a separate sheet of paper, please write specific suggestions for improving this book and anything else you'd care to share about your experience in using this book.

_____

_____

_____

_____

_____

_____

WADSWORTH
CENGAGE Learning

BUSINESS REPLY MAIL
FIRST-CLASS MAIL   PERMIT NO. 34   BELMONT CA

POSTAGE WILL BE PAID BY ADDRESSEE

Attn: Sociology

Wadsworth Cengage Learning
20 Davis Drive
Belmont, CA   94002-9801

FOLD HERE

OPTIONAL:

Your name: _____ Date: _____

May we quote you, either in promotion for *Sociology: The Essentials,* Sixth Edition, or in future
publishing ventures?

Yes: _____ No: _____

Sincerely yours,

Margaret L. Andersen
Howard F. Taylor